SUSTAINABLE FOOD AND AGRICULTURE

SUSTAINABLE FOOD AND AGRICULTURE

An Integrated Approach

Edited by

CLAYTON CAMPANHOLA

The Food and Agriculture Organization of the United Nations (FAO), Rome, Italy

SHIVAJI PANDEY

Special Advisor to the Food and Agriculture Organization of the United Nations (FAO), Rome, Italy

Published by

The Food and Agriculture Organization of the United Nations and Elsevier Inc.

Academic Press is an imprint of Elsevier
125 London Wall, London EC2Y 5AS, United Kingdom
525 B Street, Suite 1650, San Diego, CA 92101, United States
50 Hampshire Street, 5th Floor, Cambridge, MA 02139, United States
The Boulevard, Langford Lane, Kidlington, Oxford OX5 1GB, United Kingdom

British Library Cataloguing-in-Publication Data
A catalogue record for this book is available from the British Library

Library of Congress Cataloging-in-Publication Data
A catalog record for this book is available from the Library of Congress

Elsevier ISBN: 978-0-12-812134-4 FAO ISBN: 978-92-5-131137-0

For Information on all Academic Press publications
visit our website at https://www.elsevier.com/books-and-journals

Publisher: Andre Gerhard Wolff
Acquisition Editor: Nancy Maragioglio
Editorial Project Manager: Tasha Frank
Production Project Manager: Nilesh Kumar Shah
Cover Designer: Christian Bilbow

Typeset by MPS Limited, Chennai, India

Contents

33. Agrifood Systems

34. Socioeconomic Dimension of Agrifood Systems

35. Natural Resource and Environmental Dimensions of Agrifood Systems

36. Molecules, Money, and Microbes

37. Policy Orientations for Sustainable Agrifood Systems

SECTION IV

OPERATIONALIZING SUSTAINABLE FOOD AND AGRICULTURE SYSTEMS

38. Conceptual Framing of the Operationalization of Sustainable Food and Agriculture

39. Identifying the Relevant Context: Key Features of the Social-Ecological System

40. Using Evidence and Dialogue to Build Knowledge and Address Trade-Offs

List of Contributors

José Aguilar-Manjarrez Food and Agriculture Organization of the United Nations (FAO), Rome, Italy

Arwen Bailey Bioversity International, Maccarese, Italy

Jennie Barron Swedish University of Agricultural Sciences, Uppsala, Sweden; CGIAR Research Program on Water, Land and Ecosystems, Colombo, Sri Lanka

Caterina Batello Food and Agriculture Organization of the United Nations (FAO), Rome, Italy

Lorenzo Giovanni Bellù Food and Agriculture Organization of the United Nations (FAO), Rome, Italy

Zareen Pervez Bharucha Anglia Ruskin University, Cambridge, United Kingdom

Kartika Bhatia Middle East and North Africa Office, World Bank, Washington, DC, United States

Regina Birner Institute of Agricultural Sciences in the Tropics (Hans-Ruthenberg-Institute), University of Hohenheim, Stuttgart, Germany

Sally Bunning Food and Agriculture Organization of the United Nations (FAO), Santiago, Chile

Clayton Campanhola Strategic Program Leader (Sustainable Agriculture), Assistant Director-General (Economic and Social Development Department), and Special Advisor (Sustainable Agriculture), respectively, of the Food and Agriculture Organization of the United Nations (FAO), Rome, Italy

Andrea Cattaneo Food and Agriculture Organization of the United Nations (FAO), Rome, Italy

Molly Conlin CGIAR Independent Science and Partnership Council, c/o Food and Agriculture Organization of the United Nations (FAO), Rome, Italy

Valerio Crespi Food and Agriculture Organization of the United Nations (FAO), Rome, Italy

Ben Davis Food and Agriculture Organization of the United Nations (FAO), Rome, Italy

Jeroen Dijkman CGIAR Independent Science and Partnership Council, c/o Food and Agriculture Organization of the United Nations (FAO), Rome, Italy

Kumuda Dorai LINK limited, Canberra, ACT, Australia

Paul Dorosh International Food Policy Research Institute (IFPRI), Washington, DC, United States

Ehsan Dulloo Bioversity International, Maccarese, Italy

Shenggen Fan International Food Policy Research Institute (IFPRI), Washington, DC, United States

Jean-Marc Faurès Food and Agriculture Organization of the United Nations (FAO), Rome, Italy

Edith Fernández-Baca United Nations Development Program (UNDP), Lima, Peru
Cornelia Butler Flora Kansas State University, Manhattan, KS, United States
Louise O. Fresco Wageningen University and Research, Wageningen, Netherlands
Lucy Garret Food and Agriculture Organization of the United Nations (FAO), Rome, Italy
Hafez Ghanem Middle East and North Africa Office, World Bank, Washington, DC, United States
Marius Gilbert Spatial Epidemiology Lab (SpELL), Free University of Brussels, National Fund for Scientific Research, Brussels, Belgium
Meredith Giordano International Water Management Institute (IWMI), Colombo, Sri Lanka
Andrew Hall Agriculture and Food, Commonwealth Scientific and Industrial Research Organisation (CSIRO), Black Mountain, Canberra, ACT, Australia
Matthias Halwart Food and Agriculture Organization of the United Nations (FAO), Rome, Italy
Sue Hartley University of York, York, United Kingdom
Patrick P. Kalas Food and Agriculture Organization of the United Nations (FAO), Rome, Italy
Gina Kennedy Bioversity International, Maccarese, Italy
Zeyaur Khan International Centre of Insect Physiology and Ecology (ICIPE), Nairobi, Kenya
Misael Kokwe Food and Agriculture Organization of the United Nations (FAO), Lusaka, Zambia
Rattan Lal Ohio State University, Columbus, OH, United States
Wilfrid Legg Consultant and formerly OECD
Preet Lidder CGIAR Independent Science and Partnership Council, c/o Food and Agriculture Organization of the United Nations (FAO), Rome, Italy
Leslie Lipper CGIAR Independent Science and Partnership Council, c/o Food and Agriculture Organization of the United Nations (FAO), Rome, Italy
Juliana C. Lopes Food and Agriculture Organization of the United Nations (FAO), Rome, Italy
Yuelai Lu University of East Anglia, Norwich, United Kingdom
Eduardo Mansur Food and Agriculture Organization of the United Nations (FAO), Rome, Italy
Alexandre Meybeck Bioversity International (CIFOR), Maccarese, Italy
Weimin Miao Food and Agriculture Organization of the United Nations (FAO), Bangkok, Thailand
Charles Midega International Centre of Insect Physiology and Ecology (ICIPE), Nairobi, Kenya
Jamie Morrison Food and Agriculture Organization of the United Nations (FAO), Rome, Italy
William Murray Food and Agriculture Organization of the United Nations (FAO), Rome, Italy

David Neven Food and Agriculture Organization of the United Nations (FAO), Rome, Italy

Hanh Nguyen Food and Agriculture Organization of the United Nations (FAO), Rome, Italy

Ephraim Nkonya International Food Policy Research Institute (IFPRI), Washington, DC, United States

Carolyn Opio Food and Agriculture Organization of the United Nations (FAO), Rome, Italy

Shivaji Pandey Strategic Program Leader (Sustainable Agriculture), Assistant Director-General (Economic and Social Development Department), and Special Advisor (Sustainable Agriculture), respectively, of the Food and Agriculture Organization of the United Nations (FAO), Rome, Italy

Monica Petri Food and Agriculture Organization of the United Nations (FAO), Vientiane, Laos

Ugo Pica-Ciamarra Food and Agriculture Organization of the United Nations (FAO), Rome, Italy

John Pickett Rothamsted International, United Kingdom

Prabhu Pingali Tata-Cornell Institute for Agriculture and Nutrition, Cornell University, Ithaca, NY, United States

Jimmy Pittchar International Centre of Insect Physiology and Ecology (ICIPE), Nairobi, Kenya

Jules Pretty University of Essex, Colchester, United Kingdom

Ewald Rametsteiner Food and Agriculture Organization of the United Nations (FAO), Rome, Italy

Sherman Robinson International Food Policy Research Institute (IFPRI), Washington, DC, United States

Timothy Paul Robinson Food and Agriculture Organization of the United Nations (FAO), Rome, Italy

Dominic Rowland Center for International Forestry Research (CIFOR), Bogor, Indonesia; SOAS, University of London, London, UK

Harpinder Sandhu Flinders University, Adelaide, SA, Australia

Fritz Schneider Bremgarten, Switzerland

Rachid Serraj CGIAR Independent Science and Partnership Council, c/o Food and Agriculture Organization of the United Nations (FAO), Rome, Italy

Andrea Sonnino Agenzia Nazionale per le Nuove Tecnologie, L'Energia e lo Sviluppo Economico Sostenible (ENEA), Biotechnologies and Agroindustry Division, Casaccia Research Centre, Rome, Italy

Kostas Stamoulis Strategic Program Leader (Sustainable Agriculture), Assistant Director-General (Economic and Social Development Department), and Special Advisor (Sustainable Agriculture), respectively, of the Food and Agriculture Organization of the United Nations (FAO), Rome, Italy

Austin Stankus Food and Agriculture Organization of the United Nations (FAO), Rome, Italy

Henning Steinfeld Food and Agriculture Organization of the United Nations (FAO), Rome, Italy

Rudresh Kumar Sugam Formerly Council on Energy, Environment and Water, New Delhi, Delhi, India

Terry C.H. Sunderland Center for International Forestry Research (CIFOR), Bogor, Indonesia; Faculty of Forestry, University of British Columbia, Vancouver, Canada

Berhe Tekola Food and Agriculture Organization of the United Nations (FAO), Rome, Italy

James Thurlow International Food Policy Research Institute (IFPRI), Washington, DC, United States

Olcay Ünver Food and Agriculture Organization of the United Nations (FAO), Rome, Italy

Cora van Oosten Centre for Development Innovation, Wageningen University, Wageningen, Netherlands

Rob Vos International Food Policy Research Institute (IFPRI), Washington, DC, United States

Ren Wang Beijing Genomics Institute, Shenzhen, P.R. China

Keith Wiebe International Food Policy Research Institute (IFPRI), Washington, DC, United States

Wei Zhang International Food Policy Research Institute (IFPRI), Washington, DC, United States

Foreword

Today, the world still produces food mainly based on the principles of the green revolution of the 1960s and 1970s. Most of this production is based on high-input and resource-intensive farming systems at a high cost for the environment. As a result, soil, forests, water, air quality, and biodiversity continue to degrade. And this focus on increasing production at any cost has not been sufficient to eradicate hunger, despite the fact that today we produce more than enough to feed everyone. In addition, we are seeing a global epidemic of obesity. This is situation is unsustainable, and we have reached an inflection point. We need to promote a transformative change in the way we produce and consume food. We need to put forward sustainable food systems that offer health and nutritious food to everyone, and also preserve the environment.

Achieving the 2030 Agenda for Sustainable Development requires us to make this transformative change with people at its center. This publication can serve as a reference in that process. It shows what viable options look like and how various efforts have succeeded or fallen short. The data and examples provide a compelling case for an integrated approach—in which social, economic, and environmental considerations are all considered. A roadmap is sketched to highlight the policy and technology options that can make agriculture more productive both across sectors and over the long term. This text offers detailed ideas on assuring inclusive and effective governance at different levels, including suggestions for balancing trade-offs, monitoring progress, and building partnerships.

Promoting sustainable agriculture and healthy diets means taking action at each and every link in the food chain. The judicious use of technologies and innovative practices, supported by appropriate policies, institutions, and incentives can help the world meet its present needs without jeopardizing the opportunities of future generations. That is what "sustainable production and consumption" is all about.

Whether countries and communities across the globe opt for climate-smart agriculture, sustainable intensification, agroecology, biotechnologies, or any of the other approaches laid out here to make agriculture more sustainable—and to eradicate hunger and poverty—the FAO stands ready to provide expertise and support throughout the value chain. Agricultural production should go beyond increasing yields, providing new job opportunities for women, men, and youth in crop, livestock, forestry, fisheries, and aquaculture production as well as in research and innovation and after-harvest activities such as storing, processing, transporting, and marketing.

Many interrelated challenges are converging, and the choices we make at this important crossroads will determine whether we achieve a world free of poverty, hunger, disease, and want. When resource use is efficient, the environment is protected, rural livelihoods, equity, and social well-being are improved, resilience is enhanced, and responsible and effective governance mechanisms are in place in agriculture, we will have the change that is warranted, and people and the planet will benefit together.

José Graziano da Silva

Director-General

Food and Agriculture Organization of the United Nations

Preface

Currently, the world's agriculture and food systems face the future challenge of providing 50 percent more food than the total needed in 2013 to feed a population of 9.73 billion by 2050. More urbanized and affluent than ever before, that population will demand more livestock-based and processed foods and fruits and vegetables. However, even for current levels of population and consumption, our agriculture and food systems are proving inadequate. Despite systems producing enough food to meet the caloric requirements of the current population, 815 million people remain hungry, 2 billion suffer from micronutrient deficiencies, and 40% of the population above 18 years of age is overweight; the natural resources that underpin agriculture and food systems are under pressure (e.g., one-third of agricultural land is moderately to highly degraded, agriculture uses 70% of all water withdrawal and causes environmental pollution, and food production is responsible for 75% of the loss in agrobiodiversity); and the agricultural production systems, including deforestation, contribute to 29% of the global greenhouse gas emissions for climate change, which is already impacting the livelihoods of the most vulnerable, especially in rural areas.

So, how will global agriculture and food systems "meet the needs of the present without compromising the ability of future generations to meet their own needs" in 2050 and beyond? Over the years, some initiatives to enhance productivity and sustainability of different agricultural sectors have been launched (e.g., Save and Grow for the crop sector, Global Agenda for Sustainable Livestock, Sustainable Forest Management, the Code of Conduct for Responsible Fisheries, and Climate-Smart Agriculture, among others). The Food and Agriculture Organization of the United Nations recently launched a new vision for and approach to promoting sustainable food and agriculture that requires explicit consideration of cross-sectoral (e.g., crops, livestock, fisheries, aquaculture, and forestry) and multiobjective (e.g., economic, social, and environmental) policies and instruments, identifying possible synergies as well as balancing trade-offs between them. At the core of that approach are five principles: improved efficiency of the resources used in food and agriculture; direct action to conserve, protect, and enhance natural resources; protection and improvement of rural livelihoods, equity, and social well-being; enhanced resilience of people, communities, and ecosystems; and responsible and effective governance mechanisms.

In recent years, productivity and sustainability of agriculture and food systems have received increased attention at the global level. In 2015, world leaders adopted Sustainable Development Goals (SDGs) to be achieved by 2030, and 16 of the 17 SDGs have strong linkage with sustainable food and agriculture. Several other global initiatives (the Addis Ababa Action Agenda on financing for development, the 2015 Paris Agreement on coordinating action to address climate change, the Sendai Framework for Disaster Risk

Reduction, and others) also have strong connections to productivity and sustainability of agriculture and food systems.

While these developments at the global level are both important and welcome, policy coherence and coordinating efforts at the country level are more critical because sustainable food and agriculture must be achieved by the countries themselves. A holistic and integrated approach at the country level that aligns national development goals with global goals increases policy coherence and synergies across relevant ministries and agencies, enhances on- and off-farm employment and income opportunities, and raises investment in food and agriculture that would address many of the aforementioned challenges. Of particular use would be specific actions aimed at maximizing resource use efficiency, reducing food loss and waste, converting waste to wealth through reuse, and protecting, conserving, and sustainably using key natural resources that underpin the entire food system. However, no country has yet implemented such a system.

This book is the first effort to point the pathway toward more sustainable agriculture and food systems through an integrated and cross-sectoral approach. Addressing the present and future challenges of the agriculture and food systems requires better understanding of the positive and negative impacts of the agricultural sectors (crop, livestock, fisheries, aquaculture, and forestry) on the world's social, economic, and environmental themes. It also requires an understanding of the drivers of change, and linkages between the rural and urban economies. Many trade-offs emerge from production to consumption with impacts on efficiency, sustainability, and inclusiveness, and they must be balanced with policies that promote an integrated approach. Designing systems requires knowledge and experience with issues surrounding the current global food systems and the capacity to look at their future.

Consequently, this book is based on collective knowledge and experience of 78 highly knowledgeable and experienced scientists, teachers, policy experts, and leaders from 30 organizations, including universities and public, private, and international institutions/organizations around the world. The lead authors of the five sections (Rob Vos and Shenggen Fan for Section I; Jules Pretty and Zareen Bharucha for Section II; Henning Steinfeld and Tim Robinson for Section III; Leslie Lipper and Jeroen Dijkman for Section IV; and Clayton Campanhola, Kostas Stamoulis, and Shivaji Pandey for Section V) have used their years of work and experience with food systems around the world to refine the initially proposed outlines and identify additional contributors to their sections. Under such leadership and effort, section I aims at increasing understanding of factors, issues, and challenges that impact on the productivity and sustainability of the world's agriculture and food systems now and will continue to do so into the 21st century. Section II describes policies, technologies, approaches, and systems that have proven to be successful in the recent past or are being tested and promoted for addressing many of the challenges faced in different parts of the world. Section III considers the growing conflict between escalating human demand for food and other benefits, with resource degradation and scarcity, analyzes how different agriculture and food systems respond to this conflict, and suggests cross-cutting approaches for coping with their complexity and diversity. Section IV describes the importance of identifying and balancing trade-offs that occur because of tension between natural and social systems, as discussed in Section III, presents a general approach to implementing sustainable food and agriculture and provides examples of successful implementation of the approach at global and national levels. Finally, Section V focuses on research and innovation, policies and incentives, resource mobilization, and governance and institutions—the four areas considered most critical to the meaningful and needed structural transformations. The section ends with a set of recommendations that, if adapted and adopted at global, national, and local levels, would improve productivity and sustainability of agriculture and food systems.

This book is aimed at policymakers, agricultural research and extension professionals, development practitioners, and students and teachers of biological, social, and agricultural science. We hope it will inform, as well as stimulate and motivate them to consult the lead authors, contributing authors, and relevant references for additional information. Finally, it is our fervent expectation that this book will help the world transition toward more productive and sustainable agriculture and food systems for the benefit of present and future generations.

Clayton Campanhola and Shivaji Pandey

The Food and Agriculture Organization of the United Nations (FAO), Rome, Italy

Acknowledgments

We would like to express our sincere gratitude to all those who provided support, read, wrote, offered comments, and assisted in the editing, management, and proofreading of this book.

We would like to thank the FAO for enabling us to publish this book. Special thanks go to the Director-General of FAO, Prof. José Graziano da Silva, for his continuous encouragement and support. We also thank Deputy Directors General of FAO Daniel Gustafson, Maria Helena Semedo, and Laurent Thomas for their constant interest and support of this effort.

We are grateful to the lead authors of the sections: Rob Vos, Shenggen Fan, Jules Pretty, Zareen Bharucha, Henning Steinfeld, Tim Robinson, Leslie Lipper, Jeroen Dijkman, and Kostas Stamoulis for bringing all their scientific, technical, and practical expertise and experience, and for aggregating so many contributors from all over the world. We also thank 67 internationally renowned contributing authors from all over the world who contributed their own knowledge and experience to various chapters of this volume.

Thanks also go to Jana Stankova, Sonia Lombardo, Jessica Mathewson, Vanda Ferreira, and Maria Giannini for their restless support with secretarial and legal issues, permissions, edits, and for keeping our steps on the right track.

Last, but not least, we thank those who have contributed to this effort in so many ways but are not mentioned here by name.

Acronyms

AAC	African Agricultural Capital Limited
AFOLU	agriculture, forestry, and land use
AFS	agriculture and food systems
AIS	agricultural innovation system
AMR	antimicrobial resistance
AR4D	agricultural research for development
ARI	agricultural research intensity
CAAS	Chinese Academy of Agricultural Sciences
CGIAR	Consultative Group for International Agricultural Research
CFS	Committee on World Food Security
CRP	CGIAR Research Program
EAP	East Asia and Pacific
FAO	Food and Agriculture Organization of the United Nations
FDI	foreign direct investment
FFS	farmer field schools
GAFSP	Global Agriculture and Food Security Program
GDP	gross domestic product
GHG	greenhouse gases
GREL	Ghana Rubber Estates Limited
HANPP	human appropriation of net primary production
HICs	high-income countries
HLPE	high-level panel of experts
ICT	information and communication technologies
IFAD	International Fund for Agricultural Development
IFPRI	International Food Policy Research Institute
IPBES	intergovernmental science-policy platform on biodiversity and ecosystem services
IPM	integrated pest management
LIC	low-income country
LMICs	lower middle-income countries
NGO	nongovernmental organization
ODA	official development assistance
OECD	Organisation for Economic Co-operation and Development
PGRFA	plant genetic resources for food and agriculture
PPP	purchasing power parity
R&D	research and development
SDG	Sustainable Development Goal
SES	social-ecological systems
SFA	sustainable food and agriculture
SSA	sub-Saharan Africa
TFP	total factor productivity
UNFCC	United Nations Framework Convention on Climate Change
WFP	World Food Programme
WTO	World Trade Organization

SECTION I

FOOD AND AGRICULTURE AT A CROSSROADS

Lead Authors

Rob Vos and Shenggen Fan (International Food Policy Research Institute (IFPRI), Washington, DC, United States)

CHAPTER

1

Food and Agricultural Systems at a Crossroads: An Overview

1.1 THE STATE OF GLOBAL FOOD AND AGRICULTURE

Over the past century, enormous progress has been made in improving human welfare worldwide. Societies have changed radically thanks to quantum leaps in technology, rapid urbanization, and innovations in production systems. Yet, conditions today are a far cry from the world "free of fear and want" envisioned by the founders of the United Nations. Amid great plenty, billions of people still face pervasive poverty, gross inequalities, joblessness, environmental degradation, disease, and deprivation. Much of humanity's progress has come at a considerable cost to the environment. The impacts of climate change are already being felt and—if left unabated—will intensify considerably in the years ahead. While globally integrated production processes have brought many benefits, challenges in regulating those processes highlight the need to steer them toward more equitable and sustainable outcomes.

Such challenges raise concerns regarding the feasibility of achieving the sustainable development goal (SDG) of ending hunger and all forms of malnutrition while making agriculture and food systems sustainable (SDG2). Are today's food and agricultural systems capable of meeting the needs of a global population that is projected to reach almost 10 billion by mid-century? Can we achieve the required production increases, even if this implies adding pressure to already dwindling land and water resources, specifically within the context of climate change?

As these challenges are strongly interrelated, addressing them in order to achieve SDG2 and other related SDGs will require a systems approach to food and agriculture. While still critical, agricultural development alone will not be enough to secure adequate food availability and stave off hunger and famine. Food systems at large will need to be sustainable in order to address multiple development challenges.

Sustainable Food and Agriculture
DOI: https://doi.org/10.1016/B978-0-12-812134-4.00001-7

Section I of this volume contains eight chapters addressing key questions regarding the sustainability of food and agriculture systems across various dimensions. The assessments coincide in the view that current trends and policy efforts *will inadequately* address these challenges, seriously jeopardizing prospects of achieving SDG2. Significant, transformative changes in agriculture and food systems need to occur to achieve a world without hunger and malnutrition and to protect the natural resource base required for feeding present and future generations.

1.2 FOOD AND AGRICULTURE AT A CROSSROADS: CHALLENGES AND OPPORTUNITIES

Global Trends and Challenges to Food and Agriculture Into the 21st Century

Rob Vos and Lorenzo Giovanni Bellù review some of the key global trends and challenges facing agriculture and food systems through the 21st century (Chapter 2: Global Trends and Challenges to Food and Agriculture Into the 21st Century). They start by addressing the core question of whether today's agriculture and food systems are capable of meeting the needs of a global population that is projected to reach almost 10 billion by mid-century and that may peak at more than 11 billion by the end of the century. They project that global food demand will increase by 50% between 2012 and 2050. During the preceding four decades, food production more than tripled, to the extent that current systems are likely capable of producing enough food. Moving forward, the challenges will be both different and more complicated.

With accelerating urbanization and continued income growth, especially in emerging economies, dietary preferences are shifting rapidly toward increased demand for more resource-intensive food, such as animal-sourced foods, fruits and vegetables, and processed foods. Satisfying this rising and changing demand through the currently prevalent farming and food processing systems will likely put added pressure on already scarce land, soil, and water resources and further degrade the quality of these resources. Some regions, especially tropical zones, already suffer from the adverse impacts of climate change. If left unabated, climate change will significantly slow agricultural productivity growth in the coming decades. Changing dietary patterns and food systems is a double-edged sword in terms of nutritional outcomes. They have facilitated the intake of more diversified diets and improved the nutritional status of many. However, at the same time, the increased consumption of animal-sourced food and the often too salty and sugary processed foods has given rise to the spread of overweight and obesity, which in turn are associated with a rising prevalence of noncommunicable diseases. Additionally, the ease of access to low-nutrient processed foods has also led to a further spread of people suffering from micronutrient deficiencies. Consequently, as Vos and Bellù show, ending hunger and all forms of malnutrition by 2030 (and not even by 2050) will be nothing but an elusive target if current trends continue. Hence, they argue, transformative changes to agriculture and food systems are urgently needed to feed the world sustainably.

The Demographics of Rural Poverty and Sustainable Agriculture and Rural Transformations

Population, income, and urban growth have been key drivers underlying many of the changes in food and agricultural systems and will continue to pose challenges to the sustainability of these systems for decades to come. Chapter 3, Demographic Change, Agriculture, and Rural Poverty, by James Thurlow, Paul Dorosh, and Ben Davies dwells further on these drivers to spell out key challenges for employment and poverty reduction in those regions where much of the demographic dynamics will appear: South Asia and, in particular, sub-Saharan Africa. These regions have lagged in the structural transformation of their economies and as a result will feel the weight of demographic pressures threatening future economic and social progress. Structural transformation entails workers leaving less-productive agriculture and moving to more productive industries, often in urban centers. Population growth slows with development, leading to greater dependence on capital and technology rather than on labor. This was East Asia's successful pathway. Sub-Saharan Africa is also transforming, but far less than other regions and with its own distinctive features. Africa is urbanizing, but rapid population growth means that rural populations are still expanding. African workers are leaving agriculture, albeit at a slower pace than in East Asia, and they are finding work in less-productive services rather than in manufacturing. The authors argue that this "urbanization without industrialization" raises concerns about Africa's ability to create enough jobs for its urban workforce and underscores the need for continued focus on rural Africa. The chapter reviews the linkages between urbanization, agriculture, and rural poverty in sub-Saharan Africa, where most of the world's poor will soon reside. It suggests that much of the economic growth and structural change that Africa enjoyed over the past two decades, while involving a shift out of agriculture, was in fact an expansion of downstream components of the agriculture food system. Like agriculture, many downstream activities have strong linkages to poverty reduction. Governments concerned about jobs and poverty will need to raise productivity not only in agriculture, but—as also assessed in Chapter 9, Changing Food Systems: Implications for Food Security and Nutrition—also throughout the food system. Since many downstream processing and trading activities take place in towns and cities, promoting future poverty reduction will require greater alignment between agricultural and urban policies. Demographic change and rural-urban linkages will continue to be powerful drivers of global poverty reduction, but ensuring inclusive transformation will require broader development perspectives and policy coordination.

Climate Change, Agriculture, and Food Security: Impacts and the Potential for Adaptation and Mitigation

Climate change is a significant and growing threat to food security—already affecting vulnerable populations in many developing countries and expected to affect an ever-increasing number of people in more areas in the future unless decisive actions are taken at once. Chapter 4, Climate Change, Agriculture and Food Security: Impacts and the Potential for Adaptation and Mitigation, by Keith Wiebe, Sherman Robinson, and Andrea Cattaneo reviews research on climate change and its impacts on agriculture and food security at global, regional, and national scales. They summarize the International Food Policy Research Institute's latest long-term projections of the impacts of climate change on agricultural area, yields, production, consumption, prices, and trade for major crop and livestock commodities, as well as their implications for food security.

A wide range of available sustainable intensification technologies and practices can help farmers both adapt to and mitigate climate change impacts. Such technologies can also help reduce food insecurity. The model-based scenario analyses presented in the chapter suggest that under a wetter and hotter climate scenario the number of food-insecure people in developing countries could be reduced in 2050 by 12% (or almost 124 million people) if nitrogen-efficient crop varieties were widely in use; by 9% (91 million people) if no-till farming were more widely adopted; and by 8% (80 million people) if heat-tolerant crop varieties or precision agriculture were adopted.

While such innovations show considerable potential, realizing these gains requires not only increased investment in research, but also measures to overcome barriers to more widespread adoption of innovations in technologies and management practices. Wiebe, Robinson, and Cattaneo argue that the impacts of climate change on agricultural producers and consumers, and the ability to adapt and mitigate those impacts, depend critically on socioeconomic factors and conditions as much as on biophysical processes. The challenges of climate change and food security are complicated by the extent to which they emerge from the individual and collective actions of some 570 million farm households and over 7 billion consumers worldwide. The global public good nature of climate change means that mitigation requires coordinated collective action in order to succeed. Meeting the challenges of climate change adaptation and mitigation equitably and effectively thus requires well-informed, evidence-based global and local policy action to ensure the appropriate enabling environments.

Water Scarcity and Challenges to Food Security

Water availability for agriculture will become a growing constraint in areas that use a high proportion of their water resources, exposing systems to high environmental and social stress, and limiting the potential for expanding irrigated areas. In fact, the rate of expansion of land under irrigation is already slowing substantially. Meredith Giordano, Jennie Barron, and Olcay Ünver assess in Chapter 5, Water scarcity and Challenges for Smallholder Agriculture, the key global challenges to adequacy of water availability and how increasing scarcity and competition for water resources are affecting agricultural productivity, especially that of smallholder producers in Africa and Asia. They also provide evidence on the viability of alternative, improved practices of sustainable water management adapted to the needs of smallholder farmers.

Severe water scarcity is a main challenge for the many smallholders in Africa and Asia. This challenge is rooted in physical limitations to available water resources and institutional obstacles preventing smallholders from accessing available supplies. Promising agricultural water management investment options exist to address the water scarcity challenge, and the authors provide evidence of four promising opportunities tailored to different water scarcity contexts. Two of the business models focus on improving access to water resources, as well as to agricultural water management technologies, by enhancing the number and quality of irrigation service providers and increasing investments in water storage. The other two models look beyond the water sector itself to identify smart solar solutions that will improve access to water in energy-poor environments, such as in rural Ethiopia, and support the sustainable use of sparse groundwater resources, such as in western India, through incentives to the energy sector.

Forests, Land-Use Change, and Challenges to Climate Stability and Food Security

Agriculture is a significant contributor to deforestation, biodiversity loss, and greenhouse gas emissions. However, agricultural productivity is most likely to suffer adverse consequences of climate change. In Chapter 6, Forests, Land Use, and Challenges to Climate Stability and Food Security, Terry Sunderland and Dominic Rowland examine the tensions between agricultural development, food security, and forest preservation. They distinguish three roles and functions of forests. The first is the provisioning function, as a direct source of food and income and as a means for agroforestry. The second is protective, in the sense of ecosystem service provision. The third is the ecosystem restorative capacity of forests, which can be leveraged through increasing the availability of trees and forests in agricultural landscapes. Such contributions include climate change mitigation via sequestration and the restoration of degraded agricultural land. Together with these three functions, the authors raise three key questions to be addressed if trade-offs between agricultural growth, food security, and forest preservation are to be overcome: How do we increase food production on existing agricultural land while reducing environmental degradation? How do we reduce the environmental degradation and loss of ecosystem services as well as important sources of wild food and income resulting from agricultural expansion into natural habitats? How do we restore degraded, unproductive, and abandoned agricultural land and natural habitats?

The authors argue that forests and trees are an essential part of the solution to all three questions. Trees in agricultural landscapes can simultaneously increase agricultural productivity and mitigate against environmental degradation. Judicious landscape-scale land-use planning that incorporates trees and forests into productive landscapes can simultaneously conserve forests and protect the ecosystem services upon which agricultural production depends. At the same time, reforestation and regeneration of forests can restore degraded land and provide new productive landscapes on abandoned or degraded land.

Land and Water Governance, Poverty, and Sustainability

Water, land, and soils are essential resources that are fundamental to food security and for alleviating hunger and poverty. How these resources are managed are critical to the sustainability of agriculture and the natural resource base. However, global population growth, changing diets, and a changing climate have together created concerns that demand may soon outstrip available resources if we continue to use them with the current intensity and production practices. Technological innovation can help to increase production and reduce demand, but most analysts now accept that the greatest benefits will come from improving the way we use and manage existing resources. Most population growth is expected to occur in developing countries where resource governance tends to be weak and fragmented, and where people are least able to cope with the challenges accompanying sustainable usage and conservation of these resources. Chapter 7, Land and Water Governance, Poverty, and Sustainability, by Olcay Ünver and Eduardo Mansur analyzes current constraints in land use and soil quality. The authors show how the problems of land tenure and the failure to integrally manage water, land, and soil resources are conspiring against the sustainable use of those resources. The chapter proposes a fundamental reform of governance mechanisms oriented at greater coordination across those dealing with land, water, and soils.

Biodiversity and Ecosystem Services

Severe biodiversity loss is caused by ongoing deforestation, simplification of agricultural landscapes, intensive use of natural resources, and impacts of climate change. Plant genetic resources for food and agriculture form the basis of food security and consist of diversity of seeds and planting material of traditional varieties and modern cultivars, crop wild relatives, and other wild plant species. Currently, only about 30 crops provide 95% of human food energy needs, four of which—rice, wheat, maize, and potato—are responsible for more than 60% of our energy intake. Because of the dependence on a small number of crops for global food security, it will be crucial to maintain a high genetic diversity within these crops to deal with increasing environmental stress, and to provide farmers and researchers with opportunities to breed crops that can be cultivated under unfavorable conditions such as drought, salinity, flooding, poor soils, and extreme temperatures.

The loss of biodiversity in beneficial organisms in agricultural landscapes, such as natural predators of crop pests, pollinators, and soil microbes, due to habitat loss and toxic chemical pesticide use threatens the ability of ecosystems ability to maintain ecological functions and equilibrium among natural flora and fauna that underline the biophysical capacity of agricultural ecosystems. The conservation and sustainable use of plant genetic resources for food and agriculture and landscape biodiversity is necessary to ensure crop production and to meet the growing environmental challenges and impacts of climate change. The loss of these resources or a lack of adequate linkages between conservation and their use poses a severe threat to the world's food security in the long term.

In Chapter 8, Biodiversity and Ecosystem Services, Wei Zhang, Ehsan Dulloo, Gina Kennedy, Arwen Bailey, Harpinder Sandhu, and Ephraim Nkonya highlight the linkages between plant genetic resources for food and agriculture, dietary diversity, and natural habitats, as well as the roles of gender and community in conservation. Diversely structured landscapes play a crucial role in supporting environmental services that maintain the productivity and stability of agroecosystems. To ensure that the global food system remains environmentally sustainable and generates a rich array of nutrients for human health, farm landscapes must be diverse and serve multiple purposes. The authors emphasize the need for economic valuation of environmental services in both monetary and nonmonetary terms to assess the global cost of land degradation and loss of biodiversity and the benefits for food production systems to restore soil quality and biodiversity.

Changing Food Systems: Implications for Food Security and Nutrition

Food systems are changing, with growing reliance in many regions on global supply chains and large-scale distribution systems that both meet and fuel the changes in food demand and dietary preferences. While improving efficiency, the changing nature of food systems is also creating new challenges and concerns regarding the high calorie and low nutritional content of many food items, access of small-scale producers to viable markets, high levels of food loss and waste, incidences of food safety, animal and human health issues, and the increasing energy intensity and ecological footprint associated with the lengthening of food chains. Chapter 9, Changing Food Systems: Implications for Food Security and Nutrition, by Hanh Nguyen, Jamie Morrison, and David Neven argues that, in order to properly understand the implications of these challenges for future food security and nutrition, actions across a multitude of actors (from producers to consumers) will need to be coordinated from a food-systemwide perspective.

The authors note the accelerated change toward modern, industrial food systems in developing countries, which is being driven by rapid urban growth and strong economic expansions in recent decades. This has given rise to the notion of a growing "global middle class," whose size is projected to increase almost threefold between 2009 and 2030. With greater disposable income and increased exposure to imported foods and large-scale retailers, this global middle class is adopting radically different lifestyles and acquiring new food preferences. At the same time, however, although overall poverty has declined, the percentage of poor people living in urban areas is growing. While the food demand of the middle class in many developing country regions is increasingly being met through global supply chains and large-scale distribution systems, the urban poor still rely significantly on informal traditional markets as their primary food supply channel.

Rapid technological innovation, especially in information and communication technologies and renewable energy, has been another major driving force of food system changes. The judicious use of these technologies, both on-farm and off-farm, has shaped and will continue to shape the productivity and competitiveness of food systems. Finally, climate change and resource scarcities pose significant risks and can threaten the productive capacity and stability of food systems. In turn, food systems have contributed significantly to global greenhouse gas emissions (almost 30%) as well as to the degradation of natural resources. The authors further argue that the development from traditional to modern food systems has had mixed outcomes in terms of food security and nutrition, rural employment, and poverty reduction. On the one hand, modern food systems can enhance farmers' access to viable markets. Farmers who enter into contracts with companies enjoy formal employment opportunities, often alongside technical assistance to improve the efficiency of the production process and the quality of their produce. On the other hand, the concentration of market power to a few large-scale processors, wholesalers, and retailers who constantly seek to reduce production and transaction costs exerts huge pressure on small suppliers. Such development can work to the disadvantage of smallholders, who find it difficult to meet the requirements of large buyers for product uniformity, consistency, and regular supply.

The dramatic pace of food system changes over the past decades has brought about complex interactions with, as mentioned, mixed outcomes for food security and the sustainability of agriculture and food systems. Many trade-offs have emerged between food system efficiency, sustainability, and inclusiveness. The authors argue that direct interventions to address such trade-offs in one area of the system may risk exacerbating problems in another. Therefore, they argue that agriculture and food policies should take a holistic system-wide approach to be effective.

1.3 GLOBAL CHALLENGES, GLOBAL RESPONSES

The international community has recognized these challenges along with the interdependencies between them. The 2030 Agenda for Sustainable Development (2030 Agenda) provides a compelling and ambitious vision for transformative change to put economies, and agriculture and food systems with it, on a sustainable footing. SDG2 explicitly aims at ending hunger, achieving food security and improving nutrition, and promoting sustainable agriculture simultaneously by 2030. The 2030 Agenda and the Addis Ababa Action Agenda on financing for development specifically call on all countries to pursue policy coherence and establish enabling environments for sustainable development at all levels and by all actors (SDG17). The 2015 Paris Agreement on climate change reflects a global commitment for concerted actions to address the perils of climate change. The Sendai Framework for Disaster Risk Reduction also gives priority to agriculture sectors.

Despite these promising international frameworks for action, achieving policy coherence will be challenging. The 2030 Agenda and other related global agreements stress the interdependence of the challenges they must address. They also recognize the need to integrate different actions to achieve linked objectives and that doing so will pose new technical demands on policymakers, at all levels, as well as new demands on institutional arrangements and coordination at various levels of governance.

To meet these demands, more and better data and research will be needed. For instance, existing evidence on the degree and ways in which climate change is affecting agriculture, food, and nutrient availability in different settings is still surrounded by a fair amount of uncertainty and there are still many unknowns. Likewise, policies would need to take new, as yet, untested directions requiring new data collection and impact assessments to monitor their effectiveness and to provide the necessary accountability. Moreover, research will be critical to understand how food systems can address emerging social issues, including various forms of inequality, youth unemployment, migration, and conflict. Additionally, it is critical to better understand how the accelerated pace of technological innovation in information and communication technologies and biotechnologies, for example, can be made to create innovations to help make agriculture and food systems sustainable.

The purpose of Section I of this volume is not to present a menu of solutions, but rather to increase understanding of the nature of the challenges that agriculture and food systems now face and will be facing throughout the 21st century.

CHAPTER

2

Global Trends and Challenges to Food and Agriculture into the 21st Century

Rob Vos[1] *and Lorenzo Giovanni Bellù*[2]

[1]International Food Policy Research Institute (IFPRI), Washington, DC, United States

[2]Food and Agriculture Organization of the United Nations, Rome, Italy

2.1 INTRODUCTION

This chapter lays out key global trends (demographic pressures, income growth, and changing dietary patterns; natural resource constraints and climate change, technological change and productivity growth, and changing food systems) and assesses their (expected) implications to the challenges posed for sustainably provisioning both food security and nutrition.

A still-expanding world population, accelerated urbanization, climate change, and increasing scarcity of land, water, and forest resources have given rise to several fundamental questions. Are today's food and agricultural systems capable of meeting the needs of a global population that is projected to reach near 10 billion by midcentury? Can we achieve the required production increases, even if this implies increasing pressure on already dwindling land and water resources and doing so in the context of climate change?

This chapter builds upon several parts of the report: FAO (2017). *The Future of Food and Agriculture – Trends and Challenges*" Food and Agriculture Organization of the United Nations. (Available at: http://www.fao.org/3/a-i6583e.pdf.) The present authors were lead authors of the this report.

Sustainable Food and Agriculture
DOI: https://doi.org/10.1016/B978-0-12-812134-4.00002-9

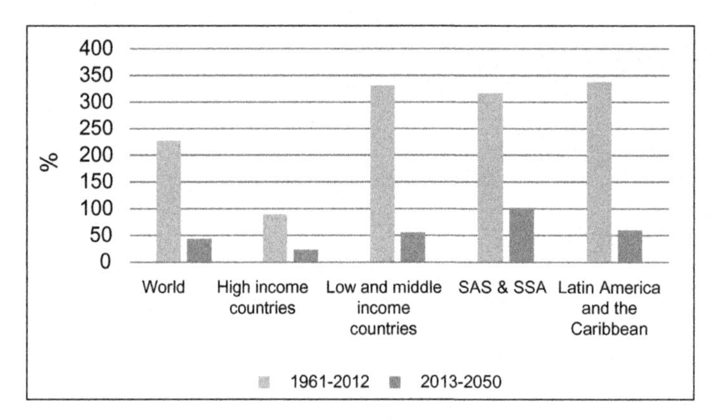

FIGURE 2.1 Agricultural output growth and demand projections, 1961–2050. Source: *Based on FAO, 2017. The Future of Food and Agriculture – Trends and Challenges. Food and Agriculture Organization of the United Nations. Available at: http://www.fao.org/3/a-i6583e.pdf: Table 5.1).*

2.2 KEY TRENDS AND CHALLENGES

Agricultural Demand is Expected to Increase Significantly

Demand for food and agricultural products is expected to rise by 50% between 2013 and 2050 (Fig. 2.1). While the expected increase is much less than in the preceding four decades (1961/63–2005/07) when global agricultural production almost tripled, it remains significant, given the much higher current volume of food demand and the much larger world population needing to be fed (Alexandratos and Bruinsma, 2012).

The rise in demand is mainly pushed by the growth of the world population, which is projected to almost 10 billion by midcentury, as well as by income growth and urbanization. More than half of the 2.3 billion people who will be added to the world population will be born in sub-Saharan Africa. Consequently, food supplies in this region would need to more than double from the present levels by 2050. The projected increase in demand assumes moderate growth of real per capita income of 1.3% per year on average for the world during the period 2005-2050, implying that average income would almost double in the coming decades to 2050: from US$7606 (in constant prices of 2005) in 2005/07 to US $13,750 in 2050. While significant, the projected income growth would be below not only historical trends, but also that assumed in other long-term global economic scenario analyses.[1] Hence, the projected agricultural demand increase should be considered moderate. All global

[1] Reference is made here to the "Shared Socioeconomic Pathways (SSPs)," which have been developed to describe alternative development futures alongside different climate change scenarios to analyze feedback between climate change and socioeconomic factors (see, e.g., O'Neill et al., 2017). As reported by the International Institute for Applied Systems Analysis (IIASA) SSP database, the second shared socioeconomic pathway (SSP2) or "middle-of-the-road" scenario, for instance, projects an increase of per capita gross domestic product (GDP) of 132% by 2050 as compared with 2005–7, implying almost 2% growth per year (similar observed global average per capita GDP growth of 1.9% per year during 1960–2005). Out of five SSPs used in the long-term forward-looking exercises for climate change analysis, only one projects lower per capita GDP than the one used here (https://secure.iiasa.ac.at/web-apps/ene/SspDb/dsd?Action = htmlpage&page = about).

projections concur, however, in projecting faster income growth for low- and middle-income countries than for high-income countries.

Higher incomes and urban lifestyles are also changing food demand toward more consumption of animal proteins and fruits and vegetables, the production of which is more resource intensive than grains. Diets change slowly over time. Over the past 50 years, the share of cereals in total apparent food-energy intake (expressed in kilo calories, kcal) in low- and middle-income countries declined from 57% to 50% (see Fig. 2.2A). The shares of animal products and fruits and vegetables, in contrast, increased from 8% to 13% and from 4% to 7%, respectively. These shifts in dietary patterns constitute, as yet, only a very mild convergence with food preferences and levels of energy and protein intake of high-income countries (Fig. 2.2A–C).

Additional demand pressure for agricultural produce is expected to be exerted by increased demand for bioenergy, both in the form of wood-based products (traditional firewood, but also and increasingly, wood pellets) and biofuels. This will depend largely on growth in biofuel demand. The estimates underlying Fig. 2.1 assume this demand will continue to grow at the current pace of 2.6% per year on average until 2050.

To Meet the Demand, Output Will Have to Expand, but Under Increasingly Tight Production Constraints

By historical standards, meeting the additional demand would not represent a challenge. Agricultural production more than tripled between 1960 and 2012 (see Fig. 2.1), due, in part, to productivity-enhancing green revolution technologies and significant expansion in the use of land, water, and other natural resources for agricultural purposes. Conditions, however, have changed and future agricultural growth will have to come from different sources.

In most regions, further expansion of arable land is limited. In the Middle East and Northern Africa and parts of Central Asia and sub-Saharan Africa, potential land expansion is constrained by water scarcity. In other parts of sub-Saharan Africa and Latin America, most of the still available land lies in remote areas, where the lack of infrastructure prevents its use for agricultural purposes, at least at current agricultural price levels. In all regions, agricultural land expansion could lead to further deforestation, which would be undesirable from the perspective of sustainability, inter alia, because of the impact on greenhouse gas emissions and biodiversity loss. Climate change may pose further limits to agricultural land expansion, as reduced or more variable rainfall and rising sea levels, may make agriculture less viable in some areas. Crop intensification can be an alternative to land expansion. However, the scope for pursuing this option while ensuring durable soil quality is relatively limited given the current state of technology (Alexandratos and Bruinsma, 2012). Under a "business-as-usual" scenario and given the already existing relative scarcity of natural resources (land and water, in particular), agricultural growth to be achieved by 2050 would mainly have to come from yield increases (Fig. 2.3).

Major resource-use efficiency improvements and conservation gains will have to be achieved globally, not only to meet food demand, but importantly, also to halt and reverse ecological degradation. This will be challenging for at least two key reasons.

First, production and productivity growth will be hampered by growing scarcity and competition for land and water resources. Projections for 2050 confirm the likelihood of

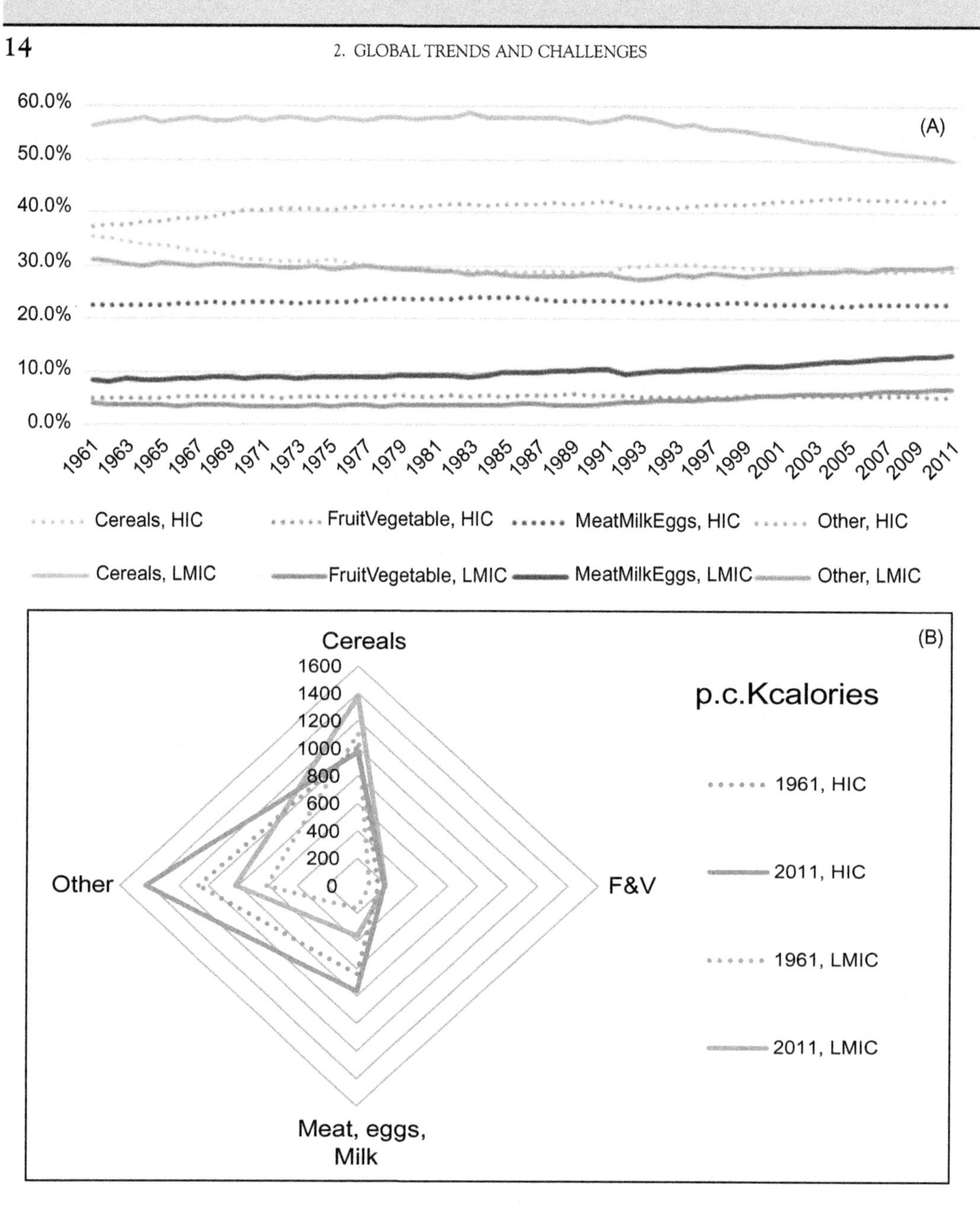

FIGURE 2.2 Dietary transition across country groups by income level. (A) Percentage share of apparent per capita food energy consumption (in kCal.); (B) Apparent food energy intake (in kCal) by main type of food, 1961 and 2011; (C) Apparent protein intake (in grams) by main type of food, 1961 and 2011. Note: *HIC*, high-income countries; *LMIC*, low- and middle-income countries; *F&V*, fruits and vegetables. Source: *Estimates based on FAOSTAT, Food balance sheets. http://www.fao.org/faostat/en/#data/FBS (accessed 11.16).*

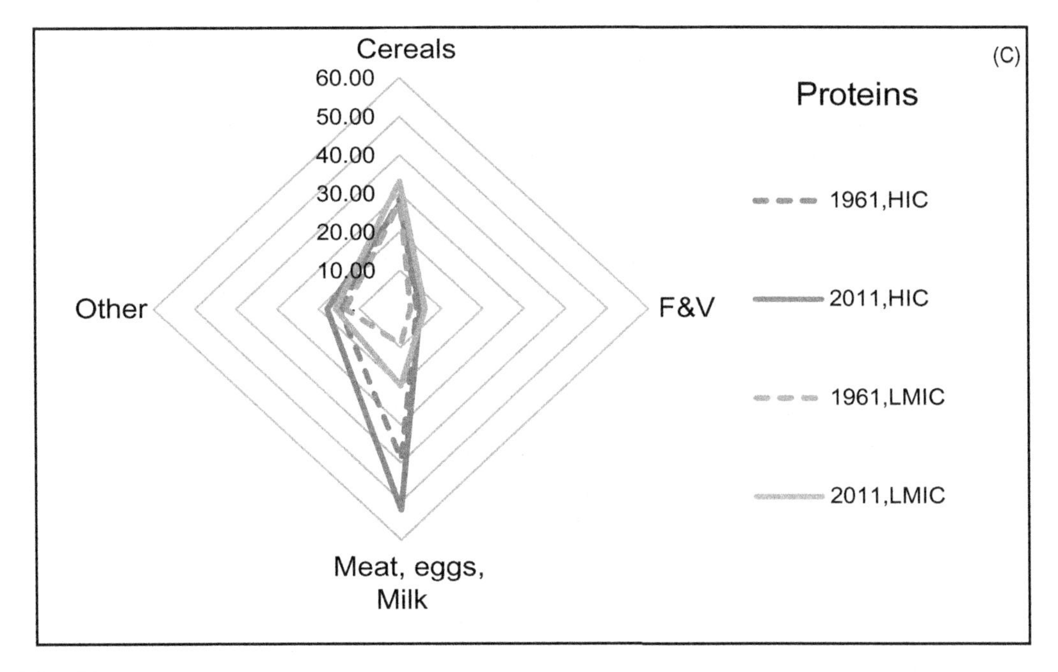

FIGURE 2.2 (Continued).

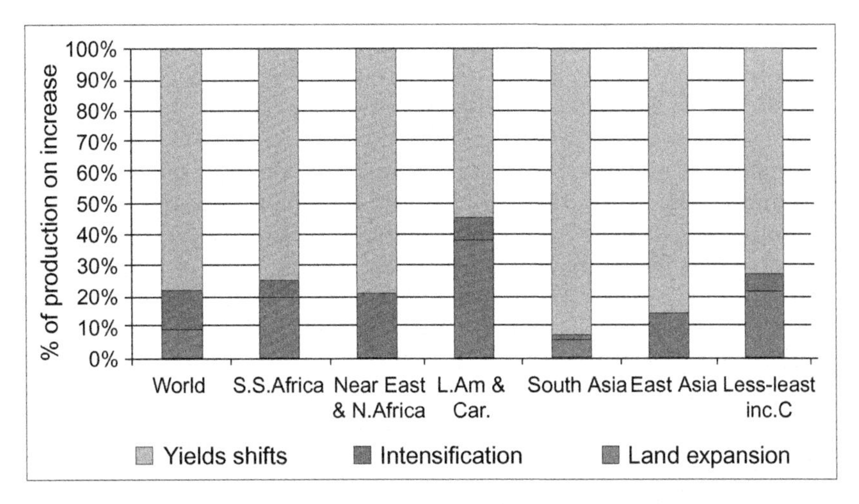

FIGURE 2.3 Future sources of agricultural output growth under a business-as-usual scenario, 2012–50. Source: *Estimates based on Alexandratos, N., and Bruinsma, J, 2012. World Agriculture Towards 2030/2050: The 2012 Revision. ESA Working Paper No. 12–03. Rome, FAO.*

growing scarcity of agricultural land, water, forest, marine capture fisheries, and biodiversity resources. Additional land requirements for agricultural production between now and 2050 are estimated at just under 0.1 billion hectares (Fig. 2.4). It is expected that demand for such land use will decrease in high-income countries, but increase in low-income

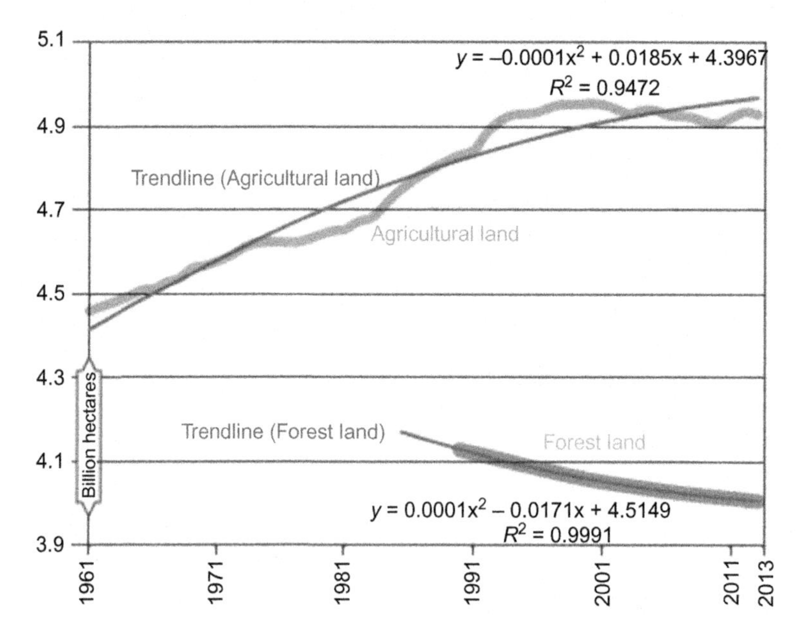

FIGURE 2.4 Changes in agricultural and forest land use, 1961–2013. *Source: FAO, 2017. The Future of Food and Agriculture – Trends and Challenges. Food and Agriculture Organization of the United Nations. Available at: http://www.fao.org/3/a-i6583e.pdf, p. 33.*

countries. This modest increase could suggest that land availability is not a constraint. In fact, the projection of increased land use for agriculture relies on the notion that most still spare land is not readily accessible, mainly because of a lack of infrastructure, physical remoteness and disconnection from markets, and/or located in disease-prone areas. Furthermore, available spare land is concentrated in a few countries only. The land availability constraint underlies the notion that agricultural production increases to meet rising food demand will mostly have to come from productivity and resource-efficiency improvements.

Increased competition for land has already emerged as a result of increased demand for bioenergy; this shift to bioenergy has severe implications for agriculture and food production. For example, in aquaculture, which provides more than 50% of all fish consumed, oilseeds are becoming a major component of fish feed, and demand for oilseeds will expand as aquaculture production methods continue to intensify. Around two-thirds of the bioenergy used worldwide involves the traditional burning of wood and other biomass for cooking and heating in low-income countries. As populations expand in these countries, increased use of such bioenergy sources will also occur. Much of this traditional wood energy is unsustainably produced and inefficiently burned, affecting the health of poor populations, and contributing to environmental degradation. At the global level, the use of woodfuel is not seen as a major contributor to deforestation and forest degradation, but in areas close to urban centers the demand for wood and charcoal for domestic needs is a serious environmental concern (FAO, 2011a).

Greater competition between food and nonfood uses of biomass has increased the interdependence between food, feed, and energy markets. This competition may risk having harmful impacts on local food security and access to land resources. Input subsidies on energy, fertilizers, and water, as well as public purchases of agricultural produce, may add unintended additional pressure on natural resources.

Water availability for agriculture will also become a growing constraint, particularly in areas that use a high proportion of their water resources, exposing systems to high environmental and social stress and limiting the potential for expanding irrigated areas. Countries are considered water-stressed if they withdraw more than 25% of their renewable freshwater resources. They approach physical water scarcity when more than 60% is withdrawn, and face severe physical water scarcity when more than 75% is withdrawn (FAO, 2016a). Water withdrawals for agriculture represent 70% of all withdrawals. The Food and Agriculture Organization of the United Nations (FAO) estimates that more than 40% of the world's rural population lives in water-scarce river basins (FAO, 2011b). In many low rainfall areas of the Middle East, North Africa and Central Asia, and in India and China, farmers use much of the available water resources, resulting in serious depletion of rivers and aquifers (Fig. 2.5). In some of these areas, about 80%–90% of the water is used for agricultural purposes. The intensive agricultural economies of Asia use about 20% of their internal renewable freshwater resources, while, in contrast, much of Latin America and sub-Saharan Africa use only a very small portion.

Given these constraints, the rate of expansion of land under irrigation is slowing substantially. The FAO has projected that the global area equipped for irrigation may increase at a relatively low rate of 0.1% annually. At that rate, it would reach 337 million hectares

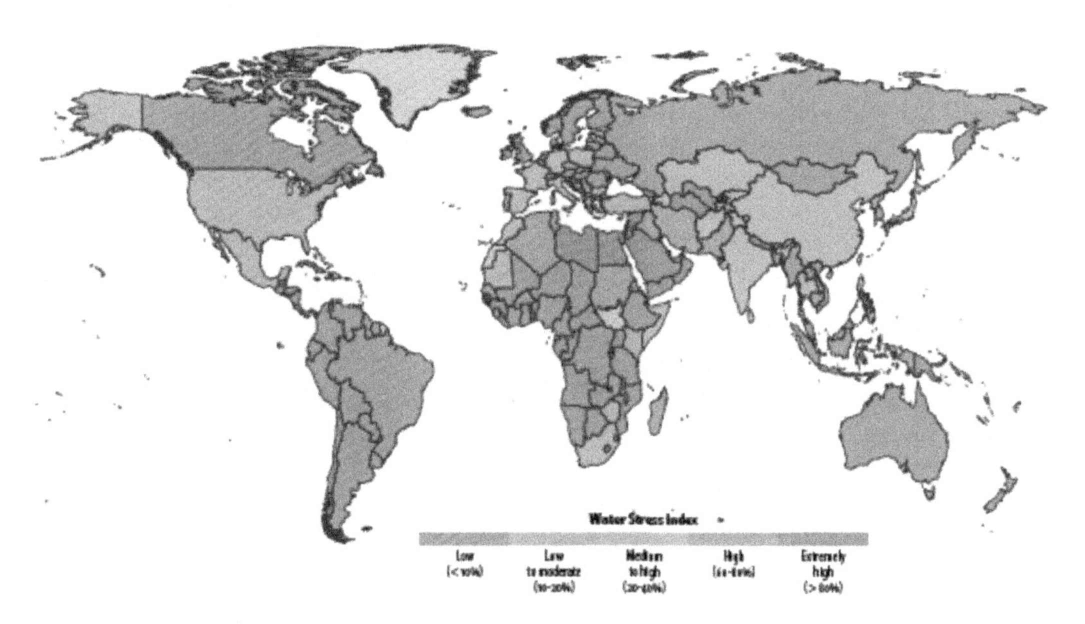

FIGURE 2.5 Freshwater withdrawals as a percentage of total renewable resources.[a] Note: a. Water withdrawals are for 2010, divided by mean available blue water, 1950–2008. Source: *FAO, 2017. The Future of Food and Agriculture – Trends and Challenges. Food and Agriculture Organization of the United Nations. Available at: http://www.fao.org/3/a-i6583e.pdf, based on FAO AQUASTAT database.*

in 2050, compared with around 317 million hectares in 2009 (FAO, 2017, p. 37). This estimation represents a significant slowdown from the period between 1961 and 2009, when the area under irrigation grew at an annual rate of 1.6% globally and more than 2% in the poorest countries. Most of the future expansion of irrigated land is projected to take place in low-income countries. Future water stress will not only be driven by changes in demand, but also by changes in the availability of water resources, arising from changes in precipitation and temperature that are driven by climate change.

Construction of dams, which interfere with fish migration, is also expected to have a negative impact on inland fisheries. While allocations of water are shifting away from agriculture to meet the needs of urban users, there is scope to exploit noncompetitive uses of water resources, such as using treated urban wastewater for irrigating crops or increasing the efficiency of water resources for inland aquaculture.

In sum, increased competition for natural resources exacerbates pressure on, and hence the degradation of, resources and ecosystems. Degradation and abandonment of natural resources can lead to increased competition over not-yet degraded natural resources and to expansion of activities into fragile and degraded areas, which then become further threatened. Reasons for diminishing resources for food and agriculture include the depletion and degradation of soil and water resources, and the loss of biodiversity and productive land for other uses. This trend is expected to continue.

Second, investments in improved technologies have been lackluster in recent decades, which has slowed agricultural yield growth to levels insufficient to meet the increases in demand. While some technological progress has been achieved, yield increases experienced in previous decades are slowing down, with increasingly evident negative side effects of high chemical inputs in crop production, thus posing serious sustainability concerns.

Yields for major crops vary substantially across regions. Estimated yield gaps—expressed as a percentage of potential yields—exceed 50% in most low-income regions and are largest in sub-Saharan Africa, at 76%, and lowest in East Asia, at 11%. Crops included in these estimates are cereals, roots and tubers, pulses, sugar crops, oil crops, and vegetables. The gap between actual farm yields and potential yields reflects the largely suboptimal use of inputs and insufficient adoption of the most productive technology. This may mean that farmers lack economic incentives to adopt more productive technologies given the constraints many farmers—especially smallholders—face to make the needed investments, such as lack of access to credit and natural resources (e.g., sufficient water or fertile land) and market constraints.

The yield gaps can be closed by increasing the quantity of inputs per unit of output, but also by improving overall efficiency in production. Total factor productivity (TFP) growth is a comprehensive measure of efficiency improvements. It expresses how much more can be produced with the same amount of inputs and factors of production. TFP growth has been the main contributor to aggregate agricultural output growth. In low-income countries, TFP growth has been slow in past decades, reflecting that most of the increase in production has been achieved by the expansion of agricultural areas (FAO, 2017, p. 50). Since 2000, however, efficiency gains have become more significant in low-income countries. From the point of view of sustainable agriculture, it is essential to use land, labor, and inputs more efficiently. This will require substantial technological progress, adoption of innovative practices, and human capital development.

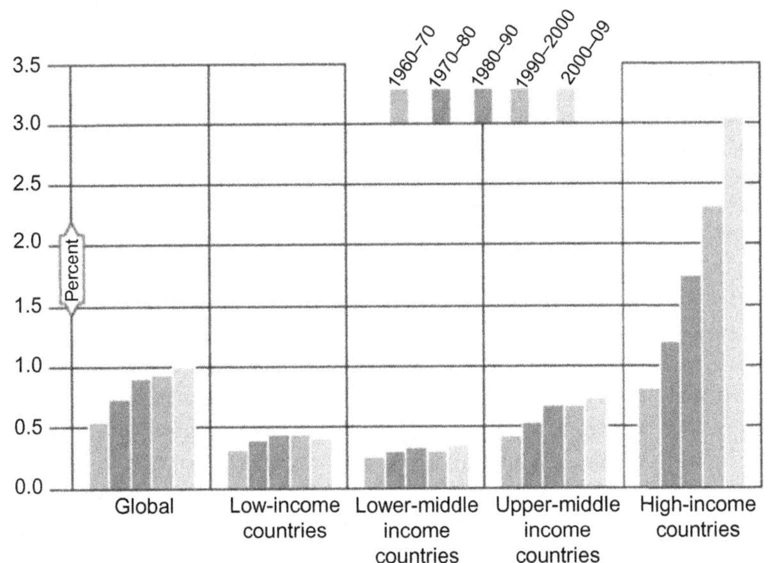

FIGURE 2.6 Averages of agricultural research intensity (ARI)[a] by country income group. Note: a. Simple average of annual agricultural research intensity (ARI), measured as the ratio of public expenditure on agricultural R&D to agricultural GDP. Source: *FAO, 2017. The Future of Food and Agriculture – Trends and Challenges. Food and Agriculture Organization of the United Nations. Available at: http://www.fao.org/3/a-i6583e.pdf, based on Pardey, P., Chan-Kang, C., and Dehmer, S., 2014. Global Food and Agricultural R, and D Spending, 1960–2009. InSTePP Report. Saint Paul, USA, University of Minnesota.*

Unfortunately, current levels of research and development (R&D) expenditures remain discomforting, especially for agricultural development in low-income countries. A commonly used indicator to assess countries' agricultural research efforts is the agricultural research intensity, which expresses national expenditure on public agricultural R&D as a share of agricultural GDP. Clearly, low-income countries are lagging behind high-income countries and are increasingly losing ground (Fig. 2.6). Increasing involvement of the private sector and use of proprietary technologies due to continued widespread poverty and climate change reinforce the importance of regulation and the strengthening of public good providers such as the Consultative Group for International Agricultural Research (CGIAR) and regional and national systems. Challenges remain, however, as the private sector mostly concentrates on fully developed commercial agriculture. Moreover, adapting new technologies, such as biotechnologies, emerging nanotechnologies to local conditions, and fully exploiting the potential of information technologies in rural areas, may be complicated by a weak regulatory system, lack of credit and insurance, and restrictions emanating from intellectual property rights. Lack of adequate extension services and insufficient attention to farmer-led research and other learning-based approaches form additional hurdles for the adoption and local adaptation of new technologies, especially among smallholder farmers.

Climate Change Adds Another Major Challenge

Climate change, along with natural and human-induced disasters, poses multiple concerns, such as damages and losses, environmental degradation of land, forests, fish stock, and other natural resources, and declining productivity growth rates; these factors all add pressure to already fragile agricultural livelihoods, food, and ecological systems.

Maintaining the capacity of the planet's natural-resource base to feed the growing world population while reducing agriculture's ecological and climate footprint is key to ensuring the welfare of current and future generations.

Climate change affects yields. It is likely that, until 2030, adverse impacts of climate trends will only modestly outweigh positive ones. Benefits derived from increased plant growth under warmer temperatures will mainly occur in temperate zones of higher latitudes, while adverse impacts will be concentrated in tropical zones at lower latitudes. Beyond 2030, adverse impacts will intensify over time with significant losses of yields in many parts of the world, no longer compensated by positive yield changes occurring in other parts (FAO, 2016b). Extreme events such as droughts and floods will intensify and become more frequent with climate change.

The Fifth Assessment Report of the Intergovernmental Panel on Climate Change (IPCC) surveyed a large number of studies projecting the impacts on crop yields at different points in time and for different geographic locations (IPCC, 2014). According to the survey, in the medium term, up to around 2030, positive and negative projections of the impacts on crop yields seem to counterbalance each other at the global level; however, after that, the balance becomes increasingly negative. Low-income countries seem to be particularly at risk of declining yields as a result of climate change. Indeed, for those countries, most projections for crop yield impacts are negative, and both the share and severity of negative outcomes increase further in the future (Fig. 2.7A). In comparison, projections for high-income countries show a much larger share of potential positive changes (Fig. 2.7B).

Climate change is also intensifying the prevalence of natural disasters. Over the past 30 years, there has been a rising trend in the occurrence of climate-related disasters worldwide, such as droughts, floods, and extreme temperatures with consequent economic damage (Fig. 2.8). Increase in weather-related events is of significant concern to crops, livestock, fisheries, and forestry given their dependence on climatic conditions. Between 2003 and 2013, agriculture sectors accounted for about 22%of damages caused by natural hazards and disasters in developing countries (FAO, 2015). The 2015–16 El Niño phenomenon was one of the strongest observed over the last 50 years and its impacts were felt worldwide. Unabated climate change is expected to exacerbate extreme weather events. While the occurrence and intensity of such extreme events are very difficult to predict, unless root causes are addressed, moving forward the global needs for humanitarian assistance will increase starkly.

Additionally, climate change and change in land cover, such as deforestation and desertification, can make plants and animals more vulnerable to pests and diseases. Changes in temperature, moisture levels, and concentrations of atmospheric gases can stimulate the growth and regeneration rates of plants, fungi, and insects, altering the interactions between pests, their natural enemies, and their hosts (FAO, 2017). Some of the most dramatic effects of climate change on transboundary animal diseases are likely to be seen among insect vectors, such as mosquitoes, midges, ticks, fleas, and sand flies, and the viruses they carry. With changes in temperatures and humidity levels, the populations of these insects may expand beyond their present geographic range and expose animals and humans to diseases to which they have no natural immunity.

Climate change, alterations in environmental conditions, and increasing pressure on land can modify not only animal production and productivity, but also disease dynamics

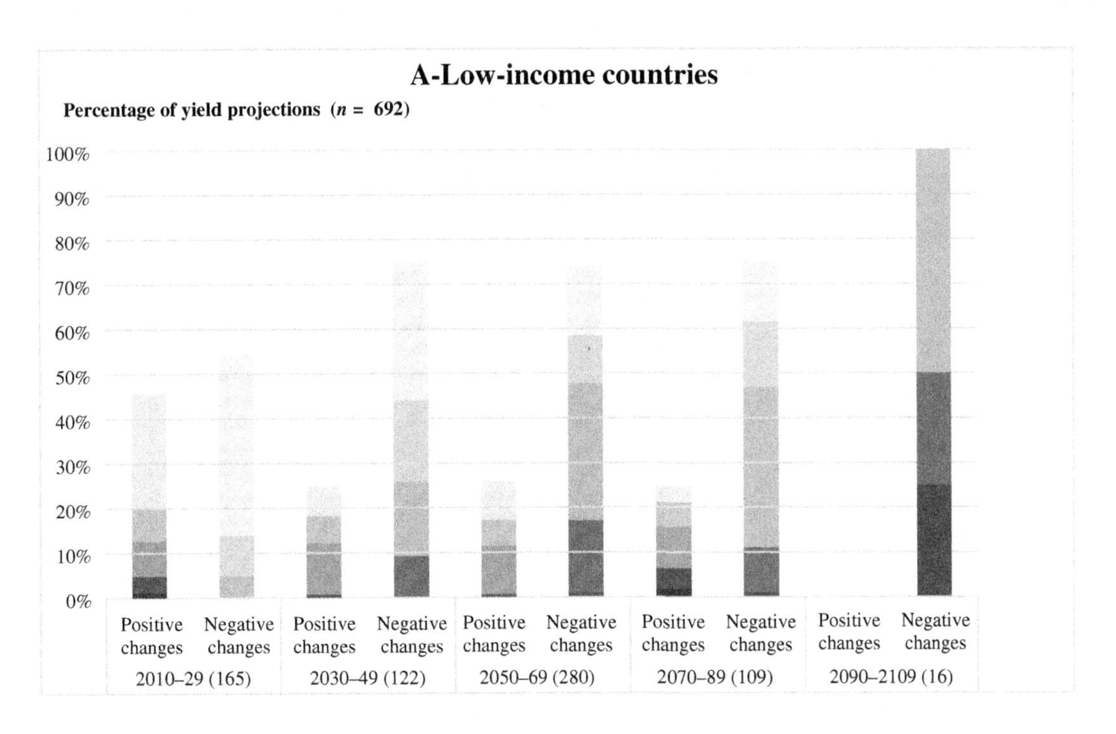

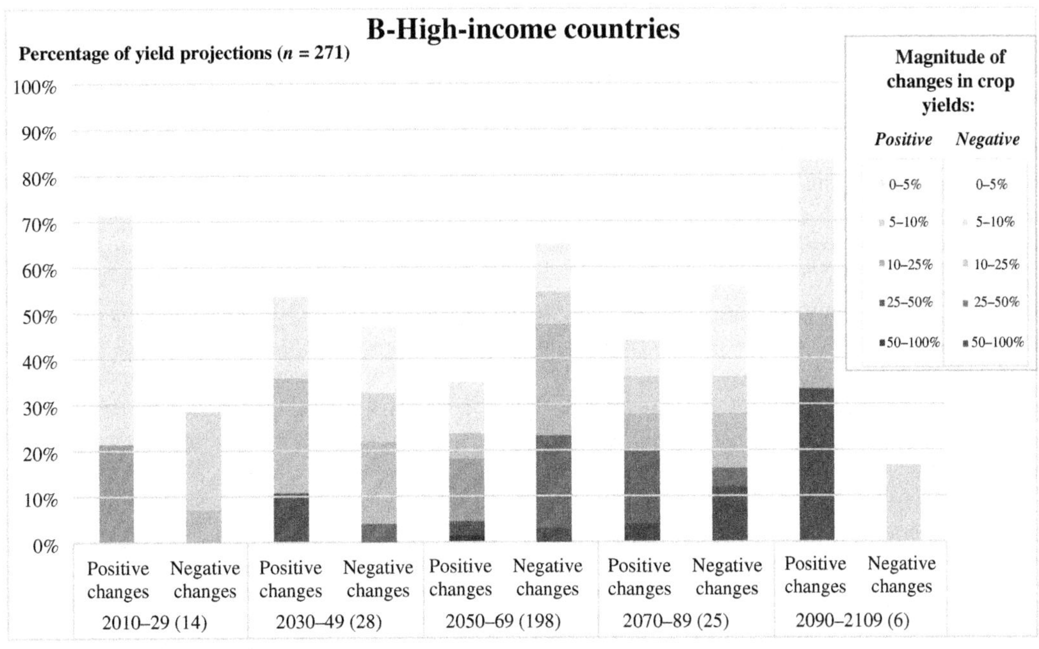

FIGURE 2.7 Projected changes in crop yields owing to climate change—percentage share of positive and negative projections. Note: Number of estimates of change in crop yield for each period is shown in parentheses. Source: *Calculations by FAO, 2016b. The State of Food and Agriculture: Climate Change, Agriculture and Food Security. FAO, Rome, p. 27, based on Porter, J.R., Xie, L., Challinor, A.J., Cochrane, K., Howden, S.M., Iqbal, M.M., et al., 2014. Food security and food production systems. In IPCC. 2014. Climate Change 2014: Impacts, adaptation, and vulnerability. Part A: Global and sectoral aspects. Contribution of Working Group II to the Fifth Assessment Report of the Intergovernmental Panel on Climate Change, Cambridge University Press, Cambridge, UK and New York, USA, pp. 485–533.*

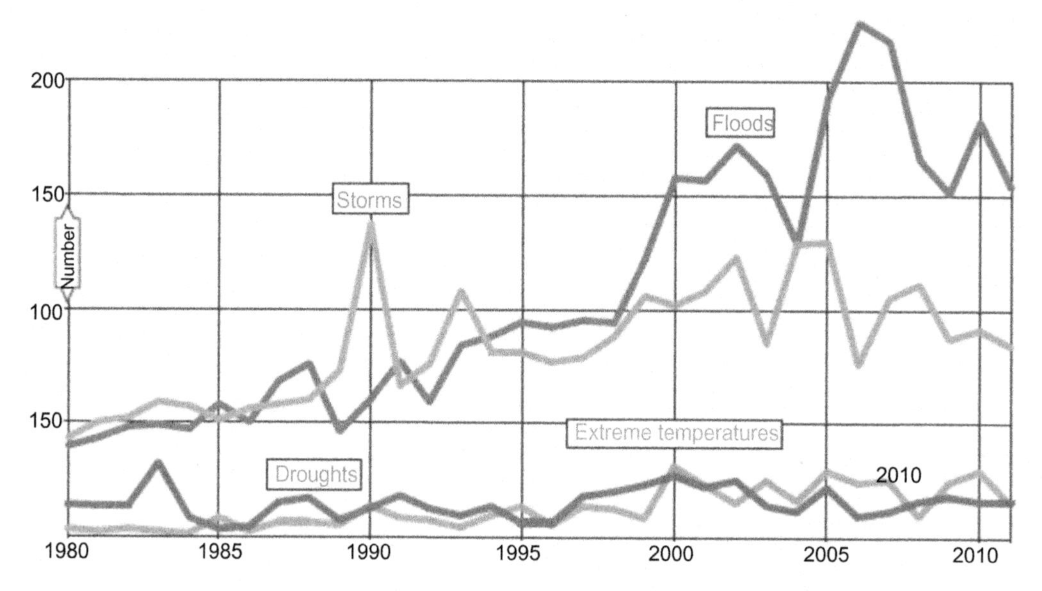

FIGURE 2.8 Number of climate-related disasters around the world, 1980–2011. *Source: FAO, 2017. The Future of Food and Agriculture – Trends and Challenges. Food and Agriculture Organization of the United Nations. Available at: http://www.fao.org/3/a-i6583e.pdf, based on: UNISDR (United Nations Office for DisasterRisk Reduction). 2016. Number of climate-related disasters around the world (1980– 2011) (dataset). (Latest updates 13 June 2012). www.preventionweb.net/files/20120613 (accessed 11.16).*

at the human–animal-ecosystem interface. They can affect the worldwide redistribution of vectors, pathogens, and infected hosts, setting off novel epidemiological patterns and driving the spread of many endemic diseases, such as bluetongue and West Nile viruses, into new areas (Kilpatrick and Randolph, 2012). In part, climate change is also responsible for the significant upsurge in transboundary plant pests and diseases (Fig. 2.9). It is modifying the dynamics of pest populations, such as locusts, and creating new ecological niches for the emergence or re-emergence and spread of pests and diseases. The effects of climate change likely will be felt in various ways, for instance in increased frequency of outbreaks, expansion of pests into new environments, the evolution of new pest strains and types, and increases in the vulnerability of plant defense mechanisms.

Because of these impacts, climate change already affects food security and, if unabated, the adverse impacts will intensify well beyond 2030. Climate change affects food availability through its increasingly adverse impacts on crop yields, fish stocks, and animal health and productivity, especially in sub-Saharan Africa and South Asia, where most of today's food insecure live. It limits access to food through negative impacts on rural incomes and livelihoods. Poor people, including many smallholder farmers and agricultural workers, are more vulnerable to the impacts of extreme events. Intensified occurrence of droughts or floods will sharply reduce incomes and cause asset losses that erode future income-earning capacity of those affected. In addition, to the extent that food supply is reduced by climate change, food prices will increase (see above). Both the urban and rural poor would

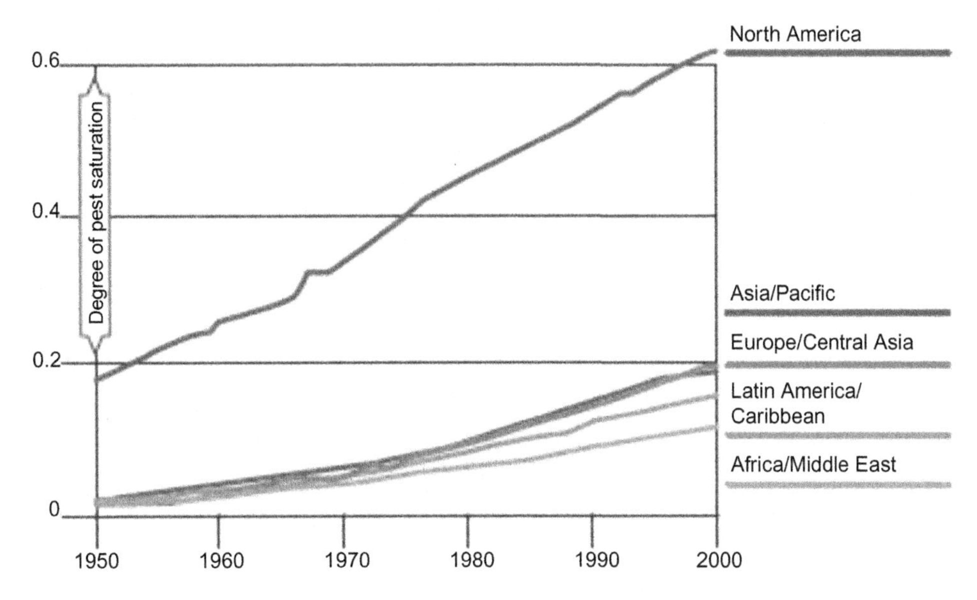

FIGURE 2.9 Global spread of crop pests and pathogens, 1950–2000. Note: The degree of pest saturation for a region is the mean of the degrees of saturation of countries in that region. The degree of saturation in a country is the number of crop pests and pathogens (CPPs) currently present divided by the number of CPPs that could occur. Source: *Bebber, D.P, Holmes, T., and Gurr, S.J., 2014. The global spread of crop pests and pathogens. Glob. Ecol. Biogeogr., 23(12): 1398–1407.*

be most affected because they spend much higher percentages of their income on food (FAO, 2016b).

All these effects are already affecting food security in the most vulnerable regions, but—at least until 2030—are only expected to mildly slow the projected downward trend of the number of the undernourished (FAO, 2016b). With unabated climate change, food security and nutrition are likely to be increasingly jeopardized beyond 2030 (see further below).

While being affected by climate change, agriculture also contributes to climate change by emitting significant amounts of the three major greenhouse gases (GHGs): carbon dioxide, methane, and nitrous oxide. About 21%of total annual GHG emissions originates in agriculture, forestry, and other land use (AFOLU), according to the IPCC classification of sources of emissions (FAO, 2016b). The bulk of current AFOLU emissions results from crop and livestock production, followed by deforestation (Fig. 2.10). Smaller amounts of emissions come from the losses of carbon in organic soils (often because of inappropriate agricultural practices) and the burning of biomass (e.g., savanna fires). Forests also mitigate climate change by removing carbon from the atmosphere through forest growth (as illustrated by the negative values in Fig. 2.8). Within the broader category of agricultural production, the main sources of emissions are enteric fermentation in ruminant livestock, the use of organic and nitrogen fertilizer, and rice production in flooded rice fields. Those emissions can be substantially reduced, however, if more sustainable production methods would be deployed.

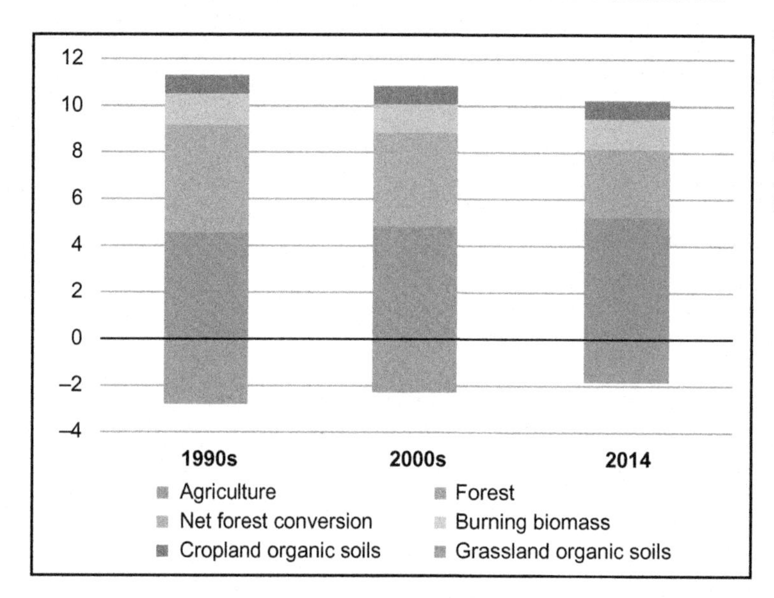

FIGURE 2.10 Annual average net emissions from AFOLU in CO_2 equivalent. Source: *FAO, 2016b The State of Food and Agriculture: Climate Change, Agriculture and Food Security. FAO, Rome, based on FAOSTAT.*

Successes in Hunger and Poverty Reduction are Undeniable, But Much More Needs to be Done

Hunger and extreme poverty have been significantly reduced globally since 1990. Between 1990–92 and 2005, the number of the undernourished fell by about 70 million. Significant achievements made in East Asia (mainly China) were offset by little or no progress in sub-Saharan Africa and South Asia, where there are still high concentrations of undernourished people (FAO et al., 2015a). Between 2005 and 2016, greater progress was made. Nearly twice as many people escaped chronic undernutrition during the past decade compared with 1990–2005. Progress in relative terms, that is, reductions in the proportion of the undernourished in the total population, has been more impressive. Between 1990 and 2015, the prevalence of undernourishment fell by almost half in low- and middle-income countries to almost meet the Millennium Development Goal target for hunger reduction. Some regions, such as Latin America, East and Southeast Asia, the Caucasus and Central Asia, and North and West Africa, have made notable progress. Less progress was made in South Asia, Oceania, the Caribbean, and Southern and Eastern Africa, where the pace was much slower. Despite the progress made, more than 800 million people were still chronically undernourished in 2016.

When extrapolated to the future, and assuming the same faster pace of progress attained over the past 10 years, the new Sustainable Development Goal target of eradicating hunger by 2030 would not be met. Based on ongoing trends and policy effort, more than 600 million people would still be undernourished in 2030 (Fig. 2.11).

The above projections do not contemplate the possible impact of climate change. Other model-based studies that simulated the possible impact of climate change suggest that by the year 2050, under a high emissions scenario, more than 40 million more people could be at risk of undernourishment than there would be in the absence of climate change

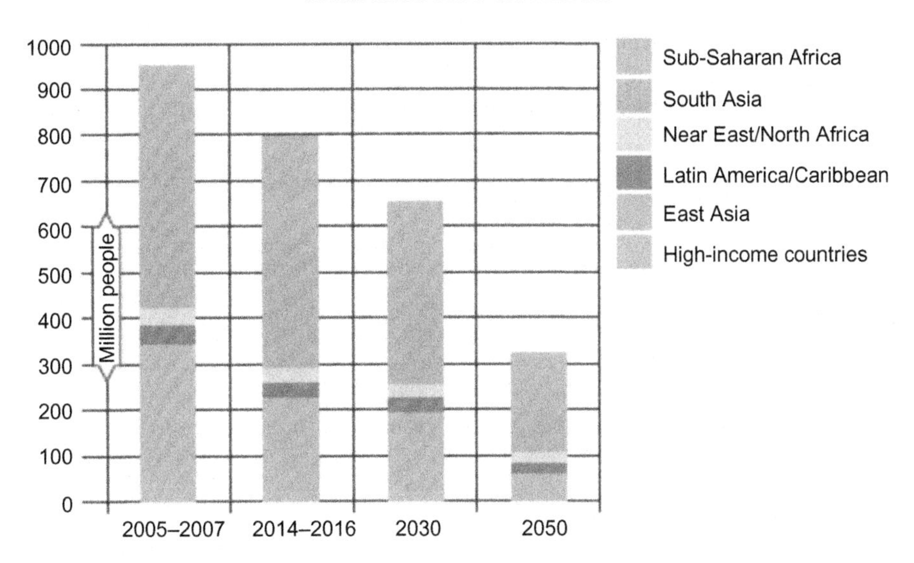

FIGURE 2.11 Undernourishment in a "business-as-usual" scenario, 2005–2050. Source: *FAO, 2017. The Future of Food and Agriculture – Trends and Challenges. Food and Agriculture Organization of the United Nations. Available at: http://www.fao.org/3/a-i6583e.pdf, based on: Years 1990–92 and 2014–16 from FAO, IFAD, and WFP. 2015a. Achieving Zero Hunger. The Critical Role of Investment in Social Protection and Agriculture. Rome, FAO; FAO, IFAD, and WFP, 2015b. The State of Food Insecurity in the World 2015. Meeting the 2015 International Hunger Targets: Taking Stock of Uneven Progress. Rome, FAO; and year 2050 from Alexandratos, N., and Bruinsma, J., 2012. World Agriculture Towards 2030/2050: The 2012 Revision. ESA Working Paper No. 12–03. Rome, FAO.*

(Wiebe et al., 2015). This may be a conservative estimate, as it is based on a "business-as-usual" assumption of economic growth and does not account for the impacts of extreme events, sea-level rise, melting glaciers, and changes in pest and disease patterns; nor does it consider other factors that are expected to change with climate, especially after 2050.

Undernourishment, defined as insufficient calorie intake, is not the only nutritional challenge. Still, widespread undernourishment coexists with a high prevalence of micronutrient deficiencies and a rapidly expanding share of the world population that is considered overweight or obese. Micronutrient deficiencies affect more than 2 billion people worldwide (Micronutrient Initiative, 2015). In 2015, for example, more than half a billion women between the ages of 25 and 45 suffered from iron deficiency. Overweight and obesity are increasing worldwide in all population groups due to increased consumption of foods that are high in energy, fats, added sugars or salt, and inadequate intake of fruits, vegetables, and dietary fiber. This "nutrition transition" reflects rapid urbanization, the increased production of processed food, and a more sedentary lifestyle. In 2014, some 40% of people aged 18 and over were overweight and, of these, 13% were obese (WHO, 2016). Almost two-thirds of the world's population live in countries where overweight and obesity kill more people than underweight (WHO, 2016).

Agriculture and food systems must meet the food and nutritional demands of people with rising incomes and changing diets, as well as the demands of a growing number of the poor and hungry. While much attention has been given to increasing farm production to meet this demand, equally critical are the supply chains that connect farmers to urban markets, along with measures such as pricing policies and social protection, which ensure consumers' access to nutritious and safe food at affordable prices. Food systems have already changed drastically in recent decades, but some of those changes have been a causal factor rather than a solution to the triple burden of malnutrition.

Food Systems are Changing, Jeopardizing many Landless and Smallholder Farmers

Urbanization increases the demand for processed food products that are easily transportable and storable. Food chains are becoming longer, more capital intensive, vertically integrated, and concentrated in fewer hands. Though this situation creates new job opportunities for some, many small-scale producers and landless households are losing out, which adds to the migration flows out of rural areas in search of jobs outside of agriculture. Where those are scarce, it is adding to international out-migration.

In Latin America, rapid urbanization has led to profound changes in food production and distribution systems, with supermarkets now accounting for more than 50% of grocery sales (Reardon et al., 2014). In much of Asia, the food purchases in urban areas are now made in supermarkets. Even in East Africa, the share of purchased food in total food consumption is just below 60% (Tschirley et al., 2015).

Although agrifood chains offer employment opportunities, the transformation of these chains in low- and middle-income countries has, in many cases, created serious barriers to the participation of smallholder producers and small-scale agroprocessors in local, national, and global markets. Since the agrifood chains are more capital intensive, agricultural production and supply processes require much less labor (Neven et al., 2009). Barriers to smallholder access to supermarket channels, combined with reduced labor requirements, may undermine the livelihoods of farm households if they cannot diversify into other rural off-farm activities.

Smallholder farmers tend to benefit from the transformation of food systems when they join vertically coordinated value chains through fair contracts with processors and traders. In recent decades, a variety of business models, national and international value chain organizations, and institutional arrangements and policies have emerged to provide incentives and support services to smallholders, with the aim of increasing sustainable food production and facilitating market access (Rao and Qaim, 2011). The innovations include institutional and market intermediaries, such as participatory guarantee systems, marketing cooperatives, training centers, private traders, and local public procurement mechanisms, which take on a wide range of roles in linking farmers to markets. These arrangements tend to be more effective in linking smallholder farmers and small-scale processors to markets, resulting in better availability of local infrastructure and stronger producer organizations and related institutions.

In addition to improving the efficiency of food delivery systems, these value chains have helped improve food quality and safety, which benefits consumer health. At the same time, they have facilitated the diversification of diets among more affluent consumers, accelerating the shift from starchy staples, which are the main foods of the poor, to livestock products, fats and oils, and fruit and vegetables. More affluent consumers tend to adopt a globally connected lifestyle that increase the demand for novel foods. However, industrialized meat production processes and higher consumption of processed foods also raise concerns related to nutrition, the environment, food security, and food safety. Large-scale food processing risks increasing the availability of cheaper foods that have a high content of fat and added sugar and salt (so-called "empty calories"). Recent studies of the links between people's diets and their food environments—that is, the food that is made available, affordable, convenient, and desirable based on consumers' tastes and education—have produced mixed findings. Sometimes, the wider availability of processed food leads to higher food consumption and greater dietary diversity. In other cases, low-income populations find it more difficult to adopt high-quality diets and are more likely to consume "empty calories." The continuing increase in overweight, obesity, and diet-related noncommunicable diseases worldwide is a clear indicator of this latter trend (Global Panel on Agriculture and Food Systems for Nutrition, 2016).

Longer food value chains may have a larger ecological footprint. As the pressure on scarce land and water resources increases, the agrifood sector must find ways of reducing its environmental impact, which includes GHG emissions, water usage, food loss, and waste, and its effects on soil health, ecosystem services, and biodiversity. Intensive production and longer food supply chains may be associated with higher GHG emissions from both production inputs (e.g., fertilizers, machinery, pesticides, veterinary products, and transport) and activities beyond farmgate (e.g., transportation, processing, and retailing). Global value chains have substantially increased the use of long-distance transport between primary production, processing, and consumption. However, there is mixed evidence of whether long-distance value chains have higher GHG emissions than the short chains. The level of overall emissions from a food production process is determined not only by transport, but also by production, processing, storage, and distribution (Kneafsey et al., 2013). In the same vein, a comparison of "local" and "nonlocal" food in terms of GHG emissions concluded that the least detrimental effect on the environment depends on the food product, the type of farm operation, transport, season, and the scale of production (Edwards-Jones et al., 2008).

Thus, the adoption of comparatively low-emission technologies in primary production phases could more than compensate for higher emissions from "long" value chains. However, if technologies that produce high levels of GHG emissions are adopted to produce food that is transported far from its origin, this will result in comparatively higher GHG emissions. For instance, farmers in Kenya supplying leaf cabbage to local supermarkets use almost twice the amount of chemicals per unit of output than farmers normally use (Neven et al., 2009). If supplying supermarkets requires more chemical fertilizer and fossil energy per unit of output, GHG emissions may increase during the transition from "traditional" to "modern" value chains. Since the production of fertilizers, herbicides, and pesticides, along with the emissions from fossil fuels used in the field, represents about 2% of total GHG emissions (HLPE, 2012), increases in the use of these inputs is likely to have significant global impacts.

2.3 KEY QUESTIONS FOR POLICIES AND GOVERNANCE TO ACHIEVE MORE SUSTAINABLE AND HEALTHY FOOD SYSTEMS

Looking ahead, the core question is whether today's agriculture and food systems are capable of meeting the needs of a global population that is projected to reach almost

10 billion by midcentury and that may peak at more than 11 billion by the end of the century. Can we achieve the required production increases, even as pressures on already scarce land and water resources and the negative impacts of climate change intensify? The common consensus is that though current systems are likely capable of producing enough food, major transformations are required to do so in an inclusive and sustainable manner.

This raises further questions. Can agriculture meet unprecedented demand for food in ways that ensure that using the natural resource base is sustainable while containing GHG emissions and mitigating the impacts of climate change? Can the world secure access to adequate food for all, especially in low-income regions where population growth is the most rapid? Can agricultural sectors and rural economies be transformed in ways that provide more and better employment and income-earning opportunities, especially for youth and women, and help stem mass migration to cities with limited labor-absorptive capacity?

Can public policies address the so-called "triple burden of malnutrition," by promoting food systems that give affordable access to food for all, eliminate micronutrient deficiencies, and redress the overconsumption of food? Can the huge problem of food losses and waste, estimated at as much as one-third of the total food produced for human consumption, be tackled? Can national and global regulatory structures protect producers and consumers against the increasing monopoly power of large, multinational, vertically integrated agroindustrial enterprises? Can the impacts of conflicts and natural disasters, both major disrupters of food security and the causes of vast migrations of people, be contained and prevented?

This leads to questions in another area: policy coherence. Can we overcome "wickedness" in policymaking, where the lack of a coherent set of well-defined goals and processes means that the response to one aspect of a problem (e.g., incentives to raise productivity) risks exacerbating others (e.g., depletion of natural resources)? Can we engage all stakeholders, including the private sector, farmer, and consumer organizations, and other civil society players in better decision-making, recognizing that more inclusive governance is essential to improving dialogue about the hard policy choices that need to be made?

The 2030 Agenda for Sustainable Development, adopted by the international community in September 2015, provides a compelling, but challenging vision on how multiple objectives can be combined to define new sustainable development pathways. The second Sustainable Development Goal (SDG 2) explicitly aims at ending hunger, achieving food security and improved nutrition, and promoting sustainable agriculture simultaneously by 2030.

The 2030 Agenda acknowledges that progress toward many other SDGs, especially the eradication of poverty and the response to climate change (SDG 13) and the sustainable use of marine and terrestrial ecosystems (SDG 14 and SDG 15), will depend on the extent to which food insecurity and malnutrition are effectively reduced and sustainable agriculture is promoted. Conversely, progress toward SDG 2 will depend on the progress made toward several of the other goals. In other words, in order to make progress on SDG 2, policymakers and all other stakeholders will need to consider interlinkages and critical interactions, both in terms of synergies and trade-offs, between SDG 2 and all other goals.

Pursuing all these objectives simultaneously will not be easy, given the sheer magnitude of the global challenges to sustainable food security and the unprecedented amount of sustained effort and policy coherence at national and global levels needed to propel the necessary transformative changes to agriculture and food systems at large. It will further be challenging given that there is still a daunting lack of knowledge of the precise impacts of climate change and other environmental constraints on food security, how these impacts will interact with demographic pressures and movements of people, and how available solutions can be scaled up in ways that will set a course toward a sustainable future for all humanity. Subsequent chapters will attempt to address several of these unknowns.

References

Alexandratos, N., Bruinsma, J., 2012. World Agriculture Towards 2030/2050: The 2012 Revision. FAO, Rome, ESA Working Paper No. 12–03.

Bebber, D.P., Holmes, T., Gurr, S.J., 2014. The global spread of crop pests and pathogens. Glob. Ecol. Biogeogr. 23 (12), 1398–1407.

Edwards-Jones, G., Canals, L.M., Hounsome, N., Truninger, M., Koerber, G., Hounsome, B., et al., 2008. Testing the assertion that 'local food is best': the challenges of an evidence-based approach. Trends Food Sci. Technol. 19 (5), 265–274.

FAO, 2011a. Land tenure, climate change mitigation and agriculture, Mitigation of Climate Change in Agriculture (MICCA) Programme, June 2011. Rome.

FAO. 2011b. The state of the world's land and water resources for food and agriculture (SOLAW). [Website]. Available at <www.fao.org/nr/solaw/solaw-home> (accessed 15.09.18.).

FAO. 2015. The impact of natural hazards and disasters on agriculture and food security and nutrition: a call for action to build resilient livelihoods, May 2015. Rome. http://www.fao.org/3/a-i4434e.pdf (accessed 15.09.18.).

FAO. 2016a. AQUASTAT [Website]. Available at www.fao.org/nr/water/aquastat/didyouknow/index2.stm. (accessed 11.16).

FAO, 2016b. The State of Food and Agriculture: Climate Change, Agriculture and Food Security. FAO, Rome.

FAO, 2017. The Future of Food and Agriculture – Trends and Challenges. Food and Agriculture Organization of the United Nations (Available at: http://www.fao.org/3/a-i6583e.pdf.).

FAO, IFAD, and WFP, 2015a. Achieving Zero Hunger. The Critical Role of Investment in Social Protection and Agriculture. FAO, Rome.

FAO, IFAD, and WFP, 2015b. The State of Food Insecurity in the World 2015. Meeting the 2015 International Hunger Targets: Taking Stock of Uneven Progress. FAO, Rome.

Global Panel on Agriculture and Food Systems for Nutrition. 2016. Food systems and diets: Facing the challenges of the 21st century. London.

HLPE. 2012. Food security and climate change. A report by the High-Level Panel of Experts on Food Security and Nutrition of the Committee on World Food Security. Rome.

IPCC, 2014. Climate Change 2014: Impacts, Adaptation, and Vulnerability. Part A: Global and Sectoral Aspects. Cambridge University Press, Cambridge, UK and New York, USA, Contribution of Working Group II to the Fifth Assessment Report of the Intergovernmental Panel on Climate Change.

Kilpatrick, A.M., Randolph, S.E., 2012. Drivers, dynamics, and control of emerging vector-borne zoonotic diseases. Lancet 380 (9857), 1946–1955.

Kneafsey, M., Venn, L., Schmutz, U., Balázs, B., Trenchard, L., Eyden-Wood, T., et al., 2013. Short Food Supply Chains and Local Food Systems in the EU. A State of Play of Their Socio-economic Characteristics. European Commission, Joint Research Centre.

Micronutrient Initiative. 2015. Micronutrient Initiative [Website]. Available at www.micronutrient.org.

Neven, D., Odera, M.M., Reardon, T., Wang, H., 2009. Kenyan supermarkets, emerging middle-class horticultural farmers, and employment impacts on the rural poor. World Dev. 37 (11), 1802–1811.

O'Neill, B.C., Oppenheimer, M., Warren, R., Hallegatte, S., Kopp, R.E., Pörtner, H.O., et al., 2017. IPCC reasons for concern regarding climate change risks. Nat. Clim. Chang. 7 (1), 28–37.

Pardey, P., Chan-Kang, C., Dehmer, S., 2014. Global Food and Agricultural R, and D Spending, 1960–2009. InSTePP Report. University of Minnesota, Saint Paul, USA.

Porter, J.R., Xie, L., Challinor, A.J., Cochrane, K., Howden, S.M., Iqbal, M.M., et al. 2014. Food security and food production systems. In IPCC. 2014. Climate Change 2014: Impacts, adaptation, and vulnerability. Part A: Global and sectoral aspects. Contribution of Working Group II to the Fifth Assessment Report of the Intergovernmental Panel on Climate Change, pp. 485–533. Cambridge, UK and New York, USA, Cambridge University Press.

Rao, E.J.O., Qaim, M., 2011. Supermarkets, farm household income, and poverty: Insights from Kenya. World Dev. 39 (5), 784–796.

Reardon, T., Tschirley, D., Dolislager, M., Snyder, J., Hu, C., White, S., 2014. Urbanization, Diet Change, and Transformation of Food Supply Chains in Asia. Michigan State University - Global Center for Food Systems Innovation, East Lansing, USA.

Tschirley, D., Snyder, J., Dolislager, M., Reardon, T., Haggblade, S., Goeb, J., et al., 2015. Africa's unfolding diet transformation: implications for agrifood system employment. J. Agribusiness Develop. Emerg. Econ. 5 (2), 102–136.

UNISDR (United Nations Office for DisasterRisk Reduction). 2016. Number of climate-related disasters around the world (1980– 2011) (dataset). (Latest updates 13 June 2012). www.preventionweb.net/files/20120613 (accessed 11.16).

WHO. 2016. Obesity and overweight. Fact sheet No. 311 [Website]. Available at www.who.int/mediacentre/factsheets/fs311.

Wiebe, K., Lotze-Campen, H., Sands, R., Tabeau, A., van der Mensbrugghe, D., Biewald, A., et al., 2015. Climate change impacts on agriculture in 2050 under a range of plausible socioeconomic and emissions scenarios. Environ. Res. Lett. 10 (08), 1–15.

CHAPTER

3

Demographic Change, Agriculture, and Rural Poverty

James Thurlow[1], *Paul Dorosh,*[1] *and Ben Davis*[2]

[1]International Food Policy Research Institute (IFPRI), Washington, DC, United States

[2]Food and Agriculture Organization of the United Nations (FAO), Rome, Italy

3.1 INTRODUCTION

Demographic change refers to shifts in the characteristics of a population, including rates of growth, urbanization, and employment status. There are well-established linkages between demographic change and economic development. As incomes rise, fertility rates and family sizes tend to fall, leading to slower overall population growth (Becker et al., 1990). This alters the underlying drivers of economic growth, with capital accumulation and productivity eventually becoming more important than labor supply. Urbanization is also associated with development, with rural people migrating into cities and towns for work (Moomaw and Shatter 1996). This usually entails a movement of workers from lower- to higher-productivity jobs, a process called "positive structural change" that is strongly associated with long-term development (McMillan et al., 2014). Faster growth in urban incomes and populations can also increase demand for agricultural products, thereby creating new opportunities for farmers (Brückner, 2012; Tschirley et al., 2015). Within rural areas, rising population density may reduce farm sizes, especially among smallholders (Chamberlin et al., 2014; Masters et al., 2013). Alternatively, this situation may encourage farmers to adopt new technologies and diversify into off-farm employment, which would lead to higher household incomes and productivity.

Although demographic change can be a driver of economic development, it can also be a constraint. Population growth makes raising per capita incomes more difficult, and higher population densities can place tremendous pressure on land and other natural resources. Declining farm sizes may reduce farmers' assets, limit their access to finance, and prevent investment in new technologies (Feder and O'Mara, 1981; Fan and Chan-Kang, 2005;

Sustainable Food and Agriculture
DOI: https://doi.org/10.1016/B978-0-12-812134-4.00003-0

Jayne et al., 2010). These developments may prompt a "premature" exit from agriculture, where smallholders abandon farming despite profitable technologies and markets. Even those who remain in agriculture may not be able to take advantage of rising urban demand, particularly farmers in more remote rural areas or where urban centers have better access to imported food (Hazell, 2005). Falling farm sizes and import competition may weaken agriculture's historically strong linkage to poverty reduction, given that most of the world's rural poor are smallholder farmers. Similarly, rapid urbanization may outpace urban economic growth and job creation, leading to greater pressure on urban infrastructure and services and a growing number of poor people living in urban areas. Finally, an influx of young job seekers into the workforce, known as a "youth bulge," can stimulate economic growth; however, it also raises concerns about an economy's capacity to create jobs (Filmer and Fox, 2014), especially the kinds of jobs that match the aspirations of younger generations (Leavy and Hossain, 2014). Conversely, some developing countries have already experienced their demographic transition and now face the prospect of an aging agricultural workforce. While demographic change can complement economic development, it can also lead to economic uncertainty and social and political instability.

This chapter considers the implications of future demographic change for agriculture and rural poverty in developing countries. Its focus is on sub-Saharan Africa (SSA) because this is the region where the world's rural poor are concentrating. We recognize that the challenge to reduce malnutrition, as opposed to income-based poverty, is most severe in South Asia (SA). By focusing on Africa, we also pay greater attention to the world's burgeoning rural youth population rather than to emerging concerns about aging rural workforces, a subject which would be more relevant for East Asia. In absolute terms, however, Africa's ongoing demographic transition is the largest in history, and because African countries are at earlier stages of development, they face a greater challenge in dealing with the implications of demographic change for agriculture and rural poverty.

Section 3.2 summarizes global demographic and economic trends, and highlights the major differences between Africa and other developing regions. One of Africa's distinguishing features is that, despite urbanization, its rural population continues to grow rapidly. This differs from other developing regions when they were at a similar stage of development and underscores one of Africa's major challenges – creating sufficient job opportunities for a young rural population. It also highlights the need to strengthen the linkages between urban growth and rural transformation.

Many African governments are targeting reductions in their agricultural workforce as a way of accelerating economic growth. Ethiopia's Second Growth and Transformation Plan, for example, aims to reduce the share of total employment in agriculture to 50% by 2025, while Uganda's Vision 2040 aims for a 31% workforce by 2040. Section 3.3 reviews the past experiences of successful developing countries to gauge how many workers Africa's nonagricultural sector can realistically absorb over the next few decades. This highlights another distinguishing feature of Africa's development pathway – that rapid economic growth over the past two decades was not driven by industrialization, but rather by (informal) services (McMillan et al., 2014). Again, this differs from Asia's experience, where economic growth and job creation was usually led by manufacturing. We find that even if Africa could emulate Asia's pace of structural transformation, agriculture will still need to play a crucial role in creating future job and income opportunities, at least for the next few decades.

Section 3.4 discusses the effects of rapid urbanization on agriculture and rural poverty. The rural-to-urban spatial transformation typically coincides with structural transformation from agriculture to nonagricultural sectors, as discussed in Section 3.3. Urbanization itself also brings about important changes in economic incentives and efficiency. Rural-urban migration provides labor to urban industry and services, and rising urban incomes can create demand for agricultural and other goods and services produced in rural areas. Migration can also generate positive agglomeration effects on total factor productivity through economies of scale, spread of technical knowledge, and the "thickening" of factor, input, and product markets (e.g., for labor and credit). This section presents dynamic economy-wide simulation analyses of these complex interactions between the rural and urban economies, focusing on public investment choices and their effects on economic growth and poverty. The final section offers conclusions and a discussion of areas for further research.

3.2 GLOBAL TRENDS AND PROJECTIONS

Population Growth and Urbanization

Both located in Asia, the world's two largest developing regions, East Asia and the Pacific (EAP) and SA, are quite different in global population trends and projections (Table 3.1).[1] More than half of EAP's population is urbanized, whereas SA is the least urbanized region; conversely, EAP's population is growing, but only at half the rate of SA's population. More importantly for this chapter, while SA's rural population continues to grow, EAP's rural population is declining in absolute terms, and this is the main reason for much faster urbanization in EAP. Because of restrictive population control policies, EAP reached its demographic transition point much earlier than SA, despite both regions starting from similar conditions in the 1960s. This is clear from today's youth dynamics. SA's youth population is expanding at close to the total population's growth rate, whereas EAP's youth population is falling in absolute terms, almost as fast as the rural population's growth rate. As a result, EAP's total population will plateau soon after 2030, whereas SA's population will continue expanding until after 2060, by which time its population will surpass that of EAP (all demographic projections are from UNDESA (2015, 2017), and use the median fertility and life expectancy projection).

Even before 2060, however, SA is likely to be surpassed by SSA as the world's most populous developing region. Indeed, SSA is unique among today's developing regions since its total population is growing rapidly and the urban population is growing even faster (from an urbanization rate that is already higher than that of SA). Even though SSA's urban population is growing twice as fast as its rural population, the latter is still growing

[1] We follow the World Bank's definition of developing countries and regions. We include all countries who were classified as either low or middle income in 2015, and we do not reclassify countries in future projections.

TABLE 3.1 Global Demographic and Economic Trends and Projections

	ECA	LAC	EAP	MENA	SA	SSA	All
Population, 2015 (millions)	411	605	2035	363	1744	1001	6159
Share of all regions (%)	6.7	9.8	33.0	5.9	28.3	16.2	100
Urban population share (%)	65.2	79.6	52.9	60.5	33.0	37.7	48.7
Poor population, 2013 (millions)	10.3	33.6	71.0	6.1	256.2	388.7	766.0
Share of all regions (%)	1.3	4.4	9.3	0.8	33.5	50.7	100.0
Poverty headcount rate (%)	2.2	5.4	3.5	1.7	15.1	41.0	12.6
GDP per capita, 2015 (US$)	17,514	14,146	12,179	11,359	5317	3486	9325
Annual growth, 1990–2015 (%)	1.0	1.5	7.1	1.5	4.3	1.3	3.2
Annual population growth, 1990–2015 (%)	0.3	1.4	1.0	1.9	1.7	2.7	1.5
Rural areas	0.0	− 0.1	− 0.7	1.1	1.3	2.1	0.6
Urban areas	0.4	2.0	3.5	2.5	2.9	4.1	2.7
Youth (15–25 years old)	− 0.3	0.9	− 0.6	1.6	1.6	2.9	0.8
Poor (1990–2013)	0.6	− 3.2	− 10.7	− 3.3	− 2.9	1.5	− 3.7
Projected population, 2030 (millions)	429	688	2206	453	2046	1471	7291
Share of all regions (%)	5.9	9.4	30.3	6.2	28.1	20.2	100
Urban population share (%)	68.0	83.0	64.0	65.0	40.0	45.0	56.0
Annual growth, 2015–30 (%)	0.3	0.9	0.5	1.5	1.1	2.6	1.1
Projected population, 2050 (millions)	429	748	2203	557	2288	2245	8470
Share of all regions (%)	5.1	8.8	26.0	6.6	27.0	26.5	100
Urban population share (%)	74.0	86.0	71.0	71.0	51.0	55.0	63.0
Annual growth, 2030–50 (%)	0.0	0.4	0.0	1.0	0.6	2.1	0.8

Note: Poverty measured using the World Bank's US$1.90 poverty line, and GDP per capita is in constant 2011 international dollars. Poverty and GDP per capita measures are adjusted for cross-country purchasing power differences. Future populations are based on the median variant fertility and life expectancy projections. Poverty estimates for the MENA region are imputed with error due to insufficient survey sample size (see World Bank, 2017b). *ECA*, Europe and Central Asia; *LAC*, Latin America and the Caribbean; *EAP*, East Asia and the Pacific; *MENA*, Middle East and North Africa; *SA*, South Asia; *SSA*, sub-Saharan Africa.

Authors' calculations using UNDESA (United Nations Department of Economic and Social Affairs), 2015. World urbanization prospects: 2014 revision. United Nations Department for Economic and Social Affairs, United Nations, New York, NY and UNDESA (United Nations Department of Economic and Social Affairs), 2017. World population prospects: 2017 revision. United Nations Department for Economic and Social Affairs, United Nations, New York, NY, and World Bank, 2017a. PovcalNet: an online analysis tool for global poverty monitoring. World Bank, Washington, DC. <http://iresearch.worldbank.org/PovcalNet> (accessed August 2017) and World Bank, 2017b. World development indicators online database. World Bank, Washington, DC. <http://databank.worldbank.org/data/reports.aspx?source = world-development-indicators> (accessed August 2017).

faster than the total populations of all other developing regions. Thus, while much attention is given to Africa's burgeoning urban centers, its rural population pressures are substantial. In fact, the pace of urbanization in Africa is similar to the past experiences of other developing regions, but it is the absolute scale of Africa's demographic change that is unprecedented (Fig. 3.1).

The left-hand panel in the figure indicates that all regions have passed their peak population growth rates and that these rates are expected to fall over the next three decades. By the 2040s, regional populations will be in absolute decline, except in SSA and the Middle East and North Africa (MENA) (not shown in the figure). EAP has experienced very fast urbanization, particularly over the last two decades. Today, SA is only as urbanized as EAP was in the 1990s, a gap between regions that is not expected to change significantly by 2050. SSA's annual population growth peaked at a higher rate than other regions, and even though a rapid slowdown is expected, SSA in the 2040s will have similar population growth rates to SA in the 2010s and EAP in the 1990s. Africa's process of demographic transition is occurring roughly four decades behind the rest of the developing world. However, SSA's population growth and urbanization trends are closely tracking other regions' historical experiences.

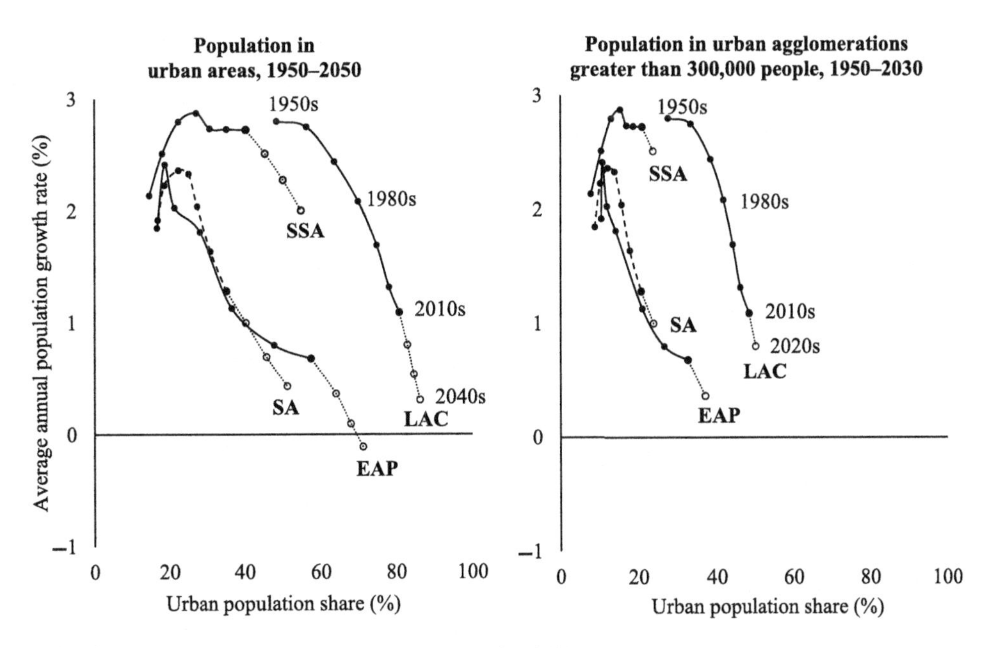

FIGURE 3.1 Population growth and urbanization, 1950–2050.
Note: Figure includes historical data (dark lines) and projections (faint lines). Future populations are based on the median variant fertility and life expectancy projections. *EAP*, East Asia and the Pacific; *LAC*, Latin America and the Caribbean; *SA*, South Asia; *SSA*, sub-Saharan Africa. Source: *Authors' calculations using population estimates for developing countries; From UNDESA (United Nations Department of Economic and Social Affairs), 2015. World urbanization prospects: 2014 revision. United Nations Department for Economic and Social Affairs, United Nations, New York, NY and UNDESA (United Nations Department of Economic and Social Affairs), 2017. World population prospects: 2017 revision. United Nations Department for Economic and Social Affairs, United Nations, New York, NY.*

Different definitions of "urban" can complicate regional comparisons (Cohen, 2004). To control for this, the right-hand panel in Fig. 3.1 only includes cities of 300,000 people or more. As we move to a stricter urban definition, the urbanization rates in the right-hand panel of Fig. 3.1 shift to the left. The magnitude of this shift indicates the share of population living in smaller cities and towns. For example, EAP's 2015 urban share falls from 58 to 33%, indicating that 25% of the region's total population live in urban centers with less than 300,000 people – this is about two-fifths of the official urban population. SSA is most affected by the stricter definition – its urban population share drops from 40 to 21%. When the standard definition is applied across regions, then we see that SSA and SA are equally urbanized. Again, it is the scale and lateness, rather than the pattern, of SSA's demographic transition that makes the region unique.

Economic Growth and Poverty

Economic growth in the developing world has led to substantial reductions in poverty over the past two and half decades. Much of this success was driven by Asia, which experienced rapid economic growth after 1990 (Table 3.1). While growth in per capita gross domestic product (GDP) in Asia was well above the global average, the rate of growth in EAP was much faster than in SA. Poverty reduction was also more pronounced in EAP. EAP's poverty headcount, i.e., the share of its total population living below the World Bank's US$1.90 per day poverty line (in purchasing power parity), fell from 60.2% in 1990 to 3.5% in 2013 (World Bank, 2017a). Today, less than 10% of the world's poor population live in EAP, down from more than half in 1990. The poverty rate also declined in SA over this period (from 44.6 to 15.1%), but the dramatic poverty reduction in EAP, coupled with faster population growth in SA, meant that SA's share of the world's poor population increased from 27.4 to 33.5%.

SSA's per capita GDP grew at only 1.3% per year during 1990–2015, almost three times slower than the developing country average. Poverty reduction in SSA was also slower. The region's poverty rate fell from 48.2% in 1990 to 41.0% in 2013. Underlying this trend is Africa's particularly poor performance during the 1990s, when economic growth barely matched population growth and poverty rates were rising. During the 2002–13 period, SSA's per capita GDP growth rate was a more impressive 3% per year, and the region's poverty rate fell by more than one percentage point per year (from 55.6% in 2002). However, despite its improved performance in recent years, SSA still lags behind the rest of the developing world, and the region's rapid population growth has meant that currently 50.7% of the world's poor live in SSA, up from 15.0% in 1990.

Economic growth and demographic change are the main drivers behind global poverty dynamics. Based on the historical trends described above, it is possible to project the future level and distribution of poverty across developing regions. We use the simple equation below to estimate the number of poor people living in each developing region r in future period t:

$$P_{rt} = (\varepsilon_r \cdot G_r)^N \cdot r_{rt} \cdot T_{rt}$$

where P is the number of poor people, ε is a poverty-growth elasticity, G is the average annual per capita GDP growth, r is the projected poverty headcount rate, T is the projected total population, and N is the number of years until period t. Given a rate of economic growth and a region's underlying poverty-growth relationship, which we assume remains constant over time, the equation estimates future poverty rates and applies these to the regional population projections from UNDESA (2017).

We estimate economic growth rates (G) using data on regional per capita GDP growth for the 20-year period 1993–2013, using data from the World Bank (2017b). A summary of the data used in our analysis is provided in Table 3.2. Given the uncertainty about future economic growth, we consider both high- and low-bound scenarios. Based on global trends, the low-growth scenario uses average growth rates from the slower-growth period 1993–2005, and high-growth scenario uses the faster-growth period 2005–13. Growth was faster in MENA and Eastern and Central Asia (ECA) regions during the period when overall global growth was slower. Given the low initial poverty rates in these regions, this does not substantially alter our poverty projections. Similarly, we estimate poverty-growth

TABLE 3.2 Projecting Global Poverty, 2013–50

	ECA	LAC	EAP	MENA	SA	SSA	All
Per capita GDP growth, 1993–2013 (%)	4.06	1.70	7.03	2.02	4.56	1.93	3.82
High growth (2005–13)	3.51	2.43	7.98	1.47	5.37	2.41	4.79
Low growth (1993–2005)	4.56	1.22	6.40	2.38	4.02	1.60	3.18
Poverty-growth elasticity, 1993–2013	− 1.71	− 2.82	− 1.79	− 2.86	− 1.16	− 0.91	− 1.49
High elasticity (2002–13)	− 2.84	− 3.40	− 2.33	− 4.69	− 1.78	− 1.02	− 1.67
Low elasticity (1993–2002)	− 0.91	− 1.98	− 1.30	− 2.11	− 0.59	− 0.80	− 1.29
Poverty headcount rate, 2013 (%)	2.15	5.40	3.54	1.71	15.09	40.99	12.44
Projected poverty rate, 2030	0.63	2.34	0.36	0.62	5.98	30.32	8.19
High growth and high elasticity	0.36	1.25	0.11	0.51	2.75	26.83	6.38
Low growth and low elasticity	1.04	3.57	0.80	0.71	10.05	32.90	10.13
Projected population, 2030 (millions)	429	688	2206	460	2046	1471	7299
Projected poor population, 2030 (millions)	3	16	8	3	122	446	598
Share of poor population, 2030 (%)	0.45	2.69	1.32	0.48	20.47	74.58	100
High growth and high elasticity	0.33	1.84	0.50	0.50	12.07	84.75	100
Low growth and low elasticity	0.61	3.32	2.40	0.44	27.80	65.43	100

Note: Poverty headcount rate measured using the World Bank's US$1.90 (purchasing power parity) poverty line, and GDP per capita is in constant 2011 international dollars. Poverty and GDP per capita measures are adjusted for cross-country purchasing power differences. Future population based on the median variant fertility and life expectancy projections. *ECA*, Europe and Central Asia; *LAC*, Latin America and the Caribbean; *EAP*, East Asia and the Pacific; *MENA*, Middle East and North Africa; *SA*, South Asia; *SSA*, sub-Saharan Africa.

Authors' calculations using UNDESA (United Nations Department of Economic and Social Affairs), 2017. World population prospects: 2017 revision. United Nations Department for Economic and Social Affairs, United Nations, New York, NY, and World Bank, 2017a. PovcalNet: an online analysis tool for global poverty monitoring. World Bank, Washington, DC. <http://iresearch.worldbank.org/PovcalNet> (accessed August 2017) and World Bank, 2017b. World development indicators online database. World Bank, Washington, DC. <http://databank.worldbank.org/data/reports.aspx?source = world-development-indicators> (accessed August 2017).

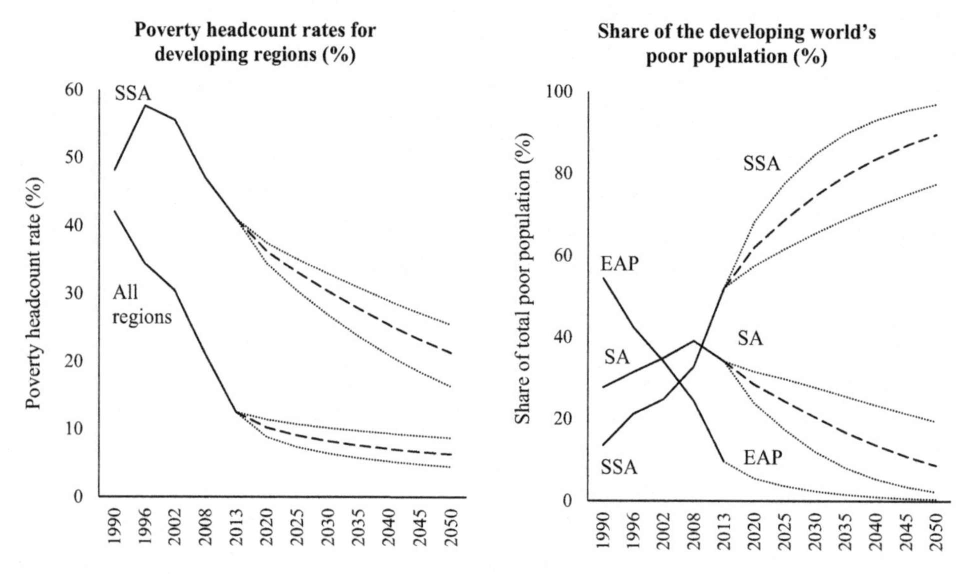

FIGURE 3.2 Past and projected distribution of global poor population, 1990–2050.
Note: Poverty headcount rate measured using a US$1.90 poverty line (purchasing power parity). Figure includes historical data (darker line) and projections (dotted and faint lines). Projections include upper and lower bounds, reflecting historical economic growth and poverty-growth elasticities for 2005–13 and 1993–2005, respectively. Projections maintain fixed list of developing countries. Annual shares do not sum to 100 because not all regions are shown. *EAP*, East Asia and the Pacific; *SA*, South Asia; *SSA*, sub-Saharan Africa.
Source: *Authors' calculations using population estimates for developing countries from UNDESA (United Nations Department of Economic and Social Affairs), 2017. World population prospects: 2017 revision. United Nations Department for Economic and Social Affairs, United Nations, New York, NY, and poverty estimates from World Bank, 2017a. PovcalNet: an online analysis tool for global poverty monitoring. World Bank, Washington, DC. <http://iresearch.worldbank.org/PovcalNet> (accessed August 2017).*

elasticities (ε) for 1993–2013, and for the two subperiods. This elasticity is defined as the percentage change in the poverty headcount rate, given a 1% change in per capita GDP. Regional poverty estimates are from the World Bank (2017b) and are based on the standard US$1.90 poverty line adjusted for differences in purchasing power across countries.[2] We again consider high and low scenarios using estimated elasticities from the subperiods 1993–2005 and 2005–13.

The main results from our poverty projections can be seen in Fig. 3.2. The left-hand panel shows past and projected poverty rates for SSA and for all developing regions. The global poverty rate is projected to fall from 12.4% in 2013 to 8.2% in 2030. This is close to the World Bank's (2015) 7.8% projected poverty rate for 2030, but is somewhat higher than projections in other studies (see, e.g., Ravallion, 2013). The lower bound estimate shown in the figure corresponds to the high growth and high elasticity combination and leads to a

[2] We impute poverty rates for the MENA region for 2005 and 2013. This is necessary because the survey sample sizes are considered too small to reliably estimate regional poverty rates. However, poverty rates are generally very low in the MENA region, and so this does not greatly affect our poverty projections.

global poverty rate of 6.4% in 2030. The figure also shows our projected poverty rate for SSA, which falls from 41.0% in 2013 to 30.3% in 2030 (and to 26.8% in the lower bound scenario). By 2030, we estimate that poverty in SSA will have fallen, even with rapid population growth and a possible return to the slow economic growth of the 1990s. However, even under more optimistic scenarios, we project that in 2030 SSA will have a poverty rate that is almost double the current SA rate.

Edward and Sumner (2014) discuss in detail the challenges of projecting poverty and the differences between approaches. In our analysis, for example, we assume constant poverty-growth elasticities, even though these normally increase over time as poverty rates fall. Similarly, we assume a continuation of the observed variation in economic growth rates across developing regions, even though we expect growth to decelerate as regions develop. That being said, and acknowledging that our analysis is relatively simple, it does generate results that are consistent with other studies' geographic distributions of poverty. The right-hand panel in Fig. 3.2 shows the three largest developing regions' projected shares of the global poor population. SA's share falls to 20% by 2030, whereas SSA's share rises to 75% (and to 85% in the lower bound scenario). More importantly, however, rapid population growth in SSA means that, while poverty rates in the region decline, the absolute number of poor people in the region is projected to increase. The number of poor people rises from 389 million in 2013 (Table 3.1) to 446 million in 2030 (Table 3.2). Under the more optimistic upper bound scenario, the number of poor people in SSA still rises to 430 million. Even if SSA could sustain the economic growth that it has enjoyed over the past decade, rapid population growth will still cause poor populations to expand in SSA over the coming decades. Tackling global poverty is rapidly becoming an African challenge.

A similar approach is used to project rural poverty. National population projections are disaggregated using urban population projections from UNDESA (2015). Rural and urban poverty-growth elasticities are estimated for developing regions using historical US$1.25 poverty rates from 58 countries, which together represent four-fifths of developing countries' total population. (Rural and urban US$1.25 – purchasing power parity – poverty rates were calculated by the World Bank for the International Fund for Agricultural Development, IFAD, 2016.) Population-weighted rural and urban elasticities are scaled proportionally to match the national elasticities used above. This assumes that the structural relationship between economic growth and rural and urban poverty reduction is similar to the US$1.25 and US$1.90 poverty lines. Thus, this approach ensures that regional poverty projections are both consistent with earlier analysis and subject to the same caveats. However, the additional assumptions required to reconcile rural and urban poverty estimates with national estimates mean that projected rural and urban poverty should be considered merely indicative (Fig. 3.3).

Poverty-growth elasticities are usually larger in urban areas. The average population-weighted rural and urban elasticities across all developing regions are −1.3 and −1.9, respectively. Only SA has a rural elasticity larger than its urban elasticity (−1.2 versus −0.8), implying that a given rate of national economic growth reduces rural poverty faster than urban poverty (although initial rural poverty rates are higher in almost all countries). Our analysis suggests that rural and urban poverty rates in SA will converge at around 6% by 2050 (down from 14.2 and 12.4% in 2015, respectively).

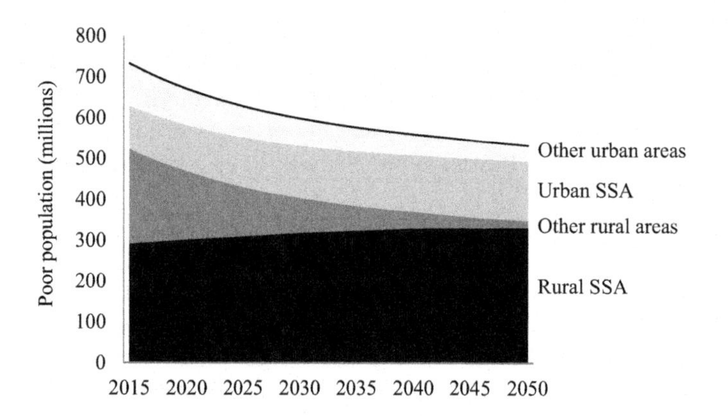

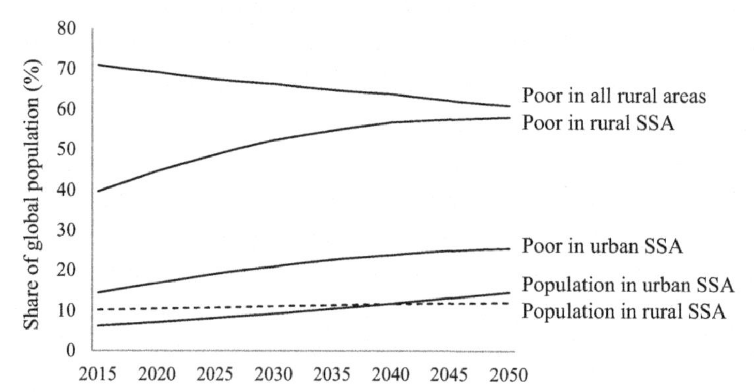

FIGURE 3.3 Projected poverty in rural and urban regions, 2015–50.

Note: Poverty headcount rate initially measured using a US$1.25 poverty line (purchasing power parity) and scaled to match the US $1.90 poverty line. Projections include historical economic growth and poverty-growth elasticities for 1993–2015. Projections maintain fixed list of developing countries. *SSA*, sub-Saharan Africa; "other" includes all developing countries outside of SSA. Source: *Authors' calculations using population estimates for developing countries from UNDESA (United Nations Department of Economic and Social Affairs), 2017. World population prospects: 2017 revision. United Nations Department for Economic and Social Affairs, United Nations, New York, NY, and poverty estimates from the World Bank (cited in IFAD, 2016. Rural development report 2016: fostering inclusive rural transformation. International Fund for Agricultural Development, Rome, Italy).*

The share of the world's poor population living in rural areas is projected to decline – from 71% in 2015 to 61% in 2050. This mainly reflects the global process of urbanization – the slower decline in urban rather than rural poverty in SA plays only a minor role. This is evident from the figure, which shows how the share of the world's poor population in rural areas falls along with the share of the world's total population living in rural areas.

Both national and rural poverty decline at the global level. By 2050, most of the world's remaining poor population will live in rural SSA (rising from 39.6% in 2015 to 58.1% in 2050).[3] This concentration of global poverty in rural SSA is in stark contrast to these areas' constant share of the global population (about one in ten people in the world live in rural SSA, a statistic that changes only slightly by 2050). In other words, the slow rate of rural poverty reduction in SSA outweighs the effects of urbanization, leaving its poor rural population growing in absolute terms. However, SSA's poor urban population expands more quickly than its poor rural population, implying a gradual "urbanization of poverty" from a relatively low base.

The challenge of reducing global poverty will increasingly focus on rural Africa. Underlying this trend are Africa's low poverty-growth elasticities. Finding ways to

[3] The share of the world's poor population living in urban SSA also increases by 11.2% points between 2015 and 2050 (compared with an 18.5% point increase for rural SSA).

enhance the structural linkages between national growth and rural poverty is crucial for accelerating the pace of global poverty reduction. The role of agricultural growth in strengthening economy-wide poverty-growth linkages is discussed later in this chapter.

Youth Bulges and Aging Farmers

Along with concerns about projected increases in total populations, African governments are increasingly focused on the age composition of the labor force. The SSA recently experienced its peak youth bulge, measured as the share of youth (15–24 years of age) in the total working age population (15–64 years). In contrast, the rest of the developing world experienced their peak youth bulges about three decades ago (Fig. 3.4). Africa only has 40–50 years before it will experience its "baby boom" transition, at which point it will face the same challenges that East Asia is starting to face, i.e., having sufficient wealth and social systems in place to support an aging population (Table 3.3).

Most of the developing world, apart from SSA, has already experienced its demographic transition, and in some regions concerns have shifted from employing youth populations to managing an aging agricultural workforce. SSA is the outlier. It has experienced its youth bulge far later than other countries have, but within only a few decades it will reach its own demographic transition point. Already, the implications of being a lagging developing region are evident, with poverty, especially rural poverty, concentrating rapidly in SSA. Future population growth is largely a given, and so Africa needs to significantly raise economic growth and enhance its effectiveness to reduce poverty. This means both increasing the pace of structural change and sectoral linkages to poverty and balancing the trade-offs between urbanization and rural development.

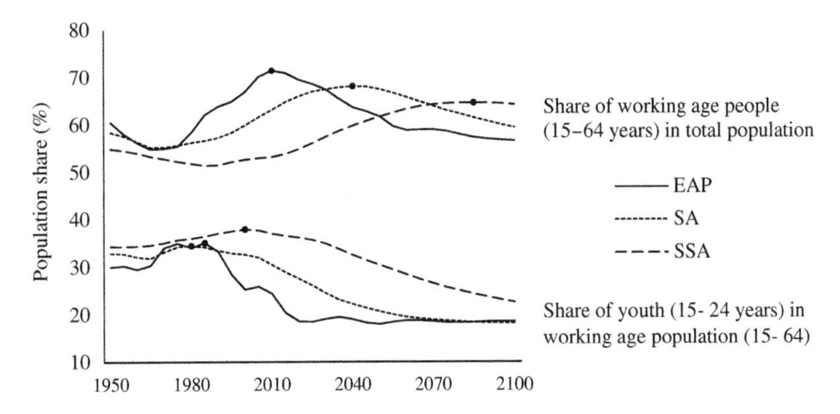

FIGURE 3.4 Onset of youth bulges and baby booms, 1950–2100.
Note:Black dots indicate years where shares peak. *EAP*, East Asia and the Pacific; *SA*, South Asia; *SSA*, sub-Saharan Africa. Source: *Own calculations using population estimates for developing countries from UNDESA (United Nations Department of Economic and Social Affairs), 2017. World population prospects: 2017 revision. United Nations Department for Economic and Social Affairs, United Nations, New York, NY.*

TABLE 3.3 Rural Youth and Older Adult Populations, 1980–2015

	ECA	LAC	EAP	MENA	SA	SSA	All
SHARE OF YOUTH (15–24) IN RURAL WORKING AGE POPULATION (15–64) (%)							
1980	16.9	19.0	19.4	18.7	19.0	18.0	18.9
1990	15.0	18.7	21.0	18.9	18.7	18.2	19.4
2005	17.3	18.7	17.4	22.8	18.9	18.9	18.5
2015	14.1	18.2	13.9	18.9	17.9	18.7	16.7
SHARE OF OLDER ADULTS (65 +) IN THE TOTAL RURAL POPULATION (%)							
1980	10.4	4.1	4.8	3.7	3.8	3.2	4.6
2015	11.8	7.2	9.1	5.4	5.6	3.5	6.6
SHARE OF GLOBAL TOTAL, 2015 (%)							
Youth in rural areas (15–24)	3.9	4.3	25.7	5.0	39.2	21.9	100
Older adults in rural areas (65 +)	8.2	4.3	42.6	3.6	30.9	10.4	100

Note: Own calculations using population estimates for developing countries from UNDESA (2017). *ECA*, Europe and Central Asia; *LAC*, Latin America and the Caribbean; *EAP*, East Asia and the Pacific; *MENA*, Middle East and North Africa; *SA*, South Asia; *SSA*, sub-Saharan Africa.
Authors' calculations using UNDESA (United Nations Department of Economic and Social Affairs), 2017. World population prospects: 2017 revision. United Nations Department for Economic and Social Affairs, United Nations, New York, NY.

3.3 GROWTH AND STRUCTURAL CHANGE

Exiting Agriculture

Economic development has historically been characterized by sustained structural change involving a shift of labor out of agriculture and into the "modern" industrial sector. The result of this process is an increased share of nonagricultural sectors in GDP, as well as in employment. In the Lewis dual-economy model (Lewis, 1954), an early theoretical formulation of this process, investment in modern sector capital drives economic growth as excess labor in agriculture (where its marginal product was zero) moves to the modern sector. For most of Asia and Latin America, this modern sector has been industry; rapid growth in Asia has also been spurred by exports of industrial products (especially in China and other parts of East Asia). In Africa in recent decades, however, structural transformation has been characterized by greater employment in (mostly informal) services rather than industry (McMillan et al., 2014).

As shown in Fig. 3.5, since the early 1990s, EAP and SA have experienced the most rapid structural change. In EAP, the share of agriculture in GDP fell by 13.5% (from 22.8 in 1991 to 9.3% in 2016), and the share of agricultural employment fell by 25.4% (from 55.3 to 29.9%). The change in SA was almost as large, with shares of agriculture in GDP and employment falling by 12.2 and 16.4%, respectively. Structural transformation in Latin America has already occurred in much of the region, and thus recent changes have been small (2.9 and 5.9% declines for shares of agriculture in GDP and employment, respectively). In contrast, SSA countries continue to have economies with very large shares of

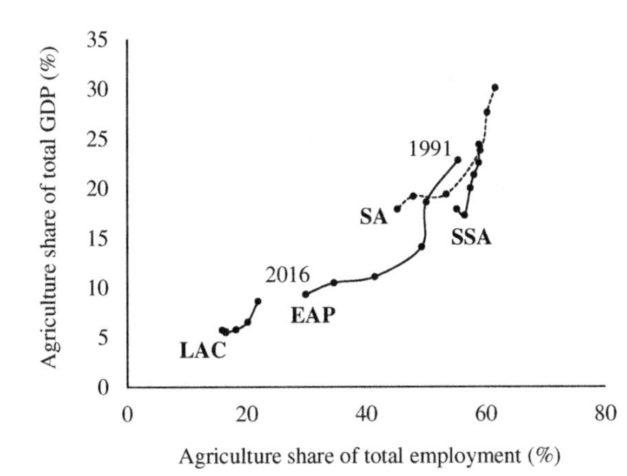

FIGURE 3.5 Agriculture and structural change by region, 1991–2016. Source: *Own calculations using employment data from ILO, 2017. Key Indicators of the Labour Market (KILM). International Labour Organization, Geneva. Available from: <http://www.ilo.org/ilostat> and GDP from the World Bank, 2017b. World development indicators online database. World Bank, Washington, DC. <http://databank.worldbank.org/data/reports.aspx?source = world-development-indicators> (accessed August 2017)).*

TABLE 3.4 Agriculture and Structural Change by Major Developing Region, 1991–2016

	EAP	LAC	SA	SSA
AGRICULTURE'S EMPLOYMENT SHARE (%)				
1991	22.8	8.6	30.1	24.4
2016	9.3	5.7	17.9	17.9
Change (%-point)	– 13.5	– 2.9	– 12.2	– 6.5
AGRICULTURE'S GDP SHARE (%)				
1991	55.3	21.8	61.6	58.8
2016	29.9	15.9	45.2	55.1
Change (%-point)	– 25.4	– 5.9	– 16.4	– 3.7

Note: *EAP*, East Asia and the Pacific; *LAC*, Latin America and the Caribbean; *SA*, South Asia; *SSA*, sub-Saharan Africa.
Calculated from FAO and World Bank data.

agricultural GDP and labor force. Moreover, structural transformation proceeds at a relatively slow structural change, as the agricultural labor force declined by only 6.5% and the share of agriculture in GDP fell by only 3.7% (Table 3.4). The rates of structural change in SSA and selected Asia and LAC countries are shown in Annex Figure 3A.1 (Fig. 3.6 and Fig 3.7).

Structural Change and Poverty Reduction

East Africa was the most successful of SSA's subregions during the period 2000–15. Rapid economic growth in this region was driven primarily by an expansion of trade and transport services, which coincided with SSA's fastest declines in national poverty. Agriculture was another important driver of growth, but this was mainly due to its large

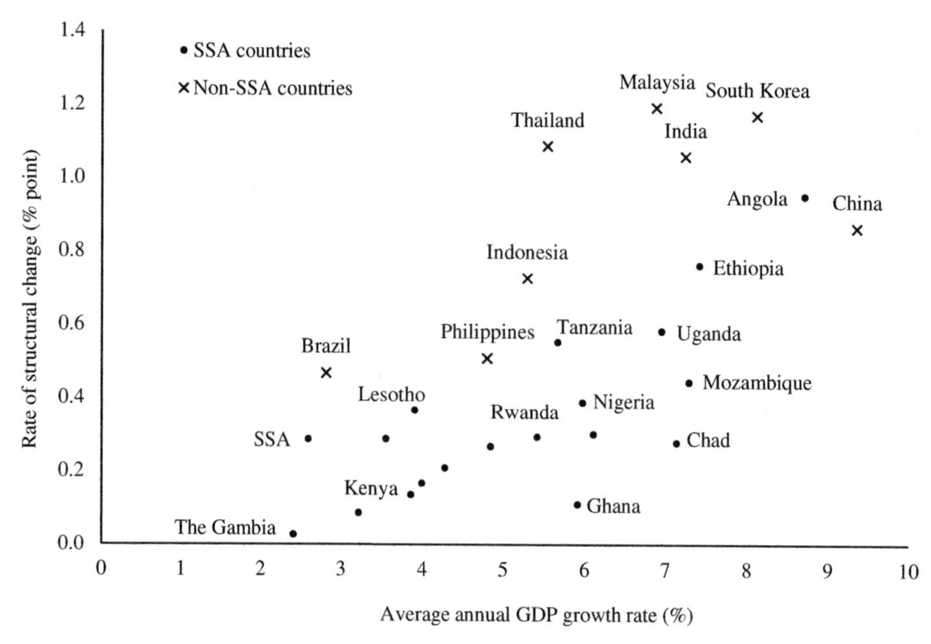

ANNEX. FIGURE 3A.1 Rates of structural change for selected countries and time periods.
Note: Rate of structural change is the annual percentage point increase in the share of nonagricultural employment in total employment. Non-SSA countries analysis is for different time periods: China (1994–2005); Indonesia (1985–2012); South Korea (1980–2000); Malaysia (1988–2001); Philippines (1991–2012); Thailand (1980–2010); India (1980–2010); and Brazil (1981–2011). All SSA countries are for 1993–2013. *SSA*, sub-Saharan Africa. Source: *Own calculations using employment data from ILO, 2017. Key Indicators of the Labour Market (KILM). International Labour Organization, Geneva. Available from: <http://www.ilo.org/ilostat>) and GDP from the World Bank (2017).*

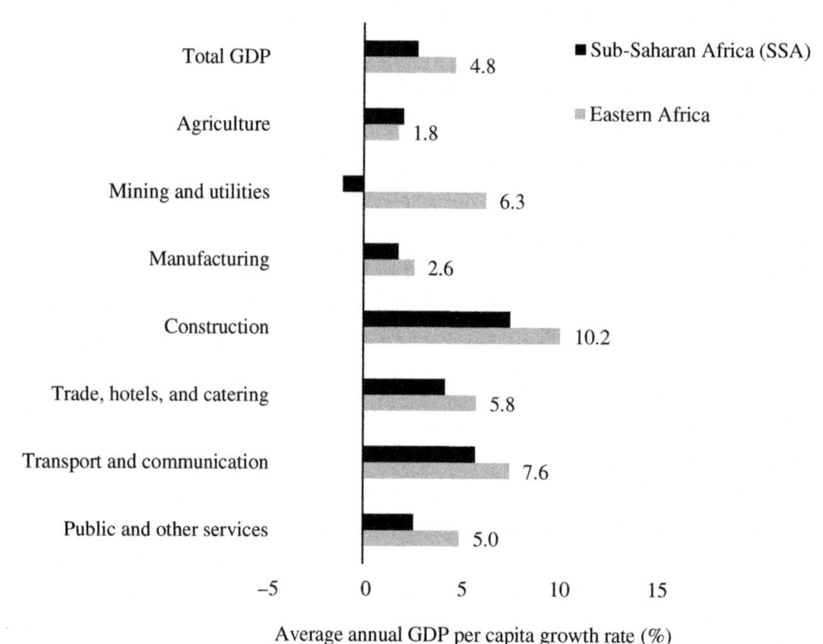

FIGURE 3.6 Sectoral GDP growth rates for selected countries in Eastern Africa, 2004–14. Source: *Own calculations national accounts data from UNSD, 2016. National accounts main aggregates database. United Nations Statistical Division, New York, NY. Available from: <http://www.unstats.un.org>.*

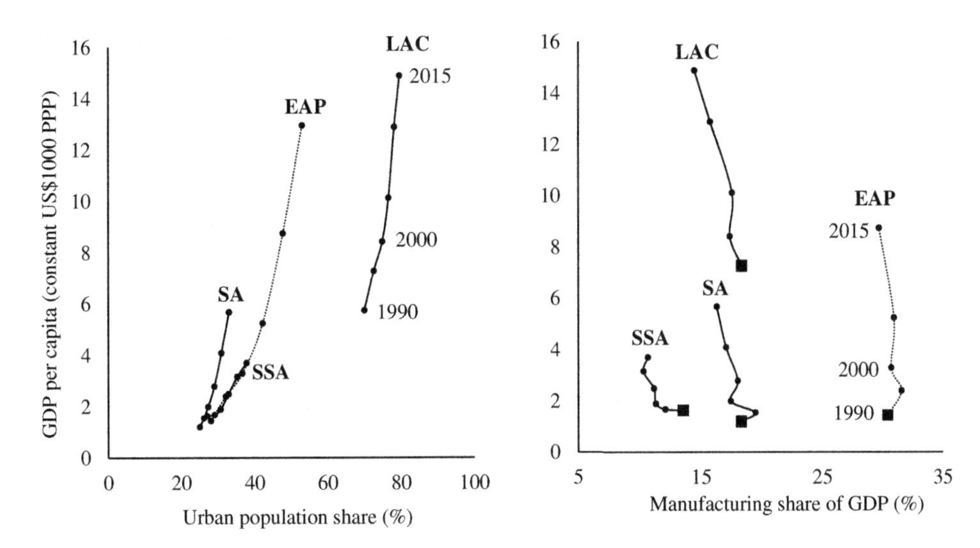

FIGURE 3.7 Urbanization, industrialization, and economic development, 1990–2015.
Note: GDP measured in constant 2011 purchasing power parity (PPP) adjusted dollars. *EAP*, East Asia and the Pacific; *LAC*, Latin America and the Caribbean; *SA*, South Asia; *SSA*, sub-Saharan Africa. Source: *Authors' calculations using populations from UNDESA (United Nations Department of Economic and Social Affairs), 2015. World urbanization prospects: 2014 revision. United Nations Department for Economic and Social Affairs, United Nations, New York, NY and UNDESA (United Nations Department of Economic and Social Affairs), 2017. World population prospects: 2017 revision. United Nations Department for Economic and Social Affairs, United Nations, New York, NY, and GDP from the World Bank, 2017b. World development indicators online database. World Bank, Washington, DC. <http://databank.worldbank.org/data/reports.aspx?source = world-development-indicators> (accessed August 2017).*

initial size. As shown by Dorosh and Thurlow (2016), both agriculture and trade services are major drivers of poverty reduction. The authors estimate poverty-growth elasticities at the sector level, and use these elasticities to compare the effectiveness of growth led by different sectors in reducing poverty. Their findings indicate that agriculture does have the highest poverty-growth elasticities, but that trade and transport services also have relatively high elasticities (Table 3.5). Even though the trade sector's elasticity is half that of agriculture, trade services accounted for about two-thirds of overall economic growth. It follows that trade and transport services probably accounted for most of East Africa's poverty reduction after 2000.

The East Africa case study suggests that nonagricultural sectors, such as trade and transport services, may be more important than agriculture for future poverty reduction in Africa. However, this finding hides the structural linkages between agriculture and Africa's nonagricultural economies. These links are particularly strong in the case of trade and transport services, both major components of Africa's agriculture and food system (AFS).

Table 3.6 provides estimates of the share of national GDP generated by the AFS in selected African countries. The analysis uses social accounting matrices, or SAMs, which are similar to input-output and supply-use tables in that they capture the upstream and downstream flows of agricultural products. Agriculture and agroprocessing (within manufacturing) simply reflect the share of GDP generated in these two sectors, as captured in national accounts data. The share of GDP from input production captures the

TABLE 3.5 Poverty-Growth Elasticities for Selected Countries in Eastern Africa

Lead growth sectors	Malawi	Mozambiq-ue	Tanzania	Uganda	Zambia	Simple Average
All sectors	− 0.51	− 0.54	− 0.60	− 0.52	− 0.52	− 0.54
Agriculture	− 0.80	− 1.27	− 1.02	− 0.94	− 0.65	− 0.94
Nonagriculture	− 0.41	− 0.36	− 0.34	− 0.45	− 0.46	− 0.40
Mining and utilities	− 0.20	n/a	− 0.20	n/a	− 0.55	− 0.32
Manufacturing	− 0.62	− 0.43	− 0.73	− 0.70	− 0.79	− 0.65
Agroprocessing	− 0.67	− 0.40	− 1.04	− 0.48	− 0.89	− 0.70
Construction	− 0.14	− 0.14	0.00	− 0.05	− 0.16	− 0.10
Utilities	n/a	− 0.41	− 0.34	n/a	− 0.48	− 0.41
Trade and transport	− 0.59	− 0.44	− 0.61	− 0.61	− 0.64	− 0.58
Finance and business	− 0.19	− 0.42	− 0.27	− 0.51	− 0.47	− 0.37
Government services	− 0.02	0.00	− 0.01	− 0.17	− 0.08	− 0.06
Other services	− 0.33	− 0.23	n/a	− 0.47	n/a	− 0.34

Note: Semielasticities are the percentage point change in the national poverty headcount rate given a 1% increase in GDP per capita. Poverty is measured using the US$1.25 poverty line (after adjusting for purchasing power).
Computable general equilibrium modeling results reproduced from Dorosh and Thurlow, 2016. Beyond agriculture versus non-agriculture: decomposing sectoral growth–poverty linkages in five African countries. World Dev. From World Bank, 2017b. World development indicators online database. World Bank, Washington, DC. <http://databank.worldbank.org/data/reports.aspx?source = world-development-indicators> (accessed August 2017).

TABLE 3.6 Estimates of the Agrifood System Gross Domestic Product (GDP) for Selected African Countries

		Share of National GDP (%)					
Country	**Year**	**Agriculture**	**Processing**	**Trade and Transport**	**Input Production**	**Prepared Meals**	**Total AFS**
Ethiopia	2011	41.7	2.1	4.4	0.8	1.2	50.2
Ghana	2013	21.7	1.6	3.9	0.7	5.4	33.3
Kenya	2013	27.8	4.0	3.9	1.3	0.5	37.7
Malawi	2010	32.3	6.0	10.8	1.7	1.2	51.9
Mozambique	2012	28.0	4.6	5.5	1.0	1.9	41.0
Nigeria	2010	24.2	4.2	10.8	0.1	0.2	39.5
Rwanda	2011	33.7	4.7	11.9	1.4	1.7	53.3
Tanzania	2007	31.8	5.6	4.9	0.8	1.5	44.7
Uganda	2007	21.7	3.5	3.8	2.5	0.4	31.9
Zambia	2007	20.2	6.0	10.6	2.6	0.7	40.0
Simple average		28.3	4.2	7.1	1.3	1.5	42.4

Note: *AFS*, agriculture and food system.
Own calculations using IFPRI social accounting matrices for the indicated years.

domestic value added that is generated when producing the intermediate inputs used in agriculture and agroprocessing. Similarly, the share of GDP attributed to the AFS is based on detailed product-level trade and transport margins for agricultural and processed products. The GDP from trade and transport therefore only captures the part of national trade and transport GDP that is actually agriculture and food related. Finally, the term "prepared meals" refers to the value added generated by restaurants and food vendors, which are an increasingly important part of the food system, especially in urban areas.

The importance of Africa's AFS goes well beyond agriculture itself. For the selected countries, agriculture accounted for 28.3% of national GDP, but the AFS accounted for 42.4% (Table 3.6). Most of the additional AFS GDP in the nonagricultural sectors comes from the value added generated in the trade and transport sectors, i.e. from moving agricultural and processed foods between farmers, firms, and consumers. SSA is rapidly shifting from subsistence to commercial agriculture, increasing demand for trade and transport services. A substantial share of Africa's expansion in trade and transport services, which has driven overall economic growth since 2000, is related to an expansion of Africa's AFS. Similarly, while manufacturing in Africa has not expanded at rates similar to those observed in EAP, much of the manufacturing that does exist is directly or indirectly related to agriculture.

Agriculture therefore still lies at the heart of much of the growth-driven poverty reduction that is taking place in Africa. Given the large size of the AFS and the growing concentration of poverty in rural Africa, agriculture will remain the major driver of both structural change and poverty reduction in the coming decades. Raising farm productivity, releasing labor to the nonfarm sector, and promoting agricultural transformation by expanding the broader AFS are crucial steps towards reducing African and, by implication, global poverty.

3.4 URBANIZATION AND AGRICULTURE

Sustaining rural employment and income growth in SSA will depend on the ability of countries to expand and transform their AFS. As mentioned above, this will not only require higher agricultural productivity, but also increased value addition and job creation in downstream agriculture-related processing, trade, and transport. Incomes from agriculture will constitute a platform for growth in the rural nonfarm economy. However, rapid urbanization and economic growth in cities and towns will play a crucial role. Rural-urban linkages have long been recognized as a key aspect of economic development (Lewis, 1954; Haggblade et al., 2007). Urban industrial and service sector growth provides employment for workers who exit agriculture, while increases in agricultural production can help avoid an increase in food prices and wages that could slow the pace of industrialization. Agricultural growth can also spur nonagricultural growth in urban areas through demand for both inputs, including fertilizer and machinery as well as consumer goods.[4]

[4] This simplified discussion essentially equates agriculture with rural and industry (and services) with the urban economy. In reality, the rural sector includes significant nonagricultural activities (including small-scale agricultural processing, trade and rural services). Likewise, some urban residents typically are involved in agricultural production in the periphery of cities or in small vegetable gardens. The simulation models described in this section include these important features of African economies.

As discussed above, structural transformation (the movement of labor from lower productivity to higher productivity sectors) generally involves a substantial out-migration of labor from rural to urban areas. To facilitate this transformation (and economic growth), many countries have emphasized public investment in infrastructure to support urban nonagricultural sectors, as well as direct investment in selected industries. Given the high share of the population, and particularly the high share of the poor in rural areas, many African countries have made agricultural investments a high priority. For example, under the Comprehensive Africa Agricultural Development Program, 40 countries have pledged to dedicate at least 10% of total public expenditures to the agricultural sector. In general, the results, in terms of agricultural growth, have been positive, with agricultural GDP increasing substantially in many countries (Benin, 2016).

Even apart from political considerations, the decision to invest in agriculture and the rural economy, on the one hand, or invest in urban industrial and service sectors, on the other, involves short-term and medium-term trade-offs between economic growth and poverty reduction outcomes. Since in general, urban sectors, particularly the industrial and formal service sector, have higher average labor productivities than the agricultural sector, promoting the migration of labor out of rural agriculture and into these urban sectors is likely to raise economy-wide labor productivity and accelerate national economic growth. However, an out-migration of working age adults from rural areas could lower rural incomes, especially if growing urban demand for agricultural and food products is met by imports rather than by domestic producers. The overall effect of urbanization on agriculture and rural poverty depends on the structural linkages between urban and rural economies, which evolve over time, at least partly in response to public investment and policy decisions.

A growing body of research evaluates the synergies and trade-offs between different urbanization pathways in SSA (Christiaensen and Todo, 2014; Dorosh and Thurlow, 2013, 2014). Economy-wide analysis for Ethiopia and Uganda suggests that, given the large shares of agriculture in these countries' GDP (48.1 and 24.2% in 2005, respectively) and high concentrations of poverty in rural areas (89 and 94%, respectively), the trade-offs between accelerating economic growth and reducing poverty do exist (Dorosh and Thurlow, 2014).[5] These authors use economy-wide models that capture structural linkages between rural and urban economies; relative-wage-driven rural-to-urban migration flows; and urban agglomeration and congestion effects arising from rising urban population densities and falling per capita public capital. Table 3.7 summarizes some of the authors' main findings.

Model simulations for 2005–25 indicate that accelerating the rate of urbanization leads to faster national economic growth. For example, a 10% increase in urbanization (from about 20% in baseline scenarios to about 30%) results in a 0.6% increase in the annual

[5] These results are based on policy simulations using neoclassical computable general equilibrium models for Ethiopia and Uganda. In the simulations, rural-urban migration is modeled as an exogenous trend (that is based on historical flows) multiplied by a constant elasticity function of relative rural and urban wage rates. Changes in the levels of public investment by sector are modeled as exogenous increases in productivity of nonagricultural public capital (for nonagricultural sectors) or increases in total factor productivity for agricultural sectors. For further details, see Dorosh and Thurlow (2013).

TABLE 3.7 Effects of Urbanization and Alternative Allocations of Public Investments (Simulation Results)

			Deviation from Baseline (% point)		
Ethiopia	**Baseline**	**Urbanization**	**Investment in Cities**	**Investment in Towns**	**Investment in Rural Areas**
Annual GDP growth	5.21	0.59	0.23	0.19	− 0.19
Annual welfare change	1.73	0.41	0.08	0.10	0.10
Poor	1.17	0.25	− 0.03	0.04	0.30
Nonpoor	1.86	0.44	0.10	0.12	0.05
Rural areas	1.38	0.90	− 0.02	0.04	0.30
Towns	0.72	− 1.22	0.14	0.19	− 0.14
Cities	1.69	− 1.95	0.29	0.15	− 0.22
UGANDA					
Annual GDP growth	5.99	1.57	0.50	0.39	− 0.24
Annual welfare change	2.45	1.31	0.31	0.20	0.01
Poor	1.50	1.05	0.16	0.08	0.29
Nonpoor	2.59	1.35	0.33	0.21	− 0.03
Rural areas	1.81	1.68	0.23	0.12	0.19
Towns	2.21	− 0.71	0.30	0.34	− 0.25
Cities	3.64	− 0.25	0.59	0.31	− 0.36

Note: *GDP*, gross domestic product.
From Dorosh and Thurlow, 2013. Agriculture and small towns in Africa. Agric. Econ. 44(3): 435–445).

GDP growth rate in Ethiopia (from 5.2% in the base to 5.8% with faster urbanization). This additional growth occurs despite the tendency of new migrants to find employment in lower-paying urban services, as opposed to higher-productivity manufacturing, a result that is consistent with observed patterns of structural change in SSA. Similarly, in Uganda, faster urbanization raises annual GDP growth by 2.3%, from 6.0 to 6.6%, as workers move to higher productivity sectors, albeit mainly into informal trade services. Moreover, declining rural labor supply caused by out-migration leads to higher real wage rates and real per capita incomes in rural areas, even without additional investments in the agricultural sector. Unfortunately, the authors find that urban migrants drive down average real wage rates in cities and towns, causing real urban per capita incomes to fall. Simply put, without supporting investments in urban infrastructure and services, an increase in urbanization may lead to higher urban poverty, despite faster urban and national economic growth. Rural areas benefit from higher urban food demand, but unsupported urbanization is not a pathway to broad-based transformation.

Urbanization requires additional public investment in urban areas to absorb and accommodate migrant workers and their families. Recent studies for Malawi and Mozambique using economy-wide modeling showed that responding to urbanization by reallocating public expenditures from agriculture to urban areas leads to higher poverty, even in urban areas (World Bank, 2016). This is because poor urban consumers are greatly affected by changes in real food prices. Reducing public spending on agriculture leads to lower agricultural production and higher food prices, which in turn reduces real incomes for poorer urban consumers. Instead, studies show that if urban areas finance their own urban investments, an increase in national economic growth results as well as faster poverty reduction in both rural and urban areas.

Other studies find similar trade-offs between rural and urban investments. Dorosh and Thurlow (2014) show that increasing investment in urban infrastructure in Ethiopia and Uganda leads to faster national economic growth, partially due to urban agglomeration effects and increased urban capital. A similar reallocation of public investment to rural sectors raises agricultural productivity, but lowers national growth, in part because agricultural productivity growth is constrained by market demand and because low density of population and economic activity in rural areas do lead to the positive agglomeration effects associated with urban development. In both countries, however, the authors find that investing in agriculture is more effective for reducing national poverty. In conclusion, rapid urbanization in Africa should not detract from the ongoing need to invest in agriculture and rural development. Rural-urban linkages mean the national AFS spans both rural and urban areas, and that broad-based investments in the AFS will benefit both rural and urban consumers, particularly those in poorer households. This conclusion is consistent with the Food and Agriculture Organization of the United Nation's recent flagship report *The State of Food and Agriculture* and the International Food Policy Research Institute's *2017 Global Food Policy Report*, which both suggest that a balanced growth strategy that includes strong agricultural growth along with growth in urban, nonagricultural sectors is needed to achieve both rapid overall growth and poverty reduction (FAO, 2017; IFPRI, 2017).

3.5 CONCLUSIONS

Both international experience and economic theory demonstrate that structural change is an essential component of long-term economic development. Urbanization is also strongly associated with structural change, primarily because a movement to urban centers is synonymous with a movement out of low-productivity agriculture. This process was the major driver of East Asian development. The region experienced rapid structural change and urbanization, along with substantial reductions in rural and urban poverty. This process is also occurring in SA, albeit at slower rates of economic growth and lower levels of urbanization. By 2050, few of the world's poor people will live in SA, and concerns about rapid population growth will shift to concerns about aging agricultural workforces and pressures on social welfare systems.

SSA, on the other hand, is not only a late-transforming region, but also faces unique challenges on an unprecedented scale. Rapid population growth means that rural populations will continue to expand in the future despite rapid urbanization. Economic growth in Africa has also not been driven by manufacturing, as it was in East Asia, but rather by low-productivity services. The movement out of agriculture has therefore led to only modest increases in labor productivity, and by implication, modest accelerations in economic growth. Rapid population growth means that not only are the benefits of structural change small in Africa, but the rate of structural change is likely to lag behind the Asian experience because Africa's urban economies are unable to absorb enough rural-to-urban migrants and their families. Together, the implication of these demographic and economic trends is that future efforts to reduce global poverty and poverty in rural Africa are fast becoming one and the same thing.

Agriculture and the rural economy will play crucial roles in overcoming Africa's future demographic and economic challenges. Economic growth in the region has accelerated over the past decade, due in some measure to structural change driven by an expansion of the agricultural and food systems. While agriculture is declining as a share of African economies, the AFS reaches well beyond the agriculture sector itself, and likely remains the major source of economic growth and poverty reduction in most African countries. Moreover, the continued shift from subsistence to commercial agriculture will create new income opportunities for rural households, including jobs in agricultural trade and transport. Urbanization, particularly in smaller towns, will also create demand for agricultural goods, including new opportunities in the food processing sectors.

Demographic change is both an opportunity and a potential constraint to agricultural growth and poverty reduction in rural Africa. Government policy is required to ensure that the opportunities created by urbanization and urban development do benefit rural households. Responding to urbanization by shifting public resources away from agriculture towards urban infrastructure, while beneficial for national economic growth, may in fact lead to higher poverty, including in urban areas. Increased public investments in both agriculture and urban infrastructure are needed, as is better alignment between agricultural and urban policy. While agriculture's importance in the overall economy declines, its role in facilitating broad-based development remains unchanged despite the demographic forces that are reshaping Africa.

While Africa's development challenges are clear, further analytical work is needed on how best to promote inclusive growth and structural transformation, and how to sustain poverty reduction in rural Africa. Among the key questions is the role of small towns and secondary cities, a major policy issue for government infrastructure investment decisions with important implications for growth and equity. Likewise, further analysis is needed regarding the new kinds of infrastructure that will be needed to transform Africa's AFSs. This includes rapid technical change in the generation and storage of electricity, particularly solar energy and improved batteries, as well as major investments in hydroelectric power that could create huge opportunities for high-value agriculture, through such measures as the establishment of cold chains linking producers of meat, dairy, and vegetables to lucrative urban and export markets. Promoting agricultural transformation will increasingly require investments outside the agricultural sector and a stronger alignment of agricultural and national development policies and investments.

References

Becker, G.S., Murphy, K.M., Tamura, R., 1990. Human capital, fertility, and economic growth. J. Polit. Econ. 98 (5), S12–S37.

Benin, S., 2016. Agricultural Productivity in Africa: Trends, Patterns, and Determinants. International Food Policy Research Institute, Washington, DC.

Brückner, M., 2012. Economic growth, size of the agricultural sector, and urbanization in Africa. J. Urban Econ. 71 (1), 26–36.

Chamberlin, J., Jayne, T.S., Headey, D., 2014. Scarcity amidst abundance? Reassessing the potential for cropland expansion in Africa. Food Policy 48 (C), 51–65.

Christiaensen, L., Todo, Y., 2014. Poverty reduction during the rural–urban transformation – the role of the missing middle. World Dev. 63, 43–58.

Cohen, B., 2004. Urban growth in developing countries: a review of current trends and a caution regarding existing forecasts. World Dev. 32 (1), 23–51.

Dorosh, P., Thurlow, J., 2013. Agriculture and small towns in Africa. Agric. Econ. 44 (3), 435–445.

Dorosh, P., Thurlow, J., 2014. Can cities or towns drive African development? Economywide analysis for Ethiopia and Uganda. World Dev. 63, 113–123.

Dorosh, P., Thurlow, J., 2016. Beyond agriculture versus non-agriculture: decomposing sectoral growth–poverty linkages in five African countries. World Dev.

Edward, P., Sumner, A., 2014. Estimating the scale and geography of global poverty now and in the future: how much difference do method and assumptions make? World Dev. 58, 67–82.

FAO, 2017. State of food and agriculture report 2017: leveraging food systems for inclusive rural transformation. Food and Agriculture Organization of the United Nations, Rome, Italy.

Fan, S., Chan-Kang, C., 2005. Is small beautiful? Farm size, productivity, and poverty in Asian agriculture. Agric. Econ. 32 (1), 135–146.

Feder, G., O'Mara, G.T., 1981. Farm size and the diffusion of green revolution technologies. Econ. Dev. Cult. Change 30 (1), 59–76.

Filmer, D., Fox, L., 2014. Youth Employment in Africa. World Bank, Washington DC.

Haggblade, S., Hazell, P.B.R., Dorosh, P.A., 2007. Sectoral growth linkages between agriculture and the rural nonfarm economy. In: Haggblade, S., Hazell, P.B.R., Reardon, T. (Eds.), Transforming the Rural Nonfarm Economy: Opportunities and Threats in the Developing World. Johns Hopkins University Press, Baltimore, MD.

Hazell, P.B.R., 2005. Is the a future for small farms? Agric. Econ. 32 (s1), 93–101.

IFAD, 2016. Rural development report 2016: fostering inclusive rural transformation. International Fund for Agricultural Development, Rome, Italy.

IFPRI, 2017. Global food policy report 2017. International Food Policy Research Institute, Washington, DC.

ILO, 2017. Key Indicators of the Labour Market (KILM). International Labour Organization, Geneva. Available from: <http://www.ilo.org/ilostat>

Jayne, T.S., Mather, D., Mghenyi, E., 2010. Principal challenges confronting smallholder agriculture in Sub-Saharan Africa. World Dev. 38 (10), 1384–1398.

Leavy, J., Hossain, N., 2014. Who wants to farm? Youth aspirations, opportunities and rising food prices. IDS Working Papers 439, 1–44.

Lewis, W.A., 1954. Economic development with unlimited supplies of labor. The Manchester School 22 (2), 139–191.

Masters, W.A., Djurfeldt, A.A., De Haan, C., Hazell, P., Jayne, T., Jirström, M., et al., 2013. Urbanization and farm size in Asia and Africa: implications for food security and agricultural research. Global Food Security 2 (3), 156–165.

McMillan, M., Rodrik, D., Verduzco-Gallo, I., 2014. Globalization, structural change, and productivity growth, with an update on Africa. World Dev. 63 (C), 11–32.

Moomaw, R.L., Shatter, A.M., 1996. Urbanization and economic development: a bias towards large cities? J. Urban Econ. 40, 13–37.

Ravallion, M. 2013. How long will it take to lift one billion people out of poverty? World Bank Policy Research Working Paper 6325. World Bank, Washington DC.

Tschirley, D., Reardon, T., Dolislager, M., Snyder, J., 2015. The rise of a middle class in East and Southern Africa: Implications for food system transformation. J. Int. Dev. 27 (5), 628–646.

UNDESA (United Nations Department of Economic and Social Affairs), 2015. World urbanization prospects: 2014 revision. United Nations Department for Economic and Social Affairs, United Nations, New York, NY.

UNDESA (United Nations Department of Economic and Social Affairs), 2017. World population prospects: 2017 revision. United Nations Department for Economic and Social Affairs, United Nations, New York, NY.

UNSD, 2016. National accounts main aggregates database. United Nations Statistical Division, New York, NY. Available from: <http://www.unstats.un.org>.

World Bank, 2015. A measured approach to ending poverty and boosting shared prosperity: concepts, data, and the twin goals. Policy Research Report.World Bank, Washington, DC.

World Bank, 2016. Malawi urbanization review: leveraging urbanization for national growth and development. World Bank, Washington, DC.

World Bank, 2017a. PovcalNet: an online analysis tool for global poverty monitoring.World Bank, Washington, DC. <http://iresearch.worldbank.org/PovcalNet> (accessed August 2017).

World Bank, 2017b. World development indicators online database. World Bank, Washington, DC. <http://databank.worldbank.org/data/reports.aspx?source = world-development-indicators> (accessed August 2017).

CHAPTER

4

Climate Change, Agriculture and Food Security: Impacts and the Potential for Adaptation and Mitigation

Keith Wiebe[1], *Sherman Robinson*[1] *and Andrea Cattaneo*[2]

[1]International Food Policy Research Institute (IFPRI), Washington, DC, United States
[2]Food and Agriculture Organization of the United Nations (FAO), Rome, Italy

4.1 CLIMATE CHANGE CHALLENGES[1]

Climate change is a significant and growing threat to food security—already affecting vulnerable populations in many developing countries and expected to affect an ever-increasing number of people across more areas in the future unless immediate actions are taken. Current scenarios for "business-as-usual" farming under climate change predict increasing food security challenges by 2050. The worst hit will be underdeveloped economic regions of the world where food security is already problematic and populations are vulnerable to shocks (Rosegrant et al., 2014). Without substantial measures that address the challenges caused by increasing temperatures and frequencies of extreme weather events, losses in crop and livestock productivity are expected to reduce the rate of

This chapter builds on IFPRI's work and scenario analysis with the IMPACT model assessing the impacts of climate change on food security and nutrition and potential for adaptation and mitigation of the agricultural sectors and food systems, as well as on FAO's work on climate-smart agriculture and findings presented in FAO's *The State of Food and Agriculture 2016*.

[1] Sections 4.1 and 4.2 are drawn in part from De Pinto et al. (2016) and from IFPRI (2017).

Sustainable Food and Agriculture
DOI: https://doi.org/10.1016/B978-0-12-812134-4.00004-2

gain from technological and management improvements (Lobell and Gourdji, 2012). Furthermore, climate change not only threatens the productivity of the world's agricultural systems, but also is expected to have consequences for a wide range of ecosystem services (Knight and Harrison, 2013).

Uncertainties in climate change scenarios make it difficult to determine the precise impacts on future agricultural productivity. Warmer temperatures and longer growing seasons may increase agricultural productivity in some high-altitude regions (Rosenzweig et al., 2014), although the expectations are mixed. On the other hand, soil quality issues in the far north and emerging pest and disease breakouts might constrain expansion and productivity. In low-lying regions, even a modest increase in maximum temperatures is expected to negatively impact agricultural production. Studies have consistently found that under the most severe scenarios of climate change, significant losses should be expected worldwide (Darwin et al., 1995, 1996; Fischer et al., 1995, 1996; Rosenzweig and Parry, 1994; Kurukulasuriya and Rosenthal, 2003; Easterling et al., 2007; Nelson et al., 2009; FAO, 2015). No matter the severity, regional differences in crop production are expected to widen over time, with the risk of widening the gap between "haves and have-nots" and increases in prices and hunger among poorer nations (Parry et al., 2004; Nelson et al., 2010). Interregional trade flows, as a result, are due to expand from mid- to high-latitude regions to low-latitude regions, although trade alone might not be a sufficient adaptation strategy to climate change (FAO, 2015).

Developing countries are expected to receive the brunt of climate change (Morton, 2007). The Intergovernmental Panel on Climate Change (IPCC) Fifth Assessment Report projects that, under more optimistic scenarios, climate change could reduce food-crop yields in parts of Africa from 10 to 20 percent, a large drop for already at-risk populations and regions. The outlook for key food crops across the African continent under climate change is mostly negative, and low productivity, together with increasing global demand, will likely drive up food prices (Jalloh et al., 2013; Waithaka et al., 2013; Hachigonta et al., 2013). Nelson et al. (2010) predict that staple food prices could rise by 42 to 131 percent for maize, 11 to 78 percent for rice, and 17 to 67 percent for wheat between 2010 and 2050 as a result of the combined effect of climate change, increasing population, and economic growth. Moreover, localized weather shocks and emerging pest and disease outbreaks are already compromising stability in crop production, highlighting the urgency for immediate and adaptable management responses (FAO-PAR, 2011).

This chapter reviews research on climate change and its impacts on agriculture and food security at global, regional, and (selected) national scales. The next section summarizes the International Food Policy Research Institute's (IFPRI) latest long-term projections of the impacts of climate change on agricultural area, yields, production, consumption, prices, and trade for major crop and livestock commodities, and their implications for food security. Results are reported at the global level and regional level for several alternative socioeconomic and climate scenarios. Subsequent sections focus on options for adaptation and mitigation, and on the critical role of institutions, governance, policy, and finance.

4.2 CLIMATE CHANGE IMPACTS ON AGRICULTURE AND FOOD SECURITY

How We Estimate Impacts

IFPRI's long-term projections of food and agriculture are based on the International Model for Policy Analysis of Agricultural Commodities and Trade (IMPACT), a multimarket model that simulates the operation of national and international markets, solving for production, demand, and prices that equate supply and demand across the globe (Robinson et al., 2015). The core model is linked to a number of "modules" that include climate models, water models (hydrology, water basin management, and water stress models), crop simulation models (e.g., Decision Support System for Agrotechnology Transfer, or DSSAT), value chain models (e.g., sugar, oils, livestock), land use (pixel-level land use and cropping patterns by regions), nutrition and health models, and welfare analysis.

When analyzing the impacts of climate change on agriculture and food in the future, we must first account for the effects of other drivers of change, including growth in population and income. To facilitate analysis and comparison across models, Shared Socioeconomic Pathways (SSPs) have been developed under the IPCC (O'Neill et al., 2014; van Vuuren et al., 2014). IMPACT baseline projections assume middle-of-the-road growth in both of these drivers, as characterized by SSP 2, but we can also explore the impacts of faster or slower growth in these factors. For example, SSP 1 assumes slower growth in population and faster growth in income based on sustainable technologies, while SSP 3 assumes faster growth in population and slower growth in income in a more fragmented world.

In the coming decades, growth in population and income will lead to increases in demand for food and other agricultural commodities (see Chapter 1.2). Analysis, using IMPACT and four other global economic models as part of AgMIP,[2] shows that at the global level and in the absence of climate change, rising demand along with improvements in agricultural technology and management practices will increase production (and consumption) of coarse grains, rice, wheat, oilseeds, and sugar by around 70 percent by 2050. These increases will come mostly from increases in yields, while changes in area harvested will play a much smaller role (as they have done over the past half century in most parts of the world). Real prices are projected to increase by 10–20 percent, while trade will increase by around 70–90 percent. Changes in yields, area, prices, and trade vary depending on the population and income growth rates assumed in the various SSPs. The effects of climate change must be analyzed in relation to these other changes that are expected in the coming decades. To do so, we combine assumptions about changes in population, income, and technology (as characterized by the SSPs) with assumptions about climate change. Here, too, standard scenarios have been developed under the IPCC to facilitate analysis and comparison, and are expressed as Representative Concentration Pathways

[2] The other four models used in this study were the ENVISAGE model of Purdue University, the MAGNET model of LEI-Wageningen University, the MAgPIE model of the Potsdam Institute for Climate Impact Research, and the FARM model of the USDA's Economic Research Service. See Wiebe et al. (2015) for more information.

(RCPs). These range from relatively slow climate change in RCP 2.6 (expressed in units of radiative forcing, in this case 2.6 watts per square meter in 2100) to relatively rapid climate change in RCP 8.5 with corresponding differences in greenhouse gas emissions, and thus in temperature and precipitation.

The same comparative model analysis explores the range of possible effects of climate change under different socioeconomic and climate scenarios by combining SSP 1 with RCP 4.5, SSP 2 with RCP 6.0, and SSP 3 with RCP 8.5. The SSP 1-RCP 4.5 reflects relatively optimistic assumptions about growth in population, income, and greenhouse gas emissions (slow, fast, and slow, respectively), and SSP 3-RCP 8.5 reflects more pessimistic assumptions (with relatively fast, slow, and fast growth in population, income, and greenhouse gas emissions, respectively); SSP 2-RCP 6.0 represents an intermediate case.

Impacts of Climate Change on Agriculture and Food Security

Projections show that, at the global scale, yields of major crops will decline by 5–7 percent relative to levels in 2050 in the absence of climate change (Fig. 4.1). This represents a loss of about a tenth of the projected growth due to improvements in technology and management practices between now and 2050, although this proportion varies significantly across crops and countries. By contrast, the area harvested in 2050 increases by around 4 percent due to climate change, nearly doubling the increase projected in the absence of climate change. Likewise, prices would increase by 10–15 percent, doubling the increases

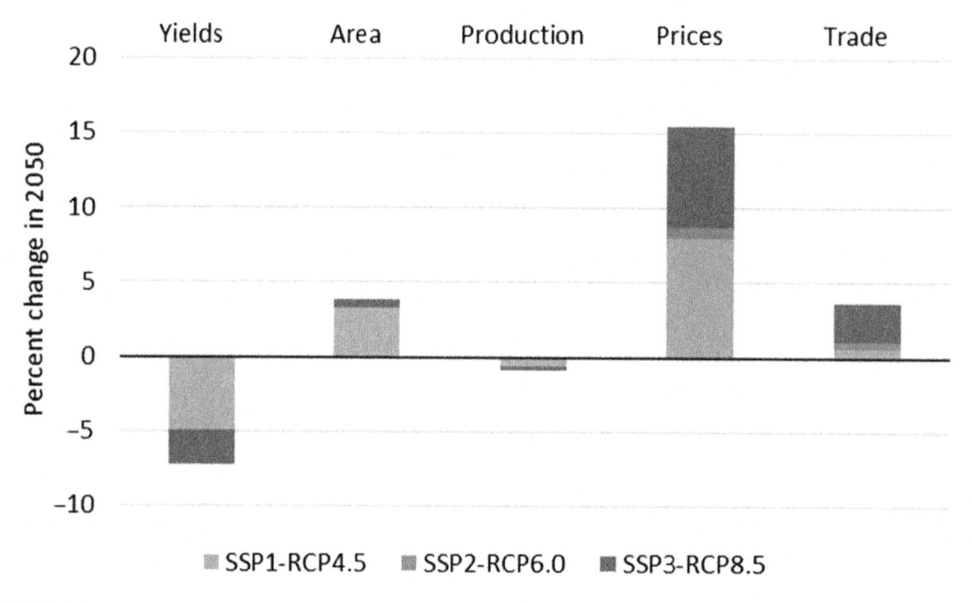

FIGURE 4.1 Climate change impacts on major crops in 2050, relative to baseline scenario.
Note: Average results for coarse grains, rice, wheat, oilseeds, and sugar from five global economic models.
Source: *Based on results from Wiebe et al., 2015. Climate change impacts on agriculture in 2050 under a range of plausible socioeconomic and emissions scenarios. Environ. Res. Lett., 10(8). doi:10.1088/1748-9326/10/8/085010. Available from <http://iopscience.iop.org/1748-9326/10/8/085010/media>.*

projected in the absence of climate change. Even though total global production (and consumption) of the five crops will change relatively little (a decline of around 1 percent relative to levels in 2050 in the absence of climate change), these climate-induced changes in harvested area and food prices are likely to have important implications for the environment and food security, respectively.

These estimated impacts are somewhat smaller than those of earlier work focusing on SSP 2 and RCP 8.5 (Nelson et al., 2014), which projected an 11 percent decrease in yields and a 20 percent increase in prices for a similar set of crops. The differences in the later study may be attributable to several factors of the study, including the set of models used, model improvements allowing greater flexibility in responding to climate change, inclusion of less extreme emissions pathways, updates in the definition of the SSP drivers used (including changes in population and income growth rates), and the inclusion of sugar, the average yields of which would increase as a result of climate change in the scenarios examined by Wiebe et al. (2015).

As growing seasons lengthen, higher latitudes tend to see smaller losses or even gains in yields due to climate change, while losses in lower latitude regions are generally greater. But impacts on cereal yields, for example, vary significantly by crop, region, and rate of climate change. Maize yields decline in most regions under most climate scenarios, with progressively greater losses under more extreme climate change scenarios. By contrast, impacts on wheat yields are small at the global level, as losses in South Asia and sub-Saharan Africa would be offset by increases elsewhere. We note that these estimates do not include the effects of carbon dioxide fertilization, which remain uncertain given the many complex interactions involved. On the other hand, these estimates also exclude other effects of climate change, such as the adverse impacts of extreme temperatures and precipitation, rising sea levels, melting of glaciers with consequent changes in availability of water for irrigation, and changes in the distribution of pests and diseases.

In the absence of climate change, most regions are projected to see declining numbers of people at risk of hunger, but these improvements are partially offset by the impacts of climate change (Table 4.1).[3] Worldwide, the number of people at risk of hunger would be expected to decline by more than half to around 406 million in 2050 in the absence of climate change, but climate change is projected to raise this number by 70 million. In sub-Saharan Africa, in particular, two-thirds of the projected reduction by 2050 in the number of people at risk of hunger is lost as a result of climate change, with roughly 40 million additional people at risk. This may be a conservative estimate, because it is based on the "middle-of-the-road" assumption of economic growth as characterized by SSP 2 and also because it does not account for the impacts of extreme events and other circumstances noted above that we expect to change with climate (and which we expect to change even more rapidly after 2050). In addition, the numbers are concentrated in sub-Saharan Africa, partly because other regions typically include at least some production in higher-latitude areas that are less affected (or even positively affected) by climate change, and partly

[3] In this analysis, the population at risk of hunger is estimated as a nonlinear function of the availability of food energy relative to requirements (see Robinson et al., 2015, for details). In general, access to food based on incomes (which may be derived from agricultural as well as nonagricultural sources) and food prices is also a key determinant of food security.

TABLE 4.1 IMPACT Projections of Food Production, Consumption, and Hunger to 2030 and 2050 With and Without Climate Change

	Aggregate Food Production (Index, 2010 = 1.00)					Per Capita Food Consumption (kcal per capita per day)					Hunger (Millions of people at risk)				
		Without Climate Change		With Climate Change			Without Climate Change		With Climate Change			Without Climate Change		With Climate Change	
	2010	2030	2050	2030	2050	2010	2030	2050	2030	2050	2010	2030	2050	2030	2050
World	1.00	1.37	1.69	1.33	1.60	2795	3032	3191	2982	3079	838.1	528.2	405.8	592.3	476.9
Developing	1.00	1.42	1.76	1.39	1.71	2683	2961	3137	2909	3020	823.3	513.3	392.2	576.7	461.1
Developed	1.00	1.24	1.47	1.15	1.29	3384	3439	3513	3406	3435	14.8	14.9	13.6	15.7	15.8
Asia and the Pacific	1.00	1.37	1.64	1.36	1.63	2656	3003	3185	2954	3072	539.8	249.8	181.8	280.9	204.6
East Asia	1.00	1.23	1.35	1.26	1.41	3009	3509	3628	3459	3516	187.2	59.2	54.7	60.3	56.8
South Asia	1.00	1.57	2.05	1.50	1.91	2361	2669	2959	2623	2848	268.5	138.3	87.7	161.6	97.0
Southeast Asia and the Pacific	1.00	1.48	1.89	1.46	1.84	2551	2852	3051	2796	2931	84.1	52.3	39.4	58.9	50.8
Africa and the Middle East	1.00	1.60	2.24	1.55	2.11	2623	2795	3002	2735	2873	238.7	229.8	185.0	258.7	227.1
Africa South of the Sahara	1.00	1.65	2.37	1.57	2.17	2358	2587	2853	2518	2713	209.5	195.7	150.5	223.0	188.7
Middle East and North Africa	1.00	1.51	2.01	1.50	2.00	3125	3250	3377	3208	3275	29.3	34.2	34.5	35.7	38.4
The Americas	1.00	1.37	1.69	1.27	1.48	3188	3290	3392	3244	3297	42.5	35.7	27.7	39.3	32.7
Latin America and the Caribbean	1.00	1.46	1.83	1.42	1.72	2878	3036	3184	2985	3081	39.5	32.1	24.0	35.8	28.7
North America	1.00	1.29	1.58	1.15	1.29	3714	3725	3735	3689	3654	3.0	3.6	3.7	3.6	4.0
Europe and former Soviet Union	1.00	1.18	1.33	1.14	1.26	3275	3390	3491	3359	3414	17.1	13.0	11.4	13.4	12.5
Former Soviet Union	1.00	1.26	1.42	1.20	1.36	3092	3321	3423	3288	3338	9.7	5.9	5.2	6.2	5.5
Europe	1.00	1.15	1.28	1.11	1.21	3370	3424	3523	3395	3450	7.4	7.0	6.2	7.3	6.9

Notes: World and regional figures include other regions and countries not reported separately. Aggregate food production is an index, by weight, of cereals, meats, fruits and vegetables, oilseeds, pulses, and roots and tubers. Per capita food consumption is a projection of daily dietary energy supply. Estimates of the number of people at risk of hunger are based on a quadratic specification of the relationship between national-level calorie supply and the share of population that is undernourished as defined by the Food and Agriculture Organization of the United Nations. Values reported for 2010 are calibrated model results. Projections for 2030 and 2050 assume changes in population and income, as reflected in the Intergovernmental Panel on Climate Change's (IPCC) Shared Socioeconomic Pathway 2. Climate change impacts are simulated using the IPCC's Representative Concentration Pathway 8.5 and the HadGEM general circulation model. Further documentation is available at http://www.ifpri.org/program/impact-model. *IMPACT*, International Model for Policy Analysis of Agricultural Commodities and Trade.

IMPACT model version 3.3 results reported in IFPRI, 2017. Global Food Policy Report. International Food Policy Research Institute, Washington, DC.

because other regions are less reliant on agriculture for their incomes and food security. Nevertheless, the overall impact of climate change to mid-century is smaller than that of the other drivers embedded in the socioeconomic scenario (including population and income).

IFPRI has collaborated with other partners to link results from IMPACT projections with models that explore implications for various indicators of environmental quality and human health. For example, Flachsbarth et al. (2015) examined implications for carbon stocks, biodiversity, and other environmental indicators in Latin America under a range of scenarios, comparing business as usual to several alternative intensification and extensification strategies. Business as usual (with and without further trade liberalization) is projected to increase carbon losses and risk of biodiversity loss. These losses are projected to accelerate under strategies that support extensification of agricultural production, but to slow under strategies that support sustainable intensification.

IFPRI also collaborated recently with health experts at Oxford University, United Kingdom, to explore the implications of climate change on diet and health (Springmann et al., 2016a, 2016b). While much research has focused on the impacts of climate change on agriculture and food security, less attention has been devoted to assessing the wider health impacts of future changes in agricultural production. The authors estimated excess mortality owing to agriculturally mediated changes in dietary and weight-related risk factors by cause of death for 155 world regions in the year 2050. IFPRI's IMPACT model was linked to a comparative risk assessment of changes in fruit and vegetable consumption, red meat consumption, and body weight for deaths from coronary heart disease, stroke, cancer, and an aggregate of other causes. The authors calculated the change in the number of deaths due to climate-related changes in weight and diets for the combination of four emissions and three socioeconomic pathways; each included six scenarios with variable climatic inputs. Underweight is the primary cause of diet-related deaths associated with climate change (additional to those expected in the no-climate-change baseline in 2050) in Africa and Southeast Asia (which in this case includes South Asia), but most diet-related deaths associated with climate change in other regions were linked to reductions in the consumption of fruit and vegetables. All other things being equal, reductions in red meat consumption, overweight, and/or obesity associated with climate change led to a reduction in diet-related deaths in all regions.

4.3 ADAPTATION OPTIONS

Addressing the new challenges posed by climate change will require innovations that strengthen the resilience of smallholder farming systems to climate change. These include enhanced resource-use efficiency through sustainable intensification of production and the adoption of agroecological production systems. Improving water resource management is another area where innovation can be effective in addressing climate change impacts. Finally, biotechnologies, both low and high tech, can assist small-scale producers in particular to be more resilient and to adapt better to climate change. While the subsections that follow focus mainly on innovation through management practices, some practices may depend on the outcomes of biotechnology, such as improved seed.

Diversification

Diversification is an important means of climate change adaptation because it helps to spread the risk of climatic variability in damaging livelihoods. First, a distinction should be made between agricultural diversification and livelihood diversification (Thornton and Lipper, 2014). Agricultural diversification means adding plant varieties and species, or animal breeds, to farms or farming communities. Livelihood diversification means farming households engaging in multiple agricultural and nonagricultural activities. Both agricultural and livelihood diversification are ways of managing climate risk.

Since climate shocks affect different farming and nonfarming activities differently, diversification can potentially reduce the impact of these shocks on income. However, if farmers diversify to low-productivity activities, such a measure may actually reduce average income, force households to sell off assets in the event of shocks, and trigger a vicious cycle of greater vulnerability and exposure to risk (Dercon, 1996). The scope of crop diversification as a means of mitigating climate risks may be limited where the risks affect equally different varieties of crops (Barrett et al., 2001). However, crop diversification may still be an option, where farm conditions are neither so marginal that they limit diversification nor sufficiently optimal for a single high-return crop (Kandulu et al., 2012).

Sustainable Intensification

Sustainable intensification raises productivity, lowers production costs, and increases the level and stability of returns from production while conserving natural resources, reducing negative impacts on the environment, and enhancing the flow of ecosystem services (FAO, 2011a). The nature of sustainable intensification strategies varies across different types of farming systems and locations. However, one of the core principles is increasing the efficiency of resource use (see mitigation section that follows).

One example is to sustainably improve productivity to deal with water scarcity or water excess, as these events may increase in frequency with climate change. In rainfed systems, which account for 95 percent of farmland in sub-Saharan Africa, better management of rainwater and soil moisture is the key to raising productivity and reducing yield losses during dry spells and periods of variable rainfall. Supplemental irrigation, using water harvesting or shallow groundwater resources, is an important but underused strategy for increasing water productivity in rainfed agriculture (HLPE, 2015; Oweis, 2014). In irrigated systems, water use efficiency can be promoted through institutional changes, such as the creation of water user associations, and infrastructural improvements, such as lining canals, more efficient drainage networks, and wastewater reuse. Water-efficient irrigation technologies such as drip emitters and better maintenance of irrigation infrastructure, combined with appropriate training to build farmers' technical knowledge, can be effective in dealing with climate change impacts on water availability and food security. However, some technologies that improve water use efficiency, such as drip irrigation, require energy.

Another area is that of better management of carbon and nitrogen cycles (see the next section), which also builds greater resilience to climate change impacts and contributes to reducing greenhouse gas (GHG) emissions (Burney et al., 2010; Wollenberg et al., 2016).

Carbon and Nitrogen Management Options

The Earth's carbon and nitrogen cycles are affected by the types of soil, nutrient, and water management practices farmers adopt, by the extent of agroforestry practiced, and by the expansion of agriculture onto nonagricultural land. Smallholders, especially, can benefit from practices that help restore soil productivity in areas where unsustainable land management has depleted soil organic carbon, natural soil fertility, and soil quality, which have reduced productivity and increased vulnerability to climate hazards such as drought, flooding, and conditions that favor pests and diseases (Stocking, 2003; Lal, 2004; Cassman, 1999; FAO, 2007).

On cropland, levels of soil organic carbon and plant-useable soil nitrogen can be improved through the adoption of such practices as agroforestry, improved fallows, green manuring, nitrogen-fixing cover crops, integrated nutrient management, minimum soil disturbance, and the retention of crop residues. On grazing lands, improving pasture management, reducing or eliminating the occurrence of fires, and introducing improved fodder grasses or legumes are important means of improving carbon management. Mixed-farming systems enhance resilience and reverse soil degradation by controlling erosion, providing nitrogen-rich residues, and increasing soil organic matter.

Context-specific climatic conditions will influence smallholders' choice of the carbon and nitrogen management options that are the most effective in improving their livelihoods. For example, the application of mineral fertilizer may generate higher yields under average climatic conditions, but lower yields under conditions of high rainfall variability or the delayed onset of rainfall. Conversely, crop rotation may produce lower yields under average climatic conditions, but higher yields and a lower probability of yield loss under conditions of high rainfall variability (FAO, 2016a).

Identifying and Addressing Barriers to Adoption

Insights into barriers smallholder farmers face in making the types of incremental changes needed for climate change adaptation are highlighted in a recent meta-analysis of the determinants of improved technology adoption in Africa (Arslan et al., 2016). The authors find that the most prominent barrier to adoption of agroforestry is access to information, primarily from extension services, which is significant in around 40 percent of the studies where it is included. Other top determinants preventing adoption of improved agroforestry practices are distance to markets, membership of farmer groups and other social capital, and tenure security. For adoption of improved agronomic practices, the main barriers were related to information access, followed by tenure security, resource endowments, and exposure to risks and shocks. The analysis also indicated a need for specifically targeting those with lower endowments, especially women farmers and woman-headed households (Arslan et al., 2016).

Explicit recognition of the costs of making changes is needed in order to adequately identify where trade-offs are possible. For example, the improvement of soil carbon stocks through improved land management and restoration carries investment costs in the form of fencing, seed and machinery, opportunity costs in the form of lost production, and operating costs in the form of annual labor inputs needed to maintain and enhance soil carbon. The costs of adopting practices that increase soil carbon can be quite significant for smallholders, particularly in the initial and transition phases. They can also outweigh the benefits to the farmers themselves, while generating benefits to others by improving landscape and watershed functions.

Addressing, in a targeted manner, issues such as access to credit, markets, and information, as well as improving tenure security and social capital, will therefore be part of the portfolio of policies needed to make agriculture more sustainable and improve resilience to climate change. The emphasis placed on the different elements will be context-dependent. Some interventions will also require collective action and coordination among stakeholders.

Policies and institutions must facilitate and support coordinated design and implementation of actions, either in a specific area, such as a watershed or forest, or in a sector, such as an entire food chain. Key areas are cross-sectoral coordination to support landscape restoration, multistakeholder dialogues for improved governance of land and water tenure systems, and the promotion of social networks.

Barriers also exist to adopting improved practices for managing climate-related risks. Overcoming such barriers requires appropriate institutions and policies. Interventions at multiple levels are needed. Several tools can help assess present and future climate change effects, including weather stations, weather and climate projection tools, environmental monitoring tools, and vulnerability assessments. Potential mechanisms for integrated risk management include national boards to coordinate with institutions for risk monitoring, prevention, and control and response at the local and global levels. Social protection programs can help vulnerable people cope with risk. Policies and institutions must also support the diversification of livelihood strategies, inter alia, by promoting improved access to credit, insurance, information, and training.

4.4 MITIGATION OPTIONS

Agricultural emissions are expected to grow along with food demand, which is being driven by population and income growth and associated changes in diets toward more animal-source products. Agriculture can contribute to mitigation by decoupling its production increases from its emissions increases through reductions in emission intensity, which is the quantity of GHG generated per unit of output. This, in turn, can be complemented by actions that reduce food losses and waste and foster changes in food consumption patterns.

Agriculture, forestry and land use (AFOLU) are responsible for about 21 percent of total greenhouse gas emissions (see Chapter 1.2). All carbon dioxide emissions from AFOLU are attributable to forestry and land use change, such as conversion of forests to pasture or crop production. The bulk of emissions of methane and nitrous oxide are attributable to agricultural practices (FAO, 2016a). Improved management of carbon and nitrogen in agriculture, therefore, will be crucial to its contribution to climate change mitigation.

Managing Carbon-Rich Forest Landscapes

Each year, forests absorb an estimated 2.6 billion tons of carbon dioxide (CIFOR, 2010), equivalent to about one-third of the carbon dioxide released from the burning of fossil fuels. However, this immense storage system, once disrupted by deforestation, becomes a major source of emissions. According to IPCC's Fifth Assessment Report, deforestation and forest degradation account for nearly 11 percent of all GHG emissions—more than the world's entire transport sector. Climate change mitigation actions in the forest sector fall into two broad categories: reducing the emission of GHG (reducing deforestation and forest degradation); and increasing removals of GHG from the atmosphere (increasing carbon density, afforestation, and off-site carbon stock in harvested wood products). The carbon mitigation potential of reduced deforestation, improved forest management, afforestation, and agroforestry differ greatly by activity, region, system boundaries, and the time horizon over which mitigation options are compared. Reductions in deforestation dominate the forestry mitigation potential in Latin America and Africa, while forest management, followed by afforestation, dominate in the countries of the Organisation for Economic Co-operation and Development (OECD), economies in transition, and Asia. Afforestation's potential contribution to mitigation ranges between 20 and 35 percent of total forestry-related potential (Smith et al., 2014: Figure 11.18). The challenge posed by most forest-related mitigation activities is the need for substantial investment before benefits and cobenefits accrue, typically over many years if not decades. The substantial mitigation potential of forestry will not materialize without appropriate financing and enabling frameworks that create effective incentives.

Livestock Emissions

There is also great potential to reduce the livestock sector's GHG emission intensity both in terms of methane and nitrous oxide. The precise potential is difficult to estimate, as emission intensities vary greatly even within similar production systems owing to the differences in agroecological conditions, farming practices, and supply chain management. Gerber et al. (2013) estimate that emissions generated by livestock production could be reduced by between 18 and 30 percent if, in each system, the practices used by the 25 percent of producers with the lowest GHG emission intensity were widely adopted. Based on six regional case studies and using a life cycle assessment model, Mottet et al. (2016) estimate that sustainable practices would lead to reductions of between 14 and 41 percent in livestock GHG emissions. In five of the case studies, mitigation resulted in increased production as well as reduced emissions, a double-win for food security and climate change mitigation.

In a modeling study by Henderson et al. (2015), improved grazing management and the sowing of legumes were the most affordable practices and therefore had the greatest economic potential. Grazing management was particularly effective in Latin America and sub-Saharan Africa, while sowing legumes appeared to work best in Western Europe.

Flooded Rice

A range of traditional and improved practices mitigate methane emissions from rice paddies, including water, straw, and fertilizer management. Stopping flooding for a few weeks saves water, and also reduces methane and GHG emissions by between 45 and 90 percent, without considering soil carbon stock increases. However, this practice can have negative impacts on yields, partly through increased weed competition. Drying early in the growing season, and then flooding, reduces emissions by 45 percent and produces yields similar to those of fully flooded rice (Linquist et al., 2015).

Nitrous-Oxide Emissions

Together with water, nitrogen is the most important determinant of crop yields (Mueller et al., 2012). However, emissions of nitrous oxide from applied fertilizer have direct negative impacts: nitrogen dioxide (NO_2) is the third most important GHG and the most significant cause of ozone depletion in the stratosphere. Sustainable nitrogen management in agriculture aims at simultaneously achieving agronomic objectives, such as high crop and animal productivity, and the environmental objectives of minimizing nitrogen (N) losses. Estimates for reduction of nitrous-oxide emissions are based on the potential for increasing N-use efficiency and/or lowering emission intensity (Oenema et al., 2014). Assumptions, based on a literature review and expert views, include improvements in crop and animal production, manure management and food utilization, and lower levels of animal protein in diets. In the business-as-usual scenario, annual nitrous-oxide emissions from agriculture increase from an estimated 4.1 million tons in 2010 to 6.4 million tons in 2030, and 7.5 million tons in 2050. Reduction strategies could potentially hold emissions at 4.1 million tons in 2030 and cut them to 3.3 million tons by 2050. Improvements in crop production, notably fertilizer use, appear to have the greatest potential.

Soil Carbon Management

In order to tap into the potential for soil carbon sequestration, sustainable soil management needs to be promoted as a system, with a range of functions that provide multiple ecosystem services (FAO and ITPS, 2015). Soil carbon sequestration rates on land in agricultural use vary in the order of 0.1 to 1 ton of carbon per hectare per year (Paustian et al., 2016). However, levels of sequestration would be relatively low at first, peak after 20 years, and then slowly decline (Sommer and Bossio, 2014). When prioritizing actions, one should also consider that carbon sequestration in agricultural soils may not be lasting. The extra soil carbon stored through improved agricultural practices is partly in unprotected forms, such that a fraction would decompose if the practices ceased.

Higher Production Efficiency, Lower Emissions, and Enhanced Food Security

Better management of the carbon and nitrogen cycles is central both to the mitigation of net GHG emissions from the agriculture, forestry, and land use sector and to increasing the efficiency of the global food system. By helping to reduce yield gaps and increase biological efficiencies, especially in developing countries, sustainable intensification of agriculture could prevent deforestation and the further expansion of agriculture into carbon-rich ecosystems, thus simultaneously enhancing food security and contributing to climate change mitigation.

Investing in Yield Improvements

Since the 1960s, the intensification of crop and livestock systems has restrained the expansion of farmland and improved the efficiency of food supply chains (Tilman et al., 2011; Gerber et al., 2013; Herrero et al., 2013). Through higher yields, agricultural intensification avoided GHG emissions between 1961 and 2005 that are estimated to total up to 161 gigatons (Gt) of carbon. Investments in productivity compare favorably, therefore, with other commonly proposed mitigation strategies because they limit agricultural land expansion and the large carbon losses associated with deforestation (Burney et al., 2010).

As agricultural and forestry efficiency have improved over the past few decades, the GHG emission intensity of many products has declined. Between 1960 and 2000, global average intensities fell by an estimated 38 percent for milk, 50 percent for rice, 45 percent for pork, 76 percent for chicken meat, and 57 percent for eggs (Smith et al., 2014). Much of the reduction in ruminant emission intensity has been due to reduced output of methane for a given amount of milk and meat (Opio et al., 2013). In both ruminants and monogastrics, improvements in feed conversion efficiency and husbandry and the selection of highly efficient animal breeds have played key roles. Strong disparities in resource-use efficiency and GHG emission intensity still exist between animal farming systems and across regions (Herrero et al., 2013), suggesting significant potential for gains.

A number of technologies can help raise production efficiency and generate cobenefits. They include the use of adapted varieties that harness genetic resources and advanced breeding, adjustments to planting dates and cropping periods, precision farming, judicious use of inorganic fertilizer in combination with organic nutrient sources and legumes, and the design of more diversified, sustainable cropping systems that also consider agroforestry approaches. Finally, reducing on-farm losses increases the efficiency of production systems.

Efficiency Improvements in Aquaculture and Fisheries

In terms of GHG reduction, there is considerable potential for lowering emissions by reducing fuel and energy use in aquaculture and fisheries. This can be achieved either directly—e.g., through more efficient fishing methods or energy use in processing—or indirectly, through a variety of actions, including energy savings along the supply and value chain and strategic waste reduction. Across the sector, the transition to more energy-efficient technologies is slow, although incentive mechanisms associated with carbon markets have shown some potential (FAO, 2013a).

Energy use in processing, storage, and transport is the main source of GHG emissions in fisheries and aquaculture. GHG outputs are usually directly related to fuel use in transport and to energy use in handling and storage. The most perishable fresh products require fast transport and energy-intensive storage. The choice of refrigerants is also important—the leakage of refrigerant gases from old or poorly maintained equipment depletes the ozone layer in the atmosphere and has significant global warming potential. More stable dried, smoked, and salted products processed in artisanal supply chains require methods of transport that are not time-sensitive and produce lower levels of GHG emissions (FAO, 2013b).

Mitigation Costs, Incentives, and Barriers

There are many feasible and promising approaches to climate change mitigation in the AFOLU sector, and the technical potential is considerable. But what are the costs and thus the economic potential of mitigation? In other words, what is the hypothetical price of carbon that would induce farmers, fisherfolk, and foresters to apply appropriate practices for sequestering carbon and reducing emissions?

Based on the combined mitigation potential of forestry and agriculture, estimated in the IPCC's Fourth Assessment Report, the IPCC suggests an economic potential in 2030 of between ≈3 and ≈7.2 Gt of carbon dioxide equivalent a year at carbon prices of US$20 and US$100 per ton, respectively (Smith et al., 2014).[4] Among regions, the largest mitigation potential for agriculture, forestry, and land use is found in Asia at all levels of carbon values.

Forestry could make a significant contribution to mitigation at all levels of carbon prices. At low prices, the contribution of forestry is close to 50 percent of the total from the AFOLU sector; at higher prices, the share of forestry is lower. Forestry represents the bulk of mitigation potential in Latin America at all levels of carbon prices. However, different forestry options have different economic mitigation potentials in different regions. Reduced deforestation dominates the forestry mitigation potential in Latin America and in the Middle East and Africa. Forest management practices, followed by afforestation, are the major options in OECD countries, Eastern Europe and Asia.

Among other mitigation options, cropland management has the highest potential at lower carbon prices of US$20 per ton. At US$100, the restoration of organic soils has the greatest potential. Also, the potential of grazing land management and the restoration of degraded lands increases at higher carbon prices (Smith et al., 2014).

These estimates of economic mitigation potential provide broad indications of how to target interventions in the most cost-effective way. However, more detailed assessments are needed in order to properly assess AFOLU's mitigation potential, the impacts on vulnerable production systems and groups, and the costs of implementation.

Beyond the Farmgate: A Food System Perspective on Emissions

Reducing food losses and waste, and promoting a transition to more sustainable diets, can also deliver emissions reductions and contribute to global food security (Bajželj et al., 2014). In low-income countries, food losses occur throughout food value chains, and result from managerial and technical limitations in harvesting, storage, transportation, processing, packaging, and marketing (HLPE, 2014). The heaviest losses are in the small- and medium-scale agricultural and fisheries production and processing sectors. Food waste in middle- and high-income countries is caused mainly by consumer behavior and by policies and regulations that address other sectoral priorities.

[4] A range of global estimates of sequestration potential at different levels of costs has been published since the IPCC's Fourth Assessment Report of 2007. The estimates differ widely. For carbon values up to US$20 a ton, they range from 0.12 to 3.03 $GtCO_2$-eq per year. For values up to US$100 per ton, they range from 0.49 to 10.6 $GtCO_2$-eq (Smith et al., 2014).

Another critical factor that needs to be considered is the energy used in modern food systems to process food and bring it to consumers. The modernization of food supply chains has been associated with higher GHG emissions from both pre-chain inputs (fertilizers, machinery, pesticides, veterinary products, transport) and post-farmgate activities (transportation, processing, and retailing). It has been estimated, using previous calculations and data from Bellarby et al. (2008) and Lal (2004), that the production of fertilizers, herbicides, and pesticides, along with emissions from fossil fuels used in the field, represented in 2005 approximately 2 percent of global GHG emissions (HLPE, 2012).

Life cycle analysis methods are needed to calculate emissions resulting from the consumption of food products. These approaches generally account for emissions from pre-chain inputs through to post-farmgate processing by including methane, nitrous oxide, and carbon dioxide emissions and fossil fuel use in food systems (Steinfeld et al., 2006; FAO, 2013b). Including post-harvest stages, around 3.4 Gt of carbon dioxide equivalent ($GtCO_2$-eq) of emissions are caused by direct and indirect energy use in the agrifood chain (FAO, 2011b). This can be compared with around 5.2 $GtCO_2$-eq generated by agriculture, and around 4.9 $GtCO_2$-eq by forestry and land use change. Food systems currently consume an estimated 30 percent of the world's available energy, with more than 70 percent of that share being consumed beyond the farmgate.

Although they are heavily dependent on fossil fuels, modern food systems have contributed substantially to improving food security. If these systems are to contribute to climate change mitigation, however, they will need to decouple future development from dependence on fossil fuel.

Further examination of demographic and social differences, including fast-growing food consumption in developing countries, is needed to inform strategies for promoting optimal diets with improved health outcomes and reduced levels of nitrate pollution and greenhouse gas emissions. Multidimensional life cycle assessments at regional and global levels are needed to estimate the adaptation and mitigation effects of different dietary transitions, including the possible trade-offs.

The Global Public Good Nature of Greenhouse Gas Mitigation and the Role of Agriculture

There is a fundamental difference between adaptation and mitigation and the incentives needed to promote them. Adaptation is something everyone will want to do in their own interest. Mitigation is something that has to be done together, in the interests of everyone. GHG mitigation is a global public good because effective controls of GHG emissions must involve all actual and potential emitters: each country's release of GHG augments the world's atmospheric stock in an additive fashion, and each country's cutback is effective only insofar as other countries also contribute to reduce their emissions. This means that collective action concerned with global climate change should be taken at the supranational level, as in the case of the United Nations Framework Convention on Climate Change's Paris Agreement of December 2015, which has set the long-term goal of holding the increase in global average temperature to "well below 2°C" above pre-industrial levels. Adaptation and mitigation objectives in agriculture, land use, land use change, and forestry figure prominently in the Intended Nationally Determined Contributions (INDCs), which, under the Paris Agreement, will guide country-level action on climate change in the coming years. They include not only targets, but also concrete strategies for addressing the causes of climate change and responding to its consequences.

An analysis of the INDCs by the Food and Agriculture Organization of the United Nations (FAO) shows that, in all regions, agriculture will play a pivotal role in accomplishing the goals related to climate change by 2030 (FAO, 2016b). Of the 188 countries that submitted INDCs, more than 90 percent included agriculture as a sector considered for mitigation and adaptation initiatives. The analysis also shows that the agriculture sectors are expected to provide the greatest number of opportunities for adaptation-mitigation synergies, as well as socioeconomic and environmental cobenefits. Around one-third of all countries acknowledge (and in some cases prioritize) actions that would create synergies between mitigation and adaptation in agriculture. There will be situations where mitigation will involve trade-offs, in which case financial incentives for such measures may be required.

4.5 INDUCING CHANGE: THE CRITICAL ROLE OF INSTITUTIONS, GOVERNANCE, POLICY, AND FINANCE

As shown in Sections 4.3 and 4.4, a wide range of sustainable intensification technologies can help farmers both adapt to and mitigate the climate change challenges introduced in Sections 4.1 and 4.2. Such technologies can also help reduce food insecurity. Rosegrant et al. (2014) linked climate, water, crop, and economic models to assess the yield and food security impacts of a broad range of agricultural technologies under varying assumptions regarding climate change and technology adoption. Under a wetter and hotter climate scenario, they found that the number of food-insecure people in developing countries could be reduced in 2050 by 12 percent (or almost 124 million people) if nitrogen-efficient crop varieties were widely in use; by 9 percent (91 million people) if no-till farming were more widely adopted; and by 8 percent (80 million people) if heat-tolerant crop varieties or precision agriculture were adopted (Fig. 4.2).

While these innovations show considerable potential, realizing these gains requires not only increased investment in research, but also measures for overcoming barriers to more widespread adoption of innovations in technologies and management practices. As previous sections have clarified, the impacts of climate change on agricultural producers and consumers, and our ability to adapt and mitigate those impacts, depend critically on socioeconomic factors and conditions as much as on biophysical processes. Key among these are access to resources, markets, information, and opportunities. These depend in turn on the nature and effectiveness of social, economic, and political institutions, policy instruments, and financial mechanisms in creating the enabling environment in which producers and consumers make choices. Examples include formal and informal mechanisms that shape land and water tenure security; exposure to shocks and risk; access to credit and insurance, as well as to extension, advisory services, and innovation systems; membership in producers' organizations and water users' associations; availability of social protection measures; and access to local, national, and international markets.

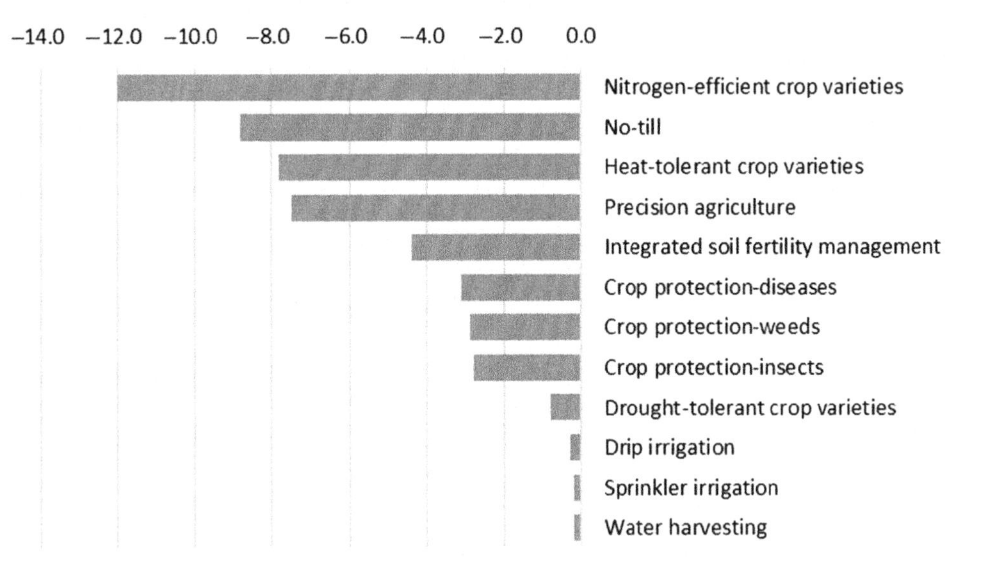

FIGURE 4.2 Percent change in the number of people at risk of hunger in developing countries in 2050, relative to the baseline scenario, under alternative agricultural technologies. Source: *IMPACT simulations and MIROC A1B climate change scenario (Rosegrant et al., 2014. Food Security in a World of Natural Resource Scarcity: The Role of Agricultural Techonologies. IFPRI, Washington, DC.).*

The challenges of climate change and food security are complicated by the extent to which they emerge from the individual and collective actions of some 500 million farm households and over 7 billion consumers worldwide. As noted above, producers and consumers have an incentive to adapt to the impacts of climate change, although they vary significantly in their ability to do so, depending on the particular set of resources and opportunities to which they have access. But the global public good nature of climate change means that mitigation requires coordinated collective action in order to succeed. Meeting the challenges of climate change adaptation equitably and the challenges of mitigation effectively thus requires well-informed, evidence-based policy actions at local, national, and global scales to ensure the appropriate enabling environment. These actions are addressed in the subsequent chapters of this book.

References

Arslan, A., Lamanna, C., Lipper, L., Rosenstock, T., Rioux, J. 2016. A Meta-Analysis on the Barriers to Adoption of Practices with CSA Potential in Africa. Mimeo, New York, NY.

Bajželj, B., Richards, K.S., Allwood, J.M., Smith, P., Dennis, J.S., Curmi, E., et al., 2014. Importance of food-demand management for climate mitigation. Nat. Clim. Change 4, 924–929.

Barrett, C.B., Reardon, T., Webb, C., 2001. Nonfarm income diversification and household livelihood strategies in rural Africa: concepts, dynamics, and policy implications. Food Policy 26 (4), 315–331.

Bellarby, J., Foereid, B., Hastings, A., Smith, P., 2008. Cool Farming: Climate Impacts of Agriculture and Mitigation Potential. Greenpeace International, Amsterdam.

Burney, J.A., Davis, S.J., Lobell, D.B., 2010. Greenhouse gas mitigation by agricultural intensification. Proc. Natl. Acad. Sci. 107, 12052–12057.

Cassman, K., 1999. Ecological intensification of cereal production systems: yield potential, soil quality, and precision agriculture. Proc. Natl. Acad. Sci. USA 96 (11), 5952–5959.

CIFOR (Center for International Forestry Research), 2010. Forests and climate change toolbox. Available from <http://www.cifor.org/fctoolbox/> (accessed 25.09.17.).

Darwin, R.F., Tsigas, M., Lewandrowski, J., Raneses, A., 1996. Land use and cover in ecological economics. Ecol. Econ. 17, 157–181.

Darwin, R.F., Tsigas, M., Lewandrowski, J., Raneses, A. 1995. World agriculture and climate change: economic adaptations, agricultural economic report number 703, United States Department of Agriculture, Economic Research Service, Washington, DC

De Pinto, A., T. Thomas, K. Wiebe, 2016. synthesis of recent ifpri research on climate change impacts on agriculture and food security, a background paper prepared by IFPRI for FAO's 2016 State of Food and Agriculture report.

Dercon, S., 1996. Risk, crop choice, and savings. Econ. Dev. Cult. Change 44, 485–513.

Easterling, W., Aggarwal, P., Batima, P., Brander, K., Erda, L., Howden, M., et al., 2007. Food, fibre and forest products. In: Parry, M.L., Canziani, O.F., Palutikof, J.P., van der Linden, P.J., Hanson, C.E. (Eds.), Climate Change 2007: Impacts, Adaptation and Vulnerability. Contribution of Working Group II to the Fourth Assessment Report of the Intergovernmental Panel on Climate Change. Cambridge University Press, Cambridge, pp. 273–313.

FAO, 2007. The State of Food and Agriculture 2007. Paying Farmers for Environmental Services. Rome.

FAO, 2011a. The State of the World's Land and Water Resources for Food and Agriculture (SOLAW)—Managing Systems at Risk. 288. Rome.

FAO, 2011b. "Energy-Smart" Food for People and Climate—An Issue Paper. Rome.

FAO, 2013a. Climate-Smart Agriculture Source Book. Rome.

FAO, 2013b. Greenhouse Gas Emissions from Ruminant Supply Chains, A Global Life Cycle Assessment. Rome.

FAO, 2015. Climate Change and Food Systems: Global Assessments and Implications for Food Security and Trade. Food Agriculture Organization of the United Nations (FAO).

FAO, 2016a. The State of Food and Agriculture 2016. Climate Change. Agriculture, and Food Security. Rome.

FAO, 2016b. The Agriculture Sectors in the Intended Nationally Determined Contributions: Analysis by Strohmaier, R., Rioux, J., Seggel, A., Meybeck, A., Bernoux, M., Salvatore, M., Miranda, J. and Agostini, A., Environment and Natural Resources Management, Working Paper No. 62. Rome.

FAO, and ITPS. 2015. Status of the World's Soil Resources (SWSR)—Main Report.FAO and Intergovernmental Technical Panel on Soils, Rome, 167.

FAO-PAR (Platform for Agrobiodiversity). 2011. Biodiversity for Food and Agriculture: Contributing to Food Security and Sustainability in a Changing World. Rome. Available from <http://agrobiodiversityplatform.org/files/2011/04/PAR-FAO-book_lr.pdf> (accessed 25.09.17.).

Fischer, G., Frohberg, K., Parry, M.L., Rosenzweig, C., 1995. Climate Change and World Food Supply, Demand and Trade. In: Climate Change and Agriculture: Analysis of Potential International Impacts. ASA Special Publication No. 59, American Society of Agronomy, Madison.

Fischer, G., Frohberg, K., Parry, M.L., Rosenzweig, C., 1996. Impacts of potential climate change on global and regional food production and vulnerability. In: Downing, T.E. (Ed.), NATO AS! Series, Vol. 137. Springer, Berlin, Climate Change and World Food Security.

Flachsbarth, I., Willaarts, B., Xie, H., Pitois, G., Mueller, N.D., Ringler, C., et al., 2015. The role of Latin America's land and water resources for global food security: Environmental trade-offs of future food production pathways. PLoS One. Available from: https://doi.org/10.1371/journal.pone.0116733.

Gerber, P.J., Hristov, A.N., Henderson, B., Makkar, H.P.S., Oh, J., Lee, C., et al., 2013. Technical options for the mitigation of direct methane and nitrous oxide emissions from livestock—a review. Animal 7, 220–234.

Hachigonta, S., G. Nelson, T.S. Thomas, L.M. Sibanda (Eds.), 2013. Southern African Agriculture and Climate Change: A Comprehensive Analysis. IFPRI Research Monograph.International Food Policy Research Institute, Washington. Available from <http://www.ifpri.org/publication/southern-african-agriculture-and-climate-change> (accessed 25.09.17.).

Henderson, B., Falcucci, A., Mottet, A., Early, L., Werner, B., Steinfeld, H., et al., 2015. Marginal costs of abating greenhouse gases in the global ruminant livestock sector. Mitig. Adapt. Strateg. Global Change 1–26.

Herrero, M., Havlík, P., Valin, H., Notenbaert, A., Rufino, M.C., Thornton, P.K., et al., 2013. Biomass use, production, feed efficiencies, and greenhouse gas emissions from global livestock systems. Proc. Natl. Acad. Sci. 110 (52), 20888–20893.

HLPE (High Level Panel of Experts), 2012. Food Security and Climate Change. A Report by the High Level Panel of Experts on Food Security and Nutrition of the Committee on World Food Security, Rome.

HLPE (High Level Panel of Experts), 2014. Food Losses and Waste in the Context of Sustainable Food Systems. A Report by the High Level Panel of Experts on Food Security and Nutrition of the Committee on World Food Security, Rome.

HLPE (High Level Panel of Experts), 2015. Water for Food Security and Nutrition. A Report by the High Level Panel of Experts on Food Security and Nutrition of the Committee on World Food Security FAO, Rome.

IFPRI (International Food Policy Research Institute), 2017. 2017 Global Food Policy Report. International Food Policy Research Institute, Washington, DC, <https://doi.org.10.2499/9780896292529>.

Jalloh, A., G.C. Nelson, T.S. Thomas, R. Zougmoré, H. Roy-Macauley, (Eds.), 2013. West African Agriculture and Climate Change: A Comprehensive Analysis. IFPRI Research Monograph.International Food Policy Research Institute, Washington, DC. Available from <http://www.ifpri.org/publication/west-african-agriculture-and-climate-change> (accessed 25.09.17.).

Kandulu, J.M., Bryan, B.A., King, D., Connor, J.D., 2012. Mitigating economic risk from climate variability in rainfed agriculture through enterprise mix diversification. Ecol. Econ. 79, 105–112.

Knight, J., Harrison, S., 2013. The impacts of climate change on terrestrial Earth surface systems. Nat. Climate Change 3, 24–29. Available from: https://doi.org/10.1038/nclimate1660 (2013).

Kurukulasuriya, P., S. Rosenthal, 2003. Climate Change and Agriculture: A Review of Impacts and Adaptations. Paper No. 91 in Climate Change Series, Agriculture and Rural Development Department and Environment Department, World Bank, Washington, DC.

Lal, R., 2004. Soil carbon sequestration impacts on global climate change and food security. Science 304, 1623–1626.

Linquist, B.A., Anders, M.M., Adviento-Borbe, M.A.A., Chaney, R.L., Nalley, L.L., Da Rosa, E.F., et al., 2015. Reducing greenhouse gas emissions, water use, and grain arsenic levels in rice systems. Global Change Biol. 21 (1), 407–417.

Lobell, D.B., Gourdji, S.M., 2012. The influence of climate change on global crop productivity. Plant Physiol. 160 (4), 1686–1697. Available from: https://doi.org/10.1104/pp.112.208298.

Morton, J.F., 2007. The impact of climate change on smallholder and subsistence agriculture. Proc. Natl. Acad. Sci. 104 (50), 19680–19685. Available from: https://doi.org/10.1073/pnas.0701855104.

Mottet, A., Henderson, B., Opio, C., Falcucci, A., Tempio, G., Silvestri, S., et al., 2016. Climate change mitigation and productivity gains in livestock supply chains: insights from regional case studies. Region. Environ. Change 1–13.

Mueller, N.D., Gerber, J.S., Johnston, M., Ray, D.K., Ramankutty, N., Foley, J.A., 2012. Closing yield gaps through nutrient and water management. Nature 490 (7419), 254–257.

Nelson, G.C., Rosegrant, M.W., Palazzo, A., Gray, I., Ingersoll, C., Robertson, R., et al., 2010. Food Security, Farming, and Climate Change to 2050: Scenarios, Results, Policy Options. International Food Policy Research Institute, Washington DC. Available from: http://dx.doi.org/10.2499/9780896291867.

Nelson, G.C., Rosegrant, M.W., Koo, J., Robertson, R., Sulser, T.B., Zhu, T., et al., 2009. Climate Change: Impact on Agriculture and Costs of Adaptation. International Food Policy Research Institute, Washington, DC, IFPRI Food Policy Report.

Nelson, G.C., Valin, H., Sands, R.D., Havlik, P., Ahammad, H., Deryng, D., et al., 2014. Climate change effects on agriculture: Economic responses to biophysical shocks. Proc. Natl. Acad. Sci. 111, 3274–3279.

Oenema, O., Ju, X., de Klein, C., Alfaro, M., del Prado, A., Lesschen, J.P., et al., 2014. Reducing nitrous oxide emissions from the global food system. Curr. Opin. Environ. Sustain. 9–10, 55–64.

O'Neill, B.C., Kriegler, E., Riahi, K., Ebi, K.L., Hallegatte, S., Carter, T.R., et al., 2014. A new scenario framework for climate change research: the concept of shared socioeconomic pathways. Clim. Change 122 (3), 387–400. Available from <http://link.springer.com/article/10.1007%2Fs10584-013-0905-2>.

Opio, C., Gerber, P., Mottet, A., Falcucci, A., Tempio, G., MacLeod, M., et al., 2013. Green House Gas Emissions from Ruminant Supply Chains—A Global Life Cycle Assessment. FAO, Rome.

Oweis, T. 2014. The Need for a Paradigm Change: Agriculture in Water-Scarce MENA Region. In: Holst-Warhaft, G., Steenhuis, T., de Châtel, F. (Eds.), Water Scarcity, Security and Democracy: A Mediterranean Mosaic. Athens, Global Water Partnership Mediterranean, Cornell University and the Atkinson Center for a Sustainable Future.

Parry, M.L., Rosenzweig, C., Iglesias, A., Livermore, M., Fischer, G., 2004. Effects of climate change on global food production under SRES emissions and socio-economic scenarios. Glob. Environ. Change A 14, 53–67. Available from: https://doi.org/10.1016/j.gloenvcha.2003.10.008.

Paustian, K., Lehmann, J., Ogle, S., Reay, D., Robertson, G.P., Smith, P., 2016. Climate-smart soils. Nature 532, 49–57.

Robinson, S., D. Mason D'Croz, S. Islam, T.B. Sulser, R.D. Robertson, T. Zhu, et al., 2015, The International Model for Policy Analysis of Agricultural Commodities and Trade (IMPACT): Model Description for Version 3. IFPRI Discussion Paper, International Food Policy Research Institute. Available from <http://www.ifpri.org/publication/international-model-policy-analysis-agricultural-commodities-and-trade-impact-model-0>.

Rosegrant, M.W., Koo, J., Cenacchi, N., Ringler, C., Robertson, R., Fisher, M., et al., 2014. Food Security in a World of Natural Resource Scarcity: The Role of Agricultural Techonologies. IFPRI, Washington DC.

Rosenzweig, C., Elliott, J., Deryng, D., Ruane, A.C., Müller, C., Arneth, A., et al., 2014. Assessing agricultural risks of climate change in the 21st century in a global gridded crop model intercomparison. Proc. Natl. Acad. Sci. USA. 111 (9), 3268–3273. Available from: https://doi.org/10.1073/pnas.1222463110.

Rosenzweig, C., Parry, M.L., 1994. Potential impact of climate change on world food supply. Nature 367, 133–138. Available from: https://doi.org/10.1038/367133a0.

Smith, P., Bustamante, M., Ahammad, H., Clark, H., Dong, H., Elsiddig, E.A., et al., 2014. Agriculture, Forestry and Other Land Use (AFOLU). In: Edenhofer, O., Pichs-Madruga, R., Sokona, Y., Farahani, E., Kadner, S., Seyboth, K., et al.,Climate Change 2014: Mitigation of Climate Change. Contribution of Working Group III to the Fifth Assessment Report of the Intergovernmental Panel on Climate Change. Cambridge University Press, Cambridge and New York, NY.

Sommer, R., Bossio, D., 2014. Dynamics and climate change mitigation potential of soil organic carbon sequestration. J. Environ. Manag. 144, 83–87.

Springmann, M., D. Mason-D'Croz, S. Robinson, T. Garnett, H.C.J. Godfray, D. Gollin, et al. 2016a, Global and Regional Health Effects of Future Food Production Under Climate Change: A Modelling Study. Lancet. Published Online: March 2, 2016. doi.org/10.1016/S0140-6736(15)01156-3. Available from <http://www.thelancet.com/journals/lancet/article/PIIS0140-6736(15)01156-3/abstract>.

Springmann, M., D. Mason-D'Croz, S. Robinson, K. Wiebe, H.C.J. Godfray, M. Rayner, et al., 2016b. Mitigation Potential and Global Health Impacts from Emissions Pricing of Food Commodities. Nat. Clim. Change. Published Online: November 7, 2016. doi.org/10.1038/NCLIMATE3155.

Steinfeld, H., Gerber, P., Wassenaar, T., Castel, V., Rosales, M., de Haan, C., 2006. Livestock's Long Shadow Environmental Issues and Options. FAO, Rome.

Stocking, M.A., 2003. Tropical soils and food security: the next 50 years. Science 302 (5649), 1356–1359.

Thornton, P., Lipper, L. 2014. How Does Climate Change Alter Agricultural Strategies to Support Food Security? IFPRI Discussion Paper 01340. International Food Policy Research Institute, Washington DC.

Tilman, D., Balzer, C., Hill, J., Befort, B.L., 2011. Global food demand and the sustainable intensification of agriculture. Proc. Natl. Acad. Sci. 108 (50), 20260–20264.

Van Vuuren, D.P., Kriegler, E., O'Neill, B.C., Ebi, K.L., Riahi, K., Carter, T.R., et al., 2014. A new scenario framework for climate change research: scenario matrix architecture. Clim. Change 122, 373–386.

Waithaka, M., G. Nelson, T.S. Thomas, M. Kyotalimye, (Eds.), 2013. *East African Agriculture and Climate Change: A Comprehensive Analysis*. IFPRI Research Monograph. International Food Policy Research Institute, Washington DC. Available from <http://www.ifpri.org/publication/east-african-agriculture-and-climate-change>.

Wiebe, K., Lotze-Campen, H., Sands, R., Tabeau, A., van der Mensbrugghe, D., Biewald, A., et al., 2015. Climate change impacts on agriculture in 2050 under a range of plausible socioeconomic and emissions scenarios. Environ. Res. Lett. 10 (8). Available from: https://doi.org/10.1088/1748-9326/10/8/085010. Available from <http://iopscience.iop.org/1748-9326/10/8/085010/media>.

Wollenberg, E., Richards, M., Smith, P., Havlik, P., Obersteiner, M., Tubiello, F.N., et al., 2016. Reducing emissions from agriculture to meet 2°C target. Global Change Biol. 22 (12), 3859–3864.

CHAPTER

5

Water Scarcity and Challenges for Smallholder Agriculture

Meredith Giordano[1], *Jennie Barron*,[2,3] *and Olcay Ünver*[4]

[1]International Water Management Institute (IWMI), Colombo, Sri Lanka [2]Swedish University of Agricultural Sciences, Uppsala, Sweden [3]CGIAR Research Program on Water, Land and Ecosystems, Colombo, Sri Lanka [4]Food and Agriculture Organization of the United Nations (FAO), Rome, Italy

5.1 INTRODUCTION

Increases in global food production are needed to eradicate hunger and secure food for a growing world population. Recent studies suggest that global food production will need to increase by nearly 50% by 2050 (up from 2013 levels) (FAO, 2017). Meeting these food production requirements will rely in large part on the world's smallholder producers. These farmers manage nearly 500 million farms worldwide and contribute an appreciable share of agricultural production and food calories in many regions of the world. Further growth in the sector promises to play an important role in poverty reduction (Christiaensen et al., 2011; Lowder et al., 2016; Samberg et al., 2016).

While agricultural production has become more efficient in recent decades, growing competition for natural resources—due to climate change, urbanization, dietary changes, and industrial development—is compromising ecological integrity and agricultural productivity. Forty percent of the world's land has already been transformed to agriculture (Ramankutty et al., 2008), and few opportunities remain globally to expand agricultural area without significant environmental, social, and economic costs (FAO, 2017). Water availability is also expected to become a growing constraint in general and for agricultural productivity in particular. Already 40% of the world's rural population live in river basins that are classified as water scarce (FAO, 2017). By 2050, nearly 4 billion people could be subject to severe water stress (Sadoff et al., 2015). Changes in water quantity, quality,

Sustainable Food and Agriculture
DOI: https://doi.org/10.1016/B978-0-12-812134-4.00005-4

and flow regimes also threaten freshwater ecosystems and the services they provide (TNC, 2016).

Agriculture is at the center of this picture. Irrigated agriculture accounts for 70% of total freshwater withdrawals globally (FAO, 2017). Recent estimates suggest that freshwater withdrawals already exceed planetary boundaries by more than 10% (Jaramillo and Destuoni, 2015). In some countries, water withdrawn for irrigation exceeds the total amount of renewable freshwater resources (Scheierling and Treguer, 2016; WWAP, 2016; FAO, 2017). The potential economic and social impacts are enormous, particularly so for low-income countries that are already facing water scarcity and/or where agriculture makes up a significant proportion of gross domestic product. Sub-Saharan Africa and South Asia are particularly vulnerable (Sadoff et al., 2015).

The importance of addressing water scarcity issues is now firmly on the global agenda. The World Economic Forum's Global Risks Report has identified water crises[1] as one of the top five global risks in terms of the impact on economies for 3 consecutive years, and one that is intrinsically linked with food crises, extreme weather events, and failure of climate-change mitigation and adaptation (WEF, 2015, 2016, 2017). Water is also a dedicated goal (Sustainable Development Goal 6), alongside zero hunger (Sustainable Development Goal 2) and climate action (Sustainable Development Goal 13) in the 2030 Agenda for Sustainable Development. The water goal aims to ensure the availability and sustainable management of water and sanitation for all sectors, including agriculture and the environment (United Nations, 2015). Complementary initiatives, such as the recently launched Global Framework for Action to Cope with Water Scarcity in Agriculture in the Context of Climate Change,[2] are supporting this aim by assisting developing countries to scale up and accelerate successful responses to the threat increasing water scarcity poses to agricultural production (IISD, 2016).

Addressing water scarcity, however, goes beyond ensuring the physical availability of water. Many people, particularly in large parts of the developing world, are unable to access and manage existing water resources for productive use because of economic or institutional barriers. This form of water scarcity, termed economic water scarcity, results from a lack of infrastructure, financial resources, appropriate institutions, and capacity that in conjunction limit the development of available water resources or that result in inequitable distribution. Economic water scarcity impacts about 1.6 billion people worldwide (Molden, 2007).

Investments are needed to support smallholder farmers to cope with water scarcity (both physical and economic) for achieving food security, livelihood benefits, and poverty reduction goals. A better understanding of the myriad issues that contribute to the resource scarcities faced by smallholder farmers will offer important insights on where and how public and private sector investors can best support and leverage this sector for greater impact (de Fraiture and Giordano, 2014; Woodhouse et al., 2017; Bjornlund et al., 2016).

[1] Defined as "a significant decline in the available quality and quantity of freshwater, resulting in harmful effects on human health and/or economic activity" (WEF, 2017, p. 62).

[2] For more information, see www.fao.org/3/a-i5604e.pdf.

This chapter assesses the key global challenges to the adequacy of water availability and how increasing scarcity and competition for water resources are affecting agricultural productivity, especially that of smallholder producers in Asia and Africa. It further analyzes alternative, sustainable water management practices that can be adapted to the needs of smallholder farmers. We provide evidence of the economic viability and potential to improve farmers' livelihoods. The opportunity to scale up the related investments for high-impact solutions is also assessed against available empirical evidence.

5.2 WATER SCARCITY DIMENSIONS, MEASURES, AND IMPLICATIONS

Water scarcity is a complex issue with multiple dimensions, both a function of the quantity and quality of water, and the ability to access sufficient supplies of sufficient quality at the right time and place. Consequently, water stress can mean different things in different contexts. To understand this complexity and the extent to which people and the environment are at risk of water scarcity, a number of methods have evolved to capture the relationship between water supply and demand.

As mentioned in Chapter 2, one common indicator of water scarcity measures water withdrawals as a percentage of total renewable water (surface and groundwater) resources. According to this method, water stressed conditions occur when withdrawals exceed 25% of renewable freshwater resources. "Physical water scarcity" exists when more than 60% of renewable water supplies are withdrawn; "severe water scarcity" exists when more than 75% of water is withdrawn (FAO, 2017). Applying this method, many countries in North Africa, the Middle East, and Central Asia face high to extremely high water stress (FAO, 2017; see also Chapter 2, Fig. 2.5).

The central role of agriculture in global and regional water scarcity becomes apparent when the above water scarcity indicator is modified to specifically isolate withdrawals for irrigated agriculture. Applying this modified indicator, countries facing medium to extreme water stress are nearly identical to those in Fig. 2.5 of Chapter 2. Countries facing extreme water stress include Saudi Arabia, where water withdrawals for irrigation are more than eight times the amount of total renewable water resources; Libya, where agricultural water withdrawals are five times greater than renewable water resources; and Egypt, where agricultural withdrawals are slightly above the total renewable water resources (Scheierling and Treguer, 2016).

Care must be taken, however, when interpreting the above results. First, water withdrawn for a particular purpose (agriculture or otherwise) is not necessarily water consumed. So-called "losses" in water conveyance and application are often captured and reused on-site or elsewhere in a hydrologic system, such as a watershed or basin, by different users for different purposes.[3] In the case of irrigated agriculture, the reuse of return flows can be significant (Scheierling and Treguer, 2016; Giordano et al., 2017).

[3] The opportunity to beneficially recycle water returning to a surface water or groundwater source may be reduced if its quantity or quality is diminished, for example, by flowing into a saltwater sink or through saline or polluted soils or groundwater (Giordano et al., 2017).

Second, these indices rely on annual national-scale data, whereas vulnerabilities to water scarcity can vary significantly over time and space. A recent analysis of water scarcity in major river basins globally, which accounts for storage capacity and seasonable changes in water supply and demand, revealed that a very different set of countries and regions are at risk of water scarcity. According to this aggregated model of water scarcity (Fig. 5.1), significant risks of water shortages exist in all continents, but northern China and parts of South Asia are particularly vulnerable to water scarcity risks. Further, the analysis suggests that the potential benefits to agriculture from mitigating such hydrologic variability are enormous, as much as US$94 billion for a single year (Sadoff et al., 2015).

Finally, in some locations actual physical water supplies are abundant relative to water use, but other factors, both socioeconomic and institutional, limit access to these resources for productive purposes (Box 5.1). Causes of economic water scarcity include inadequate infrastructure to access, store, and manage water; limited financial and/or human capital; and weak institutional structures that result in inequitable distribution of water resources. This form of water scarcity can be strongest in poor countries, countries facing physical water scarcity, and/or those that have a high dependence on agriculture (Sadoff et al., 2015). As illustrated in Fig. 5.2, much of sub-Saharan Africa and parts of South and Southeast Asia suffer from this form of water scarcity (Molden, 2007).

For both forms of water scarcity—economic and physical—the least developed regions, and the smallholders operating within them, are particularly vulnerable. The hydrologic variability present in many parts of sub-Saharan Africa and South and Southeast Asia, coupled with financial and institutional constraints limiting reliable access to existing

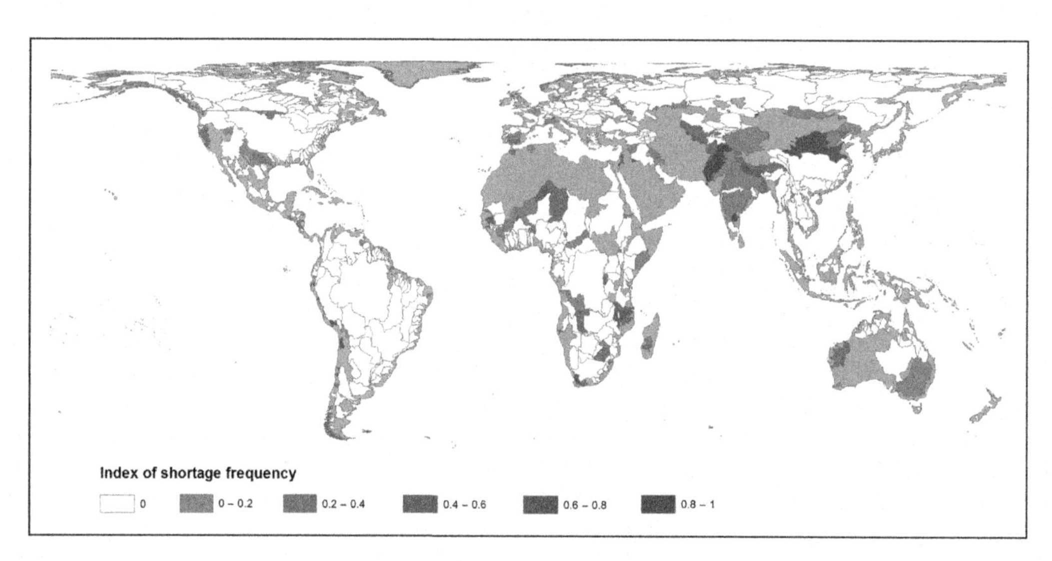

FIGURE 5.1 Index of frequency of shortages of water available for use on a month-to-month basis. Note: The index of shortage frequency denotes the frequency at which reservoir levels are predicted to fall below 20% of total storage. The 20% threshold is considered to be the average storage level at which water use restrictions may be applied. Source: *Sadoff, C.W., Hall, J.W., Grey, D., Aerts, J.C.J.H., Ait-Kadi, M., Brown, C., et al., 2015. Securing Water, Sustaining Growth: Report of the GWP/OECD Task Force on Water Security and Sustainable Growth. University of Oxford, Oxford, UK.*

BOX 5.1

DIFFERENT FORMS AND SYMPTOMS OF WATER SCARCITY

Physical Scarcity

Physical scarcity occurs when there is not enough water to meet all demands, including environmental flows. Arid regions are most often associated with physical water scarcity, but water scarcity also appears where water is apparently abundant, when water resources are overcommitted to various users owing to overdevelopment of hydraulic infrastructure, most commonly for irrigation purposes. In such cases, there simply is not enough water to meet both human demands and environmental flow needs. Symptoms of physical water scarcity are severe environmental degradation, declining groundwater, and water allocations that favor some groups over others.

Economic Water Scarcity

Economic scarcity is caused by a lack of investment in water or a lack of human capacity to satisfy the demand for water. Much of the scarcity is due to how institutions function, favoring one group over another, and not hearing the voices of various groups, especially women. Symptoms of economic water scarcity include scant infrastructure development, either small or large scale, so that people have trouble acquiring enough water for agriculture or drinking. Even where infrastructure does exist, the distribution of water may be inequitable.

Source: Water for Food Water for Life *(Molden, D. (Ed.), 2007. Water for Food, Water for Life: A Comprehensive Assessment of Water Management in Agriculture. International Water Management Institute (IWMI), London, UK: Earthscan; Colombo, Sri Lanka. 645p.)*

water resources, negatively affects farmers' ability to optimize their management of water resources for productive agriculture (Giordano et al., 2012; Sugden, 2014; Sadoff et al., 2015; Mosello et al., 2017). The water stress caused by economic and physical water scarcity also severely compromises terrestrial and aquatic habitats and the multiple ecosystem services (including the provision of food) these systems provide. Flow depletion is a key threat to biodiversity; and the lack of water-related assets and institutions in many economically water scarce countries reduces the ability to mitigate the related water and food security risks (Vörösmarty et al., 2010). The following section further defines smallholder farming, its current and potential role in supporting food security and livelihoods, and key resource scarcity constraints and challenges faced by this sector.

5.3 SMALLHOLDER AGRICULTURE: DEFINING THE SECTOR, ITS POTENTIAL FOR GROWTH, AND KEY RESOURCE SCARCITY CONSTRAINTS AND CHALLENGES

The positive food security and livelihood impacts from investments in smallholder agriculture are well documented (de Fraiture and Giordano, 2014; Giordano and de Fraiture, 2014;

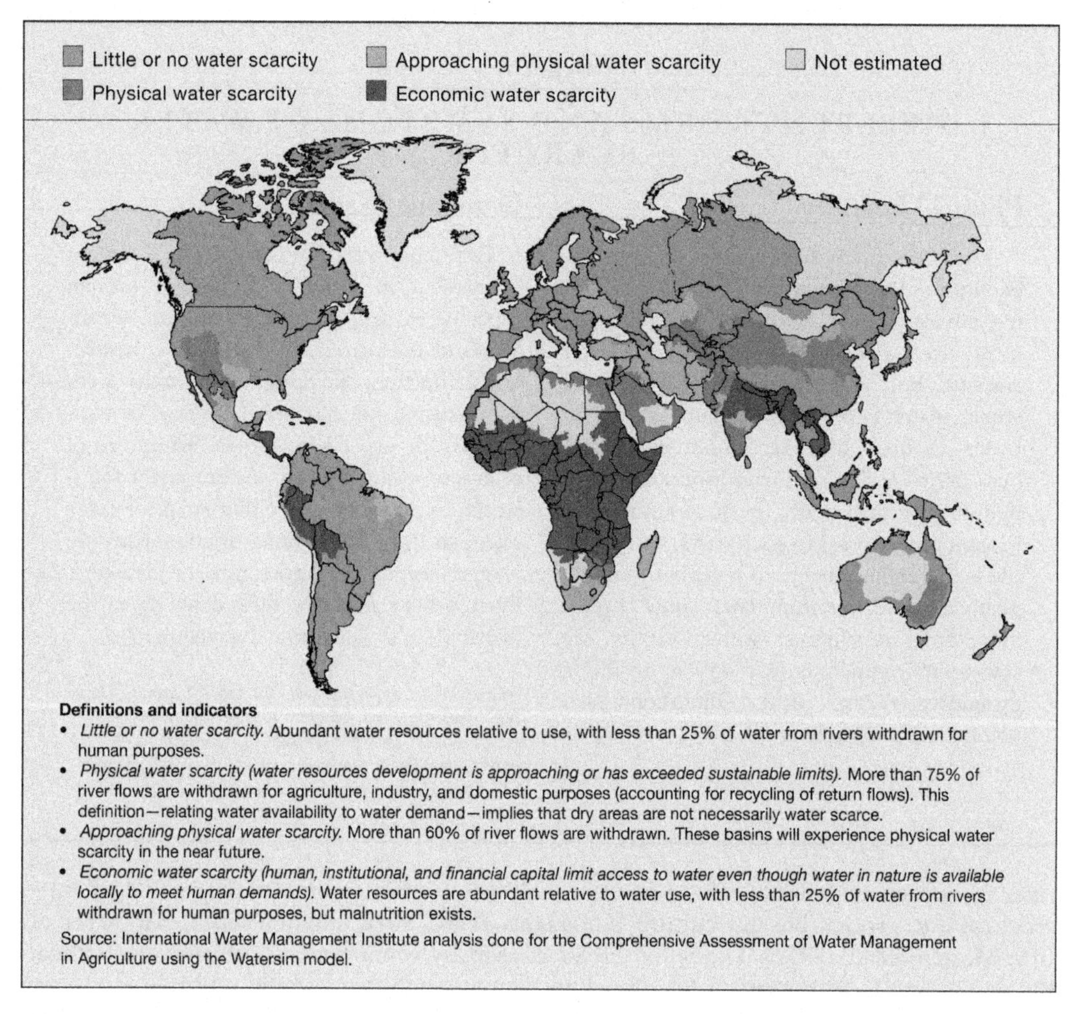
Little or no water scarcity
Approaching physical water scarcity
Not estimated
Physical water scarcity
Economic water scarcity
Definitions and indicators
• *Little or no water scarcity.* Abundant water resources relative to use, with less than 25% of water from rivers withdrawn for human purposes.
• *Physical water scarcity (water resources development is approaching or has exceeded sustainable limits).* More than 75% of river flows are withdrawn for agriculture, industry, and domestic purposes (accounting for recycling of return flows). This definition—relating water availability to water demand—implies that dry areas are not necessarily water scarce.
• *Approaching physical water scarcity.* More than 60% of river flows are withdrawn. These basins will experience physical water scarcity in the near future.
• *Economic water scarcity (human, institutional, and financial capital limit access to water even though water in nature is available locally to meet human demands).* Water resources are abundant relative to water use, with less than 25% of water from rivers withdrawn for human purposes, but malnutrition exists.
Source: International Water Management Institute analysis done for the Comprehensive Assessment of Water Management in Agriculture using the Watersim model.

FIGURE 5.2 Global physical and economic water scarcity. Source: *Molden, D. (Ed.), 2007. Water for Food, Water for Life: A Comprehensive Assessment of Water Management in Agriculture. International Water Management Institute (IWMI), London, UK: Earthscan; Colombo, Sri Lanka. 645p.*

Rockström et al., 2016). For example, in Ethiopia, earthen embankments to capture rainwater have doubled sorghum yields (Binyam and Desale, 2015). In southern Niger, farmer-managed natural land regeneration, using improved, local agroforestry practices on nearly 5 million hectares of land, has increased cereal production and improved the livelihoods of an estimated 2.5 million people (Reij et al., 2009). In India, farm ponds have allowed farmers to produce more staple crops, extend the cropping area, increase dry-season cropping, diversify activities, and increase incomes by as much as 70% (Malik et al., 2014). Achieving the sector's full livelihood and food security potential requires understanding the nature of the sector—its food and livelihood contributions—as well as overcoming resource scarcity conditions faced by smallholders and the array of factors that contribute to this scarcity.

Defining the Smallholder Agricultural Sector and Its Contributions to Food and Livelihoods

No single definition of smallholder agriculture exists. Small is a relative concept—whether a landholding is considered large or small depends on the socioeconomic and agroecological setting and can vary significantly by region. A "small" landholding in Latin America, for example, may be considered large in many countries in sub-Saharan Africa or Asia. Time scale also plays a role, with the average farm size increasing between 1960 and 2000 in many high-income countries and decreasing for many low- and medium-income countries (Lowder et al., 2016).

That said, landholding size is commonly used as a proxy to define smallholder agriculture, with 2 hectare (ha) or less often used as a threshold to define a farm as "small." A commonly cited statistic is that there are approximately 500 million small farms worldwide. Based on agricultural census data gathered from 167 countries and territories, Lowder et al. (2016) recently reviewed the evidence and estimate that conservatively there are some 570 million farms globally. Of these 570 million farms, about 84% (or 475 million farms) are less than 2 ha in size, and 72% (or 410 million farms) are less than 1 ha in size.

Beyond landholding size, other key attributes of smallholder agriculture are important to note. According to the High Level Panel of Experts on Food Security and Nutrition (2013), a smallholding, no matter its exact size, is considered "small" because "resources are scarce." Other key characteristics of the smallholder agricultural sector are: (1) it is generally practiced by family groups; (2) family groups are often headed by women; and (3) women play important roles not only in production but also in the processing and marketing of the agricultural outputs (HLPE, 2013).

The importance of the smallholder agricultural sector to livelihoods and food security is significant. Seventy percent of the world's poor are estimated to live in rural areas and depend in part or solely on agriculture for their livelihoods (HLPE, 2013; World Bank, 2016). While the exact contribution of smallholder farms to agricultural production is debated, their role is not insignificant. In low- and lower-middle-income countries, smallholder farms (with less than 2 ha of land) manage 30%–40% of agricultural land and contribute a sizeable percentage of agricultural production (Lowder et al., 2016). Farms of less than 2 ha in size contribute more than 64% of global rice production, 50% of global groundnut, and 23% of wheat production (Samberg et al., 2016).

Growth in the smallholder agriculture sector has a disproportionately higher impact on poverty alleviation than growth in other sectors (Deininger and Byerlee, 2012). Agricultural growth was found to be eleven times more effective than growth outside of agriculture in sub-Saharan Africa, and five times more effective in other resource-poor, low-income countries (Christiaensen et al., 2011). Significant additional potential exists for the sector's contribution to agricultural livelihoods, productivity, and transformational change (Deininger and Byerlee, 2012; Giordano et al., 2012; Larson et al., 2016). A recent review of over 30 field- and community-level case studies on investments in small-scale

TABLE 5.1 Potential Reach and Possible Additional Household Revenue for Selected Investments in Agricultural Water Management Interventions

Solution	Sub-Saharan Africa People Reached (Million)	Net increase HH Income (USD billions/ year)	South Asia People Reached (Million)	Net increase HH Income (USD billions/ year)
Motor pumps	185	22	40	4
Rainwater harvesting	147	9	205	6
Small reservoirs	369	20	N/A	N/A

HH, household.
Source: *Adapted from Giordano, M., de Fraiture, C., Weight, E., van der Bliek, J., 2012. Water for Wealth and Food Security: Supporting Farmer-Driven Investments in Agricultural Water Management. International Water Management Institute (IWMI), Colombo, Sri Lanka.*

irrigation (Table 5.1) found that targeted investments in agricultural land and water management and associated policy interventions could double or even triple rainfed crop yields in sub-Saharan Africa and South Asia while generating additional net household revenues for millions of people (Giordano and de Fraiture, 2014).

While the impact to date of smallholder agriculture in sub-Saharan Africa has been more localized than the transformational changes experienced during Asia's Green Revolution, a recent World Bank study (Larson et al., 2016) suggests that investments in smallholder productivity in the region, particularly in combination with policies that prepare for employment outside of the agricultural sector, is the best option to accelerate economic growth, reduce rural poverty, and support global food security needs.

However, for the sector to achieve its full potential in both Asia and Africa, a number of constraints and risks related to efficiency, equity, and long-term sustainability of water management need to be addressed. Meeting these challenges requires more than just changes in agricultural production; it requires a better understanding of the complex constraints faced by smallholder farmers, as well as the opportunities for integrated solutions—across sectors and scales—that support existing farmer-led initiatives and enable more productive, equitable, and sustainable smallholder farming systems.

Resource Scarcity Challenges Faced by Smallholder Producers

Smallholder farming systems are influenced by an array of complex ecological, social, economic, and political factors specific to the context in which they operate. Yet, research on improving smallholder agricultural water management (AWM) has tended to focus on technical solutions to improve agricultural productivity. This rather narrow view overlooks the range of factors that shape farmers' perceptions, choices, constraints, and decision-making, and the broader impacts on the landscape and supporting ecosystem services (Barron and Noel, 2011; Cordingley et al., 2015; Adimassu et al., 2015;

Snyder et al., 2016; Bjornlund et al., 2016). Consequently, many water management technologies, while beneficial in one context, may not be relevant or feasible in others (Larson et al., 2016; Brown and Nuberg, 2017).

Water management is critical to achieve future agricultural production needs. Yet, for nearly 2.8 billion people,[4] many of whom are poor, water resources are scarce due to physical, economic, or institutional limitations (de Fraiture et al., 2007). Of sub-Saharan Africa's abundant renewable water resources, for example, only 3% are withdrawn for agriculture, and just 5% of agricultural land, 6 million hectares, is equipped for irrigation (FAO, 2011a). In South Asia, lack of access to water also hinders agricultural productivity gains, even where water is relatively abundant. The Terai that spans eastern India, Bangladesh, and Nepal hosts some of the world's most abundant surface water and groundwater resources. However, not only do frequent droughts take place, but the Terai is also intensely flood-prone and subject to prolonged surface waterlogging after normal monsoons (Giordano et al., 2012).

Many technologies remain out of reach for smallholder farmers. Women, who play a substantial role in agriculture and food production in developing countries, are particularly disadvantaged (FAO, 2011b; Giordano and de Fraiture, 2014; FAO, 2017; Oates et al., 2017). Women are often confronted with poor access to natural resources and the assets to productively use these resources. In addition, women tend to have limited options to respond to the different labor inputs new AWM production systems often require. While individualized technologies could offer women significant production opportunities, women are often underrepresented in the use and ownership of small-scale irrigation equipment. Because of this gender gap in access to resources, agricultural productivity levels of women farmers are as much as 30% lower than those of men farmers (FAO, 2017).

Market inefficiencies negatively affect farmer decision-making and technology access. These inefficiencies include poorly developed supply chains; high taxes and transaction costs; lack of information and knowledge on irrigation, seeds, marketing, and equipment; and the uneven distribution of information and power in output markets (Giordano and de Fraiture, 2014; Bjornlund et al., 2016). As a result, investments in small-scale production are often financed by farmers themselves (de Fraiture and Giordano, 2014; Woodhouse et al., 2017). Surveys carried out in Ethiopia, Ghana, and Zambia, for example, found that more than 80% of all owners of small-scale irrigation equipment used their own or their family's savings (Giordano et al., 2012).

Further, the uncontrolled and unregulated spread of AWM technologies can have undesirable social and environmental consequences. If not managed within the landscape context, and with the water demands of other users in mind, accelerated investments in smallholder AWM, together with greater use of chemicals such as fertilizers, pesticides, herbicides, and fungicides, could further degrade water and soil quality and negatively impact downstream or groundwater users. In Burkina Faso, for example, declining water quality and quantity in downstream canal-based irrigation systems has been attributed to

[4] According to the Comprehensive Assessment of Water Management in Agriculture (Molden, 2007), approximately 1.2 billion people live in areas of physical water scarcity and 1.6 billion face economic water scarcity (de Fraiture and Wichelns, 2007).

vegetable farmers "informally" pumping from reservoirs upstream (de Fraiture et al., 2014). In Ethiopia, rapid expansion of pump irrigation has led to increased competition and conflict over limited water resources between motor pump users and farmers practicing traditional irrigation (Dessalegn and Merrey, 2014).

To avoid the adverse impacts of intensification, public policies should encourage investments in sustainable agricultural intensification practices and techniques, incorporating multiple aspects such as better land and soil management, adaptive water management, diversified agricultural systems, agroecology, and agroforestry. Anchoring the integrity of the ecosystems, resilience, and sustainability of the landscapes and intensification of agricultural production is fast becoming the norm, although large-scale field implementation has yet to happen (FAO, 2011c; FAO, 2014; Rockström et al., 2016). In the next section, a set of promising investment pathways are explored to address the resource scarcity constraints faced by smallholder producers and enable more productive, equitable, and sustainable smallholder farming systems.

5.4 INVESTMENT PATHWAYS TO UNLOCK THE POTENTIAL FOR SMALLHOLDER AGRICULTURE

Research clearly indicates that many diverse factors influence local water and land management decisions. Consequently, narrowly defined AWM interventions, made without consideration for the environmental and social context, will likely neither solve important, complementary issues for farmers nor address cross-scale, intersectoral synergies and trade-offs (Cumming et al., 2014; Barron et al., 2015; Snyder et al., 2016). We provide here four different business models to support smallholder farmers in addressing water resource scarcity (both physical and economic) in the diverse and complex landscapes in which they operate.

Increase Equitable Access to Agricultural Water Management Technologies[5]

Pump rental markets are a fairly widespread practice in parts of India and a developing phenomenon in sub-Saharan Africa. Smallholders who cannot afford to purchase their own pump can rent one by the day or for a season. Rental markets create opportunities for poor farmers to access water for irrigation. A related concept is that of irrigation service providers. In this model, an entrepreneur travels from farm to farm with small motor pumps to irrigate land for a fixed fee per hour. Smallholder farmers gain access to motorized pumps without the cost of ownership, operation, and maintenance, while the entrepreneur profits from leasing pumping equipment and providing irrigation services and information. The model provides employment opportunities and incomes for entrepreneurs, including rural youth, and improved access to information and AWM technologies for smallholders (Williams et al., 2012; Giordano and de Fraiture, 2014).

[5] The model presented in this section is based on the Irrigation Service Provide Business Plan detailed in de Fraiture and Clayton (2012) and summarized in Williams et al. (2012).

Based on existing experiences in India and Burkina Faso, de Fraiture and Clayton (2012) developed a business plan for investing in irrigation service providers in sub-Saharan Africa, where economic water scarcity prevails in much of the region. Three investment areas are identified in the model: (1) the establishment of business development services to recruit, train, and support pump service providers; (2) loan guarantees to help irrigation service providers access credit for business start-up; and (3) loan guarantees to help smallholder farmers access credit to acquire the needed agricultural inputs for irrigated vegetable production. As an example, to establish 10 business development services, each supporting 30 irrigation service providers, and the requisite loan guarantees for both the irrigation service providers and smallholder farmers, the estimated investment cost is US $3.8 million, which would generate annual net profits of US$6.1 million after a 3-year period.

This investment would in turn result in:

- Approximately 375 new irrigation service providers, operational after 2 years, each irrigating an average of 7.4 ha per dry season or 2775 ha of irrigated vegetable crops; and earning an average net income of US$1,235 per dry season cropping cycle (approximately US$463,000 in total annual net profit).
- Up to 7,500 smallholder farmers (including women) earning at least US$750 each in supplementary net income (approximately US$5.7 million in total net profit).

This equates to approximately US$1.60 in net profit for the service provider and smallholder farmer for every US$1.0 invested in the business model (de Fraiture and Clayton, 2012; Williams et al., 2012). As noted in Table 5.1, the overall potential for investments, through irrigation service providers or other models, in motorized pumps in sub-Saharan Africa is significant. Taking into account both physical and market limitations, Xie et al. (2014) estimate potential application area at nearly 30 million hectares, benefiting up to 185 million people and generating net revenues of US$22 billion annually.

Invest in Water Storage[6]

In India's Madhya Pradesh, farmers have made significant investments in on-farm ponds since 2006 to address declining groundwater quality and quantity and the fact that limited and erratic electricity constrained access to available water supplies. Between 2006 and 2012, more than 6,000 farmers in the state had invested in on-farm ponds. Malik et al. (2014) conducted an *ex post* impact assessment of the on-farm ponds in Dewas District, Madhya Pradesh, specifically focusing on two geologically diverse blocks: Tonkkhurd block, characterized by hard rock aquifers, and Khategaon block, characterized by soft rock aquifers interspersed with areas of hard rock.

In both blocks, farmer investments in on-farm ponds led to significant improvements in the availability of irrigation water, a revival of the agricultural economy of the region, and substantial increases in farmer incomes and livelihoods. Specifically, the ponds allowed farmers the option of dry season cultivation. Before the construction of the ponds, more

[6] The model presented in this section is based on the Decentralized Rainwater Harvesting business case detailed in Malik et al. (2014).

than 75% of cultivable area during the dry season was left fallow. The ponds reversed this situation, enabling farmers to cultivate wheat and gram during the dry season, resulting in an increase in the proportion of area cultivated from 22% to about 96% in the dry season, and an overall increase in cropping intensity from about 122% to 198% (Table 5.2). Consequently, farm incomes increased by more than 70% on average, and farmers were able to improve their livestock herds and further diversify production activities. Farmers also reported improvements in groundwater recharge due to seepage from the structures and increased density and availability of wildlife in the area (Malik et al., 2014).

The benefit-cost ratios differed between the two blocks because of the geology and related differences in capital costs of the structures. In general, without government subsidies, the benefit-cost ratio was between 1.92 in Khatagaon and 1.48 in Tonkkhurd, and farmers were able to recuperate their costs between 2.5 and 3 years. With government subsidies, the benefit-cost ratio was between 2.39 in Khatagaon and 1.72 Tonkkhurd, and the payback period reduced to between 2 and about 2.5 years (Malik et al., 2014). Taking into account both biophysical- and livelihood-based demands, an estimated 400,000 to 1.9 million hectares in Madhya Pradesh are suitable for investments in rainwater harvesting (Fig. 5.3), with the potential to benefit between 270,000 and 1.3 million households (FAO, 2012).

TABLE 5.2 Changes in Cultivated Area, Cropping Patterns, and Cropping Intensity – Khategaon and Tonkkhurd Blocks, Dewas District, Madhya Pradesh, India

		Khategaon		Tonkkhurd	
Season	**Indicator**	**Before[a] (%)**	**After[a] (%)**	**Before[a] (%)**	**After[a] (%)**
Wet	Operated area[b] allocated to:				
	Soybean	68	91	97	98
	Cotton	30	7	0	0
	Operated area left fallow	2	2	3	2
Dry	Operated area allocated to:				
	Wheat	18	47	9	53
	Gram	4	46	15	43
	Operated area left fallow	78	7	76	4
Annual	Cropping Intensity[c]	122	194	125	198

[a]*"Before" and "after" refer to periods before and after construction of water harvesting structures.*
[b]*Operated area – owned area + leased-in area – leased-out area.*
[c]*Cropping intensity is the ratio of gross cropped area to net sown area expressed as percentage.*
Khategaon and Tonkkhurd are administrative blocks within Dewas district, Madhya Pradesh. The survey was conducted in August 2010. We interviewed 59 farmers in Khategaon and 61 farmers in Tonkkhurd, for a total of 120 households. 45 households within each block adopted water harvesting structures. The above data relates to these 90 households.
Source: *Malik, R.P.S., Giordano, M., Sharma, V., 2014. Examining farm-level perceptions, costs, and benefits of small water harvesting structures in Dewas, Madhya Pradesh. Agric. Water Manage. 131, 204–211.*

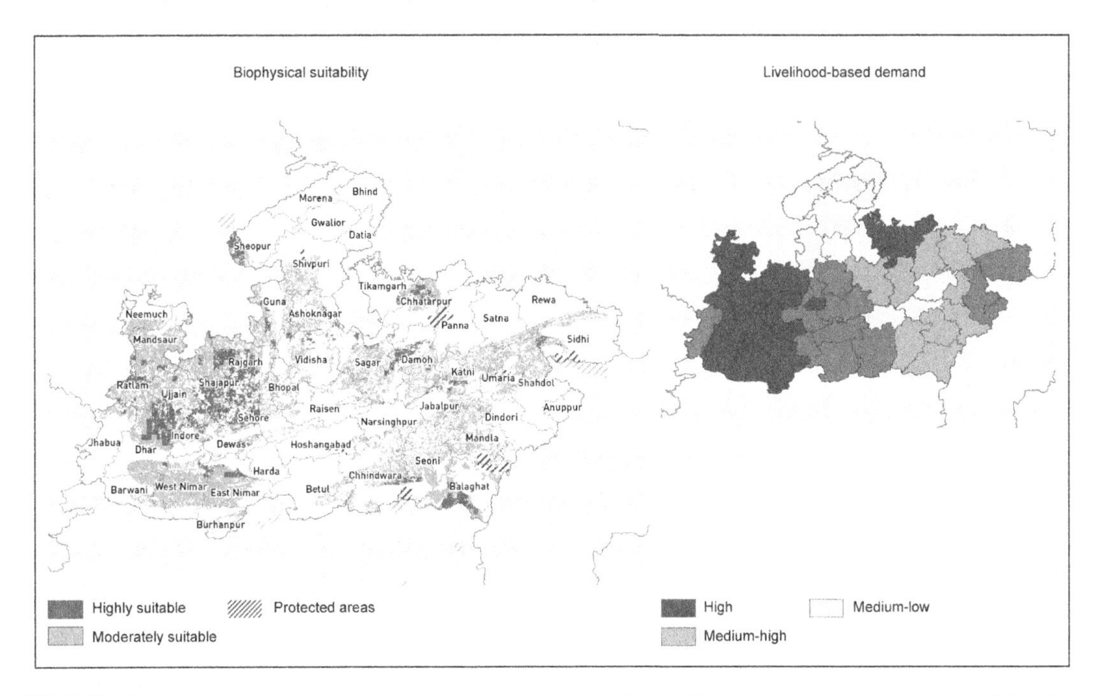

FIGURE 5.3 Suitable locations for rainwater harvesting in Madhya Pradesh. Source: *Adapted from FAO, 2012. Mapping and Assessing the Potential for Investments in Agricultural Water Management: Madhya Pradesh, India. Investment Brief, FAO Water for AgWater Solutions Project, Rome, Italy.*

Invest in Alternative Energy Solutions[7]

In Ethiopia, researchers are exploring investment scenarios to improve access to water resources for smallholder farmers living in remote rural areas without access to electricity. Between 1.1 million and 6.8 million hectares of land in Ethiopia are suitable for solar irrigation development (Fig. 5.4), and the Ethiopian Government has made tax exemptions to support investments and private-sector enterprises. Yet, the investment costs are two to three times higher than current diesel or petrol pumps, and microfinance options can be difficult or expensive to access (Otoo et al., 2018).

Making photovoltaic solar energy-based pumps a viable option for smallholder farmers in Ethiopia will require improvements in supply chains to reduce farmer costs and risks, and better access to affordable credit schemes. Otoo et al. (2018) explore three scenarios: improved rural microfinance opportunities for smallholder farmers to directly purchase solar pumps for on-farm and household use; prefinancing of solar pumps through commercial outgrower or insurance companies; or the sale or lease of pumps to farmers by pump producers or distributors. While further analysis is required, findings from an initial pilot suggest that farmers can achieve a positive return from investments in solar pump irrigation (Table 5.3).

[7] The model presented in this section is based on Solar Water Pumping for Irrigation: Business Model Scenarios and Suitability for Ethiopia in Otoo et al. (2018).

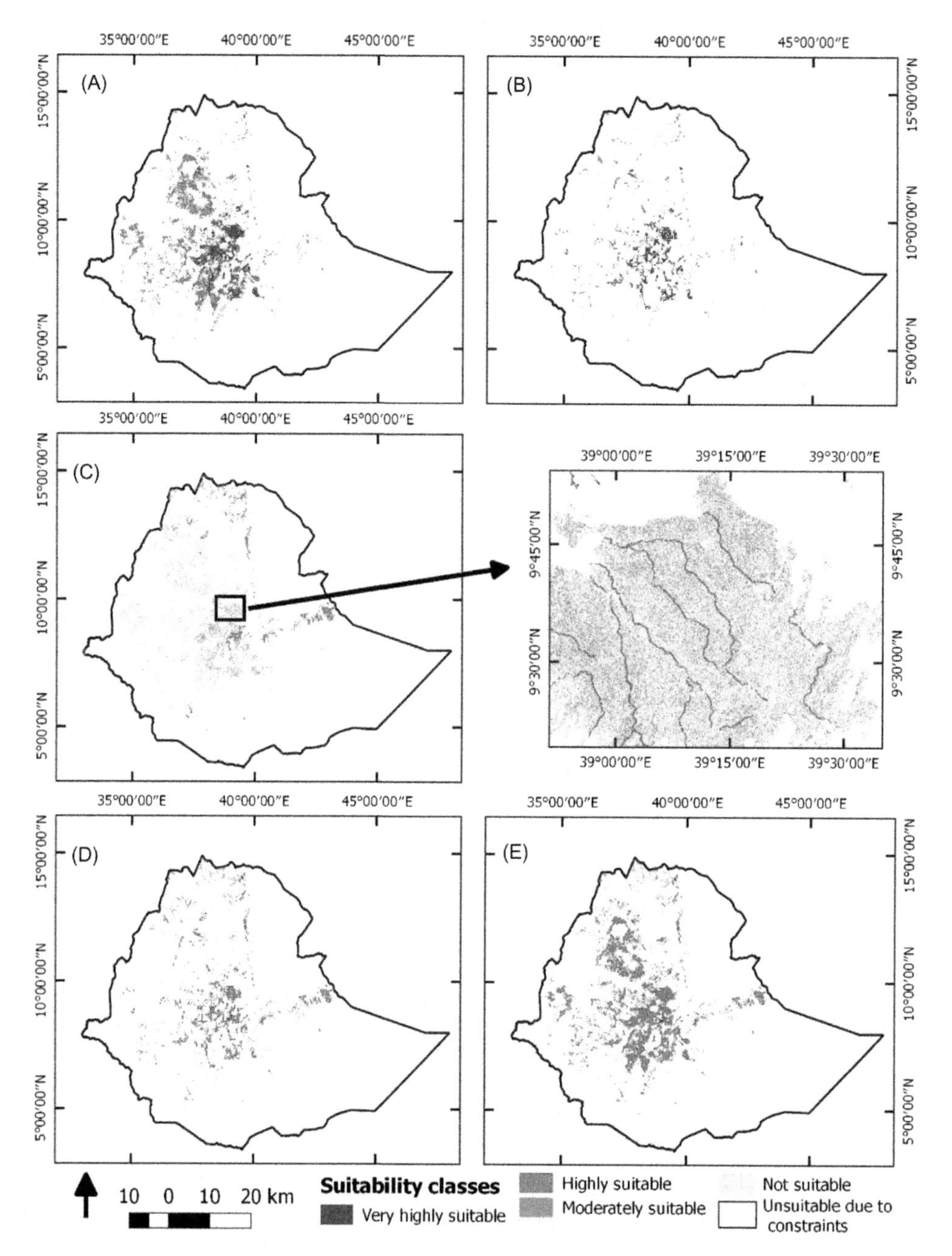

FIGURE 5.4 Suitable areas for solar pump use in Ethiopia.
Suitability maps: (A) Scenario 1: groundwater resources only (depth classes: 0–7 m, 7.1–349 25 m); (B) Scenario 2: groundwater resources only (depth class: 0–7 m); (C) Scenario 3: surface water 350 resources only; (D) Scenario 4: groundwater and surface water resources combined (groundwater 351 depth class up to 7 m); and (E) Scenario 4: with groundwater depth up to 25 m. Source: *Schmitter, P., Kibert, K., Lefore, N., Barron, J., 2018. Suitability mapping framework for solar PV pumps for smallholder farmers in sub-Saharan Africa.* Appl. Geogr. 94, 41–57.

TABLE 5.3 Benefit-Cost Analysis: Comparison of Returns from Solar Pump Irrigation, by Crop Type and Water Application System (in US$)

Variables	Drip-Pepper	Furrow-Pepper	Overhead-Pepper	Overhead-Cabbage, carrot, fodder
Labor costs/hectare (ha)	972	1399	1603	1779
Non-labor input cost/ha	1868	1687	572	427
Total cost/ha	2840	3086	2175	2206
Value of production/ha	8478	5486	5059	3819
Benefit-cost ratio	**2.985**	**1.778**	**2.326**	**1.731**

Adapted from *Otoo, M., Lefore, N., Schmitter, P., Barron, J., Gebregziabher, G., 2018. Business Model Scenarios and Suitability: Smallholder solar pump based irrigation in Ethiopia. International Water Management Institute (IWMI), Colombo, Sri Lanka. Figures in US$ converted from Ethiopian Birr.*

Create Policy Synergies[8]

Incentives for sustainable groundwater irrigation development can lie outside the agricultural water management sector. Solar-powered irrigation is a classic example that demonstrates the interconnectedness of the water, food, and energy sectors. Technological advances in solar pumps coupled with state subsidy programs are causing a major expansion of low-cost tube wells with a lower carbon footprint in India. The government of India has set an ambitious solar target of 100 GW by 2022, largely through megawatt-scale greenfield projects and urban rooftop solar systems. While such an approach could achieve the stated green energy target, detractions include the facts that large capital costs are involved, solar pump owners would be disbursed (and costly to reach), and solar farmers would remain net buyers of grid power. Moreover, without carefully designed programs, solar pumps could further threaten groundwater sustainability (Fig. 5.5) because of the dramatic reduction in the cost of pumping (Shah et al., 2016).

To maximize the productivity and livelihood benefits while minimizing potential environmental externalities, the first ever solar irrigation cooperative is being piloted in Gujarat, India. The scheme enables farmers to sell excess solar power to the utility grid, thereby supplementing their income while incentivizing them to conserve groundwater and energy use (Shah et al., 2016). Six solar pumps have been installed by the cooperative as a pilot with a formal power purchase agreement with the local power utility.

The potential benefits from this arrangement are multiple. Solarizing India's existing 15 million electric tube wells through a cooperative arrangement could allow the government of India to meet its solar capacity target of 100 GW. It has the potential to reduce carbon emissions by 4%–5% per year, and in western India to decrease groundwater overdraft 160 billion cubic meters (BCM) per year to 100–120 BCM per year. The

[8] The model presented in this section is based on the Solar Power as Remunerative Crop business case outlined in Shah et al. (2016).

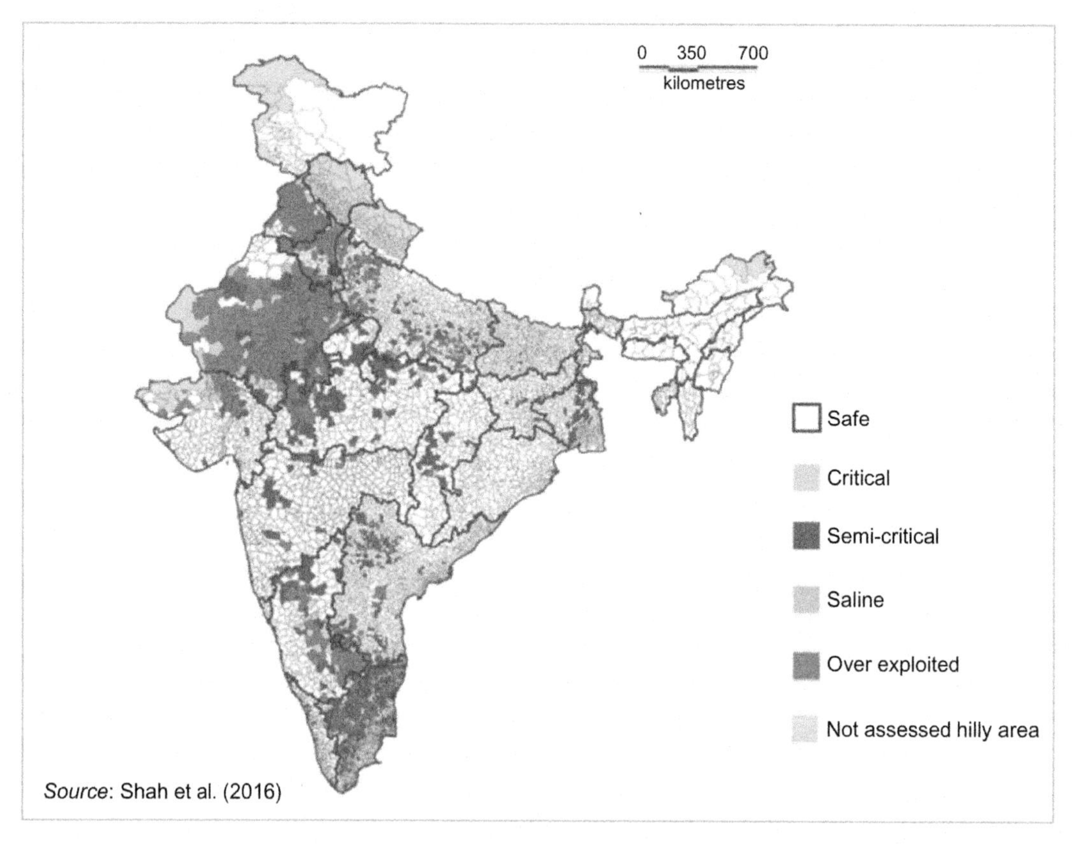

FIGURE 5.5 Groundwater development by district in India. Source: *Shah, T., Durga N., Verma, S., Rathod, R., 2016. Solar Power as Remunerative Crop. Water Policy Research Highlight. IWMI Tata Water Policy Program.*

ability to sell back excess power to the grid will generate additional net income for farmers (~US$770 per year). Researchers estimate that under the cooperative arrangement, an individual farmer with 1 ha of land, using a 7.5 kWp (kilowatt peak) solar pump that delivers 13,500 kW hours per year of on-farm energy supply, could meet his or her irrigation needs while generating additional income of about US$770 per year for selling back the excess power generated. It can also open grid capacity for alternative uses and save billions of dollars in farm power subsidies paid by the utilities and state governments (Shah et al., 2016).

5.5 CONCLUSIONS

Focused investments in agricultural sector can leverage broader socioeconomic development, especially in resource-poor economies highly dependent on agriculture. Research demonstrates that investments in smallholder agriculture can serve as a key pathway to reducing rural poverty, generating economic growth, and supporting food security needs. As the potential to expand irrigated agricultural land is limited globally, intensification of agricultural production is inevitable and the sustainable management of water resources will play an increasingly important role in this equation.

For the many smallholders in Asia and Africa, this will require addressing severe water scarcity challenges spurred by both physical limitations of the resource as well as inadequate water assets and institutions to access available supplies. A number of promising agricultural water management investment opportunities exist to support smallholder farmers to address resource scarcity constraints (and the array of factors that contribute to this scarcity) and enable more productive, equitable, and sustainable smallholder farming systems.

We provide evidence of four such promising opportunities here, tailored to different water scarcity contexts. Two of the business models focus on improving access to agricultural water management technologies and water resources through the promotion of irrigation service providers and investments in water storage. The other two models look beyond the water sector itself to identify smart solar solutions that will improve access to water in energy-poor environments, such as in rural Ethiopia, and support the sustainable use of sparse groundwater resources, such as in western India, through incentives from the energy sector.

The adoption of the United Nations Sustainable Development Goals and the Paris Agreement further strengthen the need for investments in the sustainability and resilience of agricultural development, particularly for the nearly 500 million smallholder farmers worldwide. Actions in the agricultural and food sectors will contribute to this agenda and the complementary aims of ending poverty, ensuring availability and sustainable management of water, and mitigating climate change and its impacts. The investment opportunities presented here offer emerging water management practices that can be adapted to the needs of smallholder farmers and support a transition to sustainable intensification of agricultural systems for poverty alleviation and livelihood improvements.

Acknowledgments

This research synthesis was carried out under the CGIAR Research Program on Water, Land and Ecosystems with support from CGIAR Fund Donors: www.cgiar.org/about-us/our-funders.

References

Adimassu, Z., Langan, S., Johnston, R., 2015. Understanding determinants of farmers' investments in sustainable land management practices in Ethiopia: review and synthesis. Environ. Dev. Sustain. 18 (4), 1005–1023.

Barron, J., Noel, S., 2011. Valuing soft components in agricultural water management interventions in meso-scale watersheds: a review and synthesis. Water Altern. 4 (2), 145–155.

Barron, J., Kemp-Benedict, E., Morris, J., de Bruin, A., Wang, G., Fencl, A., 2015. Mapping the potential success of agricultural water management interventions for smallholders: where are the best opportunities? Water Resour. Rural Dev. 6, 24–39.

Binyam, A.Y., Desale, K.A., 2015. Rain water harvesting: an option for dry land agriculture in arid and semi-arid Ethiopia. Int. J. Water Resour. Environ. Eng. 7, 17–28.

Bjornlund, H., van Rooyen, A., Stirzaker, R., 2016. Profitability and productivity barriers and opportunities in small-scale irrigation schemes. Int. J. Water Resour. Dev. 33 (5), 690–704.

Brown, B., Nuberg, I. 2017. Is Africa really undergoing a smallholder agricultural revolution? The Conversation. February 5, 2017. Available at: <https://theconversation.com/is-africa-really-undergoing-a-smallholder-agricultural-revolution-72100>.

Christiaensen, L., Demery, L., Kuhl, J., 2011. The (evolving) role of agriculture in poverty reduction: an empirical perspective. J. Dev. Econ. 96, 239–254.

Cordingley, J.E., Snyder, K.A., Rosendahl, J., Kizito, F., Bossio, D., 2015. Thinking outside the plot: addressing low adoption of sustainable land management in sub-Saharan Africa. Curr. Opin. Environ. Sustain. 15, 35–40.

Cumming, G.S., Buerkert, A., Hoffmann, E.M., Schlecht, E., von Cramon-Taubadel, S., Tscharntke, T., 2014. Implications of agricultural transitions and urbanization for ecosystem services. Nature 515, 50–57.

Deininger, K., Byerlee, D., 2012. The rise of large farms in land abundant countries: do they have a future? World Dev. 4 (40), 701–714.

Dessalegn, M., Merrey, D.J., 2014. Is 'Social Cooperation' for Traditional Irrigation, While 'Technology' is for Motor Pump Irrigation? International Water Management Institute (IWMI), Colombo, Sri Lanka, 37p. (IWMI Research Report 161).

FAO (Food and Agriculture Organization of the United Nations), 2011a. The State of the World's Land and Water Resources for Food and Agriculture (SOLAW) - Managing Systems at Risk. Food and Agriculture Organization of the United Nations, Rome, and London: Earthscan.

FAO, 2011b. The State of Food and Agriculture 2010–2011: Women and Agriculture Closing the Gap for Development. Food and Agriculture Organization of the United Nations, Rome, Italy.

FAO. 2011c. Save and Grow. A Policymaker's Guide to the Sustainable Intensification of Smallholder Crop Production, Food and Agricultural Organization of the United Nations, Rome, Italy.

FAO, 2012. Mapping and Assessing the Potential for Investments in Agricultural Water Management: Madhya Pradesh, India. Investment Brief. FAO Water for AgWater Solutions Project, Rome, Italy.

FAO. 2014. Building a Common Vision for Sustainable Food and Agriculture: Principles and Approaches, Food and Agricultural Organization of the United Nations, Rome, Italy.

FAO. 2017. The Future of Food and Agriculture – Trends and Challenges, Food and Agricultural Organization of the United Nations, Rome, Italy.

de Fraiture, C., Clayton, T., 2012. Irrigation Service Providers: Business Plan. International Water Management Institute (IWMI), Colombo, Sri Lanka.

de Fraiture, C., Giordano, M., 2014. Small private irrigation: a thriving but overlooked sector. Agric. Water Manage. 131, 167–174.

de Fraiture, C., Wichelns, D., Rockström, J., Kemp-Benedict, E., Eriyagama, N., Gordon, L.J., et al., 2007. Looking ahead to 2050: scenarios of alternative investment approaches. In: Molden, D. (Ed.), Water for Food, Water for Life: A Comprehensive Assessment of Water Management in Agriculture. Earthscan, London, UK, Colombo, Sri Lanka: International Water Management Institute (IWMI). 645p.

de Fraiture, C., Kouali, G.N., Sally, H., Kabre, P., 2014. Pirates or pioneers? Unplanned irrigation around small reservoirs in Burkina Faso. Agric. Water Manag. 131, 212–220.

Giordano, M., de Fraiture, C., 2014. Small private irrigation: enhancing benefits and managing trade-offs. Agric. Water Manag. 131, 175–182.

Giordano, M., de Fraiture, C., Weight, E., van der Bliek, J., 2012. Water for Wealth and Food Security: Supporting Farmer-Driven Investments in Agricultural Water Management. International Water Management Institute (IWMI), Colombo, Sri Lanka.

Giordano, M., Turral, H., Scheierling, S.M., Tréguer, D.O., McCornick, P.G., 2017. Beyond "More Crop per Drop": Evolving Thinking on Agricultural Water Productivity. International Water Management Institute (IWMI), Colombo, Sri Lanka, Washington, DC, USA: The World Bank. 53p. (IWMI Research Report 169).

HLPE (High Level Panel of Experts), 2013. Investing in Smallholder Agriculture for Food Security. A Report by the High Level Panel of Experts on Food Security and Nutrition of the Committee on World Food Security. HLPE, Rome.

IISD. 2016. COP 22 Agriculture and Climate Update: Improving Food Systems' Resilience while Mitigating Climate Change, 23 November 2016.

Jaramillo, F., Destuoni, D., 2015. Local flow regulation and irrigation raise global human water consumption and footprint. Science 350 (6265), 1248–1251.

Larson, D., Muraoka, R., Otsuka, K., 2016. On the Central Role of Small Farms in African Rural Development Strategies. World Bank Policy Research Working Paper 7710. The World Bank Group, Washington DC.

Lowder, S.K., Skoet, J., Raney, R., 2016. The number, size, and distribution of farms, smallholder farms, and family farms worldwide. World Dev. 87, 16–29.

Malik, R.P.S., Giordano, M., Sharma, V., 2014. Examining farm-level perceptions, costs, and benefits of small water harvesting structures in Dewas, Madhya Pradesh. Agric. Water Manage. 131, 204–211.

Molden, D. (Ed.), 2007. Water for Food, Water for Life: A Comprehensive Assessment of Water Management in Agriculture. Earthscan, London, UK, Colombo, Sri Lanka: International Water Management Institute (IWMI). 645p.

Mosello, B., Oates, N., Jobbins, G. 2017. Pathways for irrigation development: policies and irrigation performance in Zimbabwe. Working Paper. Pathways to resilience in semi-arid economies.

Oates, N., Mosello, B., Jobbins, G., 2017. Pathways for Irrigation Development: Policies and Irrigation Performance in Tanzania. Working Paper. Pathways to Resilience in Semi-arid Economies. FANRPAN.

Otoo, M., Lefore, N., Schmitter, P., Barron, J. and Gebregziabher, G., 2018. Business Model Scenarios and Suitability: Smallholder solar pump based irrigation in Ethiopia, International Water Management Institute (IWMI), Colombo, Sri Lanka, 67p. (IWMI Research Report 172).

Ramankutty, N., Evan, A.T., Monfreda, C., Foley, J.A., 2008. Farming the planet: 1. Geographic distribution of global agricultural lands in the year 2000. Global. Biogeochem. Cycles 22.

Reij, C., Tappan, G., Smale, M., 2009. Agroenvironmental Transformation in the Sahel: Another Kind of "Green Revolution". IFPRI Discussion Paper 00914. International Food Policy Research Institute (IFPRI), Washington, DC.

Rockström, J., Williams, J., Daily, G., Noble, A., Matthews, N., Gordon, L., et al., 2016. Sustainable intensification of agriculture for human prosperity and global sustainability. Ambio 46 (1), 1–14.

Sadoff, C.W., Hall, J.W., Grey, D., Aerts, J.C.J.H., Ait-Kadi, M., Brown, C., et al., 2015. Securing Water, Sustaining Growth: Report of the GWP/OECD Task Force on Water Security and Sustainable Growth. University of Oxford, Oxford, UK.

Samberg, L.H., Gerber, J.S., Ramankutty, N., Herrero, M., West, P.C., 2016. Subnational distribution of average farm size and smallholder contributions to global food production. Environ. Res. Lett. 11 (124010), 1–12.

Scheierling, S.M., Treguer, D.O., 2016. Enhancing water productivity in irrigated agriculture in the face of water scarcity. Choices Q. 3.

Schmitter, P., Kibert, K., Lefore, N., Barron, J., 2018. Suitability mapping framework for solar PV pumps for smallholder farmers in sub-Saharan Africa. Appl. Geogr. 94, 41–57.

Shah, T., Durga, N., Verma, S., Rathod, R., 2016. Solar Power as Remunerative Crop. Water Policy Research Highlight. IWMI Tata Water Policy Program.

Snyder, K.A., Miththapala, S., Sommer, R., Braslow, J., 2016. The yield gap: closing the gap by widening the approach. Exp. Agric. 53 (1), 1–15.

Sugden, F., 2014. Landlordism, Tenants and the Groundwater Sector: Lessons from Tarai-Madhesh, Nepal. International Water Management Institute (IWMI), Colombo, Sri Lanka, 33p. (IWMI Research Report 162).

TNC (The Nature Conservancy), 2016. Water Share: Using Water Markets and Impact Investment to Drive Sustainability. The Nature Conservancy, Washington DC.

United Nations. 2015. Transforming our world: The 2030 agenda for sustainable development. Resolution adopted by the General Assembly on September 25, 2015. A/Res/70/1.

Vörösmarty, C.J., McIntyre, P.B., Gessner, M.O., Dudgeon, D., Prusevich, A., Green, P., et al., 2010. Global threats to human water security and river biodiversity. Nature 467 (7315), 555–561.

WEF, 2015. Insight Report: The Global Risks Report 2015, tenth ed. World Economic Forum, Geneva, Switzerland.

WEF, 2016. Insight Report: The Global Risks Report 2016, eleventh ed. World Economic Forum, Geneva, Switzerland.

WEF (World Economic Forum), 2017. Insight Report: The Global Risks Report 2017, twelth ed. World Economic Forum, Geneva, Switzerland.

WWAP (United Nations World Water Assessment Programme), 2016. The United Nations World Water Development Report 2016: Water and Jobs. UNESCO, Paris.

Williams, T.O., Giordano, M., Weight, E., 2012. Investment Opportunities in Smallholder Agricultural Water Management for Improved Food Security in Sub-Saharan Africa. International Conference on Food Security in Drylands. Doha, November 14–15, 2012.

Woodhouse, P., Veldwisch, G.J., Venot, J.P., Brockington, D., Komakech, H., Manjichi, A., 2017. African farmer-led irrigation development: re-framing agricultural policy and investment? J. Peasant. Stud. 44 (1), 213–233.

World Bank. 2016. Agricultural and Rural Development Databank. <http://data.worldbank.org/topic/agriculture-and-rural-development?end = 2014&start = 2009> (last accessed 31.03.17).

Xie, H., You, L., Wielgosz, B., Ringler, C., 2014. Estimating the potential for expanding smallholder irrigation in Sub-Saharan Africa. Agric. Water Manag. 131, 183–193.

CHAPTER

6

Forests, Land Use, and Challenges to Climate Stability and Food Security

Terry C.H. Sunderland[1,2] *and Dominic Rowland*[1,3]

[1]Center for International Forestry Research (CIFOR), Bogor, Indonesia
[2]Faculty of Forestry, University of British Columbia, Vancouver, Canada
[3]SOAS, University of London, London, UK

6.1 INTRODUCTION

The current mode of global food production is characterized by negative impacts on both human and planetary health (Haddad et al., 2016; Pinstrup-Andersen, 2013; Whitmee et al., 2015; Chapter 2). Fifty years after the Green Revolution, the world still faces multiple forms of malnutrition while much of the agricultural expansion related to achieving global food security often occurs at the expense of natural systems, including forests (Gibbs et al., 2010; Leblois et al., 2017).

The environmental toll of unsustainable agriculture threatens to undermine progress toward achieving global food security (Gordon et al., 2017). Climate change, to which agriculture is a major contributing factor, threatens crop production around the world (Kent et al., 2017; Martinich et al., 2017). At the same time, loss of fertility, desertification, loss of ecosystem services, and natural habitats undermine the long-term stability of the global food system (Haddad et al., 2016). Thus, the major questions facing global sustainable production are: (1) How do we increase production on existing agricultural land while reducing environmental degradation? (2) How do we reduce the environmental degradation and loss of ecosystem services as well as important sources of wild food and income resulting from agricultural expansion into natural habitats? (3) How do we restore degraded, unproductive, and abandoned agricultural land and natural habitats?

Forests and trees are an essential part of the solution to all three questions. Trees in agricultural landscapes can simultaneously increase production and mitigate against environmental

Sustainable Food and Agriculture
DOI: https://doi.org/10.1016/B978-0-12-812134-4.00006-6

degradation. Judicious landscape-scale land use planning that incorporates trees and forests into productive landscapes can simultaneously conserve forests and protect the ecosystem services upon which agricultural production depends (Baudron and Giller, 2014; Reed et al., 2016). At the same time, reforestation and regeneration of forests can restore degraded land and provide new productive landscapes on abandoned or degraded land.

Here, we divide the roles and functions of forests into three categories. First, is the provisioning function of forests; that is, the direct provision of food and income from forests and agroforestry. Second, forests have a protective function, a term referring to the oft-neglected contributions of ecosystem service provision, and the potential of forest regeneration to reclaim degraded land and increase agricultural production and mitigate climate change. The third function relates to the restorative capacity of forests that can be leveraged through increasing the availability of trees and forests in agricultural landscapes. Such contributions include climate change mitigation via sequestration and the restoration of degraded agricultural land. The latter can occur either through secondary forest succession or deliberate planting and restoration.

This chapter begins with a brief overview and outlook on the link between agriculture, deforestation, and climate change. In the next section, we explore the direct contributions forests make to food security through the provision of food and income and as a safety net for poor and vulnerable people. In Section 6.4, we examine the protective function of forests in terms of often-neglected ecosystem services upon which current agricultural production depends. The restorative function of forests and the potential of forest and tree-based agricultural systems to provide sustainable alternatives to contemporary industrial agriculture and mitigate climate change are discussed in Section 6.5. Finally, we discuss the governance and landscape management challenges and opportunities of incorporating forests and trees into agricultural landscapes.

6.2 ENVIRONMENTAL CHALLENGES TO SUSTAINABLE AGRICULTURE

Agricultural development plays a major role in improving food security and nutrition by increasing the quantity and diversity of food, as a driver of economic development, and due to agriculture being the main source of income for most of the world's poorest people (Ruel and Alderman, 2013; Carletto et al., 2015). While global food security and the economic development of low- and middle-income countries depends on agricultural production, progress has historically come at a huge cost to the environment. The current global agricultural system is not sustainable (Haddad et al., 2016). Over 40% of the nonice, land surface of the Earth is agricultural (Ellis et al., 2010), contributing over 20% of global greenhouse gas emissions (Chapter 2; Tubiello et al., 2014), over 70% of all water withdrawals (Chapter 2), and polluting soils, marine and freshwaters with pesticides, nitrates, phosphorous, and sediments (Crist et al., 2017). Agriculture is the largest driver of global deforestation, and in tropical and subtropical regions commercial agriculture accounts for 40% of deforestation, while small-scale cultivation accounts for an additional 33% (FAO, 2016).

The environmental impact of agriculture has knock-on effects for food security. Climate change, to which industrial agriculture significantly contributes, threatens to thwart progress toward achieving global food security. In addition, agricultural expansion exerts growing pressure on natural resources, particularly forests, and biodiversity, which provide invaluable ecosystem services for agricultural production (Reed et al., 2017). These pressures have reached critical levels, if not absolute tipping points, in some countries due to the high levels of deforestation and forest degradation (FAO, 2015a). Unsustainable pressure on forests threatens the very ecosystem services upon which the global food system depends (Reed et al., 2017). Agricultural expansion must be weighed against the need to conserve forests for sustainable production in the long term, especially given the vulnerability of agriculture, fisheries, livestock, and wildlife to climate change.

Forest conversion to agriculture also presents trade-offs for agricultural households. Conversion of forests to agriculture can result in improved food security via increased rural incomes and market access, though it can lead to impoverishment of both ecosystems and of livelihoods in the long term (Deakin et al., 2016; Baudron et al., 2017).

Agriculture and Deforestation

The state of the global forest area is a net balance between forest loss and forest gain. From 1990 to 2015, the global forest area fell by 129 million hectares (ha), with a loss of 195 million ha in tropical forest and a gain of 67 million ha for temperate forest (FAO, 2015b). Despite relatively high rates of ongoing deforestation, particularly in the tropics, the overall rate of global forest loss has slowed over the past two decades: the rate of annual net loss of forest decreased from 7.3 million ha per year (0.18%) in the 1990s to 3.3 million ha per year (0.08%) between 2010 and 2015 (FAO, 2015b; Keenan et al., 2015). Between 2010 and 2015, the tropical forest area decreased at a rate of 5.5 million ha per year, only 58% of the rate in the 1990s, while temperate forest area increased at a rate of 2.2 million ha per year (Keenan et al., 2015). In Brazil, the net loss rate between 2010 and 2015 was only 40% of the rate experienced in the 1990s, while Indonesia's net loss rate has also dropped by two-thirds over the same period (Keenan et al., 2015).

If agriculture is to expand, it must find suitable land on which to do so. The Global Agro-Ecological Zones Study conducted by the International Institute for Applied Systems Analysis (IIASA) and the Food and Agriculture Organization of the United Nations (FAO) identified 7.2 billion ha of suitable rainfed agricultural land based upon ecological conditions (Fischer et al., 2008). Of this land, approximately 1.5 billion ha are already productive agricultural land. Of the remaining 5.7 billion ha, 2.8 billion ha are forest land (or already used for alternative purposes) and an additional 1.5 billion ha are of poor quality. Thus, 1.4 billion ha of suitable rainfed agricultural land are potentially available given sufficient investment. Alexandratos and Bruinsma (2012) estimate that of this 1.4 billion ha of potentially suitable land, only around 70 million ha are needed to meet rising food demand, with the rest achieved through increases in yields, reduction in fallow lengths, and multiple cropping. In theory, then, land constraints should be no barrier to increasing agricultural productivity to match demands. However, this analysis fails to account for several crucial factors. Most of the available unused "prime" agricultural land is currently not serviced by the required infrastructure and is unevenly distributed—with over 60% of the land lying within only 13 countries (Fischer et al., 2008). More importantly, it assumes that food insecurity is caused by insufficient production rather than by factors affecting access and utilization, such as poor distribution, waste, market failures, nutrition, and health. It also fails to account for feedback mechanisms between agricultural expansion and loss of productivity. Agricultural expansion risks destroying areas of natural habitat that support existing agriculture through the provision of ecosystem services (such as soil quality, pollination, and hydrological regulation) (Zhang et al., 2007). Finally, conversion of land to agricultural purposes is not optimized as this form of analysis mistakenly implies. Land is converted to agricultural purposes on the basis of local economic and political contexts and in response to global demand. This can mean that financial or political incentives can result in diminishing marginal returns for increasing yields on existing agricultural land while simultaneously providing incentives to further expand into "free" land (Phelps et al., 2013).

Of all the natural habitats under threat from agriculture, tropical forests will experience most pressure from future trends (Geist and Lambin, 2002). Tropical regions are projected to undergo rapid population growth, rises in income, per capita food consumption, and dietary transitions (Lawrence and Vandecar, 2015). These regions are also likely to experience the greatest rise in biofuel production because year-round sunlight and high rainfall provide advantageous conditions for biofuel crops.

The nutrition transition, operating in a globalized world economy, changes agricultural demand and creates new environmental challenges. For example, rising consumption of animal-sourced foods in China has historically been partially sustained by soybean feed produced in Brazil, a major driver of Amazonian deforestation (Naylor et al., 2005). Similar effects are being caused by rising global demand for vegetable oils, in particular from India and China, on deforestation and land degradation in Indonesia and Malaysia (Gill et al., 2015; Fitzherbert et al., 2008). In fact, tropical and subtropical forests are likely to be disproportionately affected by the expansion of agriculture in natural ecosystems (MEA, 2005). Such deforestation and land use changes contribute substantially to climate change and the loss of other ecosystem services, further compounding global food insecurity.

Agriculture has long been considered a major driver of deforestation, but only recently has the relative role of subsistence agriculture and commercialized agriculture been debated (Angelsen and Kaimowitz, 2001). Recent studies, however, have shown that agricultural commercialization, combined with liberalization of agricultural trade, can and does drive substantial deforestation (Leblois et al., 2017; DeFries et al., 2010). The link between increasing agricultural exports and increasing deforestation is strongest in low- and middle-income countries with large areas of intact forest—the very same countries under pressure to both reduce deforestation and reduce poverty through economic growth (Leblois et al., 2017).

Agriculture, Forests, and Climate Change

The impacts of climate change on agriculture and the sector's contribution toward global emissions are discussed in detail in Chapter 4, while climate-smart agricultural solutions are discussed in Section II of this volume. Here, we examine the contributions of the forestry sector and of forest loss and degradation to global greenhouse gas emissions, as well as the importance and potential of tree-integrated agricultural systems in mitigating and adapting to climate change.

GHG Emissions From Forest Loss, Land Use Change, and Degradation

Forests are of critical importance in relation to climate change and food security (Locatelli et al., 2015). Forests will likely play a significant future role in both mitigation and adaptation of climate change. Forests have been an essential component of all climate change mitigation strategies following the United Nations Framework Convention on Climate Change (UNFCCC) Conference of the Parties in Bali, where they were taken up on the agenda and included in the Bali Action Plan; they were less prominent in the Paris Agreement. Trees within agricultural landscapes already contribute significantly to the mitigation of climate change through carbon sequestration.

Forest loss and degradation is the second largest single source of greenhouse gas (GHG) emissions (Tubiello et al., 2014). Total emissions from deforestation and degradation amount to around 6 billion tons of carbon per year, approximately 6%–17% of global carbon dioxide (CO_2) emissions (van der Werf et al., 2009). Of the 10 billion tons of CO_2 equivalent GHG emissions from the agricultural sector in 2010, 3.8 billion tons originated from the net conversion of forests (FAOSTAT, 2016). Forest conversion is not the only source of GHG emissions from forests, as soil degradation, peat oxidation, and forest fires, particularly on degraded forests, release vast quantities of carbon dioxide to the atmosphere (Avitabile et al., 2016). Meanwhile, the forestry sector itself sequesters around 2 billion tons of CO_2 equivalent per year (FAOSTAT, 2016) and is one of the most cost-effective methods of climate change mitigation (Stern et al., 2006).

Emissions from forestry and other land uses vary dramatically by region. In most regions, especially Asia and Latin America and the Caribbean, overall emissions from the sector have declined substantially since the 1980s—concomitant with reduced deforestation in countries such as Brazil, and net afforestation in countries such as China, Viet Nam, and India (Smith et al., 2014). However, models diverge substantially when examining emissions from the Middle East and Africa, ranging from increasing emissions to net carbon sinks. Baccini et al. (2012) break down mean CO_2 fluxes in tropical countries coming from deforestation and forest management. Net emissions from managed industrial logging are relatively small, as emissions are counteracted by carbon sequestration from regenerating forests. The same principle applies to shifting cultivation, where regrowth during fallow cycles balances out emissions from deforestation.

Belowground biomass stores two to three times more carbon than above ground biomass (Vashum and Jayakumar, 2012). Data on soil organic carbon (SOC) stocks, however, are limited by coarse data, varied methodological approaches, and regional biases in data availability. In general, data from consistent data sets such as the Harmonized World Soil Database (FAO, 2009) show high levels of SOC in northerly latitudes and lower SOC in tropical latitudes at a depth of 1 m—the inverse trend of phytomass carbon (Scharlemann et al., 2014). The high levels of SOC in boreal and permafrost forests are of particular concern due to potential feedback effects through climate change. Global data sets showing SOC to a depth of 1 m may significantly underestimate carbon stocks in deep peat soil, especially in the tropics. One estimate suggests that measuring SOC up to 11 m of depth may double current SOC estimates in permafrost and tropical regions (Page et al., 2011;

TABLE 6.1 Distribution of Terrestrial Organic Carbon by the Intergovernmental Panel on Climate Change (IPCC) Climate Region in Soil, and in Above- and Belowground Phytomass Carbon Pools

	Soil				
IPCC Climate Region	**Topsoil (Pg C)**	**Subsoil (Pg C)**	**Total (Pg C) (%)**	**Phytomass (Pg C) (%)**	**Terrestrial carbon Pool (Pg C)**
Tropical wet	62.6	65.4	128.0 (47.7)	140.2 (52.3)	268.1
Tropical moist	78.6	72.3	150.9 (49.9)	151.7 (50.1)	302.6
Tropical dry	67.3	69.0	136.2 (76.2)	42.5 (23.8)	178.7
Tropical montane	29.6	26.5	56.1 (58.1)	40.5 (41.9)	96.6
Warm temperate moist	33.3	29.7	63.0 (68.7)	28.7 (31.3)	91.7
Warm temperate dry	38.9	39.6	78.5 (76.4)	24.2 (23.6)	102.7
Cool temperate moist	104.1	106.2	210.3 (88.1)	28.5 (11.9)	238.8
Cool temperate dry	52.2	50.0	102.2 (91.8)	9.1 (8.2)	111.3
Boreal moist	162.0	194.7	356.7 (93.8)	23.5 (6.2)	380.2
Boreal dry	32.0	37.0	69.1 (93.1)	5.1 (6.9)	74.2
Polar moist	30.6	21.7	52.4 (96.0)	2.2 (4.0)	54.5
Polar dry	8.0	4.3	12.3 (96.2)	0.5 (3.8)	12.8
Total	699.3	716.4	1415.7 (74.0)	496.6 (26.0)	1912.2

Pg C, petagrams of carbon.
Scharlemann, J.P., Tanner, E.V., Hiederer, R., Kapos, V., 2014. Global soil carbon: understanding and managing the largest terrestrial carbon pool. Carbon Manag. 5(1), 81–91 (2014: Table 1).

Tarnocai et al., 2009). Tropical peatland accounts for 11% of global peatlands, of which over half is found in southeast Asia. Peatlands may account for as much as 60%–75% of SOC in tropical forested countries such as Malaysia and Indonesia (Page et al., 2011). A breakdown of SOC and phytomass carbon pools taken from Scharlemann et al. (2014) is shown in Table 6.1.

Drained organic soils increase emissions through increasing decomposition rates. Globally, 250,000 km^2 of cropland and pasture have drained organic soils, releasing GHG emissions of 0.9 gigaton of carbon dioxide equivalent ($GtCO_2$-eq) per year (Smith et al., 2014). There are an estimated 500,000 km^2 of drained peatlands—much of it under forest or degraded forest land—releasing 1.3 $GtCO_2$-eq per year (Joosten, 2010). Soil carbon stocks change dramatically with altered land use. A change from native forest to plantation results in a 13% decline in soil carbon stock, and forest to crop a 42% decline (Guo and Gifford, 2002). However, the inverse is also true, offering opportunities for carbon mitigation through afforestation and agroforestry. Loss of SOC varies by region and depending on climatic, hydrological, and management conditions (Scharlemann et al., 2014). In boreal and temperate forests, a general increase in SOC was experienced between 1990 and 2007, largely as a result of natural and managed forest expansion and reduced timber harvesting. In tropical forests, however, over the same period, deforestation led to a net decline in SOC of 7.6%, even after accounting for forest regrowth (Pan et al., 2011).

Released carbon from biomass and soils is not the only concern with regard to climate change. Forests carry on multiple "complex, nonlinear interactions" with the global climate system (Bonan, 2008). For instance, a growing body of evidence exists of positive feedback mechanisms between atmospheric warming and the rate of decomposition of soil organic matter. Should such mechanisms exist, global warming may increase the rate of emissions from organic soils, further exacerbating climate change (Davidson and Janssens, 2006). Similarly, forest fires, especially on peatlands, are a major contributor to global emissions, but they are also initiated by climatic events, which are predicted to worsen with climate change. Around one quarter of emissions from fires are not compensated for by regrowth of vegetation; of this, around 60% is derived from deforestation and degradation in humid tropical forests, 20% in dry tropical forests, and 20% from tropical peat fires (van der Werf et al., 2010).

Forests and Adaption to Climate Change

Research on livelihood resilience and local people's coping strategies during times of economic or climatic shocks (Pain and Levine, 2012) suggests that households use their assets to mitigate the impact of a shock on their food and nutrition security. In rural communities, many such assets are derived from forests. In many parts of the world, livestock represent one of the most common and most valuable (and transferable) assets. In areas with forest cover, the maintenance of livestock is directly dependent on access to forest resources, either through grazing on communal or state-owned forested land or through fodder collected from uncultivated areas.

There is significant evidence that communities around the world already use forests and trees to cope with food shocks caused by climatic variables (Cotter and Tirado, 2008). Evidence reviewed in Jamnadass et al. (2015) demonstrates the diverse ways in which forests and trees contribute to household-level food security and nutrition under a variety of circumstances. These include the regular consumption of animals, birds, fish, insects, and fruits and vegetables from forest-based sources. This type of forest product consumption typically provides nutritious supplements to otherwise monotonous diets.

Agricultural systems with a diversity of cropping and land use types are more resilient to extreme weather events caused by climate change. This presents the opportunity to diversify agricultural systems, which, at the same time, will increase dietary diversity and reverse trends towards homogeneous food systems. Increasing agrobiodiversity in agricultural landscapes reduces the chances of catastrophic crop failure while simultaneously providing more diverse diets.

Diversity in agricultural practices helps mitigate risk from crop failures and environmental change. As markets increasingly dominate the global food system, those on the poverty line are increasingly susceptible to price fluctuations and especially soaring food prices, often leading to social unrest (Bush and Martiniello, 2017). Climatic events resulting in crop failures are no longer localized to the region in which they occur; rather, the shock is spread and absorbed throughout global food markets, as seen in 2008. Economic crises and food price shocks have demonstrable negative effects on nutritional outcomes and child mortality. Thus, sustainable smallholder agriculture and access to forest resources may mitigate against the extremes of global food price fluctuations.

Strengthening resilience for all vulnerable populations involves adopting practices that enable them to protect their existing livelihood systems, diversifying their sources of food and income, changing their livelihood strategies, and migrating if there is no other option.

6.3 THE PROVISIONING FUNCTION OF FORESTS AND TREES

Forests and their environs contribute to the livelihoods of an estimated 1–1.5 billion people in some way, while 60 million indigenous people worldwide are estimated to be entirely forest dependent (Chao, 2012). As well as direct food provision, these forest-based livelihoods provide crucial sources of income and nonfood forest products, such as medicines, fuelwood, and charcoal that impact upon food security and nutrition in myriad ways (Wunder et al., 2014). Millions of people are also dependent on a wide variety of tree-based production systems such as agroforestry, plantations, woodlands, and homestead trees (Jamnadass et al., 2015); thus, the influence of trees within agricultural systems are also of significant importance (Rahman et al., 2017).

Forests for Food

Convincing evidence exists that fruit and vegetable consumption is associated with increased consumption of certain micronutrients and is strongly associated with reduced risk of chronic diseases: recent World Health Organization (WHO) guidelines recommend 400 g/person per day of fruits and vegetables (WHO, 2015). Global production of fruits and vegetables is far below these requirements, while at the same time approximately 50% of the fruit that is produced globally is produced on trees (Powell et al., 2013a). From total fruit production in 2013, 64.4% consisted of only five species (bananas, apples, grapes, oranges, and mangoes) of a total of 21 species of fruits traded internationally; there are no records available for forest fruits alone. Nevertheless, there has been increasing interest in many wild fruits over the past two decades, and their products are not only consumed by forest-dependent people, but may also be found in urban markets in cities and even in international markets (WHO/CBD, 2015; Vira et al., 2015).

A recent review of wild food consumption found high levels of variation in the importance of wild foods from one site to another (Powell et al., 2015). Rowland et al. (2017) investigate the dietary contributions of wild forest foods in smallholder-dominated forested landscapes in 24 tropical countries, using data derived from the Poverty and Environment Network of doctoral research in forest-dependent communities throughout the tropics; the study estimated the contributions of micronutrient-rich forest foods to meeting dietary recommendations. The findings suggest high variability in forest food use across four forest food use site typologies to characterize the variation: forest food-dependent, limited forest food use, forest food supplementation, and specialist forest food consumer sites (Rowland et al., 2017). The results show that forest foods do not universally contribute significantly to diets, but where they are consumed in large quantities, their contribution towards dietary adequacy is substantial.

Because the majority of wild forest foods consumed are fruits and vegetables, consumption figures do not accurately represent the contribution of wild forest foods to dietary quality or diversity (Vinceti et al., 2013; Powell et al., 2013a). For example, despite low to moderate contribution to energy intake, wild foods contributed 36% of total vitamin A and 20% of iron in the diet in a study from Gabon (Blaney et al., 2009); 31% of retinol activity equivalents (vitamin A) and 19% of iron in the diet in a study from the United Republic of Tanzania (Powell et al., 2013b); and in a traditional swidden agricultural community in the Philippines, wild foods contributed 42% of calcium, 17% of vitamin A, and 13% of iron (Schlegel and Guthrie, 1973). As a result, wild foods are regarded to contribute significantly to dietary diversity and nutritional quality (Powell et al., 2015).

Animal source foods are good sources of highly bio-available micronutrients. In many rural areas, bushmeat provides much of the animal source foods consumed (even in many of the developed countries). In tropical areas where livestock production is limited due to tsetse fly and other environmental constraints, bushmeat is a particularly important source of micronutrients. Numerous case studies have identified the positive effects of animal source foods on nutritional status and health. For example, data from Madagascar have shown that the loss of access to wild bushmeat would result in a 29% increase in the number of children with anemia (Golden et al., 2011). Estimates of per capita consumption of illegally harvested bushmeat (driven largely by low dietary standards and poverty) from the Congo Basin, for instance, range from 180 g/person per day in Gabon, to 89 g/person per day in the Congo, and 26 g/person per day in Cameroon (Fa et al., 2003). Fallow forests that attract wild animals play a critical function in the food and dietary security of millions of rural families in Amazonia (Nasi et al., 2011). In addition to wild meat, trees support livestock and dairy production through the provision of fodder, which provide a low capital and labor input form of fodder and reduce the cost of production of animal source foods while raising incomes (Franzel et al., 2014).

Studies pairing satellite-based information on tree cover and dietary-intake information yield evidence of a positive link between tree cover and dietary diversity, as well as fruit and vegetable consumption (Ickowitz et al., 2014; Ickowitz et al., 2016; Johnson et al., 2013; Powell et al., 2011). To date, one of the major limitations of most of these studies is that they have not differentiated the type of forest cover associated with these forest patterns (Ickowitz et al., 2014; Powell et al., 2015). Although there are no studies that have empirically measured the associations between agroforestry and dietary quality (Powell et al., 2015), it seems very likely that the links between tree cover and dietary quality are strongest in agroforestry or swidden systems in which trees play an integral role in the production system (HLPE, 2017).

Shifting cultivation, a form of tropical agriculture that involves the rotation of fallow plots combined with slash-and-burn clearing, is widely practiced throughout the tropics. Shifting cultivation systems are highly suitable for the landscapes in which they predominate (Padoch and Sunderland, 2013). Fallow systems in forest mosaics allow for natural regeneration, nitrogen fixing, and the accumulation of biomass. Swidden systems using slash-and-burn recycle nutrients result in sustainable yields on poor soils often without a need for fertilizer, and thus reducing capital and labor input. Many wild and semi-wild plants found in fallows are used as edible plants contributing to household food supplies. Although swidden systems do not directly provide animal source foods, proximity to forests often provides opportunistic access to bushmeat. Though direct evidence of the nutrition and food security benefits of swidden cultivation is lacking, some evidence points to swiddeners consuming higher quality diets. For instance, in some provinces in Indonesia, land cover types associated with swidden and agroforestry were associated with higher consumption of micronutrient rich food groups (Ickowitz et al., 2016).

Although the number of shifting cultivators worldwide is unknown, it is estimated that in South East Asia alone, there are between 14 and 34 million shifting cultivators. According to recent case studies in Asia, most traditional people in the region still rely on shifting cultivation or swidden agriculture for their livelihoods, crucial to rural food security, particularly among indigenous societies. Shifting cultivation systems are present throughout most of the world's upland forested regions, but good data on the scale and extent are nonexistent. Swidden and fallow systems are typically not recorded in statistical data sets, and the dynamic nature of swidden systems makes remote sensing a challenge at large scales. Regional estimates for southeast Asia suggest the practice is common in most countries, but no aggregate figures have been calculated.

Forests for Income

Employment in the formal forest sector, including wood, wood-based panels, and pulp and paper, was in the order of 13.9 million workers in formal enterprise in 2011 (FAO, 2014a). An additional 4.6 million are estimated to work in furniture manufacture, most of which is wood-based. Nonformal enterprises that employ most forest workers in developing countries are estimated to provide employment for an additional 41 million workers, chiefly in woodfuel collection and charcoal manufacture (FAO, 2014b). However, because of the informal nature of these markets, estimates on employment in nontimber forest product gathering and utilization remain scanty (HLPE, 2017). In terms of direct forest income, the pantropical Poverty and Environment Network calculated that almost one-fifth (22.2%) of rural income is derived from forest and environmental resources, often equivalent to, or even outstripping that of direct income from agriculture (Angelsen et al., 2014).

Safety Net Function

The direct contribution of forests in terms of food provision is almost certainly underestimated (HLPE, 2017), primarily due to both a lack of data and the poor quality of data when available. Despite their importance to local economies, most forest resources, aside from wood products, remain under-recorded in forestry accounting. Forests supply nutritious fruits and vegetables to supplement diets during lean seasons. In Sahel ecosystems with dry spells lasting up to seven months a year, trees and shrubs are vital sources of food to supplement cereal staples (Nyong et al., 2007), as well as fodder for livestock (Franzel et al., 2014). Provision of forest-based foods can also produce a year-round supply of vital nutrients in otherwise seasonal agricultural supply owing to the variety of fruiting phenologies (Jamnadass et al., 2015).

For some communities, forest foods play an important function as a safety net during times of agricultural crop failure or seasonal downturns in agricultural production

(Shackleton and Shackleton, 2012). At these critical moments, the roles that forests play in preventing hunger can be worth much more than can be captured by the annualized and aggregated estimates of the proportions of household food consumption attributable to forest-based sources. To focus, therefore, on these forms of data is to neglect the ways in which forests supplement crop-based nutrients in household diets, and how they play especially important roles in times of scarcity and vulnerability.

The economic contributions of forestry enable households to command access to food through markets, especially during lean seasons. For example, tree foods contribute some 30% of rural diets in Burkina Faso (Thiombiano et al., 2013), and many rural people in tropical countries depend on trees for livestock fodder (Baudron et al., 2017). In West Africa, over 4 million women earn about 80% of their income from the collection, processing, and marketing of oil-rich nuts collected from shea trees that occur naturally in the forests (Boffa, 2015).

6.4 THE PROTECTIVE FUNCTION OF FORESTS AND TREES

Agriculture is highly dependent on ecosystem services provided by forests and other natural ecosystems (Foli et al., 2014; Reed et al., 2017). Positive relationships exist between forest proximity and a wide range of ecosystem services for major crops, though ecosystem disservices are also present (Mitchel et al., 2014). Trees provide shelter and habitat for a number of species that provide beneficial services at various spatial scales. Habitats for pollinators and natural enemies of pests are key examples of trees and forests enhancing ecosystems stability. Such processes occur at local scale, but inevitably have an effect on landscape and regional scales in adjacent agricultural systems. Forests and other tree cover formations regulate watersheds and water provisioning at the regional scale. At the global scale, forests contribute to climate regulation, including carbon sequestration but also the cycling of oxygen. They are also stores of genetic diversity and conserve endemic species in various hotspots (Jennings et al., 2001).

There is adequate evidence to ascertain that forests underpin agricultural systems through ecosystem services provisioning (Foli et al., 2014; Reed et al., 2017). Forests provide habitat to wild pollinators that are crucial in sustaining crop yield in animal pollinated crops (Aizen et al., 2009). This excludes the global cereal crops of maize, wheat, and rice. Approximately 70 crops are dependent on animal-mediated pollination, and globally 35% of crop production relies on animal pollinators (Winfree et al., 2011). Bees, especially the honey bee, are the backbone of agricultural pollination. Global agricultural intensification has focused on the efficiency of the single honey bee species to provide pollination services to intensively managed systems. Managed honey bees are declining because of disease die-offs, and so the attention is shifting back to native wild bees that are more resilient. Wild bee pollinators have been found to enhance fruit set in crops complementing the role of the honey bee (Garibaldi et al., 2011; Garibaldi et al., 2013). Moreover, with mysterious die-offs in honey bee colonies, native species of bees have been compensating for the deficit. Forests provide the natural habitat required by wild species.

Recognition of the importance of ecosystem services for food production and the appreciation of the complex relationship between proximity and effect, social-ecological processes and competing demands between ecosystem services at local and global scales have led to the development of viewing ecosystem services through the lens of multifunctional landscapes (Reed et al. 2017). A recent systematic review by Reed et al. (2017) found that the majority of studies reported positive effects of tree presence on agricultural yields and livelihoods, though varying strongly by region. However, there is still a lack of consistent and clear evidence of how and under what conditions trees on farms affect food production.

6.5 THE RESTORATIVE FUNCTION OF FORESTS AND TREES

Up to 40% of global arable land may be degraded because of soil erosion, loss of fertility, salination, or overgrazing (Wood et al., 2000). At the same time, degraded, secondary, and heavily logged forests may constitute as much as 60% of global forest area (FAO, 2015b). Restoring degraded land is a central challenge and a target of both the Convention on Biological Diversity (where the challenge is to restore 15% of degraded ecosystems) and the Bonn Challenge (where the challenge is to restore 150 million hectares of degraded and deforested land by 2020) (Normile, 2010). The restoration of degraded land leads to restoration of natural biodiversity and ecosystem services. However, ecosystem service provision does not automatically follow from increased biodiversity, and trade-offs may exist between different ecosystem services.

The carbon storage of agricultural systems is mainly found in soils. It is estimated that the loss of carbon from organic soils in agriculture amounts to 50 gigatons of carbon (Smith et al., 2007). Restoring degraded land and reversing the trends of erosion and desertification can reverse some of these losses depending on local contexts. At the same time, restoration of degraded agricultural land can have multiple benefits: increasing yields, reclaiming unusable land, increasing carbon sequestration, and mitigating climate change. Multiple changes to the management of agricultural land are needed to restore degraded land, including many nontree solutions such as improved agronomic practices and nutrient and water management. However, land cover change (via reforestation and afforestation) and agroforestry practices are important components of the solution.

Afforestation and Reforestation

The reversal of current trends of climate change and biodiversity loss requires drastic actions to protect forests. Forestry and agriculture combined have a total mitigation potential of ~3 to ~7.2 $GtCO_2$-eq per year (Smith et al., 2014). Included in this figure are the potential of better cropland management, restoration of organic soils, better grazing land and manure management, as well as the potential of agroforestry and regrowth on abandoned land. Though not a solution in isolation from other measures, the potential of reforestation and afforestation is substantial, although variable. Where soil carbon stocks are low, reforestation and afforestation may sequester between 1 and 1.5 gigatons of CO_2 per year, though on land with high carbon stocks (e.g., grasslands) afforestation could result in declines in carbon accumulation (Nabuurs and Masera, 2007). Globally, depending on initial carbon stocks and tree species, the potential of carbon sequestration ranges from 1 to 35 gigatons of CO_2 per hectare per year (Richards and Stokes, 2004).

Two main approaches for reforestation exist: passive restoration that utilizes natural succession and active restoration. Passive restoration is cost-effective (excluding any opportunity costs) and typically covers larger areas over a longer time period. The process, however, is slow, taking a minimum of 30–40 years for a forest structure to be restored (Chazdon, 2008). Active reforestation, in contrast, involves the deliberate planting of trees to speed up restoration of forests. Historically, passive restoration has exceeded active restoration in terms of scale because of natural processes, but increased development of plantations and reforestation programs have reversed this situation. Active restoration is now 4.9 million ha $year^{-1}$, relative to 2.9 million ha $year^{-1}$ for passive (FAO, 2015a). Active restoration sequesters more carbon, more quickly than passive restoration, though at a cost of lower biodiversity and loss of access to nontimber forest products. Active restoration, particularly when timber species are planted, may have added income benefits. A compromise between the active and passive approach has emerged over recent years, called "woodland islets," or "applied nucleation," whereby deliberate "clumps" of woodlands are planted to encourage seed dispersers such as birds (Corbin and Holl, 2012).

Restoration of forest land has numerous benefits, from global to local ecosystem service provision, carbon sequestration, and improved farm productivity (Tilman et al., 2002). Though there is reason, and some evidence, to presume that reforestation will increase the provision of ecosystem services that positively affect agricultural production (such as litter decay and nutrient cycling), quantitative estimates of their contributions to food production and security are not available (Parrotta et al., 2015) The likely benefits vary by scale and are highly dependent on context and the landscape configuration, as well as species composition and crop types. Thus, the issues of land sharing and land sparing become vital as to how and where to reforest degraded land.

Agroforestry and Low-Impact Farming

Agricultural production for food security and environmental services need not be in tension. Often high-input unsustainable intensive agriculture is less productive (and costly to farmers) than lower-input methods. In a review of 286 interventions across 57 countries, adoption of sustainable agricultural practices resulted in a net increase in yields of 79% (Pretty et al., 2006). Much of these increased yields may be due to higher labor inputs rather than industrialized agriculture through family labor (Wiggins et al., 2011). Although sustainable intensification has become a recognized process in achieving sustainable agricultural growth, alternative schemes also need to be considered, such as agroecology, agroforestry, and climate smart agriculture (Garnett and Godfray, 2012). For instance, there is a growing body of research showing that small-scale, ecologically-based, organic and even traditional peasant systems can approach, match, and even exceed the productivity of industrial systems when measured by the number of people fed per unit of land or the food biomass produced per unit area (Rosset and Martínez-Torres, 2012). These agroecosystems are usually the kinds of diverse, multilayered, and integrated systems that are most common in smallholder, traditional farming systems in the developing world, with a focus on meeting local needs, providing food for the larger communities in which they participate, and maintaining the productive capacity of the soil for the long term (FAO, 2014b).

Agroforestry systems take on a number forms. Agrosilvicultural systems incorporate trees (for food crops or nonfood services) with agricultural crops. Included under Agrosilvicultural systems are alley cropping, mixed tree gardens, multilayer tree and shrub gardens, and tree fallow systems. Silvopastoral systems include tree crops with grazing pasture or fodder production. Agrosilvopastoral systems combine domestic livestock with crops, multifunctional hedgerows, woodlots, or fodder trees. Within these broad categories, numerous varieties exist in their functional characteristics (i.e., tree and shrub components), productive functions (i.e., crops, food, fodder, and fiber produced), and protective functions (e.g., soil conservation, windbreaks, and fertility improvement) (Powell et al., 2015). Agroforestry is not a new solution. Zomer et al. (2016) estimate that 43% of global agricultural land already has more than 10% tree cover. However, converting single-crop agricultural systems to diverse agroforestry systems could have multiple benefits, including climate change mitigation, soil restoration, biodiversity conservation, and improved yields.

Agroforestry offers opportunities to improve production and increase the sustainability of agriculture while simultaneously creating carbon sinks. Albrecht and Kandji (2003) estimate that agroforestry systems could potentially sequester between 12 and 228 megagrams of carbon per hectare upon the area of land that may be suitable for agroforestry ($585-1215 \times 10^6$ ha); thus, 1.1–2.2 petagrams of carbon could be stored over a 50-year period. Despite forests having higher carbon storage potential, agroforestry compares favorably to other land uses, primarily due to the scale of available land suitable for agroforestry development (Verchot et al., 2007). The Intergovernmental Panel on Climate Change (IPCC) estimates 630 million ha of unproductive cropland could be converted to agroforestry, potentially sequestering 0.586 teragram of carbon per year (Kumar and Nair, 2011).

Carbon storage potential of different agroforestry systems differ by region and agroforestry system, as shown in Table 6.2. However, for some types of agroforestry systems, especially those utilizing nitrogen-fixing species, nitrogen dioxide emissions may be higher (Verchot et al., 2007).

Agroforestry also has the potential to improve soil quality and fertility and increase crop yields. Lack of nitrogen, and loss of nitrogen, in and from soils is a major limiting factor to agricultural productivity and thus economic development in many low- and middle-income countries, especially in sub-Saharan Africa. In Malawi, for instance, loss of nitrogen and phosphorous through erosion is estimated to cause lower crop yields equivalent to 3% of agricultural gross domestic product (Bojö, 1996). Fertilizer-tree systems and legume-based agroforestry provide vast potential for reducing this problem. Legume-based agroforestry incorporates nitrogen-fixing species being planted in intercropped, multistrata or fallow-based systems to increase the available nitrogen in the soil and reduce dependency on agricultural inputs. Tree fertilizer systems can fix nitrogen in soils where lack of nitrogen is a limiting factor to crop growth. These systems involve the planting of quick-growing nitrogen-fixing trees and shrubs, which produce nitrogen dense biomass that is released into the soils upon decomposition. Tree fertilizer systems can be intercropped or incorporated into improved fallows, or even used to generate biomass transferred to crops. Such systems have been shown to be highly successful in many

TABLE 6.2 Potential Carbon Storage for Agroforestry Systems in Different Ecoregions of the World

	Ecoregin	System	Mg C ha^{-1}
Africa	Humid tropical high	Agrosilvicutural	29–53
South America	Humid tropical low	Agrosilvicutural	39–102[a]
	Dry lowlands		39–195
Southeast Asia	Humid tropical	Agrosilvicutural	12–228
	Dry lowlands		68–81
Australia	Humid tropical low	Silvopastoral	28–51
North America	Humid tropical high	Silvopastoral	133–154
	Humid tropical low	Silvopastoral	104–198
	Dry lowlands	Silvopastoral	90–175
Northern Asia	Humid tropical low	Silvopastoral	15–18

[a]*Carbon storage values were standardized to 50-year rotation.*
Mg C ha, megagrams of carbon per hectare.
Taken from Albrecht, A., and S. Kandji. Carbon sequestration in tropical agroforestry systems. 2003. Agric. Ecosyst. Environ. 99, 1, 15–27: Table 2).

regions of sub-Saharan Africa, dramatically increasing yields of maize and wheat while reducing the cost and labor input of fertilizers to improve soils (Chaudhury et al., 2011).

Soil degradation and desertification are major global problems. Cultivation of fragile soils exposed to erosion through wind and water, combined with overgrazing, reduction in fallow size and lengths, and unsustainable use of agricultural technology, among other factors, is contributing to a global decline in soil quality and fertility. Every year, 24 billion tons of fertile soils are removed from the topsoil, and 12 million ha are lost to drought and desertification (UNCCD, 2017). Around 52% of the lands used for agricultural production are moderately or severely affected by desertification, leading to an estimated potential future loss of food production of up to 12% (UNCCD, 2017). Forests and trees can restore degraded land by fixating soils through providing windbreaks, shelterbelts, and woodlots (FAO, 2015a). Globally, an estimated 2 billion ha are suitable for restoration through reforestation, including 715 million ha in Africa, 550 million ha in Latin America, and 400 million ha in south and east Asia (UNCCD, 2017). Flagship programs such as the Great Green Wall for the Sahara and the Sahel Initiative are pushing the boundaries of restoring degraded land through both reforestation and agroforestry while providing a programming tool and opportunities to improve livelihoods and food security and reduce rural poverty.

6.6 RESOLVING TENSIONS BETWEEN FORESTS AND AGRICULTURE

While cultivated cropland may retain trees or accommodate natural tree regeneration, these alone are insufficient to provide the environmental goods and services garnered from formerly intact or largely natural forests (HLPE, 2017). Thus, forests must be protected for the sake of agriculture, as well as a multitude of other reasons.

Sustainable intensification is seen as a major component of sustainable food and agricultural production (Pretty and Bharucha, 2014). The goal of sustainable intensification is to increase food production from existing land while minimizing pressure on the key natural resources and the environment. The assumption is that as crop, livestock, and aquaculture productivity per unit of land or water is increased, less of those resources will be required to supply a given level of harvest. Some studies show reduced expansion of agricultural areas as crop yield increases. But simply increasing yield has been shown to not necessarily reduce deforestation, particularly if farmers are responding to market opportunities, and particularly in the absence of effective regulatory measures. Local deforestation rates have been shown to increase in line with increases in commodity prices (Byerlee et al., 2014).

Governance of Forests for Food Security

Historically, forests have been regarded as discrete entities within landscapes with little connection or contribution to agriculture (Sayer et al., 2013). A key challenge in relation to integrating the role of sustainable forestry into the discussion of food security and nutrition results from the compartmentalized and fragmented nature of decision-making in relation to these issues at local, national, regional, and global governance levels (Clark and Tilman, 2017). Responsibilities for agricultural production and storage, provision and distribution of food, social welfare, and safety nets are often handled in separate ministries and organizations that do not necessarily engage with the forestry or environment sectors, or see any relevance of forestry-related activities in relation to their core mandate. More problematically, in a territorial sense, priorities for forestry and other forms of land use are often perceived to be competing for valuable land resources. While there are growing calls for more integrated approaches to decision-making across these divisions (Reed et al., 2016), a systems approach towards more inclusive forestry and agriculture sector remains elusive (Fischer et al., 2008). Even in the 2030 Agenda for Sustainable Development, goals for hunger (Sustainable Development Goal 2) have targets and indicators that are separate from those that are associated with the sustainable management of forests and terrestrial ecosystems (Sustainable Development Goal 15). Line agencies that are concerned with these issues could interpret their mandate in a narrow sense, and fail to recognize the considerable overlaps and synergies that can emerge from addressing these issues in a more integrated fashion.

Despite progress in slowing the rates of deforestation, action is required to further reduce the long-term impacts of forest loss. Over the past decade, there has been a spate of innovative schemes designed to address global deforestation, such as Reducing Emissions from Deforestation and Degradation (REDD+), Payments for Ecosystem Services (PES), and biodiversity offsetting and banking. While still in their infancy, such schemes have offered potential funding mechanisms and frameworks through which reduction targets can be achieved, yet the potential benefits are argued to have been somewhat overstated (Lund et al., 2017).

In addition, considerable progress has been made to restore degraded forested lands. The Bonn Challenge created a target to restore 150 million ha by 2020, sparking a series of regional initiatives derived from it, such as the "Initiative 20 × 20" in Latin America. Most afforestation and reforestation takes place through the use of exotic, nonnative species. On the one hand, such practices encourage afforestation, since fast-growing, commercially profitable species provide significant economic returns. On the other hand, encouraging the regeneration of slower growing native species supports greater biodiversity and reduces pressure on natural forests.

Paradoxically, although intensification of agropastoral systems is seen as one way to reduce pressure on native forests, increased returns on intensified agriculture may increase incentives for further expansion of monocultures and pastures into forest frontiers (Pretty and Bharucha, 2014). Policies must therefore balance the benefits and trade-offs of agricultural intensification, expansion, and the conservation of biodiversity through appropriate and institutionalized land use planning and enforcement coupled with policy harmonization to avoid policy distortions that negate achievement of balance between benefits and costs. In the face of climate change and variability, this will be crucial from a long-term perspective.

In sum, tensions between agriculture and environmental degradation are a major driver of global biodiversity loss and climate change and represent a major threat to sustainable agricultural production and food security. However, improved agricultural practices, utilizing both trees on farms and trees and forests within agricultural landscapes, have great potential to mitigate negative effects and even reverse trends. Not only can trees and forests mitigate climate change and loss of ecosystem services, but they can also actively restore the potential of degraded and abandoned agricultural land. While forests and trees are one part of a broader movement toward sustainable and nutrition sensitive landscapes, they are also an essential part of the solution.

References

Aizen, M.A., Garibaldi, L.A., Cunningham, S.A., Klein, A.M., 2009. How much does agriculture depend on pollinators? Lessons from long-term trends in crop production. Ann. Bot. 103 (9), 1579–1588.

Albrecht, A., Kandji, S., 2003. Carbon sequestration in tropical agroforestry systems. Agric. Ecosyst. Environ. 99 (1), 15–27.

Alexandratos, N., J. Bruinsma. 2012. World agriculture towards 2030/50: the 2012 revision, ESA Working Paper 2012. FAO, Rome.

Angelsen, A., Jagger, P., Babigumira, R., Belcher, B., Hogarth, N.J., Bauch, S., et al., 2014. Environmental income and rural livelihoods: a global-comparative analysis. World Dev. 64 (1), S12–S28.

Angelsen, A., Kaimowitz, D., 2001. Agricultural Technologies and Tropical Deforestation. CABi/CIFOR.

Avitabile, V., Herold, M., Heuvelink, G., Lewis, S., Phillips, O., Asner, G., et al., 2016. An integrated pan-tropical biomass map using multiple reference datasets. Glob. Change Ecol. 22, 1406–1420.

Baccini, A., Goetz, J., Walker, W., Laporte, N., Sun, M., Sulla-Menashe, D., et al., 2012. Estimated carbon dioxide emissions from tropical deforestation improved by carbon-density maps. Nat. Clim. Change 2 (3), 182–185.

Baudron, F., Giller, K.E., 2014. Agriculture and nature: trouble and strife? Biol. Conserv. 170, 232–245.

Baudron, F., Durieux, J.-Y., Remans, R., Sunderland, T., 2017. Indirect contributions of forests to dietary diversity in Southern Ethiopia. Ecol. Soc. 22 (2), 28.

Bharucha, Z., Pretty, J., 2010. The roles and values of wild foods in agricultural systems. Philos. T. Roy. Soc. B. 365 (1554), 2913–2926.

Blaney, S., Beaudry, M., Latham, M., 2009. Contribution of natural resources to nutritional status in a protected area of Gabon. Food Nutr. Bull. 30 (1), 49–62.

Boffa J-M. 2015. Opportunities and challenges in the improvement of the shea (Vitellaria paradoxa) resource and its management. Occasional Paper 24. Nairobi: World Agroforestry Centre.

Bojö, J., 1996. The costs of land degradation in sub-Saharan Africa. Ecol. Econ. 16 (2), 161–173.

Bonan, G., 2008. Forests and climate change: forcings, feedbacks, and the climate benefits of forests. Science 320 (5882), 1444–1449.

Bush, R., Martiniello, G., 2017. Food riots and protest: Agrarian modernisations and structural crises. World Dev. 91, 193–207.

Byerlee, D., Stevenson, J., Villoria, N., 2014. Does intensification slow crop land expansion or encourage deforestation? Glob. Food Sec. 3 (2), 92–98.

Carletto, G., Ruel, M., Winters, P., Zezza, A., 2015. Farm-level pathways to improved nutritional status: introduction to the special issue. J. Dev. Stud. 51 (8), 945–957.

Chaudhury, M., Ajayi, O., Hellin, J., Neufeldt, H. 2011. Climate change adaptation and social protection in agroforestry systems: enhancing adaptive capacity and minimizing risk of drought in Zambia and Honduras. ICRAF Working Paper 137.

Chao, S. 2012. Forest peoples: numbers across the world. Moreton-in-Marsh, UK, Forest Peoples Programme. <http://www.forestpeoples.org/sites/fpp/files/publication/2012/05/forest-peoplesnumbers-across-world-final_0.pdf>.

Chazdon, R., 2008. Beyond deforestation: restoring forests and ecosystem services on degraded lands. Science 320, 1458–1460.

Clark, M., Tilman, D., 2017. Comparative analysis of environmental impacts of agricultural production systems, agricultural input efficiency and food choice. Environ. Res. Lett. 12, 064016.

Corbin, J., Holl, K., 2012. Applied nucleation as a forest restoration strategy. For. Ecol. Manage. 265, 37–46.

Cotter, J., Tirado, R., 2008. Food security and climate change: the answer is biodiversity. A Review of Scientific Publications on Climate Change Adaptation in Agriculture. Greenpeace Research Lab, University of Exeter.

Crist, E., Mora, C., Engelman, R., 2017. The interaction of human population, food production, and biodiversity protection. Science 356 (6335), 260–264.

Davidson, E., Janssens, I., 2006. Temperature sensitivity of soil carbon decomposition and feedbacks to climate change. Nature 440, 165.

Deakin, E., Kshatriya, M., Sunderland, T. (Eds.), 2016. Agrarian Change in Tropical Landscapes. Center for International Forestry Research, Bogor, Indonesia.

DeFries, R., Rudel, T., Uriarte, M., Hansen, M., 2010. Deforestation driven by urban population growth and agricultural trade in the twenty-first century. Nat. Geosci. 3, 178–181.

Ellis, E., Goldewijk, K., Siebert, S., Lightman, D., Ramankutty, N., 2010. Anthropogenic transformation of the biomes, 1700 to 2000. Glob. Ecol. Biogeogr. 19 (5), 589–606.

Fa, J., Currie, D., Meeuwig, J., 2003. Bushmeat and food security in the Congo Basin: linkages between wildlife and people's future. Environ. Conserv. 30 (1), 71–78.

FAO, 2009. International Institute for Applied Systems Analysis, International Soil Reference and Information Centre, Institute of Soil Science – Chinese Academy of Sciences, Joint Research Centre of the European Commission. *Harmonized World Soil Database (Version 1.1)*. FAO, Rome, Italy; International Institute for Applied Systems Analysis, Laxenburg, Austria.

FAO, 2014a. State of the World's Forests. Enhancing the Socio-economic Benefits from Forests. Food and Agriculture Organisation, Rome, <http://www.fao.org/3/a-i3710e.pdf>.

FAO, 2014b. Strengthening the Links Between Resilience and Nutrition in Food and Agriculture. Food and Agriculture Organisation, Rome, <http://www.fao.org/3/a-i3777e.pdf>.

FAO. 2015a. Global guidelines for the restoration of degraded forests and drylands in landscapes. Building resilience and benefiting livelihoods. Forestry Paper No. 175. Food and Agriculture Organisation, Rome. <http://www.fao.org/3/a-i5036e.pdf>.

FAO, 2015b. Global Forest Resources Assessment 2015. How are the World's Forests Changing? Food and Agriculture Organisation, Rome. Available from: http://www.fao.org/3/a-i4793e.pdf.

FAO. 2016. FAOSTAT database. Rome. Website available at: <faostat.fao.org>.

Fischer, G., Nachtergaele, F., Prieler, S., Van Velthuizen, H., Verelst, L., Wiberg, D. 2008. Global Agro-Ecological Zones Assessment forAgriculture (GAEZ 2008). IIASA, Laxenburg. Austria and FAO, Rome, Italy.

Foli, S., Reed, J., Clendenning, J., Petrokofsky, G., Padoch, C., Sunderland, T., 2014. To what extent does the presence of forests and trees contribute to food production in humid and dry forest landscapes? A systematic review protocol. Environ. Evid. 3 (1), 15.

Fitzherbert, E., Struebig, M., Morel, A., Danielsen, F., Brühl, C., Donald, P., et al., 2008. How will oil palm expansion affect biodiversity? Trends Ecol. Evol. 23 (10), 538–545.

Franzel, S., Lukuyu, C., Sinja, J., Wambugu, C., 2014. Fodder trees for improving livestock productivity and smallholder livelihoods in Africa. Curr. Opin. Environ. Sustain. 6, 98–103.

Garibaldi, L., Steffan-Dewenter, I., Kremen, C., Morales, J., Bommarco, R., Cunningham, S., et al., 2011. Stability of pollination services decreases with isolation from natural areas despite honey bee visits. Ecol. Lett. 14 (10), 1062–1072.

Garibaldi, L.A., Steffan-Dewenter, I., Winfree, R., Aizen, M.A., Bommarco, R., Cunningham, S.A., et al., 2013. Wild pollinators enhance fruit set of crops regardless of honey bee abundance. Science 339, 1608–1611.

Garnett, T., Godfray, C. 2012. Sustainable intensification in agriculture. Navigating a course through competing food system priorities. Food climate research network and the Oxford Martin Programme on the Future of Food. University of Oxford, UK.

Geist, H., Lambin, E., 2002. Proximate causes and underlying driving forces of tropical deforestation: tropical forests are disappearing as the result of many pressures, both local and regional, acting in various combinations in different geographical locations. BioScience 52, 143–150.

Gibbs, H., Ruessch, A., Achard, F., Clayton, M., Holmgren, P., Ramankutty, N., et al., 2010. Tropical forests were the primary sources of new agricultural land in the 1980s and 1990s. Proc. Natl. Acad. Sci. 107, 16732–16737.

Gill, M., Feliciano, D., MacDiarmid, J., Smith, P., 2015. The environmental impact of nutrition transition in three case study countries. Food Sec. 7, 493–504.

Golden, C., Fernald, L., Brashares, J., Rasolofoniaina, B., Kremen, C., 2011. Benefits of wildlife consumption to child nutrition in a biodiversity hotspot. Proc. Natl. Acad. Sci. 108, 19653–19656.

Gordon, I., Prins, H., Squire, G., 2017. Food Production and Nature Conservation; Conflicts and Solutions. Earthscan from Routledge, London.

Guo, L., Gifford, R., 2002. Soil carbon stocks and land use change: a meta analysis. Glob. Change Biol. 8, 345–360.

Haddad, L., Hawkes, C., Webb, P., Thomas, S., Beddington, J., Waage, J., et al., 2016. A new global research agenda for food. Nature 540, 30–32.

HLPE. 2017. Sustainable forestry for food security and nutrition. A Report by the High Level Panel of Experts on Food Security and Nutrition of the Committee on World Food Security, Rome. <http://www.fao.org/fileadmin/user_upload/hlpe/hlpe_documents/HLPE_Reports/HLPE-Report-11_EN.pdf>.

Ickowitz, A., Powell, B., Salim, M., Sunderland, T., 2014. Dietary quality and tree cover in Africa. Glob. Environ. Change 24, 287–294.

Ickowitz, A., Rowland, D., Powell, B., Salim, M., Sunderland, T., 2016. Forests, trees, and micronutrient-rich food consumption in Indonesia. PLoS One 11 (5), e0154139.

Jamnadass, R., McMullin, S., Miyuki, I., Dawson, I., Powell, B., Termote, C., et al., 2015. Understanding the roles of forests and tree-based systems in food provision. In: Vira, B., Wildburger, C., Mansourian, S. (Eds.), Forests and Food: Addressing Hunger and Nutrition Across Sustainable Landscapes. Open Book Publishers, Cambridge, UK, pp. 29–55.

Jennings, S.B., Brown, N.D., Boshier, D.H., Whitmore, T.C., do CA Lopes, J., 2001. Ecology provides a pragmatic solution to the maintenance of genetic diversity in sustainably managed tropical rain forests. Forest Ecol. Manag. 154 (1–2), 1–10.

Johnson, K., Jacob, A., Brown, M., 2013. Forest cover associated with improved child health and nutrition: evidence from the malawi demographic and health survey and satellite data. Glob. Health Sci. Pract. 1 (2), 237–248.

Joosten, H., 2010. The Global Peatland CO^2 Picture: Peatland Status and Drainage Related Emissions in All Countries of the World. Wetlands International, Wageningen, The Netherlands.

Keenan, R., Reams, G., Achard, F., de Freitas, J., Grainger, A., Lindquist, E., 2015. Dynamics of global forest area: results from the FAO global forest resources assessment 2015. For. Ecol. Manage. 352, 9–20.

Kent, C., Pope, E., Thomspson, V., Lewis, K., Scaife, A., Dunstone, N., 2017. Using climate model simulations to assess the current climate risk to maize production. Environ. Res. Lett. 12, 054012.

Kumar, M., Nair, P. (Eds.), 2011. Carbon Sequestration Potential of Agroforestry Systems: Opportunities and Challenges, Vol. 8. Springer Science, and Business Media.

Lawrence, D., Vandecar, K., 2015. Effects of tropical deforestation on climate and agriculture. Nat. Clim. Change 5 (1), 27.
Leblois, A., Damette, O., Wolfsberger, J., 2017. What has driven deforestation in developing countries since the 2000s? Evidence from new remote sensing data. World Dev. 92, 82–102.
Lund, J., Sungusia, E., Mabele, M., Schebe, A., 2017. Promising change, delivering continuity: REDD+ as a conservation fad. World Dev. 89, 124–139.
Locatelli, B., Catterall, C., Imbach, P., Kumar, C., Lasco, R., Marín-Spiotta, E., et al., 2015. Tropical reforestation and climate change: beyond carbon. Restorat. Ecol. 23, 337–343.
Martinich, J., Crimmins, A., Beach, R., Thomson, A., McFarland, J., 2017. Focus on agriculture and forestry benefits of reducing climate change impacts. Environ. Res. Lett. 12, 060301.
Millennium Ecosystem Assessment (MEA), 2005. Ecosystems and Human Wellbeing: A Framework for Assessment. Island Press, Washington, DC.
Mitchell, M.G., Bennett, E.M., Gonzalez, A., 2014. Forest fragments modulate the provision of multiple ecosystem services. Journal of Appl. Ecol. 51 (4), 909–918.
Nabuurs, G, Masera, O. 2007. Forestry. Climate Change 2007: Mitigation of Climate Change. Contributions of Working Group III to the Fourth Assessment Report of the Intergovernmental Panel on Climate Change.
Nasi, R., Taber, A., van Vliet, N., 2011. Empty forests, empty stomachs? Bushmeat and livelihoods in the Congo and Amazon Basins. Int. For. Rev. 13, 355–368.
Naylor, R., Steinfeld, H., Falcon, W., Galloway, J., Smil, V., Bradford, E., et al., 2005. Agriculture. Losing the links between livestock and land. Science 310, 1621–1622.
Normile, D., 2010. UN biodiversity summit yields welcome and unexpected progress. Science 330, 742–743.
Nyong, A., Adesina, F., Elasha, B., 2007. The value of indigenous knowledge in climate change mitigation and adaptation strategies in the African Sahel. Mitig. Adapt. Strateg. Glob. Change 12 (5), 787–797.
Padoch, C., Sunderland, T., 2013. Managing landscapes for greater food security and improved livelihoods. Unasylva 64 (241), 3–13.
Page, S., Rieley, J., Banks, C., 2011. Global and regional importance of the tropical peatland carbon pool. Glob. Change Biol. 17, 798–818.
Pain, A., Levine, S., 2012. A Conceptual Analysis of Livelihoods and Resilience: Addressing the 'Insecurity of Agency'. Overseas Development Institute, London.
Pan, Y., Birdsey, R.A., Fang, J., Houghton, R., Kauppi, P.E., Kurz, W.A., et al., 2011. A large and persistent carbon sink in the world's forests. Science 333, 988–993.
Parrotta, J., Dey de Pryck, J., Obiri-Darko, B., Padoch, C., Powell, B., Sandbrook, C., et al., 2015. The historical, environmental and socio-economic context of forests and tree-based systems for food security and nutrition. In: Vira, B., Wildburger, C., Mansourian, S. (Eds.), Forests and Food: Addressing Hunger and Nutrition Across Sustainable Landscapes. Open Book Publishers, Cambridge, UK, pp. 73–136.
Phelps, J., Carrasco, R., Webb, E., Koh, L., Pascual, U., 2013. Agricultural intensification escalates future conservation costs. Proc. Natl. Acad. Sci. 110, 7601–7606.
Pinstrup-Andersen, P., 2013. Can agriculture meet future nutrition challenges? Eur. J. Dev. Res. 25, 5–12.
Powell, B., Hall, J., Johns, T., 2011. Forest cover, use and dietary intake in the east Usambara Mountains, Tanzania. Int. For. Rev. 13, 305–317.
Powell, B., Ickowitz, A., McMullin, S., Jamnadass, R., Vasquez, M., Sunderland, T., 2013a. The Role of Forests, Trees and Wild Biodiversity for Nutrition-Sensitive Food Systems and Landscapes. FAO/WHO, Rome.
Powell, B., Maundu, P., Kuhnlein, H., Johns, T., 2013b. Wild foods from farm and forest in the East Usambara Mountains, Tanzania. Ecol. Food Nutr. 52 (6), 451–478.
Powell, B., Thilsted, S., Ickowitz, A., Termote, C., Sunderland, T., Herforth, A., 2015. Improving diets with wild and cultivated biodiversity from across the landscape. Food Secur. 7 (3), 535–554.
Pretty, J., Noble, A., Bossio, D., Dixon, J., Hine, R., Penning, F., et al., 2006. Resource-conserving agriculture increases yields in developing countries. Environ. Sci. Technol. 40, 1114–1119.
Rahman, S., Sunderland, T., Roshetko, J., Healey, J., 2017. Facilitating smallholder tree farming in fragmented tropical landscapes: challenges and potentials for sustainable land management. J. Environ. Manage. 198, 110–121.
Reed, J., van Vianen, J., Deakin, E., Barlow, J., Sunderland, T., 2016. Integrated landscape approaches to managing social and environmental issues in the tropics: learning from the past to guide the future. Glob. Change Biol. 22 (7), 2540–2554.

Reed, J., Van Vianen, J., Clendinning, J., Petrokovsky, G., Sunderland, T., 2017. Trees for life: the ecosystem service contribution of trees to food production and livelihoods in the tropics. For. Policy Econ. Available from: https://doi.org/10.1016/j.forpol.2017.01.012.

Richards, K., Stokes, C., 2004. A review of forest carbon sequestration cost studies: a dozen years of research. Clim. Change 63, 1–48.

Rosset, P., Martínez-Torres, M., 2012. Rural social movements and agroecology: context, theory, and process. Ecol. Soc. 17 (3), 17.

Rowland, D., Ickowitz, A., Powell, B., Nasi, R., Sunderland, T., 2017. Forest foods and healthy diets: quantifying the contributions. Environ. Conserv. 44 (2), 102–114.

Ruel, M., Alderman, H., 2013. Nutrition-sensitive interventions and programmes: how can they help to accelerate progress in improving maternal and child nutrition? Lancet 382, 536–551.

Sayer, J., Sunderland, T., Ghazoul, J., Pfund, J.L., Sheil, D., Meijaard, E., et al., 2013. The landscape approach: ten principles to apply at the nexus of agriculture, conservation and other competing land-uses. Proc. Natl. Acad. Sci. 110 (21), 8345–8348.

Scharlemann, J., Tanner, E., Hiederer, R., Kapos, V., 2014. Global soil carbon: understanding and managing the largest terrestrial carbon pool. Carbon Manage 5 (1), 81–91.

Schlegel, S., Guthrie, H., 1973. Diet and the tiruray shift from swidden to plow farming. Ecol. Food Nutr. 2 (3), 181–191.

Shackleton, S., Shackleton, C., 2012. Linking poverty, HIV/AIDS and climate change to human and ecosystem vulnerability in Southern Africa: consequences for livelihoods and sustainable ecosystem management. Int. J. Sustain. Dev. World Ecol. 19, 275–286.

Smith, P., Martino, D., Cai, Z., Gwary, D., Janzen, H., Kumar, P., et al., 2007. Agriculture. In climate change 2007: mitigation. Contribution of Working Group III to the Fourth Assessment Report of the Intergovernmental Panel on Climate ChangeIn: Metz, B., Davidson, O.R., Bosch, P.R., Dave, R., Meyer, L.A. (Eds.), Cambridge University Press, Cambridge, United Kingdom and New York, NY, USA.

Smith, P., Bustamante, M., Ahammad, H., Clark, H., Dong, H., Elsiddig, E.A., et al., 2014. Agriculture, forestry and other land use (AFOLU). In: Edenhofer, O., Pichs-Madruga, R., Sokona, Y., Farahani, E., Kadner, S., Seyboth, K., et al.,Climate Change 2014: Mitigation of Climate Change. Contribution of Working Group III to the Fifth Assessment Report of the Intergovernmental Panel on Climate Change. Cambridge University Press, Cambridge, United Kingdom and New York, NY, USA.

Stern, N., Peters, S., Bakhshi, V., Bowen, A., Cameron, C., Catovsky, S., et al., 2006. Stern Review: The economics of climate change, Vol. 30. HM treasury, London, p. 2006.

Tarnocai, C., Canadell, E., Schuur, G., Kuhry, P., Mazhitova, G., Zimov, S., 2009. Soil organic carbon pools in the northern circumpolar permafrost region. Glob. Biogeochem. Cycles 23, 1–11.

Thiombiano, D., Lamien, N., Castro Euler, A., Vinceti, B., Agundez, D., Boussim, I., 2013. Local communities demand for food tree species and the potentialities of their landscapes in two ecological zones of Burkina Faso. Open J. For. 3, 79–87.

Tilman, D., Cassman, K., Matson, P., Naylor, R., Polasky, S., 2002. Agricultural sustainability and intensive production practices. Nature 418, 671.

Tubiello, F., Salvatore, M., Cóndor Golec, R., Ferrara, A., Rossi, S., Biancalani, R., et al., 2014. Agriculture, Forestry and Other Land Use Emissions by Sources and Removals by Sinks. FAO, Rome, FAO Statistics Division, Working Paper Series.

UNCCD. 2017. UNCCD desertification land degradationand drought (DLDD): Some global facts and figures. Available at: <http://www.unccd.int/Lists/SiteDocumentLibrary/WDCD/DLDD%20Facts.pdf> (accessed 18.08.17).

Van der Werf, G., Morton, D., DeFries, R., Olivier, J., Kasibhatla, P., Jackson, R., et al., 2009. CO^2 emissions from forest loss. Nat. Geosci. 2, 737–738.

Van der Werf, G., Randerson, J., Giglio, L., Collatz, G., Mu, M., Kasibhatla, P., et al., 2010. Global fire emissions and the contribution of deforestation, savanna, forest, agricultural, and peat fires (1997–2009). Atmos. Chem. Phys. 10, 11707–11735.

Vashum, K.T., Jayakumar, S., 2012. Methods to estimate above-ground biomass and carbon stock in natural forests. Annu. Rev. J. Ecosyst. Ecogr. 2, 116. Available from: https://doi.org/10.4172/2157-7625.1000116.

Verchot, L., Van Noordwijk, M., Kandji, S., Tomich, T., Ong, C., Albrecht, A., et al., 2007. Climate change: linking adaptation and mitigation through agroforestry. Mitig. Adapt. Strateg. Glob. Change 12, 901–918.

Vinceti, B., Termote, C., Ickowitz, A., Powell, B., Kehlenbeck, K., Hunter, D., 2013. The contribution of forests and trees to sustainable diets. Sustainability 5 (11), 4797–4824.

Vira, C., Wildburger, Mansourian, S. (Eds.), 2015. Forests and Food: Addressing Hunger and Nutrition Across Sustainable Landscapes. Open Book Publishers, Cambridge, UK.

Wiggins, S., Argwings-Kodhek, G., Leavy, J., Poulton, C. 2011. Small farm commercialisation in Africa: reviewing the issues. Future Agricultures Consortium, Research Paper 23.

Whitmee, S., Haines, A., Beyrer, C., Boltz, F., Capon, A., de Souza Dias, B., et al., 2015. Safeguarding human health in the Anthropocene epoch: report of The Rockefeller Foundation-Lancet Commission on planetary health. Lancet 386, 1973–2028.

WHO/CBD. 2015. Connecting global priorities: biodiversity and human health: a state of knowledge review. <https://www.cbd.int/health/SOK-biodiversity-en.pdf>.

WHO (World Health Organization). 2015. Global Health Observatory data repository. <http://apps.who.int/gho/data/node.main.CODREG6?lang = en>.

Winfree, R., Bartomeus, I., Cariveau, D., 2011. Native pollinators in anthropogenic habitats. Annu. Rev. Ecol. Evol. Syst. 42 (1), 1–22.

Wood, S., Sebastian, K., Scherr, S. 2000. Pilot Analysis of Global Ecosystems: Agroecosystems. International Food Policy Research Institute and World Resources Institute, Washington, DC, 2000).

Wunder, S., Borner, J., Shively, J., Wyman, M., 2014. Safety nets, gap filling and forests: a global comparative perspective. World Dev. 64 (1), S29–S42.

Zhang, W., Ricketts, T., Kremen, C., Carney, K., Swinton, S., 2007. Ecosystem services and dis-services to agriculture. Ecol. Econ. 64 (2), 253–260 (2007).

Zomer, R., Neufeldt, H., Xu, J., Ahrends, A., Bossio, D., Trabucca, A., et al., 2016. Global tree cover and biomass carbon of agricultural land: the contribution of agroforestry to global and national carbon budgets. Sci. Rep. 6, 29987.

Further Reading

Lamb, D., Erskine, P., Parrotta, J., 2005. Restoration of degraded tropical forest landscapes. Science 310, 1628–1632.

Parris, K., 2011. Impact of agriculture on water pollution in OECD countries: recent trends and future prospects. Int. J. Environ. Anal. Chem. 27, 33–52.

Van Vianen, J, Reed, J., Sunderland, T. 2016. From commitment to action: Establishing action points toward operationalising integrated landscape approaches. InfoBrief No. 158. Centre for International Forestry Research, Bogor, Indonesia.

CHAPTER

7

Land and Water Governance, Poverty, and Sustainability

Olcay Ünver and Eduardo Mansur

Food and Agriculture Organization of the United Nations (FAO), Rome, Italy

7.1 PRESENT STATUS OF WATER, LAND, AND SOILS[1]

Over the past century, rising demand for food, feed, and fiber has been met largely by expanding the use of water, soils, and land. As shown in Chapters 2 and 5, projected increases in demand for agricultural output and resource use may well outstrip our available water, land, and soil resources if production practices continue to follow the trodden pathways. Even though at present only one-third of the global land area (or 4.8 billion hectares) is used for agricultural production or forestry, room for expansion of land for such purposes is considered extremely limited, as indicated in Chapter 2. The Food and Agriculture Organization of the United Nations (FAO) projects that only an additional 0.1 billion hectare of agricultural land is expected to be brought under cultivation by 2050 (FAO, 2017a), and available spare land is not readily accessible owing to lack of infrastructure, physical remoteness and disconnection from markets, and/or located in disease-prone areas. This projection further assumes ongoing deforestation will slow considerably. Furthermore, available spare land is only concentrated in few countries. Cultivated land area per person in low-income countries is less than half that in high-income countries. Most of the easily accessible land is already cultivated, and there is little room for expansion except in some areas of sub-Saharan Africa and South America.

The authors acknowledge the comments and suggestions provided by Rob Vos and the content and editorial support from Melvyn Kay.

[1] This section draws in part from FAO's 2011 report *The State of the World's Land and Water Resources for Food and Agriculture* (FAO, 2011a).

Sustainable Food and Agriculture
DOI: https://doi.org/10.1016/B978-0-12-812134-4.00007-8

Over the past 50 years, land expansion for agricultural use has occurred at the expense of forests, wetlands, and grassland. Agricultural production almost tripled in this period. Land expansion was merely one factor, as much of output growth is owing to shifts toward farming systems highly intensive in terms of input and water use. The total agricultural land area under irrigation has doubled since 1961. Intensification has relieved pressures on land expansion and lessened encroachment on forests. At the same time, however, more intense use of land, soils, and water have degraded the quality of these resources because of poor conservation practices (FAO, 2011a).

Rainfed agriculture still dominates most farming around the world, particularly that practiced by the hundreds of millions of smallholder family farmers. Most depend on unpredictable rainfall and soil moisture over the growing season and low-fertility tropical soils. As a result, yields are only half of the potential in many low-income countries. Furthermore, potential yields of crops cultivated under irrigation are, on average, three to four times higher than those of rainfed crops.

Most rainfed, low-productivity farms use few inputs, including limited use of fertilizers, and tend to retrieve nutrients from soils without allowing these to restore. As a result, land degradation tends to be severest on the holdings of the poorest farmers (FAO, 2011a).

Pressures on soils are increasing and are reaching critical limits. Some problems stemming from the way land is managed require immediate attention. Other problems are more gradual and long term, requiring vigilance and a sustained policy response over decades. Yet, few countries have such policies and structures in place.

Likewise, competition for water resources is increasing across uses (for agriculture, industries, and households). Estimates suggest that more than 40% of the world's rural population now live in water-scarce regions (see Chapter 5). Irrigated agriculture is the main source of global freshwater withdrawals, accounting for about 70% of the total. In areas where renewable water resources are already fully utilized, scarcity is becoming a bigger constraint to agricultural production than land availability. Groundwater use for irrigation has grown significantly over the past 20 years; currently it is the source of one-third of the water used in irrigation systems. However, groundwater use is largely unregulated in low-income countries, and declining aquifer levels and abstraction on nonrenewable groundwater are now putting this resource at risk.

Water, land, and soils already naturally function as a complex integrated system. However, the lack of clear land and water rights and weak capacity to regulate and enforce the rules only contribute to conflicts over resource access and overexploitation. It is high time for planners and decision-makers to recognize this reality, and to move from managing resources as separate entities towards thinking about the need for cooperation between land, soil, and water institutions.

This chapter explores current water, land, and soil governance, their symbiotic nature, the positive and negative impacts that each has on the other resources, and the challenges and potential benefits of both improving and coordinating governance.

7.2 CHALLENGES

Looking to the future, innovative systems are needed to protect and enhance natural resources while increasing productivity. Transformative "holistic" approaches are needed that embody agroecology, agroforestry, climate-smart agriculture, and conservation agriculture, which build upon indigenous and traditional knowledge.

One of the greatest challenges is achieving coherent, effective subnational, national, and international governance, with clear development objectives and commitment to achieving them. The United Nations 2030 Agenda for Sustainable Development embodies such a vision—one that goes beyond the divide of "developed" and "developing" countries. Sustainable development is a universal challenge and collective responsibility for all countries, requiring fundamental changes in the way all societies produce and consume.

Countries with weak and fragmented governance structures face additional challenges to effect such changes. In these contexts, many communities still rely on local, informal institutions and customary practices. This is particularly so in fragile situations (as defined by the World Bank), where formal governance may not even exist or cannot function because of significant political upheaval and insurgency.

Water, land, and soils are clearly essential resources to achieving and sustaining food security, alleviating hunger and poverty, and contributing to sustainable economic development. Two questions arise from this. First, will building and strengthening governance, as a priority, offer significant opportunities to address the balance of supply and demand in a way that is sustainable for the foreseeable future? Second, will coordinated governance of land, soil, and water resources based on a holistic approach to resource management also be beneficial for the sustainability of these resources?

Governance and Integration

In recent decades, concerns about natural resource governance and the need for an integrated approach have emerged, particularly at the international level, among those involved in planning and managing biodiversity, water, land, and soils resources. International bodies representing each resource promote an integrated approach, but continue to work within their specialized silos, meaning there is little evidence of integration on the ground.

Land managers expressed views about "[t]he new reality being one of shared dependency on limited resources" and were "seeking to address the challenges of food production, ecosystem management, and rural development by reaching across traditional sectoral boundaries to find partnerships that solve what are clearly inter-connected problems" (EcoAgriculture Partners, 2012). They advocate a "whole landscape approach," or Integrated Landscape Management. This approach begins with farmers who in effect manage most of the biodiversity, water, land, and soil resources, and thus influence the multifunctional nature of the landscape, including its ability to provide sources of freshwater for communities at large and act as a repository for biodiversity. But farmers alone cannot manage all the benefits that society now demands from the land as well as the challenges that come from resource shortages, such as synergies, trade-offs, and stakeholder interests, both of which give rise to potential conflict.

Soil scientists are active at an international level in setting up the global structures for soil governance. In 2011, the Global Soil Partnership was established to create soil awareness among decision-makers. In 2015—the International Year on Soils—the World Soil Charter (FAO, 2015) reminded everyone that soils are a core component of land resources, providing the foundation of agricultural development and ecological sustainability.

Over the past 25 years, concerns about water rather than land and soil resources have tended to dominate the international development debate. Indeed, since 2012, the World Economic Forum (WEF, 2015) put water at the top of the agenda as one of the top five greatest risks facing the world. Agriculture is now under scrutiny as the major water user. It currently consumes some 7130 km^3 of freshwater annually, 80% of which is "green" water, used in rainfed farming, and 20% "blue" water, abstracted from rivers and groundwater (IWMI, 2007). The Organisation for Economic Co-operation and Development (2012) estimates global water demand could increase by 55% by 2050 if we continue to use water at current rates.

In 1992, the idea of integrated water resources management (IWRM), as a means of making the best use of limited water resources, was adopted at the United Nations Conference on Environment and Development in Rio de Janeiro, Brazil. This idea is now enshrined in what has become known as the "Dublin Principles." In 1995, the Global Water Partnership was established to promote the IWRM concept (GWP, 2000). IWRM focuses on improving water resources management by connecting the many different water services and providing good governance, appropriate infrastructure, and sustainable financing. The advent of IWRM marked a fundamental shift away from traditional top-down, supply-led, technology-based solutions to water problems, and toward a stronger focus on more effective and sustainable water management (McDonnell, 2008). The conference also addressed the growing concerns over the links between land and water, many of which have led to land degradation and reduced water availability and quality. Some argue that IWRM should be changed to integrated land and water resources management (ILWRM) to reflect the importance and interconnectedness of land and water management. The link between land and water is now being increasingly referred to in many legal and policy instruments at both national and international levels and is the subject of many publications (FAO, 1993, 2002). However, the steps needed to put coordination into practice have yet to be taken in most countries.

The focus on water supply has led to relative neglect of land management and its impact on water use and management. To some extent, this was caused by the priority given to improving people's livelihoods and poverty reduction through the promotion of economic growth and social policies. Little thought was given to the impact this might have on both land and water resources and, more broadly, on the natural aquatic ecosystems that are an essential part of ensuring water quantity and quality and services.

Today, the idea of promoting a "silo" approach to managing water would seem archaic. Equally, it is argued that planning separately for water, land, and soil resources is equally archaic, and an integrated approach is essential to ensure the best use of available resources. In the wake of the 2007–08 food and energy crises and the related global rush by governments and the private sector to acquire farmland (often called "land grabbing"), more voices began advocating a coordinated approach to land and water governance. But some commentators see this coordination as too resource focused. They advocate a "nexus" approach, which links those who use water, such as water for food, energy, and the environment (UNESCO, 2014). Which is the better approach, IWRM or the nexus? A look at how each is defined suggests they are both advocating integration in the water and water-using sectors with similar objectives and so there is small choice between them.

In 2015, the United Nations 2030 Agenda for Sustainable Development emphasized such interdependencies across all of the 17 Sustainable Development Goals (SDGs). For the present assessment, this would imply looking integrally across the water and water-using sectors, including those reliant on land, and recognizing that the achievement of SDG 2, which aims to end hunger, achieve food security and improved nutrition, and promote sustainable agriculture, must indeed go hand in hand with pursuing goals related to water (SDG 6) and improving life on land and soil quality (SDG 15).

7.3 CHALLENGES TO LAND AND WATER GOVERNANCE

Lack of harmonization and coordination of water, land, and soil resources can adversely impact rural and urban lives, the economy, and the environment, especially where natural resources are already under intense pressure (Hodgson, 2004).

Land management influences water availability and quality. Conversely, available water quantity and quality influences how the land is used to produce food and preserve a healthy natural environment. In other words, decisions regarding the use and allocation of one resource impact directly or indirectly on the use and allocation of the other. Extensive literature exists on the physical interactions between land (including soil) and water and on the links between land tenure and water rights. However, little is known about how best to integrate the management of these resources, and what would be the cost and benefits of doing so. If integrated management is so evidently the way forward, why is it not practiced widely?

Land and Water: Distinct Regimes

We need to go back to ancient Roman times to see how two quite distinct regimes emerged with separate approaches to allocating and administering rights over land and water resources, both for perfectly rational reasons. Hodgson (2004) suggests that this practice stems largely from land rights being the important issue and land ownership being left to market forces. Historically, water rights were seen as subsidiary to land tenure and so were considered less important. However, the current concern about water availability has led to the development of modern statutory water rights regimes, which take little or no account of the form and content of land rights. Indeed, it is increasingly rare for legislation to restrict water rights to land owners. South Africa's Water Act, for example, entitles a person to use water and to convey it over land belonging to another person.

All these developments have served to further separate land governance from that for water, rather than combining them. Water rights are enshrined in water laws and policies. It is a field studied by water lawyers and water sector professionals, such as hydrologists and hydraulic engineers, with their own textbooks and literature. Likewise, land tenure is regulated by land legislation and policies, handled by quite different professionals, such as land tenure lawyers, surveyors, economists, and rural development experts. Even at a policy level there is often a disconnect, as both governments and international agencies separate land and water policies by institutions. Few, if any, mechanisms exist in law to ensure an integrated approach to allocating land tenure and water rights (Hodgson, 2004).

In 2012, FAO member governments endorsed the "Voluntary Guidelines on the Responsible Governance of Tenure of Land, Forestry and Fisheries in the Context of National Food Security" (VGGT) (FAO, 2012). Water, as a major determinant of land use, was not included in the VGGT. The High Level Panel of Experts on Food Security and Nutrition (HLPE, 2017) suggests that this omission is attributable to difficulties in the water sector in aligning with the VGGT's inclusive "people-centered" approach and emphasis on human rights.

Before the 1970s, little attention was paid to this divergence, as resources seemed plentiful and users were few. As a result, the rules for managing and sharing water were minimal and basic. As societies have become more aware of the limits to water availability and of the challenges to meet increasing demand by households, agriculture, industries, and energy sectors, water allocation has become more complex. Typically, multiple institutions have been established to allocate water among the different users. This has resulted, in most contexts, in a "silo" approach in managing water systems, with each institution looking after its own area of interest in satisfying water needs. Siloed structures become highly problematic when water shortages emerge and competition intensifies between different demands for water intensify (FAO, 2016).

Impacts of Water on Land and Land on Water

Every land use decision and practice leaves a water footprint; this, by itself, is an important reason for coordinating land and water governance. Globally, about 70% of water withdrawals from rivers and groundwater are committed to agriculture. In most regions of the world, this is expected to increase in the future. Pressures on land and water resources are so great that both are rapidly degrading, as well as the natural environment on which both resources depend (FAO, 2011a). This creates a "vicious circle" undermining sustainable food production, as changes in land use are reducing water availability and quality, affecting land productivity.

Changing plant cover affects water consumption, which in turn affects runoff from land into rivers and aquifer recharge (FAO, 2002). Reducing forest cover can increase water yield, but it may also increase peak flows, causing flooding downstream, and put steep catchments at risk of soil erosion. Soil compaction during agricultural operations reduces infiltration and increases runoff. Expanding urban areas to accommodate migration and population growth and building roads can also encourage rapid runoff and increase peak flows. Water quality also deteriorates because of soil erosion and chemical leaching caused by misuse of agricultural land. In arid regions, irrigation and drainage activities may lead to increased salinity of both surface and groundwater.

Conversely, water management influences land use. Poorly managed irrigation and drainage can make soil saline and unproductive. Some irrigated soils are prone to slaking, which reduces infiltration and increases runoff. Poor land management is also the main cause of soil erosion, which not only destroys fertile land but also pollutes watercourses.

Land use change can have lasting and, in some cases, irreversible impacts on groundwater aquifers. As indicated, agriculture is the largest user of groundwater. Consumption has grown significantly over the past 50 years thanks to the advent of affordable tube-well and pump technology in both the arid and humid regions of the world. Groundwater is also the main source of domestic water for many towns and villages, as well as the source of river flows and aquatic ecosystems. The groundwater footprint of towns and cities increases directly with urban population growth. This may lead to shortages, causing pressure on the agricultural sector to relinquish use of groundwater to meet urban demand. Therefore, to sustain livelihoods, groundwater productivity must increase—more food, feed, and fiber will need to be produced with less water (Shah, 2014).

Likewise, large-scale land acquisitions are as much about water as they are about land. Acquiring land may also involve acquiring freshwater resources whether intentional or not (Woodhouse and Ganho, 2011). This can impact smallholder farmers, downstream water users, biodiversity, and groundwater recharge. The Land Matrix Global Observatory has recorded 1,591 concluded transactions involving 49 million hectares (Land Matrix, 2018). In the United Republic of Tanzania, land allocated for biofuels caused controversy over the impact on water abstractions and the environment of rivers in Bagamoyo (Havnevik et al., 2011).

Virtual water trade is also as much about land as it is about water. Virtual water trade normally flows from water-rich countries to water-poor countries. Yet, in many countries the trade results from abundance of land rather than water. Thus, assessing future food security based purely on a water resource perspective distorts the picture and may impact decision-making on future trade. The same argument can be made from a land resource perspective (GWP, 2014).

Although governments may do so, farmers do not treat water separately from land. For them, the connection is self-evident. In a dry climate, land without water is of little use, as is access to water without land. Securing access to land can open opportunities to secure access to water, which enables farmers to invest with confidence in management practices and technologies that enable them to improve their livelihoods and use limited resources wisely (see Box 7.1).

Secure land rights can also secure women's access to water. Despite the increasing feminization of farm labor in the developing world, it is estimated that less than 5% of women have access to secure land rights (Niasse, 2013). Closing the gender gap in agricultural productivity worldwide would result in a substantial increase in crop yield on land owned by women. According to one estimate, this would benefit 100–150 million of the 925 million people estimated to be undernourished in 2010 (FAO, 2011b; GWP, 2014).

7.4 COORDINATING AND INTEGRATING LAND AND WATER GOVERNANCE

Hodgson (2004) posed the question whether it really matters if land tenure and water rights are governed through separate mechanisms? In other words, Is integrated management necessary? Although the author's answer was that by and large it is not, this conclusion was essentially derived from a legal perspective. However, in terms of sustainable resource management, this solution is not the case, as also argued above (GWP, 2014). Trade-offs between scale (area size, population served, etc.) and scope (socioeconomic

BOX 7.1

BENEFITS OF SECURE ACCESS TO LAND AND WATER.

Numerous examples show significant improvements in livelihoods when farmers and other producers acquire secure access to both land and water. Two examples (Bangladesh and Mauritania) are presented illustrating this case, and one example (the Sahel) illustrating a need for reforms that would make land and water rights more secure.

Bangladesh

Major reforms in the governance of inland waterbodies significantly improved the livelihoods of poor landless fishers. Before the reforms, poor landless fisherfolk lacked access to inland lakes. Wealthy fishers and fishing companies dominated annual leasing arrangements, leaving poor fishers little other option than to work as share catchers at low pay. These tenure arrangements provided little incentive to fisherfolk to invest in the lake maintenance and productive capacity.

A package of reforms introduced long-term lease arrangements for the lakes and decentralized resource management to fisher groups with membership favoring the poor. The new arrangements encouraged private investment in the lakes, leading to improved productivity, increased fish stock, better infrastructure, and higher incomes for fisherfolk. Women were also important beneficiaries through complementary reforms, which lifted existing constraints that women producers were facing when accessing inputs and support measures.

Mauritania

In order to improve flood recession farming in Maghama, land access needed to be renegotiated to secure equitable access to water. An agreement was brokered with landowners, which provided landless households access to flood recession agriculture in the lowlands of the river floodplain. Through this agreement—referred to as *entente foncière*, or land tenure agreement—long-term land-use rights were equally distributed between landowners and landless families.

The Sahel

In the Sahelian irrigation schemes, property rights over land and water vary substantially across countries and schemes. Broadly speaking, a recurrent problem is that farmers lack tenure security over their irrigated plots. In practice, they enjoy "conditional land use rights" and are subject to payment of water fees. In case of nonpayment, the irrigation agency may deprive farmers from access to the land. While this provides an effective sanction to ensure payment of water fees, it makes farmers more vulnerable to fluctuations in harvests and incomes, as disasters may cause inability to pay water fees, which will make farmers lose access to land.

Source: IFAD, (n.d). Land and Water Governance: IFAD experience. International Fund for Agricultural Development. http://bit.ly/2r8crMl; IFAD, 2011. Rural Poverty Report 2011 IFAD Rome; IIED, 2006. Land and water rights in the Sahel. Issue Paper 139. Intern.

aspects and sectors involved) should influence the decision between coordination and integration (Ünver, 2007).

Coordination

GWP (2014) suggests that "coordination," rather than integration, would best serve the need to bring land and water governance together (Fig. 7.1). However, there can be different degrees of coordination. Coordination may be strong, such that land and water institutions de facto combine into one system. It can also mean that land and water continue to be governed separately, but that institutions work closer together, either by sharing common governance arrangements or enabling information flow between the sectors.

The benefits of coordination differ depending on the scale of the area of competence, that is, whether it is at the local, national, river basin (including transboundary basins), or at the global level. Some of these benefits, as compared with situations where the silo approach prevails, are listed in Table 7.1.

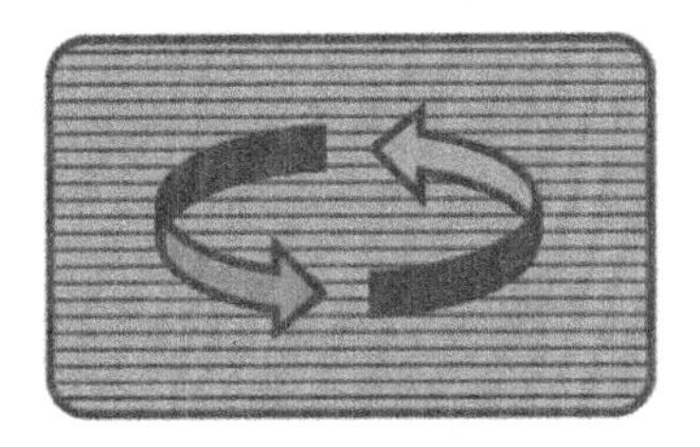

Third level:
Integrated governance (recognising that
the two resources inter-act as part of a single system)

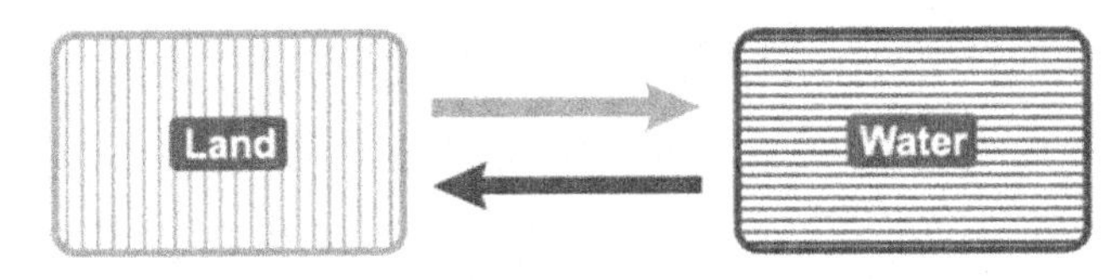

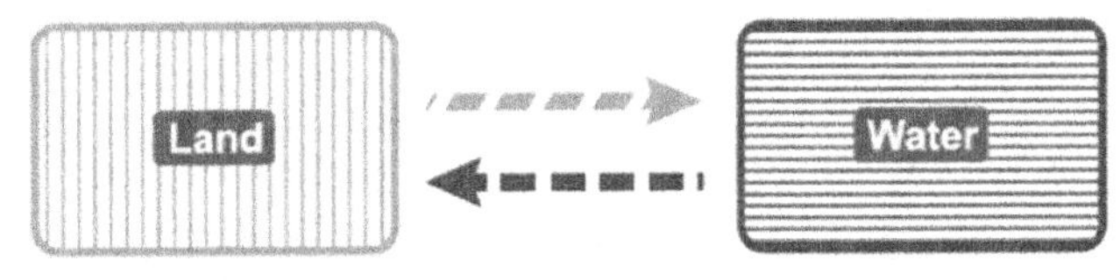

FIGURE 7.1 Three dimensions of coordinated approach to water and land governance. Source: *GWP, 2014. Coordinating Land and Water Governance. Global Water Partnership Perspective Paper.*

TABLE 7.1 Spatial Scales for Addressing Land and Water Governance for Food Security

Scale	Land/Water Governance with Silo Approach	Benefits of Coordinated Governance
Local/subnational	Inequities in access to land and/or to water; Tenure insecurity of land and/or water; Unsustainable levels of abstraction and use of land and/or water resources; Land and water degradation; Land- and/or water-related conflicts; Poor implementation and enforcement of national land and water laws; inconsistencies in land and water tenure systems	Securing access to land rights as means of ensuring greater equity in access to water; Investing in water infrastructure as means of securing land rights; Securing land rights as incentive for sustainable water conservation and for improved land productivity; Securing land and water rights to women; Factoring water access rights in land use plan, as means of preventing or resolving resource use conflicts (e.g., between farmers, pastoralists and fisherfolk); Soherency in local/community-based land use and water management plans
Nation/state	Subnational inequities in land access and tenure insecurity; formulation of national land- and/or water-related laws, strategies, action plans (rural development strategies, climate adaptation plans, plans against desertification, poverty reduction strategies) done in parallel; Inconsistencies and incoherencies in approaches to land and agrarian reform; water policies and water law reforms	Water policies and laws are informed by challenges and risks to sectors dealing with land reform and land use and vice versa; Water policies used as enabling environment for land policy objectives and vice versa; Ensure wider citizen input to water and land policy formulation, with involvement of water- and land-concerned actors; more coherent and inclusive national platform for food security and poverty reduction strategies, especially in agrarian economies
River basin/regional interstate integration groupings	Typically, undefined responsibilities of river basin organizations on the management/governance of river basin land (basin cooperation agreements typically deal only with water issues); national territorial sovereignties clashing with each other and constraining water cooperation for addressing food sovereignty; high occurrence of conflicts in transborder land and water use (cross-river or upstream-downstream water and land users)	Basin cooperation more sensitive to and supportive of agriculture and food security needs of riparian states and communities; fairer allocation of water across sectors (irrigation, energy); national decisions on land (e.g., concessions to investments) take into account interstate commitments on water
Global	Disconnected land and water discourses, theories, and normative policy processes	Cross-sectoral concerted normative response to water and land; joint collaborative efforts to address land and water challenges; improved cooperation, experience and perspectives sharing between global alliances of social movements of land- and water-concerned actors
Interplay between scales	Disconnect between scales	International and interstate commitments better informed by realities at lower levels; Land and water policy practices at lower levels are inspired by and comply with international and interstate agreements

GWP, 2014. Coordinating Land and Water Governance. Global Water Partnership Perspective Paper.

Evidence of Integration

There are promising movements toward the integration of land and water governance. EcoAgriculture Partners (2012) cite a number of land-based examples in Rwanda, the United Republic of Tanzania, and Central America. UN-Water (2012) surveyed 134 nations across the world to assess progress in reform of the water sector. An encouraging 83% had embarked on reforms to improve the enabling environment and/or come to IWRM. About 65% had developed IWRM plans, and 34% reported they were at an advanced stage of putting plans into practice. However, many developing countries confirmed that obstacles to implementation included weak and conflicting legal frameworks, inadequate or nonexistent strategic planning, and concerns about coping with "difficult hydrology" such as absolute water scarcity and severe flood risk, which increase infrastructure costs for managing and controlling water.

Shah (2016) offers evidence on the successes and failures on implementing IWRM, which has direct relevance to integrating land and water resources. The author points to early failures and blames this on integration being treated as a "blueprint" project to be implemented with a budget and time frame and as an end in itself rather than a means to an end. Rather, integration is a process that needs to evolve over time depending on local circumstances. Shah (2016) reminds us that the way in which water management mechanisms are integrated would need to be tailored to physical, social, and economic circumstances of each country and region. Referring to successful examples of integrated management in Italy, Sweden, Switzerland, the United States, the Caribbean, as well as in Central and Eastern Europe, Shah (2016) identifies two ingredients to success. First, he argues that integration must be gradual and nuanced. Forcing the pace particularly in the developing world has in fact proved counterproductive. Second, countries at different stages of development evolution have different needs and capabilities and it is essential to reflect this understanding in the approach taken. He suggests four stages of development, from fragile states to highly formed industrial economies. For each stage, Shah (2016) describes what is possible to achieve and what is needed in terms of capacity and resources (Table 7.2). There is little point, for example, in trying to introduce water demand management instruments in a fragile situation when there is a lack of basic infrastructure to harness, manage, and control water. However, such instruments for water management are likely to be successful in countries where infrastructure and strong institutions exist and the need for integration is well recognized.

Although Shah's work focuses on IWRM, land is implicit in this integration process. This approach offers a means of assessing where and how integration of land and water may best succeed.

Another promising sign comes from the food security and nutrition community. The High Level Panel of Experts on Food Security and Nutrition of the Committee on World Food Security identified a set of critical and emerging issues on food security and nutrition, which included the need for governance and calls for cross-sectoral approaches (HLPE, 2017).

7.5 CASE IN POINT: THE GOVERNANCE OF IRRIGATION AND DRAINAGE

Sixty percent of future food needs is expected to come from irrigated agriculture (FAO, 1996). Irrigation brings together water, land, and soils in a highly structured manner:

TABLE 7.2 Indicative priorities for an IWRM strategy to succeed

Indicative Priorities for an Iwrm Strategy Contextualized to Water Economies at Different Stages of Evolution

Evolutionary Stage	Stage I Completely Informal	Stage II Largely Informal	Stage III Formalizing	Stage IV Highly Formal Water Industry
% of users in formal water economy	5–15	15–35	35–75	75–95
Examples	Congo, Bhutan	Bangladesh, Tanzania	Mexico, Thailand, Turkey, China	United States, Canada, France, Australia
Capacity building	Invest in basic techno-managerial capacities for creating affordable infrastructure and service	Build capacities for efficient management of water infrastructure and water service provision	Build local capacities for catchment/river basin level water resources management	High level techno-managerial capacity for water and energy efficient water economy
Institutional reforms	Make existing institutions equitable and gender-just without emasculationg them	Create representative and participatory institutions at project or watershed levels	Integrate customary and formal user organizations and iterritorial agencies into basin organization	Modern water industry with professionally managed service providers
Policy and legal regime	Effective policies for water for livelihoods and food security; create a regulatory framework for bulk water users	Establish basic water policy and water law consistent with local institutions and customary law	Introduce policy and legal regime for a transition to basin level water governance	Policy and regulatory framework for a modern water industry and transboundary water governance
Investment priority	Establish and improve water infrastructure for consumptive and productive use by the poor and women	Invest in infrastructure modernization for improved service delivery and water use efficiency	Invest in infrastructure for basin level water allocation and management including interbasin transfers and managed aquifer recharge	Technologies and infrastructure for improvning water and energy efficiency in water economy
Managing ecosystem impacts	Create broad-based awareness of aquatic ecosystem; regulate water diversion and pollution by corporate consumers	Proactive management of water quality and ecosystem impacts at project level; invest in low-cost recycling	Focus on water quality and health management, urban waste water recycling, control of groundwater depletion	Zero or minimal discharge water economy; reduce carbon footprint
Water as a social and economic good	Minimize perverse subsidies; make subsidies smart, rationing to minimize waste	Volumetric water pricing for bulk users; partial cost recovery for retail consumers; targeted subsidies for the poor	Full financial cost recovery of water services; water supply; 90% population covered by service providers	Full economic cost recovery of water services including the costs of managing ecosystem impacts

Shah, 2016. Increasing Water Security: The Key To Implementing The Sustainable Development Goals. Global Water Partnership Stockholm. TEC Background Paper No. 22.

physically, technically, and institutionally. Its impact on water, land, and soil governance is significant.

Irrigated agriculture may have to play an important role in expanding food production, especially in sub-Saharan Africa. Increasing cropping intensities on existing irrigated farms will be a likely trend in Asia. Irrigation has the potential to remove the risks of uncertain and unreliable rainfall and so provide substantial benefits for feeding poor and disadvantaged communities. It offers opportunities to increase food security in difficult and changing climates. When properly managed, it can help protect aquatic environments and increase water security. But irrigation has a downside. If poorly managed, irreparable harm may result to cultivable land, soils, and freshwater resources.

Modern irrigation is one of the success stories of the twentieth century. It has helped to feed and improve the lives of many millions who otherwise might have continued to suffer hunger and poverty. Since the early 1900s, the global irrigated area has grown from 40 million hectares to over 340 million hectares today—an eightfold increase. Irrigated farming has significantly increased food production in many countries by pushing up yields and enabling farmers to grow multiple crops each year. Only 20% of the world's cultivated land is irrigated. Nonetheless, irrigated areas yield over 40% of the world's agricultural food, feed, and fiber production. Together, China, India, Indonesia, and Pakistan account for almost half the irrigated agricultural land area, and irrigated cropping provides for more than half their domestic food needs. The "Green Revolution" of the 1960s and 1970 dramatically expanded irrigated rice production, and this has helped prevent an imminent hunger crisis in Asia.

There is still great potential to expand irrigated farming in developing countries, particularly in sub-Saharan Africa where there are large, untapped endowments of rainfall to be harnessed using conservation farming practices and supplementary irrigation. However, progress in this direction could plant the seeds of future failure. Irrigation may cause degradation of land and soils through increased salinity, waterlogging, and soil erosion. Poor water management practices may waste water and cause water scarcity. Irrigation can damage aquatic ecosystems by overexploiting groundwater and contaminating water through misuse of agrochemicals. Many of these problems stem not only from weak and poorly integrated institutions that govern water, land, and soil resources, but also from poor engineering and agronomic and marketing support services. Although the area under irrigation is growing annually at some 3.5 million hectares, some 700,000 ha are lost each year through mismanagement.

Addressing these adverse effects from poor managed irrigation systems is not a mere technical problem. Rather, it requires better governance mechanisms able to secure the sustainability of soils, water, and land resources.

Origins of Poor Irrigation Management

Modern irrigation has developed over the past 100 years, from treating irrigation as an engineering/water distribution problem to irrigation being currently regarded as a complex mix of engineering, agriculture, socioeconomics, management, governance, politics, and environmental issues.

In many developing countries, modern irrigation practices were established during colonial times, such as in Egypt with construction of the first Aswan Dam, in Mesopotamia (Iraq and the Syrian Arab Republic), and in India. The priority was to ensure adequate food supplies. Major engineering works were needed to control and divert some of the world's biggest rivers into canal systems to distribute water over many thousands of hectares. Although initially successful, in more recent times problems have gradually emerged as many of the larger systems fell into disrepair and were poorly managed and maintained. Serious disconnects began to emerge between ministries of irrigation, set up to build schemes and allocate water, but staffed by engineers having little knowledge of agriculture, and ministries of agriculture, staffed by agricultural experts to provide support for farmers, but who possessed little knowledge of irrigation engineering, and operating and maintaining major engineering works and canal distribution systems.

During the 1970s and 1980s, governments and their development partners saw large-scale irrigation as an attractive investment for alleviating poverty and increasing food security. But many schemes suffered from poor project implementation with long maturation periods, high costs (particularly in Africa), lack of support services for operation and maintenance and farming operations, and poor productivity. All this happened at a time of falling world food prices towards the end of the twentieth century. This left a continuing perception, especially in Africa, that irrigation had failed to live up to its expectations (Inocencio et al., 2007). Too much attention was given to construction and infrastructure, and too little to how schemes would be managed and maintained, and the wider socioeconomic implications of farmer livelihoods and producing marketable food and fiber crops. The gap between engineering and agricultural institutions persisted, and still persists, in many countries. This has hindered effective irrigation management with consequent continued deterioration of the quality of land and soils, and increasingly, inefficient use of water.

In the 1990s, attention shifted towards transferring irrigation system management from governments to farmer groups. But many people saw this as means of reducing government costs and responsibilities. The strategy had limited success, as it continued to overlook the power relationships among farmers, rural elites, and irrigation bureaucracies (Yalcin and Mollinga, 2007). One positive outcome was the development of water user associations, which formed to take responsibility for managing, allocating, and distributing water.

The lessons from this era shifted interest towards small-scale irrigation, as farmers individually and in groups had greater control over both resources and, more importantly, their livelihoods. Small-scale irrigation was driven by private interests and was, and still is, largely independent of government. However, most small-scale irrigation relies on groundwater. Unregulated extraction is leading to unsustainable groundwater exploitation with consequences for sustainable production and impacts on river flows, land use and soil quality, and aquatic ecosystems. Solving this problem is an almost impossible administrative task. How does one regulate and administer the extraction of relatively small quantities of water by many millions of private smallholders? (Shah, 2014).

Renewed Interest in Irrigation

Whether for small-scale or large-scale systems, renewed interest in irrigation has been expressed by development agencies and private-sector actors alike, as concerns over water

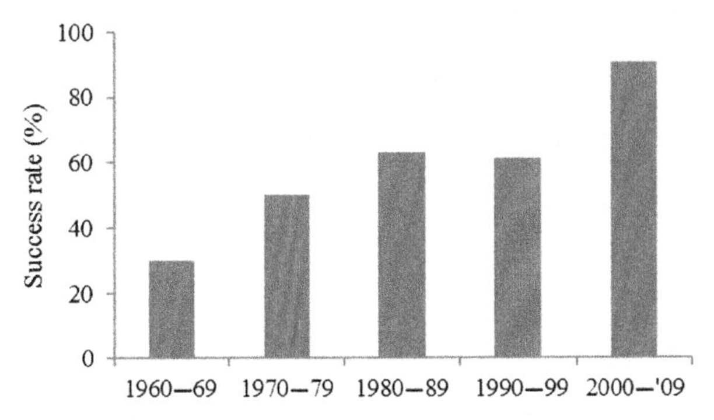

FIGURE 7.2 Irrigation scheme success rate in Southern Africa, 1960–2009. Source: *Mutiro and Lautze, 2015. Irrigation in Southern Africa: success or failure? Irrig. Drain. 64: 180–192.*

and food security have grown. As a result, evidence from southern Africa points at an increasing prevalence of successful irrigation investment projects (Fig. 7.2). In the 1960s, only 30% of schemes were determined as successful, whereas 90% were judged successful in 2000–09. Africa's heads of state signed up to the Malabo Declaration (June 2014) for transforming Africa's agriculture with targets to increase agricultural productivity and production and specifically identified increasing the area under irrigation as a key part of this strategy. Ünver et al. (2016) identify "smart" actions that could enhance the contribution of irrigation agriculture to hunger and poverty alleviation. These include facilitation of private-sector involvement in small-scale irrigation systems. Private investments in small-scale systems are already important in South Asia and Africa. Several multilateral development banks, including the World Bank and the Asian Development Bank, are exploring enhancing financing of public-private partnerships, so as to draw commercial interests into the scaling up of irrigation systems.

The lesson from past experience is that the right technology by itself, while essential, is not sufficient to guarantee successful investment. Such investments will only work well if they come with compensatory action that can handle environmental, socioeconomic, and environmental trade-offs, and put adequate institutional arrangements in place for the management of land and water resources. Huppert (2008) mapped out the changing perceptions of irrigation from simple engineering solutions to ones that recognize irrigation systems as a complex arrangement of technical, environmental, socioeconomic, and institutional factors (Fig. 7.3).

The challenge for planners and managers is to find the most effective pathways through a maze of issues that enables successful development to take place. Huppert (2008) also offers a framework to identify needs for action on the various dimensions to secure successful investment in irrigation systems (Fig. 7.4). Issues such as infrastructure development and routine maintenance, he argues, may be less complex requiring less discussion among stakeholders, whereas reforming institutions, establishing water rights and forming water user associations are highly complex issues that require more consultation and effort, especially where this requires changing attitudes and behaviors as well.

Strengthening Governance

Most analysts recognize the importance of building strong institutions and coordinated governance to successfully deal with the increasing complexity of irrigation development.

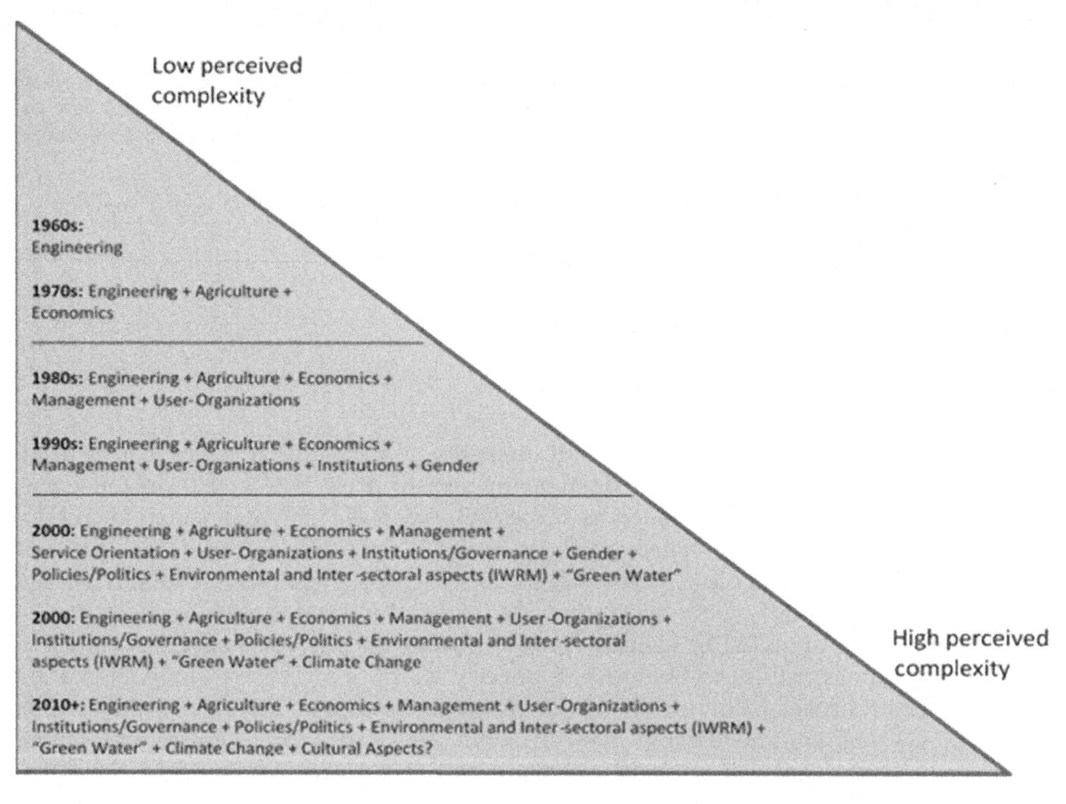

FIGURE 7.3 Chronology of perceived complexity of irrigation systems. Source: *Huppert, 2008. Coping with Complexity: Innovative Approaches to an Emerging Issue in Agricultural Water Management. IFAD, Rome.*

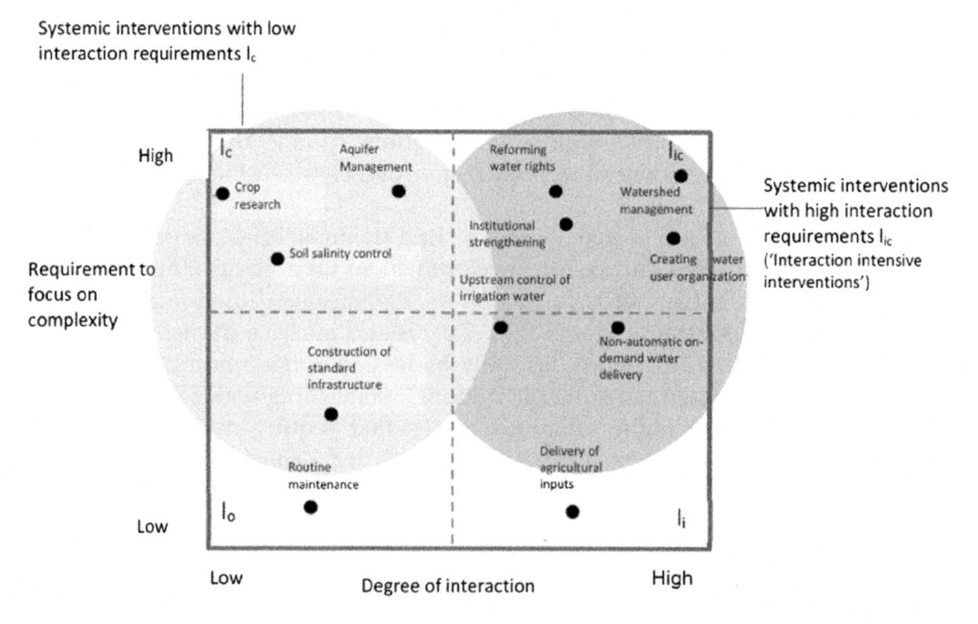

FIGURE 7.4 Framework to assess complexity of interventions in irrigation. Source: *Huppert, 2008. Coping with Complexity: Innovative Approaches to an Emerging Issue in Agricultural Water Management. IFAD, Rome.*

Institutions that support irrigation play a crucial role, providing the rules, incentives, and controls, agreements and rights, and finance which determine service provision (Rauch, 2008).

Institutions that support irrigation are both formal and informal. Large-scale schemes need formal institutions. However, an ever-present disconnect persists among the ministries of agriculture, irrigation, water resources, and environment, which continue to pursue their own agendas, often seeking to make use of the same land and water resources that others already use for other purposes.

Informal rules and regulations are common among smallholder water users, particularly in fragile situations where formal institutions are weak or nonexistent. Most are economically water scarce, war torn, having limited data on water resources and water demand, experiencing underinvestment in infrastructure, and weak institutions and organizations. As a result, households and communities build informal means to secure water supplies for domestic and productive uses (Allouche, 2014). The challenge is to build institutional and human capacity, which is essentially formal ("top-down"), but does not harm the informal and customary ("bottom-up") systems which exist, and which are vital for poor water users as they seek to sustain their livelihoods, respond to uncertainties, and mobilize around rights and resources.

Irrigation governance must also recognize the trend towards multiple-use systems—to include domestic water supplies and links to public health and livestock watering—which more closely match realities of water use by poor people (IWMI, 2006).

Change is taking place, but slowly. In many countries, significant reforms will be needed to reorient institutions. Attitudes must shift from a "top-down" system management approach to one which regulates the sector and helps farmers to take responsibility for operating and maintaining irrigation systems. Shifts must also take place towards integration and coordination with other water and land use interests, such as the aquatic environment and health. Overall, the institutional shift would need to be from managing resources for people to one which facilitates and regulates the ways in which people ensure the quality of soils and the sustainability of land and water resources. The indicative approach proposed by Shah (2016) could be the way towards successful integration in the irrigation and drainage sectors (see also Table 7.2).

7.6 SOIL GOVERNANCE

The story of soil governance has not captured international attention as have water and land. Soils are seen as implicit within land management, but in reality they are essential complements of both land and water resources. FAO describes soils as the core of land resources and the foundation of agricultural development and ecological sustainability. If water is the "life blood" of agriculture, then soils are the "body." Soils are also described as contributors to increasing food security, alleviating poverty, and sustaining development. They provide valuable services, such as regulating climate and safeguarding ecosystem services and biodiversity (FAO, 2015).

Pressures on soils are increasing and are reaching critical limits. Some problems require immediate attention and stem from the way the land is managed. Other problems are more gradual and long term. They require vigilance and a sustained policy response over decades. Few countries have such policies and structures in place.

Like the land and water communities, the soil community has been active in setting up the global structures for soil governance, albeit in relative isolation. In 2011, the Global Soil Partnership was established to create soil awareness among decision-makers. An Intergovernmental Technical Panel on Soils provides technical and scientific guidance and complements the work of the Intergovernmental Panel on Climate Change, the Intergovernmental Science-Policy Panel on Biodiversity and Ecosystem Services, and the United Nations Convention to Combat Desertification's Science-Policy Interface.

In 2015—International Year of Soils—the World Soil Charter (FAO, 2015) was adopted, setting out the case for the importance of soil and the need for its protection and conservation. Like other natural resources, soil management is influenced by socioeconomic conditions with consequent changes in the environmental services that soils provide. The Voluntary Guidelines for Sustainable Soil Management (FAO, 2017b) provide a reference for technical and policy recommendations on sustainable soil management to inform strategic and context-specific decision-making at all levels.

Soil governance concerns policies to safeguard soils and the rules and mechanisms for assessing and managing soil resources, which are all essential but tend only to receive attention at the global, regional, and national levels. This may seem remote for some individuals concerned about local soil issues. But today, the consequences of soil degradation and loss of production can be felt across the world as they interrupt international food supply chains.

The most effective responses derive from actions at a local level by individual land managers. Farmers have strong interest in soils for producing good crop yields, their suitability for irrigation and drainage, and their ease of cultivation. Understanding this interconnectedness and the consequence of actions at each level is central to effective soil governance and policy (FAO, 2015).

Soil governance is different from land and water governance. People own legal rights to land and increasingly have rights to access water, and governments have established institutions to govern them. There are no ministries of soils. Soils are seen as implicit within land, and soil centers are affiliated or incorporated within other institutions. This does not diminish the importance of soils. Soil governance is perceived to be just as important as land and water governance, at least among the cadre of soil scientists who function as the guardians of a nation's soils.

7.7 CONCLUSIONS

This chapter has explored deficiencies in current land, soil, and water governance. It has stressed the symbiotic nature of water, land, and soil resources, and the challenges to and potential benefits of both improving and coordinating governance. We began by asking questions about whether strengthening governance would address the equilibrium of supply and demand balance for resources in a sustainable manner, and if integration across land and water would support sustainable development. These are issues that are central to the 2030 Agenda for Sustainable Development and that will be essential to achieving the goals of ending hunger, ensuring food security, and making agriculture sustainable. Politicians and professionals alike require evidence-based knowledge. The limited available evidence suggests that there is much to be gained from more integral water, land, and soil management. Most of this evidence was obtained from contexts which already possess relatively strong institutions and where economic development is more advanced. More research and experimentation is needed to better grasp how institutions for coordinated and integrated water, land, and soil management could best be set up in contexts with, as yet, weak institutional and economic development. Lessons from existing practices demonstrate that improved water, land, and soil governance is more likely to succeed when developed through a gradual process tailored to local conditions rather than through a "blueprint" modeled after successful experiences elsewhere.

References

Allouche, J., 2014. The role of informal service providers in post-conflict reconstruction and state building. In: Weinthal, E., Troell, J., Nakayama, M. (Eds.), Water and Post-Conflict Peacebuilding. Earthscan, London.

EcoAgriculture Partners, 2012. Landscapes for People, Food and Nature: The Vision, the Evidence, and the Next Steps. Washington, DC.

FAO, 1993. Land and water integration and river basin management. FAO Land and Water Bulletin 1. Proceedings of an FAO Informal Workshop. Rome, Italy, 31 January–2 February 1993.

FAO, 1996. Report of the World Food Summit, 13–17 November. FAO, Rome.

FAO, 2002. Land and water linkages in rural watersheds. FAO Land and Water Bulletin 9. Proceedings of the Electronic Workshop, October 2000.

FAO, 2011a. The State of the World's Land and Water Resources for Food and Agriculture (SOLAW)—Managing Systems at Risk. Food and Agriculture Organization of the United Nations, Rome and Earthscan, London.

FAO, 2011b. The State of Food and Agriculture. Women in Agriculture. Closing the Gender Gap for Development. FAO, Rome.

FAO, 2012. Voluntary Guidelines on the Responsible Governance of Tenure of Land, Fisheries, and Forests in the Context of National Food Security. FAO, Rome.

FAO, 2015. Status of the World's Soil Resources (SWSR)—Main Report. FAO and the Intergovernmental Technical Panel on Soils, Rome, Italy.

FAO, 2016. Exploring the Concept of Water Tenure. FAO Land and Water Discussion Paper 10. Based on the work of consultant S Hodgson. Rome, Italy.

FAO, 2017a. The Future of Food and Agriculture: Trends and Challenges. FAO, Rome.

FAO, 2017b. Voluntary Guidelines for Sustainable Soil Management. FAO, Rome.

GWP, 2000. Integrated Water Resources Management, TAC Background Papers No. 4. Global Water Partnership, Stockholm.

GWP, 2014. Coordinating Land and Water Governance. Global Water Partnership Perspective Paper.

Havnevik, K., Haaland, H., Abdallah, J., 2011. Biofuel, Land and Environmental Issues: The Case of SEKAB's Biofuel Plans in Tanzania. Nordic Africa Institute, the University of Agder, Norway and Sokoine University of Agricultural Sciences, Tanzania.

HLPE, 2017. High Level Panel of Experts on Food Security and Nutrition 2nd Note on Critical and Emerging Issues for Food Security and Nutrition, Prepared for the Committee on World Food Security, 27 April 2017. http://www.fao.org/fileadmin/user_upload/hlpe/hlpe_documents/Critical-Emerging-Issues-2016/HLPE_Note-to-CFS_Critical-and-Emerging-Issues-2nd-Edition__27-April-2017_.pdf (accessed on 15.05.2017)

Hodgson, S., 2004. Land and Water—the Rights Interface. FAO Legislative Study 84.

Huppert, W., 2008. Coping with Complexity: Innovative Approaches to an Emerging Issue in Agricultural Water Management. IFAD, Rome.

IFAD (n.d.) Land and Water Governance: IFAD Experience. International Fund for Agricultural Development. <http://bit.ly/2r8crMl>

IFAD, 2011. Rural Poverty Report 2011. IFAD, Rome.

Inocencio, A., Kikuchi, M., Tonosaki, M., Mayurama, A., Merrey, D., Sally, H., et al., 2007. Costs and Performance of Irrigation Projects: A Comparison of Sub-Saharan Africa and Other Developing Regions. IWMI Research Report 109. International Water Management Institute (IWMI), Colombo, Sri Lanka.

IIED, 2006. Land and Water Rights in the Sahel. In: Cotula, L. (Ed.), Issue Paper 139. International Institute for Environment and Development.

IWMI, 2006. Taking a Multiple-Use Approach to Meeting the Water Needs of Poor Communities Brings Multiple Benefits. IWMI, Colombo. http://www.iwmi.cgiar.org/Publications/Water_Policy_Briefs/PDF/wpb18.pdf (accessed February, 2006)

IWMI, 2007. Comprehensive Assessment of Water Management in Agriculture. Water for Food, Water for Life: A Comprehensive Assessment of Water Management in Agriculture. Earthscan, and Colombo: International Water Management Institute, London, 645 pp.

Land Matrix, 2018. The online public database on land deals. http://landmatrix.org (accessed on 12 September 2018).

McDonnell, R., 2008. Challenges for integrated water resources management: how do we provide the knowledge to support truly integrated thinking? Int. J. Water Resources Dev. 24 (1), 131–143.

Mutiro, J., Lautze, J., 2015. Irrigation in Southern Africa: success or failure? Irrig. Drain 64, 180–192.

Niasse, M., 2013. Gender equality: it's smart and it's right. In: Manzi, M., Zwart, G. (Eds.), The Future of Agriculture. Oxfam Discussions Papers, pp. 51–53. <http://www.oxfam.org/sites/www.oxfamorg/files/dp-future-of-agriculture-synthesis-300713-en.pdf>

Organisation for Economic Co-operation and Development (OECD), 2012. Environmental Outlook to 2050: Key Findings on Water. OECD, Paris. <http://www.oecd.org/environment/indicators-modelling-outlooks/49844953.pdf>

Rauch, T., 2008. The New Rurality: Its Implications for a New Pro-poor Agricultural Water Strategy. InnoWat IFAD, Rome.

Shah, T., 2014. Groundwater Governance and Irrigated Agriculture. Global Water Partnership Stockholm. TEC Background Paper No. 19.

Shah, T., 2016. Increasing Water Security: The Key To Implementing The Sustainable Development Goals. Global Water Partnership Stockholm. TEC Background Paper No. 22.

UNESCO, 2014. World Water Development Report 2014, p. 69.

Ünver, O., 2007. Water-based sustainable integrated regional development. In: Mays, L.W. (Ed.), Water Resources Sustainability , McGraw-Hill, New York (Chapter 12), p. 242.

Ünver, O., Wahaj, R., Lorenzon, E., Mohammedi, K., Osias, J., Reinders, F., et al., 2016. Key and Smart Actions to Alleviate Hunger and Poverty through Irrigation and Drainage, Background Paper, World Irrigation Forum, 6–8 November 2016, Thailand.

UN-Water, 2012. Status report on the application of integrated approaches to water resources management. <http://www.unwater.org/rio2012/report/index.html>

WEF, 2015. Global Risks 2015, tenth ed. World Economic Forum, Switzerland.

Woodhouse, P., Ganho, A.S., 2011. Is water the hidden agenda of agricultural land acquisition in sub-Saharan Africa? Paper presented at International Conference on Global Land Grabbing, Institute of Development Studies and Future Agricultures Consortium, University of Sussex, UK, 6–8 April 2011.

Yalcin, R., Mollinga, P. 2007. Institutional transformation in Uzbekistan's agricultural and water resources administration: the creation of a new bureaucracy. ZEF Working Paper Series, No. 22. <http://nbn-resolving.de/urn:nbn:de:0202-20080911223>

CHAPTER

8

Biodiversity and Ecosystem Services

Wei Zhang[1], *Ehsan Dulloo*[2], *Gina Kennedy*[2], *Arwen Bailey*[2], *Harpinder Sandhu*[3] *and Ephraim Nkonya*[1]

[1]**International Food Policy Research Institute (IFPRI), Washington, DC, United States**
[2]**Bioversity International, Maccarese, Italy** [3]**Flinders University, Adelaide, SA, Australia**

8.1 INTRODUCTION

Ecosystem services are benefits humankind receives from ecosystems. They are generally categorized into four major types: provisioning services (food, fiber, freshwater); regulating services (pest regulation, pollination, water purification); supporting services (nutrient cycling, soil formation, photosynthesis); and cultural services (recreation, tourism, aesthetic values) (MEA, 2005). Ecosystem services are described as nature's contributions to people by the United Nations Intergovernmental Science-Policy Platform on Biodiversity and Ecosystem Services (IPBES), as they are the channel between nature and a good quality of life (Díaz et al., 2015; Pascual et al., 2017). Biodiversity underpins ecosystem functioning and the provision of ecosystem services is essential for human well-being (CBD, 2010). Protecting biodiversity and improving the supply of—and equitable access to—ecosystem services is a vital global interest to sustain a healthy planet and deliver benefits essential for all people. This is particularly true in lower-income countries given that biodiversity and ecosystem services constitute the wealth of the poor who depend heavily on natural resources for their livelihoods (Daily et al., 2011; Dasgupta 2010;Sandhu and Sandhu, 2014; 2015).

Covering 38% of the terrestrial, ice-free surface of the planet, agricultural ecosystems, or agroecosystems, are both the largest ecosystems of the anthropocene and a major contributor to the breaching of multiple planetary boundaries (Rockström et al., 2009; Steffen et al., 2015). The Sustainable Development Goals (SDGs) mark a momentous opportunity to improve human well-being and social equity while conserving Earth's natural resources and the vital ecological functions on which we depend (Landscapes for People, Food and Nature Initiative, 2015). Biodiversity and ecosystem services are essential for the

Sustainable Food and Agriculture
DOI: https://doi.org/10.1016/B978-0-12-812134-4.00008-X

achievement of the SDGs. According to a review of the SDG targets and goals by the International Counsel for Science (ICSU and ISSC, 2015), all SDGs benefit to some degree from ecosystem protection, restoration, and sustainable use. It is particularly important to strengthen specific ecosystem services-related targets in the poverty reduction, food security, human health, and water SDGs, as each depends heavily on ecosystems (Wood and De Clerck, 2015). Better understanding of the dependence of global food and agriculture systems on biodiversity and ecosystem services can help develop evidence-informed policies for long-term environmental and economic sustainability.

This chapter discusses biodiversity and ecosystem services in the context of sustainable food and agriculture. Following the introduction, Section 8.2 reviews key issues and challenges on plant genetic resources for food and agriculture, Section 8.3 discusses the crucial role of diversely structured agricultural landscapes in supporting ecosystem services that maintain the productivity and stability of agroecosystems, and Section 8.4 provides an overview of economic valuation for ecosystem services, illustrated by two case studies on the global cost of land degradation and the full benefits and costs of a selected food production system.

8.2 PLANT GENETIC RESOURCES FOR FOOD AND AGRICULTURE

Current Trends

Plant genetic resources for food and agriculture (PGRFA) are one component of biodiversity, whose management can impact ecosystem services. PGRFA provide provisioning services, such as food, as well as regulating, supporting and cultural services. Today's agricultural systems focus on improving productivity. In this context, a strong focus on only provisioning ecosystem services has threatened PGRFA, as uniform, improved varieties have displaced locally adapted varieties, eliminating considerable PGRFA from farmers' fields (Fowler and Mooney, 1990). By 1999, only twelve crops and five animal species made up 75% of the world's food demand (FAO, 1999). Globally, the land area devoted to high-yielding cereals has increased starkly over the past 50 years. Rice, wheat, and maize collectively increased from 66% to 79% of all cereal area between 1961 and 2013, at the expense of other cereals containing a higher micronutrient content (DeFries et al., 2015). For example, the cultivated area for sorghum, barley, oats, rye, and millet declined from 33% to 19% of the total cereal area.

Globalization has resulted in homogeneity not just in the types of crop cultivated, but also in the diets of people. While there is growing diversity on individuals' plates, the difference between national diets is decreasing and cultivation of foreign crops is increasing (Khoury et al., 2014; and Chapter 2: Global Trends and Challenges to Food and Agriculture into the 21st Century). Foreign crops represent 69% of current production systems, and thus countries are becoming more interdependent in accessing diversity of foods and nutrients (Khoury et al., 2014; 2016).

In response to the decline in diversity, scientists have been actively collecting and conserving PGRFA in ex situ facilities (Thormann et al., 2006). To date, over 7 million samples are conserved in 1,750 genebanks worldwide (FAO, 2010). However, genebanks have largely focused on the conservation of major staple crops, while nonstaple crops represent only 2% of materials stored and crop wild relatives are also poorly represented (Castañeda-Álvarez et al., 2016). To conserve more diversity, in situ approaches in which plants are managed in their natural environment either on-farm or in the wild are recommended, allowing the materials to continue to evolve in response to human and natural pressures. The benefits of ex situ and in situ conservation are each necessary, but insufficient on their own (Dulloo et al., 2010). Integrated approaches are required to link ex situ and in situ conservation efforts. For example, in Peru, links have been created between communities and genebanks, facilitating reintroduction of local varieties that were previously stored ex situ in collecting missions.

Adoption of on-farm conservation requires that farmers benefit from maintaining PGRFA. Adaptability to certain microenvironments, increasing options in the face of increasingly unpredictable weather patterns, and cultural and culinary preferences are some of the reasons for farm households to diversify crop choices (Jarvis et al., 2011, 2016; Sthapit et al., 2008). Providing more than 70% of food calories to people in Asia and sub-Saharan Africa (Samberg et al., 2016), smallholder farms play a substantial role in maintaining the genetic diversity of our food supply (Fanzo, 2017). Despite trends towards loss of valuable PGRFA (FAO, 2010), many farmers continue to maintain remarkable levels of diversity in their fields (Bellon et al., 2016; Jarvis et al., 2008, 2011).

Loss of genetic diversity is an intrinsic challenge to the future of agriculture and dietary diversity. Farmers and plant breeders need a diversity of genetic resources to equip crops to cope with more extreme weather events and other consequences of climate change. Beyond the provisioning service of food, PGRFA is, if anything, even more important for its role in human health and system resilience.

Dietary Diversity and Health

Poor diets form the biggest single factor contributing to global ill health (Global Panel on Agriculture and Food Systems for Nutrition, 2016). For example, inadequate intake of fruit and vegetables is the sixth leading cause of death in the world (Ruel et al., 2005). Many health outcomes, such as obesity, are related to changing lifestyles and the quantity of food available. Others, such as xerophthalmia and anaemia, are more closely tied to single micronutrients. Approaching malnutrition on a nutrient-by-nutrient basis does not offer sustainable improvements (IFPRI, 2016; FAO, 2016). To end multiple forms of malnutrition simultaneously, a fundamental shift is required that takes a holistic view of the entire diet of people, considered over the whole cycle of the seasons (Allen, 2008; Jacobs and Tapsell, 2007). This is not only a question of individual well-being. The economic toll of poor diets to national gross domestic product (GDP) is staggering. For example, the cost to GDP of diet-related chronic disease in Asia increased from 0.3%–2.4% in the late 1990s to 11% in the 2010s (IFPRI, 2016). Countries thus should be concerned with the economic costs of deteriorating diets.

Dietary diversity is a key determinant of nutritional outcomes (Arimond and Ruel, 2004; Pellegrini and Tasciotti, 2014). Dietary diversity and nutritional adequacy are strongly and positively correlated (Ruel, 2003). Dietary guidelines worldwide recommend a varied, diverse diet for optimal health (Gonzalez Fischer and Garnett, 2016). PGRFA underpin healthy, diverse diets as part of food-based approaches that focus on the quality of diets all year round.

On one hand, nutrient composition between species and within species can vary greatly. For example, the Cavendish banana contains almost no pro-vitamin A carotenoids, while the pro-vitamin A carotenoid content of some banana cultivars can be as high as 1412 mcg/100 g (Englberger et al., 2006). Between species, cereals such as millets and oats often contain more zinc and iron than most varieties of rice (DeFries et al., 2015). On the other, crop diversification generally leads to increased dietary diversity for individuals or households (Powell et al., 2015; Jones, 2016). The more types of food crops are grown on a holding, the more food groups appear in the diet and, by extension, the more nutritious the diet becomes. Intensive commercial production of maize, rice, and wheat has reduced hunger in many developing countries, but monotonous, staple-based diets with low diversity of other food groups have negative impacts on nutritional status (Arimond and Ruel, 2004).

PGRFA also underpins the availability of wild foods that contribute to dietary diversity, especially among rural communities. The Food and Agriculture Organization of the United Nations (FAO) estimates that around 1 billion people consume wild foods worldwide (Aberoumand, 2009; and Chapter 6: Forests, Land Use, and Challenges to Climate Stability and Food Security). The mean use of wild foods in 22 countries in Asia and Africa is 90–100 species per location. Aggregate country estimates can reach 300–800 wild edible species (CBD and WHO, 2015; Ickowitz et al., 2016).

Finally, there is a connection between landscape diversity and dietary diversity. Children in Africa who live in areas with more tree cover had more diverse and nutritious diets than those living in areas with less tree-cover density (Ickowitz et al., 2014). Populations with access to forest foods tend to consume large quantities of forest foods, which contribute substantially towards dietary adequacy (Rowland et al., 2017).

Natural habitats in landscapes contribute to maintaining the genetic diversity of important crops through conserving crop wild relatives. This repository of genetic diversity for crop improvement is estimated at more than US$120 billion a year (PwC, 2013). They are sources of traits, such as resistance to pests and diseases, salinity, flooding or drought, which are needed to increase yields under changing environmental conditions. The role of crop wild relatives is increasingly recognized (FAO, 2010), and the International Treaty on Plant Genetic Resources for Food and Agriculture refers to the promotion of in situ conservation of crop wild relatives and wild plants for food production.

Gender Roles

The role of gender in conserving and using PGRFA differs both within and across societies (Howard, 2003). In general, men grow major commodity crops, while women tend to be responsible for the nutrition and health of household members, including by gathering and gardening plant foods (Howard, 2003; IFPRI, 2011; SUN, 2016). Women's knowledge, education, social status, and control over resources are key factors that affect nutritional outcomes (Keller et al., 2006). Social and economic empowerment for women is thus an avenue to address hunger and malnutrition (IFPRI, 2011). There is a largely unmet opportunity to document, validate, strengthen, share and transmit women's knowledge on PGRFA to improve the sustainability of diets, nutrition, and health. Given the role of women in conserving and using PGRFA, gender-responsive approaches are needed, as is research into how PGRFA conservation is changing in the context of male migration and agricultural feminization.

Role of Communities

Communities can be supported through targeted interventions, such as community biodiversity management approaches, which ensure that communities gain livelihood benefits from the PGRFA they manage and are thus incentivized to maintain them (Sthapit et al., 2016; Subedi et al., 2013). Community seedbanks are increasingly used to manage and conserve PGRFA at the local level and make seeds available to farmers (Vernooy et al., 2014, 2015; Shrestha et al., 2006). Communities also can play an important role in conserving crop wild relatives, as they are the principal users of this type of PGRFA in protected areas (Hunter and Heywood, 2011). Participatory approaches, such as community biodiversity registers, have been found to be an effective way to document local genetic diversity to inventory it, but also as proof in cases of enforcing farmers' rights (Dulloo et al., 2017).

8.3 BIODIVERSITY AND ECOSYSTEM SERVICES IN AGRICULTURAL LANDSCAPES

To provide provisioning ecosystem services, such as food, fiber, and fuel, agroecosystems depend upon a variety of supporting and regulating services (e.g., soil fertility, nutrient cycling, pest regulation, and pollination), which in turn rely on how agroecosystems are managed at the site scale and on the diversity, composition, and functionality of the surrounding landscape determined by land cover and land use choices over a range of scales (Tilman, 1999). Natural habitat and diverse land use in agricultural landscapes help foster biodiversity and provide ecosystem services that are essential to farming and human well-being (Tscharntke et al., 2016), including pest suppression (Landis et al., 2000; Bianchi et al., 2006), soil conservation (Mäder et al., 2002), nutrient retention (Raudsepp-Hearne et al., 2010), crop pollination (Klein et al., 2003; Carvalheiro et al., 2010), and cultural services (van Zanten et al., 2014).

Diverse land use in agricultural landscapes underpins the provision of many ecosystem services that are important to the productivity of agriculture and which facilitate less reliance on agrochemicals. For example, arthropod predators and parasitoids suppress populations of insect herbivores (Naylor and Ehrlich, 1997). To do so, these predators and parasitoids rely on a variety of plant resources such as nectar, pollen, sap, or seeds (Wilkinson and Landis, 2005). Noncrop areas can provide habitats where beneficial insects mate, reproduce, and overwinter.

Some microorganisms in the fungal and bacterial genera are biological control agents that protect plants from disease damages. Landscape structure and biodiversity are critically important in plant disease epidemiology (Garrett, 2008). In recent years, the general trend toward greater crop homogeneity in most world regions has become a common challenge for plant disease management. This homogeneity makes it easier for plant pathogens adapted to common crop varieties to spread rapidly throughout crop populations (Garrett, 2008). A better understanding of how landscape structure and heterogeneity influence interactions between host, pathogen, and environment may suggest realistic alternatives to reduce the impacts of plant diseases (Cheatham et al., 2009).

Pollination is critical for food production and human livelihoods, and directly links wild ecosystems with agroecosystems (IPBES, 2016). Three-quarters of the crop species used for food depend on insect pollination to some degree (Klein et al., 2007). Wild pollinators can nest in fields or fly from nesting sites in nearby habitats to pollinate crops (Ricketts, 2004). Conserving wild pollinators in habitats within agricultural landscapes improves the level and stability of pollination, leading to increased yields and income (Klein et al., 2003).

Changes in land use can profoundly alter landscape patterns and ecosystem functions, compromising biodiversity and the provision of ecosystem services (Lawler et al., 2014). In most world regions, agronomic intensification has transformed many agricultural landscapes into expansive monocultures with little natural habitat, raising a concern that such landscape simplification may compromise the provision of ecosystem services (Meehan et al., 2011). Managing for biodiversity and ecosystem services in agricultural landscapes requires shifting the scale of investigations and management from the field or farm scale to the landscape scale (Plantegenest et al., 2007).

8.4 ECONOMIC VALUATION OF ECOSYSTEM SERVICES

Fig. 8.1 illustrates important ecosystem services supporting agriculture and the "disservices" or externalities of agriculture that impact society. The biophysical and ecological processes, impacts, and interdependencies within agroecosystems are still being uncovered, but substantial gaps exit in the socioeconomic field to translate relations into measurable "values" (not just economic, but anything that humans care about) for improved decision-making.

The Why and How[1]

When values of ecosystem services are evaluated and incorporated into decision-making, we can make more informed decisions about sustainably managing natural capital for human well-being and livelihoods. The values can guide between management practices, technologies, and policies to maximize social welfare and balance various societal objectives, as well as highlight changes in management that deliver enhanced ecosystem services—specifically those supporting and regulating the ecosystem services that lack markets (Swinton et al., 2015). Equally important is to quantify the impacts of varied

[1] Part of the material in this section draws from Matthews et al. (in press).

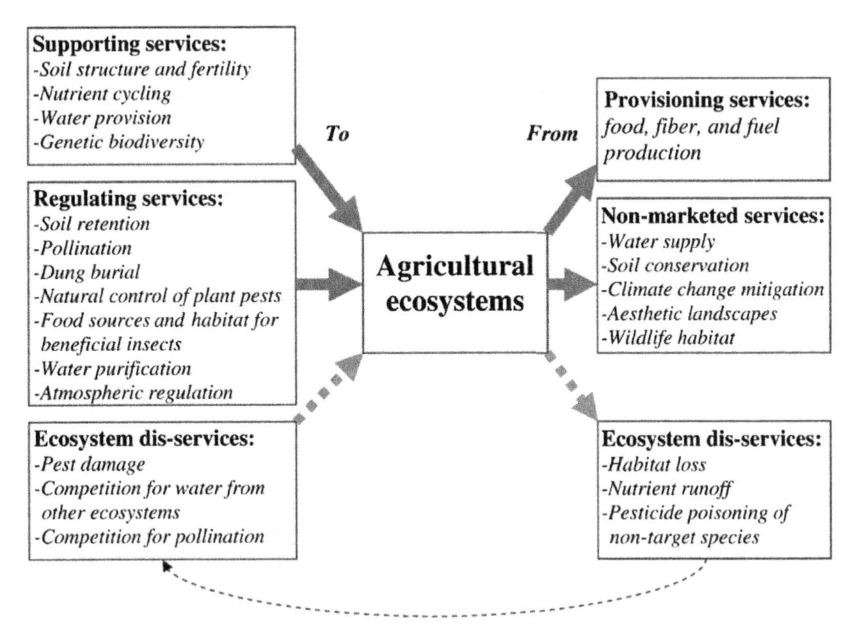

FIGURE 8.1 Ecosystem services and disservices to and from agriculture. Note: Solid arrows indicate services, whereas dashed arrows indicate disservices. Source: *Zhang, W., Ricketts, T.H., Kremen, C., Carney, K., Swinton, S. M., 2007. Ecosystem services and dis-services to agriculture. Ecol. Econ. 64, 253–260.*

production systems on ecosystem services so as to illustrate the long-term sustainability of production systems.

There are many different methods to estimate ecosystem services value (Freeman, 2003; Shiferaw et al., 2005). The methods used for agricultural ecosystem services focus on values that people obtain from use of the services: revealed preference methods capture values revealed by existing markets; production input, profitability trade-off, and stated preference methods estimate the value of changes to the status quo, such as changing current farmer cropping systems to include ecologically recommended practices (Swinton et al., 2015). When ecosystem services can substitute for an existing marketed input or contributes to a measurable marketed output, the economic value of changes in the level of the service can be readily inferred (Drechsel et al., 2005; Freeman, 2003). A widespread application of this factor input valuation method is the fertilizer replacement value to measure the value of biological nutrient cycling, for example, in cereal-legume systems (Bundy et al., 1993). Zhang and Swinton (2012) used a dynamic optimal pest control model to estimate the economic value of natural pest regulation ecosystem services for US soybeans. Other attempts at placing economic values on pest regulation ecosystem services estimated the total cost of averted pest damage due to all pest control practices and then attributed a fraction of the total to natural enemies (Losey and Vaughan, 2006; Pimentel et al., 1997).

New and innovative valuation methods attribute economic values to a suite of ecosystem services at the field, regional, and global level using a "bottom-up" approach. This approach involves a biophysical assessement of ecosystem services using field experiments, followed by an economic valuation using the direct market, avoiding cost and replacement cost methods (Ghaley et al., 2015; Porter et al., 2009; Sandhu et al., 2008, 2010; 2015). This field-based experimental approach can better reflect the economic value of all ecosystem services for decision-making and sustainability at the farm and landscape level.

An important caveat is that the economic values are only useful to the extent that they can capture the revealed or stated market and nonmarket benefits of ecosystem services with monetary metrics. Monetarization is not feasible for many important contributions of ecosystem services (e.g., social equity) that warrant consideration. A further concern is that the diversity of values of nature and its contributions to people's quality of life are associated with different cultural and institutional contexts (Brondizio et al., 2009) and are difficult to compare on the same yardstick (Gómez-Baggethun et al., 2016). This makes it necessary to expand the way society recognizes the diversity of values that need to be promoted in decision-making and to embrace pluralistic valuation approaches as adopted by the IPBES (Pascual et al., 2017).

Global Cost of Land Degradation[2]

Land degradation—the long-term loss of ecosystem services (MEA, 2005) —is a global problem, negatively affecting the livelihoods and food security of billions of people. A careful evaluation of the costs and benefits of action versus costs of inaction against land degradation is needed to support intensified efforts, more investments, and strengthend policy commitment. A Total Economic Value approach to determine the costs of land degradation goes beyond the conventional market values of only crop and livestock products lost to land degradation, and captures all major terrestrial losses of ecosystem services. Using Total Economic Value, land degradation is estimated to affect 30% of the total global land area, and around 3 billion people reside on degraded lands. Annual costs of land degradation due to land use and land cover change are about US$231 billion, or about 0.41% of the global GDP of US$56.5 trillion in 2007, of which sub-Saharan Africa accounts for the largest share of 26%.

Local tangible losses (mainly provisioning services) account only for 46% of the total cost of land degradation, while the remainder of the cost is due to the losses of biodiversity and ecosystem services, accruable largely to beneficiaries other than the local land users. This suggests that the global community, as a whole, incurs larger losses than the local communities experiencing land degradation.

The cost of soil fertility mining due to land degrading practices on maize, rice, and wheat is estimated at US$15 billion per year, or 0.07% of the global GDP, revealing the high cost of land degradation for the production of major food crops of the world. Returns to investment in action against land degradation are twice as large as the cost of inaction in the first 6 years alone. Moreover, with a 30-year planning horizon, the returns are US$5

[2] Material discussed in the section draws from Nkonya et al. (2016).

dollars for each dollar invested. This provides a strong incentive for taking action against land degradation.

True Cost Accounting for Food Production

Current metrics for agricultural performance do not account for the various environmental and social benefits and costs of food production (TEEB, 2015; Sukhdev et al., 2016). These benefits and costs are often termed as "externalities" (Tilman et al., 2001, 2002). Social capital and externalities associated with agricultural systems involve the well-being of farmers, farming families, and farm workers, and extend beyond the farmgate to rural communities and health of all animal and human beings. The role of natural capital and environmental externalities in agriculture are recognized through various ecosystem services (including disservices; Zhang et al., 2007). Different production systems either suppress social and natural capital or enhance them (Sandhu et al., 2008, 2010, 2012, 2015; Swinton et al., 2007). However, in the absence of a monetary value or quantifiable information on the positive and negative externalities associated with alternative production systems, it is impossible to account for them in farm or national accounts (TEEB, 2015). At the level of farm economy, true cost accounting can be one way to understand the full benefits and costs of food production. The framework for valuation of ecosystem services in agriculture (Fig. 8.2) captures the economic value of positive and negative externalities, grouped into production benefits, social benefits, environmental benefits, and environmental and social costs (Sandhu et al., 2015, 2016).

An application of the above method to monetize all positive and negative externalities associated with a representative corn/soybean farm in the United States is provided in Box 8.1 (Sandhu, 2016). Although data on most key externalities are included in the example, limited information regarding the public health impacts exists apart from the social cost of carbon, pesticide poisoning, and food safety. Further research is required to include monetary values of the impacts of pesticides, antibiotic resistance, and the risks of human and animal diseases. Also needed is quantifiable information on social equity, labor rights, corporate farming, and its impact on rural community while capturing the full benefits and costs of food production systems.

8.5 CONCLUSIONS

To enhance the benefits to all from biodiversity and ecosystem services, a system-based paradigm shift that both respects planetary boundary and enhances equitable human prosperity is needed. The paradigm is captured by the "planetary doughnut" framework, which depicts environmental boundaries as limits to be maintained (outer circle) and social foundations to be raised (inner circle) to secure a "safe and just operating space for humanity" (Dearing et al., 2014; Raworth, 2017). Biodiversity and ecosystem services bridge these two domains by providing a means of managing environmental processes for human well-being (DeClerck et al., 2016). The SDGs provide an opportunity to improve human well-being while conserving natural resources, but myriad challenges remain with

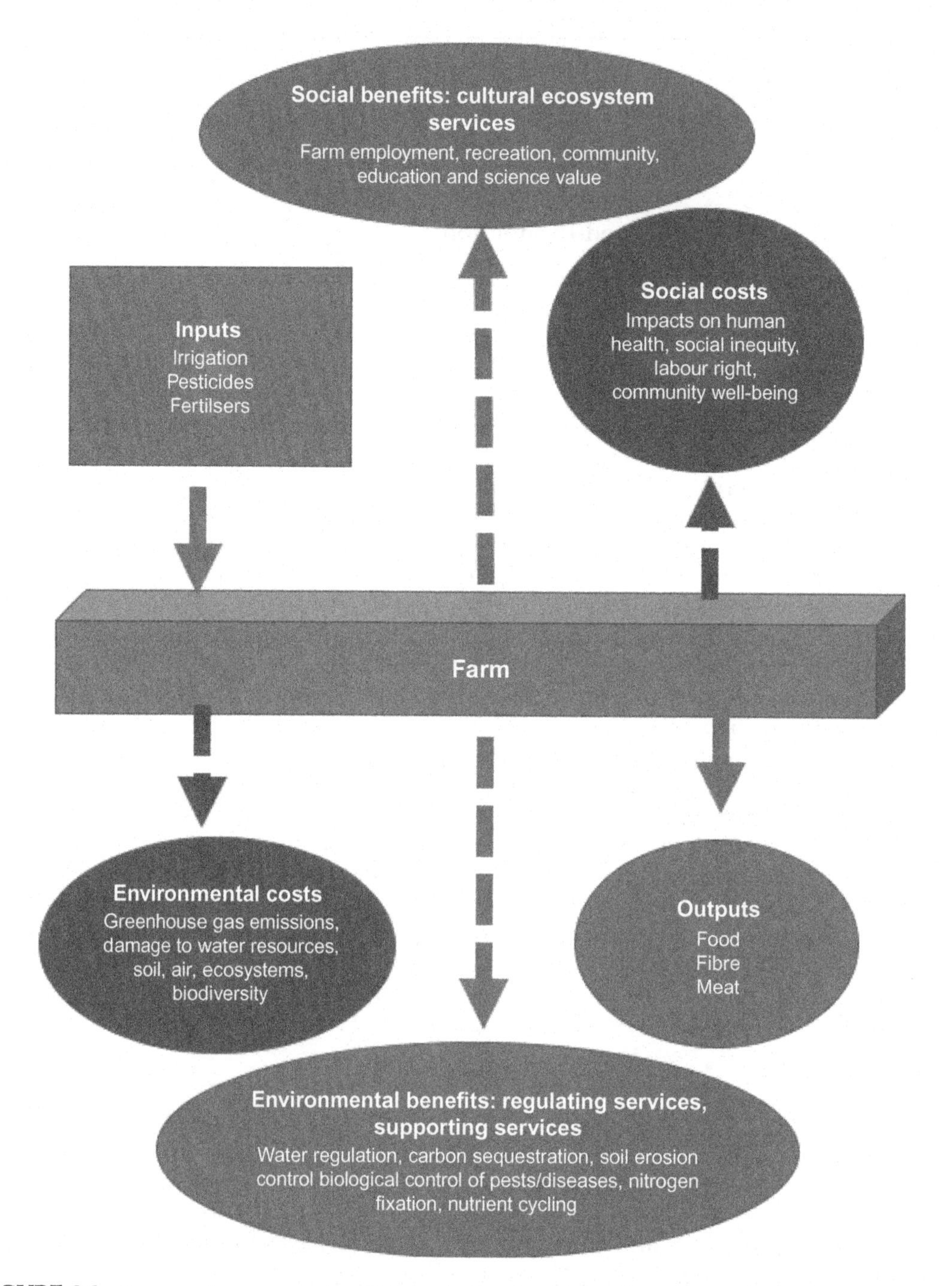

FIGURE 8.2 Framework to assess externalities at the farm level. Note: Environmental benefits comprise regulating and supporting ecosystem services, whereas social benefits include cultural ecosystem services. Environmental and social costs comprise damage to environment and human health. Arrows indicate the flow of inputs and outputs (production benefits) from the farm. Broken arrows indicate those "invisible" benefits and costs that are not accounted for in current food production systems. Source: *Sandhu H., 2016. The Future of Food and Agriculture: Quantifying the Social and Environmental Benefits and Costs of Different Production Systems. Sustainable Food Trust, UK.*

BOX 8.1

ASSESSMENT OF EXTERNALITIES IN A CORN AND SOYBEAN FARM

A family farm with 310 hectares of arable land in Minnesota, United States, produces corn and soybean in rotation using conventional farming practices with strip tillage (http://sustainablefoodtrust.org/wp-content/uploads/2013/04/The-Future-of-Food-and-Agriculture-web.pdf). Benefits and costs were estimated using true cost accounting. The total production value of corn was US$2,183 and soybean US$1,704 per hectare per year. The environmental benefits per hectare amounted to US$220 per year, which includes water regulation, annual carbon sequestration by trees on the farm, and the value of nitrogen fixation by soybean. The farm generates social benefits in terms of employment, worth US$331 per hectare per year. Environmental costs (US $541 per hectare per year) involve greenhouse gas emissions from the use of seed, strip tillage, fertilizers, pesticides, external costs associated with damage to human and environmental health, and transportation, fuel and electricity used to dry corn.

Benefits and costs of the corn and soybean farm (percent per hectare per year)

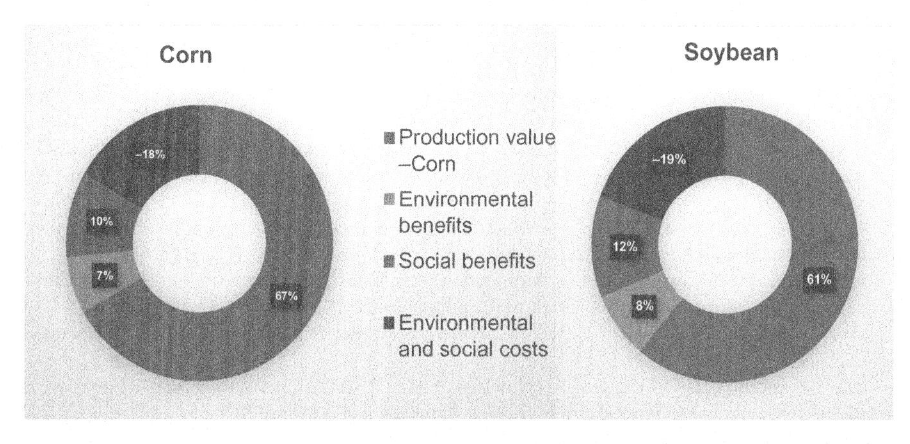

Source: Sandhu (2016).

respect to developing measurable metrics to monitor and evaluate trade-offs and outcomes. Economic valuation plays an important role in informing decision-makers about economic and environmental trade-offs when managing for multiple objectives. However, current metrics for agricultural performance do not account for the full costs or benefits from the perspective of a whole system (Sukhdev et al., 2016). A comprehensive evaluation that goes beyond monetary value and economic efficiency is needed to account for social values such as equity as well as long-term sustainability and resilience (TEEB, 2015).

Biodiversity conservation and ecosystem services management require a systems approach, scale-sensitive solutions, and collective effort. To ensure that the global food supply remains environmentally sustainable and generates a rich array of nutrients for human health, farm landscapes must be diverse and serve multiple purposes (Herrero et al., 2017). Experience has shown that effective management requires community engagement. Rapidly changing and uncertain environmental and social changes in communities require redoubling of efforts to create sustainable socioecological systems that benefit people and nature. Community-based natural resource management programs are one such scheme, but they need to overcome barriers to collaboration by creating governance structures that promote participatory decision-making (Musavengane and Simatele, 2016). There is also a great need to design effective payment for ecosystem services programs that will enhance incentives to sustainably manage natural resources. As diminishing genetic diversity across farms and decreasing land size are compounded by absolute poverty in rural places (Fanzo, 2017), key challenges lie ahead, but the benefit of actions will be monumental.

References

Aberoumand, A., 2009. Nutritional evaluation of edible *Portulaca oleraciaas* plant food. Food Analyt. Meth. 2, 204–207.

Allen, L.H., 2008. To what extent can food-based approaches improve micronutrient status? Asia Pacific J. Clin. Nutr. 17, 103–105.

Arimond, M., Ruel, M.T., 2004. Dietary diversity is associated with child nutritional status: evidence from 11 demographic and health surveys. J Nutr. 134 (10), 2579–2585.

Bellon, M.R., Ntandou-Bouzitou, G.D., Caracciolo, F., 2016. On-farm diversity and market participation are positively associated with dietary diversity of rural mothers in Southern Benin, West AfricaJ. van Wouwe, ed. PLoS ONE 11 (9), e0162535.

Bianchi, F.J.J.A., Booij, C.J.H., Tscharntke, T., 2006. Sustainable pest regulation in agricultural landscapes: a review on landscape composition, biodiversity and natural pest control. Proc. R. Soc. B 273, 1715–1727.

Brondizio, E.S., Ostrom, E., Young, O.R., 2009. Connectivity and the governance of multilevel social-ecological systems: the role of social capital. Annu. Rev. Environ. Resour. 34, 253–278.

Bundy, L.G., Andraski, T.W., Wolkowski, R.P., 1993. Nitrogen credits in soybean-corn crop sequences on three soils. Agron. J. 85, 1061–1067.

Carvalheiro, L.G., Seymour, C.L., Veldtman, R., Nicolson, S.W., 2010. Pollination services decline with distance from natural habitat even in biodiversity-rich areas. J. Appl. Ecol. 47, 810–820.

Castañeda-Álvarez, N.P., Khoury, C.K., Achicanoy, H.A., Bernau, V., Dempewolf, H., Eastwood, R.J., et al., 2016. Global conservation priorities for crop wild relatives. Nat. Plants 2, 16022.

Cheatham, M.R., Rouse, M.N., Esker, P.D., Ignacio, S., Pradel, W., Raymundo, R., et al., 2009. Beyond yield: plant disease in the context of ecosystem services. Phytopathology 99 (11), 1228–1236.

Convention on Biological Diversity (CBD). 2010. The Strategic Plan for Biodiversity 2011-2020 and the Aichi Biodiversity Targets. Decision adopted by the Conference of the Parties (COP) to the Convention on Biological Diversity at its tenth meeting.

Convention on Biological Diversity (CBD) and World Health Organization (WHO), 2015. Connecting Global Priorities: Biodiversity and Human Health. A State of Knowledge Review. Geneva.

Daily, G.C., Kareiva, P.M., Polasky, S., Ricketts, T.H., Tallis, H., 2011. Mainstreaming natural capital into decisions: theory and practice of mapping ecosystem services. In: Kareiva, P., Tallis, H., Ricketts, T.H., Daily, G.C., Polasky, S. (Eds.), Theory and Practice of Mapping Ecosystem Services. Oxford University Press, New York.

Dasgupta, P., 2010. Nature's role in sustaining economic development. Philos. Trans. R. Soc. B 365, 5–11.

Dearing, J.A., Wang, R., Zhang, K., Dyke, J.G., Haberl, H., Hossain, M.S., et al., 2014. Safe and just operating spaces for regional social-ecological systems. Glob. Environ. Change Hum. Policy Dimensions 28, 227–238.

DeClerck, F.A.J., Jones, S., Attwood, S., Bossio, D., Girvetz, E., Chaplin-Kramer, B., et al., 2016. Agricultural ecosystems and their services: the vanguard of sustainability? Curr. Opin. Environ. Sustain. 23, 92–99.
DeFries, R., Fanzo, J., Remans, R., Palm, C., Wood, S., Anderman, T.L., 2015. Metrics for land-scarce agriculture. Science 349, 238–240.
Díaz, S., Demissew, S., Carabias, J., Joly, C., Lonsdale, M., Ash, N., et al., 2015. The IPBES conceptual framework-connecting nature and people. Curr. Opin. Environ. Sustain. 14, 1–16.
Drechsel, P., Giordano, M., Enters, T., 2005. Valuing soil fertility change: selected methods and case studies. In: Shiferaw, B., Freeman, H.A., Swinton, S.M. (Eds.), Natural Resource Management in Agriculture: Methods for Assessing Economic and Environmental Impacts. CABI Publishing, Wallingford, UK.
Dulloo, M.E., Hunter, D., Borelli, T., 2010. *Ex situ*and *in situ* conservation of agricultural biodiversity: major advances and research needs. Notulae Botanicae Horti Agrobotanici Cluj-Napoca 38 (2), 123–153.
Dulloo, M.E., Rege, J.E.O., Ramirez, M., Drucker, A.G., Padulosi, S., Maxted, N., et al., 2017. Chapter 5: Conserving agricultural biodiversity for use in sustainable food systems. Mainstreaming Agrobiodiversity in Sustainable Food Systems: Scientific Foundations for an Agrobiodiversity Index. Bioversity International, Rome, Italy.
Englberger, L., Schierle, J., Aalbersberg, W., Hofmann, P., Humphries, J., Huang, A., et al., 2006. Carotenoid and vitamin content of Karat and other Micronesian banana cultivars. Int. J. Food Sci. Nutr. 57 (5–6), 399–418.
Fanzo, J., 2017. From big to small-the significance of smallholder farms in the global food system. Lancet Planet. Health 1 (1), e15–e16.
FAO, 1999. Women: Users, Preservers and Managers Of Agrobiodiversity. Rome, Italy.
FAO, 2010. The Second Report on the State of The World's Plant Genetic Resources for Food and Agriculture, Rome, Italy.
FAO, 2016. United Nations Decade of Action on Nutrition 2016–2025. Rome, Italy.
Fowler, C., Mooney, P., 1990. Shattering: Food, Politics, and the Loss of Genetic Diversity. The University of Arizona Press, Tucson, Arizona, USA.
Freeman III., A.M., 2003. The Measurement of Environmental and Resource Values: Theory and Methods, Second ed. Resources for the Future, Washington, DC, USA.
Garrett, K., 2008. Climate change and plant disease risk. In: Relman, D.A., Hamburg, M.A., Choffnes, E.R., Mack, A. (Eds.), Global Climate Change and Extreme Weather Events: Understanding the contributions to Infectious Disease Emergence: Workshop Summary. National Academy of Science, Washington D.C, Rapporteurs, Forum on Global Health.
Ghaley, B., Porter, J., Sandhu, H., 2015. Relationship between C:N/C:O stoichiometry and ecosystem services in managed production systems. PLoS One 10 (4), e0123869.
Global Panel on Agriculture and Food Systems for Nutrition. 2016. Food Systems and Diets: Facing the Challenges of the 21st Century, London, UK.
Gómez-Baggethun, E., Barton, D., Berry, P., Dunford, R., Harrison, P., 2016. Concepts and methods in ecosystem services valuation. In: Potschin, M., Haines-Young, R., Fish, R., Turner, R.K. (Eds.), Routledge Handbook of Ecosystem Services. Routledge, London and New York.
Gonzalez Fischer, C., Garnett, T. 2016. Plates, pyramids, planets. Developments in national healthy and sustainable dietary guidelin a state of play assessment. FAO/FCRN. <www.fao.org/3/a-i5640e.pdf>.
Herrero, M., Thornton, P.K., Power, B., Bogard, J.R., Remans, R., Fritz, S., et al., 2017. Farming and the geography of nutrient production for human use: a transdisciplinary analysis. Lancet Planet Health 1, e33–42.
Howard, P. (Ed.), 2003. Women and Plants: Gender Relations in Biodiversity Management and Conservation. Zed Press, and Palgrave Macmillan, London, and New York.
Hunter, D., Heywood, V. (Eds.), 2011. Crop Wild Relatives. A Manual of In Situ Conservation. Earthscan, London.
Ickowitz, A., Powell, B., Salim, M.A., Sunderland, T.C.H., 2014. Dietary quality and tree cover in Africa. Glob. Environ. Change 24, 287–294.
Ickowitz, A., Rowland, D., Powell, B., Salim, M.A., Sunderland, T., 2016. Forests, trees, and micronutrient-rich food consumption in Indonesia. PloS One 11 (5), e0154139.
ICSU and ISSC, 2015. Review of the Sustainable Development Goals: The Science Perspective. International Council for Science (ICSU), Paris. Available from: https://www.icsu.org/cms/2017/05/SDG-Report.pdf.

IFPRI, 2016. 2016 Global Nutrition Report – From Promise to Impact: Ending Malnutrition by 2030, Washington, DC.

IFPRI, 2011. Agriculture, Nutrition, Health: Exploiting the Links, Washington DC.

IPBES, 2016. Summary for policymakers of the assessment report of the Intergovernmental science-policy Platform on Biodiversity and Ecosystem services (IPBES) on pollinators, pollination and food production. Intergovernmental Science-Policy Platform on Biodiversity and Ecosystem Services, Bonn, Germany.

Jacobs, D.R., Tapsell, L.C., 2007. Food, not nutrients, is the fundamental unit in nutrition. Nutr. Rev. 65 (10), 439–450.

Jarvis, D.I., Brown, A.H.D., Cuong, P.H., Collado-Panduro, L., Latournerie-Moreno, L., Gyawali, S., et al., 2008. A global perspective of the richness and evenness of traditional crop-variety diversity maintained by farming communities. Proc. Natl. Acad. Sci. 105 (14), 5326–5331.

Jarvis, D.I., Hodgkin, T., Sthapit, B.R., Fadda, C., Lopez-Noriega, I., 2011. An heuristic framework for identifying multiple ways of supporting the conservation and use of traditional crop varieties within the agricultural production system. Crit. Rev. Plant Sci. 30 (1–2), 125–176.

Jarvis, D.I., Hodgkin, T., Brown, A.H.D., Tuxill, J., Lopez-Noriega, I., Smale, M., et al., 2016. Crop Genetic Diversity in the Field and on the Farm: Principles and Applications in Research Practices. Yale University Press, New Haven and London.

Jones, A. 2016. What matters most for cultivating healthy diets: agricultural diversification or market integration? Conference on Agri-Health Resarch ANH Academy Week, Addis Ababa, Ethiopia, 22 June 2016.

Keller, G.B., Mndiga, H., Maass, B.L., 2006. Diversity and genetic erosion of traditional vegetables in Tanzania from the farmer's point of view. Plant Genet. Resour. 3, 400–413.

Khoury, C.K., Bjorkman, A.D., Dempewolf, H., Ramirez-Villegas, J., Guarino, L., Jarvis, A., et al., 2014. Increasing homogeneity in global food supplies and the implications for food security. Proc. Natl. Acad. Sci. 111 (11), 4001–4006.

Khoury, C.K., Achicanoy, H.A., Bjorkman, A.D., Navarro-Racines, C., Guarino, L., Flores-Palacios, X., et al., 2016. Origins of food crops connect countries worldwide. Proc. R. Soc. B 283, 20160792.

Klein, A.M., Steffan-Dewenter, I., Tscharntke, T., 2003. Pollination of *Coffea canephora* in relation to local and regional agroforestry management. J. Appl. Ecol. 40, 837–845.

Klein, A.M., Vaissiere, B.E., Cane, J.H., Steffan-Dewenter, I., Cunningham, S.A., Kremen, C., et al., 2007. Importance of pollinators in changing landscapes for world crops. Proc. R. Soc. B 274, 303–313.

Landis, D.A., Wratten, S.D., Gurr, G.M., 2000. Habitatmanagement to conserve natural enemies of arthropod pests in agriculture. Annu. Rev. Entomol. 45, 175–201.

Landscapes for People, Food and Nature Initiative, 2015. Landscape Partnerships for Sustainable Development: Achieving the SDGs through Integrated Landscape Management - A White Paper to discuss the benefits of using ILM as a key means of implementation of the Sustainable Development Goals.

Lawler, J.J., Lewis, D.J., Nelson, E., Plantinga, A.J., Polasky, S., Withey, J.C., et al., 2014. Projected land-use change impacts on ecosystem services in the United States. Proc. Natl. Acad. Sci. 111 (20), 7492–7497.

Losey, J.E., Vaughan, M., 2006. The economic value of ecological services providedby insects. Bioscience 56, 311e323.

Mäder, P., Fliessbach, A., Dubois, D., Gunst, L., Fried, P., Niggli, U., 2002. Soil fertility and biodiversity in organic farming. Science 296, 1694–1697.

Matthews, N., Zhang, W., Bell, A. in press. Ecosystems at the Nexus, Chapter 4 in The Food-Energy-Water Nexus, P. Saundry editor, Springer, New York.

Meehan, T.D., Werling, B.P., Landis, D.,A., Gratton, C., 2011. Agricultural landscape simplification and insecticide use in the Midwestern United States. Proc. Natl. Acad. Sci. 108 (28), 11500–11505.

Millennium Ecosystem Assessment (MEA), 2005. Ecosystems and Human Well-Being: Synthesis. Island Press, Washington, DC.

Musavengane, R., Simatele, D., 2016. Significance of social capital in collaborative management of natural resources in Sub-Saharan African rural communities: a qualitative meta-analysis. S. Afr. Geogr. J. 99 (3), 267–282.

Naylor, R., Ehrlich, P., 1997. Natural pest control services and agriculture. In: Daily, G. (Ed.), Nature's Services: Societal Dependence on Natural Ecosystems. Washington DC.

Nkonya, E., Anderson, W., Kato, E., Koo, J., Mirzabaev, A., von Braun, J., et al., 2016. Global cost of land degradation. In: Nkonya, E., Mirzabaev, A., von Braun, J. (Eds.), Economics of Land Degradation and Improvement: A Global Assessment for Sustainable Development. Springer.

Pascual, U., Balvanera, P., Díaz, S., Pataki, G., Roth, E., Stenseke, M., et al., 2017. Valuing nature's contributions to people: the IPBES approach. Curr. Opin. Environ. Sustain. 26, 7–16.

Pellegrini, L., Tasciotti, L., 2014. Crop diversification, dietary diversity and agricultural income: empirical evidence from eight developing countries. Can. J. Dev. Stud. 35 (2), 211–227.

Pimentel, D., Wilson, C., McCullum, C., Huang, R., Dwen, P., Flack, J., et al., 1997. Economic and environmental benefits of biodiversity. BioScience 47, 47e757.

Plantegenest, M., Le May, C., Fabre, F., 2007. Landscape epidemiology of plant diseases. J. R. Soc. Interface 4, 963–972.

Porter, J., Costanza, R., Sandhu, H., Sigsgaard, L., Wratten, S., 2009. The value of producing food, energy and ecosystem services within an agro-ecosystem. Ambio 38, 186–193.

Powell, B., Thilsted, S.H., Ickowitz, A., Termote, C., Sunderland, T., Herforth, A., 2015. Improving diets with wild and cultivated biodiversity from across the landscape. Food Secur. 7 (3), 535–554.

Price waterhouse Coopers (PwC). 2013. Crop wild relatives-A valuable resource for crop development. PwC, London. <http://pwc.blogs.com/files/pwc-seed-bank-analysis-for-msb-0713.pdf>.

Raudsepp-Hearne, C., Peterson, G.D., Bennett, E.M., 2010. Ecosystem service bundles for analyzing tradeoffs in diverse landscapes. Proc. Natl. Acad. Sci. 107, 5242–5247.

Raworth, K., 2017. Doughnut Economics: Seven Ways to Think Like a 21st-Century Economist. Chelsea Green Publishing.

Ricketts, T.H., 2004. Tropical forest fragments enhance pollinator activity in nearby coffee crops. Conserv. Biol. 18, 1–10.

Rockström, J., Steffen, W., Noone, L., Persson, A., Chapin, F.S., Lambin, E.F., et al., 2009. A safe operating space for humanity. Nature 461, 472–475.

Rowland, D., Ickowitz, A., Powell, B., Nasi, R., Sunderland, T., 2017. Forest foods and healthy diets: quantifying the contributions. Environ. Conserv. 44 (2), 102–114.

Ruel, M., 2003. Operationalizing dietary diversity: a review of measurement issues and research priorities. J. Nutr. 133, 3911S–3926S.

Ruel, M., Minot, N., Smith L. 2005. Patterns and determinants of fruit and vegetable consumption in sub-Saharan Africa: a multicountry comparison. Background paper for the Joint FAO/WHO Workshop on Fruit and Vegetables for Health, 1–3 September 2004, Kobe, Japan.

Samberg, L.H., Gerber, J.S., Ramankutty, N., Herrero, M., West, P.C., 2016. Subnational distribution of average farm size and smallholder contributions to global food production. Environ. Res. Lett. 11, 124010.

Sandhu, H.S., Wratten, S.D., Cullen, R., Case, B., 2008. The future of farming: the value of ecosystem services in conventional and organic arable land. An experimental approach. Ecol. Econ. 64, 835–848.

Sandhu, H.S., Wratten, S.D., Cullen, R., 2010. The role of supporting ecosystem services in arable farmland. Ecol. Complex. 7, 302–310.

Sandhu, H., Crossman, N., Smith, F., 2012. Ecosystem services and Australian agricultural enterprises. Ecol. Econ. 74, 19–26.

Sandhu, H., Sandhu, S., 2014. Linking ecosystem services with the constituents of human well-being for poverty alleviation in eastern Himalayas. Ecol. Econ. 107, 65–75.

Sandhu, H., Sandhu, S., 2015. Poverty, development, and Himalayan ecosystems. Ambio 44, 297–307.

Sandhu, H., Wratten, S., Costanza, R., Pretty, J., Porter, J., Reganold, J., 2015. Significance and value of non-traded ecosystem services on farmland. PEERJ 3, e762.

Sandhu H. 2016. The Future of Food and Agriculture: Quantifying the Social and Environmental Benefits and Costs of Different Production Systems. Sustainable Food Trust, UK.

Sandhu, H.S., Wratten, S.D., Porter, J.R., Costanza, R., Pretty, J., Reganold, J., 2016. Mainstreaming ecosystem services into future farming. Solutions 7 (2), 40–47.

Scaling Up Nutrition (SUN). 2016. Empowering women and girls to improve nutrition: building sisterhood of success. The SUN Movement Secretariat. <https://scalingupnutrition.org/wp-content/uploads/2016/05/IN-PRACTICE-BRIEF-6-EMPOWERING-WOMEN-AND-GIRLS-TO-IMPROVE-NUTITION-BUILDING-A-SISTERHOOD-OF-SUCCESS.pdf>.

Shiferaw, B., Freeman, H.A., Navrud, S., 2005. Valuation methods and approaches for assessing natural resource management impacts. In: Shiferaw, B., Freeman, H.A., Swinton, S.M. (Eds.), Natural Resource Management in Agriculture: Methods for Assessing Economic and Environmental Impacts. CABI Publishing, Wallingford, UK.

Steffen, W., Richardson, K., Rockström, J., Cornell, S.E., Fetzer, I., Bennett, E., et al., 2015. Planetary boundaries: guiding human development on a changing planet. Science 347, 736–746.

Shrestha, P., Subedi, A., Sthapit, S., Rijal, D., Gupta, S., Sthapit, B., 2006. Community seed bank: a reliable and effective option for agricultural biodiversity conservation. In: Sthapit, B.R., Shrestha, P., Upadhyay, M.P. (Eds.), On-Farm Management of Agricultural Biodiversity in Nepal: Good Practices. NARC/LI-BIRD/Bioversity International, Nepal.

Sthapit, B., Rana, R., Eyzaguirre, P., Jarvis, D., 2008. The value of plant genetic diversity to resource-poor farmers in Nepal and Vietnam. Int. J. Agric. Sustain. 6 (2), 148–166. Available from: https://doi.org/10.3763/ijas.2007.0291.

Sthapit, B., Lamers, H.A.H., Rao, V.R., Bailey, A., 2016. Community biodiversity management as an approach for realizing on-farm management of agricultural biodiversity. In: Sthapit, B.R., Lamers, H.A.H., Rao, V.R., Bailey, A. (Eds.), Tropical Fruit Tree Diversity: Good Practices for In Situ and On-Farm Conservation. Routledge, Abingdon, Oxon UK.

Subedi, A., Shrestha, P., Upadhyay, M., Sthapit, B., 2013. The evolution of community biodiversity management as a methodology for implementing in situ conservation of agrobiodiversity in Nepal Participation in a global programme: initial steps. In: de Boef, W.S., Subedi, A., Peroni, N., Thijssen, M., O'Keeffe, E. (Eds.), Community Biodiversity Management: Promoting Resilience and the Conservation of Plant Genetic Resources. Routledge, Abidon, UK, 2013.

Sukhdev, P., May, P., Müller, A., 2016. Fix food metrics. Nature 540, 33–34.

Swinton, S.M., Lupi, F., Robertson, G.P., Hamilton, S.K., 2007. Ecosystem services and agriculture: cultivating agricultural ecosystems for diverse benefits. Ecol. Econ. 64, 245–252.

Swinton, S.M., Jolejole-Foreman, C.B., Lupi, F., Ma, S., Zhang, W., Chen, H., 2015. Economic value of ecosystem services from agriculture. In: Hamilton, S.K., Doll, J.E., Robertson, G.P. (Eds.), The Ecology of Agricultural Landscapes: Long-Term Research on the Path to Sustainability. Oxford University Press, New York, New York, USA, Chapter 3.

The Economics of Ecosystems, and Biodiversity (TEEB). 2015. TEEB for Agriculture, and Food: an interim report, United Nations Environment Programme, Geneva, Switzerland.

Thormann, I., Dulloo, M.E., Engels, J.M.M., 2006. Techniques of ex situ plant conservation. In: Henry, R. (Ed.), Plant Conservation Genetics. Lismore, Australia: Centre for Plant Conservation Genetics. Southern Cross University, The Haworth Press.

Tilman, D., 1999. Global environmental impacts of agricultural expansion: the need for sustainable and efficient practices. Proc. Natl. Acad. Sci. 96, 5995–6000.

Tilman, D., Fargione, J., Wolff, B., D'Antonio, C., Dobson, A., Howarth, R., et al., 2001. Forecasting agriculturally driven global environmental change. Science 292, 281–284.

Tilman, D., Cassman, G., Matson, P.A., Naylor, R., Polasky, S., 2002. Agricultural sustainability and intensive production practices. Nature 418, 671–677.

Tscharntke, T., Karp, D.S., Chaplin-Kramer, R., Batáry, P., DeClerck, F., Gratton, C., et al., 2016. When natural habitat fails to enhance biological pest control: five hypotheses. Biol. Conserv. 204, 449–458.

van Zanten, B.T., Verburg, P.H., Koetse, M.J., van Beukering, P.J., 2014. Preferences for European agrarian landscapes: a meta-analysis of case studies. Landsc. Urban Plan. 132, 89–101.

Vernooy, R., Sthapit, B., Galluzzi, G., Shrestha, P., 2014. The multiple functions and services of community seedbanks. Resources 3 (4), 636–656.

Vernooy, R., Sthrestha, P., Sthapit, B. (Eds.), 2015. Community Seed Banks: Origins, Evolution and Prospects. Earthscan Routledge, London, UK.

Wilkinson, T.K., Landis, D.A., 2005. Habitat diversification in biological control: the role of plant resources. In: Wackers, F.L., van Rijn, P.C.J., Bruin, J. (Eds.), Plant Provided Food and Plant–Carnivore Mutualism. Cambridge University Press, Cambridge, UK.

Wood, S.L., DeClerck, F., 2015. Ecosystems and human well-being in the sustainable development goals. Front. Ecol. Environ. 13, 123.

Zhang, W., Ricketts, T.H., Kremen, C., Carney, K., Swinton, S.M., 2007. Ecosystem services and dis-services to agriculture. Ecol. Econ. 64, 253–260.

Zhang, W., Swinton, S.M., 2012. Optimal control of soybean aphid in the presence of natural enemies and the implied value of their ecosystem services. J. Environ. Manage. 96 (1), 7–16.

CHAPTER

9

Changing Food Systems: Implications for Food Security and Nutrition

Hanh Nguyen, Jamie Morrison and David Neven

Food and Agriculture Organization of the United Nations (FAO), Rome, Italy

9.1 WHAT ARE THE ISSUES?

Food systems determine the way in which food is produced and consumed. They are also key to supporting the economic livelihoods of the majority of the world's population, important contributors to social welfare, and can significantly affect the use of natural resources. The future development of food systems will be central both to the degree and way in which several of the 17 Sustainable Development Goals (SDGs) will be achieved, that is, whether hunger can be eliminated, health and well-being improved, more responsible production and consumption fostered, decent work and employment ensured, and environmental stewardship promoted. While developments in food systems over the past decades have generated many positive outcomes in these respects, the associated rapid structural transformations, driven by significant changes in consumption patterns, the growing importance of global supply chains, and an increasing complexity of food systems, have also resulted in increasing and significant challenges with potentially wide--reaching consequences for the future of food security and nutrition.

This chapter argues that a better understanding of how food systems function is critical to ensuring that these systems develop in ways that minimize negative impacts and maximize positive contributions. The complexity of food systems requires going beyond the traditional disciplinary view toward a more holistic and coordinated approach. Systems thinking necessitates a perspective that considers all activities related to food production, aggregation, processing, distribution, and consumption, as well as the interests of and

Sustainable Food and Agriculture
DOI: https://doi.org/10.1016/B978-0-12-812134-4.00009-1

interactions between all actors involved and the socioeconomic and ecological contexts in which food systems are embedded.

The chapter begins with Section 9.2, which provides a definition of food systems and explores the linkages between food system developments and food security and nutrition, outlining some of the current concerns about how these systems are developing. Section 9.3 considers the historical evolution of food systems, which have been the main drivers at different stages, and the implications these developments have had for food security conditions and nutritional status of different groups in society. Section 9.4 discusses the possible trajectories food system developments could take in the coming decades and suggests a greater role for systems thinking in addressing future challenges. Finally, Section 9.5 explores possible actions that can be taken by different actors to shape the development of food systems toward more inclusive and sustainable pathways.

9.2 DEFINING FOOD SYSTEMS AND SUSTAINABLE FOOD SYSTEMS

Meadows (2009) emphasizes that a system has three essential aspects: elements, interconnections, and an overall function or purpose. Rather than functioning as the mere sum of its parts, a system is a complex whole that consists of various related elements. The relationships and interactions between these elements define the characteristics of the system and cause it to exhibit certain behaviors, which contribute to its purpose. While two systems can have the same set of elements, how those elements come together to serve their underlying purposes will determine their different attributes and outcomes.

Similarly, a food system comprises a number of elements that are interconnected in various ways, with the overall objective of providing food for people. The individual is the most basic element of a society. We are all part of the food system as consumers. In addition, some of us obtain incomes from interconnected activities in the food system, including production, aggregation, processing, and distribution. Value can be added or lost at each stage along the chain (FAO, 2014a). As employees or shareholders, individuals in the food system upstream from consumption are typically part of an agroenterprise (including farms) or an enterprise providing physical inputs or support services. Alternatively, they may be part of government or civil society organizations that regulate or facilitate these activities. Firms are connected to each other and to final consumers through their respective business models and via various market systems. Value chains link all the enterprises necessary to move food products from production to retailing. All of these elements and their interconnections operate within an enabling environment, comprising natural elements, such as soil, air, water, and biodiversity, which can impact the process of producing and transforming food products and societal elements, such as regulations and policies, sociocultural norms, infrastructure, and interprofessional organizations.

The system's characteristics and performance in economic, social, and environmental terms are determined through the way these various elements and actors interact with each other. Those interactions and hence food systems differ from context to context. However, in most contexts traditional, modern and postmodern supply channels coexist. These channels are shaped by the way in which actors respond to changes in consumer demand: private sector enterprises in taking advantage of new opportunities opened up by changing demand, and public sector institutions aiming to meet societal objectives. Food systems also overlap and interact with other systems and subsystems, such as energy systems, economic systems, health systems, political systems, and trade systems.

According to the High Level Panel of Experts on Food Security and Nutrition (HLPE, 2014), a food system is considered sustainable if it delivers food security and nutrition for all in such a way that the economic, social and environmental bases to generate food security and nutrition for future generations are not compromised. Therefore, the achievement of food security and nutrition for current and future generations becomes the central objective of the food system. In practice, however, and perhaps counterintuitively, food system developments have not been led with that objective in mind. It is therefore important to understand how food system developments impact food security and nutrition outcomes.

Food security outcomes are widely identified along four dimensions: food availability, food access, food utilization, and the stability of the other three dimensions over time. Nutritional outcomes are determined not only by food security, but also by other factors related to care and feeding and health and sanitation (FAO et al., 2017; Ghattas, 2014). In linking production to consumption, food systems significantly impact the availability, accessibility, affordability, and desirability of diverse, safe, and nutritious food, and consequently the food environment in which consumers make their food choices (FAO, 2013). The growth and modernization of food systems has yielded many positive results. Over the past three decades or so in developing countries, this generated many new off-farm employment opportunities as food industries developed (Reardon and Timmer, 2012). Food systems growth has also broadened food choices through supplying other items beyond traditional staple foods, as well as improving the quality and safety of food (FAO, 2017a). However, rapid transformations coupled with the complexity of the systems are creating new challenges and concerns. These include the high calorie and low nutritional value of many food items that are now widely available and consumed, limited access of small-scale producers to viable markets, high levels of food loss and waste, increased incidences of food safety, animal and human health risks, and increased energy intensity and ecological footprints associated with the lengthening of food supply chains.

9.3 HOW HAVE FOOD SYSTEMS CHANGED AND WHAT ARE THE IMPLICATIONS?

To understand the drivers of food system transformations and their implications for food security and nutrition outcomes, as well as to identify risks, opportunities, and leverage points, it is useful to review how food systems have evolved over time. Food has always been central to human civilization. From hunting and gathering to farming, from animal domestication techniques to modern processing and storing technologies, agriculture and food systems have radically transformed the way societies are fed and organized. Although the evolution of food systems has spanned thousands of years, the speed and extent to which food systems have changed over the past half century is unprecedented.

Current Trends

Most of these recent changes are the result of interlinked trends and forces that have driven food systems dynamics over time. The world is becoming increasingly globalized, urban, and crowded. The world population has grown from 1.6 billion people in 1900 to 7.6 billion in 2017, and is estimated to reach 9.8 billion in 2050 (UNDESA, 2017). Over the course of a century, urban populations have increased from 2% of the total world's population in 1900 to roughly 50% today (UNDESA, 2014). Developing countries will account for the majority of the future population growth in urban areas, especially in secondary cities. Even in countries with large rural populations, up to 70% of the food markets cater to rapidly expanding urban food demand (FAO, 2017a). Urban food markets are becoming the main outlets for farmers in Africa and Asia. To illustrate, for the last 30 years the volume of food flowing from rural to urban areas in Africa increased by 800%, while the increase in Southeast Asia was almost 1000% (Reardon and Zilberman, 2016).

Next to rapid urban growth, many low- and middle-income countries have also witnessed strong economic expansions in recent decades. This has given rise to the notion of a "global middle class," whose size is projected to increase almost threefold between 2009 and 2030 (OECD, 2012). With greater disposable income and increased exposure to imported foods and large-scale retailers, this global middle class is adopting radically different lifestyles and acquiring new food preferences. At the same time, although overall poverty has declined, the percentage of poor people living in urban areas is growing (Ravallion et al., 2007). While the food demand of the middle class in many regions is increasingly being met through global supply chains and large-scale distribution systems, the urban poor still rely on informal traditional markets as their primary food supply channel (FAO, 2013).

Rapid technological innovation, especially in information and communication technologies and renewable energy, has been another major driving force of food system changes. The judicious use of these technologies, both on-farm and off-farm, have shaped and will continue to shape the productivity and competitiveness of food systems. Finally, climate change and resource scarcities pose significant risks and can threaten the productive capacity and stability of food systems. In turn, food systems have contributed significantly to greenhouse gas emissions, at around 29% of global emissions (Vermeulen et al., 2012). They are also contributing to the degradation of natural resources. Key statistics attribute 60% of terrestrial biodiversity loss to food production (PBL Netherlands Environmental Assessment Agency, 2014); categorize over 70% of the global marine fish stocks as "significantly depleted," "overexploited," or "fully exploited" (FAO, 2014c); and 33% of soils as moderately to highly degraded (FAO and ITPS, 2015).

Evolution of Food Systems: A Typology

How did we get to this point? The world's food systems have undergone notable transitions over the past half century, from traditional to industrial, and latterly, new "alternative" systems have emerged in some parts of the world. The evolution of food systems is illustrated in Fig. 9.1, which provides a simplified typology of the main food system types corresponding to historical stages of development and builds on

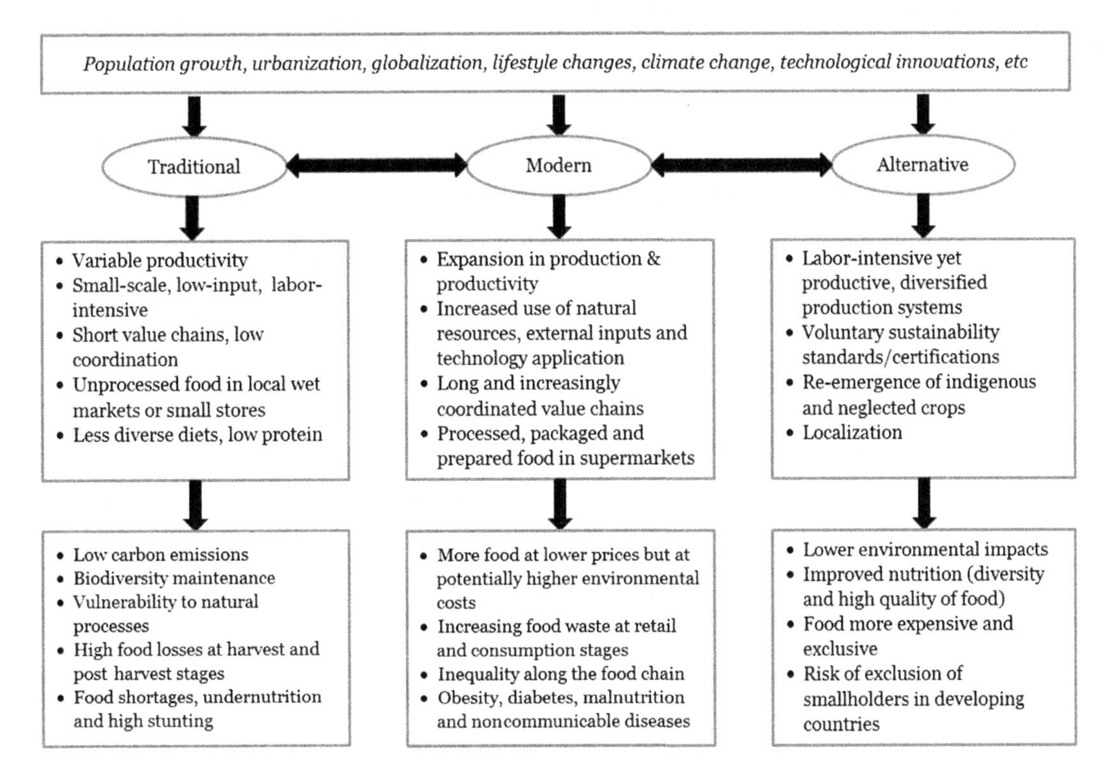

FIGURE 9.1 Evolution of food systems—a typology. Source: *Adapted from Ericksen P.J., 2008. Conceptualizing food systems for global environmental change research. Glob. Environ. Change 18, 234–245.*

previous efforts by, among others, Ericksen (2008), Reardon and Timmer (2012), and UNEP (2016). It does not attempt to capture the whole spectrum of food systems across regions, nor does it seek to promote one particular type of system over the others. Recently, food systems in many developing countries have evolved rapidly from traditional toward industrial types (FAO, 2013), while sustainability concerns and consumer preferences for ethical and ecological foods have driven a parallel move from industrial to alternative systems (Goodman et al., 2012). Food systems in most countries and regions of the world tend to combine different elements of the three types. For example, the majority of consumers in Asia still purchase unprocessed or slightly processed food from independent small shops and traditional wet markets, but growing wealth and food safety concerns have driven a drastic increase in food purchases from supermarkets. The processing and retailing sectors in this region, notably China, Indonesia, Malaysia, Thailand, India, and Viet Nam, are also undergoing substantial consolidation (Reardon et al., 2012).

The typology helps in the analysis of patterns of food systems development, such as the transition from farm-based labor-intensive systems to high capital and external input systems in many parts of the world (UNEP, 2016). These shifts are not linear and complete, and there are many intermediate situations in between, while some distinct elements of one system type can be present in another. They constantly adapt to and compete with each other. It is important to recognize that not all components of the food system have changed at the same rate or in the same direction. Food systems have evolved in a manner in which traditional, industrial, and high-end and alternative components often coexist.

Traditional Food Systems

Traditional food systems comprise short supply chains, along which food is produced and consumed within a local area or a country. Farmers and fisherfolk in traditional food systems follow well-established techniques and generally do not make extensive use of external inputs, with production units being small-scale and labor-intensive (Ericksen, 2008). The crops and livestock on one farm may be diverse because of their low level of specialization. Productivity may be variable, but often at the lower end of the spectrum compared with systems with higher levels of intensification. Food is either used for direct consumption, or is sold unprocessed or slightly processed through spot market transactions. It is then distributed to consumers through local wet markets and small independent stores (Reardon and Timmer, 2012). Traditional food systems are characterized by a low degree of coordination.

Although traditional food systems continue to play a key role in ensuring availability and access to food, particularly for the rural poor, they are subject to production shocks that can lead to localized food shortages, thus compromising food security. In many contexts, these systems have not been able to adjust to increasing pressures from population growth and may be unable to adequately meet today's standards for quality and safety for healthy and nutritious diets. These issues can be further compounded by food losses, especially at the harvest and postharvest stages (FAO, 2017b), which are caused by a lack of processing and preservation technologies and storage facilities.

Modern Industrial Food Systems

Modern or industrial food systems comprise long supply chains that span national borders. Contrary to traditional food systems that are labor-intensive, industrial food systems rely on capital to expand the scale of production, processing, and distribution through increased use of external inputs (e.g., fertilizers, pesticides, machinery) and the application of modern technologies (UNEP, 2016). Monoculture, including monocropping and high stocking density of farm animals of particular breeds, is widely practiced. Next to a high degree of specialization, industrial food systems are also characterized by close coordination and increasing vertical integration of different functions along the value chain (Neven et al., 2009; FAO, 2014b). Many commodities are grown almost exclusively for export. In consequence, their conditions for production and exchange often need to comply with contracts and standards set by global buyers. The growing power of big retailers and international traders, in turn, drives the restructuring of the procurement systems and influences the structure and practices further upstream (Reardon et al., 2003). Consumers in these systems are subject to food marketing and advertising placed by big brands, and tend to purchase processed, packaged, and prepared food products (FAO, 2013).

Modern food systems have contributed substantially to food security thanks to higher productivity, compared with traditional systems, and low prices of food products. Over the past few decades, economies of scale coupled with competition among big agribusinesses and retailers have reduced food prices, directly improving food availability and accessibility. Increased productivity can also free up on-farm labor and facilitate the transition of rural employment to nonagricultural sectors (FAO, 2017a). Provided that these sectors offer higher incomes for households to purchase food, food security can be enhanced indirectly. Yet, these improvements are made potentially at the expense of the natural environment, which can negatively affect the stability of future food supply. Greenhouse gas emissions from "agriculture, forestry, and other land use" have almost doubled within the past half century and are projected to grow by 2050 (FAO, 2014b). Expansion of monoculture can lead to land erosion and biodiversity loss. Furthermore, excessive use of external inputs can exhaust soil fertility, contaminate waterways, and reduce air quality. Environmental concerns do not lie exclusively at the production stage, but are relevant for postfarm activities as well. Food transportation, processing, and retailing in modern systems involve many water- and fossil fuel-dependent facilities, contributing appreciably to greenhouse gas emissions (FAO, 2016). High levels of food loss and waste, especially at the retail and consumption stages, also represent an enormous waste of natural resources. It is estimated that about one-third of all the food produced globally is lost or wasted each year (FAO, 2017b). For instance, cosmetic standards imposed by big supermarkets on their suppliers can result in the disposal of perfectly safe and edible food. In the United Kingdom, for example, up to 40% of fruit and vegetables produced by its farmers are wasted, as supermarkets reject those foods because of their "ugly" appearance (for more information, see www.jamieoliver.com/news-and-features/features/reclaiming-wonky-veg).

The modernization of food systems has also given rise to serious concerns about the nutritional value of food. The "nutrition transition" (Popkin, 1999) from traditional to industrial systems has witnessed a shift from starchy, low-fat diets to diets that are higher in animal protein, saturated fat, and refined sugars. Increased consumption and overconsumption of nutrient-poor processed foods, in general, contribute to accelerating obesity, micronutrient deficiency, and other degenerative diseases such as diabetes and heart diseases. Overweight and obesity are linked to more deaths than underweight (WHO, 2017). This transition has happened in industrialized and middle-income countries and is increasingly occurring in developing countries as well (Popkin et al., 2012). Furthermore, it is estimated that the increasing consumption of animal source foods in developing countries will require large increases in livestock production and its intensive use of resources (FAO, 2016). Additionally, food safety and animal and human health issues arise; for example, the excessive use of antibiotics fed to industrially farmed animals, which has been found to lead to drug-resistant strains of bacteria that could be passed to other animals, farmers, general consumers, and the environment (NHS, 2015). Intensive animal farming for mass-produced animal protein, apart from the growing concerns over animal welfare, is conducive to the spread of diseases, which is also of great concern to human health. There are also other issues, including the chemicals from packaging that can leach into food; although these minute amounts are individually benign, the effects of long-term exposure are not well understood (Muncke et al., 2014).

The development from traditional to modern food systems, furthermore, has had mixed impacts on rural transformation and inclusiveness, and thus indirectly on food security and nutrition. On the one hand, modern food systems can enhance farmers' access to viable markets. Farmers who enter into contract with companies enjoy formal employment opportunities, often alongside technical assistance to improve the efficiency of the production process and the quality of their produce. On the other hand, the concentration of market power in the hands of a few large-scale processors, wholesalers, and retailers (supermarkets) that constantly seek to reduce production and transaction costs exerts huge pressures on small suppliers (Neven et al., 2009). Such development can work to the disadvantage of smallholders, who find it difficult to meet big buyers' requirements for product uniformity, consistency, and regular supply (Timmer, 2014).

Emerging Alternative Food Systems

Recently, in high-income countries and in different parts of Asia and Latin America, alternative food systems have emerged, often in response to the impacts of industrial food systems on consumer health and the environment. Alternative systems consist of diversified farming systems, and include organic farming, conservation farming, biodynamic agriculture, and permaculture, which can draw on certain elements of traditional practices (FAO, 2016). These systems involve the limited use of external inputs and are labor-intensive yet remunerative (due to higher product value). They make use of many varieties of crops and edible wild plants. Consumers can purchase organic and whole foods in specialized independent stores, health-food retail outlets, and specialized departments in supermarket chains.

There are different variants of alternative food systems. The global variants work within the framework of global value chains and seek to address sustainability and fairness issues of industrial food systems. Agroecological and ethical markets that make use of voluntary standards and third-party certifications (notably organic and fair trade) related to sustainability, quality, and traceability belong to this strand. Coordination between different functions of the value chains is high, as consumers want to know more about the origin, production, and procurement processes of their food. The local variants of alternative food systems comprise short value chains and the consumption of "slow" local foods (Kilometre Zero, family farms, box delivery schemes, urban agriculture). They emphasize the direct connection of local producers and consumers, advocate for strong farmer organizations, and avoid a fully commercial orientation (DuPuis and Goodman, 2005). Adding to this evolution of food systems, indigenous food systems that promote the so far "neglected" crops, such as chia, quinoa, and argan, are gaining prominence and influencing mainstream markets (FAO, 2016).

Alternative food systems that follow sustainable practices in production, processing, and distribution may become an increasing force in reducing carbon emissions and mitigating other environmentally degrading effects associated with intensive farming and the lengthening of the supply chains. Such systems may also enhance crop diversity and thus complement industrial food systems that rely on a few staple crops, increasing the availability of more nutritious and diverse foods. Nevertheless, accessibility does not automatically follow. As agroecological food products are still sold at high premiums, their consumers mostly belong to niche and high-end market strata. Furthermore, while agroecological markets offer potentially higher returns for participating producers, smallholders in developing countries can easily be excluded from participation in differentiated value chains because of the high costs of documentation and standards compliance (Dauvergne and Lister, 2012). Powerful downstream buyers can also dictate the formulation and implementation of private sustainability standards to be followed by smallholders (Clapp and Fuchs, 2009). As a result, alternative food systems can also affect food security and nutrition negatively through market and income pathways.

9.4 POSSIBLE FUTURE PATHWAYS FOR FOOD SYSTEMS AND THE NEED FOR A FOOD SYSTEMS APPROACH

The evolution of food systems has been driven by a combination of changing consumer demands and preferences, latterly increasingly expressed through growing consumer concerns about health and sustainability, and the agenda of social movements, among others. Changes have been shaped further through business actions and investments, which seek to take advantage of the opportunities created by shifting consumer demands, and through government interventions generally seeking to ensure that societal objectives are not compromised.

Although it is impossible to predict how these trends, actions, and reactions will combine to determine the future evolution of food systems in specific contexts, some insights into the potential roles of the different categories of actors can be gained by considering possible evolutions in each element and their implications for food security and nutrition.

Changing Consumer Preferences

At an aggregate level, there have been significant changes in consumption patterns over the past decades, with a shift toward diets higher in sugar, salt, and fat content, an increase in the consumption of animal products, and a greater reliance on processed foods, particularly in the more advanced developing countries. These trends are also now observable in other developing regions as incomes increase.

However, there have been some incipient countermovements from small groups of consumers in developed countries, as well as groups of urban middle-income consumers in developing countries, who have shifted preferences toward organic, natural, slow, and ethical foods. The counter-reaction has come as a response to a mix of concerns about the ways in which food is produced (environmental sustainability, animal welfare, food safety, etc.) and over the poor nutritional value of many food items. Indications are that this countermovement is likely to grow in the coming decades.

Changing Behavior of Private Sector Actors

Aggregate level trends in food consumption in many parts of the world are likely to drive further consolidation as private sector operators strive to deliver higher quantities of food to growing and increasingly urbanized populations, particularly in developing countries. These mainstream food systems are focused on efficiency improvement and commercial viability and tend to be characterized by deeper globalization, more consolidation, more vertical integration, and ever-tighter margins.

However, as mentioned, consumer concerns about the origin, safety, and nutritional value of foods have led to the emergence and widening of agroecological markets, ethical markets, as well as indigenous and local food chains and urban agriculture. Currently, an estimated 800 million people around the world are engaged in urban agriculture, 200 million of whom are doing so as a commercial activity, creating an estimated 150 million full-time jobs (Zezza and Tasciotti, 2010). Various forms of social movements focused on localism and small-scale, organic agriculture have also emerged and gained considerable ground. Food Sovereignty, Slow Food, Community Supported Agriculture, and Fair Trade have expanded their social base on the grounds of democracy, ecology, and quality, and in turn engendered niche, differentiated food supply chains (DuPuis and Goodman, 2005). These developments can potentially lead to more positive food security and nutrition outcomes. They can also lower greenhouse gas emissions commonly associated with long food chains and place less burden on natural resources. However, there is a risk that access to food might be compromised as food becomes more expensive and thus exclusive.

Changing priorities of public sector regulators

As noted in Section 9.2, there have been several unintended consequences of food system development that are causing increased concerns among policymakers. These include the growing incidence of malnutrition and the associated burden of noncommunicable diseases, increased inequality of access for both producers and consumers, concerns related to food safety and transboundary diseases, and about the resilience of food systems to the impacts of climate change. These concerns are likely to provide a focus for the types of interventions that the public sector makes in relation to food systems over the coming years.

While the difficulty of understanding how these different elements might interact has been elaborated above, there have been attempts to think through possible future scenarios for food systems. One recent attempt is the scenario analysis by the World Economic Forum's System Initiative on Shaping the Future of Food Security and Agriculture (2017). The analysis helpfully sets out four contrasting scenarios as to how food systems could evolve by 2030 (see Box 9.1), explaining the trends, actions and, more often than not, inactions that result in these scenarios.

BOX 9.1

WORLD ECONOMIC FORUM SCENARIOS

The World Economic Forum report defines four potential future worlds:

- Survival of the richest, in which resource-intensive consumption coexists with disconnected markets, resulting in a sluggish global economy and stark divisions between the "haves" and "have-nots."
- Unchecked consumption, in which resource-intensive consumption associated with strong market connectivity results in high gross domestic product growth to the detriment of the environment.
- Open-source sustainability, where resource-efficient consumption in the context of highly connected markets contributes to increased international cooperation and innovation, but with some groups in society left behind.
- Local is the new global, in which resource-efficient consumption and fragmented local markets combine to deliver local foods for resource-rich countries, but hunger risks for import-dependent regions.

Source: WEF (2017).

Though the world already witnesses the symptoms of each of these scenarios, the first two predominate. Whether those will continue to be principal features of future food systems is hard to predict, but it is rather unlikely that the desirable elements of the "open-source sustainability" and "local is the new global" scenarios will prevail unless major efforts are made in that direction.

The scenarios also clearly point to the many trade-offs that will need to be addressed in the pursuit of more sustainable food systems and in identifying actions that are required to ensure improved food security and nutrition outcomes. In the coming decades, stakeholders and decision-makers will need to confront key competing priorities of the food systems: poverty reduction, increased agricultural productivity, improved nutrition, and environmental sustainability. Crucial questions about the future of food systems have emerged: How do we shape the development of food systems that meet the needs of our changing world without harming the environment? How do we provide nutritious and sustainably produced food to a world population projected to reach almost 10 billion by 2050?

The Need for an Integrated Food Systems Approach in Guiding Future Actions

Addressing these challenges requires integrated actions committed by all stakeholders across multiple fronts—not just agriculture, but also health, environment, gender, trade, education, and so on. Traditional food security programs have focused on increasing food supply through agricultural productivity growth. Today, however, inadequate food production is a main cause of food insecurity in only a few parts of the world.

The dramatic pace of food system changes over the past decades has brought about complex interactions and feedback loops that impact food security and nutrition in many different ways. Direct interventions in one area of the system risk creating or exacerbating problems in another. Therefore, a holistic, system-wide view is essential. "Silo" thinking likely will exacerbate rather than address food system challenges. A food systems approach considers the food system in its totality, taking into account all the elements of a food system, its relationships, and related effects (Ingram, 2011). It is not confined to one single sector or discipline, and thus broadens the framing and analysis of a particular issue as the result of an intricate web of interlinked activities. By encouraging us to see the bigger picture, a systems approach helps integrate analyses of the full set of food system activities encompassing production, food security, and nutrition outcomes, as well as ensures multistakeholder collaboration to address future challenges. While there will clearly be trade-offs to be made, opportunities will also emerge to accomplish multiple objectives. An integrated systems approach can help in the identification of such synergies and further the coordination needed to achieve them.

9.5 WHAT ACTIONS NEED TO BE TAKEN TO ENSURE IMPROVED FOOD SECURITY AND NUTRITION OUTCOMES AND BY WHOM?

It is clear that a systems approach to the development of more sustainable, inclusive, and resilient food systems will require a broader, yet more targeted and coordinated set of

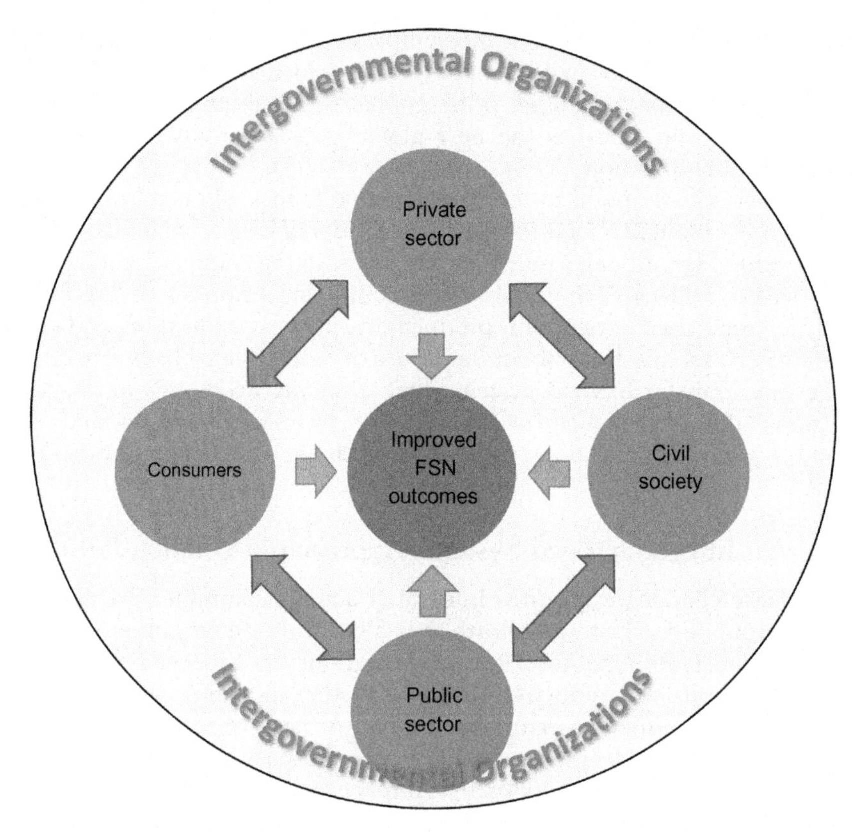

FIGURE 9.2 Improved food security and nutrition (FSN) as the collective outcome of the interrelated actions by main food system actors. Source: *Author's own elaboration.*

actions, implemented by a wider range of actors (Fig. 9.2). Rather than incremental changes with localized impacts, there is a need for systemic changes that have impacts at scale (WEF, 2017). Such an approach will require better performance measurement tools and analyses that focus on the drivers of behavior, root causes of problems, leverage points in the system, and solutions that tackle all critical constraints simultaneously.

Public Sector Action

The public sector can push food system developments toward more sustainable outcomes and improved food security and nutrition through a multitude of pathways. It can create a strong supporting enabling environment for businesses that focuses on the production and distribution of sustainable and nutritious foods through fiscal, legal, and policy measures (e.g., territorial investment in infrastructure linked to new urban market food systems or environmental regulations to reflect the real costs of the food system). It can engage in public–private partnerships based on common interests, such as commercial demands that overlap with public goods demands, and direct its institutional food procurement strategies to stimulate local development, especially targeting small farms and firms. School feeding programs can be promoted based on both innovative institutional procurement and education strategies. Governments can also influence consumer demands through new food labeling requirements and consumer education and awareness programs.

For the development of all of these pathways, collaborations between various government ministries are essential. The strategies of ministries related to agriculture, forestry and fisheries, health, environment, education, trade, finance, infrastructure, rural development, and so on, are often contradictory and thus can be aligned to create incentives for sustainable food systems development. Intersectoral collaborations can highlight trade-offs that may need to be made between different goals, such as between more diverse and nutritious food supplies and reduction of ecological footprints. A food systems approach, in essence, would necessitate the development of an overarching food policy that informs the development of related sectoral policies, for example, farm policies.

Civil Society Action

Civil society organizations (nongovernmental organizations, consumer groups, religious communities, etc.) and community leaders, including those with many followers linked through social networks, can play a key role in awareness-raising to foster greater demand for nutritious and socially and environmentally responsible foods. Such organizations and individuals wield more political and market power than individual consumers, and are typically important cultural influencers to change consumers' perception of desirable foods and eating habits. They can also put pressure on businesses and governments to ensure that the social and environmental impacts of food production, processing, and distribution are factored strongly in their food security and nutrition-related agendas.

Consumer Action

Consumers are central actors in the food systems, as—by voting with their "money"—they determine which channels in the food system will grow and which ones will shrink. This consumer choice is influenced by a great number of elements that make up the food environment; these elements include, among others, the range and nature of products available to consumers along with information about these products to which they are exposed, such as the marketing campaigns of firms and retailers, public sector education for healthy diets, and information provided through consumer associations, magazines, and other sources. Within the set of food products accessible to them, consumers will furthermore make choices based on their preferences in terms of taste, convenience, appearance, household needs, cultural traditions, and so on.

Consumer action in support of more sustainable food systems can thus take on two interrelated forms: (1) consumers can choose to buy food that is more nutritious, socially responsible, or greener when these types of food are accessible, affordable, and sufficiently appealing to them; and (2) they can join consumer associations, networks, or forums that expose unsustainable food products, or advocate for measures from the public or private sector side to promote more sustainable food products.

Private Sector Action

The private sector, from farm to retail outlet, is the main supplier of foods and as such is directly responsible for actions that determine the economic, social, and environmental impacts of how food flows to the consumer. These impacts include not only the direct impact on food availability in markets, but also the indirect impact on the broader set of food products that households can access through income pathways (salaries, farms, and small business incomes). Private sector behavior is driven by a complex set of factors, including their need to compete in the marketplace, the regulations, policies, and laws they must adhere to, the needs and wants of consumers, the intricacies of their collaborative networks, and their internal risk management and social responsibility strategies. They have to balance short-term financial objectives with long-term sustainability goals. While changing consumer wants and needs constantly creates new market opportunities, including ones for more sustainable food products, these markets are often small initially and risky to establish a foothold. At the same time, there are win-win strategies that could improve firm competitiveness and profitability, generate positive social and environmental impacts, and improve food security and nutrition in a society (e.g., improved technologies that reduce food loss). Within the private sector, systemic solutions to food system challenges will require new partnership formats, both vertically (e.g., contract farming, inclusive business models) and horizontally (e.g., farmer cooperatives, industry associations).

Catalytic Role of Intergovernmental Organizations

Intergovernmental organizations and platforms, such as the United Nations Committee on World Food Security (CFS, 2016), can play a key catalytic role in driving more sustainable food systems. Through studies and analyses, these organizations can provide policy advice and knowledge on food system transformation and its impacts on food security and nutrition. Furthermore, they offer technical support and capacity development linked to investment and policy processes that affect food system developments. They are also neutral actors that can facilitate the interministerial, multistakeholder collaborations needed to move beyond traditional "silo" structures and toward multidisciplinary and systemic approaches.

Coordinated Action Through a Food Systems Approach

Taking a food systems approach enables integrated responses and better-aligned programs through collaboration between different groups of actors and across sectors. It also facilitates the move from short-term to long-term visions and from reactive to proactive policies. Many food system development efforts often focus on a particular issue within the food system, such as developing and providing inputs for farmers, improving market access, or enhancing nutrition for consumers. Solutions to these issues are often not direct, and the solution base is not restricted to a certain area of focus. In many cases, there will be multiple binding constraints that need to be addressed simultaneously to create changes and make an impact. To illustrate, nutrition outcomes are not solely improved through interventions like food fortification or nutrition education, but will also depend on other factors, such as the conditions determining incomes and allowing people to purchase more nutritious foods, or available infrastructure to facilitate adequate storage of food and preservation of nutrients. Effective solutions are never merely technical and should consider political and sociocultural dimensions. Greater gender equality, for instance, has been found to contribute positively to nutrition, as women tend to direct more of the resources they manage toward food, education, health, and care than men do (FAO, 2017c).

By looking at the system as a whole, a food systems approach can foster more effective cooperation between different sectors and disciplines to create synergies and balance trade-offs. This calls for integrated interventions in multiple sectors involved in the development of sustainable food systems, such as agriculture, trade, health, and education rather than a series of single interventions within one sector. Embracing a food systems approach will therefore require commitment and action from all interested parties, regionally, nationally, and internationally.

References

Clapp, J., Fuchs, D.A., 2009. Corporate Power in Global Agrifood Governance. The MIT Press, Cambridge, US.

Committee on World Food Security (CFS). 2016. Inclusive value chains for sustainable agriculture and scaled up food security and nutrition outcomes. Forty-third session, Rome. Also available at <http://www.fao.org/3/a-mr587e.pdf>.

Dauvergne, P., Lister, J., 2012. Big brand sustainability: governance prospects and environmental limits. Glob. Environ. Change 22 (1), 36–45.

DuPuis, E.M., Goodman, D., 2005. Should we go "home" to eat? Toward a reflexive politics of localism. J. Rural Stud. 21 (3), 359–371.

Ericksen, P.J., 2008. Conceptualizing food systems for global environmental change research. Glob. Environ. Change 18, 234–245.

FAO, 2013. The State of Food and Agriculture 2013. Food systems for better nutrition. Rome. 114pp. Also available at <http://www.fao.org/docrep/018/i3300e/i3300e00.htm>.

FAO, 2014a. Developing Sustainable Food Value Chains. Guiding Principles. Rome. 75pp. Also available at <http://www.fao.org/3/a-i3953e.pdf>.

FAO, 2014b. The State of Food and Agriculture 2014. Innovation in family farming. Rome. 161pp. Also available at <http://www.fao.org/3/a-i4040e.pdf>.

FAO, 2014c. The State of World Fisheries and Aquaculture. Opportunities and Challenges. Rome. 243pp. Also available at <http://www.fao.org/3/a-i3720e.pdf>.

FAO, 2016. The Future of Food and Agriculture. Trends and Challenges. Rome. 180pp. Also available at <http://www.fao.org/3/a-i6583e.pdf>.

FAO, 2017a. The State of Food and Agriculture 2017. Leverage food systems for inclusive rural transformation. Rome. Also available at <http://www.fao.org/3/a-I7658e.pdf>.

FAO, 2017b. Food loss and food waste [online]. Rome. [Cited 29 July 2017]. <http://www.fao.org/food-loss-and-food-waste/en/>.

FAO, 2017c. Nutrition-sensitive agriculture and food systems in practice: options for intervention. Rome. Also available at <http://www.fao.org/3/a-i6983e.pdf>.

FAO, IFAD, UNICEF, WFP, WHO, 2017. The State of Food Security and Nutrition in theWorld 2017. Building resilience for peace and food security. Rome, FAO. 132pp. Also available at <http://www.fao.org/3/a-I7695e.pdf>.

FAO and ITPS, 2015. Status of the World's Soil Resources (SWSR). Main Report by Intergovernmental Technical Panel on Soils (ITPS). Rome, FAO. 648pp. Also available at <http://www.fao.org/3/a-i5199e.pdf>.

Ghattas, H., 2014. Food security and nutrition in the context of the nutrition transition. Technical Paper. Rome, FAO. 21pp. Also available at <http://www.fao.org/3/a-i3862e.pdf>.

Goodman, D., Dupuis, E.M., Goodman, M.K., 2012. Alternative Food Networks: Knowledge, Practice, and Politics. Routledge, New York.

HLPE. 2014. Food losses and waste in the context of sustainable food systems. A report by the High Level Panel of Experts on Food Security and Nutrition of the Committee on World Food Security. HLPE Report No.8. Rome. 117pp. Also available at <http://www.fao.org/3/a-i3901e.pdf>.

Ingram, J., 2011. A food systems approach to researching food security and its interactions with global environmental change. Food Secur. 3 (4), 417–431.

Meadows, D., 2009. Thinking in Systems: A Primer. Earthscan, London.

Muncke, J., Myers, J.P., Scheringer, M., Porta, M., 2014. Food packaging and migration of food contact materials: will epidemiologists rise to the neotoxic challenge? J. Epidemiol. Commun. Health 68 (7), 592–594.

Neven, D., Odera, M.M., Reardon, T., Wang, H., 2009. Kenyan supermarkets, emerging middle-class horticultural farmers, and employment impacts on the rural poor. World Dev. 37 (11), 1802–1811.

NHS. 2015. Antibiotic use in farm animals 'threatens human health'. [online]. London. [Cited 29 July 2017]. <http://www.nhs.uk/news/2015/12December/Pages/Antibiotic-use-in-farm-animals-threatens-human-health.aspx>.

OECD. 2012. An emerging middle class. [online]. Paris. [Cited 25 May 2017]. <http://oecdobserver.org/news/fullstory.php/aid/3681/An_emerging_middle_class.html>.

PBL Netherlands Environmental Assessment Agency. 2014. How sectors can contribute to sustainable use and conservation of biodiversity. CBD Technical Series No. 79. The Hague. Also available at <https://www.cbd.int/doc/publications/cbd-ts-79-en.pdf>.

Popkin, B.M., 1999. Urbanization, lifestyle changes and the nutrition transition. World Dev. 27 (11), 1905–1916.

Popkin, B.M., Adair, L.S., Ng, S.W., 2012. Global nutrition transition and the pandemic of obesity in developing countries. Nutr. Rev. 70 (1), 3–21.

Ravallion, M., Chen, S., Sangraula, P., 2007. New Evidence on the Urbanization of Global Poverty. Policy Research Working Paper No. 4199, World Bank. Washington DC.

Reardon, T., Timmer, P., 2012. The economics of the food system revolution. Ann. Rev. Resour. Econ. 4, 225–264.

Reardon, T., Timmer, P., Minten, B., 2012. Supermarket revolution in Asia and emerging development strategies to include small farmers. Proc. Natl. Acad. Sci. 109 (31), 12332–12337.

Reardon, T., Timmer, P., Barrett, C., Berdegué, J., 2003. The rise of supermarkets in Africa, Asia, and Latin America. Am. J. Agric. Econ. 85 (5), 1140–1146.

Reardon, T., Zilberman, D., 2016. Climate smart food supply chains in developing countries in an era of rapid dual change in agrifood systems and the climate. In: Lipper, L., McCarthy, N., Zilberman, D., Asfaw, S., Branca, G. (Eds.), Climate Smart Agriculture. Building Resilience to Climate Change. FAO, Rome, Natural Resource Management and Policy No. 52.

Timmer, P., 2014. Managing structural transformation: A political economy approach. UNU-WIDER Annual Lecture 18. Helsinki, UNU-WIDER (United Nations University - World Institute for Development Economics Research).

United Nations Environment Programme (UNEP). 2016. Food Systems and Natural Resources. A Report of the Working Group on Food Systems of the International Resource Panel. Nairobi, Kenya. Also available at <http://apps.unep.org/publications/index.php?option = com_pubandtask = downloadandfile = 012067_en>.

United Nations, Department of Economic and Social Affairs (UNDESA), Population Division. 2014. World urbanization prospects: The 2014 Revision. New York. Also available at <https://esa.un.org/unpd/wup/publications/files/wup2014-highlights.Pdf>.

United Nations, Department of Economic and Social Affairs (UNDESA), Population Division. 2017. World Population Prospects: The 2017 Revision, Key Findings and Advance Tables. New York. Also available at <http://www.un.org/en/development/desa/news/population/2015-report.html>.

Vermeulen, S.J., Campbell, B.M., Ingram, J.S., 2012. Climate change and food systems. Annu. Rev. Environ. Resour. 37, 195–222.

World Economic Forum (WEF). 2017. Shaping the future of global food systems: A scenarios analysis. Cologny, Switzerland. Also available at <http://www3.weforum.org/docs/IP/2016/NVA/WEF_FSA_FutureofGlobalFoodSystems.pdf>.

World Health Organization (WHO). 2017. Obesity and Overweight Fact Sheet.[online]. Geneva, Switzerland. Updated October 2017. [Cited29 July 2017]. <http://www.who.int/mediacentre/factsheets/fs311/en/>.

Zezza, A., Tasciotti, L., 2010. Urban agriculture, poverty, and food security: empirical evidence from a sample of developing countries. Food Policy 35 (4), 265–273.

SECTION II

CURRENT APPROACHES TO SUSTAINABLE FOOD AND AGRICULTURE

Lead Authors

Jules Pretty (University of Essex, Colchester, United Kingdom) and Zareen Pervez Bharucha (Anglia Ruskin University, Cambridge, United Kingdom)

Co-authors

Caterina Batello (Food and Agriculture Organization of the United Nations (FAO), Rome, Italy), José Aguilar-Manjarrez (Food and Agriculture Organization of the United Nations (FAO), Rome, Italy), Edith Fernández-Baca (United Nations Development Program (UNDP), Lima, Peru), Cornelia Butler Flora (Kansas State University, Manhattan, KS, United States), Valerio Crespi (Food and Agriculture Organization of the United Nations (FAO), Rome, Italy), Matthias Halwart (Food and Agriculture Organization of the United Nations (FAO), Rome, Italy), Sue Hartley (University of York, York, United Kingdom), Zeyaur Khan (International Centre of Insect Physiology and Ecology (ICIPE), Nairobi, Kenya), Rattan Lal (Ohio State University, Columbus, OH, United States), Wilfrid Legg (Consultant and formerly OECD), Weimin Miao (Food and Agriculture Organization of the United Nations (FAO), Bangkok, Thailand), Charles Midega (International Centre of Insect Physiology and Ecology (ICIPE), Nairobi, Kenya), William Murray (Food and Agriculture Organization of the United Nations (FAO), Rome, Italy), John Pickett (Rothamsted International, United Kingdom), Jimmy Pittchar (International Centre of Insect Physiology and Ecology (ICIPE), Nairobi, Kenya), Ewald Rametsteiner (Food and Agriculture Organization of the United Nations (FAO), Rome, Italy), Fritz Schneider

(Bremgarten, Switzerland), Austin Stankus (Food and Agriculture Organization of the United Nations (FAO), Rome, Italy), Rudresh Kumar Sugam (Formerly Council on Energy, Environment and Water, New Delhi, Delhi, India), Berhe Tekola (Food and Agriculture Organization of the United Nations (FAO), Rome, Italy), Yuelai Lu (University of East Anglia, Norwich, United Kingdom), Monica Petri (Food and Agriculture Organization of the United Nations (FAO), Vientiane, Laos), Sally Bunning (Food and Agriculture Organization of the United Nations (FAO), Santiago, Chile), Cora van Oosten (Centre for Development Innovation, Wageningen University, Wageningen, Netherlands), Jean-Marc Faurès (Food and Agriculture Organization of the United Nations (FAO), Rome, Italy), and Lucy Garret (Food and Agriculture Organization of the United Nations (FAO), Rome, Italy).

CHAPTER

10

Context for Sustainable Intensification of Agriculture

The Food and Agriculture Organization of the United Nations (FAO) recently launched a new vision and approach for promoting sustainable food and agriculture (SFA) that addresses the challenges of transition and defining sustainability as a process rather than as an end point (FAO, 2014a). The SFA approach requires explicit consideration of cross-sectoral (e.g., crops, livestock, fishery, forestry) and multiobjective (e.g., economic, social, and environmental) policy objectives, identifying possible synergies between them, as well as balancing trade-offs. The approach builds on efforts to identify transition pathways adapted to specific agroecological, political, and socioeconomic conditions. Accordingly, the aim is to build the enabling conditions for sustainability, tailored to fit the dimensions of location-specific sustainability processes.

Building on the concepts and practice of sustainable intensification of agricultural systems, the core of the SFA approach is comprised of five supporting principles: (1) improved efficiency of the resources used in food and agriculture; (2) direct action to conserve, protect, and enhance natural resources; (3) protection and improvement of rural livelihoods, equity, and social well-being; (4) enhanced resilience of people, communities, and ecosystems; and (5) responsible and effective governance mechanisms. Since the launch of the SFA approach in 2014, sustainability of food and agriculture has received greater attention at the global level. In 2015, world leaders adopted 17 Sustainable Development Goals (SDGs) to be achieved by 2030, wherein SFA is key to SDG 2 and plays a contributing role in most of the remaining SDGs.

Despite significant recent progress, total food production will need to grow again before the world population stabilizes. The desire for agriculture to produce more food without causing environmental harm, and even making positive contributions to natural and social capital, has been reflected in calls for a wide range of different types of more sustainable agriculture: for a doubly green revolution, alternative agriculture, an evergreen revolution, agroecological intensification, green food systems, greener revolutions, *agriculture durable*, and EverGreen Agriculture.

Sustainable Food and Agriculture
DOI: https://doi.org/10.1016/B978-0-12-812134-4.00010-8

The concept of sustainable intensification is an attempt to explicitly unify the interlinked goals of agricultural productivity and social-ecological well-being. "Intensification" has largely been synonymous with a type of agriculture characterized as causing harm while simultaneously producing food. At the same time, the term "sustainable" was often used to describe all that could be good about agricultural processes. In combination, the terms attempt to indicate that desirable ends (more food, better environment) could be achieved by a variety of means. Sustainable intensification is defined as a process or system where yields are increased without either adverse environmental impact or the cultivation of more land. The concept is thus relatively open, in that it does not articulate or privilege any particular vision of agricultural production. It emphasizes ends rather than means, and does not predetermine technologies, species mix, or particular design components. Sustainable intensification can be distinguished from earlier conceptions of intensification because of its explicit emphasis on a wider set of environmental and social outcomes than solely productivity enhancement.

Since many pathways toward agricultural sustainability exist, no single configuration of technologies, inputs, and ecological management is more likely to be widely applicable than another. Agricultural systems with high levels of social and human assets are able to innovate in the face of uncertainty, and farmer-to-farmer learning has been shown to be particularly important in implementing the context-specific, knowledge-intensive, and regenerative practices of sustainable intensification.

In general, pathways toward sustainability in agroecosystems combine technical interventions, investments, and enabling policies and instruments, and involve a variety of actors, operating at different scales. Successful transitions need to be specific both to national and local contexts and particular scales and time periods. Tracing pathways to intensified practice among rural smallholders in particular shows that households develop an array of different strategies. Thus, even within small communities, there is no "one size fits all." Diversification both within and beyond cropping systems is a common theme, but actual levels and types of intensification vary greatly.

A vital need persists for coherence and integration among agriculture, economic, nutrition, education, and health policies at the national level, and for the improvement of international coordination across these sectors. Pathways can be grounded on very different narratives, each of which drives a selection of options.

CHAPTER

11

Sustaining Capital Assets for Agroecosystems

Agricultural systems are amended ecosystems with a variety of important properties (Table 11.1). Sustainable agroecosystems, by contrast, seek to shift some of these functions back toward natural systems without significantly trading off productivity. In affluent economies, modern agroecosystems have tended toward high through-flow systems, with energy supplied by fossil fuels directed out of the system (either deliberately for harvests or accidentally through side effects). For a transition toward sustainability, renewable sources of energy need to be maximized, and some energy flows directed toward internal trophic interactions (e.g., to soil organic matter or to nonagricultural biodiversity for farmland birds) in order to maintain other ecosystem functions. These properties suggest a role for agroecological redesign of systems in producing both food and environmental assets.

What makes agriculture unique as an economic sector is that it directly affects many of the very assets on which it relies for success. Agricultural systems at all levels rely on the value of services flowing from the total stock of assets that they influence and control, and five types of assets—natural, social, human, physical, and financial capital—are recognized as important (Pretty, 2008).

As agricultural systems shape the very assets on which they rely for inputs, a vital feedback loop occurs from outcomes to inputs. Sustainable agroecosystems are those tending to have a positive impact on natural, social, and human capital, while unsustainable ones feed back to deplete these assets, leaving fewer for the future. The concept of sustainability does not require that all assets be improved at the same time. One agricultural system that contributes more to these assets than another can be said to be more sustainable, but there are still likely to be trade-offs, with one asset increasing as another falls.

As most contemporary, intensive agroecosystems are considerably more simplified than natural ecosystems, some natural properties need to be designed back into systems to decrease losses and improve efficiency. For example, loss of biological diversity (to improve crop and livestock productivity) results in the loss of some ecosystem services, such as pest and disease control. For sustainability, biological diversity needs to be increased both to recreate natural control and regulation functions and to manage pests

Sustainable Food and Agriculture
DOI: https://doi.org/10.1016/B978-0-12-812134-4.00011-X

TABLE 11.1 Properties of Natural Ecosystems Compared with Recent Agroecosystems Typical of Affluent Economies and Sustainable Agroecosystems

Property	Natural Ecosystem	Recent Agroecosystem Typical of Affluent Economies	Sustainable Agroecosystem
Productivity	Medium	High	Medium (possibly) high)
Species diversity	High	Low	Medium
Output stability	Medium	Low-medium	High
Biomass accumulation	High	Low	Medium-high
Nutrient recycling	Closed	Open	Semi-closed
Trophic relationships	Complex	Simple	Intermediate
Natural population regulation	High	Low	Medium-high
Resilience	High	Low	Medium
Dependence on external inputs	Low	High	Medium
Human displacement of ecological processes	Low	High	Low-medium
Sustainability	High	Low	High

Pretty, J. and Bharucha, Z.P., 2018. The Sustainable Intensification of Agriculture. Routledge, London., drawing on Gliessman, S.R., 2005. Agroecology and agroecosystems. In: Pretty J., (Ed.), The Earthscan Reader in Sustainable Agriculture. London: Earthscan.

and diseases rather than seeking to eliminate them. Modern agricultural systems have come to rely on synthetic nutrient inputs obtained from natural sources but requiring high inputs of energy, usually from fossil fuels. These nutrients are often used inefficiently, and result in losses in water and air as nitrate, nitrous oxide, or ammonia. To meet the principles of sustainability, such nutrient losses need to be reduced to a minimum, recycling and feedback mechanisms introduced and strengthened, and nutrients diverted for capital accumulation (Thomson et al., 2012). Mature ecosystems are now known to be in a state of dynamic equilibrium that buffers against large shocks and stresses. Because modern agroecosystems by contrast have weak resilience, transitions toward sustainability will need to focus on structures and functions that improve resilience while meeting the primary goal of producing agricultural commodities.

Ecosystem health is a prerequisite for productive and sustainable agriculture (MEA, 2005; NRC, 2010; Foresight, 2011; NEA, 2011). Agriculture is both a driver and recipient of the impacts of global environmental change. Meeting projected demands for food, fodder, and fiber will necessitate finding ways to navigate the legacy of past environmental degradation while building natural capital. In future, it is likely that many crops will have to be produced under less favorable climatic and economic conditions than those that enabled yield increases during the past century (Glover and Reganold, 2010).

CHAPTER

12

The Term "Sustainable Intensification"

The desire for agriculture to produce more food without environmental harm, or even positive contributions to natural and social capital, has been reflected in calls for a wide range of different types of more sustainable agriculture: for a "doubly green revolution" (Conway, 1997), for "alternative agriculture" (NRC, 1989), for an "evergreen revolution" (Swaminathan, 2000), for "agroecological intensification" (Milder et al., 2012), for "green food systems" (Defra, 2012), for "greener revolutions" (Snapp et al., 2010), and for "EverGreen Agriculture" (Garrity et al., 2010). All center on the proposition that agricultural and uncultivated systems should be considered interdependent. In light of the need for the sector to also contribute directly to the resolution of global social-ecological challenges, there have also been calls for nutrition-sensitive (Thompson and Amoroso, 2011), climate-smart (FAO, 2013), and low-carbon agriculture (Norse, 2012).

Sustainable agricultural systems exhibit a number of key attributes (Pretty, 2008; Royal Society, 2009). They should:

1. utilize crop varieties and livestock breeds with a high ratio of productivity to use of externally and internally derived inputs;
2. avoid the unnecessary use of external inputs;
3. harness agroecological processes, such as nutrient cycling, biological nitrogen fixation, allelopathy, predation, and parasitism;
4. minimize use of technologies or practices that have adverse impacts on the environment and human health;
5. make productive use of human capital in the form of knowledge and capacity to adapt and innovate, and social capital to resolve common landscape-scale or systemwide problems (such as water, pest, or soil management); and
6. minimize the impacts of system management on externalities, such as greenhouse gas emissions, clean water, carbon sequestration, biodiversity, and dispersal of pests, pathogens, and weeds.

Sustainable Food and Agriculture
DOI: https://doi.org/10.1016/B978-0-12-812134-4.00012-1

Agricultural systems emphasizing these principles tend to display a number of broad features that distinguish them from the processes and outcomes of conventional systems. First, these systems tend to be multifunctional within landscapes and economies (Dobbs and Pretty, 2004; MEA, 2005; IAASTD, 2008). They jointly produce food and other goods for farmers and markets while contributing to a range of valued public goods, such as clean water, wildlife and habitats, carbon sequestration, flood protection, groundwater recharge, landscape amenity value, and leisure and tourism opportunities. In their configuration, they capitalize on the synergies and efficiencies that arise from complex ecosystems and social and economic forces (NRC, 2010).

Second, these systems are diverse, synergistic, and tailored to their particular social-ecological contexts. Agricultural sustainability implies the need to adapt these factors to the specific circumstances of different agricultural systems (Horlings and Marsden, 2011). Challenges, processes, and outcomes will also vary across agricultural sectors: in the United Kingdom, Elliot et al. (2013) found that livestock and dairy operations transitioning toward sustainability experienced particular difficulties in reducing pollution while attempting to increase yields.

Third, these systems often involve more complex mixes of domesticated plant and animal species and associated management techniques, requiring greater skills and knowledge by farmers. To increase production efficiently and sustainably, farmers need to understand under what conditions agricultural inputs (such as seed, fertilizer, and pesticide) can either complement or contradict biological processes and ecosystem services that inherently support agriculture (Settle and Hama Garba, 2011; Royal Society, 2012). In all cases, farmers need to see for themselves that added complexity and increased knowledge inputs can result in substantial net benefits to productivity.

Fourth, these systems depend on new configurations of social capital, comprising relations of trust embodied in social organizations, horizontal and vertical partnerships between institutions, and human capital comprising leadership, ingenuity, management skills, and capacity to innovate. Agricultural systems with high levels of social and human assets are able to innovate in the face of uncertainty (Pretty and Ward, 2001; Wennink and Heemskerk, 2004; Hall and Pretty, 2008; Friis-Hansen, 2012), and farmer-to-farmer learning has been shown to be particularly important in implementing the context-specific, knowledge-intensive, and regenerative practices of sustainable intensification (Pretty et al., 2011; Settle and Hama Garba, 2011; Rosset and Martínez-Torres, 2012).

Some conventional thinking about agricultural sustainability has assumed that it implies a net reduction in input use, thus making such systems essentially extensive (requiring more land to produce the same amount of food). Organic systems often accept lower yields per area of land in order to reduce input use and increase their positive impact on natural capital. However, such organic systems may still be efficient if management, knowledge, and information are substituted for purchased external inputs. Recent evidence shows that successful agricultural sustainability initiatives and projects arise from shifts in the factors of agricultural production (e.g., from use of fertilizers to nitrogen-fixing legumes; from pesticides to emphasis on natural enemies; from ploughing to zero tillage). A better concept than that of extensive systems centers on the sustainable intensification of resources, thus making better use of existing resources (e.g., land, water, biodiversity) and technologies (IAASTD, 2008; Royal Society, 2009; NRC, 2010; Foresight, 2011; Tilman et al. 2011).

Compatibility of the terms "sustainable" and "intensification" was hinted at in the 1980s (e.g., Raintree and Warner, 1986; Swaminathan, 1989), and then first used in conjunction in a paper examining the status and potential of African agriculture (Pretty, 1997). Until this point, "intensification" had become synonymous with a type of agriculture that inevitably caused harm while producing food (e.g., Collier, Wiradi, and Soentoros, 1973; Poffenberger and Zurbuchen, 1980; Conway and Barbier, 1990). Equally, "sustainable" was seen as a term to be applied to all that could be good about agriculture. The combination of the terms was an attempt to indicate that desirable ends (more food, better environment) could be achieved by a variety of means. The term was further popularized by its use in a number of key reports: *Reaping the Benefits* (Royal Society, 2009), *The Future of Food and Farming* (Foresight, 2011), and *Save and Grow* (FAO, 2011a, 2016).

In 2011, FAO also called for a paradigm shift in agriculture, to sustainable, ecosystem-based production. FAO's new model, Save and Grow, aims to address today's intersecting challenges: raising crop productivity and ensuring food and nutrition security for all while reducing agriculture's demands on natural resources, its negative impacts on the environment, and its major contribution to climate change (FAO, 2011a, 2016). The Save and Grow approach recognizes that food security will depend as much on environmental sustainability as it will on raising crop productivity. It seeks to achieve both objectives by using improved varieties, drawing on ecosystem services—such as nutrient cycling, biological nitrogen fixation, and pest predation—and minimizing the use of farming practices and technologies that degrade the environment, deplete natural resources, add momentum to climate change, and harm human health.

Sustainable intensification is defined as a process or system where yields are increased without either adverse environmental impact or the cultivation of more land. The concept is thus relatively open, in that it does not articulate or privilege any particular vision of agricultural production. It emphasizes ends rather than means, and does not predetermine technologies, species mix, or particular design components. Sustainable intensification can be distinguished from former conceptions of agricultural intensification as a result of its explicit emphasis on a wider set of drivers, priorities, and goals than solely productivity enhancement.

CHAPTER

13

Agroecological Approaches to Sustainable Intensification

It is increasingly recognized that increasing agricultural production without the use of more land, water, and fertilizers and pesticides is unlikely to be achievable unless the ecosystem services provided by the biodiversity in agroecosystems can be far better harnessed than is currently the case (Power, 2010). The value of better ecological management is clear in areas such as integrated pest management, soil health, and pollination services, but these benefits have often proved hard to achieve, not least because success requires a shift in our thinking: we need to be focused on ecosystem processes rather than relying on artificial inputs (Altieri et al., 2017). An overemphasis on intensifying production at the expense of ecosystem health risks reduces the resilience of agroecosystems (Huxham et al., 2014; Truchy et al., 2015). Adding urgency to the drive for new approaches is the threat associated with climate change, the impacts of which are already becoming apparent on crop yields (Lobell et al., 2011).

One of the key areas where better understanding of ecological processes could be transformative for agricultural production is soil health. Soil characteristics, biota, and functions are central to resistance and adaptation to climatic and other perturbations, nutrient and water acquisition by crops, and reducing losses due to soil-borne pests and diseases. To date, farmers have primarily focused on the benefits of improvements to physical and chemical soil parameters, not least because of the catastrophic losses of soil through erosion, loss of structure, and a decline in soil organic carbon (Banwart, 2011), but it is increasingly apparent that the diversity of soil microbes, particularly those closely associated with plant roots—the so-called root microbiome—is key to increasing the resilience of crops to both biotic and abiotic stresses and making agricultural practices more sustainable.

Microbial symbionts such as fungal endophytes can protect forage grasses against insect pests and increase plant tolerance of heat and salt stress (Hartley and Gange, 2009; Redman et al., 2011); plant growth-promoting rhizobacteria are increasingly recognized for their contributions to primary productivity through promotion of growth and triggering of systemic plant defense mechanisms (Hol et al., 2013); while arbuscular mycorrhizal fungi

Sustainable Food and Agriculture
DOI: https://doi.org/10.1016/B978-0-12-812134-4.00013-3

not only increase uptake of essential crop nutrients such as nitrogen and phosphorus, but also improve resistance to pests and drought (Hartley and Gange, 2009). Exploiting partnerships with fungi to reduce crop dependence on fertilizer inputs and increase both nutrient and water use efficiency is an active area of current research (Cameron et al., 2013). There is overwhelming evidence that the rhizosphere microbiome is critical for crop health (Berendsen et al., 2012), and focus has shifted to the potential for manipulating the microbiome to increase its ability to supress soil diseases, activate host plant defense mechanisms, and promote plant growth, a prospect that has been more tractable with the use of high-throughput sequencing approaches (Edwards et al., 2015; Dessaux et al., 2016).

The soil can also be a source of defense against pests and diseases for crops: many of them accumulate silicon, which is then deposited on the leaf surface as abrasive bodies (so-called phytoliths) or in the epidermal cells or in leaf spines and hairs (Cooke and Leishman, 2011; Hartley et al., 2015). This deposition forms an effective defense against insect pests (Massey and Hartley, 2009; Hartley and DeGabriel, 2016; Reynolds et al., 2016) and pathogens (Guérin et al., 2014), both above and below ground (Johnson et al., 2016). There is increasing interest in the use of silicon in agriculture as a natural means of crop protection—it is routinely added to crops in the United States and in many countries in Asia, including China (Guntzer et al., 2012) —but our knowledge of the molecular basis of silicon uptake (Ma et al., 2006) could make it a target for crop breeding. Silicon also increases yield and resistance to abiotic stresses such as drought (Guntzer et al., 2012; Meharg and Meharg, 2015), so a better understanding of the functional roles of silicon in plants and its impact on ecological interactions in agricultural systems could bring real benefits (Cooke et al., 2016).

Considerable research effort has now gone into sustainable intensification and various mechanisms for achieving it, but progress has been slow and patchy (Barnes and Thomson, 2014; Barnes, 2016). One reason may be that the focus has centered on "tinkering" with the current agricultural system to try and make it more efficient rather than redesigning farming systems around genuinely agroecological principles. In other words, emphasis must be placed on redesigning farming systems away from a trajectory of intensification and toward one that is more ecological (Bommarco et al., 2013; Huxham et al., 2014), but such an endeavor may encounter numerous obstacles.

Agroecology is the science of applying ecological concepts and principles to the design and management of sustainable food systems (Gliessman, 2014), capitalizing on interactions among plants, animals, humans, and the environment, and applying solutions that harness ecosystem services and conserve biodiversity using both scientific and traditional knowledge.

Agroecological approaches bring science, practice, and social movements into dialogue (Wezel et al., 2009; Tomich et al., 2011). Since the 1980s, community-led social movements have been crucial to the dissemination of agroecological practices worldwide. These have been driven by farmer associations and nongovernmental organizations, who have engaged in advocacy and promotion, as well as developing new forms of markets, distribution systems, and farmer-consumer relationships.

How do we maximize the benefits of such ecological approaches? One obvious enabling factor is ensuring sufficient research to deliver the innovations and technologies required (Khan et al., 2014); Bommarco et al., 2013), but perhaps of even greater importance is delivering the results of that research to farmers. A particularly acute problem is that academic research on crop improvement and agricultural practices has often failed to translate to the field or reach practitioners. Knowledge transfer and farmer engagement strategies need urgent improvement (Green et al., 2016). Improved investment in extension services, together with more participatory approaches focusing on codesign with stakeholders, is the way forward (Malézieux, 2012). This methodology is gaining traction, but better ways of spreading best practice are urgently needed, as the lack thereof can limit the uptake of more agroecological approaches.

One example is biofumigation, a sustainable method of pest and pathogen control, which involves the incorporation into soil of brassicaceous plants, usually mustards, which produce secondary metabolites—glucosinolates, whose breakdown products are toxic (Kirkegaard and Matthiessen, 2005). Biofumigation has been employed, particularly for nematode control, in several regions of the world, but efficacy is variable and inconsistent and there is a lack of research aimed at understanding the range of factors (such as seeding density of the biofumigant, timing of incorporation, and soil conditions) that ensure optimum biofumigant production, release, and efficacy. This lack of detailed data on deployment under a range of agronomic situations prevents the effective, widespread uptake of this sustainable pest control technique (Matthiessen and Kirkegaard, 2006; Motisi et al., 2010).

Best practice for any agroecological technique will reflect local and regional conditions: ecological techniques that could increase yields sustainably will reflect ecosystem-specific biophysical factors, as well as socioeconomic and biological ones (Godfray et al., 2010; Poppy et al., 2014). Thus, the ability of a particular patch of agricultural land to deliver a given ecosystem service is hugely variable both spatially and temporally, which complicates the management of habitats for ecological service provision (Nelson et al., 2009). However, agricultural management practices are key to realizing the benefits of ecosystem services. Although there have been several recent advances in estimating the value of various ecosystem services related to agriculture, and in analyzing the potential for minimizing trade-offs and maximizing synergies, more work is needed to understand the spatial and temporal complexities (Power, 2010; Gunton et al., 2016). Future research will need to confront these challenges in spatially and temporally explicit frameworks.

13.1 TRADE-OFFS OR SYNERGIES: ARE THERE WIN–WINS?

Farmers may not implement these approaches, even if they know about them, as they may not have confidence in their impact on yield. At this point, it is worth noting that yield is not actually the bottom line: less intensive farm management may indeed reduce yield, but if the inputs are less time- and cash-demanding, on-farm profits may actually increase. In any case, increasing evidence indicates that trade-offs between the provisioning services of agricultural production and other ecosystem services are not inevitable and that "win-win" scenarios are possible (Badgley et al., 2007; Power, 2010). Indeed, yields can even increase in response to management of additional ecosystem services, at least in developing countries (Pretty et al., 2006; Pretty, Toulmin, and Williams, 2011), arguably where these approaches are most vital.

However, a global quantitative assessment of the yields obtained from conventional farming compared with organic farming showed that yields from organic farms were between 5% and 34% lower (Seufert et al., 2012). In certain crop types, and when best organic practices were used, organic crop systems almost matched conventional ones, but under less favorable site and system characteristics, the yield "penalty" was higher. However, this analysis does not account for the economic and environmental benefits of the organic system. The challenge, then, is to be able to manage agricultural landscapes for ecological diversity, but in a way that is compatible with real-world agriculture and that maintains yields in addition to providing other benefits (Box 13.1). Recent research on pollination services provides clear examples of where this is practicable: harnessing the services provided by native biodiversity-enhanced crop production (Hoehn et al., 2008; Bommarco et al., 2012; Pywell et al., 2015) and integrating pest management (Pretty and Bharucha, 2015).

BOX 13.1

ECOSYSTEM SERVICE SYNERGIES OR TRADE-OFFS?

Increasing the amount of a provisioning service (such as food) may involve enhancing the delivery of other services (a synergy) or reducing them (a trade-off). Here, the first two examples illustrate synergies, while the last describes a trade-off. The examples lie along a scale of intensification, from a largely unmanaged natural system to an intensively managed and artificial one. Trade-offs are more likely at the intensive end of this scale.

Case 1: Increasing Fisheries Provisioning From Mangroves

Mangrove forests provide a wide range of services, such as provisioning of timber, medicines, fish, and shellfish. Good catches of fish and shellfish, especially crabs and shrimp, rely on good ecosystem health, including proper flushing by seawater, dense and productive stands of trees, and ecological connections with adjacent ecosystems such as seagrass meadows. Hence, managing mangrove forests to enhance fisheries provisioning implies enhancing ecosystem health, with concomitant benefits for other services such as coastal protection and carbon sequestration. Positive interactions between individual trees raise the prospect of synergies for timber production and other services as well (Huxham et al., 2010).

Case 2: Increasing Crop Yields in Mixed Maize Systems in Malawi

African agriculture suffers from a large "yield gap"—the difference between potential production and achieved harvests. Productivity of maize farming in Malawi has recently been dramatically improved following a subsidy for nitrogen fertilizer, allowing millions of farmers to boost yields. However, increasing inorganic nitrogen inputs is expensive and vulnerable to future

BOX 13.1 (*cont'd*)

hikes in prices. In a large-scale experiment, Snapp et al. (2010) compared monoculture maize treatments with maize intercropped with nitrogen-fixing shrubby legumes. They found that the legume treatment produced the same (enhanced) yield of maize as a fertilized monoculture, but with only half the amount of expensive fertilizer. In addition, the yield was more stable and other ecosystem services, such as soil condition, biodiversity, and carbon content showed signs of improvement. Agroforestry initiatives in Malawi and other African countries have demonstrated how the use of nitrogen-fixing "fertilizer trees" can result in increased maize yields along with new services from trees, including honey and fuelwood (Asaah et al., 2011; Pretty et al., 2011).

Case 3: Increasing Timber Yields From Monoculture Plantations

While forests may bring multiple ecological and social benefits, plantations managed for maximum productivity often cause problems. Jackson et al. (2005) describe the trade-offs between monoculture tree plantations (often planted for fast production of wood) and hydrological services. Using a global data set, they show how plantations reduce river flows; this may prove a benefit in wet, flood-prone areas, but constitute a threat in many drier areas. Plantations also tend to salinize and acidify soils. The authors report how these effects are less pronounced or absent in natural forests.

13.2 WHERE IS THE BIODIVERSITY GOING TO COME FROM?

Perhaps the most significant constraint to increasing the use of agroecological approaches is that they rely on biodiversity, but the diversity of many agricultural landscapes, particularly intensive ones, has been critically denuded (Tscharntke et al., 2005). Not merely has this occurred in the case of the obvious limited diversity of crop monocultures, or very limited rotations (Robinson and Sutherland, 2002), but this is possibly particularly acute below ground. For example, it has long been known that the abundance and diversity of mycorrhizal fungi is far lower in agricultural systems than in other ecosystems (Helgason et al., 1998). How can this be reversed? This is where agroecological approaches can have a significant impact, as already well-established approaches exist for increasing on-farm diversity with significant agroecological benefits (Altieri et al., 2017).

Although the relationship between species diversity and ecosystem service provision is recognized as context-dependent and more complex than suggested by the early studies on model communities (Hooper et al., 2005), there is evidence that increased species richness benefits crop yields (Bommarco et al., 2012). Pest control and pollination (Hoehn et al., 2008) services are particularly good examples, and interestingly, the evidence for the importance of biodiversity for effective ecosystem functioning tends to strengthen when assessed over larger spatio-temporal scales (Cardinale et al., 2007; Truchy et al., 2015). On-farm management practices, particularly those that emphasize crop diversity through the use of polycultures, cover crops, crop rotations, and agroforestry, can significantly enhance biodiversity in agroecosystems and the services it provides (Tscharntke et al., 2005; Power, 2010). The challenge is to apply this knowledge to redesign agricultural landscapes for multiple benefits.

Agroecosystems need to be redesigned as multifunctional landscapes to deliver a range of ecosystem services, including not only food production, but also water and soil conservation, soil carbon storage, nutrient recycling, and pest control, all of which underpin and impact on food production. Therefore, sustainable intensification cannot be considered as shorthand for "increasing agronomic efficiency"; a broader view of agroecosystems to deliver sustainable agricultural management practices is needed (Power, 2010; Loos et al., 2014). Recently, Gunton et al. (2016) proposed the following approach: "Sustainable intensification means changes to a farming system that will maintain or enhance specified kinds of agricultural provisioning while enhancing or maintaining the delivery of a specified range of other ecosystem services measured over a specified area and specified time frame." This perspective encapsulates the importance of appropriate temporal and spatial scales, and emphasizes the need to redress the balance between sustainability and intensification by moving nonproduction ecosystem services center stage.

Delivering such rebalancing of the agenda away from intensification and toward agroecological approaches will require innovative thinking, cooperation between disciplines, and a willingness to be open to radical solutions by engaging effectively with a range of stakeholders. In sum, genuine sustainability that harnesses the benefits of ecological approaches effectively requires an understanding of, or even transformation of, the relationships between the social, the ecological, and the technological aspects of agriculture.

CHAPTER

14

Measuring Impacts

There have been a number of meta-level analyses where multiple farms, projects, and initiatives of different types have been analyzed for their impact. There are three key questions: Can the sustainable intensification of agricultural systems work? Can sustainable intensification, at the first and production stage of food chains, produce more food, fiber, and other valued products while improving natural capital? Is it possible to produce more while not trading off harm to key renewable capital assets?

Because "sustainable intensification" is an umbrella term that includes many different agricultural practices and technologies, and because it is more an approach than a distinct set of technologies and processes, the precise extent of existing sustainable intensification practice is also unknown. Milder et al. (2012) estimated that globally some 200 million hectares (Mha) of agricultural land are cultivated under some form of agroecological regime. At the same time, smallholder production is particularly dependent on healthy ecosystems on and around farms (IFAD and UNEP, 2013), and it has been estimated that half the world's smallholders practice some form of sustainable intensification of agriculture (Altieri and Toledo, 2011; IFAD and UNEP, 2013).

However, documenting and evaluating evidence from sustainable intensification is complex and sometimes contentious. Conceptual diversity and the inclusivity of the approach mean that it is difficult to bound evaluations. Agroecological approaches involve multiple practices, adapted from place to place depending on farmer and community needs. There may be no clear conceptual, methodological, or practical dividing line between a new or alternative practice and a conventional or normal practice. Depending on need and ability, farmers may apply agroecological principles to industrial farms, or introduce the mechanization and inorganic inputs into otherwise agroecologically managed farms (Milder et al., 2012). Where studies seek to demonstrate simultaneous improvements to yields and environmental outcomes, results are sometimes sensitive to the variables and parameters selected to capture environmental improvements, the timescales involved, and any weightings used (Elliot et al., 2013).

Some assessments have been found to suffer from methodological flaws (Milder et al., 2012). First, despite the heterogeneity of practices involved in any intensification strategy, assessments often focus on yields from specific, labeled approaches, such as conservation agriculture, agroforestry, or integrated pest management. Analysis of distinct approaches

Sustainable Food and Agriculture
DOI: https://doi.org/10.1016/B978-0-12-812134-4.00014-5

is also difficult. For example, evidence is mixed on outcomes from conservation agriculture and the system of rice intensification, and debate on the general applicability and scalability of these approaches has been "high profile, sustained, and at times acrimonious and emotive" (Sumberg et al., 2013). Second, syntheses, meta-analyses, and overviews have so far focused primarily on yield increases rather than on multiple outcomes and benefits (but see Pretty et al., 2006; Pretty, Toulmin, and Williams, 2011; Milder et al., 2012). Finally, there is not yet sufficient data on how different sustainable intensification strategies (e.g., using agroecological methods) might meet aggregate regional and global goals for food security and economic success.

A number of syntheses have highlighted increased yields (among other positive social-ecological outcomes) as a result of the application of agroecological methods and system redesign. These again have emphasized the beneficial outcomes of "both-and" rather than "either-or" approaches. Outcomes are key; pathways differ. Yields, though, can be a crude measure of the successful outputs or impacts of agricultural systems, particularly where more sustainable systems are expected to have positive impacts on the natural components of both agricultural and wild systems and habitats. Analyses tend to be of two types: involving temporal measures of impact (changes over time at the same location, sometimes called longitudinal) and spatial measures (changes measured at the same point but with different treatments, sometimes called latitudinal).

Farmers adopting sustainable intensification approaches have been able to increase food outputs by sustainable intensification in two ways. The first is multiplicative—by which yields per hectare have increased by combining the use of new and improved varieties with changes to agronomic-agroecological management. The second is improved food outputs by additive means—by which diversification of farms resulted in the emergence of a range of new crops, livestock, or fish that added to the existing staples or vegetables already being cultivated. These additive system enterprises include aquaculture for fish raising (in fish ponds or concrete tanks); small patches of land used for raised beds and vegetable cultivation; rehabilitation of formerly degraded land; fodder grasses and shrubs that provide food for livestock (and increase milk productivity); raising of chickens and zero-grazed sheep and goats; new crops or trees brought into rotations with staple (e.g., maize, sorghum) yields not affected, such as pigeonpea, soybean, and indigenous trees; and adoption of short-maturing varieties (e.g., sweet potato, cassava) that permit the cultivation of two crops per year instead of one.

CHAPTER

15

Impacts on Productivity

Some of the most significant progress toward sustainable intensification has taken place in developing countries within the past two decades (Pretty et al., 2006; Pretty et al., 2011; Pretty and Bharucha, 2014).

The largest study to date analyzed 286 projects in 57 countries. In all, some 12.6 million farmers on 37 Mha (about 3% of the total cultivated area in developing countries) were engaged in redesign transitions involving sustainable intensification. In 68 randomly resampled projects from the original study, there was a 54% increase over the subsequent 4 years in the number of farmers, and a 45% increase in the number of hectares (Pretty et al., 2006; Pretty, 2008). These resurveyed projects comprised 60% of the farmers and 44% of the hectares in the original sample of projects. For the 360 reliable yield comparisons from 198 of the projects, the mean relative yield increase was 79% across the very wide variety of systems and crop types. However, there was a wide spread in results. While 25% of projects reported relative yields of more than 2.0 (i.e., 100% increase), half of all the projects had yield increases of between 18% and 100%. Though the geometric mean is a better indicator of the average for data with a positive skew, this still shows a 64% increase in yield for eight different crop groups (Fig. 15.1).

Table 15.1 summarizes changes in yields, along with numbers of farmers and hectares, for eight different smallholder farming systems: irrigated, wetland rice, rainfed humid, rainfed highland, rainfed dry/cold, dualistic mixed, coastal artisanal, and urban-based and kitchen garden. The mean farm size for these 12.6 million farmers was 2.9 hectares.

It is clear that African countries and farmers have not benefited from previous agricultural improvements in the same way as have those in Asia and South America. The United Kingdom Government Office for Science Foresight Program commissioned reviews and analyses from 40 projects in 20 African countries where sustainable intensification had been developed or practiced in the 2000s (Pretty, Toulmin, and Williams, 2011; Pretty et al., 2014). The cases comprised crop improvements, agroforestry and soil conservation, conservation agriculture, integrated pest management, horticultural intensification, livestock and fodder crops integration, aquaculture, and novel policies and partnerships. By early 2010, these projects had recorded benefits for 10.4 million farmers and their families on approximately 12.75 Mha. Across the projects, crop yields rose on average by a factor of 2.13 (i.e., slightly more than doubled) (Table 15.2). The timescale for these

Sustainable Food and Agriculture
DOI: https://doi.org/10.1016/B978-0-12-812134-4.00015-7

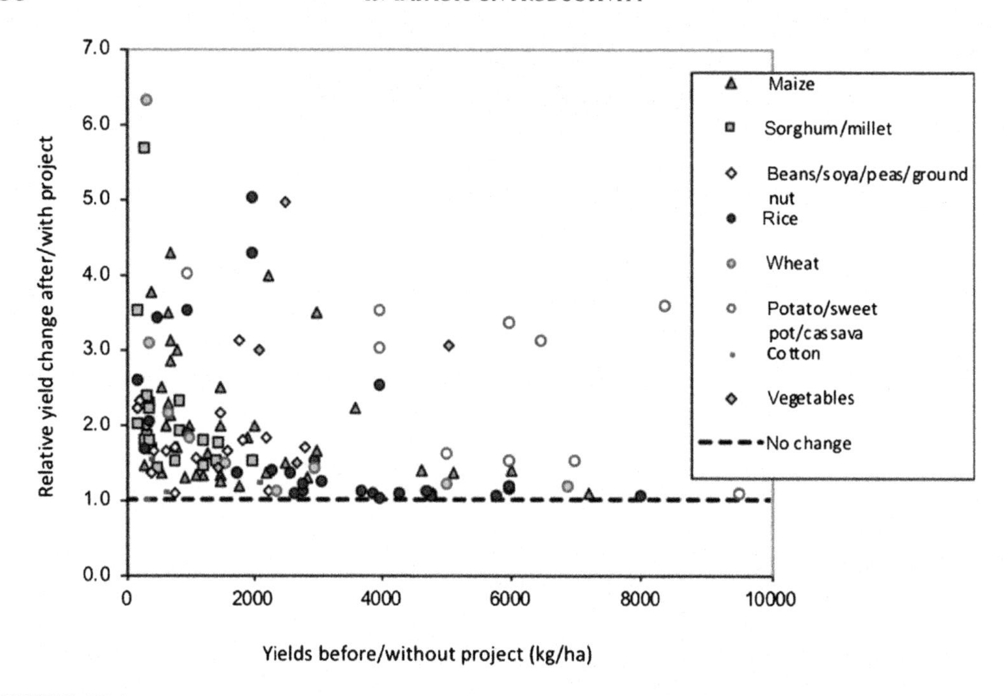

FIGURE 15.1 Relative changes in crop yields under sustainable intensification, 198 projects, 57 countries. Source: *Pretty J., Noble A.D., Bossio D., Dickson J., Hine R.E., Penning de Vries F.W.T., et al., 2006. Resource-conserving agriculture increases yields in developing countries. Environ. Sci. Technol. 40, 1114–1119.*

TABLE 15.1 Summary of Adoption and Impact of Agricultural Sustainability Technologies and Practices on 286 Projects in 57 Countries

FAO Smallholder Farm System Category[a]	Number of Farmers Adopting	Number of Hectares Under Sustainable Intensification	Average % Increase in Crop Yields[b]
1. Irrigated	177,287	357,940	129.8 (± 21.5)
2. Wetland rice	8,711,236	7,007,564	22.3 (± 2.8)
3. Rainfed humid	1,704,958	1,081,071	102.2 (± 9.0)
4. Rainfed highland	401,699	725,535	107.3 (± 14.7)
5. Rainfed dry/cold	604,804	737,896	99.2 (± 12.5)
6. Dualistic mixed	537,311	26,846,750	76.5 (± 12.6)
7. Coastal artisanal	220,000	160,000	62.0 (± 20.0)
8. Urban-based and kitchen garden	207,479	36,147	146.0 (± 32.9)
All projects	12,64,774	36,52,903	79.2 (± 4.5)

[a] *Smallholder farm categories from Dixon, J., A Gulliver, and D. Gibbon (2001).*
[b] *Yield data from 360 crop-project combinations; reported as percentage increase (thus, a 100% increase is a doubling of yields). Standard errors in brackets.*

TABLE 15.2 Summary of Productivity Outcomes From Sustainable Intensification Projects in Africa

Thematic Focus	Area Improved (ha)	Mean Yield Increased (Ratio)
Crop variety and system improvements	391,060	2.18
Agroforestry and soil conservation	3,385,000	1.96
Conservation agriculture	26,057	2.20
Integrated pest management	3,327,000	2.24
Horticulture and small-scale agriculture	510	Nd
Livestock and fodder crops	303,025	Nd
Novel regional and national partnerships and policies	5,319,840	2.05
Aquaculture	523	Nd
Total	**12,753,000**	**2.13**

Nd, no data—largely because horticulture, livestock, and aquaculture are additive components to systems, increasing total food production, but not necessarily yields.
Milder et al. (2012) undertook a broad review of five sets of agroecological systems, examining their contribution to yields, as well as nine distinct ecosystem services that were relevant to both on- and off-farm beneficiaries. These systems were in both developing and industrialized countries.
Foresight, 2011. The Future of Global Food and Farming. Final Project Report. London: Government Office for Science London.; Pretty J., Bharucha Z.P., Hama Garba M., Midega C., Nkonya E., Settle W., et al., 2014. Foresight and African agriculture: innovations and policy opportunities. Report to the UK Government Foresight Project. <https://www.gov.uk/government/uploads/system/uploads/attachment_data/file/300277/14-533-future-african-agriculture.pdf (31.08.14)>.

improvements varied from 3 to 10 years. It was estimated that this resulted in an increase in aggregate food production of 5.79 Mt per year, equivalent to 557 kg net per farming household (in all the projects). The next chapter shows how some of these changes were achieved.

Milder et al. (2012) undertook a broad review of five sets of agroecological systems, examining their contribution to yields, as well as nine distinct ecosystem services that were relevant to both on- and off-farm beneficiaries (Table 15.3). These systems were in both developing and industrialized countries.

In 1989, the US National Research Council (NRC) published the seminal *Alternative Agriculture*. Partly driven by increased costs of fertilizer and pesticide inputs, in addition to the growing scarcity of natural resources (such as groundwater for irrigation) and continued soil erosion, farmers had been adopting novel approaches in a wide variety of farm systems. The NRC noted that alternative agriculture was not a single system of farming practices but several; these often diversified systems were compatible with large and small farms. Such alternative agricultural systems used crop rotations, integrated pest management, soil and water conserving tillage, animal production systems that emphasized disease prevention without antibiotics, and genetic improvement of crops to resist pests and disease and use nutrients more efficiently. Well-measured alternative farming systems "nearly always used fewer synthetic chemical pesticides, fertilizers and antibiotics per unit of production than comparable conventional farms" (NRC, 1989). They also required "more information, trained labor, time and management skills per unit of production."

TABLE 15.3 Global Extent of Five Agroecological Systems

System	Extent Under Agroecological Intensification Systems (Mha)	Total land under Analogous Production Systems (Mha)	Proportion of Land Under Agroecological Intensification (%)
Conservation agriculture	116	2098	6
Holistic grazing management	40	3200	1.25
Organic agriculture	37	2459	1.5
Precision agriculture	Nd	2098	Nd
System of rice intensification	>1.5	153	>1

Nd, no data.
Milder J.C., Garbach K., DeClerck F.A.J., Driscoll L., Montenegro M., 2012. An Assessment of the Multi-Functionality of Agroecological Intensification. A report prepared for the Bill& Melinda Gates Foundation.

The NRC (1989) commissioned 11 detailed case studies of 14 farms as exemplars of effective and different approaches to achieving similar aims: economically successful farms with a positive impact on natural capital. The NRC (2010) later conducted follow-up studies in 2008 on 10 of the original farms. These included integrated crop-livestock enterprises, fruit and vegetable farms, one beef cattle ranch, and one rice farm. After 22 years, common features of farms included:

(1) All farms emphasized the importance of maintaining and building up their natural resource base and maximizing the use of internal resources.
(2) All farmers emphasized the values of environmental sustainability and the importance of closed nutrient cycles.
(3) The crop farms engaged in careful soil management, the use of crop rotations, and cover crops; the livestock farms continued with management practices that did not use hormones or antibiotics.
(4) More farmers participated in nontraditional commodity and direct sales markets (via farmers markets and/or the Internet); some sold at a premium with labeled traits and products (e.g., organic, naturally-raised livestock).
(5) Most farms relied heavily on family members for labor and management.
(6) The challenges and threats centered on rising land and rental values associated with urban development pressure, the availability of water, and the spread of new weed species.

In France, the Institut de l'Agriculture Durable (2011) called for a new European agriculture based around maintaining healthy soil, biodiversity, appropriate fertilization, and appropriate plant protection techniques. Deploying these helps protect the environment while producing more yield in better and viable ways. Testing 26 indicators categorized into 7 themes (economic viability, social viability, input efficiency, soil quality, water quality, greenhouse gas emissions, and biodiversity) across 160 different types of farms, the authors found that positive ecological externalities can be both achieved and measured. In the United Kingdom, Elliot et al. (2013) explored outcomes across 20 farms, of which 4 appeared to have achieved yield increases alongside environmental improvements, using technologies and practices such as improved genetics and precision farming, zero tillage and improved water management, diversification (the installation of small-scale energy generation), and application of available agrienvironmental schemes.

CHAPTER

16

Improving Environmental Externalities with Integrated Pest Management

Pathogens, weeds, and invertebrates cause up to 30% losses in particular locations for particular crops (Oerke and Dehne, 2004; Flood, 2010). Viewed in terms of food security, crops lost to pests represent the equivalent of food required to feed over 1 billion people (Birch et al., 2011). Over the coming decades, pest damage may worsen due to the response of pest species to global climate change (Birch et al., 2011) and the distribution of pests and pathogens will change (Bebber et al., 2013). Resistance to synthetic pesticides poses a continuing challenge. Since the discovery of triazine resistance in common groundsel in the late 1960s, herbicide resistance in weeds has grown rapidly. Globally, some 220 species of weeds have evolved herbicide resistance. These pose a particular challenge to cereals and rice (Heap, 2014). Five species groups (*Avena* spp., *Lolium* spp., *Phalaris* spp., *Setaria* spp., and *Alopecurus myosuroides*) infest over 25 Mha of cereal crops globally (Heap, 2014). For insect pests, over 600 cases of insecticide resistance have become apparent since the 1950s (Pretty, 2005; Head and Savinelli, 2008).

Overall, the impact of synthetic pesticides has been mixed. While these prevent a significant amount of potential losses, especially in the short term, they pose threats to human and ecosystem health, and their effectiveness declines as pest and weed resistance develops (Pretty, 2005). Thus, overreliance on agrichemicals as the sole form of plant protection is not sustainable (Shaner, 2014). Moreover, with the development of poorly regulated generics, especially in developing countries, cheap alternatives that do not meet international quality standards lock in the use of obsolete pesticides (Popp et al., 2012) and result in associated risks for ecosystems and human health (de Bon et al., 2014).

Effective integrated pest management (IPM) centers on the principle of deploying multiple complementary methods for pest, weed, and disease control. IPM has been defined as a "decision-based process for coordinating multiple tactics for control of all classes of pests in an ecologically and economically sound fashion" (Ehler, 2006). This incorporates the simultaneous management and integration of tactics, the regular monitoring of pests

Sustainable Food and Agriculture
DOI: https://doi.org/10.1016/B978-0-12-812134-4.00016-9

and natural enemies, and the use of thresholds for decisions, and spans methods from pesticide product management/substitution to whole agroecosystem redesign. This broad range of options allows for many interpretations of IPM (Parsa et al., 2014; Gadanakis et al., 2015).

IPM approaches can be classified into four main types (Table 16.1). These vary along a spectrum from targeted or changed use of pesticide compounds to habitat and agroecological design.

IPM provides some of the best evidence on redesigning systems to improve environmental externalities. Although IPM is said to have become the dominant crop protection paradigm, its adoption rate remains low (Parsa et al., 2014). It has also not yet led to a reduction in total pesticide use, nor has it eliminated negative externalities. However, evaluations contain multiple reasons for optimism. Existing IPM-based projects have seen reductions in synthetic pesticide use. For example, in cotton and vegetables in Mali, pesticide use fell from an average of 4.5 to 0.25 L/ha (Settle and Hama Garba, 2011). In some cases, biological control agents have been introduced to replace pesticides, or habitat design has led to effective pest and disease management (Royal Society, 2009; Khan et al., 2011).

TABLE 16.1 Typology of IPM

Type	Example Applications
1a. Substitution of pesticidal products by other compounds	• Synthetic pesticide with high toxicity substituted by another product with low toxicity • Use of agrobiologicals or biopesticides (e.g., derived from neem)
1b. Management of application of pesticides	• Targeted spraying • Threshold spraying prompted by decision making derived from observation/data on pest, disease, or weed incidence
2. Crop or livestock breeding	• Deliberate introduction of resistance or other traits into new varieties or breeds (e.g., recent use of genetic modification for insect resistance and/or herbicide tolerance)
3a. Release of antagonists, predators, or parasites to disrupt or reduce pest populations	• Sterile breeding of male pest insects to disrupt mating success at population level • Identification and deliberate release of parasitoids or predators to control pest populations
3b. Deployment of pheromone compounds to move or trap pests	• Sticky and pheromone traps for pest capture
4. Agroecological habitat design	• Seed and seed bed preparation • Deliberate use of domesticated or wild crops/plants to push–pull pests, predators, and parasites • Use of crop rotations and multiple cropping to limit pest, disease, and weed carryover across seasons or viability within seasons • Adding host-free periods into rotations • Adding stakes to fields for bird perches

Source: Pretty, J., Bharucha, Z.P., 2015. Integrated pest management for sustainable intensification of agriculture in Asia and Africa. Insects 6 (1):152–182.

The most significant innovation for IPM has been the development and deployment of farmer field schools (FFSs), beginning in the 1980s (Kenmore et al., 1984). The aims of the schools are education, colearning, and experiential learning so that farmers' expertise is improved to provide resilience to current and future challenges in agriculture. FFSs are not just an extension method: they increase knowledge of agroecology, problem-solving skills, group building, and political strength. FFS play a central role in training farmers in IPM, and have been shown to result in improved outcomes (van den Berg and Jiggins, 2007; Pretty et al., 2011; Settle et al., 2014).

Thus, complementary and alternative modes of pest control, relying on the manipulation of pest ecologies, have been gaining increased attention as part of a suite of diversified strategies to control pest populations and boost crop resilience to infestation. The preservation and strategic use of on- and off-farm biodiversity is an overarching principle in new forms of pest management. A key principle is that biodiverse agroecosystems demonstrate less pest damage and more natural pest enemies than those lacking biodiversity (Letourneau et al., 2011).

Various IPM programs were spurred onward due to a number of pest outbreaks and, concomitantly, growing awareness of the potential health concerns associated with the use of conventional inorganic pesticides. Over time, IPM strategies have transitioned from individual field-based practice to coordinated, community-scale decision making covering wider landscapes (Brewer and Goodell, 2012). While this improves the effectiveness of pest control, it presents a significant obstacle to wider adoption (notably in developing country contexts) by presenting a collective-action dilemma (Pretty, 2003).

Relatively few cross-country evaluations exist regarding the effectiveness of IPM (van den Berg, 2004; van den Berg and Jiggins, 2007; Pretty, 2008). In a 2015 meta-analysis, Pretty and Bharucha analyzed 85 IPM projects from 24 countries of Asia and Africa implemented over a 25-year period (1990–2014) in order to assess outcomes on productivity and reliance on pesticides. Projects had been implemented in Bangladesh, Cambodia, China, India, Indonesia, Japan, Laos, Nepal, Pakistan, the Philippines, Sri Lanka, Thailand, and Vietnam, as well as in Burkina Faso, Egypt, Ghana, Kenya, Malawi, Mali, Niger, Senegal, Tanzania, Uganda, and Zimbabwe. An estimated 10–20 million farmers used IPM methods and approaches on 10–15 Mha, involving substantial redesign of agroecosystems. The mean yields across projects and crops increased by 40.9% (standard deviation 72.3), combined with a decline in pesticide use to 30.7% (standard deviation 34.9) of the original pesticide use. A total of 35 of 115 (30.4%) crop combinations resulted in a transition to zero pesticide use, and 18 of 115 (15.7%) crop combinations resulted in no changes in yields but did reduce costs (Fig. 16.1).

In principle, there are four possible trajectories an agroecosystem can take with the implementation of IPM (see Fig. 16.1):

1. both pesticide use and yields increase (PY);
2. pesticide use increases but yields decline (Py);
3. both pesticide use and yields fall (py); and
4. pesticide use declines, but yields increase (pY).

The conventional assumption is that pesticide use and yields are positively correlated, suggesting that only trajectories into PY or py are likely. A shift into Py should be against

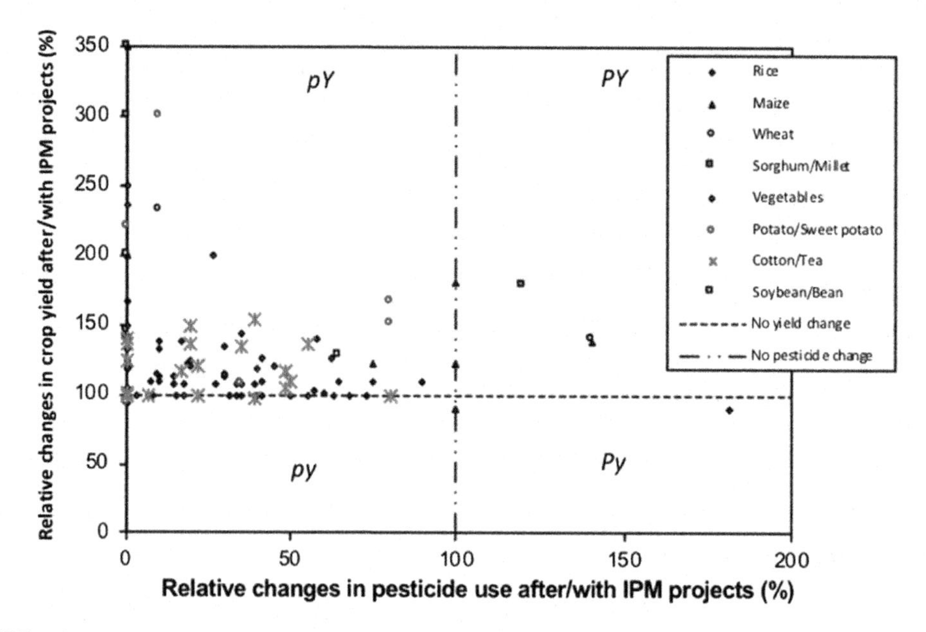

FIGURE 16.1 Impacts of integrated pest management (IPM) projects and programs on pesticide use and crop yields (data from 115 crop combinations, 85 projects, 24 countries in Africa and Asia). Source: *Pretty, J., Bharucha, Z.P., 2015. Integrated pest management for sustainable intensification of agriculture in Asia and Africa. Insects 6 (1):152–182.*

economic rationale, as farmers' returns would be lowered, and thus there should be incentives to change practices. A shift into pY would suggest that current pesticide use is inefficient and could be amended. In the above meta-analysis, only one Py case (increased pesticide use combined with reduced yield) was found in the recent literature, reported by Feder et al. (2004). This evaluation considered the impact of FFS training in Indonesian rice cultivation. A number of py (declines in both pesticide use and yield) cases have been reported elsewhere for transitions in industrialized systems typical in Europe following adoption of integrated farming systems. Here, large reductions in pesticide use were accompanied by up to 20% falls in yields (Röling and Wagemakers, 1997). PY contains examples that have involved the adoption of conservation and no-till agriculture, where the reduced tillage improves soil health, reduces erosion, and improves surface water quality, but also requires increased use of herbicides for weed control. The rapid growth in herbicide use in Argentina and Brazil has followed the widespread adoption of zero-tillage systems, and these show a transition to PY (though this is not inevitable, as there are organic zero-tillage systems in Latin America: Petersen et al., 2000). The majority of cases reported here show transitions to the pY sector (pesticide use falls, yields increase).

While pesticide reductions with IPM may be expected, yield increases induced by IPM are more complex. IPM may, for example, reduce the incidence of severe-loss years, though not increase yields in a normal year, thus increasing mean production across years. Many IPM projects involve interventions focused on more than just pest management (e.g., push–pull systems where legumes fix soil nitrogen as well as help manage parasitoids and depress weeds); or if they involve a significant component of farmer training (e.g., through FFS), then farmers' capabilities at innovating in a number of areas of their agroecosystems will have increased, such as in soil and water management (Settle et al., 2014). Yield improvements may also be accompanied by new income streams that open up with increased flows of ecosystem goods and services. Push–pull IPM systems, for example, may provide improved livestock feed, enabling smallholders to diversify into dairying and poultry. This in turn provides increased manure for farms (helping soil health and yields), and also translates into increased income and better nutrition. Farmers may also be able to invest cash savings made from reducing pesticide use into better seeds and fertilizers, both of which would increase yields. These arrays of improvements are examples of some of the synergistic cycles that can result from sustainable intensification.

CHAPTER

17

Push–Pull Redesign: Multifaceted Innovation Systems in Africa

The push–pull technological innovation, developed by the International Centre of Insect Physiology and Ecology, Rothamsted Research in the United Kingdom, and partners in east Africa, addresses smallholder agricultural constraints, food insecurity, environmental degradation, and loss of biodiversity (Khan et al., 2014; Midega et al., 2015a).

Push–pull is a polycropping innovation that holistically combines multifunctional resource-conserving integrated pest management (IPM) with integrated soil fertility management approaches while making efficient use of natural resources to increase farm productivity by addressing most aspects of smallholders' constraints (Cook et al., 2007; Hassanali et al., 2008). Developed for minimum tillage-based systems, push–pull maintains continuous soil cover with perennial cover crops and plant residue in a diversified cereal–legume–fodder intercropping practice. The perennial intercrop provides live mulching, thus improving aboveground and belowground arthropod abundance, agrobiodiversity, and the food web of natural enemies of stemborers (Khan et al., 2006; Midega et al., 2015a). The technology effectively controls the lepidopteran stemborers and the devastating parasitic *Striga* weeds, both of which can cause severe yield loss to cereals. Furthermore, it improves soil health and conserves soil moisture. The technology involves use of intercrops and trap crops in a mixed cropping system (Khan et al., 2006). These companion plants release behavior-modifying stimuli (plant chemicals) to manipulate the distribution and abundance of stemborers and beneficial insects for management of the pests (Fig. 17.1). The system is well suited to African socioeconomic conditions, as its efficacy is not based on high external inputs, but rather upon biological management of local resources.

The main cereal crop is planted with a repellent intercrop such as *Desmodium* (push) and an attractive trap plant such as *Brachiaria* or Napier grass (pull) planted as a border crop around this intercrop. Gravid stemborer females are repelled from the main crop and are simultaneously attracted to the trap crop (Cook et al., 2007). The companion plants are valuable themselves as high-quality animal fodder, thereby facilitating livestock integration.

Sustainable Food and Agriculture
DOI: https://doi.org/10.1016/B978-0-12-812134-4.00017-0

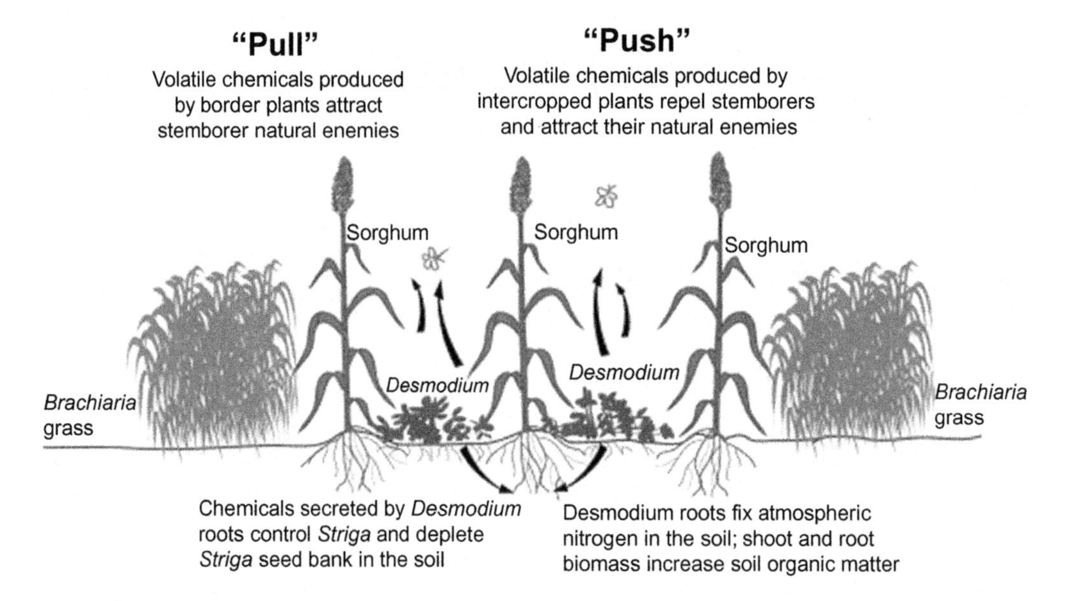

FIGURE 17.1 Push–pull technology. Source: *Adapted from Khan Z.R., Midega C.A.O., Pittchar J.O., Murage A. W., Birkett M.A., Bruce T.A.J. and et al. (2014) Achieving food security for one million sub-Saharan African poor through push–pull innovation by 2020. Philos. Trans. R. Soc. B Biol. Sci. 369: 20120284.*

17.1 BIOLOGICAL CONTROL MECHANISMS

Trap crops produce higher levels of volatile cues than maize or sorghum (Birkett et al., 2006), thus attracting stemborer moths for egg-laying. Larvae that hatch on the grass have a very low survival rate of around 20% (Khan et al., 2006) since in response to feeding by the larvae, Napier grass tissues produce sticky sap, which traps and kills them. Simultaneously, the intercrop, legumes in the *Desmodium* genus, produce repellent volatile chemicals that push away the stemborer moths. Additional benefits of *Desmodium* include effective control of *Striga* weed, increased availability of nitrogen, and soil shading. Several African adapted *Desmodium* spp. Have been evaluated and are currently being used as intercrops in maize, sorghum, and millets (Midega et al., 2015a).

Stemborers and *Striga* were effectively controlled by the technology under farmers' conditions; farmers also reported additional benefits, such as increased soil fertility, up to threefold increases in grain yields, and improved availability of animal fodder resulting in increased milk production (Khan et al., 2008a,b; Midega et al., 2014, 2015b).

The technology has been shown to provide other ecosystem services, including reduced soil erosion and improved water retention, lowered production costs, and climate change adaptability and mitigation. Its reliance on locally adapted companion plants and reduction of the dependence on inorganic fertilizer reduces its carbon footprint while enhancing both the environmental and economic sustainability of the system.

In a detailed economic analysis utilizing data from over seven cropping seasons, returns to investment for the basic factors of production under push–pull were significantly higher compared to those from maize-bean intercropping and maize monocrop systems (Khan et al., 2008c). Positive total revenues ranged from $351/ha in low potential areas to $957/ha in high potential areas, with general increases in subsequent years. The returns to labor that were recovered within the first year of establishment of push–pull ranged from $0.5/person day in low potential areas to $5.2/person day in higher potential areas, whereas in the maize monocrop, this was negligible or even negative. Furthermore, the net present value from push–pull was positive and consistent over several years. De Groote et al. (2010) used discounted partial budget and marginal analysis to corroborate these findings and concluded that push–pull earned the highest revenue compared to other soil fertility management technologies, including green manure rotation.

Thus far, on-farm uptake of push–pull has confirmed that the technology is very effective and has significant impacts on food security, human and animal health, soil fertility, conservation of agrobiodiversity, provision of agroecosystem services, empowerment of women, and income generation for resource-poor farmers.

17.2 CLIMATE-SMART TECHNOLOGICAL ADAPTATION

Adaptation of the push–pull technology for hotter and drier agroclimatic conditions has provided the resilience and adaptability that farmers need to deal with the additional problems associated with climate change (Khan et al., 2014). The trap crops and intercrops used in conventional push–pull were rainfall and temperature limited, as the initial system was developed under average rainfall (800–1200 mm) and moderate temperatures (15–30°C). New drought-tolerant trap (*Brachiaria* cv. Mulato) and intercrop (drought-tolerant spp. Of *Desmodium*, e.g., *D. intortum*) plants have been tested under controlled conditions, validated under farmers' conditions, and incorporated into the push–pull technology.

In addition to being able to control stemborers, *Brachiaria* cv. Mulato II is a source of fodder and improves on-farm biodiversity and soil organic matter. The commercial availability of its seed was also assumed to facilitate more rapid adoption.

17.3 FIELD IMPLEMENTATION OF CLIMATE-SMART PUSH–PULL TECHNOLOGY

Some 61,800 smallholder farms in drier parts of Kenya, Tanzania, Uganda, and Ethiopia have reported effective control of stemborers and *Striga* weed, resulting in significant increases in grain yields of both maize and sorghum (Midega et al., 2015b). The rate of adoption of the climate-smart push–pull by women farmers (62% of all adopters) is growing faster than that of men farmers because of the technology's labor-saving advantages. The *Brachiaria* grass is easier for women to manage than the Napier grass used in the previous conventional push–pull system.

Validation of gross returns from the climate-smart push–pull shows a marginal rate of return of 109% and 143% for sorghum and maize, respectively, implying that the net increase in benefits of climate-smart push–pull outweighed the net increase in costs compared with farmers' practices (Murage et al., 2015). Additionally, over 80% of the interviewed farmers ($n = 350$) reported effective control of *Striga* and stemborers, improvement in soil fertility, and improved grain yields. Other benefits mentioned included increases in fodder and milk production.

Smallholder farmers in sub-Saharan Africa typically practice multiple cropping, where cereals are intercropped with food legumes. Therefore, the technology has been adapted where edible beans are planted either in the same hole with maize or in between maize plants within a row. This has increased the technology's appeal to farmers, as it guarantees an additional protein source in the diet (Khan et al., 2014), resulting in higher technology adoption rates in the region.

17.4 MAXIMIZING MULTIPLE BENEFITS OF PUSH–PULL TECHNOLOGY

On-farm uptake of conventional and climate-adapted push–pull by over 132,400 farmers in eastern Africa has confirmed that it has been widely accepted, and has significantly impacted on food security, human and animal health, soil fertility, income generation, empowerment of women, and conservation of agrobiodiversity and provision of agroecosystem services. The climate-smart push–pull is spreading these benefits to a wider range of farming systems, conferring the benefits upon additional crops and agroecologies.

In addition to the pest control benefits, push–pull systems generate a number of positive social–ecological externalities. Companion plants ensure availability of high-quality fodder at the farms. This improves the productivity of livestock, while surplus fodder is preserved as silage or hay. Both *Desmodium* and *Brachiaria* have benefits for soil health. Recently compiled evidence indicates that long-term use of the push–pull intercropping system dramatically increases total soil carbon and nitrogen stocks, which builds soil fertility and improves soil health, drainage, and water-holding capacity while also reducing soil erosion. Improvements in soil health are, in turn, the basis for increased drought tolerance. Furthermore, the use of biological nitrogen fixation to support crop production in place of synthetic nitrogen fertilizer reduces fossil fuel dependency and greenhouse gas emissions.

Once established, the technology reduces the drudgery of digging and weeding, a task performed most often by women, freeing up their time and labor for more productive tasks such as selling milk or starting other enterprises. Diversified farm income means there is more money available to buy medicines, household goods, and other essentials. Stall-feeding dairy cattle also frees the women and children from the task of herding cattle to graze.

17.5 DELIVERY MECHANISMS AND SCALING UP APPROACHES

The adoption of push–pull technologies in sub-Saharan Africa has used innovative communication and institutional arrangements tailored to the needs of different types of smallholder users. Different farmer-learning approaches have been field-tested. Novel communication strategies and cost-effective scaling-up models have been used to achieve faster learning and adoption of technology, leading to food security, nutrition, and environmental impacts at scale. Ranked in terms of their increasing resource- and knowledge-intensiveness from our previous research (Khan, 2008b; Murage, et al., 2015), these include print materials with local stockists, field days, farmer teachers, and farmer field schools. Given the nature of push–pull technology as a knowledge-intensive technology, these pathways have been used either singly or in combination to enhance the quality of information received. The following delivery mechanisms have been used:

- Print material gives the learner clear and effective understanding of the technology's functioning, components, and recommended agronomic practices. Print material is used in creating awareness and stimulating farmers' interest in the technology. Once stimulated, farmers may gain enough interest to seek additional information. Print material has the added advantage of preserving the information and can therefore be used for a long time as a permanent reference.
- The farmer teachers, also referred to in the "innovative farmer approach," involves selection of progressive early adopters of new technologies. The method capitalizes on local social networks. The concept of farmer trainers is based on the hypothesis that there are always farmers who have above average skills, knowledge, and talents in farm management. These farmers motivate other farmers, help them to improve their skills, share their know-how, and are therefore trained to train other farmers. Farmer trainers are identified based on their knowledge and understanding of the technology, and motivated by the recognition they receive in society. Their training is typically hands-on, initially at the trainer's farm and later at the trainee's farm, with monitoring and evaluation visits by the trainer to ensure quality.
- Field days are day-long events commonly used in rural agricultural extension. Farmers are invited to a particular field or plot and specific information about the technology is demonstrated and discussed. A field day ranges from structured presentations to more informal events where participants walk through the field plot at their own pace to view the demonstrations. Farmers are able to interact with the facilitators as well as with other farmers and exchange ideas and experiences. Hands-on training and physical participation of the farmers is usually encouraged. In this approach, farmers are actively engaged in planning and training activities. At the end of a field day, participants are interviewed to assess the effectiveness of the field day as a knowledge transfer model.
- Described earlier for IPM, farmer field schools offer greater learning opportunities for resource-poor or marginalized members such as women or youth than less interactive or supported approaches (Phillips, Waddington, and White, 2014; Ramisch et al., 2006). Farmers experiment and solve their problems independently, and are able to adapt the technologies to their own specific environmental and cultural needs. Participants are encouraged to share their knowledge with other farmers and trained to teach the courses themselves, thus reducing the need for external support.
- New media: innovative approaches have been integrated in the technology dissemination package, including the use of participatory video, cartoon books, and drama where farmers share their own experiences in local contexts, embedded within

strategic partnership platforms. Audiovisual methods are more effective than printed material in disseminating knowledge-intensive technologies to farmers with low literacy levels. Moreover, farmers more readily share knowledge and their own findings if they use participatory video as part of their work (Ouma et al., 2014; Gandhi et al., 2009). Participatory communication functions as a catalyst for action and as a facilitator of knowledge acquisitions and knowledge sharing among people (Nair, 1994).

Interactive methods are now recognized as important knowledge communication strategies for developing capacities of resource-poor farmers (Berdegué and Escobar, 2001). They are an improvement on the traditional top-down linear dissemination approaches that achieved limited success in catalyzing adoption of technological innovations (Röling, 1996; Rivera et al., 2001; Gandhi et al., 2009), shifting the communication strategy from instrumental to more participatory and decentralized approaches (Leeuwis and van den Ban, 2004). The new media, used in combination with social organization and mediated learning, are complementing traditional dissemination methods and resulting in improved knowledge acquisition, retention and correct application, cost-efficiency, local relevance, and stimulation of real adoption.

CHAPTER

18

Crop Variety Improvements

Varietal improvements, particularly focused on increased yield and pest resistance, have long been at the forefront of agricultural intensification. Yield improvements in key agricultural staples—wheat (208%), paddy rice (109%), maize (157%), potato (78%), and cassava (36%) —between 1960 and 2000 (Pingali and Raney, 2005) were key to reducing protein-energy malnutrition (undernourishment) in the developing world (Gómez et al., 2013) by increasing output and reducing food prices. The key technological development came from conventional plant breeding (crossing plants with different genetic backgrounds and selecting individuals with desirable characteristics). An international network of public sector bodies, the Consultative Group for International Agricultural Research (CGIAR), played a central role. This provided the dominant source of improved germplasm, particularly for rice, wheat, and maize, especially in developing countries. Even in the 1990s, some 36% of all varietal releases were based on CGIAR crosses and 26% of all modern varieties had some CGIAR content (Evenson and Gollin, 2003). Crucially, global benefits from conventional plant breeding in the mid- to late 20th century were realized as being the result of the international spread of germplasm. This enabled countries to make strategic decisions about how much they needed to invest in their own plant-breeding capacity and enabled developing countries to capture the spillover effects of international investment in crop improvement (Pingali and Raney, 2005).

From the 1990s onward, varietal improvement has also focused on the methods of biotechnology and genetic modification. Since they were first commercialized in 1996, there has been a hundredfold increase in global hectarage of genetically modified crops. As of 2013, some 18 million farmers sowed approximately 175 Mha in 27 countries (ISAAA, 2013). Just over half of the global hectarage under genetically modified crops was in Latin America, Asia, and Africa, the primary crops being soybean, cotton, maize, and canola (oilseed rape). To date, commercialized varieties mainly express two traits: herbicide tolerance and resistance to specific insect pests. Some other traits have been developed to deliver nutritional benefits (e.g., golden rice with high levels of vitamin A), but have not yet been cultivated commercially.

This gene revolution has catalyzed a substantial break between private and public sector involvement in varietal improvement (Pingali and Raney, 2005; Gray and Dayananda, 2014). Globally, private investment in agricultural research has significantly overtaken

Sustainable Food and Agriculture
DOI: https://doi.org/10.1016/B978-0-12-812134-4.00018-2

public investment (Gray and Dayananda, 2014). In 2005, Pingali and Raney reported that the world's top 10 multinational bioscience corporations had a collective annual expenditure on agricultural research and development of $3 billion. The CGIAR, by contrast, spent just under $300 million a year on plant improvement research and development. Public investment in agricultural biotechnology in the developing world has been led by China, followed by Brazil and India (Pray and Naseem, 2003). Bangladesh is soon to follow. In China, varietal improvement through hybridization in rice has been credited with feeding some 60 million additional people a year and reducing the land allocated to rice by 14% since 1978 (Li et al., 2010). Many other developing countries have been increasing capacity to adopt and adapt innovations developed elsewhere (Pingali and Raney, 2005).

Adoption has been relatively scale-neutral, with both large and small farmers using genetically modified crops. Some 90% (15 million) of farmers cultivating genetically modified crops were small and resource-poor farmers in developing countries (ISAAA, 2013). In a 2006 review, Raney concludes that: "The economic evidence available to date does not support the widely held perception that transgenic crops benefit only large farms; on the contrary, the technology may be pro-poor. Nor does the available evidence support the fear that multinational biotechnology firms are capturing all the economic value created by transgenic crops."

To develop varieties that are adapted to ecosystem-based agriculture and to climate change, plant breeders need continual access to the widest possible sources of desirable traits, which are found in cereal accessions held in germplasm collections, in wild relatives, and in landraces in farmers' fields (see also Chapter 8: Biodiversity and Ecosystem Services). Crop varieties need to be better adapted to ecosystem-based agriculture. Work on this is in progress, with proposals for a breeding program for wheat cultivars suited to zero tillage (Trethowan et al., 2009). Among traits that have been identified are more rapid establishment and faster root growth (Kohli and Fraschina, 2009). Other research institutions are responding to growing water scarcity by identifying wheat genotypes with higher water-use efficiency (Solh and van Ginkel, 2014).

The breeding of more productive, efficient, and nutritious crops needs to be matched by seed systems that ensure rapid multiplication of the seed and its supply to smallholder farmers. Seed production is especially critical for hybrids of cross-pollinated crops such as maize. In many countries, the lack of such systems has prevented farmers from adopting new varieties and improved farming practices. Many farmers continue to use seed from harvests in their fields. This "saved seed" is often of poor quality and its repeated use can result in low yields.

Alongside these developments, there have also been calls for increased attention to "orphan crops" in developing countries (Varshney et al., 2012a). These are crops that are "valued culturally, often adapted to harsh environments, nutritious, and diverse in terms of their genetic, agroclimatic, and economic niches" (Naylor et al., 2004). They are important for nutritional security in the world's most vulnerable regions. Many orphan cereal crops and legumes are also coming to attention as a result of international research collaborations and advances in sequencing and genotyping technologies (Varshney et al., 2012b; Bohra et al., 2014). Orphan legumes are better adapted to climatic extremes than major crops and demonstrate high tolerance to drought and seasonal extremes (Cullis and Kunert, 2016).

Sweet potato is particularly important in Uganda, where it underpins farm-level food security as an important source of starch, calcium, and riboflavin. Ugandan yields for sweet potato are around 4 tons/ha (compared with an average global yield of 14 tons/ha) (Naylor et al., 2004). Conventional breeding for sweet potato is relatively slow because it is a vegetatively propagated perennial. Biotechnology-based approaches on the other hand could confer pest resistance to weevils and viruses and improved starch/dry matter ratios, and it has been estimated that effective resistance could raise yields to 7 tons/ha (Naylor et al., 2004). Transfer of the technology across half the sweet potato area in sub-Saharan Africa would result in a gross annual benefit of $121 million (Naylor et al., 2004). In Kenya, it has been estimated that effective resistance to viruses and weevils could result in income increases of 28%–39% (Qaim, 1999). Other approaches to sweet potato improvements have centered on conventional breeding to improve vitamin A content and to shorten time to harvest (Mwanga and Ssemakula, 2011). Emerging evidence demonstrates the value of participatory varietal development. Promising models of participatory varietal development have been demonstrated in Ethiopia for teff (Assefa et al., 2011) and white pea bean (Assefa et al., 2013), in Ghana for cassava (Manu-Aduening et al., 2014), and in Uganda for sweet potato (Mwanga et al., 2011).

Varietal improvements will increasingly need to focus on improved nutritional content, better resource use efficiency, tolerance to biotic and abiotic constraints, and the reduction of greenhouse gas emissions (Bouis, 1996; Tilman et al., 2002). As annuals cover nearly 70% of global cropland (Pimentel et al., 1997), there is scope for increasing the share of perennials in the global crop mix. These offer several advantages for the preservation of ecosystem services and cost-savings to farmers (Pimentel et al., 1997; Dewar, 2007; Glover and Reganold, 2010; Kell, 2011; Crews and Brookes, 2014). Beneficial traits may include the ability to be grown on resource-poor and "marginal" lands and the ability to sustain more production per unit of land than most annual crops grown on fertile lands. Varietal development is also needed in order to produce perennial alternatives that express desired traits—larger seed size, stronger stems, improved palatability, and higher seed yield (Glover and Reganold, 2010). Breeding plants with deeper and bushy root systems may also offer improved soil structure, carbon sequestration, water and nutrient retention, and higher yields (Kell, 2011). Estimates suggest that between 5 and 10 kg/m^2 (50–100 tons/ha) of carbon could be sequestered through increased root mass, resulting in globally significant sequestration of atmospheric carbon (Kell, 2011). Advances in biotechnologies, especially those related to genome editing (e.g., Clustered Regularly Interspaced Palindromic Repeats, or CRISPR), promise to complement traditional plant-breeding methods to further sharpen the focus and increase the efficiency of crop varietal improvement.

CHAPTER

19

High-Yielding Hybrids Help Adapt to Climate Change

Climate change necessitates new priorities for varietal improvement. Where climate-resilient varieties are absent, farmers may switch crop types, or be forced to bear losses.

In Asia, many rice farmers have maintained year-round production by growing either wheat or rice in the dry winter season after the monsoon rice crop. Over the past two decades, however, rice–maize farming systems have expanded rapidly throughout Asia, driven by a strong demand for maize and by the development of maize hybrids suited to areas with insufficient water for continuous rice cultivation (Timsina et al., 2011). At last count, rice–maize systems were practiced on more than 3.3 Mha of land in Asia, with the largest production areas found in Indonesia (1.5 Mha), India (0.5 Mha), and Nepal (0.4 Mha). Recent expansion of the area under rice–maize rotation has been most rapid in Bangladesh, where farmers began growing maize to sell as feed to the country's booming poultry industry. Between 2000 and 2013, maize production increased from just 10,000 tons to 2.2 million tons, and the harvested area from 5000 to 320,000 ha.

Maize grows well in Bangladesh's fertile alluvial soils and yields there are among the highest in the region. The crop is sown at the start of the cool Rabi season, which runs from November to April, after the harvesting of the rice crop grown during the July–December Aman monsoon season. While Rabi maize is generally cultivated as the sole crop, many farmers have begun to intercrop it with potatoes and with early maturing vegetables, such as red amaranth, spinach, radish, coriander, and French beans. Peas are also intercropped with maize because they do not compete for sunlight, nutrients, or space (Ali et al., 2009).

Farmers typically use high-yielding hybrid maize, which requires significant inputs of nutrients. The cost of maize production is actually higher than that of other traditional winter cereals and, as a result, poorer farmers often plant maize on only a small area of land. However, the gross margin from maize sales, per hectare, is 2.4 times greater than that of wheat or rice. Maize also has fewer pest and disease problems. The diversification to maize could also be a good strategy for climate change adaptation because maize is more tolerant of high temperatures, which are a growing problem for wheat, and is less

Sustainable Food and Agriculture
DOI: https://doi.org/10.1016/B978-0-12-812134-4.00019-4

thirsty for water. In Bangladesh, for example, 850 L of water produce a kilogram of maize grain, compared with 1000 L/kg of wheat and more than 3000 L for the same amount of rice. By reducing the extraction of groundwater for irrigation, maize production also helps reduce arsenic contamination of the soil, a severe problem in many areas of Bangladesh.

Farmers and agronomists in Bangladesh have noted that grain yields tend to decline in fields where maize has been cultivated as a dry season crop for five or more years. To ensure the sustainability of rice–maize systems, farmers need to carefully time the planting and harvesting of each crop, improve their soil and water management practices, and use quality seed.

The soil requirements of rice and maize are very different, which makes the timing of maize planting tricky. Transplanted Aman rice is usually grown in well-puddled, wet clay soils, while maize grows best in well aerated loamy soils (Ali et al., 2009). After the rice harvest, therefore, conventional preparation of fields for maize often involves three to five passes of a rotary tiller behind a two-wheel tractor. Ploughing requires considerable investments of time, fuel, and labor, and farmers need to wait for up to three weeks before the rice fields are dry enough to be tilled. Late planting of maize, in turn, can reduce yields by as much as 22 percent (Ali et al., 2009).

Conservation agriculture practices are reducing the need for ploughing and, with it, delays in maize planting. Establishing rice and maize on untilled permanent beds, using straw as mulch, has produced higher grain yields using fewer inputs than crops sown on ploughed land. Increased productivity has been attributed to higher levels of soil nitrogen and generally better soil conditions. In India, research has shown that permanent beds not only produce higher yields than ploughed land, but do so using up to 38% less irrigation water (Gathala et al., 2014). In Bangladesh, saving water is crucial during the dry months from February to May, when shallow tube wells often run dry.

The Bangladesh Agricultural Research Institute and the International Maize and Wheat Improvement Center (CIMMYT) have adapted and promoted drill seeders originally developed for wheat, so that they can be used to sow maize and rice without tillage. In northwest Bangladesh, farmers using these seeders obtained rice yields similar to those of transplanted rice, but using less water and labor, and were able to harvest the crop 2 weeks earlier. A study in Bangladesh compared yields and profitability under ploughing and zero tillage. With permanent bed planting of maize, the combined productivity of rice and maize was 13.8 tons per ha, compared to 12.5 tons on tilled land. The annual costs of rice–maize production on permanent beds was $1532 per ha, compared with $1684 under conventional tillage (Gathala et al., 2014).

Hybrid maize requires large amounts of nitrogen to produce high yields, but Bangladesh's reserves of natural gas, which is used to produce urea fertilizer, are small, finite, and nonrenewable. One promising solution to soil nutrient depletion is the application of poultry manure, which is becoming abundant; Bangladesh's poultry sector now produces about 1.6 million tons of manure every year (Ali et al., 2009). Good maize yields have been obtained by replacing 25% of the mineral fertilizer normally applied with poultry manure. Soil nitrogen can also be partially replenished by growing legumes, such as mung beans, after the maize harvest. In tropical monsoon climates, a summer mung-bean crop also mops up residual nitrogen and prevents nitrate pollution of aquifers (Borlaug Institute for South Asia, 2015).

Planting short duration rice varieties would allow farmers to plant maize earlier. However, those rice varieties produce lower yields, and farmers are generally unwilling to sacrifice the production of their main food crop. The Bangladesh Rice Research Institute is, therefore, developing higher yielding, shorter duration Aman rice varieties. The future of sustainable rice–maize farming in South Asia also hinges on the development of high-yielding maize hybrids that mature quickly and tolerate both waterlogging and drought.

Maize farming in Bangladesh is still new territory for many farmers; it will take time for them to fully integrate it into cropping systems that optimize production and improve soil health. Critical to the rapid and widespread adoption of sustainable maize production is the training of farmers in the precise timing of sowing and more effective management of irrigation and mineral fertilizer (Hasan et al., 2007; CIMMYT, 2009). Domestic maize production has reduced Bangladesh's dependence on imports. The shift to maize has also provided farmers with a means of diversifying their income and their diets. Many farmers do not sell their entire maize harvest: they feed some to their own poultry and sell eggs and meat at local markets. Increasingly, maize is being consumed as food, not just fed to poultry. As the price of wheat flour has increased, many families are mixing maize flour into wheat flour for their *chapatis* (flatbreads).

CHAPTER

20

Conservation Agriculture

A variety of measures to mitigate soil erosion, improve water-holding capacity, and increase soil organic matter help to improve soil health and boost crop yields. A key feature is the elimination or reduction of soil disturbance through tilling. Zero tillage involves no ploughing prior to sowing. Conservation agriculture (CA) consists of a group of management strategies to minimize soil disturbance, maintain soil cover, and rotate crops. This seeks to maintain an optimum environment in the root zone in terms of water availability, soil structure, and biotic activity (Kassam et al., 2009).

CA evolved partially as a response to the severe soil erosion that devastated the Midwest in the United States in the 1930s. Currently, CA systems are practiced successfully across a range of agroecological conditions, soil types, and farm sizes (Derpsch et al., 2010). At present, CA practices cover just over 8% of global arable cropland, but are estimated to be spreading globally by some 6 Mha per year to a total of 155 Mha in 2014 (Kassam et al., 2014). Adoption varies greatly by region. CA covers some 69% of arable cropland in Australia and New Zealand, 57% of arable cropland across South America, and 15% in North America (Jat et al., 2014). By contrast, adoption across Europe and Africa is low, covering only 1% of arable cropland in each of the two continents (Jat et al., 2014). A successful example of use of CA for rice–wheat systems in the Indo-Gangetic Plains is presented in Box 20.1.

Increased productive potential has resulted in yield differences ranging from 20% to 120% for CA compared with conventional tillage systems (Kassam et al., 2009). Beneficial impacts in terms of resource efficiency include reduced need for fertilizer application over time, lower runoff, and increased resilience to pest and disease. All these result in significant savings, which, combined with yield increases may translate to significant financial benefits for farmers relative to conventional ploughing practice (Sorrenson, 1997). A recent European Conference on Green Carbon highlighted the importance of soil carbon content as a marker and enabler of sustainability in agroecosystems, and concluded that CA presents a good strategy to maintain and improve soil carbon levels. Comparisons between conventional tillage and reduced or no-till systems have found higher soil organic carbon, lower

Sustainable Food and Agriculture
DOI: https://doi.org/10.1016/B978-0-12-812134-4.00020-0

BOX 20.1

CONSERVATION AGRICULTURE FOR RICE–WHEAT SYSTEMS IN THE INDO-GANGETIC PLAINS

The Indo-Gangetic Plains are both the rice bowl and breadbasket of 1.8 billion people across South Asia, with a rice–wheat system covering 13.5 million hectares and producing about 80 million tons of rice and 70 million tons of wheat annually (Ladha et al., 2009). Around one in three farm households in the plains has adopted at least one resource-conserving technology as of 2009, with the highest rates—of almost 50%—in the northwest.

Land management: By the early 2000s, some 300,000 ha of wheat were under zero tillage, contributing to yield increases in the range of 6%–10% (Singh et al., 2008) and savings of $50 to $70 per ha on water (Erenstein and Laxmi, 2008). In some areas, irrigation water productivity has improved by 65% relative to conventional practice (Sharma et al., 2010). On the western Indo-Gangetic Plains, the adoption of zero tillage in wheat production has reduced farmers' costs per hectare by 20% and increased net income by 28% while reducing greenhouse gas emissions (Aryal et al., 2015). Laser-assisted land leveling can reduce water applications by as much as 40%, improving the efficiency of fertilizer and boosting rice and wheat yields by 5%–10%. Importantly, this agricultural method is affordable by smallholders and profitable on all farm sizes (IRRI, 2009). In northwest India, zero-tillage seed drills were the most common item of agricultural machinery after tractors. Their high adoption rate was made possible by the ready availability of seed drills developed by the private sector, with strong support from state and local governments (Erenstein, 2009). To reduce the wasteful use of fertilizer, the Rice–Wheat Consortium for the Indo-Gangetic Plains promoted "needs-based" nitrogen management by introducing leaf color charts indicating the best times for fertilization. The charts were originally designed for rice, but were spontaneously adapted to wheat by farmers (Singh et al., 2009). Using the charts, farmers have reduced fertilizer applications by up to 25% with no reduction in yield.

Planting and seeding: Broadcasting or drum seeding presoaked wheat seeds, without tillage, is a low-cost technology particularly suited to smallholders (Gupta et al., 2015), reducing costs (IRRI, 2009) and avoiding yield losses due to late sowing. For rice, a change to short-season cultivars and direct dry seeding has saved time and resources by eliminating the need for transplanting and reducing water use, energy costs, and labor requirements. Farmers also use practices such as raised-bed planting, alternate wetting and drying, and aerobic rice, where seeds are sown directly into the dry soil and then irrigated. Several strategies have been proposed to increase the uptake of dry seeding in rice production, including intercropping with *Sesbania*, which reduces weed infestations and boosts yields in unpuddled rice fields. However, large-scale adoption of dry seeding is hindered by lack of farmer access to suitable equipment. A decisive shift to CA practices in rice—particularly the retention of crop residues—would create positive synergies in the production of the two

BOX 20.1 (*cont'd*)

cereals. While many farmers have adopted the drill seeding of wheat into residues of the preceding rice crop, most continue to burn rice straw after the harvest, which leads to serious air pollution (Fischer et al., 2014). To discourage burning off and encourage mulch-based zero tillage, the governments of Punjab and Haryana states are now upscaling a new technology, the "Happy Seeder," which can drill wheat seed through heavy loads of rice residues (Gathala et al., 2013). Finally, in the northwest, sugar cane, mung beans, mint, maize, and potatoes are now cultivated as part of rotations in the rice–wheat system. On the eastern plains, where winters are shorter, there is a growing trend towards replacing wheat entirely with potato and maize, which offer higher economic returns. Farmers have also introduced new crop rotations that disrupt the lifecycles of insect pests and weeds and promote soil health. In Pakistan's Punjab province, smallholder farmers rotate rice with berseem clover, a fodder that improves soil fertility and suppresses weeds that might otherwise infest subsequent cereal crops (Hussain et al., 2012). On the eastern plains, where fields generally remain fallow for 80 days after the wheat harvest, a summer mung-bean crop planted on zero-tilled soil produces 1.45 tons/ha, worth $745.

emissions, and improved soil quality under the former (Brenna et al., 2013 in Italy; Franzluebbers, 2005, 2010, for the southeastern United States; Spargo et al., 2008 in Virginia, United States), especially on certain types of soils and with the addition of cover cropping.

While CA offers much potential for sustainable intensification, scientific debate highlights a number of difficulties and contentions. Each component in the CA "package" requires interpretation (Stevenson et al., 2014), and the applicability and scalability of CA to smallholder systems has been questioned, especially for developing country contexts (Giller et al., 2009, 2011; Stevenson et al., 2014). While the applicability of CA to sub-Saharan contexts has been called into question, some case studies nevertheless show remarkable social–ecological outcomes. Comparisons between CA and conventional plots demonstrate yield increases in the former, as well as reduced soil loss, increased soil carbon content, improved soil structure, and water productivity (Marongwe et al., 2011). Collectively, recent evidence from the African Union (Pretty et al., 2011) shows that the adoption of CA principles has led to improvements on 26,000 ha, with a mean yield increase ratio of 2.20 and annual net multiplicative yield increases in food production of some 11,000 tons per year. In addition, a number of positive externalities, with cost-saving or income-boosting effects were also reported, including reduced soil erosion, increased resilience to climate-related shocks, increased soil carbon, improved water productivity, reduced debt, livelihood diversification, and improved household-level food security (Marongwe et al., 2011; Owenya et al., 2011; Silici et al., 2011).

Nevertheless, there have been calls for improvements to the evidence base on CA (Brouder and Gómez-MacPherson, 2014). Meta-analyses and reviews across cases also show that the evidence on yield impacts and carbon sequestration potential is mixed (Stevenson et al., 2014). This may, in part, reflect the context sensitivity of CA, where outcomes depend on the precise combination of practices used, and differ by crop type. There is some evidence that zero tillage may result in yield penalties in the short term (Brouder and Gómez-MacPherson, 2014). Such evidence may hinder adoption among smallholders, who may "attribute more value to immediate costs and benefits than those incurred in the future," as they must navigate precarious and pressing concerns over food and livelihood security (Giller et al., 2011, p. 3). However, other studies find yield increases to be stable over time. For example, Derpsch (2008) compared conventional and CA systems over a decadal scale and found yield decreases over the period in the former and yield increases over the period under CA, in addition to lower use of inputs. There is also a need for more evidence on the precise implications of improved land management across agricultural sectors and agroecosystems (e.g., Morgan et al., 2010) using appropriate soil sampling strategies (Baker et al., 2007) and standardized methodologies (Derpsch, 2014). An example of successful use of zero tillage is found in the Kazakh Steppe of Central Asia (Box 20.2).

BOX 20.2

ZERO TILLAGE IN CENTRAL ASIA, KAZAKH STEPPE

In the spring of 2012, as farmers across the semiarid steppes of northern Kazakhstan were sowing their annual wheat crop, the region entered one of its worst droughts on record. In many areas, no rain fell between April and September and daily summer temperatures rose to several degrees above normal (CIMMYT, 2013). That year, many farmers lost their entire crop and Kazakhstan's wheat harvest, which had reached 23 million tons in 2011, plummeted to less than 10 million tons. Some farmers, however, did not lose their crop. They were among the growing number of Kazakh wheat growers who have fully adopted CA, including zero tillage, retention of crop residues on the soil surface, and crop rotation. Those practices have increased levels of soil organic carbon and improved soil structure in their fields, allowing better infiltration and conservation of moisture captured from melting winter snow (Nurbekov et al., 2014). As a result, some farmers in Kostanay region achieved yields of 2 tons/ha in 2012, almost double the national average of recent years (CIMMYT, 2013). Overall, the adoption of CA in Kazakhstan has enabled an increase in annual wheat production of almost 2 million tons, sufficient to feed some 5 million people.

Around 2 million of Kazakhstan's 19 million hectares of cropland are under full CA. On 9.3 million hectares, farmers have adopted minimal tillage, which uses narrow chisel ploughs at shallow depths (Karabayev et al., 2014). State policy actively promotes CA, and designates the development and dissemination of water-saving technologies as the top priority in agricultural research. In 2011, Kazakhstan introduced subsidies on CA equipment that are three to four times higher than those on conventional technologies (Nurbekov et al.,

BOX 20.2 (*cont'd*)

2014). Government support has encouraged farmers in northern Kazakhstan to invest an estimated $200 million to equip their farms with zero-tillage machinery. In 2000, the International Maize and Wheat Improvement Center (CIMMYT) and the Food and Agriculture Organization of the United Nations (FAO), together with Kazakh scientists and farmers, launched a program to introduce CA in rainfed areas and raised-bed planting of wheat under irrigation in the south of the country (Karabayev and Suleimenov, 2010). Trials in the north showed that zero-tilled land produced wheat yields 25% higher than ploughed land, while labor costs were reduced by 40% and fuel costs by 70%. The trials demonstrated the advantages of growing oats in summer instead of leaving land fallow. With an oat crop, the total grain output from the same area of land increased by 37%, while soil erosion was much reduced.

Land management: Today, Kazakhstan ranks among the world's leading adopters of zero tillage, rising from zero in 2000 to 1.4 million hectares by 2008, mainly driven by high adoption rates on large farms and those with rich black soils, where high returns provide the capital needed for investment. Many farmers have found that combining zero tillage with permanent soil cover also helps to suppress weeds. Retaining crop residues increases the availability of water to the wheat crop while reducing or even eliminating erosion, and thus addressing desertification and land degradation that costs Central Asian countries an estimated $2.5 billion annually. On-farm research has found that the use of residues to capture snow, along with zero tillage, can increase yields by 58%.

Planting and seeding: Farmers are taking advantage of available—and sometimes abundant—rainfall to grow oats, sunflower, and canola. Studies have shown the high potential of other rotational crops, including field peas, lentils, buckwheat, and flax (Karabayev, 2012). Further increases will be possible with the development of high-yielding wheat varieties better suited to zero tillage and the north's harsh winters and increasingly hot summers. That option is being explored through a program with CIMMYT, which, in Mexico, crosses local Kazakh wheat varieties with Mexican, Canadian, and US cultivars.

CHAPTER

21

Climate-Smart Soil Redesign

21.1 CARBON SEQUESTRATION IN AGROECOSYSTEMS

Meeting the climate change challenge to agriculture requires good soil health. Physical, chemical, biological, and ecological soil health helps build resilience to climatic variation. Additionally, soil is fundamental to mitigating agricultural emissions of methane, nitrous oxide, and carbon dioxide.

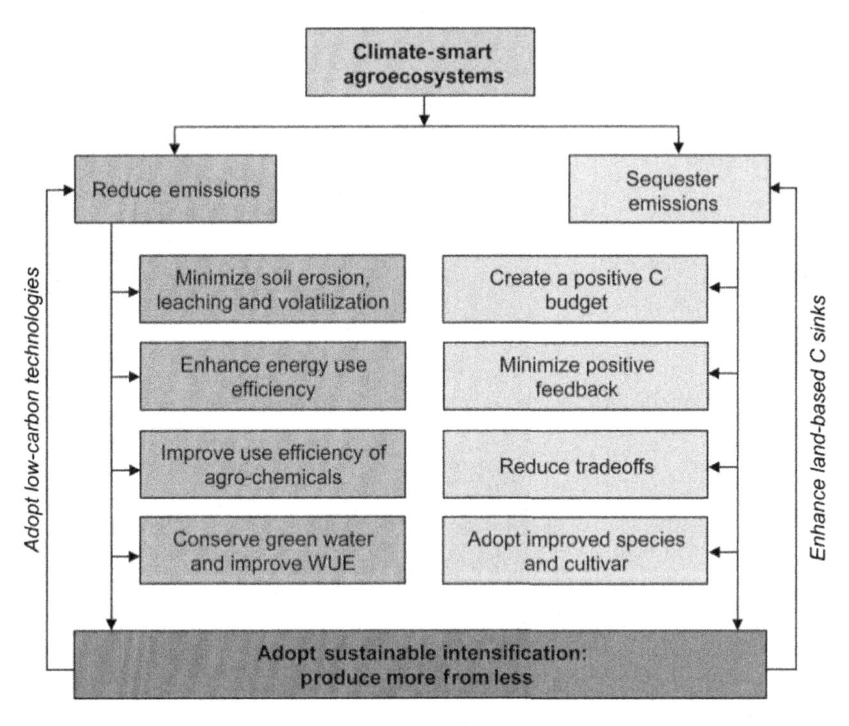

FIGURE 21.1 Basic principles of reducing and sequestering emission of carbon dioxide in agroecosystems. Source: *Authors.*

Sustainable Food and Agriculture
DOI: https://doi.org/10.1016/B978-0-12-812134-4.00021-2

In recognition of the importance of soils, the Paris COP 21 climate summit adopted the "4 per Thousand" program for advancing food security and mitigating climate change. The strategy is to sequester organic carbon in the world's soils at the rate of 0.4%/year to 40 cm depth, an amount equivalent to ~3.6 petagrams of carbon (PgC) per year. More recently, the Fiji-Bonn summit in 2017 (COP 23) builds on the initiatives of COP 21 and COP 22 by developing programs that exploit the carbon storage capacity of land-based sinks. As with other measures to improve sustainability, this has a number of cobenefits for people and ecosystems, including improved water quality, higher biodiversity, and increased input efficiency.

Two distinct but closely interrelated paths to achieving climate-smart agroecosystems are reducing emissions and sequestering emissions (Fig. 21.1). Strategies of reducing

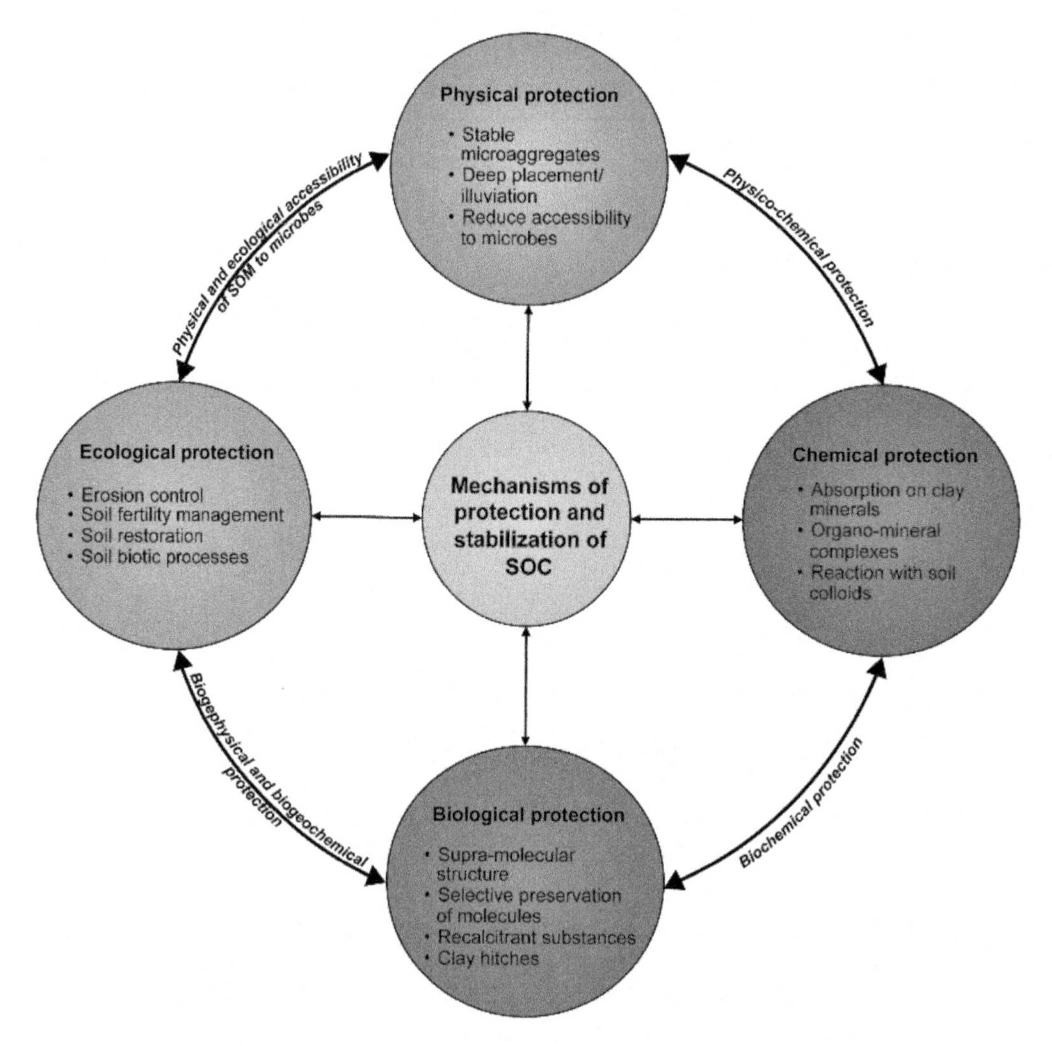

FIGURE 21.2 Mechanisms of protection of soil organic matter to reduce access and prolong the mean residence time. Source: *Authors.*

emissions involve adoption of land use and soil management options that: (1) minimize losses of soil organic carbon stocks by accelerated erosion, leaching, and volatilization/mineralization; (2) maximize energy use efficiency by reducing tillage and other vehicular traffic; (3) improve use efficiency of fertilizers, pesticides, and other agrochemicals; and (4) conserve green water supply in the root zone and enhance water use efficiency. The second pathway is to sequester emissions by creating a positive soil/ecosystem carbon budget (Fig. 21.1), specifically through minimizing positive feedback to climate change, reducing trade-offs, and adopting improved species and cultivars. The overall strategy is to increase efficiency, thus producing more from less, adopt low-carbon technologies, and enhance sand/soil-based carbon sinks (Fig. 21.1).

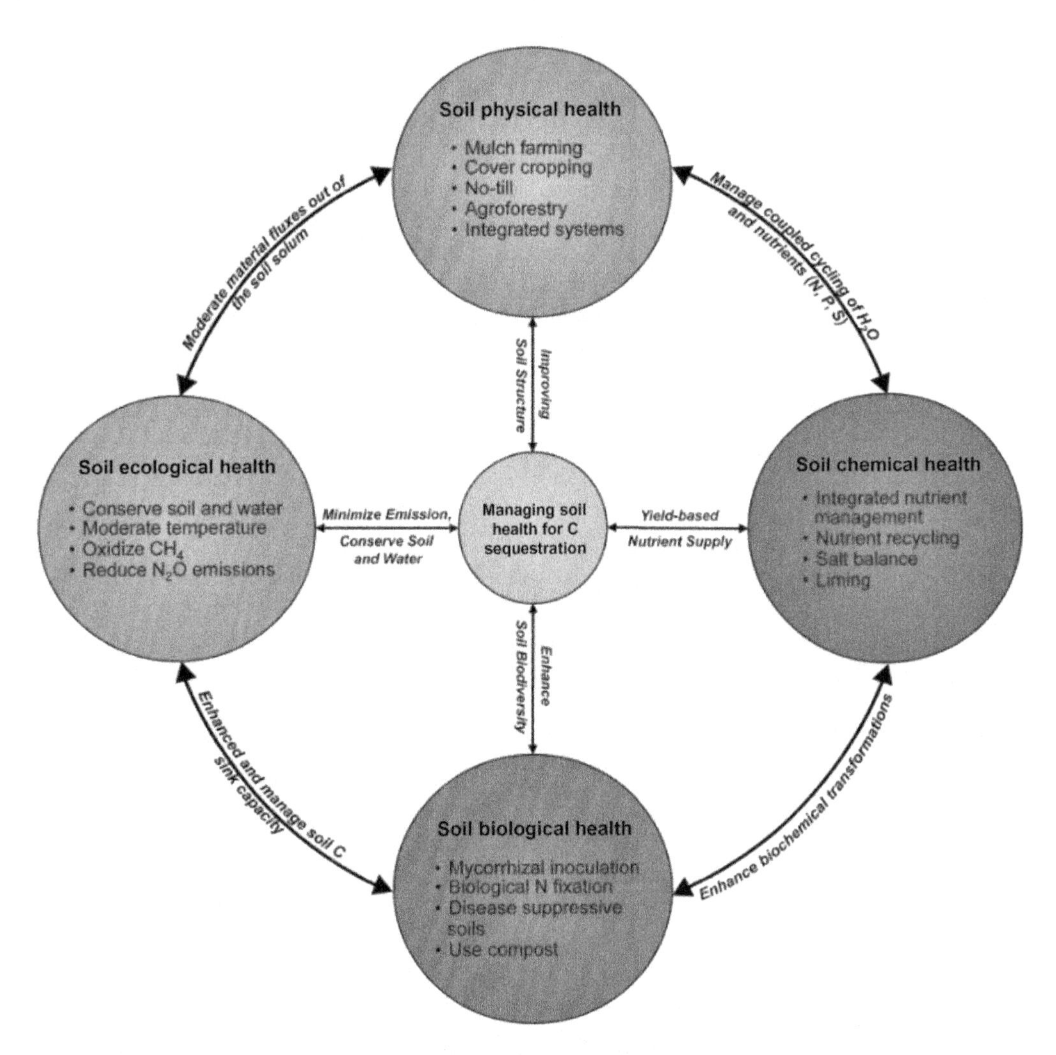

FIGURE 21.3 Managing soil for carbon sequestration through improvements of physical, chemical, biological, and ecological aspects of soil health. Source: *Authors.*

Once sequestered, soil carbon must be stabilized and protected by a variety of means (Fig. 21.2).

Sequestration of carbon in soil (both as soil organic carbon and as secondary carbonates) also necessitates the judicious management of soil health (Fig. 21.3).

21.2 AGRICULTURAL PRACTICES FOR CARBON SEQUESTRATION

A variety of basic strategies help to sequester soil carbon (Table 21.1). There is no one-size-fits-all agricultural practice for all of the 300,000 soil series of the world. Context matters, and site-specific best-management practices have to be identified and fine-tuned.

21.3 GLOBAL TECHNICAL POTENTIAL OF SOIL CARBON SEQUESTRATION IN AGROECOSYSTEMS

The technical potential refers to the upper limit (the maximum soil carbon sink capacity) of carbon sequestration. In general, it is equivalent to the historic loss of soil organic carbon caused by prior land use and soil, crop, and animal management practices. In addition, soil carbon sink capacity also depends on texture (clay and silt contents), nature of clay minerals (2:1 type and expanding lattice), depth of soil solum, available water capacity, and nutrient reserves (nitrogen, phosphorus, and sulfur). The technical potential of

TABLE 21.1 Strategies and Practices of Carbon Sequestration in Soils

Strategy	Agricultural Practices
1. Conserve soil and water	Conservation agriculture, agroforestry, complex farming systems, land forming, contour hedges, controlled grazing
2. Enhance green water supply	Water harvesting, drip subirrigation, condensation irrigation, deep-rooted crops, using treated greywater, reducing losses by runoff and evaporation
3. Soil fertility management	Integrated nutrient management, recycling agricultural by-products, precision farming, balanced nutrient application, yield-based fertilization, fertigation, biological nitrogen fixation, mycorrhizal inoculation
4. Managing soil compaction	Deep-rooted crops, enhancing activity of earthworms and other soil biota, precision tillage, guided traffic, dual tires, reducing vehicular traffic
5. Weed and pest management	Cover cropping, complex rotations, disease suppressive soils through inoculation and composting, selective herbicides
6. Sustainable intensification	Enhancing productivity, adopting improved species and varieties, reducing losses, improving use efficiency of inputs
7. Creating a positive soil carbon budget	Recycling biomass (crop residues, animal manure, compost, sawdust), reducing losses by erosion and runoff, oxidizing methane by improved aeration, minimizing nitrous oxide by using slow release formulations and split application, reducing tillage and traffic

Source: Authors.

soils is estimated to be 0.4–1.2 PgC/year for croplands and 0.3–0.5 PgC/year for grasslands/pasture lands. The technical potential of restoring eroded and desertified soils is 0.2–0.7 PgC/year and that of reclaiming salt-affected soils is 0.3–0.7 PgC/year. The total technical potential of soils of managed and degraded or eroded ecosystems is 1.2–3.1 PgC/year, with an average of 2.15 PgC/year (Table 21.2) (Lal, 2010).

In comparison, anthropogenic emissions (by fossil fuel combustion, cement production, and land use conversion) are estimated to be 9.9 PgC/year for 2015. Thus, the technical potential of carbon sequestration in soils of managed ecosystems can offset about 22% of

TABLE 21.2 Technical Potential of Carbon Sequestration in Soils of Managed and Degraded Ecosystems

Land Use	Technical Potential (Gigaton of Carbon Per Year)
Cropland	0.4–1.2
Grasslands/grazing lands	0.3–0.5
Restoration of eroded desertified soils	0.2–0.7
Restoration of salt-affected soils	0.3–0.7
Total	1.2–3.1 (2.15)

Source: Adapted from Lal, R. 2010. Managing soils and ecosystems for mitigating anthropogenic carbon emissions and advancing global food security. BioScience. 60 (9): 708–721.

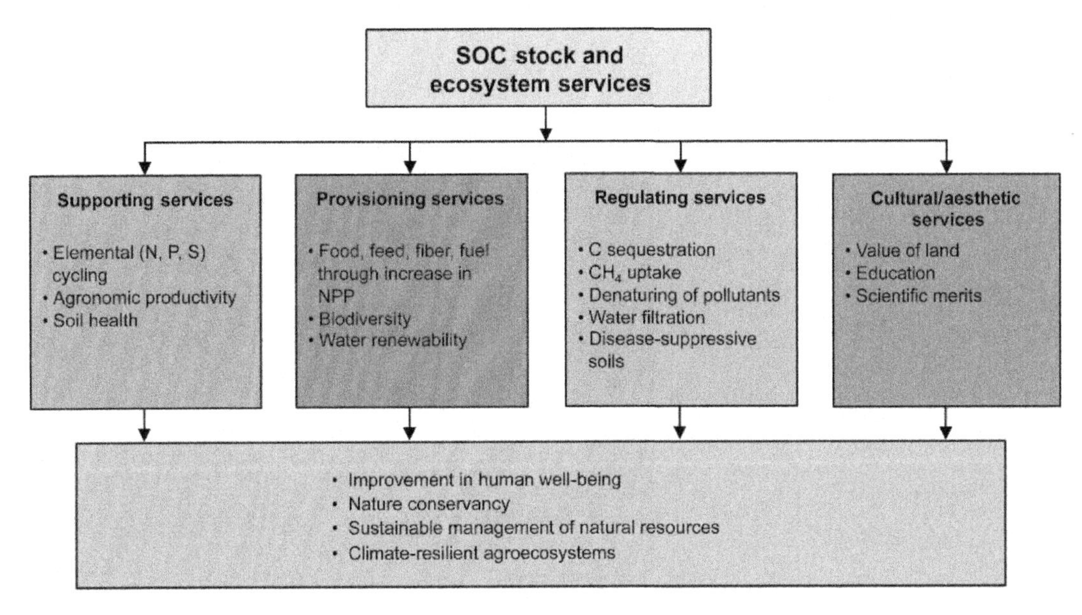

FIGURE 21.4 Ecosystem services provisioned by soil carbon sequestration and soil health management. Source: *Authors.*

the anthropogenic emissions under a best-case scenario. Despite this finite potential, the strategy of soil organic carbon sequestration is a cost-effective option and has numerous cobenefits through provisioning of ecosystem services (Fig. 21.4). Thus, even the incremental drawdown of atmospheric carbon dioxide through soil organic carbon sequestration is of global significance.

CHAPTER

22

Sustainable Livestock and Animal-Sourced Food

The livestock sector is a powerful driver of major economic, social, and environmental changes in food systems worldwide, and provides an entry point for understanding the issues around sustainable agricultural development as a whole. The sector is particularly dynamic and complex. Though critical issues are global in nature, the ways in which they manifest themselves or manner in which they can be dealt with take diverse forms in different livestock systems and across countries.

Livestock accounts for around 43% of agricultural gross domestic product globally, reaching up to 80% in countries such as Sudan and New Zealand. The sector has implications for animal-feed demand, for market concentration in agricultural supply chains, for the intensification of production at the farm level, for farm income, for land use, and for human and animal nutrition and health. Improving the sustainability of livestock and animal-sourced food production is essential to meeting both climate targets and other human development goals. Traditional livestock systems support the livelihoods of around 70% of the world's poor, while intensive, large-scale operations cater to the growing demand for meat, milk, and eggs worldwide. It is this growth in demand that has caused livestock to set the speed of change in agriculture in recent decades. As a result, the sector is a key driver of global environmental change. It is the largest user of land resources; permanent meadows and pastures represent 26% of global land area and feed crops account for one-third of global arable land. Globally, manure supplies up to 12% of nitrogen for crops and up to 23% in mixed crop-livestock systems in developing countries. As a driver of deforestation, demand for feed, and transportation and processing infrastructure, the livestock sector is directly and indirectly responsible for 14.5% of greenhouse gas emissions. At the same time, some livestock systems are among the most vulnerable to climate change (particularly those in dry areas) and to newly emerging environment-related diseases. If the best existing practices in a given system and region can be shared and learned from more widely, the livestock sector has huge potential for improvement.

Sustainable Food and Agriculture
DOI: https://doi.org/10.1016/B978-0-12-812134-4.00022-4

Going forward, it is estimated that consumption of animal-sourced food is expected to rise until 2050, and will do so faster in developing countries. Much of the increased crop demand in the period to 2050 will be for feedstuffs for livestock.

The livestock sector plays a crucial economic role for around 60% of rural households in developing countries, including smallholder farmers, agropastoralists, and pastoralists. It contributes to the livelihoods of about 1.7 billion people. In these communities, livestock is integral to the cultural and traditional practices, values, and landscapes. The sector contributes significantly to household income, nutritional security, and resilience. The relative importance of each of these different roles varies by livestock species, agroecological zone, production system, and sociocultural context, and livestock often assumes several roles simultaneously. Examples of improved and sustainable silvopastoral production system in Colombia is provided in Box 22.1, and of goat production system in Kenya in Box 22.2.

Contributions of livestock include the provision of services (e.g., draught/hauling power, insurance, and savings), food (e.g., meat, milk, and eggs), nonfood products (e.g., wool, hides, and skins), and less tangible benefits such as status and inclusion in social networks. Livestock products such as milk, meat, and eggs provide essential micronutrients, such as calcium, zinc, iron, vitamin A, vitamin B12, and vitamin B2 (riboflavin).

Small stock such as poultry, pigs, sheep, and goats are particularly important assets for rural women, whose role in agriculture is often unappreciated. With livestock activities, women can earn income that remains under their control, with implications for the intrahousehold allocation of food and resources. Women's roles within livestock production systems differ from region to region, and the distribution of ownership of livestock between men and women is strongly related to social, cultural, and economic norms. Too often, however, women face multiple forms of discrimination, from lack of access to education and productive resources to discriminatory political and legal systems that in conjunction limit their ability to benefit from the livestock sector. At present, not enough gender-disaggregated data are available to fully understand the specific challenges faced by women in this sector.

As one of the fastest growing sectors in agriculture, livestock production offers employment opportunities along numerous animal production value chains. It also generates jobs in related sectors, including transport, trade, feed, and input provision, as well as veterinary services. Gainful and productive employment opportunities can also be derived from value-enhancing activities in livestock. Due to limited requirements in capital investment and land ownership, short-cycle species production (small ruminants, poultry, dairy, etc.) in particular presents unique opportunities for the rural poor, especially women and youth, to benefit from surging demands for animal products.

Finally, livestock is important for its links with other sectors, particularly crop production. The sector uses crops and crop byproducts as feed, and crops are often reliant on animals for fertilizer (manure) and power (animal traction). Livestock also enhance labor productivity, reduce drudgery and dependence on hand tools, and smoothen labor demand across agricultural seasons. An important role is converting organic material not suited for human nutrition into high-value food and nonfood products (Otte et al., 2012). Though numbers vary significantly by region and species, globally almost 86% of total animal feed comes from nonedible matters.

BOX 22.1

SUSTAINABLE SILVOPASTORAL SYSTEMS IN COLOMBIA

In Colombia, extensive cattle ranching, with a population equivalent to 0.62 livestock units per capita (Vera, 2006), has caused soil degradation and deforestation and in dry areas has hastened desertification. This type of ranching provides limited feed quality and is vulnerable to seasonal extremes of temperature and drought due to limited shade, poor soil quality, and access to water.

Silvopastoral systems have been proposed to increase efficiency while reducing the environmental burdens of extensive ranching systems. Intensive silvopastoral systems (ISPS) are a type of silvopastoral system that combine high-density cultivation of fodder shrubs (4,000–40,000 plants per hectare) with: (1) improved tropical grasses; and (2) tree species or palms at densities of 100–600 trees per hectare. These systems are managed under rotational grazing with high stocking rates and brief grazing periods interspersed with long recovery periods (Calle et al., 2012). ISPS have the potential to deliver much more feed of higher quality. The additional plant matter, in addition to root density and biodegradable material, can increase soil quality and water retention as well as carbon content in the soil. By using animal breeds adapted to tropical environments, the ISPS have the potential to achieve high levels of production from local feed sources in pasture-based environments. This measure maintains good health, natural behavior, and ease of animal management.

A study carried out on the establishment of ISPS on three farms showed higher feed production and profit: La Luisa, a beef finishing farm in the Cesar valley with four groups of beef animals and a total of 500 cattle on the farm; Petequí, a dairy farm in the Cauca valley with around 70 crossbred dairy animals; and El Hatico, a dairy farm in the Cauca valley rearing Lucerna breed animals. The herd was divided into five groups, ranging from preparturition cows to high-, medium- and low-lactation cows.

The results showed that ISPS are more productive and profitable than extensive cattle ranching systems, while productivity goes hand in hand with animal welfare. Their success is based on good management, extension, and access to capital, which build farmers' long-term capacity to deliver efficient and increasingly productive beef and dairy production, while responsible investment in sustainable environmental management delivers potential climate mitigation benefits. The study provided evidence for the ability of ISPS to create "triple-win" solutions for sustainable livestock production: productivity and profitability gains; environmental improvements; and animal welfare benefits. The knowledge developed in these farms is used by the project Colombia—Mainstreaming Biodiversity in Sustainable Cattle Ranching, led by FEDEGÁN-FNG and in partnership with the Center for Research on Sustainable Farming Systems (CIPAV), the Nature Conservancy, and Fondo Acción. The project is administered by the World Bank, with funding from the Global Environment Facility and the United Kingdom's Department of Energy and Climate Change, and aims at establishing 10,000 ha of ISPS and an additional 40,000 ha of other silvopastoral systems in the country. As ISPS are management-intensive, capacity building via extension and advisory services is a key component of successful delivery. Targeted investment early in the establishment of the silvopastoral system, and an effective capacity-building program, tailoring knowledge development to individual farmers' needs, can provide increased potential for success.

BOX 22.2

IMPROVED GOAT PRODUCTION IN KENYA

Small livestock are important sources of food and income for smallholders, and make a particularly important contribution to women and children in such households. Goats are well adapted to small farms, as they are able to efficiently convert unused vegetation into milk and income, and thus contribute to the food and livelihood security of vulnerable families who are unable to keep more expensive livestock such as cows (Peacock, 1996).

In Meru district, Kenya, Farm Africa developed a model to improve smallholder goat enterprises mostly managed by women and assist them to develop local and regional goat milk markets. The model is suitable for mixed-farming smallholders with up to 2 ha, at least 500 mm/annum rainfall, and which grow a variety of crops (Farm Africa, 2007). It includes the establishment of intensive dairy goat enterprises with housed goats, on-farm fodder development and conservation, crossbreeding of local goats with an improved dairy breed, and mapping out market opportunities and developing linkages to those that are viable. In the model, farmer groups, community-based private providers, and local nongovernmental organizations manage all the necessary support services and inputs. For example, replacement bucks are bred locally at group-managed breeding units. In the scaled-up intervention, 120,000 goat milk enterprises were targeted for assistance over 10 years. The model has resulted in tenfold increases in milk production and income within 5 years. As a result, families have been able to benefit from increased availability of milk and meat, reduced incidence of animal diseases, and increased availability of manure (Peacock and Hastings, 2011).

There are also strong linkages with aquaculture, particularly in traditional systems in Asia. Forests and pastures are closely linked in land use change dynamics. By focusing on linkages with other sectors, rural development and investment in crops, livestock, forestry, fisheries, and aquaculture can turn these sectors into powerful tools to end poverty and hunger within overall sustainable development. There has been a tendency for agricultural development analysts to focus solely on the developmental impact of crop production when, in fact, the subsectors of livestock, fishery, and forestry also have an important bearing on productivity and human well-being in nonindustrial countries.

Greater resource efficiency in livestock production is needed to maintain production systems within critical planetary limits; preserve the ecosystem services on which agricultural production relies; and reduce land degradation, biodiversity loss, and pressure on water use and quality. Resource efficiency can be improved through different technical means, including improving livestock management, careful breeding, and health and feed efficiency; closing the nutrient cycle; and reducing food losses and waste.

In addition to these general challenges, each livestock system is also confronted with specific challenges. These, along with key recommendations, are outlined in Table 22.1.

TABLE 22.1 Challenges and Recommendations for Four Livestock Systems

System Type	Challenges	Recommendations
Smallholder mixed-farming systems	• Limited access to resources, markets, and services • Variable resource efficiency • Large yield gaps • Vulnerability to rapid structural transformation	• Enhance economic viability and access to markets and prioritize fairer markets and measures to overcome obstacles faced, especially by women and marginalized and vulnerable groups engaged in small-scale livestock. • Create an enabling environment for collective organizations and actions of smallholders and invest in market information and infrastructure. • Strengthen security, tenure, and title of customary lands, property rights, and governance of common natural resources, building on the Committee on World Food Security Voluntary Guidelines on the Responsible Governance of Tenure of Land, Fisheries and Forests and other relevant instruments in the international legal framework. • Increase potential of livestock for sustainable livelihoods in smallholder. mixed farming systems.
Pastoral systems	• Limited access to resources, markets, and services • Variable resource efficiency • Large yield gaps • Vulnerability to rapid structural transformation • Conflicts over land and water • Economic and political exclusion • Gender inequity • Poor animal health • Vulnerability to losses from zoonotic diseases	• Strengthen the role of local pastoralist organizations in adaptive land management and governance in order to increase the resilience of pastoral systems and households, in particular for climate change, conflicts, and protracted crises, as well as price volatility. • Consider the use of innovative financing mechanisms to invest in basic services adapted to the needs and ways of life of pastoralists, including culturally appropriate education, health, communications, drinking water and sanitation services, and renewable energy systems. • Improve the connection of pastoralists to local, national, and international markets. • Strengthen security, tenure, and title of customary lands, property rights, and governance of grazing resources, building on the Committee on World Food Security Voluntary Guidelines on the Responsible .Governance of Tenure of Land, Fisheries and Forests. • Enable the mobility of pastoralists, including transboundary passage, through appropriate infrastructure, institutions, agreements, and rules.
Commercial grazing systems	• Land degradation • Competition for land and resources from other sectors • Poor labor conditions • Technical inefficiencies	• Support sustainable management of livestock, pastures, and feed in order to minimize harmful environmental externalities, methods that include promoting models of production that preserve biodiversity and ecosystem services and that reduce greenhouse gas emissions. • Explore context-specific technical and policy initiatives for integration of plants and animals at diverse scales, such as for instance, agrosilvopastoral systems. • Promote practices that enhance resource efficiency and resilience of commercial grazing systems.

(*Continued*)

TABLE 22.1 (Continued)

System Type	Challenges	Recommendations
Intensive livestock systems	• Environmental challenges resulting from intensification • Harm to human health arising from antimicrobial resistance • The emergence of new diseases • Social consequences of intensification, including rural abandonment, poor working conditions, low wages, and vulnerability of migrant labor • Economic vulnerabilities caused by dependence on external inputs • Market concentration • Price volatility • Inequitable distribution of value-added products • Difficulty of internalizing externalities in prices	• Ensure that the working and living conditions of workers—especially women and other vulnerable workers, including temporary and migrant workers—at all stages of production, transformation, and distribution meet international standards and are protected by domestic laws. • Undertake a lifecycle assessment along the complete food chain to identify options for increasing production while minimizing negative environmental impacts and excessive use of energy, water, and nitrogen. • Improve technical efficiency by monitoring the individual performance of herds and animals. • Support and improve animal health and welfare by promoting good practices and by establishing and enforcing robust standards for different species in intensive systems, building upon the World Organization for Animal Health guidelines and private sector initiatives. • Explore and implement approaches for the reduction of antimicrobial use in livestock production. • Develop innovative approaches with farmer organizations at diverse scales to facilitate the use of manure as organic fertilizer, and to promote the use of crop coproducts or residues and waste as feed, including through technical innovations.

Source: HLPE, 2016. Sustainable agricultural development for food security and nutrition: what role for livestock? A report by the high level panel of experts on food security and nutrition of the Committee on World Food Security. Rome.

22.1 MULTISTAKEHOLDER INITIATIVES FOR SUSTAINABLE LIVESTOCK

Sustainable livestock systems require coordinated action across a number of different sectors, bringing together sustainable practice in agronomies of feed crops, feed production, animal husbandry, and food production. Multistakeholder initiatives are essential to govern for sustainability in this context (Barling et al., 2002). In recognition of this, a number of platforms are aiming to bring together stakeholders from the private, public, and third sectors to coordinate sustainability transitions in the livestock sector. These include FAO's Global Agenda for Sustainable Livestock (elaborated upon in Section IV), the Livestock Environmental Assessment and Performance Partnership (LEAP), the Global Roundtable for Sustainable Beef (GRSB), and a number of location- or sector-specific platforms (e.g., the Ethiopia Fodder Roundtable). These initiatives play a supportive role towards continuous improvements in policy and practice by facilitating dialogue among stakeholders and building consensus, conducting and supporting joint analysis, and promoting innovation and supporting investments. However, a key challenge is to initiate dialogue and push transitions in an existing context where monocentric governance and short planning horizons remain dominant (Breeman et al., 2015).

22.2 PRIVATE SECTOR INITIATIVES

During the last decade, a number of private initiatives have been developed to improve sustainability along the livestock supply chain to reduce harmful environmental impacts while improving animal welfare and nutritional attributes from increasing production. Some of the main umbrella organizations bringing together national sustainability initiatives, mainly in intensive and commercial grazing systems, are the International Meat Secretariat, representing the global meat and livestock sector, the International Dairy Federation, the International Poultry Council, the International Egg Commission, and the International Feed Industry Federation. are the main umbrella organizations bringing together national sustainability initiatives, mainly in intensive and commercial grazing systems.

Initiatives generally involve reporting and sharing information on best practices drawn from evidence-based scientific studies, pilot schemes on representative farms and enterprises, benchmarking, and developing indicators to assess progress. In some cases, livestock farmers are offered certification by independent organizations for their sustainability and animal welfare practices. Supermarkets, food outlets, and livestock processors often require livestock farmers to enter into contracts whereby they adhere to codes and standards of practice in order to sell their livestock products.

The Global Roundtable for Sustainable Beef is the largest global, multistakeholder initiative to advance continuous improvement in sustainability of the global beef value chain. GRSB consists of producers and producer associations, the trading and processing sector, retail companies, civil societies, and national and regional roundtables. The major beef-producing countries—Australia, Brazil, Canada, New Zealand, and the United States—are represented. GRSB does not set standards or create a certification program, but rather provides a common baseline understanding of sustainable beef that national roundtables and other initiatives can use while recognizing the diversity of beef production systems around the world.

The Global Dairy Agenda for Action, launched in 2009, is the international dairy sector's platform to develop a common vision of sustainability for the dairy sector and to proactively collaborate in solving the challenges, recognizing the diversity of production systems and of priorities at a local level. The global Dairy Sustainability Framework, a program of the Global Dairy Agenda for Action, was launched in 2013 as a tool to shape, assess, and monitor the continuous efforts of the international dairy industry to progress toward sustainability. The framework defines 11 key sustainability criteria unique to the dairy value chain and covering environmental, social, and economic aspects of sustainability.

CHAPTER

23

Sustainable Forest Management

Globally, around 129 Mha of forest, almost the size of South Africa, have been lost since 1990 (FAO, 2015a). Eighty percent of forest loss is driven by agriculture, which in turn represents approximately 11% of total global greenhouse gas emissions. Forecasts indicate that pressure on forests and agriculture will remain high, with the need for a 50% increase in food production as well as a 27%–63% increase in energy supply by 2050 (WEC, 2013).

Of the total 129 Mha of deforestation between 1990 and 2015, about 58 Mha were cleared for pastures and 69 Mha were directly or indirectly cleared for cropland. Six agricultural crop commodities caused half of the 69 Mha of deforestation associated with cropland expansion: soybeans (13 Mha, mostly in Argentina, Bolivia, Brazil, and Paraguay); maize (8 Mha); oil palm (6 Mha, mostly in Indonesia and Malaysia); wood products (5 Mha); rice (4 Mha); and sugar cane (3 Mha). To a lesser degree, commodities such as rubber, sugar, cocoa (particularly in West Africa), and coffee also contributed to deforestation.

Palm oil (African oil palm, *Elaeis guineensis*) recently surpassed soybeans (*Glycine max*) as the most widely consumed vegetable oil in the world. The bulk of the increase in palm oil production has come from area expansion rather than improvements in yield. Increased demand for palm oil and limited availability of land in Southeast Asia has led to expansion in other continents. Latin America has more than doubled its output since 2000, particularly in Colombia, Ecuador, and Honduras. Based on a sample of 342,032 ha of oil palm plantations across Latin America, Furumo and Aide (2017) reported that 79% replaced previously farmed lands (e.g., pastures, croplands, banana plantations), primarily cattle pastures (56%). The remaining 21% came from areas that were classified as woody vegetation (e.g., forests), most notably in the Amazon and the Petén region in northern Guatemala.

Healthy forests and trees provide a range of services relevant for agricultural production and family livelihoods, including food, feed for livestock, fuel, material for construction, fiber, and other uses. Equally important, and essential for sustainable agriculture, trees or forests provide key ecosystem services that help regulate ecosystem processes. They host a major part of biodiversity, including pollinators such as insects and birds (Klein et al., 2007), pest control agents (Karp et al., 2013), and provide habitat for living organisms that are key for soil fertility, including bacteria and fungi. They play a role in

Sustainable Food and Agriculture
DOI: https://doi.org/10.1016/B978-0-12-812134-4.00023-6

soil protection (Pimentel and Kounang, 1998), nutrient recycling (Power, 2010), water regulation (de Groote et al., 2002), microclimate regulation, and climate change mitigation (Foli et al., 2014; Reed et al., 2017).

Forest fragments influence the provision of multiple ecosystem services in adjacent agricultural fields, but quantitative evidence is scarce. Mitchell et al. (2014) looked at six ecosystem services (crop production, pest regulation, decomposition, carbon storage, soil fertility, and water quality regulation) in soybean fields at different distances from adjacent forest fragments that differed in isolation and size across an agricultural landscape in the province of Quebec, Canada. The study showed significant effects of distance-from-forest, fragment isolation and fragment size on crop production, insect pest regulation, and decomposition. Distance-from-forest and fragment isolation had unique influences on service provision for each of the ecosystem services measured. For example, pest regulation was maximized adjacent to forest fragments (within 100 m), while crop production was maximized at distances of 150–300 m from the forest (Mitchell et al., 2014).

Proximate tree cover can be beneficial to crop yield in agroforestry systems, but can also have unintended negative outcomes for crop yield. This includes the harboring of pests within adjacent forests or acting as incubation zones for plant diseases that can be transferred to crop plants. Trees also directly compete for water, nutrients, and light where their niches overlap with food crops. Wildlife animal populations encroach on to human-claimed land, causing problems such as crop damage, livestock predation, and increased risk of certain diseases. However, the design of integrated systems, such as agroforestry systems, take such factors explicitly into account, for example, limiting competition through adequate shade and root architectures of companion species.

The socioeconomic benefits and contributions of forests and trees to livelihoods are diverse, but their quantification is often impeded by lack of data. Both the cash and non-cash contributions to livelihoods of forests, in particular for rural families and marginalized groups, are often underestimated. Wood and fiber make a significant contribution to the shelter of at least 1.3 billion people, or 18% of the world's population (FAO, 2014a), and some 50% of the fruit consumed globally comes from trees, much of this collected by women and children (Vira et al., 2015).

23.1 MANAGING FORESTS FOR SUSTAINABILITY

New management regimes that take account of the key roles of forests have meant that landscapes are being managed for a much more diverse set of purposes (Ribot et al., 2006). Hence, about two-thirds of the global forest area is explicitly designated to serve purposes other than production. Thirty-one percent of the forest area is specifically managed for protection of soil and water, and 13% for biodiversity conservation. Some countries have established and many have amended their government compensation schemes to provide public goods by forests unrecognized by markets. While some countries use payment for ecosystem services through markets, it is also being explored and piloted by several others, particularly for recreation, water, and carbon.

Where large-scale commercial agriculture is the principal driver of land use change, effective regulation of land use change, with appropriate social and environmental safeguards, is needed. A range of governmental initiatives has been formulated to address illegal harvesting of wood and related trade and illegal land conversion through changing trade rules, including the US Lacey Act, the Australian Illegal Logging Prohibition Act, and the European Union Timber Regulation. Initiatives on "forest law enforcement, governance, and trade" ensure that imported timber is legally harvested and traded. Verification of the legality of harvested timber is slowly expanding, enhancing the role of the private sector in strengthening sustainable forest management.

Private governance initiatives led to the formation of the Roundtable on Sustainable Palm Oil and the Round Table on Responsible Soy. Around 14 major voluntary sustainability standards now operate in agriculture and forestry-related fields, including for wood, palm oil, soybeans, beef, bananas, cocoa, coffee, cotton, and tea. These programs aim to provide premium access to markets, price premiums, and other benefits to producers for adhering to environmental and labor practices established by the certifying entities. While they are growing, their share of total land area is still small. Also, certification programs have difficulties in adequately addressing smallholder producers.

By far, the largest share of certified area of agriculture and forestry land is covered by forest certification and two major forest certification schemes—the Forest Stewardship Council and the Programme for the Endorsement of Forest Certification—covering a combined global total of 429 Mha (11%) of the global forest area as of May 2017 (UNECE and FAO, 2017). However, only 15% of the total certified area is located in Africa, Latin America, Asia, and Oceania, where the bulk of deforestation and degradation occurs.

A range of initiatives has mobilized commitments to stamp out deforestation from commodity chains responsible for much of the tropical forest destruction, especially cattle, palm oil, soybeans, and wood products (EDF, 2017). An example of such an effort in Brazil is presented in Box 23.1. Mobilizing action in livestock seems to be more difficult, yet it is important. For example, in countries with high rates of deforestation, beef

BOX 23.1

SUSTAINABLE BEEF VALUE CHAINS AND DEFORESTATION

Brazil's three largest meatpacking companies (JBS, Marfrig, and Minerva) signed an agreement with the government in 2009, and later with Greenpeace, under which they committed to buy only from suppliers that promised to reduce deforestation to zero. Unfortunately, the agreements left room for cattle laundering through which the ranchers who are not signatories to the agreement could raise and fatten their cattle on properties not covered by the pacts. Then, they could sell the animals to ranchers who are direct suppliers. That loophole was closed with a publicly available database that covered all animals and the areas where they were being transported, leading to about 60% of the suppliers registering, and compliance reaching 96% by 2013.

production accounts for more than twice as much deforestation as that caused by the production of soybeans, palm oil, and wood products combined.

Moving forward, much remains to be done to address deforestation and forest degradation driven by agriculture. Both policy and development practice still have a long way to go toward explicitly integrating the direct and indirect benefits and services provided by forests or trees in food production or development initiatives. Forest policymakers often struggle to make the case for sustaining and investing in forests to provide ecosystem benefits in the face of alternative land uses that promise higher short-term or more visible and direct economic returns.

CHAPTER

24

Agroforestry

Agroforestry is a form of intercropping where perennial trees or shrubs are intercropped with annual herbaceous crops. In a first attempt at estimating the global extent of agroforestry initiatives, just under half of all agricultural land had over 10% tree cover, 27% of agricultural land had over 20% tree cover, and about 7% of global agricultural land had over 50% tree cover. Trees were also found to be an integral part of the agricultural landscape in all regions except North Africa and West Asia (Zomer et al., 2009).

A wide variety of agroforestry systems are practiced worldwide. Lorenz and Lal (2014) estimate that worldwide tree intercropping covers some 700 Mha, multistrata systems 100 Mha, protective systems 300 Mha, silvopasture 450 Mha, and tree woodlots 50 Mha. Over 100 distinct agroforestry systems have been recorded (Atangana et al., 2014), and modern agroforestry systems have a number of precursors, such as traditional systems across India (Murthy et al., 2013), the Sahel (Garrity et al., 2010), or the Javanese systems of *pekarangan* (combining agricultural and tree crops with livestock) and *kebun-talun* (rotational cultivation of agricultural and tree crops) (Christanty et al., 1986).

The inclusion of trees within cultivated landscapes delivers a range of social–ecological benefits, including for crop yields, income, and resilience (Reed et al., 2017; Pramova et al., 2012). In addition, trees improve soil structure and water filtration, which makes farms, especially rainfed systems, more resilient to drought and the effects of climate change. Moreover, they can play an important role in climate change mitigation. Conservation agriculture with trees sequesters from 2 to 4 tons of carbon per hectare per year, compared to 0.2–0.4 tons without trees (FAO, 2015a). In addition, by increasing maize production and the supply of fuelwood, farming systems that integrate trees with maize can reduce the need for converting forests to farmland, which is a major source of greenhouse gas emissions. In Sahelian countries, such as Burkina Faso and Niger, agroforestry has been shown to improve yields of other cereals, such as millet and sorghum. With further research and farmer engagement, conservation agriculture with trees could be expanded into a much broader range of food cropping systems.

Trees planted in an agroforestry system generate additional benefits for the farmer. In Niger, Larwanou and Adam (2008) estimate that each additional tree generates an average of $1.40/year through improved soil fertility, fodder, fruit, firewood, and other produce. This would result in additional production value of at least $56 ha/year, and a total

Sustainable Food and Agriculture
DOI: https://doi.org/10.1016/B978-0-12-812134-4.00024-8

production value of $280 million. In India, Murthy et al. (2013) review carbon sequestration within agrosilvicultural and silvopastoral systems and estimate that these could hold a soil stock of 390 million tons of carbon (excluding soil organic carbon stocks). Ickowitz et al. (2013) found that in 21 African countries, dietary diversity among children increases with tree cover after controlling for relevant confounding variables. Under EverGreen Agriculture systems, trees planted within fields of annual crops replenish soil fertility, and provide food, fodder, timber, and fuelwood. Overall, this portfolio of benefits provides "an overall value greater than that of the annual crop within the area that they occupy per m^3" (Garrity et al., 2010, p. 199).

Agroforestry does not require large financial investment. In fact, low-income farmers are often quicker to embrace it than wealthier farmers (FAO, 2015a). Although more labor is required during the initial shift to a maize-forestry system, farm labor can be used more efficiently once farmers master new practices. However, incorporating trees into crop production is a knowledge-intensive activity. Policy support, continued research, and rural advisory services that engage smallholder farmers are thus crucial for the long-term expansion of these farming systems.

Legume tree-based farming systems offer an important route to increasing the availability of nitrogen (N) while avoiding synthetic fertilization. This has led to the use of the term "fertilizer tree" (Garrity et al., 2010). Nitrogen fixation depends on tree species and soil status, and can range from 5 to over 300 kg N/ha/year (Rosenstock et al., 2014). The use of *Gliricidia sepium* in improved fallows resulted in a 55% increase in sorghum yield over two cropping seasons (Hall et al., 2006). Sileshi et al. (2012) compared yields across three systems: maize-*Gliricidia* (the agroforestry cohort), conventional monoculture (with fertilization), and regular practice (absent any external input). Yields in the agroforestry system were comparable to those achieved via synthetic fertilization, but 42% higher than nonfertilized fields. They were also more stable over time than yields achieved through synthetic fertilization.

Another promising fertilizer tree, *Faidherbia albida*, presents one of the few examples of intercropping being practiced at scale (Montpellier Panel, 2013), resulting in what has been called the Great Green Wall of the Sahel (Reij and Smaling, 2008). *F. albida* is a nitrogen-fixing acacia indigenous to Africa and the Middle East. The tree is particularly suited to intercropping with maize, as it sheds its leaves during the monsoon season when maize is sown. This prevents it from competing with maize seedlings for light and nutrients, while falling leaves provide nutrients (Barnes and Fagg, 2003). Thanks to the decaying leaves, the soil under the trees contains up to twice as much organic matter and nitrogen as soil outside the canopy. There is also a marked increase in soil microbiological activity and an increase in water-holding capacity.

These mixed tree-crop systems address two pressing vulnerabilities limiting smallholder maize yields: fluctuating rainfall and temperatures, and low soil fertility, particularly low nitrogen levels caused by decades of intensive cultivation. Integrating trees provides affordable, high-quality, nitrogen-rich residues that improve soil fertility, boost yields, and provide new sources of income. In Malawi and Zambia, maize yields average a low 1.2 tons/ha. Only about one in four smallholder farmers in Zambia and one in five in Malawi grow enough maize to sell in markets. In Malawi, a drought in 2004/05 caused average maize yields to drop to just 0.76 tons/ha, and 5 million Malawians, nearly 40% of the population, needed food aid (Garrity et al., 2010). In response, both countries promote *F. albida* as part of conservation agriculture systems that offer smallholder farmers a means of increasing maize productivity and earning higher incomes from sales. National recommendations are to grow 100 trees/ha in a $10 \times 10\ m^2$ grid pattern; in some cases, farmers have been able to establish most of the *F. albida* stands simply by assisting the natural regeneration of tree seedlings on their land.

In Zambia, the planting of *F. albida* within maize fields is practiced on over some 300,000 ha (Garrity et al., 2010). In Malawi, there are about half a million farms with the trees, and *G. sepium*, *Sesbania sesban*, and *Tephrosia* spp. are additionally intercropped with maize. *G. sepium* takes relatively less time to establish than *F. albida* and is now promoted by the World Agroforestry Centre as an intercrop in maize fields. Trees are pruned back two or three times a year, and the leaves mixed in with the soil. Findings from a decade-long study indicate that on unfertilized plots where *G. sepium* is intercropped with maize, yields average 3.7 tons/ha, and reach 5 tons/ha in good years. On unfertilized plots without *G. sepium*, average yields were only 0.5–1.0 tons/ha (Garrity et al., 2010).

Yield increases for maize grown in association with *F. albida* are also shown in numerous studies, particularly where soil fertility has been low. Cereal yields of 3 tons/ha without the application of additional fertilizer have been reported. Additional benefits include significant carbon sequestration, weed suppression, increased water filtration, and increased adaptability to serious droughts (Garrity et al., 2010). In Zambia, maize planted outside the tree canopy produced average yields of 1.9 tons/ha, compared to 4.7 tons/ha when the crop was grown under the canopy; in Malawi, maize yields increased by 100%–400% when the crop was grown with *F. albida*. However, it is important to note that although *F. albida* is one of the fastest growing acacia species, it is not a quick fix for low soil fertility. In a survey of 300 Zambian farmers, one-third said that yields increased over a period of 1–3 years, while 43% said that it took up to 6 years before they saw benefits in production.

In areas where landholdings are larger than 1 ha, growing leguminous shrubs such as *S. sesban* on fallow fields is another option for revitalizing the soil and increasing maize yields. Leguminous trees and shrubs add from 100 to 250 kg of N/ha to the soil in fields that are left fallow for 2 or 3 years. Even though fields are unproductive for 2 out of every 5 years, overall production and returns on investment are higher when maize is grown in rotation with nitrogen-fixing shrubs and trees. In eastern Zambia, one study found that the average net profit was $130 per ha when farmers cultivated maize without fertilizer; $309 when it was grown in rotation with *S. sesban*; and US$327 when it was intercropped with *G. sepium*. For each unit of investment, farmers who integrated trees with maize benefited from higher returns than those who used mineral fertilizer, subsidized or unsubsidized, for continuous maize production (Ajayi et al., 2009). The study confirmed that maize production in agroforestry-based systems is both socially profitable and financially competitive compared to maize production using only mineral fertilizer.

Niger has experienced the most remarkable "regreening" as a result of farmer-managed regeneration of trees in fields. As a result of the relaxation of restrictive policies prohibiting farmers from managing the trees on their own lands, agricultural landscapes in Niger now harbor significant densities of *F. albida* (Garrity et al., 2010). It has been estimated that regreening has resulted in an annual increase of 500,000 tons of food (Reij et al., 2009).

Maize, sorghum, millet, groundnuts, and cotton have all shown increased yields, even without additional fertilizer application, when integrated with *F. albida* (Garrity et al., 2010). In Burkina Faso, *Combretum glutinosum* and *Piliostigma reticulatum* are included to generate additional biomass and provide fodder and increased yields (Garrity et al., 2010).

In East and Southern Africa, one barrier to overcome for adoption of conservation agriculture is the lack of crop residues to maintain a constant soil cover. Because most African smallholders also raise livestock, they often use crop residue biomass as animal fodder. With trees growing on their farms, there is now enough biomass both to meet livestock needs and improve maize yields. The trees also provide fuel for rural households; in Zambia, farmers were able to gather 15 tons of fuelwood per hectare after the second year of fallow with *S. sesban* and 21 tons after the third year (Garrity et al., 2010). However, pressure on young parkland needs to be well monitored and managed. In the Senegalese groundnut basin, for example, regenerated *Acacia* parkland is under strong pressure from fodder collection, as livestock intensification has occurred alongside.

Legume-based agroforestry systems could result in increased nitrous oxide emissions (Hall et al., 2006), depending on species, soil type, climatic conditions, and management practices. However, based on a review of evidence from agroforestry systems in tropical regions and improved fallows in sub-Saharan Africa, Rosenstock et al. (2014) conclude that "Legume-based agroforestry is unlikely to contribute an additional threat to increasing atmospheric N_2O concentrations, by comparison to the alternative (e.g., mineral fertilizers)." Legume-based tree systems also sequester and accumulate carbon and may affect methane exchanges. Overall, there is a need for further research on the precise impacts of changes to carbon and nitrogen cycling, and an exploration of other environmental trade-offs such as increased leaching of nitrogen into local water supplies. Given these potential negative externalities, further context-appropriate research is needed to identify "win–win–win" systems that deliver yields, climate resilience, and water quality. This will require a "fundamental departure" from conventional research on the subject, which has focused on single outcomes (e.g., yield) or single media (e.g., outcomes for water, or air, or carbon or nitrogen in isolation) (Rosenstock et al., 2014).

CHAPTER

25

System of Rice Intensification in Asia

Rice farmers are adopting crop, soil, and water management practices that in conjunction produce more rice and income using less water, less fertilizer, and less seed. Traditionally, rice has been cultivated in most of Asia in the following way: fields are first flooded then ploughed to create soft, muddy soil often overlying a dense, compacted layer that restricts downward loss of water. Rice seedlings 20–60 days old are then transplanted to the fields in clumps of two to four plants, randomly distributed or in narrowly spaced rows. To suppress weeds, the paddy is continuously flooded with 5–15 cm of water until the crop matures (Africare et al., 2010; Berkhout et al., 2015). That system has enabled the cultivation of rice for millennia at relatively stable yields. When the modern Green Revolution introduced high-yielding varieties, mineral fertilizer, and chemical pest control, productivity in many Asian rice fields doubled in the space of 20 years.

A set of crop, soil, and water management practices known as the System of Rice Intensification (SRI) takes a strikingly different approach. Seedlings 8–15 days old are transplanted singly, often in grid patterns with spacing of $25 \times 25\ cm^2$ between plants. To promote moist, but aerated soil conditions, intermittent irrigation is followed by dry periods of 3–6 days. Weeding is done at regular intervals, and compost, farmyard manure, and green manure are preferred to mineral fertilizer. Once the plants flower, the field is kept under a thin layer of water until 20 days before the harvest (Uphoff, 2008; Berkhout et al., 2015).

Since SRI was first developed in Madagascar in the 1980s, the system has been found to improve grain yields above those obtained under flooded systems by 40% in India and Iraq and almost 200% in the Gambia. Compared to trials with current improved practices in China, SRI methods increased rice yields by more than 10% (Wu et al., 2015). Rice grown using SRI consumed 25%–47% less water than flooded systems in India and China (Zhao et al., 2011), and required 10%–20% less seed than traditional systems in Nepal.

The governments of Cambodia, China, Indonesia, and Vietnam—where much of the world's rice is produced—have endorsed SRI methods in their national food security programs, and millions of rice farmers have adopted SRI practices. More than 1 million Vietnamese rice farmers are reported to be applying SRI; their per hectare incomes have increased by an average of $110 thanks to a 40% reduction in production costs. Farmers

Sustainable Food and Agriculture
DOI: https://doi.org/10.1016/B978-0-12-812134-4.00025-X

who were trained in site-specific nutrient management in Vietnam benefited from additional annual income of up to $78 per hectare (Nga et al., 2010).

The system has the potential to reduce emissions of methane from irrigated systems (Gathorne-Hardy et al., 2013). At present, more than 90% of the world's rice is harvested from flooded fields, which emit methane totaling some 625 million tons of carbon dioxide equivalent annually. Emissions could be reduced by almost one-sixth if all continuously flooded rice fields were drained at least once during the growing season. SRI involves draining several times during the growing season (Uphoff, 2008).

Scientists are seeking rigorous explanations for SRI's lower resource use and higher productivity, as well as examining the ways in which farmers follow SRI guidelines. Possible explanations include increased populations of beneficial soil bacteria in the root zone due to use of compost and intermittent irrigation (Anas et al., 2011; Lin et al., 2011) and better absorption of sunlight by singly planted seedlings, permitting larger root systems. A more general explanation may be that SRI methods allow the rice plant to fully exploit its genetic potential.

Assessments are notoriously contentious. Reviews of SRI's reported high yields have to contend with the diversity in SRI practices, making it difficult to draw general conclusions about the impact of SRI as a "singular technological package." As with many other agroecological packages, SRI may demand innovation in methods for evaluation and assessment of impacts.

Another focus of debate around SRI is the potential for SRI to involve increased demand for labor. In the Gambia, labor costs of transplanting were two to three times higher than those of conventional flooded rice. A recent study in India found that because it was labor intensive, the system carried much higher production costs and was "really uneconomical" (Borlaug Institute for South Asia, 2015). However, proponents of SRI respond that it generates employment. In Tamil Nadu, India, SRI production was found to be the most suitable option for employing otherwise idle family labor during the dry season. The labor requirements of SRI cultivation could be lowered by use of seedling trays that simplify seedling preparation and transplanting and direct seeding of rice.

CHAPTER

26

Intercropping, Multicropping, and Rotations

Intercropping, multicropping, and rotations involve the addition of system components that increase yields, add resilience, and generate cobenefits for people and ecosystems. A recent meta-analysis of intercropping in African systems found that on average it increased crop yields by 23% and brought gross income increases of $170/ha (Himmelstein et al., 2016). In Asia, intercropping rice and water spinach also had a dramatic effect on productivity and income. Growing wet rice with water spinach reduced rice blast and sheath blight disease, increased rice tiller numbers, and raised net photosynthesis. Net income increased fivefold (Liang et al., 2016).

26.1 WHEAT AND LEGUME ROTATIONS

Growing legumes can be a very good investment in its own right, but intercropping with cereals can boost yields while saving costs. Wheat farmers grow legumes to improve the health of soil and provide a natural source of nitrogen, which boosts yields. Since they derive 70%–80% of their nitrogen needs from the atmosphere, through biological nitrogen fixation in their root nodules, grain and forage legumes generally do not require nitrogen fertilizer to achieve optimum yields (Dong et al., 2003). Grain legumes, such as lentils, are high in protein, dietary fiber, vitamins, minerals, antioxidants, and phytoestrogens and can be sold to generate income. Forage legumes, such as alfalfa, can be used on the farm to feed livestock.

When grown before wheat, legumes produce another significant benefit—nitrogen in legume residues reduces the need to apply nitrogen fertilizer to the wheat crop. It is estimated that globally, some 190 Mha of grain legumes contribute around 5–7 million tons of nitrogen to soils (Peoples et al., 2009). Wheat grown after legumes produces higher grain yields and has higher protein content than wheat grown after another wheat crop. The high productivity of wheat-legume rotations has long been recognized by wheat farmers since as far back as 2000 years ago in Western Asia and North Africa. Typical rainfed

Sustainable Food and Agriculture
DOI: https://doi.org/10.1016/B978-0-12-812134-4.00026-1

wheat-based rotations include grain legumes such as chickpeas, lentils, and faba beans, and forage legumes such as vetch, berseem clover, and *Medicago* species.

Choosing the right legume for a specific wheat farming system is extremely important, as different legume species and varieties growing in the same location can differ significantly in dry matter production, nitrogen fixation and accumulation, and residue quality. Residual nitrogen values from grain legumes vary greatly, but can cover between 20% and 40% of wheat's nitrogen needs (Evans et al., 2001). While grain legumes can add from 30 to 40 kg of N/ha to the soil, legumes grown as green manure crops or as forage for livestock build up nitrogen much more quickly and can fix as much as 300 kg of N/ha (Fischer et al., 2014).

Legumes also enhance wheat's uptake of other nutrients. Wheat grown after legumes tends to have a healthier root system than wheat-after-wheat, making it better able to use other available nutrients. The roots of chickpeas and pigeon peas secrete organic acids, which can mobilize fixed forms of soil phosphorus and make it more readily available. Legumes also release hydrogen gas into the soil, at rates of up to 5000 L/ha per day. Hydrogen is oxidized by soil microbes surrounding the root system of the legume plant, leading to changes in the soil biology that improve the development of the wheat plant. Deep-rooting legumes such as pigeon peas, lablab, and velvet beans help build soil structure and biopores, which improve drainage and aeration.

Wheat sown in the autumn and followed by a summer fallow is the predominant production system in dry areas. In the Middle East and North Africa, fields are commonly left fallow owing to the lack of sufficient moisture to sustain reliable production of rainfed summer crops. However, with the development of early maturing legume varieties, farmers can now replace long fallows with legume crops, which make more productive use of land. Growing food legumes in summer not only helps enhance soil fertility and water-use efficiency, but also boosts yields of the subsequent wheat crop (Gan et al., 2014). In the highlands of Ethiopia, pulses are grown in rotation with cereals, or as intercrops, to spread out the risks of drought and to improve soil fertility (Amanuel et al., 2000).

Managing legumes to achieve "win–win" outcomes—a profitable legume and maximum benefits for the subsequent wheat crop—is a complex practice for many farmers. Legumes are generally seen as more risky to grow than wheat or other cereals. This is partly because legumes are often more susceptible to biotic and abiotic stresses, which can reduce yields and plant biomass. If the legume fails to produce enough biomass to yield well and also leaves residual nitrogen in straw and root residues, the smallholder loses income in one growing season without realizing compensation in the next. In addition, prices for grain legumes are often more volatile than for cereals. Due to their shorter growing season, some legume crops not only do not remove as much soil water as wheat, but also leave more residual moisture for the wheat crop. However, this moisture can easily be lost if the legume residues are heavily grazed or removed for other purposes. It is recommended, therefore, that residues are left as a surface cover and wheat is drill seeded with minimum soil disturbance.

To manage risk, farmers are advised to plant legumes only when there is sufficient moisture stored in the soil profile, or available as irrigation. While early planting enhances biomass production and nitrogen fixation, it can also increase susceptibility to pathogens. To realize the full benefits of wheat-legume rotation, residues should be retained on the soil surface, and both legumes and wheat crops should be established with zero tillage to conserve soil structure, soil water, and soil nutrients.

26.2 MAIZE AND LEGUME MULTIPLE CROPPING

The high productivity of maize-legume systems makes them especially suitable for smallholders. Maize-legume systems come in three basic configurations. One is intercropping, in which maize and legumes are planted simultaneously in the same or alternating rows. Another approach is relay cropping, where maize and legumes are planted on different dates and grow together for at least a part of their life cycle. Maize and legumes may also be grown as monocultures in rotation, with maize being planted in the same field after the legume harvest. Such systems are common throughout the developing world. Commonly planted legumes include beans, pigeon peas, cowpeas, groundnuts, and soybean, which are grown mainly for food, and nonedible legumes such as velvet beans and jack beans, which are used as feed for livestock. All fix nitrogen in the soil and are useful as sources of residues that can be retained on the soil surface as mulch.

Maize-bean intercropping is a traditional practice among smallholder farmers in Latin America, especially in the land-scarce highlands. In Peru, practically all beans, and in Ecuador about 80%, are planted along with maize. In areas of Central America where land is limited and rainfall low, maize is often intercropped with field beans. When maize and beans are intercropped, their yields are generally lower than those of maize or beans grown in monoculture. Studies have found that maize yielded 5.3 tons/ha when monocropped, 5.2 tons when intercropped with bush beans, and 3.7 tons when intercropped with climbing beans (Francis, 1981). However, under intercropping, production costs per unit of output are usually lower and, because beans sell for up to four times the price of maize, farmers' incomes are higher and more stable.

Being drought tolerant, pigeon pea is often intercropped with cereals in smallholder farming systems in Asia, Africa, and the Caribbean. Pigeon pea is also deep-rooting, so it does not compete with maize for water, and is slow growing in its early stages, which allows maize to establish properly. As with maize and beans, both maize and pigeon pea, when planted together, yield slightly less than they do when cultivated alone. However, the overall yield from intercropping exceeds that which would have been produced by the corresponding monocrops—a comprehensive study of maize-pigeon pea intercropping in South Africa found that the system was nearly twice as productive as monocrops per unit of area (Mathews et al., 2001). In maize-pigeon pea systems in India and Sri Lanka, planting four rows of maize to two rows of pigeon pea provided the highest net returns (Marer et al., 2007).

A 3-year study conducted in central Malawi found that intercropping maize and pigeon pea under conservation agriculture produced almost double the vegetative biomass, and in drier years 33% more maize grain than conventionally tilled maize monocropping (Ngwira et al., 2012). In Mozambique, long-term maize-legume intercropping and zero tillage improved rainfall infiltration five times over due to good quality biomass production, which provided mulch (Rusinamhodzi et al., 2012). In Panama, planting maize on jack bean mulch saved farmers 84 kg N/ha in applications (FAO, 2016a).

Relay cropping is practiced in Brazil, Colombia, and Central America, where maize is planted in May-June and beans are sown between the maize plants in August-September. This allows the maize to develop sufficiently to provide a support for the climbing beans (Francis, 1981). In northern Ghana, planting fields with cowpeas, from three to six weeks before maize, yields nutritious food at a time when other crops are not yet mature and, with the retention of residues, provides nitrogen for the soil. Maize-legume rotations also help to maintain soil fertility. For example, in Mexico, smallholder farmers have developed a system that grows velvet beans in the maize "off season," and leads to significantly higher levels of soil pH, organic matter, and nitrogen. That, in turn, contributes to a 25% increase in the yield of the subsequent maize crop (Ortiz-Ceballos et al., 2015).

In East and Southern Africa, a CIMMYT-led program for the sustainable intensification of maize-legume cropping systems found that under conservation agriculture, the highest maize yield increases were achieved when the cereal was rotated with legumes such as beans, cowpeas, and soybeans. In Malawi, farmers' normal practice obtained maize yields of 3.7 tons/ha; with conservation agriculture, yields rose to 3.9 tons; and with conservation agriculture and after soybeans, yields reached 4.5 tons (FAO, 2015b).

A highly productive maize-soybean rotation system is practiced in Nigeria. Planted before maize, the soybeans reduce *Striga* infestations by inducing premature germination of its seeds. The soybeans produce about 2.5 tons of grain and 2.5 tons of forage per hectare, and residues that supply 10 to 22 kg of N/ha. Nitrogen is utilized by the subsequent maize crop, which produces yields up to 2.3 times higher than those under monocropping. Nigerian farmers' soybean output rose from less than 60,000 tons in 1984 to 600,000 tons in 2013, encouraged by gross income that is 50%–70% higher than that obtained from continuous maize. Increases in legume area of just 10% in the northern Guinea savanna in Nigeria, and increases in yield of 20%, have resulted in additional fixed nitrogen valued annually at $44 million (Sanginga et al., 2003).

Soybeans are often rotated with maize in Brazil. In the southern states of Mato Grosso and Paraná, maize is a second-season crop that is planted on the mulch of early maturing soybeans, which improves moisture availability for the maize and reduces soil erosion. The rotation allows two harvests from the same field and alleviates pest pressure on both crops, leading to more sustainable production and improvements in farmers' incomes and livelihoods.

While the benefits of maize-legume systems are well known, smallholder farmers who rely on food crops for household food security, especially in Africa, are often reluctant to occupy their fields with nonedible legumes for a half or a full year, regardless of the long-term benefits. Adoption of these systems in Africa is also constrained by dysfunctional markets for rotational crops, the unavailability of seed, and the farmer's perception of risk (Thierfelder et al., 2012). Governments may invest in the development of smallholder maize-legume systems as a means of ensuring food security, improving farmer incomes, and improving soil health. Since nonedible legumes such as velvet bean have very high carbon sequestration potential, climate change mitigation funding may be available to encourage smallholder adoption.

Maize and legume varieties that produce high yields in monoculture generally also have high yields when intercropped. However, differences in the suitability of certain varieties for maize-legume systems have been observed. Breeding efforts should exploit productive interactions, such as strong-stalk maize that supports higher weights of beans. Generally, maize-legume systems also exhibit considerable site specificity. Therefore, the system and its variations require extensive validation in farmers' fields.

26.3 INTEGRATING LIVESTOCK AND CROPS WITH NUTRIENT PUMPS IN LATIN AMERICA

Livestock production is particularly important in smallholder farming systems on the savanna grasslands of Latin America. However, output per animal unit in tropical areas is far below that achieved in temperate regions. A major constraint is the quantity and quality of forage, a key feed source in ruminant systems. Overgrazing, farming practices that deplete soil nutrients, and a lack of forage species that are better adapted to biotic and abiotic stresses all contribute to low productivity. Improving pasture forage quality and productivity would help to boost production of meat and milk.

Many livestock farmers in Latin America have adopted a sustainable livestock production system that integrates forages with cereals. A key component of the system is *Brachiaria*, a grass native to sub-Saharan Africa, which grows well in poor soils, withstands heavy grazing, and is relatively free from pests and diseases. Due to its strong, abundant roots, *Brachiaria* is very efficient in restoring soil structure and helps prevent soil compaction, which reduces rainwater infiltration and stifles root growth. It also has the ability to convert residual soil phosphorus into organic, readily available forms for a subsequent maize crop (Resende et al., 2007). Brazilian farmers have integrated *Brachiaria* in a direct-seeded maize system that is replacing soybean monocropping. Recent research by Centro Internacional de Agricultura Tropical (CIAT) has identified another special characteristic of *Brachiaria*: a chemical mechanism found in the roots of one *Brachiaria* species inhibits emissions from the soil of nitrous oxide, which is derived mainly from mineral fertilizer and is one of the most potent of the greenhouse gases causing climate change (CGIAR, 2013). The versatile grass is now grown on an estimated 80 Mha in Latin America. While the adaptation of *Brachiaria* to low-fertility soils has led to its use for extensive, low-input pastures, it is also suitable for intensively managed pastures (Rao et al., 2014). In Mexico and Central America, the productivity of animals feeding on *Brachiaria* pastures is up to 60% higher than those feeding on native vegetation. The value of the additional production has been estimated at $1 billion a year (Holmann et al., 2004). In Brazil, annual economic benefits have been put at $4 billion (CIAT, 2013).

Rotation of annual crops with grazed pasture is increasing in the Cerrado ecoregion of Brazil, where beef cattle are a major source of income for many farmers. Years of poor herd management, overgrazing, and lack of adequate soil nutrient replacement have led to declining productivity and reduced profitability in traditional livestock production systems (Rao et al., 2014). Where natural ecosystems have been replaced by intensive soybean monoculture, much of the soil is compacted and susceptible to erosion from heavy rainfall. Under those conditions, traditional techniques of soil erosion control, such as contour planting, have proved ineffective (Pacheco et al., 2013).

In response, many farmers have adopted zero-tillage systems, which increase soil cover and bring other environmental benefits. In the early 1990s, less than 10% of the Cerrado was under zero tillage; by 1996, this figure had risen to 33%, where more than 4 Mha were cultivated using diversified direct seeding mulch-based cropping (DMC) systems, replacing inefficient, tillage-based soybean monoculture. A typical sequence is maize (or rice), followed by another cereal such as millet or sorghum, or the grass *Eleusine* intercropped with a forage species such as *Brachiaria* (Scopel et al., 2004). It has been estimated that around 50% of the total cropped area in Brazil is under DMC systems, which usually support three crops a year, all under continuous direct seeding.

The forages work as "nutrient pumps," producing large amounts of biomass in the dry season that can be grazed or used as green manure. Combining maize and *Brachiaria* at the end of the rainy season taps soil water from levels deeper than 2 m and promotes active photosynthesis later during the dry season. It results in vigorous vegetative regrowth after the first rains of the following season, or after rain during the dry season, thus ensuring permanent soil cover.

Because *Brachiaria* provides excellent forage, farmers can then choose to convert the area into pasture, or keep it in grain production for another year. Such systems are found under irrigation and in wetter regions with frequent heavy rains that recharge deep water reserves. In the best DMC systems, total annual dry matter production, above and below the soil, averages around 30 tons/ha compared to the 4–8 tons found under monocropping. To reduce crop competition, novel intercropping systems have been developed. In the "Santa Fé" system for maize and *Brachiaria*, developed in Brazil, the grass is made to germinate after the maize crop, either by delaying its planting or by planting it deeper. The young *Brachiaria* plants are shaded by the maize and provide little competition to the cereal. At maize harvest, however, shading is reduced and the established pasture grows very quickly over the maize residues.

This tight integration between forage and grain crops leads to a better use of the total farm area and a more intensive use of the pastures, with less pasture degradation. Similar DMC systems are being tested in other parts of the world, including sub-Saharan Africa.

CHAPTER

27

Patch Intensification

The use of small plots of land to cultivate crops or rear fish, poultry, or small livestock near places of human settlement represents the oldest form of agriculture (Niñez, 1984). These plots of land represent perhaps "one of the last frontiers for increasing food production in the struggle against world hunger and malnutrition" (Niñez, 1984, p. 35) and are among the most diverse and productive (per unit area) cultivation systems in the world (Conway, 1997).

They provide several significant nutritional, financial, and ecosystem benefits, including pollination; gene-flow between plants inside and outside the garden; control of soil erosion and improvements to soil fertility; improved urban air quality; and carbon sequestration and temperature control through the creation of microclimates. Patches contribute directly to household food and nutritional security by increasing the availability, accessibility, and utilization of nutrient-dense foods (Galhena et al., 2013), including wild edible species and traditional varieties no longer cultivated on a commercial scale (Galluzzi et al., 2010). Intensification on patches often comprises an additive change to productivity of agricultural and farm systems.

Tropical home gardens are typically multilayered environments with multiple trophic levels and management zones, and can include fruit trees, shaded coffee, residentials, ornamentals with shade trees, multipurpose trees, herbaceous crops, ornamentals with vine-crop shade, grass, space for working and storage, and ornamentals. Such agroforestry home gardens can host over 300 plant species (Méndez et al., 2001). In Peru, four kinds of small food production system have been documented: fenced gardens near homes, plots planted as gardens near fields, field margins cropped with vegetables, and intercropping of the outer rows of staple fields with climbing vegetables (Niñez, 1984). In Java, village agroforestry gardens (*pekarangan*) provide a safeguard against crop failure, and Tanzanian *chagga* gardens produce 125 kg beans, 275 bunches of banana, and 280 kg of parchment coffee annually on plots of just over half a hectare, insuring against crop failure and supporting poultry and small livestock (Niñez, 1984).

Patch cultivation at varying scales is also important for urban food security. An estimated 800 million people practice some form of urban food production around the world, most on relatively small patches of land cultivated for subsistence in many countries (Lee-Smith, 2010). In Peru and Brazil, urban home gardens increased the availability of

Sustainable Food and Agriculture
DOI: https://doi.org/10.1016/B978-0-12-812134-4.00027-3

staple and nonstaple foods to slum dwellers (Niñez, 1985; WinklerPrins, 2003; WinklerPrins and de Souza, 2005). Small patches are also important for household resilience during lean seasons, in conditions of political instability and turmoil, for marginalized households, in degraded or highly populated areas with few endowments of land and materials, in disaster areas, and in conflict and postcrisis situations. Examples include the use of gardens for food during the Tajik Civil War (Rowe, 2009); the ethnic riots in Sri Lanka (Niñez, 1984); the use of "relief" and "victory" gardens in the United States and the United Kingdom during the World Wars; and the use of gardens to ride over political and economic crises in Cuba (Pretty, 2003). Food growing on neglected patches of city land along highways "often represent the only green spots in abandoned and neglected city parks" (Niñez, 1984, p. 28).

In recognition of these benefits, home gardening and the cultivation of small patches around fields have been promoted in development initiatives to improve food security and incomes. Across Africa, an important strategy has been the construction of raised beds to improve the retention of water and organic material (Pretty et al., 2011). In Kenya and Tanzania, the Farm Africa project has encouraged the cultivation of African indigenous vegetables by 500 participating farmers who have used 50% less fertilizer and 30% less pesticide than in conventionally grown vegetables (Muhanji et al., 2011). In Bangladesh, Homestead Food Production has involved 942,040 households (some 5 million beneficiaries) between 1988 and 2010 (Iannotti et al., 2009). Through home gardening and small-animal husbandry, the project has achieved notable success in increasing the production of fruit, vegetables, and eggs relative to controls and increased income from the sale of produce.

Research on home gardens has focused more on tropical home gardens in the developing world rather than on temperate gardens (Galhena et al., 2013; but see Vogl and Vogl-Lukasser, 2003, and Calvet-Mir et al., 2012, for studies from Austria and Spain). Agricultural extension and research have also largely ignored patches that fall outside regular field boundaries (Pretty et al., 2011), despite calls for more attention to be given to home gardens, kitchen gardens, and other small-scale, subsistence-oriented enterprises in agricultural research, development, and extension (Niñez, 1984). There remains a lack of research on urban agriculture and home gardening in industrialized countries, and studies remain largely descriptive (Taylor and Lovell, 2012a,b).

CHAPTER

28

Integrated Aquaculture and Aquaponics

Aquaculture, especially when integrated within crop cultivation, is also a form of patch management. Brummett and Jamu (2011) reviewed Integrated Aquaculture–Agriculture (IAA) systems in Malawi and Cameroon, describing the outcomes of a project undertaken by the WorldFish Center, supported by various donors and partnered by international and national research and rural development agencies as well as local farmers. Where successful, these partnerships have resulted in increased agricultural productivity, per capita farm income, and per capita fish consumption. There has also been "a 40% improvement in farming system resilience (i.e., defined by the ability to maintain positive cash flows through drought years), a 50% reduction in nitrogen loss and improved nitrogen-use efficiency."

In Nigeria, the rise of peri-urban aquaculture "brings into focus several innovative 'firsts' in African aquaculture development" (Miller and Atanda, 2011). Aquaculture has progressed from a subsistence-focused activity to one connected to a growing market. Miller and Atanda (2011) report "a remarkable 20% increase in growth per year for the past 6 years, with high growth in small- to medium-scale enterprises and a number of large-scale intensively managed fish farms." Aquaculture here is primarily focused on the African catfish (*Clarias gariepinus*), and is characterized by a highly developed production chain of suppliers, processors, and marketers, supported by private-sector technical staff, a growing competence with public extension services, government support via livelihood programs, and credit support from lenders. Nigerian catfish aquaculture is estimated to provide some 17% of the country's total domestic fish production. Producers have been able to diversify into further animal husbandry, keeping cows, pigs, goats, and sheep. Aquaculture also supports crop cultivation via the supply of irrigation and fertilization from fish ponds.

28.1 RICE–FISH SYSTEMS IN ASIA

A field of rice in standing water is more than a crop—it is an ecosystem teeming with life, including ducks, fish, frogs, shrimp, snails, and dozens of other aquatic organisms.

Sustainable Food and Agriculture
DOI: https://doi.org/10.1016/B978-0-12-812134-4.00028-5

For thousands of years, rice farmers have harvested that wealth of aquatic biodiversity to provide their households with a wide variety of energy- and nutrient-rich foods. The traditional rice–fish agroecosystem supplied micronutrients, proteins, and essential fatty acids that are especially important in the diets of pregnant women and young children (Halwart, 2013). Farmers benefit from lower costs, higher yields, and improved household nutrition. Combining rice and aquaculture also makes more efficient use of water. However, rice–fish farming requires about 26% more water than rice monoculture, and therefore these systems are not recommended where water supplies are limited. Nevertheless, an estimated 90% of the world's rice is planted in environments that are suitable for the culture of fish and other aquatic organisms.

During the 1960s and 1970s, traditional farming systems that combined rice production with aquaculture began to disappear, as policies favoring the cultivation of modern high-yielding rice varieties—and a corresponding increase in the use of agrochemicals—transformed Asian agriculture. As the social and environmental consequences of this transformation have become more apparent, there is renewed interest in raising fish in rice fields (FAO, 2014c). There are two main rice–fish production systems. The most common is concurrent culture, where fish and rice are raised in the same field at the same time; rotational culture, where the rice and fish are produced at different times is less common. Both modern short-stem and traditional long-stem rice varieties can be cultivated, as can almost all the important freshwater aquaculture fish species and several crustacean species (FAO, 2004, 2014c).

In China, rice farmers raise fish in trenches up to 100 cm wide and 80 cm deep, which are dug around and across the paddy field and occupy about 20% of the paddy area. Bamboo screens or nets prevent fish from escaping. While fish in traditional rice–fish systems feed on weeds and by-products of crop processing, more intensive production usually requires commercial feed. With good management, a 1 ha paddy field can yield from 225 to 750 kg of finfish or crustaceans a year while sustaining rice yields of 7.5–9 tons/ha (FAO, 2007).

The combination of different plant and animal species makes rice–fish systems productive and nutritionally rich. Equally important are the interactions among plant and animal species, which improve the sustainability of production. Studies in China found that the presence of rice stemborers was around 50% less in rice–fish fields. A single common carp can consume up to 1000 juvenile golden apple snails every day; the grass carp feeds on a fungus that causes sheath and culm blight. Weed control is generally easier in rice–fish systems because the water levels are higher than in rice-only fields. Fish can also be more effective at weed control than herbicides or manual weeding. By using fish for integrated pest management, rice–fish systems achieve yields comparable to, or even higher than, rice monoculture while using up to 68% less pesticide. This safeguards water quality as well as biodiversity.

The interactions among plant and animal species in rice–fish fields also improve soil fertility. The nutrients in fish feed are recycled back into fields through excreta and made immediately available to the rice crop. Reports from China, Indonesia, and the Philippines indicate that rice–fish farmers' spending on fertilizer is lower. Cultivating fish reduces the area available for planting rice. However, higher rice yields, income from fish sales, and savings on fertilizer and pesticide lead to returns higher than those of rice monoculture.

Profit margins may be more than 400% higher for rice farmers culturing high-value aquatic species (FAO, 2014c).

Raising fish in rice fields also has community health benefits. Fish feed on the vectors of serious diseases, particularly mosquitoes that carry malaria. Field surveys in China found that the density of mosquito larvae in rice–fish fields was only a third of that found in rice monocultures. In one area of Indonesia, the prevalence of malaria fell from 16.5% to almost zero after fish production was integrated into rice fields (FAO, 2004). In China, aquaculture in rice fields has increased steadily over the last two decades, and production reached 1.2 million tons of fish and other aquatic animals in 2010. New opportunities for diversifying production are opening up in Indonesia, where the *tutut* snail (*Pila ampullacea*), a traditional item in rural diets, is becoming a sought-after health food for urban consumers. The resurgence in rice–fish farming is being actively promoted by the Government of Indonesia, which recently launched the "One Million Hectare Rice–Fish Program" (Box 28.1). While there is compelling evidence of the social, economic, and environmental benefits of aquaculture in rice farming systems, its rate of adoption remains low outside of China. Elsewhere in Asia, the area under rice–fish production is only slightly more than 1% of the total irrigated rice area. Interestingly, the rice–fish farming area is proportionally largest outside Asia, in Madagascar, at nearly 12%.

Yet, there is much scope for wider uptake of these systems. Improved awareness of its benefits is required. Where smallholders can access low-cost pesticides, these may lock in rice monocultures. Farmers may also have limited access to credit for investment in fish production. Overcoming these barriers is difficult because multisectoral policymaking is involved. Rice–fish farming needs to be championed by agricultural policymakers and agronomists who recognize the benefits of integrating aquaculture and rice and who can deliver that message to rice-growing communities. Just as agricultural development strategies once promoted large-scale rice monoculture, they can now help to realize the potential of intensive, but sustainable, rice–fish production systems.

Another highly innovative initiative on rice–fish farming is being implemented in Sleman district of Yogyakarta province in Indonesia (Box 28.1).

28.2 INTEGRATED AGRICULTURE–AQUACULTURE IN ARID LANDS, ALGERIA

The increasing scarcity of water in dry areas is a well-recognized problem, particularly in the Near East and North Africa region, where water shortage and needs are increasing, and the competition for water among urban, industrial, and agricultural sectors is becoming more intensive. Growth of human population and the continuous exploitation of land and water resources, particularly in arid lands, will require the application of new strategies to ensure adequate food production (animal protein and vegetables) by populations living in remote areas.

The most suitable aquaculture practice in desert and arid lands is an aquaculture system fully integrated with agriculture. This system, if correctly conceived, allows the production of different crops (fish, agriculture, and livestock) using the same quantity of water. Water used to grow fish is then used to water agriculture crops, which benefit from

BOX 28.1

INNOVATIVE RICE–FISH FARMING IN SLEMAN DISTRICT, YOGYAKARTA PROVINCE, INDONESIA

The Government of Indonesia started its national campaign to promote rice–fish farming in 2011. It evolved later as the national One Million Hectare Rice–Fish Program.

In order to support the national rice–fish farming program in Indonesia, in 2015, FAO's Regional Initiative on Blue Growth supported the Directorate General of Agriculture of the Ministry of Marine Affairs and Fisheries to demonstrate the innovative rice–fish farming system through a holistic approach in Sleman district of Yogyakarta province and Lima Puluh Kota district, West Sumatra province. In Sleman district, 151 rice farmers participated in the demonstration with a total rice–fish area of 25 ha. The main fish cultured in the demonstration was tilapia.

To address the deficiencies of the former rice–fish farming demonstration activities, the demonstration project included the following major innovations in addition to achieving symbiosis between fish and rice in the usual rice–fish system (e.g., integrated pest and weed control, natural food for fish in the rice field, and fertilization for rice with fish feces):

- The project employed a farmer group approach and emphasized capacity building. In Sleman district of Yogyakarta province, all the demonstration farmers joined five farmer groups, which were used as the platform for technical training, experience sharing, and joint production activities.
- A new technical manual for innovative rice–fish farming was developed, which included improved stocking (large-size fingerlings at appropriate density and good handling), standard rice-plot and fish-trench preparation, and improved water management and feeding practices. Technical training was provided for the demonstration farmers.
- The project also emphasized a value chain approach, supported the establishment of local large-size fish seed production capacity, and organized marketing of products through farmer groups.
- The project effectively engaged other government ministries, particularly the Ministry of Agriculture. The Ministry of Marine Affairs and Fisheries and the Ministry of Agriculture jointly promoted the scaling up of innovative rice–fish farming in the country.

The FAO-supported demonstration project achieved significantly improved results compared with the demonstration previously implemented by the Directorate General of Agriculture:

- Rice production from integrated rice–fish farming was normally 20% higher than rice farming alone. The average yield of rice reached 7.2 tons/ha.
- The average fish yield achieved by the 151 demonstration farmers was 1.2 tons/ha. The highest fish yield from an individual demonstration farmer was 2.5 tons/ha. The survival rate of fish reached 82% on average. The harvesting size of the fish reached 150 g on average, which was significantly larger than the fish from earlier demonstration activities of the Directorate General of Agriculture.

BOX 28.1 (*cont'd*)

- The economic profit from the demonstration of innovative rice–fish farming averaged $1390/ha and the benefit–cost ratio 1.50.

Important lessons from the demonstration of innovative rice–fish farming include:

- Innovative rice–fish farming can greatly improve the productivity from the rice system. Fish can overtake rice as the main source of income to the farmer, or at least be an equally important source. Significantly improved incomes of farmers can contribute greatly to sustained rice production.
- A group approach is the key for sustainable development of rice–fish farming, which facilitates the value chain development, effective sharing of experiences and lessons and collective production activities, and supports the farmers in continuing the practices after the financial support is withdrawn.
- The technical package of rice–fish farming should consider the characteristics of the rice–fish system and aim to maximize the benefit of the integrated rice–fish system. It cannot simply apply the pond fish farming practices to the rice–fish system. Fish seed for the rice–fish system should be much larger than for pond culture. The water should be retained in the rice–fish system to develop natural food and keep the nutrients for rice. The feeding rate applied for fish in the rice–fish system should be much lower than in ponds to maximize use of natural food. The use of inorganic fertilizer should be minimized.

The demonstration project achieved satisfactory results in general, but there was variation among the performance of individual demonstration farmers. Continuous efforts are required to support the farmers to master the good rice–fish farming practices for reduced feed/fertilizer cost, improved survival, and growth of fish.

nutrient-rich and fertilized water. Agriculture production has shown increasing performance growth since crops have been irrigated using water from fish ponds.

Water in arid lands is available from dams, ponds, irrigation canals, salt and brackish-water lakes, temporary rivers, rainfed reservoirs, but mainly from ground fossil water (aquifers). In recent years, through the technical assistance of FAO, the Ministry of Fisheries and Fish Resources of Algeria has carried out pilot projects to assess the development constraints and opportunities for the expansion of aquaculture in five *wilayas* (provinces), namely Ouargla, Ghardaïa, Laghouat, El Oued, and Biskra.

The aquaculture stakeholders range from small-scale farmers (i.e., those owning 1–2 ha of land and 1 earthen pond with an average surface area of 150 m^2 and 1 m depth mainly used for irrigation), to medium- and large-scale farmers (i.e., those with farms covering a surface area between 10 and 20 ha with several ponds of an average surface area of 100–150 m^2).

The main cultured species are Nile tilapia (*Oreochromis niloticus*), red tilapia (hybrid—*O. mossambicus* × *O. niloticus*), and common carp (*Cyprinus carpio*) reared in integrated aquaculture systems; and North African catfish (*C. gariepinus*) and whiteleg shrimp (*Litopenaeus vannamei*) farmed in intensive closed recirculating aquaculture systems. There are a number of governmental hatcheries for production and distribution of seed. Farmers mainly employ imported commercial feed, but in some cases, farm-made feed is produced by using selected local ingredients. Fish production varies from 2–10 tons/year on small-scale farms to 20–100 tons/year on medium-scale farms, and 500–1000 tons/year on intensive large-scale farms (catfish and shrimp).

These results show that there is a high potential for aquaculture development in arid lands across the country. The five *wilayas* cover an agricultural area of about 30,000 ha, with some 13,700 farmers and 6605 irrigation ponds. Current and future developments of aquaculture in desert and arid lands depend on the proper use of water-adopting farming practices that ensure the smart use of this limited resource. However, the engagement and support of governments, as in the case of Algeria, seems to be the right approach for promoting and strengthening this subsector.

28.3 AQUAPONICS

Aquaponics is a symbiotic integration of two food production disciplines: (1) aquaculture, the practice of farming aquatic organisms; and (2) hydroponics, the cultivation of plants in water without soil. Aquaponics combines the two within a closed recirculating system. A standard recirculating aquaculture system filters and removes the organic matter ("waste") that builds up in the water, keeping the water clean for the fish. However, an aquaponic system filters the nutrient-rich effluent through an inert substrate containing plants. Here, bacteria metabolize the fish waste and plants assimilate the resulting nutrients, with the purified water then returning to the fish tanks. The result is value-added products such as fish and vegetables as well as lower nutrient pollution into watersheds.

Indonesia has developed a particular type of aquaponics, known as Yumina–Bumina, that has been adopted in many districts. Yumina–Bumina uses simple and locally available materials, and has proved to be an efficient and low-energy demanding system with which to build resilience and self-sufficiency in single households or communities and to provide effective diet diversification. Aquaponics has the potential for higher yields of produce and protein with less labor, less land, and fewer chemicals, and takes up a fraction of the water usage. Being a strictly controlled system, it combines a high level of biosecurity with a low risk of disease and external contamination without the need for fertilizers and pesticides. Moreover, it is a potentially useful tool for overcoming some of the challenges of traditional agriculture in the face of freshwater shortages, climate change, and soil degradation. Aquaponics works well in places where the soil is poor and water is scarce, for example, in urban areas, arid climates, and on low-lying islands.

However, commercial aquaponics is not appropriate in all locations, and many start-ups have failed. Before investing in large-scale systems, operators need to consider several factors carefully, especially the availability and affordability of inputs (i.e., fish feed, building, and plumbing supplies), the cost and reliability of electricity, and access to a significant market willing to pay premium prices for locally produced, pesticide-free vegetables. Aquaponics combines the risks of both aquaculture and hydroponics, and thus expert assessment and consultation are essential.

Indonesia is a strong supporter of aquaponics, where a socioeconomic survey of 33 practitioners using aquaponics has been conducted noting the technical, social, and economic characteristics, as well as a nutrient composition study of aquaponic vegetables. Interviews with the Ministry of Agriculture and the Ministry of Marine Affairs and Fisheries officials reveal that aquaponics has blossomed, starting from around 2011, although data on the total number of aquaponic farms in Indonesia are not available. Aquaponics production takes place in nine production centers located in Jakarta, West Java, Central Java, East Java, Yogyakarta, South Sumatra, South Sulawesi, and Central Kalimantan. Experts estimate that around 8000–10,000 households are exposed to aquaponic extension activities, though only about 10% of those are actively practicing aquaponics.

During the primary data collection survey in Indonesia, several questions regarding the social and economic impacts of aquaponics at the household level were asked. The respondents identified the greatest advantages of the aquaponic system to be improved community status, improved diet, improved education opportunities for youth, increased income, and reduced budget for food. Generally, aquaponics affected and/or changed the diet of respondent families through eating more fish and a greater diversity of vegetables, with 29 of the 33 respondents (88%) recognizing this as a key result. The nutrient composition study supported claims that no significant differences exist between water spinach grown in aquaponics and that grown with traditional practices, though this needs to be studied further. In addition to affecting dietary consumption patterns, 21 of the 33 respondents reported selling the fish and 28 reported selling the vegetables.

However, negative effects were also reported from the adoption of the aquaponic systems, which included tension among household members over control/usage of the aquaponic system (52%), increased overall workload (33%), and loss of money (5%). The biggest opportunity with aquaponics was the ability to save water and create a cost-effective business, while the biggest challenges were the higher initial costs, the need for more information and education, and the difficulties in commercializing as a profitable business.

One of the most effective strategies in Indonesia was found to be working with community women's groups, often associated with schools and education, as these groups maintain an important role in the village and neighborhoods. Individual extension agents and volunteer extension or village champions are useful to support and follow up with extension activities. Communications, in all forms, need to improve to support consumer awareness of the value of aquaponics.

As demonstrated in Indonesia, aquaponics has the technical potential to become an additional means of addressing the global challenge of food supply. It is still, however, a relatively new enterprise with inadequate technology transfer, enterprise-oriented capacity development, locally inspired innovations, and organized awareness-raising campaigns.

CHAPTER

29

Landscape Approaches for Sustainable Food and Agriculture

Landscape approaches include a "set of concepts, tools and methods deployed in a bid to achieve multiple economic, social, environmental objectives (multifunctionality) through processes that recognize, reconcile and synergize interests, attitudes and actions of multiple actors" (Minang et al., 2015).

The mitigation of global environmental change requires change and cooperation across landscapes. This is also the scale at which work must integrate social, economic, and environmental objectives. Landscape approaches allow issues to be addressed in a multifaceted way, integrating domains, involving stakeholders, and working at different scales. They recognize that each locality or territory has a set of physical, environmental, human, financial, institutional, and cultural resources that jointly constitute its asset endowment and development potential.

A recent FAO–Intergovernmental Panel on Climate Change Expert Meeting (FAO and IPCC, 2017) concluded that food security under climate change requires integrated frameworks and approaches combining scientific findings with socioeconomic and institutional assessments. There is also need to manage production systems and natural resources across an area large enough to produce vital ecosystem services and small enough to be managed by the people using the land and producing those services (FAO, 2017). This requires long-term collaboration among different groups of land managers and stakeholders to achieve their multiple objectives and expectations (LPFN, 2016). Land use planning is central to successful landscape approaches. When conducted in a way that allows for the participation of all stakeholders, it helps in reducing conflicts over the use of resources, therefore leading to sustainable outcomes for the long-term resilience of populations in the face of unpredictable and extreme weather events (IPCC, 2014).

Overall, the rationale for applying integrated approaches at the landscape scale is threefold: landscape approaches offer a comprehensive platform across sectors and increase the likelihood of successful and sustainable outcomes of development interventions by addressing and negotiating the externalities that occur beyond traditional interventions at the farm and community level. Also, these approaches contribute toward building resilience

Sustainable Food and Agriculture
DOI: https://doi.org/10.1016/B978-0-12-812134-4.00029-7

in social–ecological systems by enhancing their capacity to withstand stresses and shocks, including from the likely future impacts of climate change.

Landscape approaches include, *inter alia*, watershed management and restoration, forest conservation, water resources management, marine and coastal management, and reversing land and soil degradation. All these approaches share common characteristics: they focus on the sustainable use, capacity, and resilience of natural resources, on the joint management of shared resources, and on addressing the needs of farmers, other resource managers, and local communities at large within a broader governance framework.

Landscape governance can be defined as the interactions, decision-making, policies, and partnerships within or related to a landscape. It has a substantive element, which is the area and its nature–human interaction, and a process element, which relates to the process of interacting and decision making (van Oosten et al., 2014). Governing at the landscape level requires an explicit unpacking of the relations between a landscape's spatial features (rivers, slopes, forests, rural–urban relations, value chains), the people living and producing within it, and the various levels at which spatial decisions are taken (individual land users, villages, sectors, districts, nations). Landscape governance aims to combine different objectives into one integrated framework based on spatial planning. When the right institutional mechanisms are in place, landscape governance can provide a basis for long-term collaboration and shared development targets among stakeholders (governments, private sector, communities, and civil society), and for reconciling conflicts between food security, productivity, landscape restoration, and socioeconomic development.

To achieve successful outcomes, the people who have a stake in the landscape must be in a position to jointly plan and negotiate practices and management actions that they consider acceptable. Ensuring the participation of all stakeholders in the decision process is key to enhancing ownership and commitment to managing landscapes and scaling up sustainable agriculture. Enabling negotiation among stakeholders necessitates involving different actors with varying degrees of power, including local authorities, local leaders, landowners, land users, central government institutions, and private entrepreneurs.

Often, stakeholders have dissimilar visions and understanding of landscape planning and goals, and diverse entry points and priorities (land use system, risk aversion, productivity increase, etc.). Setting up a successful negotiation process involves considering all stakeholders' interests in the formulation of land use/resources management plans, solving conflicts, or addressing trade-offs. Such a negotiated process should follow procedures and rules that are agreed upon by the stakeholders and enforced by a credible and legitimized third party (FAO, 2012). The Green Negotiated Territorial Development (GreeNTD) approach, developed by FAO, is an example of multistakeholder engagement to foster a progressive consensus leading to a comprehensive, multiscale, and negotiated vision. It promotes a concerted decision-making method contributing to leveling the power asymmetries among different stakeholders, particularly women, minorities, youth, and other marginalized groups.

Landscape interventions need to be supported by dedicated financial mechanisms. Farmers and land users play a key role in managing natural resources across the landscape. The adoption of more sustainable practices often requires additional effort from farmers. However, upfront financing, land, and labor during establishment, inputs, seeds, technology, and risk of poor performance may all deter farmers from making long-term sustainable investments.

Incentives can help certain categories of land users overcome the barriers that prevent them from adopting practices that benefit others in the landscape, or the environment as a whole. The spectrum of available incentive options range from policy driven to voluntary, sourced from existing public programs, private sector investment, or civil society initiatives (FAO, 2016b). These incentives can be financial or nonfinancial in nature. Financial incentives include payments for ecosystem services, which is a mechanism to compensate farmers and farming communities for the lost opportunity cost of maintaining ecosystem services. Payments for ecosystem services can be used as a market-based innovation to scale up sustainable land management practices or restoration activities. A wide range of nonfinancial incentives also exist, such as capacity development, knowledge building, provision of materials, and the development of alternative livelihoods. Improving access to higher value markets, such as through certification for specific commodities, provides additional incentives for investments in sustainable agricultural initiatives. Improved coordination of existing incentives across sectors can provide a package of actions to support short-term transitional needs and the long-term sustainability of agriculture systems.

Examples of incentive mechanisms exist, such as the Upper Tana-Nairobi Water Fund (The Nature Conservancy) in Kenya; the Rio Rural Partnership in Brazil; the Pro-poor Rewards for Environmental Services in Africa program (World Agroforestry Centre—ICRAF); and the Rewards for, Use of and Shared Investment in Pro-poor Environmental Services (RUPES) in Asia.

A combination of incentives can help rehabilitate or protect ecosystem services. Examples include helping farmers to get organized so as to facilitate their access to markets, or negotiate compensation for hydrological services that are provided at the landscape scale. As many of these sustainable practices have benefits beyond the farm, these linkages become clearer in a landscape approach and can capture funding from other sectors to support sustainable agriculture choices.

CHAPTER

30

Social Capital and Redesign

Many encouraging innovations for sustainable intensification have emerged in recent years, in both small as well as large farms. Central to all these are, of course, farmers and their families. In many cases, the shift from substitution to redesign can only occur if changes are made across whole landscapes. Save for the largest of farms and rangeland properties, individual decision making for farming occurs at a lower system level than many environmental services and assets operate. A core principle of sustainable intensification centers on the positive contributions that agricultural production can make to the natural capital of soil, water, biodiversity, and atmospheric quality. In most systems, this implies the need for collaboration and cooperation between farmers. One farmer can change the world through leadership and demonstration; it takes many farmers in a particular landscape to regenerate ecosystems.

There are four central features of social capital: relations of trust; reciprocity and exchanges; common rules, norms, and sanctions; and connectedness, networks, and groups.

Trust lubricates cooperation and thus reduces the transaction costs between people. Instead of having to invest in monitoring others, individuals are able to trust them to act as expected. Besides saving money and time, this also creates a social obligation, as trusting someone also engenders reciprocal trust. There are two different types of trust between individuals: one being the trust we have in individuals we know, and the other being the trust we have in those we do not know, but which arises because of our confidence in a known social structure. Trust takes time to build, but is easily damaged; when a society is pervaded by distrust, cooperative arrangements are very unlikely to emerge or persist.

Reciprocity and regular exchanges increase trust, and so are also important for social capital. Reciprocity comes in two forms: specific reciprocity is the simultaneous exchange of items of roughly equal value, while diffuse reciprocity refers to a continuing relationship of exchange that at any given time may be unrequited, but which over time is repaid and balanced. Again, both contribute to the development of long-term obligations between people, which is an important part of achieving positive sum gains for natural capital.

Common rules, norms, and sanctions are the mutually agreed or handed-down norms of behavior that place group interests above those of individuals. They give individuals the confidence to invest in collective or group activities, knowing that others will do so, too. Individuals can take responsibility and thus ensure their rights are not infringed.

Sustainable Food and Agriculture
DOI: https://doi.org/10.1016/B978-0-12-812134-4.00030-3

Mutually agreed sanctions ensure that those who break the rules know they will be punished. These rules of the game, sometimes called the internal morality of a social system or the cement of society, reflect the degree to which individuals agree to mediate their own behavior. Formal rules are those set out by authorities, such as laws and regulations, while informal ones shape our everyday actions. Norms, by contrast, indicate how we should act (when driving, norms determine when we let other drivers into the traffic queue; rules tell us on which side of the road to drive).

Connectedness, networks, and groups are the fourth key feature of social capital. Connections are manifested in many different ways, such as trading of goods, exchange of information, mutual help, provision of loans, and common celebrations and rituals. They may be one-way or two-way, and may be long established, and thus not respond to current conditions, or subject to regular update. Connectedness is institutionalized in different types of group at the local level, from guilds and mutual aid societies to sports clubs and credit groups, from forest, fishery, or pest management groups to literary societies and mothers' groups. High social capital also implies the likelihood of multiple membership of organizations and good links persisting between groups.

For farmers to invest in collective action and social relations, they must be convinced that the benefits derived from joint approaches will be greater than those from going it alone. External agencies, by contrast, must be convinced that the required investment of resources to help develop social and human capital, through participatory approaches or adult education, will produce sufficient benefits to exceed the costs. As Elinor Ostrom (1998, p. 18) puts it: "participating in solving collective-action problems is a costly and time-consuming process. Enhancing the capabilities of local, public entrepreneurs is an investment activity that needs to be carried out over a long-term period." For initiatives to persist, the benefits must exceed both these costs and those imposed by any free-riders in collective systems.

One mechanism to develop the stability of social connectedness is for groups to work together by federating to influence district, regional, or even national bodies. This can open up economies of scale to create even greater economic and ecological benefits. The emergence of such federated groups with strong leadership also makes it easier for government and nongovernmental organizations to develop direct links with poor and formerly excluded groups, though if these groups were dominated by the wealthy, the opposite would be true. This can result in greater empowerment of poor households, as they can far easily draw on public services. Such interconnectedness between groups is more likely to lead to improvements in natural resources than regulatory schemes alone.

Social capital is thus seen as an important prerequisite to the adoption of sustainable behaviors and technologies over large areas. Three types of social capital are commonly identified (Hall and Pretty, 2008). These are the ability to work positively with those closest to us who share similar values (referred to as bonding social capital); working effectively with those who have dissimilar values and goals (bridging social capital); and finally, the ability for positive engagement with those in authority whether to influence their policies or to garner resources (which is termed linking social capital). Linking social capital encompasses the skills, confidence, and relationships that farmers employ to create and sustain rewarding relationships with staff from government agencies. To gain the most from social capital, individuals and communities require a balanced mixture of bonding, bridging, and linking relationships.

30.1 FARMER FIELD SCHOOLS

A remarkable social innovation arose in the 1980s when Peter Kenmore, Kevin Gallagher, and colleagues at FAO undertook groundbreaking research to show that in irrigated rice systems, increased pesticide use resulted in increased damage from pests. They realized that insecticides were killing beneficial insects and arthropods. They also realized that farmers would not be aware of this. Detailed entomological knowledge is rarely a feature of local indigenous knowledge systems. Their first hypothesis: Could irrigated rice management be amended to reduce insecticide use, and could the beneficial insects do the work of pest management? Their second: Could they create a system of learning to allow farmers to demonstrate to themselves that this really works—that they would, in short, not lose all their crops and starve? The first farmer field schools (FFS) were established in Indonesia and the Philippines and within the space of a generation have spread to many countries and benefited large numbers of farmers.

The aims of FFS, often called schools without walls, are education, colearning, and experiential learning so that farmers' expertise is improved to provide resilience to current and future challenges in agriculture (Table 30.1). FFS are not just an extension method: they increase knowledge of agroecology, problem-solving skills, group building, and political strength. FFS have also been recently complemented by modern methods of extension involving video, radio, market stalls, popups, and songs (Bentley, 2009). These can be particularly

TABLE 30.1 The Principal Elements of Farmer Field Schools

- Each farmer field school (FFS) consists of a group of 25–30 farmers working in small subgroups of about five people. The training is field-based and season-long, usually with one meeting per week.
- The season starts and ends with a "ballot box" pretest and posttest, respectively, to assess trainees' progress.
- Each FFS has one training field divided into two parts: one managed by integrated pest management (IPM) (management decisions decided on by the group, not a fixed formula); the other with a conventional treatment regime, either as recommended by the agricultural extension service or through consensus of what farmers feel to be the usual practice for their area.
- In the morning, the trainees go into the field in groups of about five to make careful observations on the growing stage and condition of crop plants, weather, pests and beneficial insects, diseases, soil, and water conditions. Interesting specimens are collected, put into plastic bags, and brought back for identification and further observation.
- On returning from the field to the meeting site (usually near the field, under a tree or other shelter), drawings are made of the crop plant, which depict plant condition, pests and natural enemies, weeds, water, and anything else worth noting. A conclusion about the status of the crop and possible management interventions is drawn by each subgroup and noted under the drawing (agroecosystem analysis).
- Each subgroup presents its results and conclusions for discussion to the entire group. As well as in the preceding field observations, the trainers remain as much as possible in the background, avoiding lecturing, not answering questions directly, but stimulating farmers to think for themselves.
- Special subjects are introduced in the training, including maintenance of "insect zoos" where observations are made on pests, beneficial insects, and their interactions. Other subjects include leaf removal experiments to assess pest compensatory abilities, lifecycles of pests and diseases, and in recent years the expansion of topics away from just IPM.
- Sociodynamic exercises serve to strengthen group bonding in the interest of post-FFS farmer-to-farmer dissemination.

Source: Settle W., Hama Garba M. 2011. Sustainable crop production intensification in the Senegal and Niger River Basins of francophone West Africa. Int. J. Agric. Sustain. 9(1): 171–185.

effective where there are simple messages or heuristics that research demonstrates will be effective if adopted. FFS have now been used for soil management, biodiversity, livestock, and as of 2005, have benefited 10–20 million farmers (Braun and Duveskog, 2009).

How effective have the FFS been? In Sri Lanka, 610 FFS were conducted between 1995 and 2002 on farms of a mean size of 0.9 ha, and on which paddy rice yields improved slightly from 3.8 to 4.1 tons/ha, while insecticide applications fell from an average of 3.8–1.5 per season (Tripp et al., 2005). More than a third of farmers eliminated pesticide use completely, and an average farmer could name four natural enemies compared to 1.5 by those who had not attended an FFS. In Benin, Burkina Faso, and Mali, 116,000 farmers have been trained in 3500 FFS for vegetables, rice, and cotton (Settle and Hama Garba, 2011).

Pesticide use has been cut to 8% (previously 19 compounds were found in the Senegal River, 40% of which were 100 or more times the maximum tolerance), biopesticides and neem use has increased by 70%–80%, and there have been substantial increases in yields (e.g., rice in Benin up from 2.3 to 5 tons/ha). In Mali, cotton farmers participating in FFS reduced pesticide use by just over 90% compared with pre-FFS use and a control group (Settle et al., 2014).

FFS and integrated pest management (IPM) have had an impact at the macro level in some countries (Ketelaar and Abubakar, 2012). Between 1994 and 2007, rice farmers in the Philippines reduced pesticide application frequency and applications per hectare by 70%, increased yields by 12%, and increased the inter-year stability of yields. Over this period, national rice production rose from 10.5 to 16.8 million tons (FAOSTAT, 2015a). In Bangladesh, pesticide use has increased substantially in recent years, and in some districts (e.g., Natore) results in 40–50 applications per season for beans and 150–200 times for *brinjal* (aubergine), sometimes on a daily basis (Bentley, 2009). Many farmers only spray crops for market, and keep those for home consumption unsprayed. In other regions of Bangladesh, there are severe shortages of rural labor where the burgeoning garment industry has attracted young people; here, FFS have helped farmers to grow rice while ensuring that fish and frogs are protected.

Evaluation of two IPM–FFS programs in Sichuan, China, found that yields slightly increased while pesticide applications fell by 40%–50% (Mangan and Mangan, 1998). In Indonesian wet rice agroecosystems, 64% of the insects and spiders were found to be predators and parasitoids, 19% neutral detritivores, and only 17% rice pests. Beneficials were extremely effective at controlling pests, until pesticides were used.

Nonetheless, it is difficult to overcome the fears that farmers have; often these have been encouraged by the pesticide industry. Farmers need to overcome the notion that insects always cause harm, that insect pests will transfer from sprayed to unsprayed fields and farms, and that crop loss will occur (Palis, 2006). In some cases, farmers have experienced anxieties about maintaining good social relations, spraying their crops secretly at night. Yet where farmers did join FFS, they had the confidence to make dramatic reductions in pesticide applications from 1.9 to 0.3 per season (van den Berg and Jiggins, 2007). There is also much scope for increasing farmers' participation in associations. In Cameroon, for example, less than half of the farmers interviewed for a study on sustainable practices were members of farmer organizations, impeding the widespread uptake of improved practices.

Field farmer schools have now been used in 90 countries, including in central and eastern Europe, the United States, and Denmark (Braun and Duveskog, 2009). Between 10 and 20 million farmers have graduated from FFS, including 1.1 million in Indonesia, 930,000 in

BOX 30.1

PARTICIPATORY IRRIGATION MANAGEMENT AND WATER USER GROUPS

Although irrigation is a vital resource for agriculture, water is—rather surprisingly—rarely used efficiently. Without regulation or control, water tends to be overused by those who have first access to it, resulting in shortages for tailenders, conflicts over water allocation, and waterlogging, drainage, and salinity problems. However, where social capital is well developed, water user groups with locally developed rules and sanctions are able to make more of existing resources than individuals working alone or in competition. The resulting impacts, such as in the Philippines and Sri Lanka, typically involve increased rice yields, increased farmer contributions to the design and maintenance of systems, dramatic changes in the efficiency and equity of water use, decreased breakdown of systems, and reduced complaints to government departments. More than 60,000 water user groups have been set up in the past decade or so in India, Nepal, Pakistan, the Philippines, and Sri Lanka.

It has become clear that the inefficiencies of public administration produced market failures in managing irrigation water, thus demanding new principles of organization. Participatory irrigation management and associated water user associations (WUAs) or groups emerged as both concepts and practice that have spread substantially. In Mexico, 2 million of the 3.2 million hectares of government-managed systems have been transformed by WUAs, and half of the systems in Turkey have been turned over to local groups (Groenfeldt and Sun, 1997). In China, a quarter of all villages have WUAs, and these have reduced maintenance expenditure while improving the timeliness of water delivery and fee collection. Farm incomes have improved while water use has fallen by 15%–20%. In Bali, there are 1800 long-standing self-organizing irrigation groups that cover nearly 20% of the land area (Kulkarni and Tyagi, 2012).

Up to 80% of India's farmers are small and marginal producers. The majority work in the rainfed drylands and have low capacity to invest due to the relatively high cost of technology per patch of land. Farmers are also constrained by insufficient access to transport and storage facilities, hindering their ability to bring produce to market even if surpluses were to be produced. Collective resource management, particularly of water, and the development of human capital through peer-to-peer learning have shown that even in this challenging context remarkable transformations are possible.

The Andhra Pradesh Farmer-Managed Groundwater Systems Project was set up in seven districts in the state of Andhra Pradesh. Here, farmers have no access to surface irrigation infrastructure, nor the capacity to invest in borewells. The project thus instituted a novel arrangement whereby groups of farmers were given access to borewells collectively managed through associations. Farmers were trained in the concept of a hydrological unit, raising awareness about the water source being tapped, as well as being given information on improved water management practices. Farmers once trained were now well versed

BOX 30.1 *(cont'd)*

in relevant concepts, such as recharge rates, evaporation losses, soil moisture content, crop water requirements, and appropriate water-saving cropping patterns. Farmer associations thus formed and trained had the added benefit of being able to use their social capital to bring in additional government support, access development schemes, and develop new marketing solutions to bring crops to the market (Mittra et al., 2014).

Community-led watershed development has also demonstrated excellent potential in India, with globally lauded examples, such as the cases of Ralegan Siddhi and Hiware Bazaar villages in the state of Maharashtra. Here, landscapes that were previously extremely vulnerable to drought and recurrent dry spells have seen long-term transformations as a result of participatory soil and water conservation, coupled with the management of village commons and forests and the creation of new rules around crop water budgeting. Together, these initiatives have had a dramatic impact on farm incomes, with per capita income in the village of Hiware Bazaar rising from \$13 to \$462.

Similar community groups have the potential to increase farmers' access to sustainable energy (e.g., via solar-powered irrigation pumps); sustainable irrigation infrastructure (e.g., drip irrigation sets); and relatively expensive equipment (e.g., tractors, harvesters, and storage facilities). Farmer groups also have the potential to improve incomes directly through the creation of marketing cooperatives and selling groups. Indian smallholders are often at a major disadvantage in the market, as small surpluses are sold to intermediaries who appropriate up to 70% of the value of the produce.

There are estimated to be 73,000 WUAs in India, covering 15 million hectares, but still only 12% of irrigated area (Sinha, 2014). Many of these are thought to exist only on paper, and in some areas they have been subject to variable performance, elite capture, and irrigation department control (Reddy and Reddy, 2005). In some contexts, rights transfers to landowners and tenant farmers have led to landless and fishing families losing out. In all collective management, the distribution of winners and losers remains a challenge: in irrigation management, this centers particularly on differential benefits for farms at the head, middle, and tail of systems. Nonetheless, there is also strong evidence from India that WUAs can lead to increases in area under irrigation (more efficient); greater equity (improved benefits for tailenders); and greater recovery of water charges (a measure of improved yields) (Sinha, 2014). In Turkey, 10 years of WUAs increased cropping intensity and increased yields by 53% (Karahan Uysal and Atış, 2010).

Vietnam, 650,000 in Bangladesh, 500,000 in the Philippines, 250,000 in India, and 90,000 in Cambodia. By the early 2000s, 20,000 FFS graduates were running FFS for other farmers, having graduated from farmers to expert trainers.

Three other examples of successful initiatives based on the principles outlined include: participatory irrigation management and water user groups (Box 30.1); advocacy coalition in response to climate change in the Peruvian Andes (Box 30.2); and Agreco farmers' organization in Brazil (Box 30.3).

BOX 30.2

SOCIAL CAPITAL AND THE CREATION OF AN ADVOCACY COALITION IN RESPONSE TO CLIMATE CHANGE IN THE PERUVIAN ANDES

Agricultural communities in developing countries are particularly vulnerable to climate change relative to communities with more diversified livelihoods or urban communities. To what degree is it possible for community-based organizations to utilize bridging and bonding social capital to join with other groups and institutions to increase the health of the ecosystem, increase social inclusion, and provide economic security in a time of rapid external changes? Using bridging and bonding social capital, vulnerable groups can utilize social capital to enhance all the capitals that impact their community. A key element is how they find allies to use existing policies, rules, and regulations to enhance their collective well-being (Flora et al., 2006; Flora and Flora, 2004). Social capital is critical in making that shift.

The high plains of the Andes are a potentially fertile but fragile ecosystem. People and animals coevolved with the plant and soil microorganisms to provide not only economic security, but also surplus that supported the rise of a variety of specialized livelihoods, from those of governors to warriors to priests. Traditional agroecosystems here depended on a combination of root crops grown at high altitudes and Andean camelids, particularly alpacas (*Vicugna pacos*). These camelids take advantage of the lean pastures from the "Puna" with greater feed conversion efficiency than the foreign animal species imported by the Spanish, including sheep, goats, hogs, horses, and cattle. For example, alpaca digestion of pasture is 22% more efficient compared to that of sheep (Lay-Lisung, interview with PROMPEX, 2007), making it a preferred species at high Andean altitudes. As its meat and fiber are highly appreciated by rural communities, alpacas have been bred to be docile and cooperative around humans. Yet even with this biological advantage, alpaca producers have been experiencing losses in their flocks due to the dramatic swings in temperature, increasingly variable rainfall, unusually high winds, and a higher incidence of hailstorms, as well as arbitrary changes to market standards.

These challenges have driven the development of community-led adaptation coalitions from 2006 onward. As part of the collaborative research support program—the Sustainable Agriculture and Natural Resource Management Collaborative Research Support Program, funded by the United States Agency for International Development—local research teams (EILs—*equipos de investigación local)* were formed by community members and trained in interview techniques. These teams worked with local institutions and communities to identify activities that could be collectively carried out to increase the quality of alpaca flocks by ensuring feed through improved pasture, including cut-and-carry feed provision during unexpected ice storms and freezes, classification of each animal in the herd by color and fiber quality, and selective breeding for high demand characteristics to increase the fineness of the fiber on

BOX 30.2 (*cont'd*)

each alpaca. This strategy enabled higher incomes with fewer alpaca. The research committee then sought out other villages to form a collection center in order to have a place to gather and categorize the fiber, while at the same time selecting their own animals for finer fiber and seeking progenitors with fine fiber.

An evaluation of this approach found it to have:

1. Strengthened human capital at the community level by increasing data-handling and processing skills.
2. Strengthened social capital at two levels: (a) with actors in the area through the strategy of visits of coordination with institutions in Ilave, a district capital, and in Puno, the regional capital, signed agreements with CONACS (Consejo Nacional de Camélidos Sudamericanos—National Council of South American Camelids), and initiated some of the activities that came from the agreements; and (b) at the community level, increased their collective ability to interact and articulate proposals and present them to the community assembly for their discussion and to make decisions after the meeting based on community priorities.
3. Woven a social web between the community and civil society actors. These were valued by the community, as demonstrated by the following testimonies by community members: "The Community is grateful to the SANREM project because thanks to you now we dare to go to the institutions where we did not before. We have that we must always be prepared and sure of what we should achieve [sic], that we need to be united as we were, and that such unity functions and that together we will achieve many things for the community." [Statement of a community member of Apopata who participated in the local research team]

> ...Now we can say to our young people that they must participate because finally they are those who will be here with all of this to do many things for the community because this way we are a group and united we can do a lot. Yes, we can do many things because of what we have achieved through our alliances. ***Testimony of a member of the Apopata community and member of the local research team, as recorded by Dr. Edith Fernandez-Baca.***

This initiative has also shown that by collectively building relationships with outside institutions (bridging social capital) and sharing those relationships with the whole community, internal relations (bonding social capital) were also strengthened. Furthermore, the need for volume led to sharing of processes and learning with neighboring communities (linking social capital) to create a center for bringing the alpaca fiber, centralizing record-keeping, and sharing knowledge learned. As the new mindset around new standards changed the way that the alpaca producers engaged with the value chain, the narratives they shared focused on their appreciation of fiber quality. An important lesson in building social capital is that those new networks between the community and the institutions required that the community become the researchers and not simply the beneficiaries of institutional largesse. The discernment process of the community

BOX 30.2 (*cont'd*)

in identifying the appropriate institutions and the initiative in setting up interviews with them was empowering and moved the locus of control of the interaction from the institution to the community. Productivity increased as the measure of productivity shifted due to changes in convention, and the advocacy coalition research process gave the community the courage—the cultural capital—to change not only what they produced and how they produced it, but how they marketed it in a global value chain.

BOX 30.3

THE CASE OF AGRECO FARMERS' ORGANIZATION, SOUTHERN BRAZIL

The Ecological Farmers Association in the Hillsides of Santa Caterina State (Agreco) is an ecological farmers' association based in the eastern hillsides of the state of Santa Catarina that has expanded its activities to ecotourism, school meals, and links to wider economic development. The organization was set up in the mid-1990s by a dozen families near Santa Rosa de Lima, and began organic production of legumes, honey, grain, and fruit. Though there are now 300 families involved in the network, Agreco does not intend to expand further, preferring to help others to set up in the same way. Produce is now sold directly to urban consumers in mixed baskets, to selected supermarkets, and to schools for children's meals. An innovation has been the emphasis on agritourism. It has adopted the French *accueil paysan* approach, with accommodation in farm buildings and families actively seeking to exchange and share life experiences with visitors.

Agreco has formed two regional forums: one to link public and private agencies, which has already had success with energy supply, rural transport, and school meals; and another to link urban with rural development. This alternative development model links an ecological approach to farming, organized small businesses, connections to food consumers, and wider economic development. Sérgio Pinheiro, a researcher in Santa Catarina state explains: "Initial resistance may be explained because individualism is normal for most people. The ability and enthusiasm to work in groups has increased among farmers, and participation and trust have grown too." This resistance is common elsewhere, as farmers who are not organized often feel that they will lose something by collaboration. Oddly, such cooperation has been fundamental to all agricultural and resource management systems throughout history.

CHAPTER

31

The Way Forward: Supporting Greener Economies

It is difficult to say how much of the world's agricultural lands already operate under forms of more sustainable agricultural production. There has been significant progress both for reducing negative impacts on natural capital and environmental services and for creating systems with the potential to improve all forms of renewable capital assets (natural, social, and human). However, it is clear that considerably more food will need to be produced as populations continue to grow and food consumption patterns converge on the diets typical to affluent countries and societies, regardless of how much progress is made on reducing waste in food systems (Foresight, 2011; Pretty, 2013).

The sustainable intensification of agricultural systems should thus be seen as part of a wide range of initiatives and efforts to create greener economies. Green growth and the greener economies have become important targets for national and international organizations, including the Organisation for Economic Co-operation and Development (2011), the United Nations Environment Programme (2011), the World Bank (2012), the Rio + 20 conference (UNCSD, 2012), and the Global Green Growth Institute (2012). The United Nations Environment Programme (2011) defines the green economy as "resulting in human well-being and social equity, while significantly reducing environmental risks and ecological scarcities." Deep political commitment is rare, even though Stern (2007) pointed to the economic value of early action with respect to climate change: the cost of stabilizing all greenhouse gases was a "significant but manageable" 1% of global gross domestic product (GDP), but a failure to reduce emissions would result in annual costs of 5%–20% of GDP.

There are policies actively promoting greener agendas in countries such as China, Denmark, Ethiopia, South Africa, and South Korea: such a pursuit of greener economies could lead to a new industrial revolution (Stern and Rydge, 2012) and promote further sustainable intensification of agriculture. China has invested $100 billion since 2000 in eco-compensation schemes, mostly in forestry and watershed management. It has developed and implemented a policy framework for sustainable intensification (Box 31.1). A total of 65 countries have implemented feed-in tariffs to encourage renewable energy generation (Renewables, 2012), and by 2010, renewable energy sources had grown to supply 16.7% of

Sustainable Food and Agriculture
DOI: https://doi.org/10.1016/B978-0-12-812134-4.00031-5

BOX 31.1

A POLICY FRAMEWORK FOR SUSTAINABLE INTENSIFICATION IN CHINA

The challenge of agricultural intensification in China is centered on its unique confluence between demand and resource availability: the Chinese agrifood sector meets most of the food needs of 20% of the world's population and produces 25% of the world's grain using less than 9% of the world's arable land, with per capita landholdings among the lowest in the world. In 2015, China's grain output reached a historic record of 621.4 million tons following 12 years of relatively continuous growth, even though there were serious regional or seasonal droughts in some years. All major crop and livestock products have experienced significant growth over the past three and a half decades since the reforms began in 1978, with some subsectors in agriculture growing faster than others. From 1978 to 2015, the output of grain (rice, wheat, maize, beans) and tubers more than doubled, from 304.8 million tons to 621.4 million tons. Meat (pork, beef, lamb) output increased more than sevenfold, from 12.05 million tons in 1980 to 86.25 million tons in 2015. Vegetable and fruit output also increased at a very quick pace, and there was a consequent boost in the amount of food available for everyone in China. Per capita availability of grain in China increased from 319 kg in 1978 to 453 kg in 2015, meat from 9 to 48 kg, and aquatic products from 4.6 to 49 kg despite the population increasing from 987 million to 1.37 billion in the same period.

Yet, existing intensification has come at a cost: groundwater pollution from the overuse of nitrogen fertilizers, pollution from intensive livestock production, degraded soil, and low efficiency of input use and therefore low competiveness. At the same time, food demand is increasing apace, driven by the change to a more protein-based diet. Rapid economic growth, urbanization, and market development are key factors underlining the changes in agricultural structure. Rising incomes and urban expansion have boosted the demand for meat, fruit, and other nonstaple foods. These changes have stimulated sharp shifts in the structure of agriculture. One significant change has been the rapid growth in livestock and fishery production. Total meat output increased from 12.05 million tons in 1980 to 86.25 million tons in 2015; aquatic products increased from 4.5 million tons in 1980 to 65 million tons in 2014. In terms of contribution to total value of agricultural outputs, the share of crops declined from 76% in 1980 to 54% in 2015, while the share of livestock increased from 18% in 1980 to 28% in 2015. The share of fisheries increased most rapidly, from just 1.7% in 1980 to 10% in 2015.

The major drivers of agricultural production in the past three and half decades include: (1) encouraging policies that mobilize farmers' motivation in agricultural production; (2) technological progress; (3) income growth and urbanization as drivers of both qualitative and quantitative changes in agricultural production; and (4) agricultural productivity growth and the increasing use of agricultural inputs (Norse and Ju, 2015). However, these driving forces are also bringing about certain constraints to

BOX 31.1 (*cont'd*)

the further development of agricultural production. The introduction of the household responsibility system has been a major driver of increased agricultural production in the early stage of rural reform. However, small-scale and fragmented household farming plots have become a barrier for further improvement in resource use efficiency, mechanization, and market competitiveness (Huang and Ding, 2016; Ju et al., 2016). Intensive land and water resource use with high input has caused degradation of natural resource bases and environmental externalities (Norse and Ju, 2015; Lu et al., 2015). In recognition of these challenges, the Chinese Government has started implementing a comprehensive strategy to modernize China's agriculture while increasing efficiency and improving environmental outcomes: the zero-growth policy in fertilizer use by 2020 is one of the important components of the strategy, as is land consolidation (Lu, 2016).

Five key concepts inform China's policy framework for development over the next 5 years include: innovation, coordination, greening, opening up, and sharing. The 2016 No. 1 Central Document, released on January 27, 2016 (Xinhua, 2016), gives a broad picture of China's strategy for sustainable intensification. This covers the following key aspects:

1. Consolidating the foundation for modern agriculture, enhancing the quality, efficiency, and competitiveness of agriculture. China will improve the quality and competitiveness of its agricultural products through high-quality farmland and professional farmers, catering to the demands of modern agriculture. This will entail developing high-quality farmland, advancing irrigation, strengthening innovation and extension systems (including that of the seed industry), coordinating use of resources and markets at home and abroad, and making full use of large family farm operations, including by professionalizing family farms.
2. Protecting resources and the ecosystem and promoting green agricultural development. This will be achieved by strengthening actions to protect resources and increase efficiency in their utilization, accelerating the pace at which environmental problems are tackled, and finally by instituting a food safety strategy.
3. Promoting the integration of primary, secondary, and tertiary industries and raising farm incomes by promoting logistics, markets, and the profit-sharing mechanism in rural regions, including the processing of agricultural products, rural tourism, and agritourism.
4. Promoting integrated rural–urban development by accelerating the development of rural infrastructure, raising the level of public services, improving rural and agricultural insurance services, and encouraging financial institutions to extend credit to agricultural businesses.

global energy consumption, with the fastest growing sector being photovoltaic systems (solar PV). This alone could have a significant impact on remote rural communities, and thus lead to changes in agricultural and food systems.

The revenue of many poorer countries is absorbed by the costs of oil imports: for example, India, Kenya, and Senegal spend 45%–50% of their export earnings on energy imports. Investing in renewable energy benefits these three countries by saving export earnings, increasing self-reliance, and improving domestic natural capital. Kenya has introduced feed-in tariffs on energy generated from wind, biomass, hydro, biogas, solar, and geothermal sources from 2008 (UNEP, 2011). In this way, a greener economy that dramatically changes aspirations and consumption patterns by increasing consumption by the current poor and reducing that of the affluent, increases well-being, and protects natural capital is not likely to greatly resemble the current economy. Relevant to all sectors of economies will be important questions about material consumption, in particular how modes of consumption based on "enough not more" can be created, resulting in mass behaviors of "enoughness" (O'Neill et al., 2010). In this way, the sustainable intensification of agricultural systems could both promote transitions toward greener economies as well as benefit from progress in other sectors.

As noted by the NRC (2010) "sustainability is best evaluated not as a particular end state, but rather a process that moves farming systems along a trajectory towards greater sustainability." This suggests that no single policy instrument, research output, or institutional configuration will work to maximize sustainability and productivity over spatially variable conditions and over time. The NRC (2010) made a series of recommendations regarding public research for public goods and integration of agencies to address multidisciplinary challenges in agriculture. It was recommended that the national (in this case, the US Department of Agriculture) and state agricultural institutions should continue publicly funded research and development of key farming practices for improving sustainability and productivity, and that federal and state agricultural research and development programs should deliberately pursue integrated research and extension on farming systems, with a focus on whole agroecosystems. It was further suggested that all agricultural and environmental agencies, universities, and farmer-led organizations should develop a long-term research and extension initiative to understand and shape the aggregate effect of farming at landscape scale. Researchers were encouraged to adopt farmer-participatory research and farmer-managed trials as critical components of their research. At the national level, there should finally be investment in studies to understand how market structures, policies, and knowledge institutions provide opportunities or barriers to expanding sustainable practices in farming. This is particularly important in enabling farmers to navigate the complex and evolving trade-offs between resource conservation and increasing farmers' incomes through participation in markets. Policies designed to conserve resources and stabilize resource availability and farmers' incomes may not work well over time given market imperatives to maximize resource use and incomes (see, for example, Bharucha et al., 2014).

In the context of developing countries, the 30 African cases of sustainable intensification (Pretty et al., 2011, 2014) have illustrated key lessons regarding policy challenges. These projects contained many different technologies and practices, yet had similar approaches to working with farmers, involving agricultural research, building social infrastructure, working in novel partnerships, and developing new private sector options. Only in some of the cases were national policies directly influential. These projects indicated that there were seven key requirements for such scaling up of sustainable intensification to larger numbers of farmers:

1. Scientific and farmer input into technologies and practices that combine crops and livestock with appropriate agroecological and agronomic management.
2. Creation of novel social infrastructure that results in both flows of information and builds trust among individuals and agencies.
3. Improvement of farmer knowledge and capacity through the use of farmer field schools, videos, and modern information communication technologies.
4. Engagement with the private sector to supply goods and services (e.g., veterinary services, manufacturers of implements, seed multipliers, milk and tea collectors) and development of farmers' capacity to add value through their own business development.
5. A focus particularly on women's educational, microfinance, and agricultural technology needs, and building of their unique forms of social capital.
6. Ensuring that microfinance and rural banking are available to farmer groups.
7. Ensure public sector support to lever up the necessary public goods in the form of innovative and capable research systems, dense social infrastructure, appropriate economic incentives (subsidies, price signals), legal status for land ownership, and improved access to markets through transport infrastructure.

However, no single project or program will be able to address all seven of these requirements at once, and thus a generic need persists for an integrated approach that seeks positive synergies over time. Despite great progress having been made, and now the emergence of the term "sustainable intensification" and its component parts, much remains to be done to ensure agricultural systems worldwide not only increase productivity fast enough, but also guarantee only positive impacts on natural and social capital.

References

Africare, Oxfam America, WWF-ICRISAT Project, 2010. More Rice for People, More Water for the Planet. WWF-ICRISAT Project, Hyderabad, India.

Ajayi, C., Akinnifesi, F., Sileshi, G., Kanjipite, W., 2009. Labour inputs and financial profitability of conventional and agroforestry-based soil fertility management practices in Zambia. Agrekon 48, 246–292.

Ali, M.Y., Waddington, S.R., Hodson, D., Timsina, J., Dixon, J., 2009. Maize-rice cropping systems in Bangladesh: status and research opportunities. Working Paper. CIMMYT. Mexico DF.

Altieri, M.A., Toledo, V.M., 2011. The agroecological revolution in Latin America: rescuing nature, ensuring food sovereignty and empowering peasants. J. Peasant. Stud. 38, 587–612.

Altieri, M.A., Nicholls, C.I., Montalba, R., 2017. Technological approaches to sustainable agriculture at a crossroads: an agroecological perspective. Sustainability 9, 349–362.

Amanuel, G., Kühne, R.F., Tanner, D.G., Vlek, P.L.G., 2000. Biological nitrogen fixation in faba bean (*Vicia faba* L.) in the Ethiopian highlands as affected by P fertilization and inoculation. Biol. Fertil. Soils 32, 353–359.

Anas, I., Rupela, O.P., Thiyagarajan, T.M., Uphoff, N., 2011. A review of studies on SRI effects on beneficial organisms in rice soil rhizospheres. Paddy Water Environ. 9, 53–64.

Aryal, J.P., Sapkota, T.B., Jat, M.L., Bishnoi, D., 2015. On-farm economic and environmental impact of zero-tillage wheat: a case of north-west India. Exp. Agric. 51, 1–16. Available from: https://doi.org/10.1017/S001447971400012X. Cambridge University Press 2014.

Asaah, E.K., Tchoundjeu, Z., Leakey, R.R.B., Takousting, B., Njong, J., Edang, I., 2011. Trees, agroforestry and multifunctional agriculture in Cameroon. Int. J. Agric. Sustain. 9 (1), 110–119.

Assefa, K., Aliye, S., Belay, G., Metaferia, G., Tefera, H., Sorrells, M.E., 2011. Quncho: the first popular tef variety in Ethiopia. Int. J. Agric. Sustain. 9 (1), 25–34.

Assefa, T., Sperling, L., Dagne, B., Argaw, W., Tessema, D., Beebe, S., 2013. Participatory plant breeding with traders and farmers for white pea bean in Ethiopia. J. Agric. Educ. Extens. 20 (5), 497–512.

Atangana, A., Khasa, D., Chang, S., Degrande, A., 2014. Major agroforestry systems of the humid tropics. In: Atangana, A., Khasa, D., Chang, S., Degrande, A. (Eds.), Tropical Agroforestry. Springer 978-94-007-7723-1.

Badgley, C., Moghtader, J., Quintero, E., Zakem, E., Chappell, M., Aviles-Vazquez, K., et al., 2007. Organic agriculture and the global food supply. Renew. Agric. Food Syst. 22, 86–108.

Baker, J.M., Ochsner, T.E., Venterea, R.T., Girffis, T.J., 2007. Tillage and soil carbon sequestration – what do we really know? Agric. Ecosyst. Environ. 118, 1–5.

Banwart, S., 2011. Save our soils. Nature 474, 151–152.

Barling, D., Lang, T., Caraher, M., 2002. Joined-up food policy? The trials of governance, public policy and the food system. Soc. Policy Admin. 36 (6), 556–574.

Barnes, A.P., 2016. Can't get there from here: attainable distance, sustainable intensification and full-scale technical potential. Reg. Environ. Change 16 (8), 2269–2278.

Barnes, R., Fagg, C., 2003. Faidherbia albida. Monograph and Annotated Bibliography. Tropical Forestry Papers No 41, Oxford. Forestry Institute, Oxford, UK. 281pp.

Barnes, A.P., Thomson, S.G., 2014. Measuring progress towards sustainable intensification: how far can secondary data go? Ecol. Indicators 36, 213–220.

Bebber, D.P., Ramotowski, M.A.T., Gurr, S.J., 2013. Crop pests and pathogens move polewards in a warming world. Nat. Clim. Change 3, 985–988.

Bentley, J.W., 2009. Impact of IPM extension for smallholder farmers in the tropics. In: Peshin, R., Dhawan, A.K. (Eds.), Integrated Pest Management: Dissemination and Impact. Springer, Berlin.

Berdegué, J.A., Escobar, G. 2001. Agricultural knowledge and information systems and poverty reduction. AKIS discussion paper No. 69. World Bank. Washington.

Berendsen, R.L., Pieterse, C.M.J., Bakker, P.A.H.M., 2012. The rhizosphere microbiome and plant health. Trends Plant Sci. 17, 478–486.

Berkhout, E., Glover, D., Kuyvenhoven, A., 2015. On-farm impact of the System of Rice Intensification (SRI): evidence and knowledge gap. Agric. Syst. 132, 157–166.

Bharucha, Z., Smith, D., Pretty, J., 2014. All paths lead to rain: explaining why watershed development in India does not alleviate the experience of water scarcity. J. Dev. Stud. 50 (9). Available from: https://doi.org/10.1080/00220388.2014.928699.

Birch, A.N.E., Begg, G.S., Squire, G.R., 2011. How agro-ecological research helps to address food security issues under new IPM and pesticide reduction policies for global crop production systems. J. Exp. Biol. 62 (10), 3251–3261.

Birkett, M.A., Chamberlain, K., Khan, Z.R., Pickett, J.A., Toshova, T., Wadhams, L.J., et al., 2006. Electrophysiological responses of the lepidopterous stemborers Chilo partellus and Busseola fusca to volatiles from wild and cultivated host plants. J. Chem. Ecol. 32 (11), 2475–2487.

Bohra, A., Pandey, M.K., Jha, U.C., Singh, B., Singh, I.P., Datta, D., et al., 2014. Genomics-assisted breeding in four major pulse crops of developing countries: present status and prospects. Theor. Appl. Genet. 127 (6), 1263–1291.

Bommarco, R., Marini, L., Vaissiere, B.E., 2012. Insect pollination enhances seed yield, quality and market value in oilseed rape. Oecologia 169, 1025–1103.

Bommarco, R., Kleijn, D., Potts, S.G., 2013. Ecological intensification: harnessing ecosystem services for food security. Trends Ecol. Evol. 28, 230–238.

Borlaug Institute for South Asia, 2015. Major accomplishments 2012–2014. BISA Report Series 1. New Delhi, p. 13.

Bouis, H., 1996. Enrichment of food staples through plant breeding: a new strategy for fighting micronutrient malnutrition. Nutr. Rev. 54 (5), 131–137.

Braun, A., Duveskog, D., 2009. The Farmer Field School Approach – History, Global Assessment and Success Stories. IFAD, Rome.

Breeman, G., Dijkman, J., Termeer, C., 2015. Enhancing food security through a multistakeholder process: the global agenda for sustainable livestock. Food Secur. 7, 245–435.

Brenna, S., Rocca, A., Sciaccaluga, M., 2013. Stock di carbonio organico e fertilità biologica. Il ruolo dell'Agricoltura Conservativa nel bilancio del carbonio – progetto Agricoltura. Regione Lombardia, pp. 53–74, QdR n. 153/2013, cap. 3.1.

Brewer, M.J., Goodell, P.B., 2012. Approaches and incentives to implement integrated pest management that addresses regional and environmental issues. Annu. Rev. Entomol. 57, 41–59.

Brouder, S.M., Gómez-MacPherson, H., 2014. The impact of conservation agriculture on smallholder agricultural yields: a scoping review of the evidence. Agric. Ecosyst. Environ. 187, 11–32.

Brummett, R.E., Jamu, D.M., 2011. From researcher to farmer: partnerships in integrated aquaculture-agriculture systems in Malawi and Cameroon. Int. J. Agric. Sustain. 9 (1), 282–289.

Calle, Z., Murgueitio, E., Chará, J., 2012. Integrating forestry, sustainable cattle-ranching and landscape restoration. Unasylva 63 (1), 31–40.

Calvet-Mir, L., Gómez-Baggethun, E., Reyes-García, V., 2012. Beyond food production: ecosystem services provided by home gardens. A case study in Vall Fosca, Catalan Pyrenees, Northeastern Spain. Ecol. Econ. 74, 153–160.

Cameron, D.D., Neal, A.L., Van Wees, S.C.M., Ton, J., 2013. Mycorrhiza-induced resistance: more than the sum of its parts? Trends Plant Sci. 18, 539–545.

Cardinale, B.J., Wright, J.P., Cadotte, M.W., Carroll, I.T., Hector, A., Srivastava, D.S., et al., 2007. Impacts of plant diversity on biomass production increase through time because of species complementarity. Proc. Natl. Acad. Sci. USA 104, 18123–18128.

CGIAR (Consultative Group for International Agricultural Research), 2013. Grassroots action' in livestock feeding to help curb global climate change. Research Program on Livestock and Fish. Available at: <http://livestockfish.cgiar.org/2013/09/14/bni/>.

Christanty, L., Abdoellah, O.S., Marten, G.G., Iskander, J., 1986. Traditional agroforestry in WestJava: the pekarangan (homegarden) and kebun-talun (annual-perennial rotation) cropping systems. In: Marten GG, (Ed.), Traditional Agriculture in Southeast Asia: A Human Ecology Perspective. ISBN: 0-8133-7026-4.

CIAT, 2013. The Impacts of CIAT's Collaborative Research. Cali, Colombia.

CIMMYT, 2009. Maize motorizes the economy in Bangladesh. CIMMYT E-News, vol. 6, no. 5, August 2009. Available at: <http://www.cimmyt.org/en/what-we-do/socioeconomics/item/maize-motorizes-the-economy-in-bangladesh>.

CIMMYT, 2013. Water-saving techniques salvage wheat in drought-stricken Kazakhstan. In: *Wheat research, Asia*. 21 March 2013. Available at: <http://www.cimmyt.org/en/what-we-do/wheat-research/item/water-saving-techniques-salvage-wheat-in-drought-stricken-kazakhstan>.

Collier, W.L., Wiradi, G., Soentoros, 1973. Recent changes in rice harvesting methods. Some serious social implications. Bull. Indonesian Econ. Stud. 9 (2), 36–45.

Conway, G.R., 1997. The Doubly Green Revolution. Penguin, London.

Conway, G.R., Barbier, E.B., 1990. After the Green Revolution. Earthscan, London.

Cook, S.M., Khan, Z.R., Picket, J.A., 2007. The use of push-pull strategies in integrated pest management. Annu. Rev. Entomol. 52, 375–400.

Cooke, J., Leishman, M.R., 2011. Is plant ecology more siliceous than we realise? Trends Plant Sci. 16, 61–68.

Cooke, J., DeGabriel, J.L., Hartley, S.E., 2016. The functional ecology of plant silicon: geoscience to genes. Funct. Ecol. 30, 1270–1276.

Crews, T.E., Brookes, P.C., 2014. Changes in soil phosphorus forms through time in perennial versus annual agroecosystems. Agric. Ecosyst. Environ. 184, 168–181.

Cullis, C., Kunert, K.J., 2016. Unlocking the potential of orphan legumes. J. Exp. Bot. Available from: https://doi.org/10.1093/jxb/erw437.

De Bon, H., Huat, J., Parrot, L., Sinzogan, A., Martin, T., Malézieux, E., et al., 2014. Pesticide risks from fruit and vegetable pest management by small farmers in sub-Saharan Africa. A review. Agron. Sustain. Dev. Available from: https://doi.org/10.1007/s13593-014-0216-7.

De Groot, R.S., Wilson, M.A., Boumans, R.M.J., 2002. A typology for the classification, description and valuation of ecosystem functions, goods and services. Ecol. Econ. 41, 393–408.

De Groote, H., Vanlauwe, B., Rutto, E., Odhiambo, G.D., Kanampiu, F., Khan, Z.R., 2010. Economic analysis of different options in integrated pest and soil fertility management in maize systems of Western Kenya. Agric. Econ. 41 (5), 471–482.

DEFRA (Department for Environment, Food and Rural Affairs), 2012. Green Food Project Conclusions. DEFRA, London.

Derpsch, R., 2008. No-tillage and conservation agriculture: a progress report. In: Goddard, T., Zoebisch, M.A., Gan, Y.T., Ellis, W., Watson, A., Sombatpanit, S. (Eds.), No-Till Farming Systems. Special Publication No. 3. World Association of Soil and Water Conservation, Bangkok.

Derpsch, R., 2014. Why do we need to standardize no-tillage research? Soil Tillage Res. 137, 16–22.

Derpsch, R., Friedrich, T., Kassam, A., Hongwen, L., 2010. Current status and adoption of no-till farming in the world and some of its main benefits. Int. J. Agric. Biol. Eng. 3 (1), 1–25.

Dessaux, Y., Grandclément, C., Faure, D., 2016. Engineering the rhizosphere. Trends Plant Sci. 21, 266–278.

Dewar, J.A., 2007. Perennial Polyculture Farming: Seeds of Another Agricultural Revolution? RAND Corporation, Santa Monica.

Dixon, J., Gulliver, A., Gibbon, D., 2001. Farming Systems and Poverty: Improving Farmers' Livelihoods in a Changing World. FAO and World Bank, Rome and Washington, DC.

Dobbs, T.L., Pretty, J., 2004. Agri-environmental Stewardship schemes and "Multifunctionality". Rev. Agric. Econ. 26 (2), 220–237.

Dong, Z., Wu, L., Kettlewell, B., Caldwell, C., Layzell, D., 2003. Hydrogen fertilization of soils – is this a benefit of legumes in rotation? Plant Cell Environ. 26, 1875–1879.

EDF, 2017. Collaboration Toward Zero Deforestation. Aligning Corporate and National Commitments in Brazil and Indonesia. Environmental Defense Fund, New York, USA.

Edwards, J., Johnson, C., Santos-Medellína, C., Luriea, E., Podishetty, N.K., Bhatnagar, S., et al., 2015. Structure, variation, and assembly of the root-associated microbiomes of rice. Proc. Natl. Acad. Sci. USA 112, E911–E920.

Ehler, L.E., 2006. Integrated pest management (IPM): definition, historical development and implementation, and the other IPM. Pest Manage. Sci. 62, 787–789.

Elliot, J., Firbank, L.G., Drake, B., Cao, Y., Gooday, R. 2013. Exploring the Concept of Sustainable Intensification. <http://www.snh.gov.uk/docs/A931058.pdf> (31.08.14).

Erenstein, O., 2009. Reality on the ground: integrating germplasm, crop management, and policy for wheat farming system development in the Indo-Gangetic Plains in. 2009. In: Dixon, J., Braun, H., Kosina, P., Crouch, J. (Eds.), Wheat Facts and Futures 2009. CIMMYT, Mexico, DF.

Erenstein, O., Laxmi, V., 2008. Zero tillage impacts in India's rice–wheat systems. Soil Tillage Res. 100, 1–14.

Evans, J., McNeill, A.M., Unkovich, M.J., Fettell, N.A., Heenan, D.P., 2001. Net nitrogen balances for cool-season grain legume crops and contributions to wheat nitrogen uptake: a review. Aust. J. Exp. Agric. 41, 347–359.

Evenson, R.E., Gollin, D., 2003. Assessing the impact of the green revolution: 1960–1980. Science 300, 758–762.

FAO, 2004. Culture of fish in rice fields. In: Halwart, M., Gupta, M. (Eds.), Rome.

FAO, 2007. Analysis of feeds and fertilizers for sustainable aquaculture development in China. Miao, W.M., Mengqing, L. 2007. In: Hasan M., Hecht T., De Silva S. (Eds.), Study and Analysis of Feeds and Fertilizers for Sustainable Aquaculture Development. FAO Fisheries Technical Paper 497. Rome.

FAO, 2011a. Save and Grow: A Policymaker's Guide to the Sustainable Intensification of Smallholder Crop Production. FAO, Rome.

FAO, 2011b. State of Land and Water. FAO, Rome.

FAO, 2012. Voluntary Guidelines on the Responsible Governance of Tenure of Land, Fisheries and Forests in the Context of National Food Security. Food and Agriculture Organization of the United Nations, Rome.

FAO, 2013. Climate-Smart Agriculture Sourcebook. FAO, Rome.

FAO, 2014a. Building a Common Vision for Sustainable Food and Agriculture. FAO, Rome, Italy.

FAO, 2014b. Conservation agriculture for irrigated areas in Azerbaijan, Kazakhstan, Turkmenistan and Uzbekistan. Project GCP/RER/030/TUR Terminal report, Annex 4. Rome (personal communication).

FAO, 2014c. Aquatic biodiversity in rice-based ecosystems: studies and reports from Indonesia, LAO PDR and the Philippines. In: Halwart M., Bartley D., (Eds.), The Asia Regional Rice Initiative: Aquaculture and Fisheries in Rice-Based Ecosystems. Rome.

FAO, 2015a. FAOSTAT. Online statistical database: Production.

FAO, 2015b. Traditional system makes more productive use of land. Save and Grow FactSheet 7. <http://www.fao.org/3/a-i5310e.pdf>.

FAO, 2016a. Pulses and Biodiversity. <http://www.fao.org/fileadmin/user_upload/pulses-2016/docs/factsheets/Biodiversity_EN_PRINT.pdf>.

FAO, 2016b. Incentives for Ecosystem Services. Web site. <http://www.fao.org/in-action/incentives-for-ecosystem-services/resources/en/>.

FAO, 2017. Landscapes for Life – Approaches to Landscape Management for Sustainable Food and Agriculture. FAO, Rome.

FAO and IPCC, 2017. Expert Meeting on Climate Change, Land Use and Food Security, January 23–25, 2017, FAO HQ Rome. <http://www.fao.org/climate-change/events/detail-events/en/c/465791/>.

Farm Africa, 2007. The goat model: a proven approach to reducing poverty among smallholder farmers in Africa by developing profitable goat enterprises and sustainable support services. Farm Africa Working Paper No. 9. <https://www.farmafrica.org/downloads/resources/WP9%20The%20Goat%20Model.pdf>.

Feder, G., Murgai, R., Quizon, J.B., 2004. Sending farmers back to school: the impact of Farmer Field Schools in Indonesia. Rev. Agric. Econ. 26 (1), 45–62.

Fischer, R.A., Byerlee, D., Edmeades, G.O., 2014. Crop Yields and Global Food Security: Will Yield Increase Continue to Feed the World? ACIAR Monograph No. 158. Australian Centre for International Agricultural Research, Canberra, Canberra.

Flood, J., 2010. The importance of plant health to food security. Food Secur. 2, 215–231.

Flora, J.L., Flora, C.B., 2004. Building community in rural areas of the Andes. In: Atria, R., Siles, M. (Eds.), Compilers. Social Capital and Poverty Reduction in Latin America and the Caribbean: Towards a New Paradigm. Economic Commission for Latin America and the Caribbean and Michigan State University, Santiago Chile, pp. 523–542.

Flora, J.L., Flora, C.B., Campana, F., García Bravo, M., Fernández-Baca, E., 2006. Social capital and advocacy coalitions: examples of environmental issues from Ecuador. In: Rhoades, R.E. (Ed.), Development with Identity: Community, Culture and Sustainability in the Andes. CABI Publishing, pp. 287–297. Available at Mountain Forum Online Library: http://www.mtnforum.org/oldocs/969.pdf.

Foli, S., Reed, J., Clendenning, J., Petrokofsky, G., Padoch, C., Sunderland, T., 2014. To what extent does the presence of forests and trees contribute to food production in humid and dry forest landscapes?: a systematic review protocol. Environ. Evid. 3, 15.

Foresight, 2011. The Future of Global Food and Farming. Final Project Report. Government Office for Science London, London.

Francis, C.A., 1981. Development of plant genotypes for multiple cropping systems. In: Frey, K.J. (Ed.), Plant Breeding II. The Iowa State University Press, Ames, 497pp.

Franzluebbers, A.J., 2005. Soil organic carbon sequestration and agricultural greenhouse gas emissions in the southeastern United States. Soil Tillage Res. 83 (1), 120–147.

Franzluebbers, A.J., 2010. Achieving soil organic carbon sequestration with conservation agricultural systems in the southeastern United States. Soil Sci. Soc. Am. J. 74, 347–357.

Friis-Hansen, E., 2012. The empowerment route to well-being: an analysis of farmer field schools in East Africa. World Dev. 40 (2), 414–427.

Furumo, P.R., Aide, T.M., 2017. Characterizing commercial oil palm expansion in Latin America: land use change and trade. Environ. Res. Lett. 12, 024008.

Gadanakis, Y., Bennett, R., Park, J., Areal, F.J., 2015. Evaluating the sustainable intensification of arable farms. J. Environ. Manage. 150 (0), 288–298.

Galhena, D.H., Freed, R., Maredia, K.M., 2013. Home gardens: a promising approach to enhance food security and wellbeing. Agric. Food Secur. 2, 8. Available from: http://www.biomedcentral.com/content/pdf/2048-7010-2-8.pdf.

Galluzzi, G., Eyzaguirre, P., Negri, V., 2010. Home gardens: neglected hotspots of agro-biodiversity and cultural diversity. Biodivers. Conserv. 19, 3635–3654.

Gan, Y.T., Liang, C., Chai, Q., Lemke, R.L., Campbell, C.A., Zentner, R.P., 2014. Improving farming practices reduce the carbon footprint of spring wheat production. Nat. Commun. 5, Article number: 5012.

Gandhi, R., Veeraraghavan, R., Toyama, K., Ramprasad, V., 2009. Digital green: participatory video for agricultural extension. Inf. Technol. Int. Dev. 5 (1), 1–15.

Garrity, D.P., Akinnifesi, F.K., Ajayi, O.C., Weldesemayat, S.G., Mowo, J.G., Kalinganire, A., et al., 2010. Evergreen Agriculture: a robust approach to sustainable food security in Africa. Food Secur. 2, 197–214.

Gathala, M.K., Kumar, V., Sharma, P.C., Saharawat, Y.S., Jat, H.S., Singh, M., et al., 2013. Optimizing intensive cereal-based cropping systems addressing current and future drivers of agricultural change in the northwestern Indo-Gangetic Plains of India. Agric. Ecosyst. Environ. 177, 85–97.

Gathala, M.K., Timsina, J., Islam, Md. S., Rahman, Md. M., Hossain, Md. I., Harun-Ar-Rashid, Md, et al., 2014. Conservation agriculture based tillage and crop establishment options can maintain farmers' yields and increase profits in South Asia's rice–maize systems: evidence from Bangladesh. Field Crops Res. 172, 85–98.

Gathorne-Hardy, A., Narasimha Reddy, D., Venkatanarayana, M., Harriss-White, B., 2013. A life-cycle assessment (LCA) of greenhouse gas emissions from SRI and flooded rice production in S.E. India. Taiwan Water Conserv. 61, 110–125.

Giller, K.E., Witter, E., Corbeels, M., Tittonell, P., 2009. Conservation agriculture and smallholder farming in Africa: the heretics' view. Field Crop. Res. 114, 23–34.

Giller, K.E., Corbeels, M., Nyamangara, J., Triomphe, B., Affhoder, F., Scopel, E., et al., 2011. A research agenda to explore the role of conservation agriculture in African smallholder farming systems. Field Crops Res. 124 (3), 468–472.

Gliessman, S.R., 2005. Agroecology and agroecosystems. In: Pretty, J. (Ed.), The Earthscan Reader in Sustainable Agriculture. Earthscan, London.

Gliessman, S.R., 2014. Agroecology: The Ecology of Sustainable Food Systems. CRC Press.

Global Green Growth Institute. 2012. <www.gggi.org>.

Glover, J.D., Reganold, J.P., 2010. Perennial grains: food security for the future. Issues Sci. Technol. 26 (2), 41–47.

Godfray, H.C.J., Beddington, J.R., Crute, I.R., Haddad, L., Lawrence, D., Muir, J.F., et al., 2010. Food security: the challenge of feeding 9 billion people. Science 327, 812–818.

Gómez, M.I., Barrett, C.B., Raney, T., Pinstrup-Andersen, P., Meerman, J., Croppenstedt, A., et al., 2013. Post-green revolution food systems and the triple burden of malnutrition. Food Policy 42, 129–138.

Gray, R., Dayananda, B., 2014. Structure of public research. In: Smyth, S., Phillips, P.W.B., Castle, D. (Eds.), Handbook on Agriculture, Biotechnology and Development. Edgar Elgar, Cheltenham.

Green, H., Broun, P., Cakmak, I., Concon, L., Fedoroff, N., Gonzalez-Valero, J., et al., 2016. Planting seeds for the future of food. J. Sci. Food Agric. 96, 1409–1414.

Groenfeldt, D., Sun, P., 1997. Demand management of irrigation systems through users participation. Water: Economics, Management, and Demand, pp. 304–312.

Guérin, V., Cogliati, E.E., Hartley, S.E., Belzile, F., Menzies, J.G., Bélanger, R.R., 2014. A zoospore inoculation method with Phytophthora sojae to assess the prophylactic role of silicon on soybean cultivars. Plant Dis. 98, 1632–1638.

Gunton, R.M., Firbank, L.G., Inman, A., Winter, D.M., 2016. How scalable is sustainable intensification? Nat. Plants 2, 1–4.

Guntzer, F., Keller, C., Meunier, J.D., 2012. Benefits of plant silicon for crops: a review. Agronomy Sustain. Dev. 32, 201–213.

Gupta, R., Jat, R.K., Sidhu, H.S., Singh, U.P., Singh, N.K., Singh, R.G. et al., 2015. Conservation Agriculture for sustainable intensification of small farms. Compendium of Invited Papers presented at the XII Agricultural Science Congress 3–6 February 2015, ICAR-National Dairy Research Institute, Karnal, India, p. 15.

Hall, J., Pretty, J.N., 2008. Then and now: Norfolk farmers' changing relationships and linkages with government agencies during transformations in land management. J. Farm Manage. 13 (6), 393–418.

Hall, N.M., Bocary, K., Janet, D., Ute, S., Amadou, N., Tobo, R., 2006. Effect of improved fallow on crop productivity, soil fertility and climate-forcing gas emissions in semi-arid conditions. Biol. Fert. Soils 42, 224–230.

Halwart, M., 2013. Valuing aquatic biodiversity in agricultural landscapes. In: Fanzo, D., Hunter, T., Borelli, Mattei, F. (Eds.), Diversifying Food and Diets – Using Agricultural Biodiversity to Improve Nutrition and Health. Bioversity International, pp. 88–108.

Hartley, S.E., DeGabriel, J.L., 2016. The ecology of herbivore-induced silicon defences in grasses. Funct. Ecol. 30, 1311–1322.

Hartley, S.E., Gange, A.C., 2009. Impacts of plant symbiotic fungi on insect herbivores: mutualism in a multitrophic context. Annu. Rev. Entomol. 54, 323–342.

Hartley, S.E., Fitt, R.N., McLarnon, E.L., Wade, R.N., 2015. Defending the leaf surface: intra- and inter-specific differences in silicon deposition in grasses in response to damage and silicon supply. Front. Plant Sci. 6, 35. Available from: https://doi.org/10.3389/fpls.2015.00035.

Hasan, M.M., Waddington, S.R., Haque, M.E., Khatun, F., Akteruzzaman, M., 2007. Contribution of whole family training to increased production of maize in Bangladesh. Progress. Agric.(Bangladesh) 18 (1), 267–281.

Hassanali, A., Herren, H., Khan, Z.R., Pickett, J.A., Woodcock, C.M., 2008. Integrated pest management: the push–pull approach for controlling insect pests and weeds of cereals, and its potential for other agricultural systems including animal husbandry. Phil. Trans. R. Soc. B 363, 611–621.

Head, G., Savinelli, C., 2008. Adapting insect resistance management programs to local needs. In: Onstad, D. (Ed.), Insect Resistance Management: Biology, Economics and Prediction. Elsevier, Amsterdam.

Heap, I., 2014. Global perspective on herbicide-resistant weeds. Pest Manage. Sci. Available from: https://doi.org/10.1002/ps.3696.

Helgason, T., Daniell, T.J., Husband, R., Fitter, A.H., Young, J.P.W., 1998. Ploughing up the wood-wide web? Nature 394, 431.

HLPE, 2016. Sustainable agricultural development for food security and nutrition: what role for livestock? A report by the high level panel of experts on food security and nutrition of the Committee on World Food Security. Rome.

Hoehn, P., Tscharntke, T., Tylianakis, J.M., Steffan-Dewenter, I., 2008. Functional group diversity of bee pollinators increases crop yield. Proc. R. Soc. Ser. B 275, 2283–2291.

Hol, W.H.G., Bezemer, T.M., Biere, A., 2013. Getting the ecology into interactions between plants and the plant growth-promoting bacterium Pseudomonas fluorescens. Front. Plant Sci. 4, 81. Available from: https://doi.org/10.3389/fpls.2013.00081.

Holmann, F., Rivas, L., Argel, P., Pérez, E., 2004. Impact of the adoption of *Brachiaria*grasses: Central America and Mexico. Livestock Res. Rural Dev. 16 (12), 2004.

Hooper, D.U., Chapin, F.S., Ewel, J.J., Hector, A., Inchausti, P., Lavorel, S., et al., 2005. Effects of biodiversity on ecosystem functioning: a consensus of current knowledge. Ecol. Monogr. 75, 3–35.

Horlings, L.G., Marsden, T.K., 2011. Towards the real green revolution?Exploring the conceptual dimensions of a new ecological modernisation of agriculture that could 'feed the world'. Glob. Environ. Change 21, 441–452.

Huang, J., Ding, J., 2016. Institutional innovation and policy support to facilitate small-scale farming transformation in China. Agric. Econ. 47 (S1), 227–237.

Hussain, I., Hassnain Shah, M., Khan, A., Akhtar, W., Majid, A., Mujahid, M., 2012. Productivity in rice-wheat crop rotation of Punjab: an application of typical farm methodology. Pak. J. Agric. Res. 25 (1), 1–11.

Huxham, M., Kumara, M.P., Jayatissa, L.P., Krauss, K.W., Kairo, J., Langat, J., et al., 2010. Intra- and interspecific facilitation in mangroves may increase resilience to climate change threats. Philos. Trans. R. Soc. Lond. B. Biol. Sci. 365, 2127–2135. Available from: https://doi.org/10.1098/rstb.2010.0094.

Huxham, M., Hartley, S.E., Pretty, J., Tett, P., 2014. No Dominion Over Nature: Why Treating Ecosystems Like Machines Will Lead to Boom and Bust in Food Supply. Friends of the Earth England and Wales, London UK.

IAASTD, 2008. Agriculture at a Crossroads. International Assessment of Agricultural Knowledge, Science and Technology for Development, Island Press, Washington, DC.

Iannotti, L., Cunningham, K., Ruel, M., 2009. Improving Diet Quality and Micronutrient Nutrition: Homestead Food Production in Bangladesh. IFPRI Discussion Paper 00928. International Food Policy Research Institute, Washington DC.

Ickowitz, A., Powell, B., Salim, M.A., Sunderland, T.C.H., 2013. Dietary quality and tree cover in Africa. Glob. Environ. Change 24, 287–294.

IFAD and UNEP, 2013. Smallholders, Food Security and the Environment. IFAD. Available from: https://www.ifad.org/documents/10180/666cac24-14b6-43c2-876d-9c2d1f01d5dd.

Institut de l'Agriculture Durable, 2011. Agriculture 2050 Starts Here and Now. Paris.

IPCC, 2014. Summary for policymakers. In: Field, C.B., Barros, V.R., Dokken, D.J., Mach, K.J., Mastrandrea, M.D., Bilir, T.E., et al.,Climate Change 2014: Impacts, Adaptation, and Vulnerability. Part A: Global and Sectoral Aspects. Contribution of Working Group II to the FifthAssessment Report of theIntergovernmental Panel on Climate Change. Cambridge University Press, Cambridge, United Kingdom and New York, NY, USA, pp. 1–32.

IRRI, 2009. Revitalizing The Rice-Wheat Cropping Systems of the Indo-Gangetic Plains: Adaptation and Adoption of Resource-Conserving Technologies in India, Bangladesh, and Nepal. International Rice Research Institute, Los Baños (Philippines), Final report submitted to the United States Agency for International Development.

ISAAA, 2013. Global Status of Commercialized Biotech/GM Crops: 2013. Brief 46-2013. <http://isaaa.org/resources/publications/briefs/46/executivesummary/default.asp> (31.08.14).

Jackson, R.B., Jobagy, E.G., Avissar, R., Roy, S.B., Barrett, D.J., Cook, C.W., et al., 2005. Trading water for carbon with biological carbon sequestration. Science 310, 1944–1947.

Jat, R.A., Sahrawat, K.L., Kassam, A.H., Friedrich, T., 2014. Conservation agriculture for sustainable and resilient agriculture: global status, prospects and challenges. In: Jat, R.A., Sahrawat, K.L., Kassam, A.H. (Eds.), Conservation Agriculture: Global Prospects and Challenges. CABI, Wallingford and Boston, 2014.

Johnson, S.N., Erb, M., Hartley, S.E., 2016. Roots under attack: contrasting plant responses to below- and above-ground insect herbivory. New Phytol. 210, 413–418.

Ju, X., Gu, B., Wu, Y., et al., 2016. Reducing China's fertilizer use by increasing farm size. Glob. Environ. Change 41, 26–32.

Karabayev, M., 2012. Conservation agriculture adoption in Kazakhstan. A Presentation Made in WIPO Conference on Innovation and Climate Change, 11–12 July 2011. Geneva.

Karabayev, M. and Suleimenov, M., 2010. Adoption of conservation agriculture in Kazakhstan. In: Lead Papers 4th World Congress on Conservation Agriculture: Innovations for Improving Efficiency, Equity and Environment, 4–7 February 2009. New Delhi.

Karabayev, M., Morgounov, A., Braun, H.-J., Wall, P., Sayre, K., Zelenskiy, Y., et al., 2014. Effective approaches to wheat improvement in Kazakhstan: breeding and conservation agriculture. J. Bahri Dagdas Crop Res. (1–2), 50–53. 2014.

Karahan Uysal, Ö., Atış, E., 2010. Assessing the performance of participatory irrigation management over time: a case study from Turkey. Agric. Water Manage. 97 (7), 1017–1025.

Karp, D.S., Mendenhall, C.D., Sandí, R.F., Chaumont, N., Ehrlich, P.R., Hadly, E.A., et al., 2013. Forest bolsters bird abundance, pest control and coffee yield. Ecol. Lett. 16, 1339–1347.

Kassam, A., Friedrich, T., Shaxson, F., Pretty, J., 2009. The spread of conservation agriculture: justification, spread and uptake. Int. J. Agric. Sustain. 7 (4), 292–320.

Kassam, A., Friedrich, T., Shaxson, F., Bartz, H., Mello, I., Kienzle, J., et al. 2014. The spread of Conservation Agriculture: policy and institutional support for adoption and uptake. FACTS Reports (in press).

Kell, D.B., 2011. Breeding crop plants with deep roots: their role in sustainable carbon, nutrient and water sequestration. Ann. Bot. 108, 407–418.

Kenmore, P., Carino, F., Perez, G., Dyck, V., 1984. Population regulation of the rice brown plant hopper Nilaparvata lugens (Stal) within rice fields in the Philippines. J. Plant Protect. Trop. 1, 19–38.

Ketelaar, J.W., Abubakar, A.L., 2012. Sustainable intensification of agricultural production. In: Kim, M., Diong, C. H. (Eds.), Biology Education for Social and Sustainable Development. Sense Springer Nature, pp. 173–181.

Khan, Z.R., Pickett, J.A., Wadhams, L.J., Hassanali, A., Midega, C.A.O., 2006. Combined control of striga and stemborers by maize-Desmodium spp. intercrops. Crop Prot. 25, 989–995.

Khan, Z.R., Midega, C.A.O., Amudavi, D.M., Hassanali, A., Pickett, J.A., 2008a. On-farm evaluation of the 'push–pull' technology for the control of stemborers and striga weed on maize in western Kenya. Field Crops Res 106, 224–233.

Khan, Z.R., Amudavi, D.M., Midega, C.A.O., Wanyama, J.M., Pickett, J.A., 2008b. Farmers' perceptions of a 'push–pull' technology for control of cereal stemborers and striga weed in western Kenya. Crop Prot. 27, 976–987.

Khan, Z.R., Midega, C.A.O., Njuguna, E.M., Amudavi, D.M., Wanyama, J.M., Pickett, J.A., 2008c. Economic performance of 'push–pull' technology for stemborer and striga weed control in smallholder farming systems. Crop Prot. 27, 1084–1097.

Khan, Z.R., Midega, C.A.O., Pittchar, J., Pickett, J.A., Bruce, T., 2011. Push-pull technology: a conservation agriculture approach for integrated management of insect pests, weeds and soil health in Africa. Int. J. Agric. Sustain. 9, 162–170.

Khan, Z.R., Midega, C.A.O., Pittchar, J.O., Murage, A.W., Birkett, M.A., Bruce, T.A.J., et al., 2014. Achieving food security for one million sub-Saharan African poor through push–pull innovation by 2020. Philos. Trans. R. Soc. B Biol. Sci. 369, 20120284.

Kirkegaard, J.A., Matthiessen, J.N., 2005. Developing and refining the biofumigation concept. Agroindustria 3, 5–11.

Klein, A.M., Vaissiere, B.E., Cane, J.H., Steffan-Dewenter, I., Cunningham, S.A., Kremen, C., et al., 2007. Importance of pollinators in changing landscapes for world crops. Proc. R. Soci. Lond. B Biol. Sci.s 274 (1608), 303–313.

Kohli M.M., Fraschina J., 2009. Adapting wheats to zero tillage in maize-wheat-soybean rotation system. 4th World Congress of World Agriculture. <http://www.fao.org/ag/Ca/doc/wwcca-leadpapers.pdf#page = 218>.

Kulkarni, S.A., Tyagi, A.C., 2012. Participatory irrigation management: understanding the role of cooperative culture. Int. Commiss. Irrig. Drain. 1–8. http://www.un.org/waterforlifedecade/water_cooperation_2013/pdf/ICID_Paper_Avinahs_Tyagi.pdf.

Ladha, J., Yadvinder-Singh, E.O., Hardy, B., 2009. Integrated Crop and Resource Management in the Rice-Wheat System of South Asia. International Rice Research Institute, Los Baños (Philippines).

Lal, R., 2010. Managing soils and ecosystems for mitigating anthropogenic carbon emissions and advancing global food security. BioScience. 60 (9), 708–721.

Larwanou, M., Adam, T., 2008. Impacts de la régénération naturelle assistée au Niger: Etude de quelques cas dans les Régions de Maradi et Zinder. Synthèse de 11 mémoires d'étudiants de 3ème cycle de l'Université AbdouMoumouni de Niamey, Niger.

Lee-Smith, D., 2010. Cities feeding people: an update on urban agriculture in equatorial Africa. Environ. Urban. 22, 483–499.

Leeuwis, C., van den Ban, A., 2004. Communication for Rural Innovation: Rethinking Agricultural Extension, Third ed. Blackwell publishing and CTA.

Letourneau, D.K., Armbrecht, I., Rivera, B.S., Lerma, J.M., Carmona, E.J., Daza, M.C., et al., 2011. Does plant diversity benefit agroecosystems? A synthetic review. Ecol. Appl. 21, 9–21.

Li, J., Xin, Y., Yuan, L., 2010. Hybrid rice technology development: ensuring China's food security. In: Spielman, D.J., Pandya-Lorch, R. (Eds.), Proven Successes in Agricultural Development: A Technical Compendium to Millions Fed. International Food Policy Research Institute, Washington, DC, pp. 271–293.

Liang, K., Yang, T., Zhang, S., Zhang, J., Luo, M., Fu, L., et al., 2016. Effects of intercropping rice and water spinach on net yields and pest control: an experiment in southern China. Int. J. Agric. Sustain. 14 (4), 448–465.

Lin, X., Zhu, D., Lin, X., 2011. Effects of water management and organic fertilization with SRI crop practices on hybrid rice performance and rhizosphere dynamics. Paddy Water Environ. 9, 33–39.

Lobell, D.B., Schlenker, W., Costa-Roberts, J., 2011. Climate trends and global crop production since 1980. Science 333, 616–620.

Loos, J., Abson, D.J., Chappell, M.J., Hanspach, J., Mikulca, F., Tichit, M., et al., 2014. Putting meaning back into "sustainable intensification. Front. Ecol. Environ. Available from: https://doi.org/10.1890/130157.

Lorenz, K., Lal, R., 2014. Soil organic carbon sequestration in agroforestry systems: a review. Agron. Sustain. Dev. 34 (2), 443–454.

LPFN, 2016. Landscape for people, food and nature. Integrated land management definition. <http://peoplefoodandnature.org/about-integrated-landscape-management/>.

Lu, Y., 2016, China's agricultural modernisation – policy framework, SAIN Information Sheet, No 6.

Lu, Y., Chadwick, D., Norse, D., Powlson, D., Shi, W., 2015. Sustainable intensification of China's agriculture: the key role of nutrient management and climate change mitigation and adaptation. Agric. Ecosyst. Environ. 209, 1–4.

Ma, J.F., Tamai, K., Yamaji, N., Mitani, N., Konishi, S., Katsuhara, M., et al., 2006. A silicon transporter in rice. Nature 440, 688–691.

Mangan, J., Mangan, M.S.A., 1998. Comparison of two IPM training strategies in China: the importance of concepts of the rice ecosystem for sustainable insect pest management. Agric. Hum. Values 15, 209–221.

Manu-Aduening, J.A., Peprah, B., Bolfrey-Arku, G., Aubyn, A., 2014. Promoting farmer participation in client-oriented breeding: lessons from participatory breeding for farmer-preferred cassava varieties in Ghana. Adv. J. Agric. Res. 2 (002), 008–017.

Marer, S.B., Lingaraju, B.S., Shashidhara, G.B., 2007. Productivity and economics of maize and pigeonpea intercropping under rainfed condition in northern transitional zone of Karnataka. Karnataka J. Agric. Sci. 20, 1–3.

Marongwe, L.S., Kwazira, K., Jenrich, M., Thierfelder, C., Kassam, A., Friedrich, T., 2011. An African success: the case of conservation agriculture in Zimbabwe. Int. J. Agric. Sustain. 9 (1), 153–161.

Massey, F.P., Hartley, S.E., 2009. Physical defences wear you down: progressive and irreversible impacts of silica on insect herbivores. J. Anim. Ecol. 78, 281–291.

Mathews, C., Jones, R.B., Saxena, K.B., 2001. Maize and pigeonpea intercropping systems in Mpumulanga, South Africa. Int. Chickpea Pigeonpea Newsl. 8, 53.

Matthiessen, J.M., Kirkegaard, J.A., 2006. Biofumigation and enhanced biodegradation: opportunity and challenge in soil-borne pest and disease management. Crit. Rev. Plant Sci. 25, 235–265.

MEA (Millennium Ecosystem Assessment), 2005. Ecosystems and Well-being. Island Press, Washington DC.

Meharg, C., Meharg, A.A., 2015. Silicon, the silver bullet for mitigating biotic and abiotic stress, and improving grain quality, in rice? Environ. Exp. Bot. 120, 8–17.

Méndez, V.E., Lok, R., Somarriba, E., 2001. Interdisciplinary analysis of homegardens in Nicaragua: micro-zonation, plant use and socioeconomic importance. Agrofor. Syst. 51, 85–96.

Midega, C.A.O., Salifu, D., Bruce, T.J.A., Pittchar, J., Pickett, J.A., Khan, Z.R., 2014. Cumulative effects and economic benefits of intercropping maize with food legumes on Striga hermonthica infestation. Field Crops Res. 155, 144–152.

Midega, C.A., Bruce, T.J., Pickett, J.A., Khan, Z.R., 2015a. Ecological management of cereal stemborers in African smallholder agriculture through behavioural manipulation. Ecol. Entomol. 40 (S1), 70–81.

Midega, C.A.O., Bruce, T.J., Pickett, J.A., Pittchar, J.O., Murage, A., Khan, Z.R., 2015b. Climate-adapted companion cropping increases agricultural productivity in East Africa. Field Crops Res. 180 (2015), 118–125.

Milder, J.C., Garbach, K., DeClerck, F.A.J., Driscoll, L., Montenegro, M., 2012. An Assessment of the Multi-Functionality of Agroecological Intensification. A report prepared for the Bill& Melinda Gates Foundation.

Miller, J.W., Atanda, T., 2011. The rise of peri-urban aquaculture in Nigeria. Int. J. Agric. Sustain. 9 (1), 274–281.

Minang, P.A., van Noordwijk, M., Freeman, O.E., Mbow, C., de Leeuw, J., Catacutan, D. (Eds.), 2015. Landscapes: Multifunctionality In Practice. World Agroforestry Centre (ICRAF), Nairobi, Kenya, 444 pp. http://asb.cgiar.org/-landscapes/digital-edition/resources/_Landscapes-LR.pdf.

Mitchell, M.G.E., Benett, E.M., Gonzalez, A., 2014. Forest fragments modulate the provision of multiple ecosystem services. J. Appl. Ecol. 51, 909–918.

Mittra, S., Sugam, R., Ghosh, A., 2014. Collective Action for Water Security and Sustainability. Preliminary Investigations. Council on Energy, Environment and Water, URL: http://ceew.in/pdf/CEEW-2030-WRG-Collective-Action-for-Water-Security-and-Sustainability-Report-19Aug14.pdf.

Morgan, J.A., Follet, R.F., Allen, L.H., Del Grosso, S., Derner, J.D., Dijkstra, F., et al., 2010. Carbon sequestration in agricultural lands of the United States. J. Soil Water Conserv. 65 (1), 7A–13A.

Motisi, N., Dore, T., Lucas, P., Montfort, F., 2010. Dealing with the variability in biofumigation efficacy through an epidemiological framework. Soil Biol. Biochem. 42, 2044–2057.

Muhanji, G., Roothaert, R.L., Webo, C., Stanley, M., 2011. African indigenous vegetable enterprises and market access for small-scale farmers in East Africa. Int. J. Agric. Sustain. 9 (1), 194–202.

Murage, A.W., Midega, C.A.O., Pittchar, J.O., Pickett, J.A., Khan, Z.R., 2015. Determinants of adoption of climate-smart push-pull technology for enhanced food security through integrated pest management in eastern Africa. Food Sec. 7, 709–724.

Murthy, I.K., Gupta, M., Tomar, S., Munsi, M., Tiwari, R., Hedge, G.T., et al., 2013. Carbon sequestration potential of agroforestry systems in India. Earth Sci. Clim. Change 4, 1. Available from: http://dx.doi.org/10.4172/2157-7617.1000131.

Mwanga, R.O., Ssemakula, G., 2011. Orange-fleshed sweet potatoes for food, health and wealth in Uganda. Int. J. Agric. Sustain. 9 (1), 42–49.

Mwanga, R.O., Niringiye, C., Alajo, A., Kigozi, B., Namukula, J., Mpembe, I., et al., 2011. 'NASPOT 11', a sweetpotato cultivar bred by a participatory plant-breeding approach in Uganda. HortScience 46 (2), 317–321.

Nair, K.S., 1994. Participatory video for rural development: a methodology for dialogic message design. Unpublished Paper. Food and Agriculture Organization. Rome.

Naylor, R.L., Falcon, W.P., Goodman, R.M., Jahn, M.M., Sengooba, T., Tefera, H., et al., 2004. Biotechnology in the developing world: a case for increased investments in orphan crops. Food Policy 29 (1), 15–44.

NEA, 2011. Synthesis of the Key Findings. NEP-WCMC, Cambridge, UK.

Nelson, E., Mendoza, G., Regetz, J., Polasky, S., Tallis, H., Cameron, D., et al., 2009. Modeling multiple ecosystem services, biodiversity conservation, commodity production, and tradeoffs at landscape scales. Front. Ecol. Environ. 7, 4–11.

Nga, N., Rodriguez, D., Son, T., Buresh, R.J., 2010. Development and impact of site-specific nutrient management in the Red River Delta of Vietnam. In: Palis, F.G., Singleton, G.R., Casimero, M.C., Hardy, B. (Eds.), Research to Impact. Case Studies for Natural Resource Management for Irrigated Rice in Asia. International Rice Research Institute, Los Baños, Philippines, pp. 317–334.

Ngwira, A., Aune, J., Mkwinda, S., 2012. On-farm evaluation of yield and economic benefit of short term maize legume intercropping systems under conservation agriculture in Malawi. Field Crops Res. 132 (2012), 149–157.

Niñez, V.K., 1984. Household gardens: theoretical considerations on an old survival strategy. Potatoes in Food Systems Research Series Report No. 1. International Potato Center. <http://pdf.usaid.gov/pdf_docs/PNAAS307.pdf> (31.08.14).

Niñez, V.K., 1985. Working at half-potential: constructive analysis of homegarden programme in the Lima slums with suggestions for an alternative. <http://archive.unu.edu/unupress/food/8F073e/8F073E02.htm> (31.08.14).

Norse, D., 2012. Low carbon agriculture: objectives and policy pathways. Environ. Dev. 1, 25–39.

Norse, D., Ju, X., 2015. Environmental costs of China's food security. Agric. Ecosyst. Environ. 209, 5–14.

NRC, 1989. Alternative Agriculture. National Academies Press, Washington DC.

NRC, 2010. Towards Sustainable Agricultural Systems in the 21st century. Committee on Twenty-First Century Systems Agriculture. National Academies Press, Washington DC.

Nurbekov, A., Akramkhanov, A., Lamers, J., Kassam, A., Friedrich, T., Gupta, R., et al., 2014. Conservation agriculture in Central Asia. In: Jat, R., Sahrawat, K., Kassam, A. (Eds.), Conservation Agriculture: Global Prospects and Challenge. CAB International.

O'Neill, D.W., Dietz, R., Jones, N. (Eds.), 2010. Enough is Enough. Ideas for a Sustainable Economy in a World of Finite Resources. Report of the Steady State Economy Conference. Centre for the Advancement of the Steady State Economy and Economic Justice for All, Leeds.

OECD (Organisation for Economic Co-operation and Development), 2011. Towards Green Growth. OECD, Paris.

Oerke, E.C., Dehne, H.W., 2004. Safeguarding production – losses in major crops and the role of crop protection. Crop Prot. 23 (4), 275–285.

Ortiz-Ceballos, A., Aguirre-Rivera, J., Salgado-Garcia, S., Ortiz-Ceballos, G., 2015. Maize-velvet bean rotation in summer and winter *milpas*: a greener technology. Agronomy J. 107 (1), 330–336.

Ostrom, E., 1998. Social Capital: A Fad or a Fundamental Concept? Centre for the Study of Institutions, Population and Environmental Change. Indiana University, Bloomington.

Otte, A., Costales, J., Dijkman, U., Pica-Ciamarra, T., Robinson, V., Ahuja, C., et al., 2012. Pro-Poor Livestock Policy Initiative, A Living from Livestock. FAO, Rome.

Ouma, M.A., Onyango, C.A., Ombati, J.M., Khan, Z.R., Midega, C.A.O., Pittchar, J., 2014. Effectiveness of participatory video in learning and dissemination of Push-pull technology among smallholder farmers in Rachuonyo North sub-County in Kenya. IJRAS 1 (3), 189–193.

Owenya, M.Z., Mariki, W.L., Kienzle, J., Friedrich, T., Kassam, A., 2011. Conservation agriculture (CA) in Tanzania: the case of the Mwangaza B CA farmer field school (FFS), Rhotia Village, Karatu District, Arusha. Int. J. Agric. Sustain. 9 (1), 145–152.

Pacheco, A.R., de Queiroz Chaves, R., Lana Nicoli, C.M., 2013. Integration of crops, livestock, and forestry: a system of production for the Brazilian Cerrados. In: Hershey, C.H., Neate, P. (Eds.), Eco-Efficiency: From Vision to Reality (Issues in Tropical Agriculture Series). Centro Internacional de Agricultura Tropical (CIAT), Cali, Colombia, pp. 51–60. , 2013.

Palis, F.G., 2006. The role of culture in farmer learning and technology adoption: a case study of farmer field schools among rice farmers in central Luzon, Philippines. Agric. Hum. Values. Available from: https://doi.org/10.1007/s10460-006-9012-6.

Parsa, S., Morse, S., Bonifacio, A., Chancellor, T.C.B., Condori, B., Crespo-Perez, V., et al., 2014. Obstacles to integrated pest management adoption in developing countries. PNAS 111 (10), 3889–3894.

Peacock, C., 1996. Improving Goat Production in the Tropics: A Manual for Extension Workers. Oxfam/FARM-Africa, Oxford.

Peacock, C., Hastings, T., 2011. Meru dairy goat and animal healthcare project. Int. J. Agric. Sustain. 9 (1), 203–211.

Peoples, M.B., Brockwell, J., Herridge, D.F., Rochester, I.J., Alves, B.J.R., Urquiaga, S., et al., 2009. The contributions of nitrogen-fixing crop legumes to the productivity of agricultural systems. Symbiosis 48, 1–17.

Petersen, P., Tardin, J.M., Marochi, F., 2000. Participatory development of non-tillage systems without herbicides for family farming: the experience of the center-south region of Parana. Environ. Dev. Sustain. 1, 235–252.

Phillips, D., Waddington, H., White, H., 2014. Better targeting of farmers as a channel for poverty reduction: a systematic review of Farmer Field Schools targeting. DSR 1, 113–136.

Pimentel, D., Kounang, N., 1998. Ecology of soil erosion in ecosystems. Ecosystems 1 (5), 416–426.

Pimentel, D., Wilson, C., McCullum, C., Huang, R., Dwen, P., Flack, J., et al., 1997. Economic and environmental benefits of biodiversity. Bioscience 47 (11), 747–757.

Pingali, P., Raney, T., 2005. From the Green Revolution to the Gene Revolution. How Will the Poor Fare? ESA Working Paper No. 05-09. November 2005. Agricultural and Economics Division, FAO. <ftp://ftp.fao.org/docrep/fao/008/af276e/af276e00.pdf> (31.08.14).

Poffenberger, M., Zurbuchen, M.S., 1980. The Economics of Village Bali: Three Perspectives. The Ford Foundation, New Delhi.

Popp, P., Petõ, K., Nagy, J., 2012. Pesticide productivity and food security. A review. Agronomy Sustain. Dev. Available from: https://doi.org/10.1007/s13593-012-0105-x.

Poppy, G.M., Chiotha, S., Eigenbrod, F., Harvey, C.A., Honzak, M., Hudson, M.D., et al., 2014. Food security in a perfect storm: using the ecosystem services framework to increase understanding. Philos. Trans. R. Soc. B Biol. Sci. 369, 20120288.

Power, A., 2010. Ecosystem services and agriculture: tradeoffs and synergies. Philos. Trans. R. Soc. B Biol. Sci. 365, 2959–2971.

Pramova, E., Locatelli, B., Djoudi, H., Somorin, O.A., 2012. Forests and trees for social adaptation to climate variability and change. Wiley Interdisciplinary Rev. Clim. Change 3 (6), 581–596.

Pray, C.E., Naseem, A., 2003.The Economics of Agricultural Biotechnology Research. ESA Working Paper No. 03-07. June 2003. Agriculture and Economics Development Analysis Division, FAO.

Pretty, J., 1997. The sustainable intensification of agriculture. Nat. Resour. Forum 21 (4), 247–256.

Pretty, J., 2003. Social capital and the collective management of resources. Science 302, 1912–1915.

Pretty, J. (Ed.), 2005. The Pesticide Detox: Towards a More Sustainable Agriculture. Earthscan, London.

Pretty, J., 2008. Agricultural sustainability: concepts, principles and evidence. Phil. Trans. R. Soc. Lond. B 363 (1491), 447–466.

Pretty, J., 2013. The consumption of a finite planet: well-being, convergence, divergence and the nascent green economy. Environ. Resour. Econ. Available from: https://doi.org/10.1007/s10640-013-9680-9.

Pretty, J., Bharucha, Z.P., 2014. Sustainable intensification in agricultural systems. Ann. Bot. 205, 1–26.

Pretty, J., Bharucha, Z.P., 2015. Integrated pest management for sustainable intensification of agriculture in Asia and Africa. Insects 6 (1), 152–182.

Pretty, J., Bharucha, Z.P., 2018. The Sustainable Intensification of Agriculture. Routledge, London.

Pretty, J., Ward, H., 2001. Social capital and the environment. World Dev. 29 (2), 209–227.

Pretty, J., Noble, A.D., Bossio, D., Dickson, J., Hine, R.E., Penning de Vries, F.W.T., et al., 2006. Resource-conserving agriculture increases yields in developing countries. Environ. Sci. Technol. 40, 1114–1119.

Pretty, J., Toulmin, C., Williams, S., 2011. Sustainable intensification in African agriculture. International. J. Agric. Sustain. 9 (1), 5–24.

Pretty, J., Bharucha, Z.P., Hama Garba, M., Midega, C., Nkonya, E., Settle, W., et al., 2014. Foresight and African agriculture: innovations and policy opportunities. Report to the UK Government Foresight Project. <https://www.gov.uk/government/uploads/system/uploads/attachment_data/file/300277/14-533-future-african-agriculture.pdf> (31.08.14).

Pywell, R.F., Heard, M.S., Woodcock, B.A., Hinsley, S., Ridding, L., Nowakowski, M., et al., 2015. Wildlife friendly farming increases crop yield: evidence for ecological intensification. Proc. R. Soc. Ser. B 282, 20151740.

Qaim, M., 1999. The Economic Effects of Genetically Modified Orphan Commodities: Projections for Sweet potato in Kenya. ISAA Briefs No. 13. ISAA: Ithaca, N.Y and ZEF: Bonn. <http://www.isaaa.org/resources/publications/briefs/13/download/isaaa-brief-13-1999.pdf>.

Raintree, J.B., Warner, K., 1986. Agroforestry pathways for the intensification of shifting cultivation. Agrofor. Syst. 4, 39–54.

Ramisch, J.J., et al., 2006. Strengthening "Folk Ecology": community-based learning for integrated soil fertility management, western Kenya. Int. J. Agric. Sustain. 4 (2), 154–168.

Raney, T., 2006. Economic impact of transgenic crops in developing countries. Curr. Opin. Biotechnol. 17, 1–5.

Rao, I., Peters, M., van der Hoek, R., Castro, A., Subbarao, G., Cadisch, G., et al., 2014. Tropical forage-based systems for climate-smart livestock production in Latin America. Rural 21, 12–15.

Reddy, V.R., Reddy, P.P., 2005. How Participatory Is Participatory Irrigation Management? Water Users' Associations in Andhra Pradesh. Econ. Polit. Wkly. 5587–5595.

Redman, R.S., Kim, Y.O., Woodward, C.J.D.A., Greer, C., Espino, L., Doty, S.L., et al., 2011. Increased fitness of rice plants to abiotic stress via habitat adapted symbiosis: a strategy for mitigating impacts of climate change. PLOS One 6, e14823.

Reed, J., Vianena, J., Foli, S., Clendenning, J., Yang, K., MacDonald, M., et al., 2017. Trees for life: the ecosystem service contribution of trees to food production and livelihoods in the tropics. For. Policy Econ. 84, 62–71.

Reij, C.P., Smaling, E.M.A., 2008. Analyzing successes in agriculture and land management in Sub-Saharan Africa: is macro-level gloom obscuring positive micro-level change? Land Use Policy 25, 410–420.

Reij C., Tappan G., Smale M., 2009. Agroenvironmental Transformation in the Sahel: Another Kind of "Green Revolution". IFPRI Discussion Paper 00914. International Food Policy Research Institute. Washington DC.

Renewables. 2012. Global Status Report. REN21. <http://www.ren21.net/ren21activities/globalstatusreport.aspx> (31.14).

Resende, Á.V., Furtini Neto, A.E., Alves, V.M.C., Curi, N., Muniz, J.A., Faquin, V., et al., 2007. Phosphate efficiency for corn following *Brachiaria*grass pasture in the Cerrado Region. Better Crops. 91 (1), 17–19.

Reynolds, O.L., Padula, M.P., Zeng, R., Gurr, G.M., 2016. Silicon: potential to promote direct and indirect effects on plant defense against arthropod pests in agriculture. Front. Plant Sci. 7, 00744.

Ribot, J.C., Agrawal, A., Larson, A.M., 2006. Recentralizing while decentralizing: how national governments reappropriate forest resources. World Dev. 34, 1864–1886.

Rivera, W.M., Qamar, M.K., Van Crowder, L., 2001. Agricultural and Rural Extension Worldwide: Options for Institutional Reform: Research, Extension and Training Division. Food and Agricultural Organization, FAO, Rome Italy.

Robinson, R.A., Sutherland, W.J., 2002. Post-war changes in arable farming and biodiversity in Great Britain. J. Appl. Ecol. 39, 157–176.

Röling, N., 1996. Towards an interactive agricultural science. Eur. J. Agric. Educ. Extens. 2, 3–48.

Röling, N.G., Wagemakers, M.A.E. (Eds.), 1997. Facilitating Sustainable Agriculture. Cambridge University Press, Cambridge.

Rosenstock, T.S., Tully, K.T., Arias-Navarro, C., Neufeldt, H., Butterbach-Bahl, K., Verchot, L.V., 2014. Agroforestry with N2-fixing trees: sustainable development's friend or foe? Curr. Opin. Environ. Sustain. 6, 15–21.

Rosset, P.M., Martínez-Torres, M.E., 2012. Rural social movements and agroecology: context, theory, and process. Ecol. Soc. 17 (3), 17.

Rowe, W.C., 2009. "Kitchen gardens" in Tajikistan: the economic and cultural importance of small-scale private property in a post-soviet society. Hum. Ecol. 37 (6), 691–703.

Royal Society, 2009. Reaping the Benefits: Science and the Sustainable Intensification of Global Agriculture. Royal Society, London.

Royal Society, 2012. People and the Planet. The Royal Society, London.

Rusinamhodzi, L., Corbeels, M., Nyamangarad, J., Giller, K., 2012. Maize–grain legume intercropping is an attractive option for ecological intensification that reduces climatic risk for smallholder farmers in central Mozambique. Field Crops Res. 136 (2012), 12–22.

Sanginga, N., Dashiell, K.E., Diels, J., Vanlauwe, B., Lyasse, O., Carsky, R.J., et al., 2003. Sustainable resource management coupled to resilient germplasm to provide new intensive cereal-grain-legume-livestock system in the dry savanna. Agric. Ecosyst. Environ. 100, 305–314.

Scopel, E., Triomphe, B., dos Santos Ribeiro, M. de F., Séguy, L., Denardin, J.E., Kochhann, R.A., 2004. Direct seeding mulch-based cropping systems (DMC) in Latin America. In: R.A. Fischer, (Ed.), New Directions for a Diverse Planet. Proceedings of the 4th International Crop Science Congress. Brisbane, Australia.

Settle, W., Hama Garba, M., 2011. Sustainable crop production intensification in the Senegal and Niger River Basins of francophone West Africa. Int. J. Agric. Sustain. 9 (1), 171–185.

Settle, W., Soumaré, M., Sarr, M., Hama Garba, M., Poisot, A., 2014. Reducing pesticide risks to farming communities: cotton farmer field schools in Mali. Philos. Trans. R. Soc. (B) 369, 20120277. Available from: https://doi.org/10.1098/rstb.2012.0277.

Seufert, V., Ramankutty, N., Foley, J.A., 2012. Comparing the yields of organic and conventional agriculture. Nature 485, 229–232.

Shaner, D.L., 2014. Lessons learned from the history of herbicide resistance. Weed Sci. 62 (2), 427–431.

Sharma, B.R., Amarasinghe, U., Cai, X., de Condappa, D., Shah, T., Mukherji, A., et al., 2010. The Indus and the Ganges: river basin under extreme pressure. Water Int. 35, 493–521.

Sileshi, G.W., Debusho, L.K., Akinnifesi, F.K., 2012. Can integration of legume trees increase yield stability in rainfed maize cropping systems in southern Africa? Agronomy J. 104 (5), 1392–1398.

Silici, L., Ndabe, P., Friedrich, T., Kassam, A., 2011. Harnessing sustainability, resilience and productivity through conservation agriculture: the case of likoti in Lesotho. Int. J. Agric. Sustain. 9 (1), 137–144.

Singh, S., Sharma, R.K., Gupta, R.K., Singh, S.S., 2008. Changes in rice-wheat production technologies and how rice-wheat became a success story: lessons from zero-tillage wheat. In: Singh, Y., Singh, V., Chauhan, B., Orr, A., Mortimer, A., Johnson, D., Hardy, B. (Eds.), Direct Seeding of Rice and Weed Management in the Irrigated Rice-Wheat Cropping System of the Indo-Gangetic Plains. International Rice Research Institute, and Pantnagar, India, Directorate of Experiment Station, G.B. Pant University of Agriculture and Technology, Los Baños (Philippines).

Singh, R., Erenstein, O., Gatdala, M., Alam, M., Regmi, A., Singh, U., et al., 2009. Socioeconomics of integrated crop and resource management technologies in the rice-wheat systems of South Asia: site contrasts, adoption, and impact using village survey findings. In: Ladha, J., Yadvinder-Singh, O.E., Hardy, B. (Eds.), Integrated Crop and Resource Management in the Rice-Wheat System of South Asia. International Rice Research Institute, Los Baños (Philippines).

Sinha P., 2014. Status of participatory irrigation management (PIM) in India. National Convention of Presidents of Water Users Associations, Ministry of Water Resources India NPIM, New Delhi, 7–8 November, 2014. <http://wrmin.nic.in/writereaddata/PIM02.pdf>.

Snapp, S.S., Blackie, M.J., Gilbert, R.A., Bezner-Kerr, R., Kanyama-Phiri, G.Y., 2010. Biodiversity can support a greener revolution in Africa. PNAS 107 (48), 20840–20845.

Solh, M., van Ginkel, M., 2014. Drought preparedness and drought mitigation in the developing world's drylands. Weather Clim. Extremes 3, 62–66.

Sorrenson W.J., 1997. Financial and Economic Implications of Non-Tillage and Crop Rotations Compared to Conventional Cropping Systems. FAO Investment Centre Occasional Paper Series No. 9. <http://www.fao.org/docrep/007/ae368e/ae368e00.HTM> (31.08.14).

Spargo, J.T., Alley, M.M., Follet, R.F., Wallace, J.V., 2008. Soil carbon sequestration with continuous no-till management of grain cropping systems in the Virginia coastal plain. Soil Tillage Res. 100 (1-2), 13–140.

Stern, N., Rydge, J., 2012. The new energy-industrial revolution and an international agreement on climate change. Econ. Energy Environ. Policy 1, 1–19.

Stevenson, J.R., Serraj, R., Cassman, K.G., 2014. Evaluating conservation agriculture for small-scale farmers in Sub-Saharan Africa and South Asia. Agric. Ecosyst. Environ. 187, 1–10.

Sumberg, J., Thompson, J., Woodhouse, P., 2013. Why agronomy in the developing world has become contentious. Agric. Hum. Values 30 (1), 71–83.

Swaminathan, M.S., 1989. Agricultural Production and Food Security in Africa. In: d'Orville H, ed. The Challenges of Agricultural Production and Food Security inAfrica. A report of the proceedings of an international conference organized by the Africa Leadership Forum 27–30 July, 1989. Ota, Nigeria. <http://www.africaleadership.org/rc/the%20challenges%20of%20agricultural.pdf#page = 23> (31.08.14).

Swaminathan, M.S., 2000. An evergreen revolution. Biologist 47 (2), 85–89.

Taylor, J.R., Lovell, S.T., 2012a. Urban home food gardens in the Global North: research traditions and future directions. Agric. Hum. Values 31 (2), 285–305.

Taylor, J.R., Lovell, S.T., 2012b. Mapping public and private spaces of urban agriculture in Chicago through the analysis of high-resolution aerial images in Google Earth. Landscape Urban Plann. 108 (1), 57–70.

The Montpellier Panel, 2013. Sustainable Intensification: A New Paradigm for African Agriculture. Agriculture for Impact, London.

Thierfelder, C., Cheesman, S., Rusinamhodzi, L., 2012. Benefits and challenges of crop rotation in maize-based conservation agriculture (CA) cropping system of Southern Africa. Int. J. Agric. Sustain. 703894, 1–17. Available from: https://doi.org/10.1080/14735903.2012.

Thompson, B., Amoroso, L., 2011. FAO's Approach toNutrition-Sensitive Agricultural Development. FAO, Rome, http://www.fao.org/fileadmin/user_upload/agn/pdf/FAO_Approach_to_Nutrition_sensitive_agricultural_development.pdf (31.08.2014).

Thomson, A.J., Giannopoulos, G., Pretty, J., Braggs, E.M., Richardson, D.J., 2012. Biological sources and sinks of nitrous oxide and strategies to mitigate emissions. Phil. Trans. R. Soc. (B) 367 (1593), 1157–1168.

Tilman, E., Cassman, K.G., Matson, P.A., Naylor, R., Polasky, S., 2002. Agricultural sustainability and intensive production practices. Nature 418, 617–677.

Tilman, D., Balzer, C., Hill, J., Befort, B.L., 2011. Global food demand and the sustainable intensification of agriculture. Proc. Natl. Acad. Sci. USA 108, 20260–20264.

Timsina, J., Buresh, R.J., Dobermann, A., Dixon, J., 2011. Rice-Maize Systems in Asia: Current Situation and Potential. International Rice Research Institute and International Maize and Wheat Improvement Center, Los Baños (Philippines), pp. 7–26, and 161–171, 232pp.

Tomich, T.P., Brodt, S., Ferris, H., Galt, R., Horwath, W.R., Kebreab, E., et al., 2011. Agroecology: a review from a global-change perspective. Annu. Rev. Environ. Resourc. 36, 193–222.

Trethowan, R., Manes, Y., Chattha, T., 2009. Breeding for improved adaptation to conservation agriculture improves crop yields. Plenary Sess. 207.

Tripp, R., Wijeratne, M., Piyadasa, V.H., 2005. What should we expect from farmer field schools? A Sri Lanka case study. World Dev. 33 (10), 1705–1720.

Truchy, A., Angeler, D.G., Sponseller, R.A., Johnson, R.K., McKie, B.G., 2015. Linking biodiversity, ecosystem functioning and services, and ecological resilience: towards an integrative framework for improved management. Adv. Ecol. Res. 53, 56–96.

Tscharntke, T., Klein, A.M., Kruess, A., Steffan-Dewenter, I., Thies, C., 2005. Landscape perspectives on agricultural intensification and biodiversity: ecosystem service management. Ecol. Lett. 8, 857–874.

UNCSD, 2012. United Nations Conference on Sustainable Development. At <http://www.uncsd2012.org/rio20/>.

UNECE/FAO, 2017. Forest Products Annual Market Review 2016-2017. ECE/TIM/SP/41. United Nations Economic Commission forEurope/ Food and Agriculture Organization of the United Nations, Geneva, Switzerland.

UNEP (United Nations Environment Programme), 2011. Towards a Green Economy: Pathways to Sustainable Development and Eradication of Poverty. UNEP, Nairobi.

Uphoff, N., 2008. Farmer innovations improving the System of Rice Intensification (SRI). Available at <http://www.future-agricultures.org/farmerfirst/files/T1a_Uphoff.pdf>.

Van den Berg, J., 2004. IPM Farmer Field Schools: A Synthesis of 25 Impact Evaluations. Global IPM Facility, Rome.

Van den Berg, H., Jiggins, J., 2007. Investing in farmers – the impacts of farmer field schools in relation to integrated pest management. World Dev. 35 (4), 663–686.

Van Oosten, C., Gunarso, P., Koesoetjahjo, I., Wiersum, F., 2014. Governing forest landscape restoration: cases from Indonesia. Forests 2014 5, 1143–1162.

Varshney, R.K., Close, T.J., Singh, N.K., Hoisington, D.A., Cook, D.R., 2012a. Orphan legume crops enter the genomics era!. Curr. Opin. Plant Biol. 12 (2), 202–210.

Varshney, R.K., Ribaut, J., Buckler, E.S., Tuberosa, R., Rafalski, J.A., Langridge, P., 2012b. Can genomics boost productivity of orphan crops? Nat. Biotechnol. 30, 1172–1176.

Vera, R.R., 2006. Country Pasture/Forage Resource Profiles: Colombia. FAO, Rome.

Vogl, C.R., Vogl-Lukasser, B., 2003. Tradition, dynamics and sustainability of plant species composition and management in homegardens on organic and non-organic small scale farms in Alpine Eastern Tyrol. Austria. Biol, Agric. Hortic. 21, 149–166.

WEC. 2013. World Energy Scenarios: Composing Energy Futures to 2050. World Energy Council, London, UK.

Wennink B., Heemskerk W., 2004. Building social capital for agricultural innovation: experiences with farmer groups in Sub-Saharan Africa. Bulletin 368. Royal Tropical Institute (KIT): Amsterdam. <http://www.iscom.nl/publicaties/buildingsocialcapitalagriculture.pdf> (31st Aug 2014)

Wezel, A., Bellon, S., Doré, T., Francis, C., Vallod, D., David, C., 2009. Agroecology as a science, a movement and a practice. A review. Agronomy Sustain. Dev. 29 (4), 505–515.

WinklerPrins, A.M.G.A., 2003. House-lot gardens in Santarem, Pará, Brazil: linking rural with urban. Urban Ecosyst. 6 (1-2), 43–65.

WinklerPrins, A.M.G.A., de Souza, P., 2005. Surviving the city: urban home gardens and the economy of affection in the Brazilian Amazon. J. Latin Am. Geogr. 4 (1), 107–126.

World Bank, 2012. Inclusive Green Growth. The Pathway to Sustainable Development. World Bank, Washington DC.

Wu, W., Ma, B.-L., Uphoff, N., 2015. A review of the system of rice intensification in China. Plant Soil 393 (1), 361–381.

Xinhua, 2016. CPC and State Council Guide Opinion on Using New Development Concepts to Accelerate Agricultural Modernisation and Realise Moderate Prosperity Society. Available online: <http://news.xinhuanet.com/fortune/2016-01/27/c_1117916568.htm>.

Zhao, L., Wu, L., Wu, M., Li, Y., 2011. Nutrient uptake and water use efficiency as affected by modified rice cultivation methods with reduced irrigation. Paddy Water Environ. 9, 25–32.

Zomer, R.J., Trabucco, A., Coe, R., Place, F. 2009. Trees on Farm: Analysis of Global Extent and Geographical Patterns of Agroforestry. ICRAF Working Paper no. 89. Nairobi: World Agroforestry Centre.

SECTION III

UNDERSTANDING SUSTAINABLE AGRI-FOOD SYSTEMS

Lead Authors

Henning Steinfeld and Timothy Paul Robinson (Food and Agriculture Organization of the United Nations (FAO), Rome, Italy)

Co-authors

Carolyn Opio, Ugo Pica-Ciamarra and Juliana C. Lopes (Food and Agriculture Organization of the United Nations (FAO), Rome, Italy), and Marius Gilbert (Spatial Epidemiology Lab (SpELL), Free University of Brussels, National Fund for Scientific Research, Brussels, Belgium).

CHAPTER

32

Introduction

32.1 OVERALL FRAMING OF SUSTAINABILITY

Sustainability is the endurance of systems and processes. While the ideal state of full sustainability will never be reached, the process of persistently working toward and approaching it gives rise to sustainable systems. Following the definition in the Brundtland Report (World Commission on Environment and Development, 1987) and adopted by the 2030 Agenda for Sustainable Development (UN, 2015), sustainability is understood as a problem of both intra- and intergenerational equity. The Brundtland Commission described sustainable development as: ".... development that meets the needs of the present without compromising the ability of future generations to meet their own needs." This entails simultaneously addressing both the needs of those who are poor and hungry today and those of future generations, otherwise described as "cooperating with the future" (Hauser et al., 2014).

Concerns about the sustainability of food and agriculture as it currently exists are outlined in a recent FAO report—*Building a Common Vision for Sustainable Food and Agriculture* (FAO, 2014) —which describes sustainability in terms of the following five principles:

- Improving efficiency in the use of resources is crucial to sustainable agriculture.
- Sustainability requires direct action to conserve, protect, and enhance natural resources.
- Agriculture that fails to protect and improve rural livelihoods, equity, and social well-being is unsustainable.
- Enhanced resilience of people, communities, and ecosystems is key to sustainable agriculture.
- Sustainable food and agriculture (SFA) requires responsible and effective governance mechanisms.

Sustainable Food and Agriculture
DOI: https://doi.org/10.1016/B978-0-12-812134-4.00032-7

For the purpose of this section, we frame sustainability in terms of the conflict between escalating human needs and demands on one side, and increasing resource scarcity and environmental degradation on the other. While this conflict is not new, the decline of environmental conditions and of productive capacity has sharply accelerated recently, notably in the form of climate change. The opposition between the Earth's natural system and the human system is mediated by food and agriculture, engaged in a coevolution with both sides. We define agrifood systems (AFS) broadly to embrace natural resources and the environment, the benefits and costs they provide to humankind, and the transformation through crops, livestock, forestry, fisheries, and aquaculture of natural resources into human benefits (Fig. 32.1).

Diverse AFS transform resources into human benefits and waste products that flow back to the environment. A much-simplified model of food and agriculture is adopted that reduces them to "agricultural production" (up to farmgate) and "agro-industries" (including processing, storage, retailing, and input supply), giving our AFS four simple nodes: (1) natural resources and the environment; (2) agricultural production; (3) agro-industries; and (4) human benefits.

Natural resources and the environment include soil, air, water, climate, and organisms. In terms of the condition of the resource, such as soil fertility and water quality, there are qualitative aspects to natural resources and the environment. Aggregate aspects also exist, for example, environmental services and biodiversity. Agricultural production refers to cultivation of crops, raising of animals, hunting, gathering, and harvesting of natural stocks, for example, of fish and forests. Agro-industries include all services that support

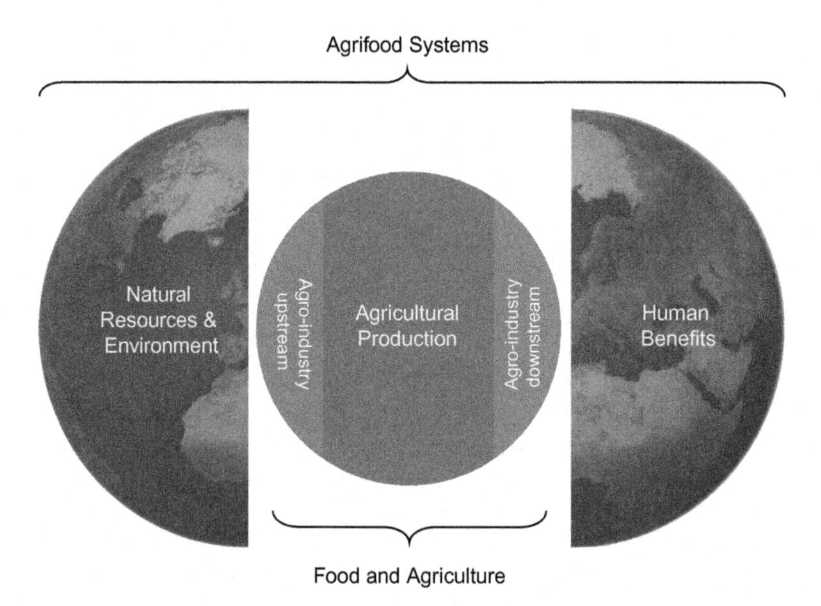

FIGURE 32.1 Diagram of agrifood systems with natural resources and the environment (green) represented on the left and human benefits (red) on the right. Food and agriculture, comprising agricultural production (blue) and the upstream and downstream agro-industries (grey) sit between the natural system and the human system, linking them in many complex ways. Source: *For interpretation of the references to color in this figure legend, the reader is referred to the web version of this book.*

agriculture both upstream and downstream of production. Upstream, agro-industry provides inputs such as seeds and breeding stocks, fertilizers, pesticides, farm machinery, feed processing, extension services, financial services, operation of government administrations, regulatory bodies, and agricultural research. Downstream of agricultural production, the agro-industry caters to the activities of handling, processing, preserving, transporting, and marketing agricultural food and nonfood products. The human benefits that are provided by agricultural production and agro-industries include food, nutrition, health, employment, livelihoods, energy, materials, and economic growth, as well as stability, security, culture, and heritage. Human benefits also include costs such as diseases related to unhealthy diets and overconsumption, zoonotic diseases, antimicrobial resistance (AMR), and hazards to which agricultural workers are exposed.

32.2 BACKGROUND AND RATIONALE

Generally, the world has been successful in feeding its growing population. Globally, the number of hungry people has been reduced from just over 1 billion in 1990–92 to less than 800 million in 2014–16 (see also Section I). This corresponds to a decline in the proportion of hungry people from 19% of the total to 11%. In East and Southeast Asia and in Latin America and the Caribbean, the percentage of hungry people has more than halved (FAO, 2016). Over the last 25 years, only sub-Saharan Africa has seen growth in the total number of hungry people, but even in this case the share of hungry people in the total population has decreased by one-third (FAO, 2016). These gains have resulted from improvements in AFS, from policies better targeted at the poor and vulnerable, and from more effective forms of social protection. Unfortunately, some of this progress has been recently undone. Regional and local armed conflicts and political instability have increased the number of hungry people—serving as a reminder of the fragility of the status quo and the value of the gains.

With peace, though, zero hunger seems within reach. The FAO, IFAD, and WFP (2015) estimate that, in a business-as-usual scenario, food insecurity will continue to decline to 653 million people by 2030, less than 8% of the global population. With better policies, even larger reductions are possible, reaching pockets of protracted food insecurity. For the first time in human history, zero hunger appears to be a realistic prospect.

Most of the hungry are also poor. Poverty is a predominantly rural phenomenon; the rural poor largely depend on AFS for food, income, and other services. Between 1990 and 2013, poverty fell from over one-third of the global population to just over one-tenth. Although poverty declined dramatically in East Asia, with South Asia following suit, it remains stubbornly high in sub-Saharan Africa (World Bank, 2014).

Not only has the number of poor people declined, but also incomes have generally risen, resulting in a burgeoning global middle class across all developing regions. As incomes rise, people move away from staples providing calories, to richer and more diversified diets and more protein, in particular, animal products such as meat, milk, eggs, and fish, and more fats and sugar. Richer diets are exacting a growing toll on human health (see also Section I). Human bodies are growing not only in high-income countries, but also in middle- and low-income countries. The proportion of overweight people now

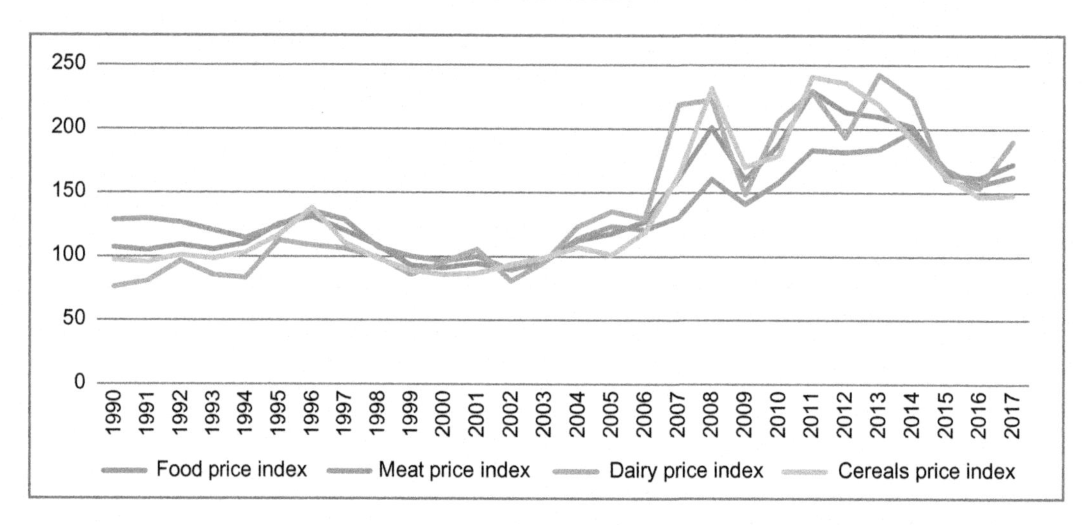

FIGURE 32.2 Annual food price indices (2002–04 = 100). Source: *FAO, 2017. World Food Situation. FAO Food Price Index. FAO, Rome.*

exceeds half of the population, with the biggest increase observable in children. In sub-Saharan Africa and Asia, the proportion of overweight people has now reached about one-quarter of the total population (WHO, 2014).

Recent improvements in food security have been helped by decreasing food prices following their extended spike over the years 2008–14 (FAO, 2017; Fig. 32.2). During those years, the food price crisis pushed millions into food insecurity and caused civil strife in numerous countries. The food price crisis has also been identified as a main cause of rebellion against autocratic regimes in the Near East—the short-lived Arab Spring.

However, while current food prices appear comfortable, and trends in food security are broadly positive, both past experiences and future projections indicate that the calm may not last and that disruption looms ahead. There is a significant increase in the incidence of extreme weather events associated with rising average temperatures. Examples include the 2003 European heat wave, the 2010 Pakistan flood and Russian heat wave, the 2011 Texas heat wave, and the recent floods in Europe (Mann et al., 2017). As temperatures continue to rise in most agricultural production areas, so does the likelihood of extreme weather events and of supply-chain disruptions.

Environmental change, together with greatly increased movements of people and agricultural products, is creating enhanced opportunities for pathogens to move and spread quickly, with sometimes devastating consequences for agricultural production. Examples are repeated flare-ups of various strains of avian and swine influenza in livestock, and the recent expansion of the fall armyworm (*Spodoptera frugiperda*) affecting mostly maize in Africa and beyond. Emerging diseases, such as Ebola in west Africa (2013–15), also create disruptions when they directly affect humans, wreaking havoc on rural lives and economies and severely compromising food security.

Geopolitics is shifting toward a world of multiple power poles and spheres of influence. Armed conflicts and political instability in many regions of Africa, the Near East, and Asia have created a series of protracted humanitarian crises. In 2016, the number of forced migrants reached 65.6 million (UNHCR, 2016) —an all-time high. Alongside this trend also is a resurgence of nationalism and protectionism in many countries, and a loss of confidence in multilateralism. A long period of globalization has created a large group of countries that are structurally food deficit and rely on open international food markets for food security. Currently, 950 million people depend on international trade to cover their food needs, with Near East and Andean countries being most dependent (Fader et al., 2013). The 2007–8 food-price crisis was exacerbated by trade restrictions imposed by traditional exporting countries. The possibility of disruptions caused by failing infrastructure (choke points) is another growing threat (Chatham Report, 2017). Any tightening of food supply, such as crop failure in major producing regions, could lead to similar and perhaps more severe supply disruptions resulting from export restrictions.

While major short-term disruption may be averted, the long-term prospects point to a collision between growing human needs and demands on the declining environmental and natural resource base, both at the local and global levels.

The demands for food and other goods and services from AFS are escalating. The global human population is projected to reach 9.7 billion by mid-century and will surpass 11.2 billion by 2100 (UN, 2017). Sub-Saharan Africa alone, today's poorest and most food-insecure region, will grow from 1 billion currently to 4 billion in 2100, accounting for more than half of the increase (UN, 2017). Urbanization will grow from 54% in 2015 to 66% in 2050, meaning that an additional 2 billion people will live in cities and depend on food mostly produced elsewhere (UN, 2014). Per capita incomes continue to grow in most parts of the world, leading to more diversified and richer diets, with increasing proportions of animal products typically associated with higher resource use. At the same time, failing AFS, often a combined result of fragile environments and conflicts, cause loss of livelihoods and displacements and dependence on food aid and other transfers.

Mean global temperatures have surpassed preindustrial levels by more than 1.1°C in the last 3 years (WMO, 2018), putting the goal of the Paris Agreement (1.5–2.0 degrees increase in the long term) plainly out of reach. More water, land, energy, and nutrients are used for food and agriculture than ever before, and rates of resource degradation, pollution, and depletion continue to rise in most countries. Human modifications of the environment, trade, and travel are exacerbating disease pressures that affect wildlife, AFS, and public health. Large production areas face stagnating, or even declining yields and diminishing returns to input use, which raises the costs of production and contributes to displacing local populations. Climate change and resource competition and depletion threaten the productive capacity and stability of AFS, putting local and global food security at risk.

Human modification of local and global ecosystems has reached such an extent that our present times are being referred to as the "anthropocene" (Crutzen, 2002a, 2002b) —a new epoch whose beginnings have been placed variably at the beginning of the industrial revolution (c. 1780) or the advent of nuclear fallout (1945). The anthropocene follows the holocene, whose relatively benign, stable, and moderate climate allowed the advent of agriculture (around 12,000 years ago) and human civilization. The anthropocene is characterized by human-induced climate change and mass extinctions of species and other geochemical signatures that will remain in the geological record of the planet.

The realization that human activities have become the main driver of environmental change has raised the issue of "planetary boundaries" (Rockstrom et al., 2009). These boundaries represent thresholds or tipping points that, once passed, face a risk of "irreversible and abrupt environmental change" (Scheffer et al., 2009). The idea of boundaries has also found application in the form of regional or local ecosystem thresholds to reflect the peculiarities and resource constraints of smaller geographical units (Steffen et al., 2015). Nine critical Earth system processes have been proposed to have planetary boundaries. Among those, critical thresholds have already been passed for biochemical flows (phosphorus, nitrogen) and biosphere integrity (genetic diversity). Boundaries are being rapidly approached for two others: climate change and land system change. The capacity of the land to produce biomass is considered one such planetary boundary (Running, 2012). Human appropriation of net primary production (HANPP) has doubled in the last 100 years, from 13% to 25% (Krausmann et al., 2013). If the world can maintain the past trends in efficiency gains, HANPP might only increase to 27%–29% by 2050. In a less sustainable future, however, HANPP might grow to 44% by 2050 (Krausmann et al., 2013). Such concerns have given rise to the concept of "planetary health" (Whitmee et al., 2015).

Considering these boundaries and Earth processes, AFS appear as key components and collectively are the largest driver of human-induced environmental change. Sustainability of AFS must be considered against the background of resource finiteness and local and global ecosystem functioning. AFS need to provide food security and other goods and services within an essentially closed system of planetary life support, driven by energy from the sun.

32.3 OVERVIEW OF THE CHAPTERS

This section tackles the growing tension between human demands and a declining environment, explaining first how food and agriculture respond to this conflict and then what can be done to improve outcomes.

Chapter 33, Agrifood Systems, describes the subsectors that make up AFS (crops, livestock, forestry, aquaculture, and fisheries) and breaks down the complexity of AFS into more distinct subsets. Based on the relative abundance of land, labor, and capital, AFS can be classified as land-intensive, labor-intensive, and capital-intensive systems. These are defined using a combination of biophysical and socioeconomic characteristics. The main structures and functions of these systems are described and mapped globally for crops, livestock, and forests, as examples. Such differentiation allows us to outline pathways of sustainable intensification that are specific to each system.

Individual farmers combine land, labor, and capital in different ways, based on relative cost. Efforts to raise productivity tend to focus on the scarcest production factor, a concept known as induced innovation. Across all AFS, intensification is a substitution process, whereby physical inputs (energy, land, nutrients) are replaced by knowledge. If this lowers the rates of extraction and degradation while increasing human benefits, we can think of them as sustainable.

Chapter 34, Socioeconomic Dimension of Agrifood Systems, focuses on the right-hand side of Fig. 32.1: the human benefits derived from AFS. These include food, human nutrition and health, social benefits such as poverty reduction and gender equity, employment, economic growth, and cultural and aesthetic values. In this chapter, we consider the contributions AFS make to growing human needs and demands and where they are failing.

Chapter 35, Natural Resource and Environmental Dimensions of Agrifood Systems, focuses on the left-hand side of Fig. 32.1: the biophysical foundations and processes that underpin AFS. It covers biomass and energy, land, water, nutrients, environmental health, and climate change. The major issues addressed relate both to the extent and nature of resource use by AFS and the impacts of AFS on the natural resource base and on ecological processes, also including resource degradation and depletion, ecosystem functioning, land cover and land use change, greenhouse gas emissions, and climate change. Given the alteration of the environment by AFS, the question then becomes how do a modified resource base and altered ecological processes affect AFS?

Chapter 36, Molecules, Money, and Microbes, refers back to Fig. 32.1 in order to address the flows and linkages between and among the biophysical and socioeconomic dimensions over time and space. It uses the dairy subsector to illustrate the externalities, synergies, and trade-offs associated with these flows. It adopts a dynamic view of the AFS—looking at its geophysical, socioeconomic, and biological aspects—taking as examples the interactions and flows of nutrients (specifically nitrogen), value (cash), and microbes over space and time. We consider the flows and linkages connecting the four compartments: the natural resource base, agricultural production, the agro-industry, and the human benefits.

A lifecycle approach is used to determine the fate of nitrogen in dairy AFS, a value chain approach to examine value flows, and a One Health approach to describe transmission of microbes and their interactions with other organisms. Accounting for both temporal and spatial dimensions of these technical strings around the same basic model helps us to understand trade-offs and identify opportunities for systems integration and points of possible intervention.

Chapter 37, Policy Orientations for Sustainable Agrifood Systems, concludes the section by combining the approaches in the preceding chapters to address the five sustainability principles described above (FAO, 2014). The first principle, on efficiency in the use of resources, we describe as managing flows. The second, to conserve, protect, and enhance natural resources, we interpret as managing stocks. The third principle, to protect and improve rural livelihoods, we equate to managing human benefits. The fourth principle, on resilience, we discuss in terms of managing risks. Lastly, we introduce issues of institution building and governance, developed in Sections IV and V.

References

Chatham Report, 2017. Annual Review 2016–17 - Sovereignty and Interdependence Geopolitics and Instability Delivering Global Public Goods. Chatham House - The Royal Institute of International Affairs, London.

Crutzen, P.J., 2002a. Geology of mankind. Nature 415, 23-23.

Crutzen, P.J., 2002b. The "anthropocene". J. Phys. (Iv) 12, 1–5.

Fader, M., Gerten, D., Krause, M., Lucht, W., Cramer, W., 2013. Spatial decoupling of agricultural production and consumption: quantifying dependences of countries on food imports due to domestic land and water constraints. Environ. Res. Lett. 8, 15.

FAO, 2014. Building a Common Vision for Sustainable Food and Agriculture, Principles and Approaches. FAO, Rome.

FAO, 2016. Food Security Indicators. FAO, Rome.

FAO, 2017. World Food Situation. FAO Food Price Index. FAO, Rome.

FAO, IFAD and WFP, 2015. Achieving Zero Hunger: The Critical Role of Investments in Social Protection and Agriculture. FAO, Rome.

Hauser, O.P., Rand, D.G., Peysakhovich, A., Nowak, M.A., 2014. Cooperating with the future. Nature 511, 220.

Krausmann, F., Erb, K.H., Gingrich, S., Haberl, H., Bondeau, A., Gaube, V., et al., 2013. Global human appropriation of net primary production doubled in the 20th century. Proc. Natl. Acad. Sci. US A 110, 10324–10329.

Mann, M.E., Rahmstorf, S., Kornhuber, K., Steinman, B.A., Miller, S.K., Coumou, D., 2017. Influence of anthropogenic climate change on planetary wave resonance and extreme weather events. Sci. Rep. 7, 10.

Rockstrom, J., Steffen, W., Noone, K., Persson, A., Chapin, F.S., Lambin, E.F., et al., 2009. A safe operating space for humanity. Nature 461, 472–475.

Running, S.W., 2012. A measurable planetary boundary for the biosphere. Science 337, 1458–1459.

Scheffer, M., Bascompte, J., Brock, W.A., Brovkin, V., Carpenter, S.R., Dakos, V., et al., 2009. Early-warning signals for critical transitions. Nature 461, 53–59.

Steffen, W., Richardson, K., Rockstrom, J., Cornell, S.E., Fetzer, I., Bennett, E.M., et al., 2015. Planetary boundaries: guiding human development on a changing planet. Science 347, 11.

UN, 2014. 2014 Revision of World Urbanization Prospects. Population Division of the Department of Economic and Social Affairs of the United Nations, New York.

UN, 2015. Transforming Our World: The 2030 Agenda for Sustainable Development. United Nations, New York.

UN, 2017. 2017 Revision of World Population Prospects. Population Division of the Department of Economic and Social Affairs of the United Nations Secretariat, New York.

UNHCR, 2016. Global Rrends - Forced Displacement in 2016, Geneva.

Whitmee, S., Haines, A., Beyrer, C., Boltz, F., Capon, A.G., Dias, B., et al., 2015. Safeguarding human health in the Anthropocene epoch: report of The Rockefeller Foundation-Lancet Commission on planetary health. Lancet 386, 1973–2028.

WHO, 2014. Global Health Observatory (GHO) Data. Obesity and overweight, Geneva.

WMO, 2018. Press: WMO confirms 2017 among the three warmest years on record. <https://public.wmo.int/en/media/press-release/wmo-confirms-2017-among-three-warmest-years-record>.

World Bank, 2014. World Development Indicators. Poverty Rates, Washington.

World Commission on Environment and Development, 1987. Our Common Future: Report of the World Commission on Environment and Development. Oxford University, New York.

CHAPTER

33

Agrifood Systems

33.1 INTRODUCTION

Agricultural production more than tripled between 1960 and 2015. The bulk of that production expanse has occurred through productivity-enhancing technologies and a major growth in the use of irrigation and chemical inputs. To a smaller extent, land expansion contributed to agricultural growth, with large variations between regions and ecosystems. The same period witnessed a rapid process of industrialization and globalization of food and agriculture. Food supply chains have lengthened dramatically as the physical distance from farm to fork has increased; the consumption of processed, packaged, and prepared foods backed up by compliance schemes has become commonplace. New consumption patterns have created new opportunities for value addition and income generation.

The engines of such rapid growth and widespread change are a broad range of agrifood systems (AFS) that are usually categorized into the subsectors of crops, livestock, forestry, fisheries, and aquaculture. AFS transform natural resources into human benefits using a variety of living organisms and different combinations of land, labor, and capital. Many of these systems are undergoing rapid structural and functional changes that span multiple geographic locations and extend into faraway places. Not all systems are successful. Resource shortage and degradation, climate change, disease, market volatility, and conflict act as disruptors, exacerbating resource pressures and human suffering.

33.2 AGRIFOOD SYSTEMS

The world's AFS operate in diverse landscapes, responding to prevalent agroecological conditions and heavily shaped by the market forces to which they are exposed. These, in turn, depend on access to consumers, their purchasing power, and their preferences. AFS adapt in response to changing contexts, such as resource availability, input costs, policy and institutional environment, and consumer preferences. Evaluation of the impact of evolving AFS, in the context of sustainability, requires an analytical framework relevant to the diverse agricultural subsystems described above—both land-based (crop, livestock,

Sustainable Food and Agriculture
DOI: https://doi.org/10.1016/B978-0-12-812134-4.00033-9

and forest) and water-based (fisheries and aquaculture) —and that cuts across disciplines and embraces agroecology, demographics, and economics.

A typology that addresses agroecological, demographic, and economic determinants of AFS lends itself to an approach recognizable from the classical economics literature, one that is based on the "production factors" of land, labor, and capital stock. In knowledge-based economies, "human capital" is often considered as a fourth production factor (Samuelson, 2010). This refers to the knowledge, talents, skills, abilities, experience, intelligence, training, judgment, wisdom, and entrepreneurship possessed collectively by individuals in a population (Krugman and Wells, 2012). Beyond human capital, the roles of institutions are also recognized in new institutional economic theory as critical factors for production. This includes aspects of land rights and access to markets and services.

The classical production factors are the building blocks of the economy; the human benefits that are derived from AFS have been generated by combinations of land, labor, and capital, which can be defined as follows.

Land comprises any natural resource connected with it, used to produce goods and services. Land resources occur in the context of agroecology, provide the raw materials for production, and are renewable (e.g., forests), or nonrenewable (e.g., fossil fuels). In the case of fisheries and aquaculture, water is equivalent to the "land" resource. In the present context, we refer to the availability and productive potential of the natural resource base for agricultural production. The income that resource owners earn in return for land resources is referred to as "rent."

Labor refers to the agricultural workforce, which includes those in full- or part-time employment, whether subsistence or commercial. For an AFS (as opposed to a farming system), this includes value-chain workers upstream and downstream of production itself. The income earned by labor is referred to as "wages."

Capital stock refers to any asset used for agricultural production. This includes items such as irrigation installations, greenhouses, tractors, and other farm machinery, animal housing, aquaculture ponds, fishing vessels, and logging machinery and infrastructure. Livestock and trees are also capital stocks, which can be liquidated at any time, but in the present context these are included under the "land" resource. The income earned by owners of capital resources is termed "interest."

Human capital facilitates the combination and conversion of the other production factors—land, labor, and capital—into "profit," which refers to the payment for entrepreneurship.

The relative abundance and cost of the three classical production factors determines the direction and extent of technological change, a process termed "induced innovation" (Hicks, 1932), which has been applied to the development of agricultural systems (Ruttan and Hayami, 1984; Ruttan et al., 1980; Hayami and Ruttan, 1971). As demand for agricultural products increases, the prices of production factors for which supplies are scarcest (i.e., those that are inelastic) will rise relative to the prices of more abundant inputs. Farmers can therefore be expected to substitute the scarcer and less-responsive factor(s) of production as they become costlier relative to more abundant ones. Technical innovations increase the productivity of a factor of production, lowering costs, and enhancing profits. In other words, it is the scarcest production factor that most attracts technological innovation and becomes intensified to ease the constraints on growth imposed by inelastic supplies or scarcities. The natural progression substitutions can also be influenced by external factors. de Haan et al. (1997), for example, applied the concept of induced innovation to livestock-environment interactions, referring to how subsidies and environmental regulations can distort the normal progression. AFS are shaped by the availability of land, labor, and capital, and their relative costs. The dynamics of these production factors, in the face of changing demand, capacity to innovate, and institutional context determine how AFS change over time. The task at hand is to steer this development along pathways of enhanced sustainability.

The optimal combination of production factors is defined by equating the marginal productivity gained by an additional unit of each factor equal to its price, namely wage for labor, interest for capital, and rent for land. As one factor price increases in relation to others, there is a substitution effect: a shift to the production factors that are relatively cheaper. Typically, mechanical innovations are motivated by an increase in wages (labor-saving incentives) and biological innovations are motivated by land rent increase (land-saving incentives), although the distinction is not clear-cut and the two often go hand in hand (e.g., a biological innovation is often needed to make the most of a mechanical innovation, or vice versa). Innovations tend to be capital intensive, labor and land being substituted by capital.

The production factors of land, labor, and capital can be represented as a triangle where the vertices represent three broad AFS: extensive (land-intensive) systems, labor-intensive systems, and capital-intensive systems. This is shown in Fig. 33.1 (adapted from Herlemann and Stamer, 1958). Because AFS extend beyond agricultural production, the type of demand for agricultural products and the supply of inputs also need to be embraced by any typology. These three broad AFS types also have characteristic input supply and market and associated consumption patterns.

Extensive systems are relatively abundant in land, and inputs of labor and capital are limited. These systems tend to have a low biomass harvest and are typically areas of low agricultural suitability and natural habitats. Extensive AFS generally supply subsistence,

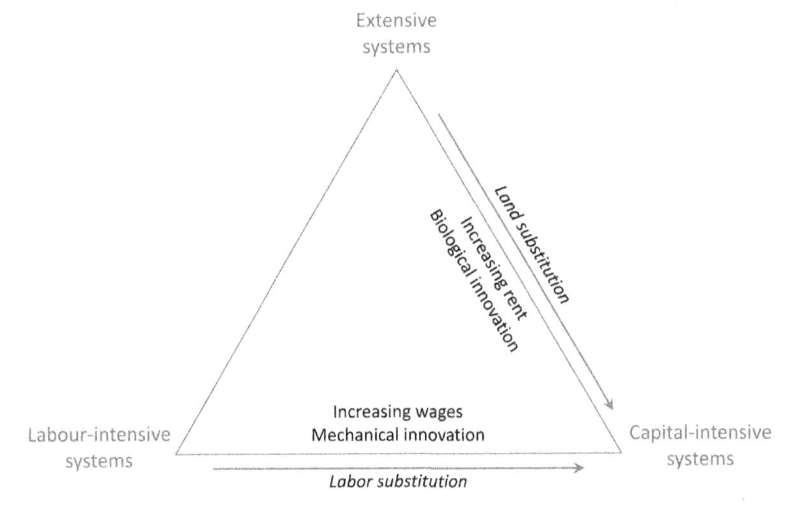

FIGURE 33.1 A broad typology of agrifood systems represented as a triangle where extreme high values of each production factor (land, labor, and capital), accompanied by low values of the other two, are indicated as the vertices. The process of induced innovation is indicated with land substitution occurring as rents increase and labor substitution occurring as wages rise. Source: *Herlemann, H.H., Stamer, H., 1958. Produktionsgestaltung und Betriebsgrosse in der Landwirtschaft unter dem Einfluss der wirtschaftlich - technischen Entwicklung. Institut fuer Weltwirtschaft an der Universitaet Kiel, Kiel.*

but can also serve distant and specialized markets, such as the export of sheep from the Horn of Africa to the Arabian Peninsula for the celebration of the Haj.

Labor-intensive systems tend to occur in less wealthy areas where labor costs are low and capital investment is relatively limited. These tend to be quite productive areas (in the absence of overexploitation) because of the natural historical tendency for people to have settled in areas of high productivity. Increased labor input is the major form of intensification; mechanization is limited and holdings are often small. These AFS typically supply subsistence farming, but there is generally a surplus of production that feeds into markets of various types depending on the context and commodity. Local trading for cash or other products and services is an important outlet, supplying local, informal markets. Some labor-intensive AFS, however, such as milk and tea, feed into more formal and often international markets, and this usually requires a high level of institutional support and organization.

Capital-intensive AFS predominate in high-income countries and are growing rapidly in middle-income countries, and also occur in discrete locations and situations in low--income countries, such as greenhouse production of asparagus in Peru and flowers in Kenya, and intensive poultry production in many countries. In these systems, the capital input per unit of land is increased through machinery, land improvements, and buildings, and through specialization, typically responding to high land prices. Capital-intensive systems tend to be heavily managed and occur in modified environments. Capital substitutes for labor use; farms often occupy little space but have high throughput. Capital-intensive systems are highly market-oriented and rapidly respond to demand created by urban consumption. Commodities produced in these systems are often traded internationally.

The AFS described above can be seen as "endpoints" and many systems fall between these extremes, often in transition from one to another, or outside them, where none of the production factors dominate. For example, ranching and pastoralism are similar in that they exploit a seasonally low-value resource to raise cattle, but pastoral areas have a higher agricultural population density compared to rangelands in which labor has been substituted by capital investment, such as fencing and quad bikes or helicopters for herding cattle. Other capital inputs to rangelands might include pasture improvement, higher yielding breeds, animal health interventions, and the provision of water and supplementary feed. Examples of typical endpoint systems are given in Table 33.1, and some falling outside are given in Table 33.2.

Framing AFS within these three dimensions, and the dynamics of transition—the interplay among land, labor, and capital—can be useful in addressing sustainable food and agriculture.

33.3 A TYPOLOGY OF AFS

Whittlesey (1936) recognized that a successful agricultural systems classification ".... must take account of the relative abundance of land, labor and capital." Varying greatly across ecosystems, AFS depend upon, or have available to them, different combinations of these production factors. These production factors also represent aspects of agroecology

TABLE 33.1 Examples of Typical Endpoint Agrifood Systems

	Land-Based Systems			Water-Based Systems	
System	Crop	Livestock	Forest	Fisheries	Aquaculture
Extensive (land-intensive)	Shifting cultivation; fallow systems	Pastoralism	Protective area management; honey harvesting; hunting	Off-shore line fishing	Low-input mangrove aquaculture; extensive pond farming in Central and Eastern Europe
Labor-intensive	Smallholder tillage—often integrated with livestock or aquaculture	Smallholder dairy—usually integrated with cropping	Charcoal making	Inland river or lake fishing with nets; artisanal coastal fishing	Aquaculture integrated into smallholder systems
Capital-intensive	Horticulture in Europe; irrigated monocropping	Intensive dairy, pig, and poultry farming; beef feedlot	Commercial plantation forestry	Trawling in exclusive economic zones	Industrial aquaculture in tanks

TABLE 33.2 Examples of Agrifood Systems Falling Between or Outside EndPoint Systems

	Land-Based Systems			Water-Based Systems	
System	Crop	Livestock	Forest	Fisheries	Aquaculture
Extensive and capital-intensive		Ranching		Off-shore trawling	
Labor-intensive and capital intensive	Market gardening; seed production				
Extensive and labor-intensive		Agropastoralism; silvopastoral production	Nonwood forest products collection (berries/mushrooms, honey, medicinal plants) and ecotourism		

(land), demography (labor), and economics (capital). Embracing diverse "drivers of change," such as land use, climate, demographics, and demand, they thus lend themselves to exploring trajectories of change under different scenarios. At the same time, individual farmer "decision making" revolves around optimizing the use of these production factors. Farmers will respond to changes in the cost or availability of land or labor, for example, in adapting to changing situations. The production factors also embrace the different dimensions of the sustainability of AFS.

Drivers of Change

While recent years have seen considerable progress in mapping some components of agricultural systems, this has not been matched by similar progress in understanding both the dynamics and sustainability of these systems in different environments and the drivers of change. Three important drivers that impact strongly on agricultural production systems are climate change, demographic change, and socioeconomic change. Potentially, changing climate has a major impact on agricultural supply, influencing the yields of staple crops (Thornton et al., 2010). Ample evidence exists for these effects already occurring in the Central American "dry corridor" related to El Niño (FAO, 2016). Climate warming and drying in Africa may reduce crop yields overall by 10%–20% (Jones and Thornton, 2009), but the localized impacts are likely to be highly variable depending on agroecology and the production system. The mixed crop-livestock systems are the mainstay of agriculture in Africa (Thornton and Herrero, 2015), and in many areas there are options for adaptation to climate change (Thornton and Herrero, 2014), mostly involving some form of intensification. The more marginal areas, however, are predicted to become so unsuitable for arable agriculture as to force more fundamental changes in agricultural systems: abandonment or pushing mixed crop-livestock systems to more livestock-oriented systems (Jones and Thornton, 2009). A major impact of climate change is to be seen in the increase in variability of production conditions and the subsequent market volatility.

Demographic and socioeconomic changes are closely related. Demographic changes include population growth in low- and middle-income countries (LMICs), particularly in Africa (Gerland et al., 2014). Increasing wealth—per capita gross domestic product (GDP) —is another strong trend in many LMICs and this is further associated with urbanization (Chen et al., 2014). These trends lead to an increasingly affluent urban population and a declining rural population, except in Africa (see also Section I). Together, these factors affect both demand and supply of agricultural products. Changing dietary preferences, in response to increasing wealth and a larger proportion of consumers living in cities, result in increased demand for animal-source foods in areas where levels of their consumption have been previously low. With diminishing rural populations disconnected from the growing markets, intensive production meets an increasing share of this demand. For example, a positive relationship has been shown between the proportion of chicken and pig production in intensive (vs. backyard) systems and national per capita GDP (Gilbert et al., 2015).

Decision Making

Innovation changes how these "inputs" are combined and optimizes the response to resource constraints and market forces, making use of novel techniques and forms of organization. While we are proposing to use proxies of the factors of production to map agricultural systems globally, these are the very factors that influence decisions made by individual farmers on the ground and other players at different stages of the AFS. If the farmgate price for a farmer's produce increases, for example, he or she may decide to invest more in its production. If land rents increase, farmers may decide to produce more per unit of land by irrigating, choosing higher yielding varieties, or stall-feeding livestock, for example. If the cost of labor increases, and cheap labor is no longer available at the time of harvest, a farmer may decide to invest in or rent some mechanized, labor-saving device to assist with the harvest. The decisions that farmers and other AFS actors make with regards to optimizing these production factors reflect their capacity to innovate, and that capacity can be strengthened with promising new approaches (see Section IV).

Sustainability of AFS

Another reason for evaluating AFS in terms of agroecology, demographics, and economics is that it helps us to conceptualize the sustainability of agricultural systems in terms of the five principles of sustainable food and agriculture adopted by FAO: (1) efficiency in the use of resources; (2) conservation, protection, and enhancement of natural resources; (3) the protection and improvement of rural livelihoods, equity, and social well-being; (4) enhancing the resilience of people, communities, and ecosystems; and (5) supporting responsible and effective governance mechanisms. In the proposed typology, AFS are defined in terms of their present values of land, labor, and capital, and systems transitions are characterized by observed or expected changes in these values. Different combinations of these production factors, and different trajectories of change, imply different issues in terms of sustainability. This issue of sustainability will be expanded on in the concluding Chapter 37, Policy Orientations for Sustainable Food and Agriculture of this section.

33.4 MAPPING THE FACTORS OF PRODUCTION

The production factors of land, labor, and capital can be plotted in a 3 parameter space, where each dimension goes from a low value to a high value, as illustrated in Fig. 33.2, which is color-coded so that each axis is assigned one of the primary colors of red (R-capital), green (G-land), and blue (B-labor), creating an "RGB color space" whereby each combination of values of the three production factors results in a unique color.

"Extensive" systems tend to occupy a darkish green zone in the parameter space shown in Fig. 33.2, with the land factor having intermediate values, since productivity tends to be relatively low in these systems, and labor (blue) and capital (red) both having low values. With capital (red) generally very low, labor (blue) high, and land (green) also relatively high, "labor-intensive" systems correspond to cyan in this parameter space. These systems tend toward blue in areas where, while the agricultural population density remains high, the land resource becomes scarcer. "Capital-intensive" systems, where both land (green) and capital (red) are high and where the agricultural labor force (blue) is low appear yellow in the parameter space. In wealthy areas where the land factor is more limited, the color will tend toward magenta. Our aim is to create a spatial representation of this parameter space, to map proxies of these production factors in their respective colors, and to combine them to produce AFS maps for systems based on crops, livestock, and forestry as examples.

The systems maps are presented as color composite images, a color-rendering method whose origins lie in remote sensing, where it is widely used to provide visual contrast to multispectral satellite images. When these three colors are combined in various

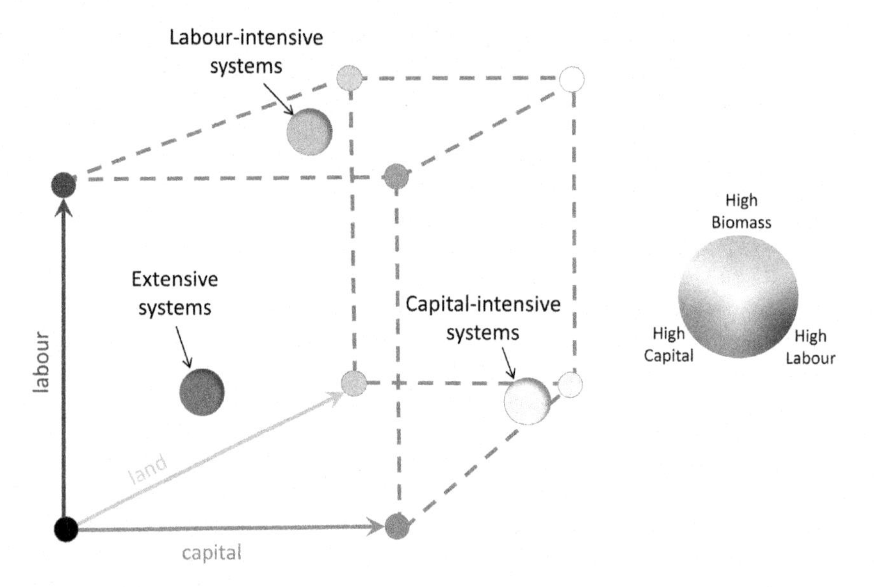

FIGURE 33.2 3D parameter space showing the interplay between land (green), labor (blue), and capital (red). The color wheel shows how different combinations of land (green), labor (blue), and capital (red) give rise to new, unique colors. Source: *For interpretation of the references to color in this figure legend, the reader is referred to the web version of this book.*

proportions, dependent on the values of the input variables, they produce different colors in the visible spectrum.

Labor

Labor in this context refers to the people available to work in a particular AFS. Geographic information system (GIS) maps of total population are readily accessible from a number of sources and are becoming increasingly detailed and accurate. LandScan, Gridded Population of the World (GPW), and WorldPop are the best known; we have produced a hybrid comprising the most reliable of each of these sources from different world regions. Total population can be converted to rural population by masking out urban areas, although definitions and boundaries of urban areas are not clear-cut and peri-urban areas could provide considerable amounts of labor. In low-income countries, rural population serves as a reasonable proxy for agricultural population, but this does not hold true in high-income countries, as in much of Europe, where rural populations are high but are not active in agriculture.

At the national level, FAOSTAT and the World Bank provide data on the proportion of the population engaged in agriculture, which they define as "... activities in agriculture, hunting, forestry and fishing" (UNSTAT, https://unstats.un.org/unsd/cr/registry/regcst.asp?Cl = 27). These data are missing for some countries, however, so a simple model was developed to predict agricultural population as a function of rural population and per capita GDP, accounting for the fact that high-income countries have very low proportions

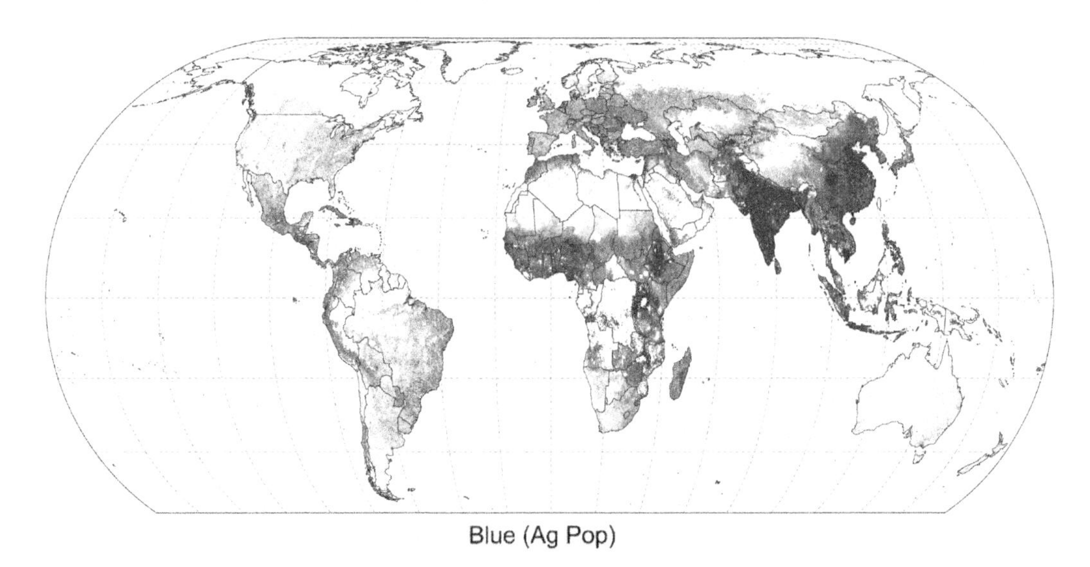

FIGURE 33.3 Global map of agricultural population density (log scale). Source: *For interpretation of the references to color in this figure legend, the reader is referred to the web version of this book.*

of agricultural population compared with low-income countries. Another model allocated agricultural populations subnationally according to human population density, such that a maximum density of agricultural population is assigned in areas with < 100 people km^{-2} (i.e., most rural areas), and a minimum density of agricultural population is assigned in areas with > 1000 people km^{-2} (i.e., urban areas). Intermediate densities are assigned to areas with population densities between these values, depending on the population density. These models were applied to the hybrid population map to give the global distribution of agricultural population shown in Fig. 33.3. From an AFS perspective, we would ideally include the labor involved in the agro-industries at all stages of the value chain, including production inputs and the processing and marketing of agrifood products. As such data are not available, we are restricted to using the labor input to production as a proxy.

Fig. 33.3 shows the very high density agricultural populations in India and China in particular, but also in other parts of southeast Asia, in the highlands of east Africa, and in west Africa, south of the Sahel. A clear gradient in agricultural population density is observable going from western to eastern Europe.

Capital

Capital stock, in the context of agricultural systems, refers to investments into agricultural equipment and infrastructure. Specifically, it refers to the inputs that are needed to modify production environments and to intensify production. Since the level of intensification at a country level, at least in the livestock sector, is closely related to per capita GDP (Gilbert et al., 2015), this can be used as a proxy for capital stock. Per capita GDP estimates are usually only available nationally, however, so this results in quite a different spatial resolution from the agricultural population.

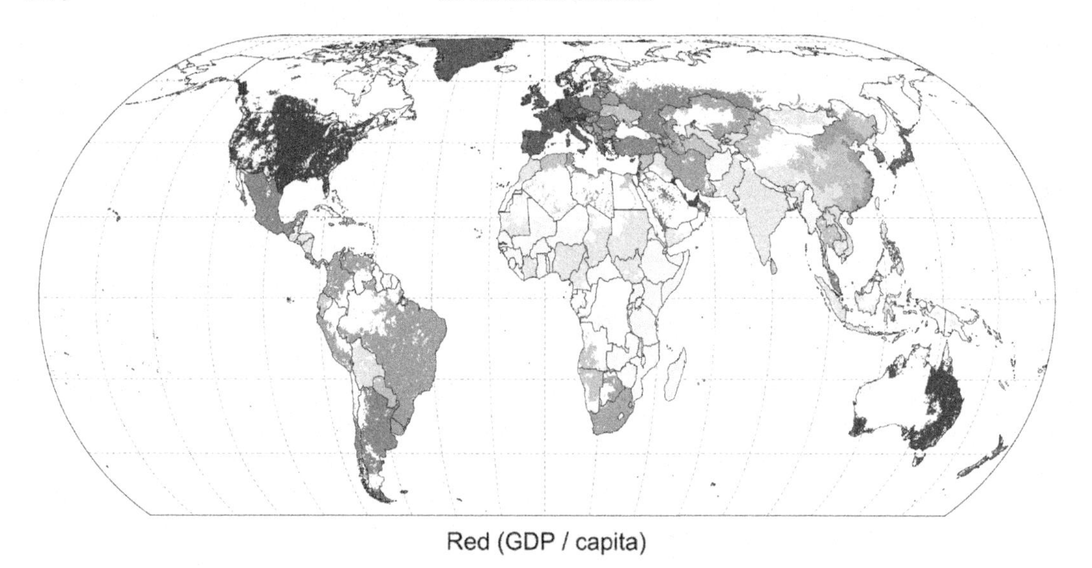

FIGURE 33.4 Global map of GDP per agricultural person (log scale). Source: *For interpretation of the references to color in this figure legend, the reader is referred to the web version of this book.*

Subnational estimates of GDP have been produced at a spatial resolution of a degree square (Nordhaus, 2006), but there are questions raised as to the validity of spatially disaggregating GDP using spatial units that are not linked to functional administrative boundaries. Numerous explorations discuss the use of remotely sensed nighttime lights as indicators of economic activity (Doll et al., 2000, 2006; Chen and Nordhaus, 2011) with promising results, though more recent analyses (Addison and Stewart, 2015; Mellander et al., 2015) are more cautious and indicate areas where the relationship seems to break down. Based on preliminary analyses, we use the robust national GDP estimates and disaggregate them by estimating the amount of GDP available per person engaged in agriculture and then using the agricultural population to assign GDP per unit of area. The resulting map is shown in Fig. 33.4.

Fig. 33.4 shows the stark contrast in GDP per agricultural person between wealthy countries of North America, Western Europe, and Australasia compared to much poorer countries of Africa and Asia, with Eastern Europe and Latin America lying in between. The gradient moving from east to west Europe is also clear.

Land

The representation of land as a production factor in agricultural systems is complicated by different concepts of availability and productivity and by the diversity of commodities involved. In many remote or unproductive areas, land is relatively abundant and prices are low. In other areas, where land value is higher, there is often competition from sectors other than agriculture, such as industrial or residential use. There are also the twin issues of land cover and land use; some land is forested, some under cultivation, and some available for livestock or aquaculture production. This also raises the question of how to represent the "land" production factor for aquaculture and for inland and offshore fisheries. Clearly, no single definition of land or surface will apply to all types of agriculture. As we did for GDP, we have normalized the relevant land resource by expressing it relative to the agricultural population. For crop agriculture, therefore, we take the average number of hectares of cultivated land per agricultural person; for livestock agriculture, we take the average number of livestock units per agricultural person; and for forestry, we take the average number of hectares of forest per agricultural person. In the case of aquaculture, and even more so for that of fisheries, this simple land-based model cannot be applied, necessitating the development of alternative approaches.

Crop-Based Agriculture

Numerous initiatives have been undertaken to map land cover, globally, often with the objective of distinguishing agricultural land. Some of the most widely used land cover maps include the LandScan land cover maps (Loveland et al., 1991), the Global Land Cover 2000 classification (Bartholomé and Belward, 2005; Mayaux et al., 2004), the MODIS land cover map (Morisette et al., 2002), and GlobCover (Bicheron et al., 2008). Because of inconsistencies in the estimated distributions of cropland among these various maps, we have used a composite cropland layer (Fritz et al., 2015) to estimate crop cover. This is expressed as the average number of hectares of cropland per agricultural person in Fig. 33.5.

Fig. 33.5 shows the greatest amount of area under cultivation per agricultural person, that is, the largest farms, occur in North America and in southwest and eastern Australia. The cultivated areas of Latin America and Europe are intermediate, and the lowest amounts, that is, the smallest farms, occur in the highly populated rural areas of Africa and Asia.

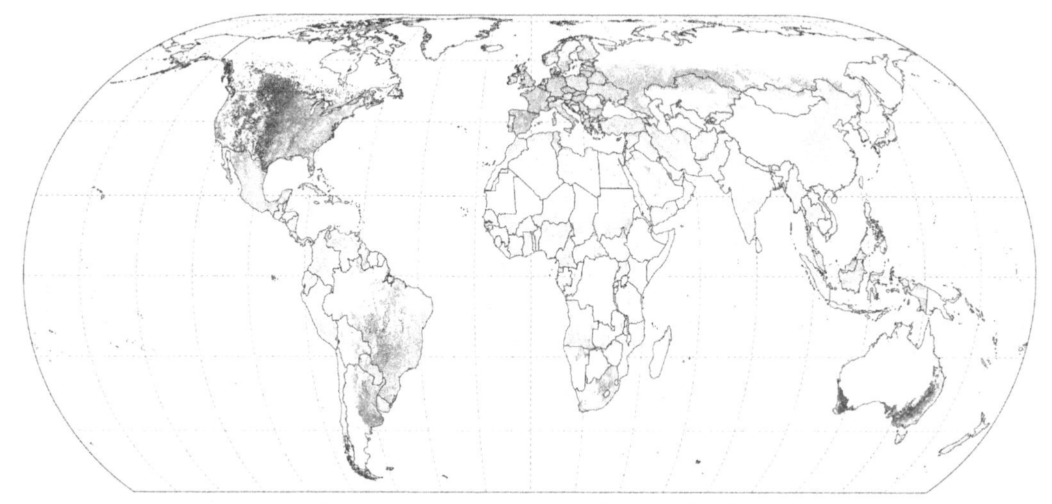

FIGURE 33.5 Global map of land for crop-based agriculture—hectares of cropland per agricultural person. Source: *For interpretation of the references to color in this figure legend, the reader is referred to the web version of this book.*

Livestock-Based Agriculture

The term "livestock geography" was coined in the 1990s by J. Slingenbergh and G.R.W. Wint, who produced the first continental livestock distribution map—a digital map of cattle densities in Africa. This was extended to global coverage and multiple species in the 2007 Gridded Livestock of the World (GLW) (Wint and Robinson, 2007). The predicted densities were based on a stratified regression analysis between reported subnational statistics on livestock numbers and a suite of spatial predictor variables relating to agroecology, demography, land cover, and terrain. Further developments (Prosser et al., 2011; Van Boeckel et al., 2011) have resulted in GLW 2 (Robinson et al., 2014). Combining different livestock species into a single aggregate estimate of livestock amount requires the different species to be converted into biomass, expressed in livestock units. Fig. 33.6 shows the global distribution of livestock units per agricultural person.

Fig. 33.6 shows the greatest livestock biomass per agricultural person to be in North America, Australasia, parts of Latin America, and South Africa. Europe has intermediate levels, and a relatively low livestock biomass per agricultural person is observed in the highly populated rural areas of Africa and Asia.

Forest-Based Agriculture

Forest classes are available in all the major land cover maps, but the same issues of inconsistency occur as with the crop-cover classes. We have taken the Hansen et al. (2013) map as the most reliable and up-to-date estimate of forest cover and expressed this as the average number of hectares of forest cover per agricultural person in Fig. 33.7.

Fig. 33.7 shows the greatest amount of area under forest per agricultural person to occur in North America, particularly in Canada and the northwest United States, and in the

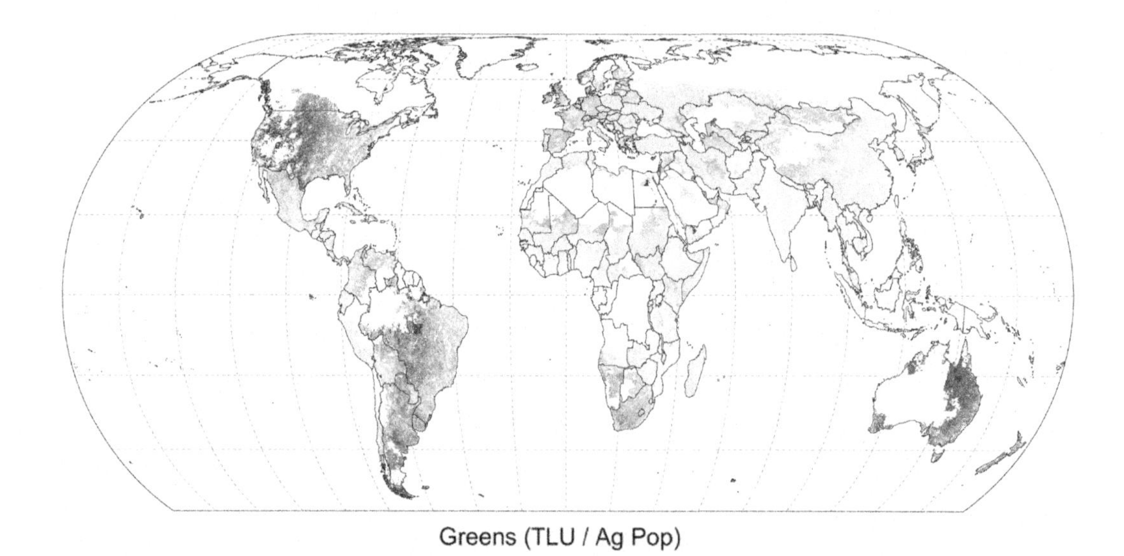

FIGURE 33.6 Global map of "land" for livestock-based agriculture—number of livestock units per agricultural person. Source: *For interpretation of the references to color in this figure legend, the reader is referred to the web version of this book.*

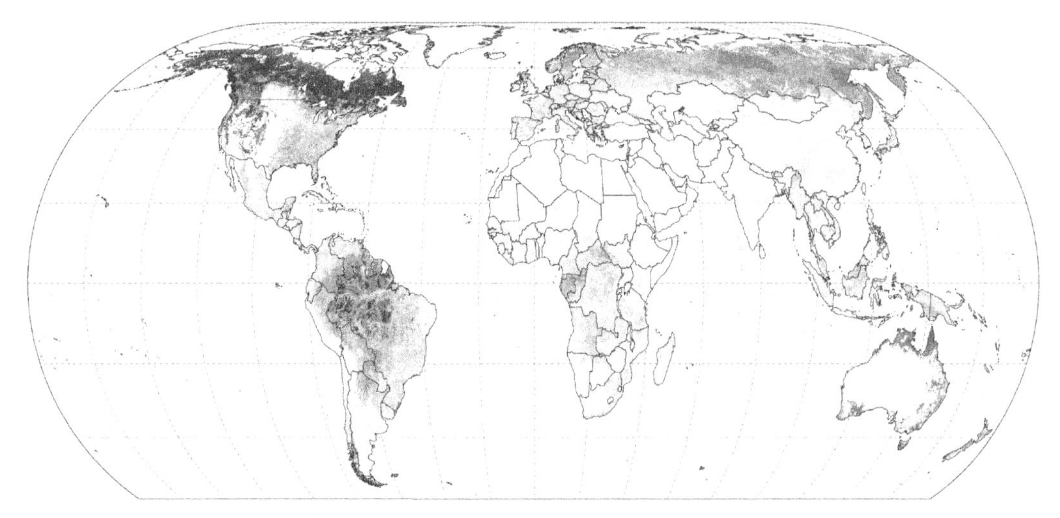

FIGURE 33.7 Global map of land for forest-based agriculture—hectares of forest per agricultural person. Source: *For interpretation of the references to color in this figure legend, the reader is referred to the web version of this book.*

Amazonian area of Latin America. Relatively high values also occur in northern Europe and in parts of Australasia. India has very little forested area per agricultural person. Forest-based agriculture is somewhat different from that based on crops or livestock because much forest is publically owned. In 2010, 82% of forests globally were publicly owned and 18% privately owned. Private companies hold management rights to around 14% of public forests (FAO, 2015).

Aquaculture and Fisheries

Global GIS data layers of aquaculture and fisheries production do not exist in the same way that these are available for crop, livestock, and forest distributions. While the same principles apply in theory, and we can recognize examples of systems in the same categories, as shown in Tables 33.1 and 33.2, determining proxy values for production factors of land, labor, and capital is more complex for aquatic systems and is therefore not done here.

33.5 COMBINING THE PRODUCTION FACTORS INTO A SYSTEMS TYPOLOGY

Having mapped proxies for each of the factors of production, the next step is to combine these in a meaningful way to provide a typology of AFS. Given the objective of creating a generic classification that allows across-sector comparison, we have retained common estimates of the labor and capital factors of production and varied only that of land—to correspond to the different types of terrestrial agriculture. The frequency

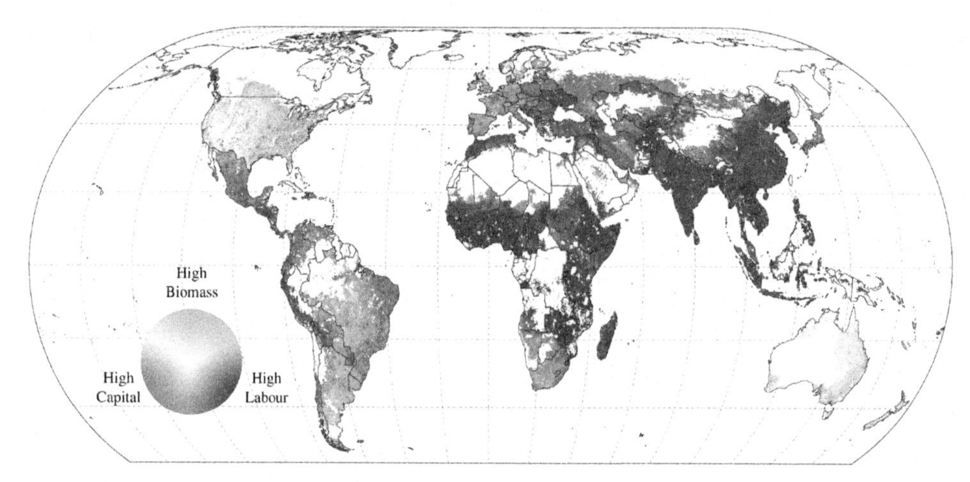

FIGURE 33.8 Color composite systems map for livestock-based agricultural systems. Land (green) is proxied by the number of livestock units per agricultural person; labor (blue) by agricultural population density; and capital (red) by GDP per agricultural person. Source: *For interpretation of the references to color in this figure legend, the reader is referred to the web version of this book.*

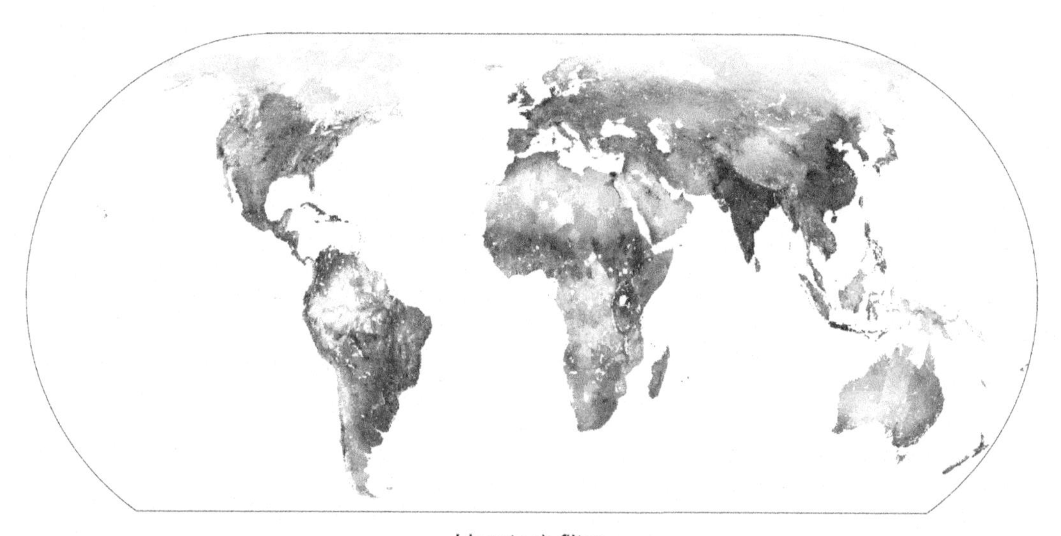

FIGURE 33.9 Transparency filter based on the density of livestock units applied to a homogenous blue background. Areas of low livestock density are relatively opaque, so the blue color is paled out; areas of high livestock density are very transparent, so the blue shines through intensely. Source: *For interpretation of the references to color in this figure legend, the reader is referred to the web version of this book.*

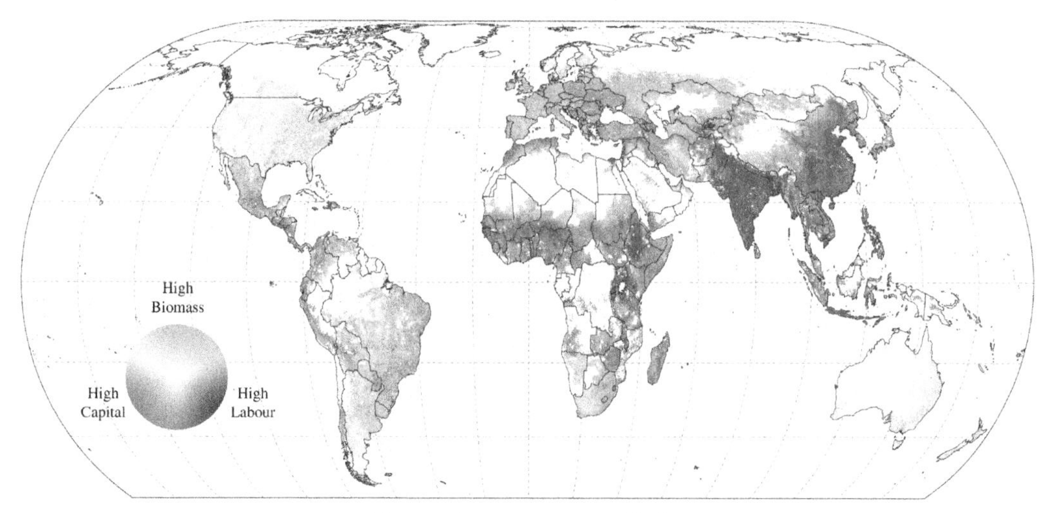

FIGURE 33.10 Color composite systems map for livestock-based agricultural systems. Land (green) is proxied by the number of livestock units (LU) per agricultural person; labor (blue) by agricultural population density; and capital (red) by GDP per agricultural person. The map has been adjusted to account for the density of LUs in absolute terms (see text). Source: *For interpretation of the references to color in this figure legend, the reader is referred to the web version of this book.*

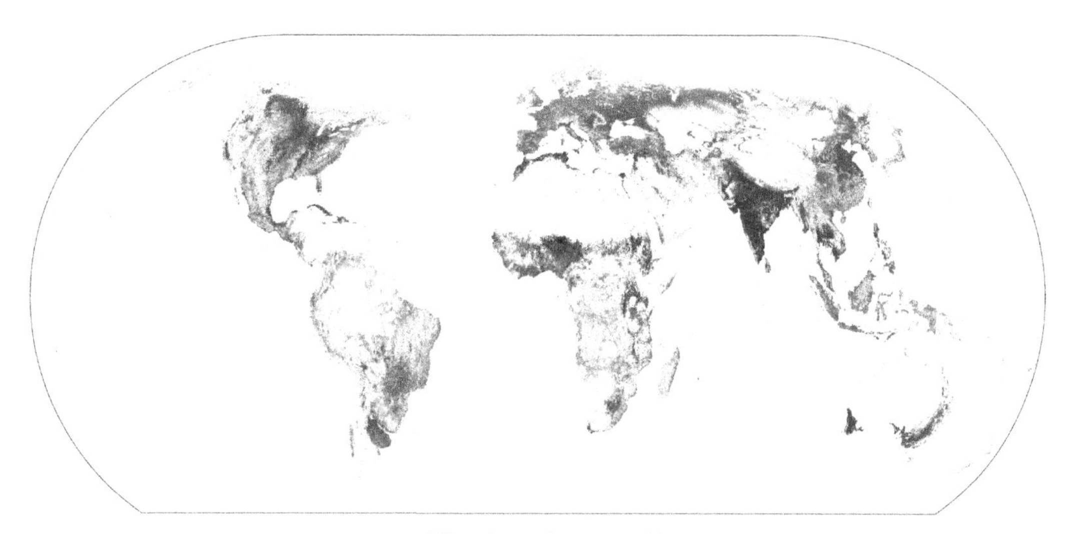

FIGURE 33.11 Transparency filter based on the percentage of crop cover applied to a homogenous blue background. Areas of low crop cover are relatively opaque, so the blue color is paled out; areas of high crop cover are very transparent, so the blue shines through intensely. Source: *For interpretation of the references to color in this figure legend, the reader is referred to the web version of this book.*

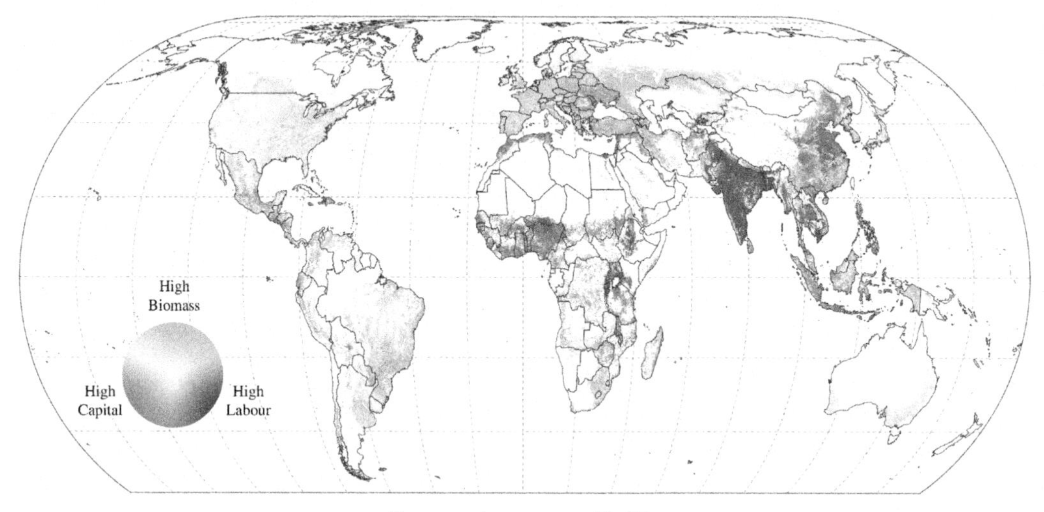

FIGURE 33.12 Color composite systems map for crop-based agricultural systems. Land (green) is proxied by the number of hectares of cropland per agricultural person; labor (blue) by agricultural population density; and capital (red) by GDP per agricultural person. The map has been adjusted to account for percentage crop cover in absolute terms (see text). Source: *For interpretation of the references to color in this figure legend, the reader is referred to the web version of this book.*

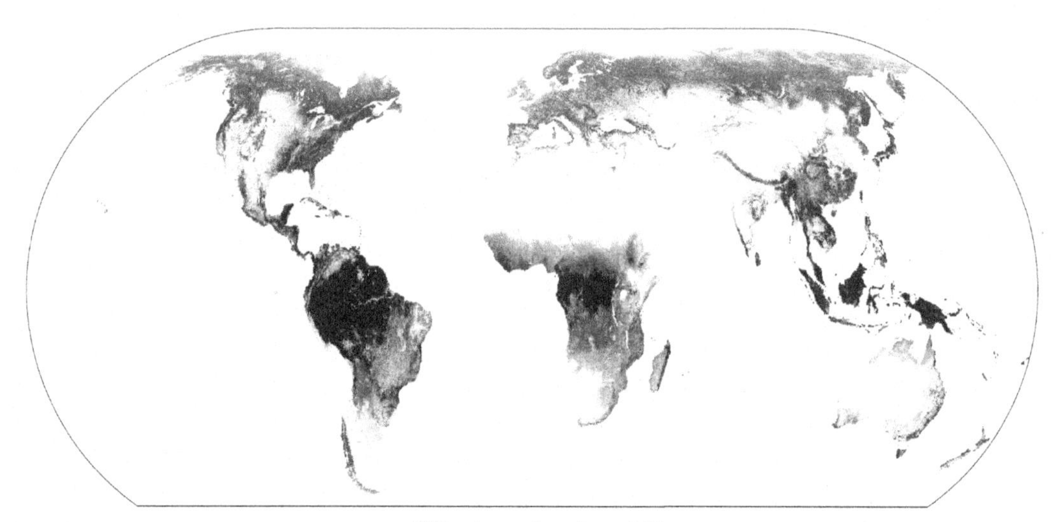

FIGURE 33.13 Transparency filter based on the percentage of forest cover applied to a homogenous blue background. Areas of low forest cover are relatively opaque, so the blue color is paled out; areas of high forest cover are very transparent, so the blue shines through intensely. Source: *For interpretation of the references to color in this figure legend, the reader is referred to the web version of this book.*

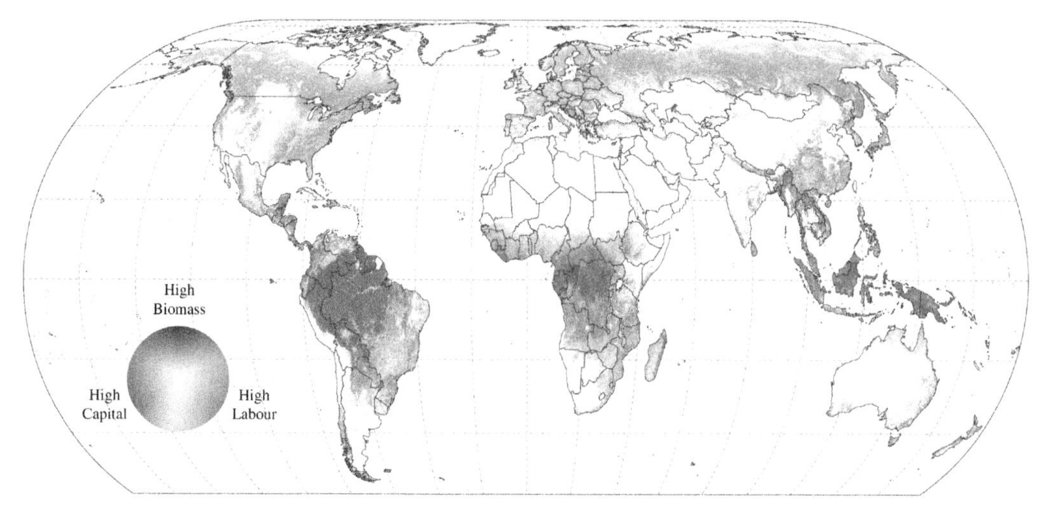

FIGURE 33.14 Color composite systems map for forest-based agricultural systems. Land (green) is proxied by the number of hectares of forest per agricultural person; labor (blue) by agricultural population density; and capital (red) by GDP per agricultural person. The map has been adjusted to account for percentage forest cover in absolute terms (see text). Source: *For interpretation of the references to color in this figure legend, the reader is referred to the web version of this book.*

distributions of most of these variables are highly skewed, so we used appropriate transformations and scaled them so that they were evenly spread across the visual range displayed. For each terrestrial agricultural type, the relevant variables have been combined into AFS maps, as presented in Figs. 33.8–33.14.

In each of the maps we can identify the three broad categories of AFS described in the introduction and shown in Fig. 33.1 and Fig. 33.2: (1) extensive systems; (2) labor-intensive systems; and (3) capital-intensive systems.

Fig. 33.8 presents the color composite map for livestock-based AFS. The most capital-intensive livestock systems appear as yellow in the western United States and in eastern Australia. The capital-intensive, but smaller livestock holdings in the eastern United States and western Europe appear as magenta. These systems tend toward blue in eastern Europe as capital decreases and agricultural population increases. The labor-intensive systems of Africa, Asia, and Central America appear as blue and tend toward magenta in areas with higher GDP, such as South Korea, the eastern coast of China, and Thailand. Extensive livestock-producing areas appear as green, with relatively low use of capital and low densities of agricultural population. These are abundant in much of Latin America, southern Africa, East Africa, and the Sahel. These areas become bluer as agricultural population densities increase, for example, in east Africa.

One particular limitation with this visualization is that it does not differentiate areas where the land resource, in this case for livestock, is abundant from those where it is scarce. This is because it is expressed per agricultural person so low agricultural population densities can effectively push up the "land" variable. A way to overcome this is to determine the color in the RGB space by the relative values of each production factor, but for the saturation of colors to be determined depending on the abundance of the land resource. In order to achieve this, we have created a transparency filter whose transparency is determined by the value of the land resource in each pixel, allowing us to show a fourth dimension in these systems maps. In the case of livestock systems, the higher the density of livestock units, the more transparent the overlay.

In Fig. 33.9, the transparency layer is shown for livestock, overlaid on a blue homogeneous background. The blue appears pale in areas where livestock are scarce and intense where livestock are abundant.

Fig. 33.10 shows the livestock systems map with the transparency filter applied. This reveals some important systems differences, for example, distinguishing the densely populated labor-intensive AFS, which remain intense blue, from the more sparsely populated extensive systems, which appear paler. The utility of this is clearly seen in the Horn of Africa, where the high-density smallholder areas in the highlands of Kenya and Ethiopia remain bright blue, but the lower-density pastoral areas—which are extensive systems—become much paler blue.

The same approach has been taken to mapping the crop- and forest-based AFS, using the percentage of crop cover (Fig. 33.11) and percentage of forest cover (Fig. 33.13), respectively, to determine the transparency masks.

Fig. 33.12 shows the crop-based AFS map with the transparency filter applied. Similar considerations apply to those for livestock, though the locations vary somewhat. For example, in North America the capital-intensive cropping zone, the "corn-belt," is further to the east of the United States. The heavily cropped and high population density areas of the east African highlands, Malawi, Nigeria, India, Pakistan, and Bangladesh appear as bright blue, and these areas tend toward magenta as GDP increases, for example, in Central America, eastern China, and Thailand.

Fig. 33.14 shows the forest-based agricultural systems map with the transparency filter applied. Heavily forested areas in the high-income countries appear yellow and these predominate in Canada, northern Europe, and Australasia. Extensive forestry systems, that is, forestry areas with limited population density to manage or utilize forests are widespread in the Amazon region of Latin America and in southern Chile, in parts of central Africa, and in southeast Asia, notably Myanmar, Cambodia, parts of Indonesia, and Papua New Guinea. Forested areas with abundant actual or potential labor availability for managing them, appearing dark blue in Fig. 33.14, dominate in central America, central Africa, eastern Madagascar, and much of Indonesia and Papua New Guinea. The eastern part of the United States, western Europe, Japan and South Korea, and parts of Australasia are characterized by a pink/magenta speckle, indicating high GDP and intermediate agricultural population densities and amounts of forested area. Note that the map depicts potential overall intensity of management of the forest-based agricultural system, not actual intensity of use. The latter is determined by additional factors, such as ownership or management and use rights of forestland, management purpose (which includes things beyond extractive production, such as biodiversity protection), and availability of infrastructure such as roads or suitability of terrain for harvesting and transportation of wood.

Examples of AFS

We can identify some examples of AFS falling into different parts of the systems typology described above. Since we have developed the systems maps only for the terrestrial agricultural subsectors, the focus is on these, and mostly on crops and livestock; however, the same structure is used briefly to discuss aquatic systems, illustrating that the same typology also has relevance for these but needs to be further developed.

Crop-Based Agriculture (Fig. 33.12)

An example of extensive crop-based agriculture is shifting cultivation, or "slash-and-burn." This ancient AFS is still practiced in the humid tropics and typically involves slashing and burning relatively small forest plots, which are cultivated for two to three years until nutrient depletion and weeds render them no longer suitable. After this, they are left for 5–40 years to regenerate.

Brady (1996) reported 300–500 million people engaged in this type of agriculture, largely centered on the tropical rain forests of Latin America, central and west Africa, and southeast Asia, stating that: "Some 240 M ha of closed forest and 170 M ha of open forest are thought to be involved in some sort of shifting cultivation." More recent estimates suggest far smaller numbers of people are still dependent on shifting cultivation.

Shifting cultivation in its traditional sense is often linked with hunting and other types of forest harvesting for providing protein and other goods and services. In its original form, shifting cultivation is associated with high levels of indigenous knowledge, is efficient (if unproductive), and is considered to be socially and environmentally sustainable, supporting population densities of up to 30 people per square kilometer. However, with population pressure forcing plots to be larger and closer together, cultivation periods to be extended, and fallow periods shortened, this method is highly unstable and probably a major cause of deforestation, land degradation, and losses to biodiversity. Nowadays, migrant populations with few other options carry out much of the slash-and-burn in a poorly controlled manner. In parts of the Amazon basin, central Africa, and southeast Asia, however, shifting cultivation has been practiced by indigenous populations for centuries and has important agricultural heritage value.

Labor-intensive crop-based agricultural systems are typified by the smallholder systems of central America, Africa, and Asia. These are usually mixed systems, combining crop production with livestock and/or aquaculture. Most labor-intensive systems are family farms and focus on producing staple foods for subsistence with surpluses sold or exchanged locally. In some cases, though, there are strong links to national and international markets, where institutions are in place that facilitate collection and marketing of a product, such as tea production in East Africa and South Asia.

In LMICs, very large numbers of people are engaged in smallholder crop or mixed farming, so these are important AFS in terms of livelihood provision. These farming households are often extremely vulnerable and plagued by low returns to labor and high levels of poverty, especially where population density is high and plots are small, having been repeatedly divided through the generations. In these cases, production is typically inefficient: they sit at the dip of a U-shaped productivity curve, moving from extensive systems through labor-intensive systems to capital-intensive systems. The systems are often characterized by nutrient mining and land degradation. Where soils are not degraded, there is considerable potential for increased production through efficiency gains.

In some circumstances, smallholder farming is highly productive, remunerative, and competitive with developed world crop agriculture, playing a major role in alleviation of rural poverty and development in many countries (see also Section I).

Highly capital-intensive crop-based agricultural systems are evident in the Corn Belt in the mid-western United States, centered on the states of Nebraska, Minnesota, Iowa, Illinois, and Indiana, which together account for 64% of national production. The farming system is very intensive and heavily mechanized with high external inputs and high outputs.

These systems are some of the great breadbaskets of the world, generating vast surpluses of grain and soybean to be exported for food and feed. As such, they are of fundamental importance for global food security.

While they are highly efficient, they are characterized by low species and management diversity, high use of fossil energy, and agrichemicals such as fertilizers and pesticides. They also dig deep into water resources in aquifers for irrigation. These factors combine to result in a heavy environmental footprint, but there is considerable potential for reducing the environmental impact of these capital-intensive cropping systems. In recent decades, improved practices such as precision agriculture, reduced or zero tillage, and agroecological adaptations have shown important progress in mitigating environmental damage.

Livestock-Based Agriculture (Fig. 33.10)

The extensive livestock systems are typically pastoralist systems, with ruminant livestock grazing large expanses of relatively, or at least seasonally, low-productivity rangelands. These can be cold and dry, such as in Mongolia where 80% of the land area is covered by grassland, giving home to about 35 million horses, cattle, sheep, goats, and camels. Half of the country's 2.7 million people depend on livestock production, which contributes more than 20% of the country's GDP. Or the lands can be hot and dry, such as the vast rangelands of Africa, which are home to 50 million herders, who own a third of the livestock and half the small ruminants in the countries where they occur.

The number of people who benefit directly from these systems is relatively small, about 200 million globally (IUCN ESARO, 2011), though in situations where the systems are linked to distant markets, such as exports from the Horn of Africa to the Arabian Peninsula, the beneficiaries are significantly increased. With few inputs beyond basic animal health care, these systems are highly efficient low in productivity. They tend to occur in areas that are unsuitable for crop growth and therefore do not compete with food production, but instead make productive use of an environment that would otherwise not contribute to feeding the world's population. As livestock are integral to the social fabric of pastoralist societies, the benefits serve many social and financial roles. Pastoralist societies are important in terms of agricultural heritage. The diversity of people and cultures is vast, ranging from the Kazakh people of the Altai Mountains, who supplement their pastoral livelihood by hunting with Golden Eagles, to the Samburu and Turkana people of northern Kenya, and the Dinka of South Sudan.

Already living at climatic extremes, pastoralist people are extremely vulnerable to climate change, with harsh winters or extended droughts causing high levels of mortality among their livestock. While generally inefficient in terms of greenhouse gas emissions—with relatively high emissions per unit of output—they account only for a small proportion of overall agricultural emissions. One of the major environmental issues for these systems is land degradation; it is estimated that globally 10%–20% of the rangelands are severely degraded (Reynolds et al., 2007).

Labor-intensive livestock systems almost invariably occur as part of mixed crop-livestock farms, sometimes including aquaculture and/or tree crops. Much of the above discussion about labor-intensive cropping systems therefore also applies. There are some quite specialized subsectors, like smallholder dairy.

In LMICs, there are many people whose livelihoods depend on livestock, many of them poor. Of the 770 million people surviving on less than US$1.90 per day, about half depend directly on livestock for their livelihoods and the vast majority are in these systems (Robinson et al., 2011). For the livestock keepers, livestock are an important source of nutritious food, including income in some cases, and provide many social and other functions, as detailed in Chapter 34, Socioeconomic Dimension of Agrifood Systems. Livestock also provide an important source of food for people outside of the farming households, particularly in dairy production, where 80% of milk in developing countries comes from smallholder farmers, as well as creating value for agro-industries.

As with crops, the production of livestock in these systems is generally inefficient compared to extensive, or capital-intensive systems, if measured simply by yields. But there are other products and services that must be accounted for, particularly in relation to nutrient cycling, giving value to crop residues and provision of draft power. Many opportunities exist to close productivity gaps through efficiency gains, but the right incentives have to be provided for farmers to invest in intensifying production.

Capital-intensive systems of beef production are concentrated in Texas, Oklahoma, Kansas, and Nebraska of the United States, which are home to large ranches and feedlots. Mega-dairy farms are also becoming more prevalent, which are highly mechanized and extremely productive. Half of the milk in the United States now comes from farms with more than 1000 cows, and this proportion increases each year. Monogastric livestock species—pigs, chickens, and ducks—particularly lend themselves to industrial production. These are the mainstay in high-income countries but are becoming much more prevalent in LMICs where they contribute to feeding the growing urban populations.

Capital-intensive systems tend to occur in highly modified environments. While very few people benefit directly as farmers from these systems, many benefit from a regular supply of clean, affordable, nutritious food. That said, industrial systems also fuel the fast-food industry, which is heavily implicated in many issues of ill health. They also provide employment and value-addition opportunities for the agro-industry. Capital-intensive pig, chicken, and dairy systems consume almost three times the amount of antimicrobials consumed in human medicine (Van Boeckel et al., 2017) and probably contribute substantially to the global problem of antimicrobial resistance. Additionally, capital-intensive systems provide conditions that give rise to the emergence of new diseases and the development and amplification of pathogenic prevalence. For example, the emergence of avian influenza H5N1 and H7N9 in China has been linked to rapid intensification of the poultry sector (Gilbert et al., 2017).

Capital-intensive livestock systems are highly efficient, but they rely on large amounts of inputs, particularly in terms of grown feed, which often has telecoupled effects in locations distant from the production unit, including deforestation, disruption of nutrient cycles, and pollution with pesticides and fertilizers. Resulting from so many animals in such high densities, excess nutrients from manure is a growing problem and a major source of soil and water pollution.

Forest-Based Agriculture (Fig. 33.14)

Important for the conservation of endangered species, extensive forest management is characterized by minimal differences between managed and natural forests. This includes protected area management, such as the tiger reserves of Bangladesh and India, protective forest management against avalanches and water-loss and soil erosion, and hunter-gathering, such as that carried out in the rainforests of central Africa. Hunter-gathering would also often be associated with some degree of shifting cultivation in cleared patches of forest. The conservation of biodiversity is the primary management objective for 13% of the world's forests. The area of forests designated for protection of soil and water represents 31% of the forest area (FAO, 2015).

Very small numbers of individuals benefit directly from extensive forest use, but the societal benefits resulting from ecosystems services provided by forest conservation are considerable. Apart from the provision of food, water, and wood to direct beneficiaries, the system also provides climate regulation, particularly through carbon sequestration, pollution control, soil protection and formation, cycling of nutrients, biodiversity protection, water regulation and supply, disturbance regulation, and recreation.

Labor-intensive forms of forest use, beyond small-scale harvesting of wood, include practices such as charcoal harvesting in east and southern Africa, collection of medicinal plants in Nepal, berry and mushroom picking in Finland, and ecotourism in Costa Rica.

Relatively small numbers of people benefit directly from these types of systems, but for those who do, these systems often represent important livelihoods options. Many of these practices would be sustainable at low levels of use, but problems arise when population densities create unsustainable demand on the natural resource base. Charcoal production exemplifies the type of destruction that can arise through overexploitation of the resource base. In parts of Zambia, for example, large areas of forest and woodland have been decimated due to very high demand for charcoal for cooking and poorly regulated charcoal harvesting.

Capital-intensive forestry operations include logging in semi-natural forests, widely practiced in Canada, Sweden, and Finland, for example, or in plantation forestry, which is led by countries such as Brazil, China, and the United States. In 2015, about 31% of the world's forests were primarily designated as production forest, and close to 28% of the forest area was designated for multiple use (FAO, 2015).

Globally, capital-intensive forest use is the primary source of wood, and the proportion of that which is based on plantations is growing rapidly. Forest plantations are typically monocultures of uniform age and composition, managed to optimize the yield of wood from a site. Clear-felling and replanting is the most common system, although, where appropriate, coppicing is used as a means of restocking. These practices raise concern that many of the sites on which trees are planted may be incapable of sustaining their productivity. Export of nutrients, physical damage to soil structure, and greater risk from pests and diseases have all been proposed as possibilities for the inherent unsustainability of intensive plantation forestry.

Aquaculture-Based Agriculture

Extensive forms of aquaculture are uncommon and very basic, involving the netting off of mangrove swamps and harvesting of whatever happens to be retained. Labor-intensive farms are very common, particularly in east and southeast Asia where they often comprise an important component of mixed farming systems, integrated with crop and livestock production. As demand for aquaculture products grows, however, these are increasingly being replaced by capital-intensive systems, which operate in more highly controlled environments and at far greater scales. Capital-intensive aquaculture is probably the fastest growing AFS of all. Aquaculture production makes a significant and growing contribution to global food security. However, concerns for sustainability include heavy use of antimicrobials in some systems, and the discharge of waste, pesticides, and other chemicals directly into ecologically fragile coastal waters, destroying local ecosystems. There are further concerns around the sources of feed both in terms of crop-based feed production and in terms of using feed originating from wild-caught fish.

Fisheries-Based Agriculture

Extensive fisheries-based agriculture would include open ocean line fishing. Labor-intensive systems would include artisanal coastal fishing and inland fisheries. Capital-intensive systems would include the deployment of large fleets of trawlers. The contribution that capture fisheries make to feeding the world leveled off in the 1980s and is decreasing in proportion as aquaculture takes over, having surpassed wild caught fish supply in 2014. The critical issues for sustainability with fisheries are the depletion of stocks and the unintended destruction of nontarget species.

33.6 CONCLUSIONS

This chapter has explored the possibility of using the factors of production to develop a crosscutting typology for agricultural production systems and has used proxies for these production factors to map agricultural systems globally—distinguishing extensive, labor-intensive, and capital-extensive systems as endpoints. It is proposed that this approach can help to anticipate the impact of drivers, such as increasing wealth and demographic change, explore the pressures that farmers will come under to make decisions, and predict the impacts such decisions will have on the sustainability of agricultural systems.

It is beyond the scope of the present book to model the likely trajectories of change and the consequences for sustainability, but there is scope for taking an example to illustrate how this might be achieved. In many LMICs, two powerful trends are increasing: wealth and urbanization. Increasing wealth results in an increase in the production factor of capital, and urbanization results in a decrease in the production factor of labor—as the rural population is drawn to urban areas. Thus, labor becomes scarcer while capital becomes more abundant and, following the concept of induced innovation, we would expect labor to be substituted for by capital investment. Considering the position of labor-intensive systems in Fig. 33.2, shown as the cyan sphere, we would expect this to move down the blue scale (less available labor) and up the red scale (more available capital) toward the yellow sphere, representing the capital-intensive systems. Such a transition will bring about a change in the issues relating to sustainability, both in terms of human benefits and in terms of natural resources and the environment. For example, while the labor-intensive systems provide livelihoods for many, the capital-intensive systems provide livelihoods for relatively few. Not only would efficiency gains be expected from such a transition, but also a greater reliance on inputs such as water, fertilizers, and pesticides for food crop and feed production, or antimicrobials in livestock and aquaculture production. The speed of the transition and the specific trajectory taken may also have a bearing on sustainability, as will the policy and institutional context in which the transition occurs. Projections are available for population growth, urbanization, and GDP growth, so changes in the production factors of land and capital could be mapped and areas identified where systems transitions can be expected to take place.

Mapping AFS combining social, economic, and environmental parameters, such as in the method presented, allows for better targeting of interventions, and helps reduce complex problems to more manageable and actionable issues. Mapping and spatial analysis has proved useful in many other spheres of food and agriculture. Robinson and Pozzi (FAO, 2011), for example, mapped growth in demand and supply of animal-source foods to 2030, highlighting areas of deficit. Many publications have mapped risk of particular pests or diseases, and a recent publication (Osgood-Zimmerman et al., 2018) has mapped stunting, wasting, and underweight in Africa from 2000 to 2015. As this publication reveals, generalized national figures can hide vast inequalities and failures in particular areas where problems are deeply ingrained. The same applies with the mapping of AFS in the context of sustainability. Such maps will help us to locate problems and opportunities, draw up baselines, and target interventions more precisely to where they are most needed.

As we have discussed, different AFS have diverse impacts both on human benefits provided and on the impact on the natural resource base and the environment. A detailed discussion of these impacts constitutes the subject matter for Chapter 34, Socioeconomic Dimension of Agrifood Systems, and Chapter 35, Natural Resource and Environmental Dimensions of Agrifood Systems.

References

Addison, D., Stewart, B., 2015. Nighttime Lights Revisited: The Use of Nighttime Lights Data as a Proxy for Economic Variables. World Bank, Washington DC.

Bartholomé, E., Belward, A.S., 2005. GLC2000: a new approach to global land cover mapping from Earth observation data. Int. J. Remote Sens. 26, 1959–1977.

Bicheron, P., Defourny, P., Brockmann, C., Schouten, L., Vancutsem, C., Huc, M., et al., 2008. GlobCover: Products Description and Validation Report. runn, Toulouse.

Brady, N.C., 1996. Alternatives to slash-and-burn: a global imperative. Agric. Ecosyst. Environ. 58, 3–11.

Chen, X., Nordhaus, W.D., 2011. Using luminosity data as a proxy for economic statistics. Proc. Natl. Acad. Sci. 108, 8589–8594.

Chen, M.X., Zhang, H., Liu, W.D., Zhang, W.Z., 2014. The global pattern of urbanization and economic growth: evidence from the last three decades. PLoS One 9, 15.

de Haan, C.H., Steinfeld, H., Blackburn, H., 1997. Livestock and the environment: Finding a balance. WRENmedia, Suffolk, UK, p. 115.

Doll, C.N.H., Muller, J.-P., Elvidge, C.D., 2000. Night-time imagery as a tool for global mapping of socioeconomic parameters and greenhouse gas emissions. Ambio 29, 157–162.

Doll, C.N.H., Muller, J.-P., Morley, J.G., 2006. Mapping regional economic activity from night-time light satellite imagery. Ecol. Econ. 57, 75–92.

FAO, 2011. Mapping supply and demand for animal-source foods to 2030, by T.P. Robinson & F. Pozzi. Animal Production and Health Working Paper, Rome.

FAO, 2015. Global Forest Resources Assessment 2015. FAO, Rome.

FAO, 2016. Dry corridor Central America. Situation Report - June 2016. FAO, Rome.

Fritz, S., See, L., McCallum, I., You, L., Bun, A., Moltchanova, E., et al., 2015. Mapping global cropland and field size. Glob. Change Biol. 21, 1980–1992.

Gerland, P., Raftery, A.E., Ševčíková, H., Li, N., Gu, D., Spoorenberg, T., et al., 2014. World population stabilization unlikely this century. Science 346, 234–237.

Gilbert, M., Conchedda, G., Van Boeckel, T.P., Cinardi, G., Linard, C., Nicolas, G., et al., 2015. Income disparities and the global distribution of intensively farmed chicken and pigs. PLoS One 10, e0133381.

Gilbert, M., Xiao, X., Robinson, T.P., 2017. Intensifying poultry production systems and the emergence of avian influenza in China: a 'One Health/Ecohealth' epitome. Arch. Public Health 75, 48.

Hansen, M.C., Potapov, P.V., Moore, R., Hancher, M., Turubanova, S.A., Tyukavina, A., et al., 2013. High-resolution global maps of 21st-century forest cover change. Science 342, 850–853.

Hayami, Y., Ruttan, V.W., 1971. Induced Innovation in Agricultural Development. Center for Economic Research, Department of Economics, Univers ity of Minnesota, Minneapolis, Minnesota 55455, p. 47.

Herlemann, H.H., Stamer, H., 1958. Produktionsgestaltung und Betriebsgrosse in der Landwirtschaft unter dem Einfluss der wirtschaftlich - technischen Entwicklung. Institut fuer Weltwirtschaft an der Universitaet Kiel, Kiel.

Hicks, J.R., 1932. The Theory of Wages. Macmillan, London.

IUCN ESARO, 2011. Supporting Sustainable Pastoral Livelihoods: A Global Perspective on Minimum Standards and Good Practices. Second Edition March 2012: published for review and consultation through global learning fora ed. IUCN ESARO office, Nairobi, Kenya, p. vi + 34pp.

Jones, P.G., Thornton, P.K., 2009. Croppers to livestock keepers: livelihood transitions to 2050 in Africa due to climate change. Environ. Sci. Policy 12, 427–437.

Krugman, P., Wells, R., 2012. Microeconomics. Palgrave Macmillan.

Loveland, T.R., Merchant, J.W., Ohlen, D.O., Brown, J.F., 1991. Development of a land-cover characteristics database for the conterminous U.S. Photogrammetric Eng. Remote Sens. 57, 11.

Mayaux, P., Bartholomé, E., Fritz, S., Belward, A., 2004. A new land-cover map of Africa for the year 2000. J. Biogeogr. 31, 861–877.

Mellander, C., Lobo, J., Stolarick, K., Matheson, Z., 2015. Night-time light data: a good proxy measure for economic activity? PLoS One 10.

Morisette, J.T., Privette, J.L., Justice, C.O., 2002. A framework for the validation of MODIS Land products. Remote Sens. Environ. 83, 77–96.

Nordhaus, W.D., 2006. Geography and macroeconomics: new data and new findings. Proc. Natl. Acad. Sci. US A 103, 3510–3517.

Osgood-Zimmerman, A., Millear, A.I., Stubbs, R.W., Shields, C., Pickering, B.V., Earl, L., et al., 2018. Mapping child growth failure in Africa between 2000 and 2015. Nature 555, 41.

Prosser, D.J., Wu, J., Ellis, E.C., Gale, F., Van Boeckel, T.P., Wint, W., et al., 2011. Modelling the distribution of chickens, ducks, and geese in China. Agric. Ecosyst. Environ. 141, 381–389.

Reynolds, J.F., Stafford Smith, D.M., Lambin, E.F., Turner, B.L., Mortimore, M., Batterbury, S.P.J., et al., 2007. Global desertification: building a science for dryland development. Science 316, 847–851.

Robinson, T.P., Thornton, P., Franceschini, G., Kruska, R., Chiozza, F., Notenbaert, A., et al., 2011. Global :ivestock Production Systems. Food and Agriculture Organisation of the United Nations, Rome.

Robinson, T.P., Wint, G.R.W., Conchedda, G., Boeckel, T.P.V., Ercoli, V., Palamara, E., et al., 2014. Mapping the global distribution of livestock. PLoS One 9, e96084.

Ruttan, V.W., Binswanger, H.P., Hayami, Y., 1980. Induced innovation in agriculture. In: Bliss, C., Boserup, M. (Eds.), Economic Growth and Resources: Natural Resources. Palgrave Macmillan, UK, London, pp. 162–189.

Ruttan, V.W., Hayami, Y., 1984. Toward a Theory of Induced Institutional Innovation. Center for Economic Research, Department of Economics, Univers ity of Minnesota, Minneapolis, Minnesota, p. 41.

Samuelson, P.A., 2010. Economics. Tata McGraw-Hill Education.

Thornton, P.K., Herrero, M., 2014. Climate change adaptation in mixed crop–livestock systems in developing countries. Global Food Secur. 3, 99–107.

Thornton, P.K., Herrero, M., 2015. Adapting to climate change in the mixed crop and livestock farming systems in sub-Saharan Africa. Nat. Clim. Change 5, 830–836.

Thornton, P.K., Jones, P.G., Alagarswamy, G., Andresen, J., Herrero, M., 2010. Adapting to climate change: agricultural system and household impacts in East Africa. Agric. Syst. 103, 73–82.

Van Boeckel, T.P., Prosser, D., Franceschini, G., Biradar, C., Wint, W., Robinson, T., et al., 2011. Modelling the distribution of domestic ducks in Monsoon Asia. Agric. Ecosyst. Environ. 141, 373–380.

Van Boeckel, T.P., Glennon, E.E., Chen, D., Gilbert, M., Robinson, T.P., Grenfell, B.T., et al., 2017. Reducing antimicrobial use in food animals. Science 357, 1350–1352.

Whittlesey, D., 1936. Major agricultural regions of the Earth. Ann. Assoc. Am. Geogr. 26, 199–240.

Wint, G.R.W., Robinson, T.P., 2007. Gridded Livestock of the World, 2007. Food and Agriculture Organization of the United Nations, Rome.

CHAPTER

34

Socioeconomic Dimension of Agrifood Systems

34.1 INTRODUCTION

Humankind, by combining land, labor, and capital with different levels of intensity, has been hugely successful in deriving benefits from agriculture. These include food, textiles, energy and construction material, on- and off-farm employment, environmental services, healthy environments, and social identity, among many others. The world's population increased from about 250 million people in the 1st century (Livi Bacci, 2012) to about 7.5 billion today. Levels of malnutrition and food insecurity have declined dramatically. Analysis of skeletons of ancient wealthy societies, such as the Greeks in southern Italy between 600 and 250 BC, suggests that more than three-quarters of the population suffered from serious disease or undernutrition (Henneberg et al., 1992). Today, such percentages are not found even in worn-torn societies, with only three countries in the world recording a stunting prevalence higher than 50% (IFPRI, 2016).

Technological and institutional innovations—from threshing machines to sharecropping, from dredges to contract farming, from drip irrigation to trade agreements—have contributed to increasing productivity in agriculture, reducing food shortages and malnutrition, and making food increasingly available for humans over the course of centuries (de Janvry and Dethier, 1985; Koppel, 1994). Increased agricultural productivity in all sectors and systems, whether they are extensive, labor-intensive, or capital-intensive, has been pivotal to broader societal development through the transfer of surplus to other sectors of the economy. Productivity increases in agriculture lowered the price of food, reduced the relative cost of labor, and enabled industrial development in 18th-century England (Toynbee, 1884) and elsewhere. Since then, there has been a constant decline of agricultural prices relative to other goods, which has been feeding the development of the nonagricultural sectors (Mundlak and Larson, 1990). In industrialized economies, households spend less than 10% of their budget on food, which allows the purchase of nonagricultural goods and services; in less-developed countries, the average household expenditures on food is often over 40% (Gray, 2016).

Sustainable Food and Agriculture
DOI: https://doi.org/10.1016/B978-0-12-812134-4.00034-0

In spite of these advances in agricultural production, the global agricultural system fails in many important ways. The coexistence of malnutrition and stunting alongside growing overweight and obesity has dramatic health implications at each end of the spectrum. Our agrifood systems (AFS) also lead to the emergence and spread of zoonotic diseases and hamper our capacity to treat human and animal diseases with antimicrobial drugs. AFS contribute to the deterioration of life-support systems through soil and water pollution, greenhouse gas emissions, agriculture-driven deforestation, and biodiversity loss. At the same time, the systems do not adequately serve the rural poor, the majority of whom depend on agriculture for their livelihoods (FAO, 2013, 2016a). With the human population expected to increase from 7.5 billion today to 9.7 billion in 2050 (UN, 2017) —in combination with increasing purchasing power for expenditure on food and rising temperatures across the globe—it will be an increasingly daunting challenge for AFS to continue providing adequate benefits to society.

This section provides a snapshot of the benefits human beings derive from agriculture, and highlights some of the key challenges the sector will face in the coming decades to continue playing its fundamental role in human society, which will need to be addressed in a way that is socially and environmentally sustainable.

34.2 AGRIFOOD SYSTEMS, FOOD SECURITY, NUTRITION, AND HEALTH

Regular intake of sufficient, safe, and nutritious food is required for an active and healthy life (FAO et al., 2015). Four elements (food availability, food access, food utilization, and stability of supply) determine if and how much human beings benefit from food, that is, whether they are well nourished or malnourished (FAO, 2016b).

Food availability refers to the supply of food, that is, the quantity of food available to a given population, which at the national level is determined by national production, national stock, and net trade. The Committee on World Food Security (CFS) recommends an initial set of five actionable indicators to measure food availability: (1) average dietary supply adequacy; (2) average value of food production; (3) share of dietary supply derived from cereals, roots, and tubers; (4) average protein supply; and (5) average supply of protein of animal origin (FAO, 2011a).

Food access is the ability of households to access food, which depends on both economic and physical dimensions. It is closely associated with the "entitlement approach," pioneered by Sen (1976), and represents the ability of people to command food through any legal means available in society. The CFS (FAO, 2011a) suggests eight indicators to assess food accessibility, which include factors such as gross domestic product (GDP) per capita, the domestic food price index, road density, and the percentage of paved roads over total roads.

Food utilization refers to the ability of households to use food properly. This includes a diverse range of factors, such as the employment of appropriate food processing and storage techniques, access to clean water sources and sanitation facilities, knowledge of nutrition, and child-care practices. For example, handwashing with soap before eating could reduce the risk of diarrhoeal diseases by 42%–47% and save about 1 million lives per year in developing countries (Curtis and Cairncross, 2003). The CFS proposes a variety of indicators to assess food utilization (FAO, 2011a): examples include access to improved water sources; access to improved sanitation facilities; percentages of children under 5 years of age affected by wasting, stunting and underweight.

Finally, stability of supply is the temporal dimension of both food availability and food access—the capacity of people to have a consistent and regular healthy diet. The 2011 report by the High Level Panel of Experts on Food Security and Nutrition reads: "price volatility has a strong impact on food security because it affects household incomes and purchasing power. Simply put, it can transform vulnerable people into poor and hungry people" (CFS, 2011). The CFS recommends a multitude of indicators to assess food stability, indicators such as cereal import dependency ratio, domestic food price volatility, per capita food production variability, and percentage of arable land equipped with irrigation.

Figs. 34.1–34.4 present indicators that provide an assessment of the current levels of food supply, access, and utilization in different regions of the world. Fig. 34.1 shows the supply of food as measured by the average dietary energy supply adequacy, which expresses the dietary energy supply as a percentage of the average dietary energy requirement. It shows that in all regions the current level of food supply is sufficient to meet the population's energy requirements. In sub-Saharan Africa, the share is 12% greater than 100, while in North America the share is 44% greater than 100. Thus, agriculture today produces more food in terms of calories than is required by the current population.

In spite of that, about 11% of the world's population consumes an amount of food that is insufficient to cover their minimum energy requirements, that is, they are undernourished (Fig. 34.2). The regions of the world most affected by undernourishment are sub-Saharan Africa (22.2% of the population), Near East and North Africa (20.2%), East and Southeast Asia (18.0%), and South Asia (15.8%). Worse still is the fact that undernourishment is often a chronic rather than a temporary condition. Fig. 34.3 shows the proportion

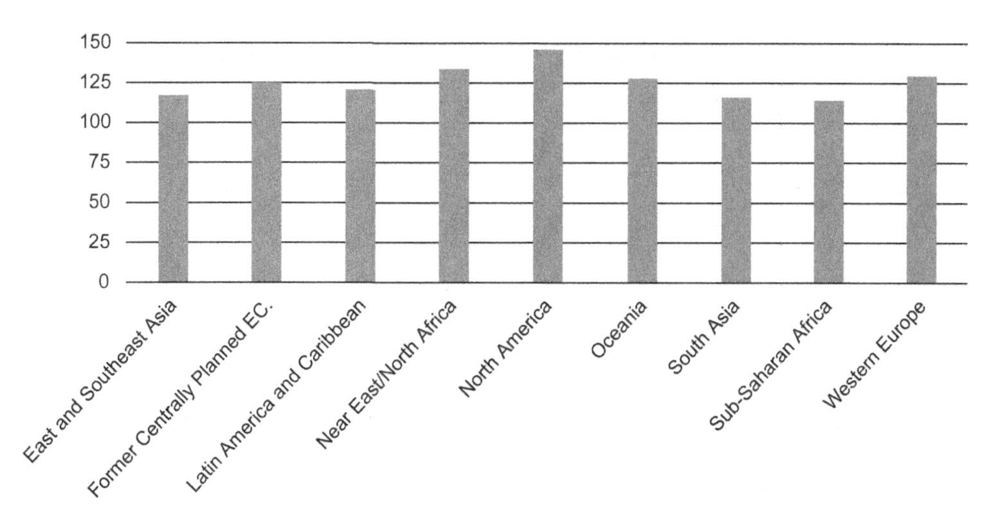

FIGURE 34.1 Average dietary energy supply adequacy (%) by world's region. Source: *FAO, 2016a. Food Security Indicators. FAO, Rome.*

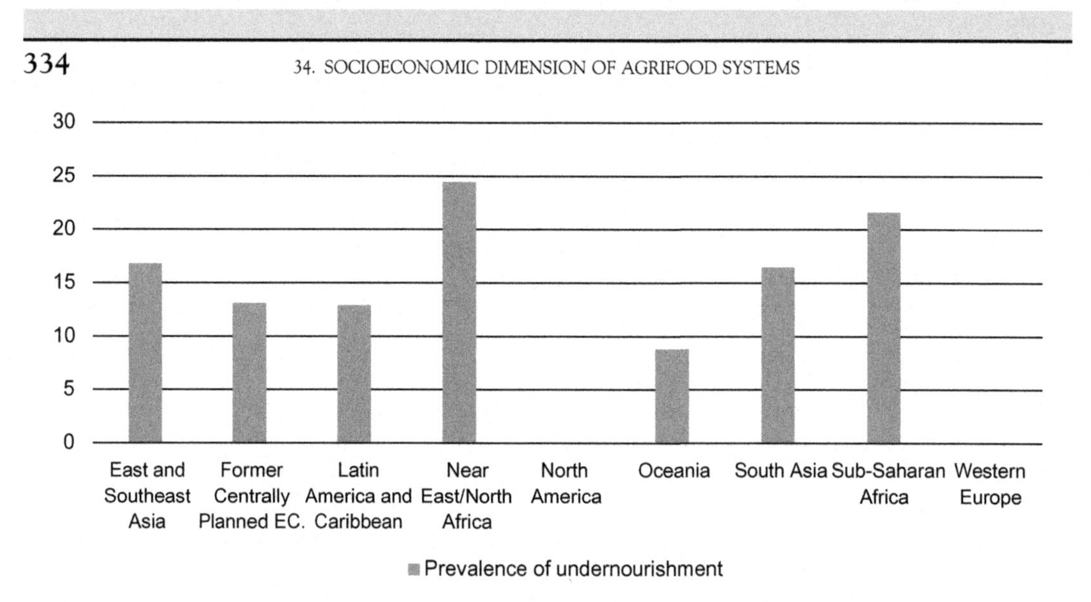

FIGURE 34.2 Undernourishment by world's region. Source: *FAO, 2016a. Food Security Indicators. FAO, Rome.*

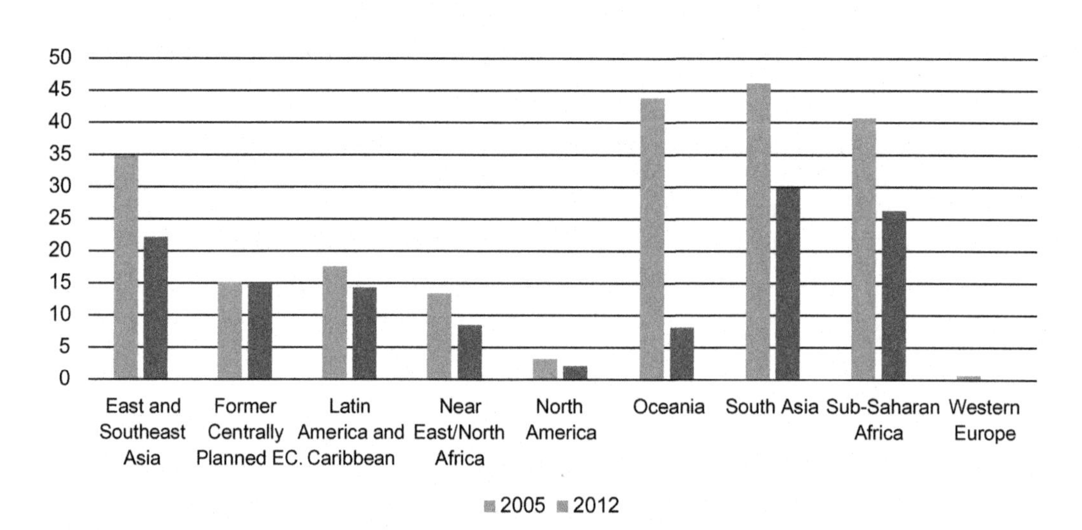

FIGURE 34.3 Proportion of stunted children under 5 years of age by world's region. Source: *FAO, 2016a. Food Security Indicators. FAO, Rome.*

of children under 5 years of age who are stunted. Stunting, defined as having a low height for age, reflects the cumulative effects of undernutrition and infections since birth, and is an indication of poor living conditions and/or long-term restriction of a child's actual growth with respect to potential growth. Stunting particularly affects children in south Asia (29.9%), sub-Saharan Africa (26.6%), and east and southeast Asia (22.1%). At the other extreme of stunting are overweight and obesity, defined as having a high body weight relative to height. These disorders denote an excessive intake of nourishment. Fig. 34.4 shows

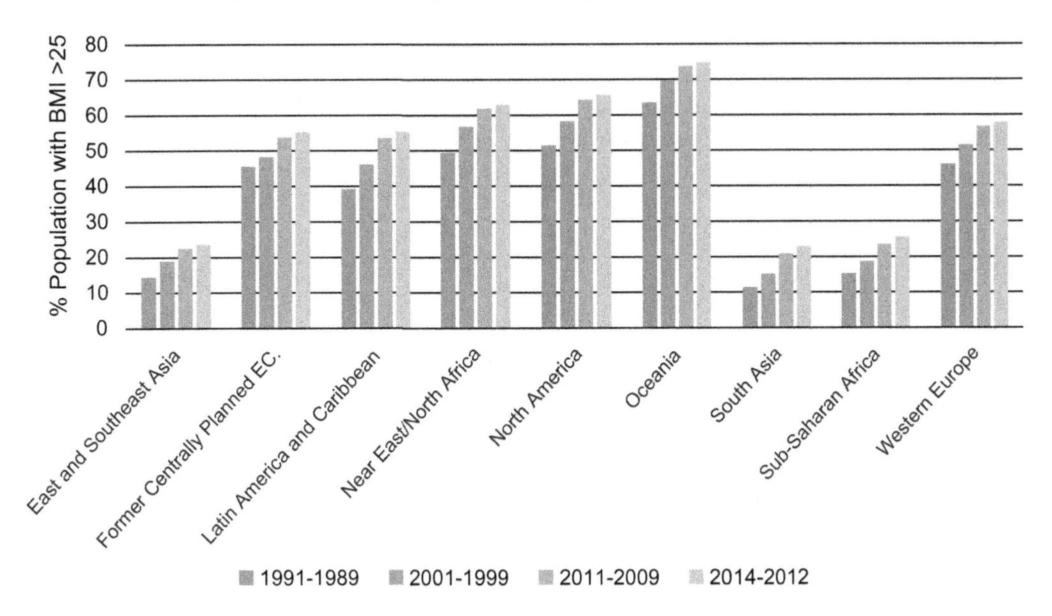

FIGURE 34.4 Proportion of overweight people by world's region. Source: *WHO, 2014. Global health observatory (GHO) data. Obesity and overweight, Geneva. WHO (2014).*

that overweight, as measured by a body mass index (BMI) greater than 25, affects large shares of the population throughout the world, from a minimum of 23.0% in south Asia up to 74.7% in Oceania. Many of the overweight are obese (BMI $>$ 30).

In spite of being to a large extent successful in providing food to a growing human population, and particularly so in the last two centuries, AFS need vast improvements if they are to fulfil food security and healthy nutrition in a satisfactory manner. The triple burden of malnutrition—a shift toward the coexistence of food insecurity, undernutrition, and overweight within the same geographies—is an emerging challenge and will become increasingly difficult as both populations and incomes grow and become concentrated in urban areas.

There are three considerations for policymakers to address when tackling the triple burden of malnutrition in order to ensure that AFS satisfy the nutritional needs of the growing human population. First, agriculture today produces more than is needed to meet the nutritional requirements of the world's population: the economic and institutional bottlenecks that prevent people from accessing food need to be given as much attention as production-related issues and constraints. Second, the development narrative around agriculture and food focuses more on undernutrition than on malnutrition, which also includes overweight and obesity. Changing this narrative, which will include stressing the differences between food demand and food needs, is fundamental for healthy diets in the future. Finally, one-third of the food produced for human consumption, about 1.3 billion tons per year, is estimated to be either lost or wasted along the value chain (FAO, 2011b). Addressing this issue is critical for AFS to properly fulfil their principal role in the coming decades.

Furthermore, AFS affect human health well beyond nutrition. Unsafe food containing harmful bacteria, viruses, parasites, or chemical substances causes more than 200 diseases, from diarrhea to cancers. About 600 million people are estimated to suffer from foodborne diseases and 420,000 die every year (WHO, 2015). Most recently, the overuse and misuse of antibiotics in veterinary medicine has been associated with the emergence of resistant bacteria, a development which makes it challenging to treat infectious diseases in both humans and animals (FAO, 2016c).

34.3 AGRIFOOD SYSTEMS, ECONOMIC GROWTH, AND LIVELIHOODS

A large share of the world's population fully or partly depends on agriculture for its livelihoods: whether for subsistence, self-employment, or wage-employment, and whether on the farm or in upstream or downstream value chains. The International Labour Organization estimates that about 30% of the world's population of working age is employed in agriculture, equivalent to 950 million people. Table 34.1 shows the global distribution of the agricultural population. In sub-Saharan Africa, 56% of the active population is engaged in agricultural production, and in South Asia and East and Southeast Asia, 46% and 30%, respectively. North America and Western Europe, in stark contrast, sees only 2% and 3%, respectively, of the active population engaged in agriculture. Asia is home to almost 70% of the global agricultural population, with another 20% living in sub-Saharan Africa.

TABLE 34.1 Employment in Primary Agricultural Production (Number and Percentage of People Employed in Agriculture; Value Added Per Agricultural Worker)

	Employment in Primary Agriculture Production (Thousands)	Total Workers Employed in Agriculture (%)	Value Added Per Agricultural Worker (US$/capita/year)
East and Southeast Asia	352,956	30	7703
Former Centrally Planned Economies	28,052	15	14,874
Latin America and Caribbean	43,373	16	11,346
Near East/North Africa	23,563	19	22,743
North America	2667	2	75,721
Oceania	2942	16	68,388
South Asia	300,044	46	5182
Sub-Saharan Africa	192,693	56	2470
Western Europe	5282	3	52,367

ILO, 2015. International Labour Organization: key indicators of the labour market (KILM). Geneva, Switzerland. ILO (2015); World Bank, 2015. Agriculture & Rural Development Indicators, Washington, DC.

Among those engaged in primary agriculture are both self-employed entrepreneurs, which includes subsistence farmers, and on-farm employees. Self-employed farmers represent the largest share of those employed in agriculture. Lowder et al. (2016) estimate the total number of farms in the world at 570 million. Most farms are located in East and Southeast Asia and South Asia, comprising over 40% and 35% of global farms, respectively. Next is sub-Saharan Africa, accounting for 9% of the world's farms. Globally, there are also some 57 million people engaged in capture fisheries and aquaculture: 75% are located in Asia, 10% in Africa, and 4% in Latin America and the Caribbean (FAO, 2016d). Self-employment in agriculture is particularly high in labor-intensive smallholder systems, and less common in extensive and capital-intensive systems. Labor-intensive smallholder systems are mostly reliant on family labor and characterized by a small but diversified resource base, including a mixture of human, social, capital, financial, and natural assets. Because of the overall low asset base, it is only through achieving a high total factor productivity in these systems that smallholders are able to generate a level of subsistence and income that meets their basic needs and livelihood support (HLPE, 2013).

Agricultural labor productivity, as measured by agricultural GDP per worker, is a common proxy for assessing how much laborers benefit from agriculture. In Asia and Africa, the labor productivity of agricultural workers is below US$10,000 per year, whereas in Western Europe and North America it is about US$55,000 and US$75,000 per year, respectively. Labor productivity, however, does not include many of the non tradable benefits agriculture provides to farmers, particularly in developing countries. Livestock farmers, for example, benefit from draft power, hauling services, and financial, insurance, and the social value of their livestock assets. Depending on the agricultural system and the institutional and legal context, farmers can use trees, animals, land parcels, or their harvest as collateral for loans, thereby deriving additional benefits from agriculture. Animal dung is a fertilizer for crops and fish ponds, and in some places has value as fuel or construction material. Agricultural by-products are a major source of fuel for cooking, heating, and lighting (RECOFT, 2015; Pica-Ciamarra et al., 2015). At the same time, however, about 70 million children (aged 5–17) are involved in hazardous work in agriculture, including exposure to pesticides and other chemicals, use of dangerous machinery, heavy loads, long hours, and hostile environments (FAO, 2017a).

On-farm wage employees represent the second category of people who directly benefit from agricultural production. Systematic global statistics on their number are not available, but some inference is possible. First, most of the 570 million farms are small, family run, and labor-intensive: 84% of farms are smaller than 2 ha and operate about 12% of the world's farmland (Lowder et al.,2016). With such a small asset base, farm operators usually have neither the resources nor the need to hire labor. The average number of permanent hired workers is in fact very small: less than one full-time worker per farm for the 55 countries for which census data are available (Lowder et al.,2016). Second, mega-farms and agroholdings of 500,000 ha or more are few and currently found only in a small number of countries, including Argentina, Australia, Brazil, China, Czech Republic, Germany (east), Poland, Romania, Russia, South Africa, and the United States (Hermans et al., 2017). Mega-farms are capital-intensive and rarely employ a significant number of people. In the United States, for example, there are about 2.1 million farms covering 370.2 million hectares, employing less than 310,000 farmworkers (USDA, 2014; BLS, 2017).

Nonfarm workers engaged in AFS—those employed upstream and downstream along the agricultural value chain—also derive direct benefits from agriculture. Available evidence indicates that in industrialized economies the number of jobs provided by agro-industries, such as in the provision of farming inputs, in food processing, marketing and retail, are considerably more than those provided by agricultural production itself. Conversely, in developing countries more people are employed in agricultural production than in agro-industries. In the United States, for example, agricultural production contributes about 1.9% to total employment, while employment in agro-industries accounts for 9.2% of total employment. In Uganda, where the active population in agricultural production amounts to about 13 million, only about 51,000 people are employed in agro-industries (UBOS, 2011). In China, there are about 200 million small farms (>0.07 ha) (Lowder et al.,2016), engaging at least 200 million people directly in agricultural production. The 2015 China Statistical Yearbook reports that around 215,000 people were employed in the wholesale of agricultural, forestry, and livestock products; about 1 million in the wholesale of food, beverages, and tobacco; 1.4 million people in supermarkets; and over 330,000 in "special retail of food, beverages and tobacco" (CSY, 2015). This totals about 3 million jobs in the post-production value chain. However, while off-farm employment along the value chain is in general beneficial, there is evidence of several safety and health hazards in agro-industry. For example, the United States Bureau of Labor Statistics reports injury and illness rates for the meat-packing industry as being 2.5 times higher than the national average. This is associated with exposure to high noise levels, dangerous equipment, slippery floors, musculoskeletal disorders, and hazardous chemicals (USDL, 2017).

Farmers in labor-intensive smallholder systems represent the largest share of those depending on agriculture for income and other benefits, with those employed in off-farm activities making up only a small percentage. As economic development progresses and agriculture intensifies, the number of people employed in agricultural production dramatically drops, with relatively few additional jobs created in agro-industry. A major challenge in the coming decades concerns the provision of sufficient jobs to meet the increased supply of labor associated with a growing human population and with people moving out of agriculture as an occupation.

There are two important considerations for policymakers in addressing this challenge. First, the intensification of agriculture must be accompanied by policies that support growth in demand for labor in economic sectors other than agriculture. The second consideration is that technological advances also provide labor-saving interventions to other sectors, such as manufacturing and service-provision through, for example, the increased use of robots and artificial intelligence. This stresses the importance of thoroughly rethinking labor market policies in the coming decades, well beyond agriculture.

Another important function of agriculture is the supply of energy. Agricultural biomass—including wood (e.g., residues from thinning), forests (fast growing trees), crops and their residues and other outputs such as farm waste —can generate biofuel, including biogas, bioethanol, and biodiesel (Awasthi et al., 2015). Estimates suggest that, by using residues and bio-organic waste, biofuel production could replace one-fourth of the global consumption of fossil fuels for transportation (Johansson et al., 2010). In the European Union, the amount of crop residues available for bioenergy represents about 3.2% of total energy consumption (Scarlat et al., 2010).

Agrifood systems also provide wood as construction material and for scaffolding, as well as for house interiors (windows, doors, floors) and furniture: available data suggests that forest products contribute between 1% and 2% of GDP, including in industrialized and developing economies (Lebedys, 2015). Today, wood is still the most important single source of renewable energy: estimates suggest that about 2.4 billion people use wood for cooking, boiling water, and heating, which makes wood the most decentralized form of energy in the world. Overall, wood fuel provides 40% of today's global renewable energy supply, as much as solar, hydroelectric, and wind power combined (FAO, 2017b). Finally, AFS provide fibers and textiles for clothing, home, and industrial use. Natural fibers represent 45% of the world's fiber consumption, with cotton being the most important (82% of all natural fiber consumption), followed by wool (6%). Other natural fibers include flax, hemp, and jute. Developing countries account for about 57% of the world's fiber consumption, with almost half being natural fibers and the rest synthetic fibers (FAO and ICAC, 2013).

34.4 AFS SOCIAL AND POLITICAL FUNCTIONS

Though AFS contribute food, raw materials, and livelihoods to human beings, many of the other benefits remain unappreciated. Social identity and cultural and community practices are often associated with food production and consumption. These are a few examples of the myriad human benefits that are linked to, but go well beyond food and livelihoods.

The poor benefit more from agricultural growth than from the growth of other sectors of the economy, though the power of agriculture to reduce poverty diminishes as economic development progresses (e.g., Bresciani and Valdés, 2007; Ligon and Sadoulet, 2007; Montalvo and Ravallion, 2009). DFID (2004) identified four major channels through which agriculture reduces poverty: (1) direct impact on farm income; (2) direct impact on people's livelihoods through the provision of affordably priced foods; (3) generation of nonfarm economic opportunities and associated supply of labor, as better-off farmers increase their demand for goods and services produced by sectors other than agriculture; and (4) promotion of economic transformation through making human and financial capital available for other sectors of the economy to develop.

People are more likely to be in good health and receive education when the agricultural sector thrives. Through the provision of calories, proteins, and micronutrients, agriculture helps the formation of human capital in early life and throughout youth. Undernutrition, including during pregnancy, can result in permanent and irreversible negative effects on brain development (Horton, 2008; Nyaradi et al., 2013), while good nutrition increases schooling performance and raises skills, both in developing and industrialized economies. A positive association between a good diet and academic performance, for example, has been found in adolescents from Canada, Chile, Iceland, the Netherlands, Norway, Sweden, and the United Kingdom (Glewwe and Miguel, 2008; Correa-Burrows et al., 2016).

Women, girls, and children directly benefit from agriculture. As about 15% of all agricultural holders are women, agriculture contributes to their well-being and helps reduce gender inequality. In some countries, such as Cape Verde, over half of all farm households are headed by women; in others, such as Jordan or Mali, less than 5% of agricultural households are women-headed (FAO, 2015a). In some countries, women are responsible for the production of over 80% of food crops. Fisheries is often a joint man-woman business, with men investing in fishing vessels, nets, and other gear and doing the fishing, and women being responsible for processing and sales. Women are often responsible for tending small ruminants (sheep and goats), poultry, and other small animals, as well as for milking dairy cows (World Bank, 2009). In many cases, however, women are disadvantaged with respect to men in accessing land, financial services, and markets. Endowing women with the same access to resources as men would increase agricultural production by 2.5%–4% in developing countries and reduce undernourishment (FAO, 2011c).

People value the environment, and agriculture provides a variety of environmental services. As evidenced with all human activities, agriculture also negatively impacts the environment. However, good farming practices generate environmental benefits, including through protecting and maintaining the quantity and quality of soil, water, air (climate change), and biodiversity. For example, leguminous plants help improve fertility and health of soils by improving nitrogen content, an important benefit as soils are essentially a nonrenewable resource because their formation is an extremely slow process. Forest and trees stabilize the soil, prevent erosion, and enhance the land's capacity to store water, moderate air and soil temperatures, and support biodiversity. Plants, through photosynthesis, remove carbon dioxide and add oxygen to the atmosphere, thereby contributing to maintaining the ozone layer (FAO, 1992; 2016e; Peoples et al., 2009). People also value culture, customs, and traditions, which are often associated with food. Agricultural landscapes have cultural heritage value. Over the centuries, human beings have created, shaped, and maintained unique agricultural systems and landscapes in different geographies using locally adapted management practices. These agricultural systems not only reflect the evolution of humankind, the diversity of knowledge, and the relationship with nature, but also demonstrate the human capacity to create resilient ecosystems that support the livelihoods of farmers, herders, forest-dependent people, and fishers in a sustainable manner (Koohafkan and Altieri, 2011; FAO, 2015b). FAO estimates that there are about 500 million hectares around the world that can be considered agricultural heritage systems, such as the ancient Ifugao rice terraces in the Philippines; the upland forestry system in the Usambara and Pare Mountains in northern Tanzania; the Oases System in the High Atlas Mountains of Morocco; and the potato system in the Chiloé archipelago in Chile (FAO, 2015b). Because of the multitude of derived services, both monetary and nonmonetary, it is difficult to assess the value for humanity of these systems. Available evidence, however, indicates people would be willing to pay fees, such as US$50 and US$33 per person per year, for ecosystem services that preserve the Chiolé archipelago and restore and preserve the Upper Paraná River floodplain, respectively (Barrena et al., 2014; Carvalho, 2007).

Society evolves and develops through better combining the different factors of production, including labor. A functional agricultural system supports a productivity-enhancing movement of labor within and across countries and continents by tackling one of the determinants of forced migrations: the lack of food (FAO, 2016a). In many cases, rural to urban migration, occurring within and across countries, is associated with a poor performance of the agricultural sector. For example, in India, the rate of rural migration is high across Bihar, Uttar Pradesh, and Rajasthan, where agricultural productivity is low and nonfarm employment opportunities are limited (Singh et al., 2011). In Nigeria, the large gap in living conditions between urban and rural areas is one of the determinants of the fast rate of urbanization (Iruonagbe, 2009). In Rwanda, the youth migrate from rural to urban areas in search of job opportunities, access to social services, and education opportunities (Mutandwa et al., 2011). Among the 185 countries with available data in 2013, 80% had policies in place to reduce rural to urban migration (UN, 2013). Official development assistance has been increasingly channeled to rural areas with the specific objective to tackle the root causes of migration, such as the European Union Emergency Trust Fund for Africa (EU, 2015).

Societies evolve better when they are politically stable and peaceful. A functional agricultural system facilitates political stability and peace. In Syria, four years of drought between 2007 and 2010 destroyed the agricultural sector and prompted massive migration to urban areas, which contributed to the outbreak of the civil war in 2011 (CEIP, 2015). In 2007/08, rises in food prices resulted in street demonstrations and riots in more than 40 countries across the world. The L'Aquila Food Security Initiative, launched by the G8 and the G20 in 2009, supported the design and implementation of programs to increase agricultural productivity across the world, including by assisting developing countries in formulating and implementing food security strategies, thereby supporting political stability (Simmons, 2017; G8, 2009).

Considering that the benefits from AFS go beyond food, nutrition, and income, this implies that sustainable investments in agriculture generate returns that accrue to society as a whole, and not just to those engaged directly in AFS. However, governments and institutions are structured around productive sectors or technical domains: industry, agriculture, environment, trade, health, transport, and so on. Each of these has very specific mandates, making it difficult for society to channel investments to optimize the multiple benefits that agriculture can provide. For example, the ministries in charge of agriculture have no mandate to reduce poverty, improve education, or promote social stability. At the same time, neither the ministry responsible for social affairs nor that in charge of the environment has responsibilities for supporting the development of agriculture.

The combination of escalating demand and environmental decline will make the nonfood and non income roles of agriculture increasingly important in the coming decades. Policy makers should make institutions accountable for these nonfood and nonincome benefits that AFS provide. This will avoid that major investments in agriculture narrowly target increased production and productivity, but explicitly address the interactions between agriculture, the environment, and public health—while also targeting poverty, empowerment of women and youth, orderly migration, stability, and other socially desirable goals.

34.5 PUTTING A VALUE ON AGRICULTURE

In valuing agriculture, it is important to appreciate its relevance for people vis-à-vis other sectors of the economy. The global value of agricultural production was estimated at US$5214 billion for the period 2010–15, contributing over 6% to the world's GDP. Eastern and Southern Asia contribute the most to global agricultural GDP, accounting, respectively, for 24% and 15% of the global value of agricultural production. Oceania contributes the least, at 0.2%. Within geographies, agriculture contributes the most to GDP in sub-Saharan Africa and South Asia (18.5% and 17.3%, respectively), with the lowest contributions being 1.5% and 1.1% in Western Europe and North America, respectively (Fig. 34.5).

Not considering fishery and forestry products, about 59% of the value of agricultural production originates in the livestock sector, with cereals coming second (32%). This holds true for all of the world's regions: the livestock sector accounts from about three-quarters of agricultural value addition in Western Europe and Latin America to 42% in Oceania. East and Southeast Asia are the most important producers in value terms of both cereals and livestock products (36.5% and 27.2% of world production, respectively). Oceania and South Asia are the largest producers, in value terms, of fruits and vegetables (34.8%) and fibers (31.9%), respectively (Fig. 34.6).

34.6 THE AGRO-INDUSTRY

The agro-industry (see Chapter 1: Food and Agricultural Systems at a Crossroads: An Overview and also Section I) includes all of the activities surrounding agricultural production, both upstream and downstream of the production unit, such as a farm, forest, or waterbody (see Fig. 32.1). Upstream, agro-industry provides agricultural inputs, such as seeds and breeding stocks; the provision and application of water; the development, manufacture provision, and marketing of fertilizers, pesticides, and veterinary drugs and vaccines; the production of farm machinery, equipment, and infrastructure; provision of extension services and plant and animal health services; provision of financial services; and the operation of government administrations, regulatory bodies, and agricultural societies and corporations.

Agricultural products are in many cases transformed before consumption, which generates additional value. This is the downstream role of the agro-industry, which includes handling, processing, preserving, transporting, and marketing agricultural food and nonfood products. It includes, for example, the manufacture of food, beverages, textiles and clothing, wood products and furniture, paper and paper products, and rubber and rubber products. There are no systematic data on the value of agro-industry across countries (Dubey and Vander Donckt, 2015), but the available evidence shows that agro-industry contributes variedly to GDP. This ranges from 2% in high-income economies, to 8% in middle-income countries, and to over 15% in low-income countries. In early stages of development, the value of primary agricultural production is higher than that of agro-industry, but as economic development progresses, agro-industry reaches comparable value to agricultural production, if not more so. While the so-called supermarket revolution is a common feature of development, it is often associated with shifting market power from primary agricultural producers to agro-industry companies, such as large processors and large retailers. Simultaneously, as most of agro-industry is characterized by low levels of technological innovation, the relative contribution of agro-industry in the economy decreases as development progresses (de Janvry, 2009; Rankin et al., 2016; UNIDO, 2013).

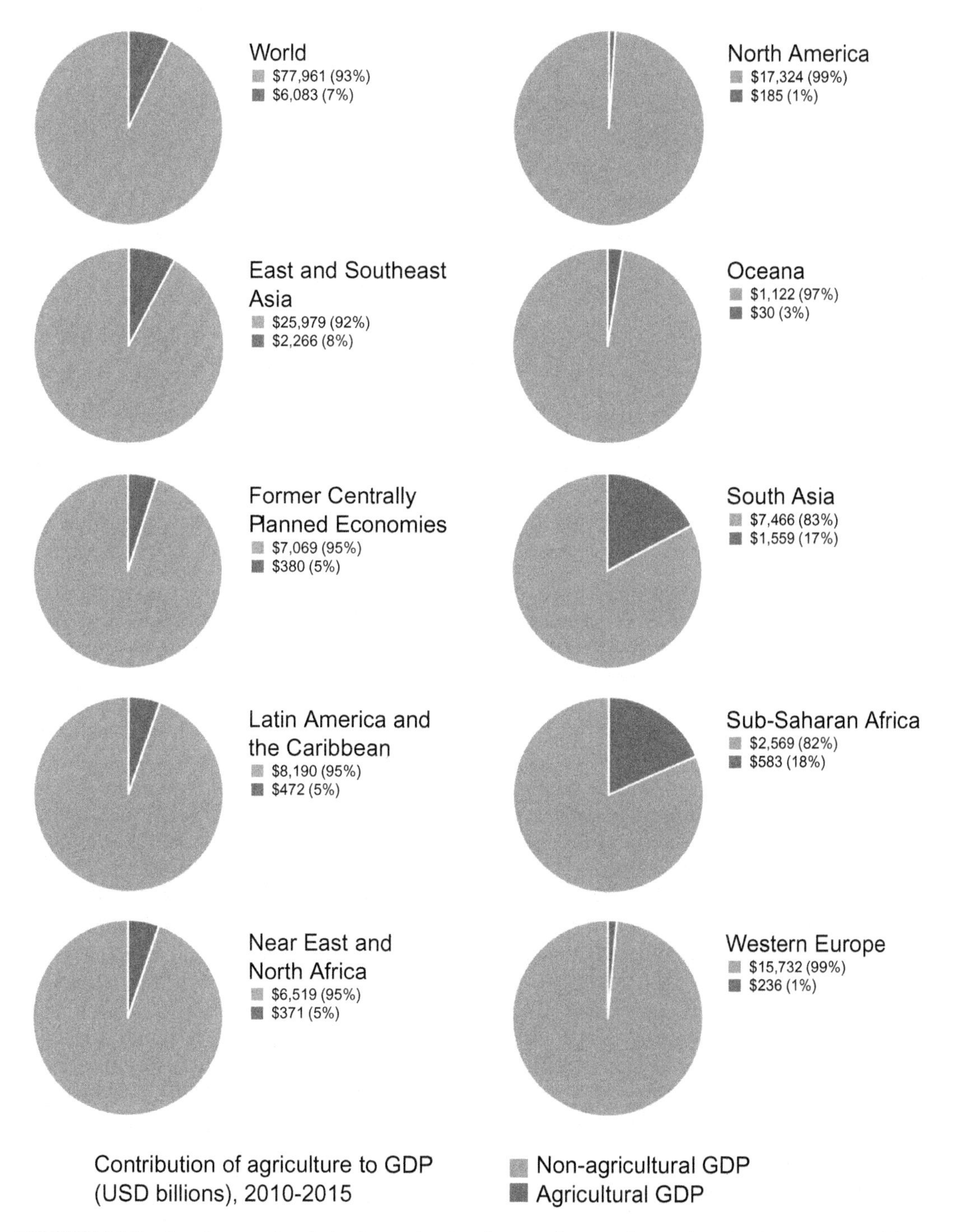

FIGURE 34.5 Set of proportional pie charts showing the contribution of agriculture to GDP in the world and in different geographies (with indication of absolute values). Source: *World Bank, 2015. Agriculture & Rural Development indicators, Washington, D.C. World Bank (2015).*

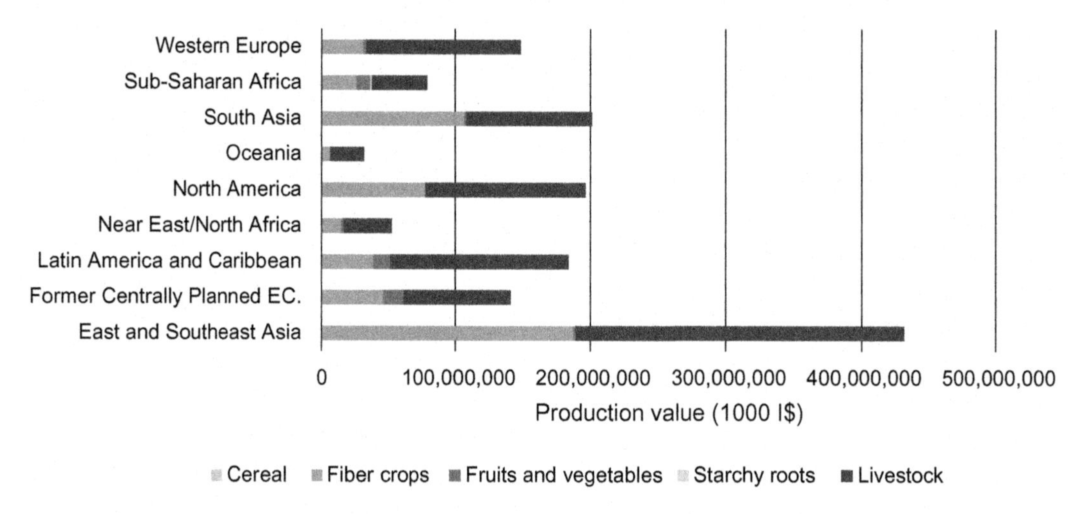

FIGURE 34.6 Set of bar charts showing the contribution of different agricultural subsectors to agricultural value added in the world and in different geographies. Source: *FAOSTAT, 2010. FAOSTAT Database. FAO Rome. FAOSTAT (2010).*

Agricultural products are traded, which generates value. The annual value of agricultural trade—including food, vegetables, animal products, oils, fats, tobacco, and beverages—amounts to around US$1.5 trillion: close to 10% of the value of all international trade (UNCTAD, 2015). Raw products account for about 45% of the value of agricultural trade, while semiprocessed and processed products account for about 30% and 25%, respectively. Asia and the Pacific, and Africa and the Near East are net importers of all major food items, while North America, Europe, and Latin America are net exporters. Trade flows, therefore, are critical to support food security and development in several developing countries and will prove increasingly so in the coming decades (FAO, 2015c). However, trade is often distorted because of subsidies to agricultural production as well as trade barriers, largely in the form of tariffs, but also in the form of sanitary and other standards. Members of the World Trade Organization (WTO), therefore, since 2000, have been engaging in talks to reform agricultural trade. In 2015, at the 10th WTO Ministerial Conference in Nairobi, Kenya, WTO members agreed on a landmark decision to eliminate agricultural export subsidies, the most important reform of international trade rules in agriculture since the WTO was established. Export subsidies will be eliminated by developed countries immediately, except for a handful of agriculture products, while developing countries have longer periods to phase out subsidies. At the conference, it was also agreed to continue negotiations on a special safeguard mechanism that would allow developing countries to temporarily raise tariffs on agriculture products in cases of import surges or price falls. These decisions will help to level the playing field for farmers around the world, particularly those in poor countries who cannot compete with rich countries that artificially boost their profits through subsidies.

The value addition of primary agricultural production and agro-industry represents the monetary well-being that human beings derive from agriculture. Basically, it is calculated as the output in value terms of the sector minus the value of all intermediate inputs, without making deductions for depreciation of fixed assets and depletion/degradation of natural resources. It does not consider the non monetary contribution of agriculture to household livelihoods, as well as the many other benefits the sector provides to society. In other words, the value of agriculture, as calculated in national accounts, only represents a small share of the benefits that humankind derives from agriculture.

A major challenge for policymakers is to better value the current and potential contribution of agriculture to society. Returns on investments, in fact, are one of the criteria used to allocate public (and private) resources. Such returns, in the case of agriculture, are largely underestimated, which ultimately reduces the allocation of resources to the sector and lowers its contribution to societal well-being. This will become a major issue in the coming decades when a larger and growing affluent and urbanized population will rely increasingly on both the tradable and nontradable benefits of agriculture in order to ensure a sustainable future for society and to achieve the Sustainable Development Goals.

References

Awasthi, P., Shrivastava, S., Kharkwal, A., Varma, A., 2015. Biofuel from agricultural waste: a review. Int. J. Curr. Microbiol. Appl. Sci. 4, 470–477.

Barrena, J., Nahuelhual, L., Baez, A., Schiappacasse, I., Cerda, C., 2014. Valuing cultural ecosystem services: agricultural heritage in Chiloe Island, southern Chile. Ecosyst. Serv. 7, 66–75.

BLS, 2017. US Bureau of Labour Statistics.

Bresciani, F., Valdés, A., 2007. Beyond Food Production: The Role of Agriculture in Poverty Reduction. FAO, Rome.

Carvalho, A.R., 2007. An ecological economics approach to estimate the value of a fragmented wetland in Brazil (Mato Grosso do Sul state). Braz. J. Biol. Rev. Bras. Biol. 67, 663–671.

CEIP, 2015. Food Insecurity in War-Torn Syria: From Decades of Self-Sufficiency to Food Dependence.

CFS, 2011. The Price Volatility and Food Security. A report by The High Level Panel of Experts on Food Security and Nutrition. FAO, Rome.

Correa-Burrows, P., Burrows, R., Blanco, E., Reyes, M., Gahagan, S., 2016. Nutritional quality of diet and academic performance in Chilean students. Bull. World Health Organ. 94, 185–192.

CSY, 2015. China Statistical Yearbook-2015. China Statistics Press.

Curtis, V., Cairncross, S., 2003. Effect of washing hands with soap on diarrhoea risk in the community: a systematic review. Lancet Infect. Dis. 3, 275–281.

de Janvry, A., 2009. Agriculture for Development - Implications for Agro-industries, Agro-Industries for Development. FAO and UNIDO, pp. 252–270.

de Janvry, A., Dethier, J.-J., 1985. Technological Innovation in Agriculture: The Political Economy of Its Rate and Bias. World Bank, Washington D.C.

DFID, 2004. Agriculture, Growth and Poverty Reduction. DFID, London.

Dubey, S., Vander Donckt, M., 2015. The FAO-UNIDO Agro Industry Measurement (AIM) Database, Presentation delivered at the FAO-UNIDO Expert Group Meeting on Agro-Industry Measurement. 23–24 November, Rome.

EU, 2015. Emergency Trust Fund for stability and addressing root causes of irregular migration and displaced persons inAfrica, in: Document, S.O. (Ed.), Bruxelles, EU.

FAO, 1992. Forests, Trees and Food. FAO, Rome.

FAO, 2011a. Committee on World Food Security (CFS) 37th Session. FAO, Rome.

FAO, 2011b. Global Food Losses and Food Waste. FAO, Rome.

FAO, 2011c. The State of Food and Agriculture, 2010–2011. FAO, Rome.

FAO, 2013. World Livestock 2013 - Changing Disease Landscapes. FAO, Rome.

FAO, 2015a. Accenting the 'Culture' in Agriculture. FAO, Rome.

FAO, 2015b. Gender and Land Rights Database. FAO, Rome.
FAO, 2015c. The State of Agricultural Commodity Markets Trade and Food Security: Achieving a Better Balance Between National Priorities and the Collective Good. FAO, Rome.
FAO, 2016a. Food Security Indicators. FAO, Rome.
FAO, 2016b. Migration, Agriculture and Rural Development. Addressing the Root Causes of Migration and Harnessing its Potential for Development. FAO, Rome.
FAO, 2016c. State of the World Fisheries and Aquaculture. Contributing to Food Security and Nutrition for All, State of the World Fisheries and Aquaculture. FAO, Rome.
FAO, 2016d. The FAO Action Plan on Antimicrobial Resistance 2016–2020. FAO, Rome.
FAO, 2016e. The State of Food and Agriculture - Climate Change, Agriculture and Food Security, Rome.
FAO, 2017a. FAO Guidance Note: Child Labour in Agriculture in Protracted Crises, Fragile and Humanitarian Contexts. FAO, Rome.
FAO, 2017b. Forests and Energy. Infographics. FAO, Rome.
FAO, IFAD, WFP, 2015. The State of Food Insecurity in the World 2015, Meeting the 2015 International Hunger Targets: Taking Stock of Uneven Progress. FAO, Rome.
FAO/ICAC, 2013. World Apparel Fiber Consumption Survey. FAO, Rome, International Cotton Advisory Committee, Washington D.C.
FAOSTAT, 2010. FAOSTAT Database. FAO, Rome.
G8, 2009. L'Aquila" Joint Statement on Global Food Security. L'Aquila Food Security Initiative (AFSI), L'Aquila, 10 July 2009.
Glewwe, P., Miguel, E.A., 2008. The impact of child health and nutrition on education in less developed countries. Handb. Dev. Econ. 49, 3561–3606.
Gray, A., 2016. Which countries spend the most on food? This map will show you. World Economic Forum, Agriculture, Food and Beverage website.
Henneberg, M., Henneberg, R., Carter, J.C., 1992. Health in Colonial Metaponto. Natl. Geogr. Res. Explor. 8, 446–459.
Hermans, F.L.P., Chaddad, F.R., Gagalyuk, T., Senesi, S., Balmann, A., 2017. The emergence and proliferation of agroholdings and mega farms in a global context. Int. Food Agribusiness Manage. Rev. 20, 175–185.
HLPE, 2013. Investing in smallholder agriculture for food security. A report by The High Level Panel of Experts on Food Security and Nutrition of the Committee on World Food Security, Rome.
Horton, R., 2008. Maternal and child undernutrition: an urgent opportunity. Lancet 371, 179-179.
IFPRI, 2016. Global Nutrition Report 2016: From Promise to Impact: Ending Malnutrition by 2030. International Food Policy Research Institute (IFPRI), Washington, D.C.
ILO, 2015. International Labour Organization: key indicators of the labour market (KILM). Geneva, Switzerland.
Iruonagbe, C.T., 2009. Rural-urban migration and agricultural development in Nigeria. Arts Soc. Sci. Int. Res. J. 1, 28–49.
Johansson, K., Liljequist, K., Ohlander, L., Aleklett, K., 2010. Agriculture as provider of both food and fuel. Ambio 39, 91–99.
Koohafkan, P., Altieri, M.A., 2011. Globally Important Agricultural Heritage Systems. A Legacy for the Future. FAO, Rome.
Koppel, B., 1994. Induced Innovation Theory and International Agricultural Development: A Reassessment. Johns Hopkins University Press, Baltimore.
Lebedys, A., 2015. Forest products contribution to GDP. Presentation delivered at the 37th Joint Working Party on Forest Statistics, Economics and Management, Geneva, Switzerland.
Ligon, E., Sadoulet, E., 2007. Estimating the effects of aggregate agricultural growth on the distribution of expenditures. World Bank, Washington, D.C., pp. 1–24.
Livi Bacci, M., 2012. A Concise History of World Population. John Wiley & Sons, Chichester.
Lowder, S.K., Skoet, J., Raney, T., 2016. The number, size, and distribution of farms, smallholder farms, and family farms worldwide. World Dev. 87, 16–29.
Montalvo, J.G., Ravallion, M., 2009. The Pattern of Growth and Poverty Reduction in China, Policy Research Working Paper. World Bank, Washington, D.C.
Mundlak, Y., Larson, D.F., 1990. On the Relevance of World Agricultural Prices, Policy, Research, and External Affairs Working Papers. World Bank, Washington, D.C., p. 36.

Mutandwa, E., Kanuma Taremwa, N., Uwimana, P., Gakwandi, C., Mugisha, F., 2011. An analysis of the determinants of rural to urban migration among rural youths in northern and western provinces of Rwanda. Rwanda J. 22.

Nyaradi, A., Li, J., Hickling, S., Foster, J., Oddy, W.H., 2013. The role of nutrition in children's neurocognitive development, from pregnancy through childhood. Front. Hum. Neurosci. 7, 97.

Peoples, M.B., Brockwell, J., Herridge, D.F., Rochester, I.J., Alves, B.J.R., Urquiaga, S., et al., 2009. The contributions of nitrogen-fixing crop legumes to the productivity of agricultural systems. Symbiosis 48, 1–17.

Pica-Ciamarra, U., Tasciotti, L., Otte, J., Zezza, A., 2015. Livestock in the household economy: cross-country evidence from microeconomic data. Dev. Policy Rev. 33, 61–81.

Rankin, M., Kelly, S., Galvez-Nogales, E., Dankers, C., Ono, T., Pera, M., et al., 2016. The transformative power of agrifood industry development: policies and tools for restructuring the agricultural sector towards greater added value and sustainable growth, ESA Conference on Rural Transformation, Agricultural and Food System Transition: Building the evidence base for policies that promote sustainable development, food and nutrition security and poverty reduction. FAO, Rome, Italy (2016-09-19 - 2016-09-20).

RECOFT, 2015. Trees as Loan collateral: valuation methodology for smallholder teak plantations, Working Paper. The Centre for People and Forest, Bankgok.

Scarlat, N., Martinov, M., Dallemand, J.F., 2010. Assessment of the availability of agricultural crop residues in the European Union: potential and limitations for bioenergy use. Waste Manage. 30, 1889–1897.

Sen, A., 1976. Famines as failures of exchange entitlements. Econ. Polit. Wkly. 11, 1273. 1275, 1277, 1279-1280.

Simmons, E., 2017. Recurring Storms. Food Insecurity, Political Instability, and Conflict. Center for Strategic & International Studies, Washington D.C.

Singh, N.P., Singh, R.P., Kumar, R., Padaria, R.N., Singh, A., Varghese, N., 2011. Labour migration in Indo-Gangetic plains: determinants and impacts on socio-economic welfare. Agric. Econ. Res. Rev. 449–458.

Toynbee, A., 1884. The industrial revolution. The Beacon Press, Boston.

UBOS, 2011. Uganda Bureau of Statistics Report on the Census of Business Establishments, 2010/11. Uganda Bureau of Statistics (UBOS), Kampala.

UN, 2013. World Population Policies 2013. Population Division of the Department of Economic and Social Affairs of the United Nations Secretariat, New York.

UN, 2017. Revision of World Population Prospects. Population Division of the Department of Economic and Social Affairs of the United Nations Secretariat, New York.

UNCTAD, 2015. Key Statistics and Trends in International Trade 2015. United Nations, Geneva.

UNIDO, 2013. The structure and growth pattern of agro-industry of African countries. Development Policy, Statistics and Research Branch Working Paper 9/2012. UNIDO, Vienna.

USDA, 2014. Census of Agriculture. Preliminary Report Highlights. USDA, Washington D.C, 2012.

USDL, 2017. Safety and Health Topics: Meatpacking. Occupational Safety and Health Administration. United States Department of Labor, Washington D.C.

WHO, 2014. Global health observatory (GHO) data. Obesity and overweight, Geneva.

WHO, 2015. Food Safety Fact Sheet. WHO, Geneva.

World Bank, 2009. Gender in Agriculture Sourcebook. The World Bank, Food and Agriculture Organization, and International Fund for Agricultural Development, Washington, DC.

World Bank, 2015. Agriculture & Rural Development indicators, Washington, DC.

CHAPTER

35

Natural Resource and Environmental Dimensions of Agrifood Systems

35.1 INTRODUCTION

Resource use and environmental impacts of agrifood systems (AFS) are significant. In general, of all economic activities, the agrifood sector makes the greatest demands on natural resources and is the most important driver of global environmental change. This chapter assesses the current status and dynamics of natural resource use and environmental impacts of AFS.

AFS are critically dependent on natural resources and functioning of global and local ecosystems and they cannot function if ecological integrity, the ability to maintain healthy functioning, is compromised beyond certain thresholds. The capacity to produce food is only one of many food and agriculture-related services; they also provide a range of regulating, supporting, and cultural ecosystem services (MA, 2005). They also rely on services provided by natural ecosystems, including pollination, biological pest control, maintenance of soil structure and fertility, nutrient cycling, and hydrological services (see also Chapter 34: Socioeconomic Dimension of Agrifood Systems and Section I). Food production is critically dependent on climate, land, energy, water, genetic resources, and minerals. Many of these resources are, in principle, renewable and, given proper management, can be used for centuries or more as they are naturally replenished or regenerated. Others, such as minerals and fossil fuels, are not renewable and the potential of those resources to provide a basis for AFS is finite. Renewable natural resources are subject to biological and physical thresholds beyond which irreversible changes in benefit provision occur (Wentworth, 2011).

Many of the natural resources on which AFS depend—including climate, energy, land, water, nutrients, biodiversity, and genetic diversity—are being depleted and degraded at growing rates. The 20th century was characterized by an unprecedented growth in

Sustainable Food and Agriculture
DOI: https://doi.org/10.1016/B978-0-12-812134-4.00035-2

population and in the size of the global economy; global population quadrupled to 6.4 billion and economic output, as measured by GDP, grew more than 20-fold (Maddison, 2001). The amount of materials extracted, harvested, and consumed worldwide increased by 60% since 1980, reaching nearly 62 billion million tons per year in 2008 (OECD, 2015; Fig. 35.1). From 1980 to 2005, both the world population and the extraction of agricultural biomass for food and feed increased by 50% (Krausmann et al., 2013).

The use of natural resources in production and consumption processes has many environmental, economic, and social consequences that extend beyond geographical borders and affect future generations. Large environmental pressures are associated with the extraction, processing, transport, use, and disposal of materials (e.g., pollution, waste, habitat loss), and their effects on environmental quality (e.g., air, climate, water, soil, biodiversity, landscape), ecosystem services, and human health. An estimated 60% of the ecosystem services that support life on Earth are being degraded or used unsustainably, with many of these changes caused in part by current and past management of land for food, fiber, and timber (MA, 2005). Estimates point to the extent of human transformation of the terrestrial surface and associated ecosystem functions that have already crossed planetary boundaries for climate, biodiversity, and biochemical cycles. Four of the nine planetary boundaries that have been exceeded relate to food systems, including climate change, loss of biosphere integrity, land-system change, and altered biogeochemical cycles (Steffen et al., 2015; Box 35.1).

Competition for natural resources with nonagricultural sectors further aggravates resource scarcity. Salinization and pollution of water courses and bodies, and degradation of water-related ecosystems are rising. In many large rivers, only 5% of former water

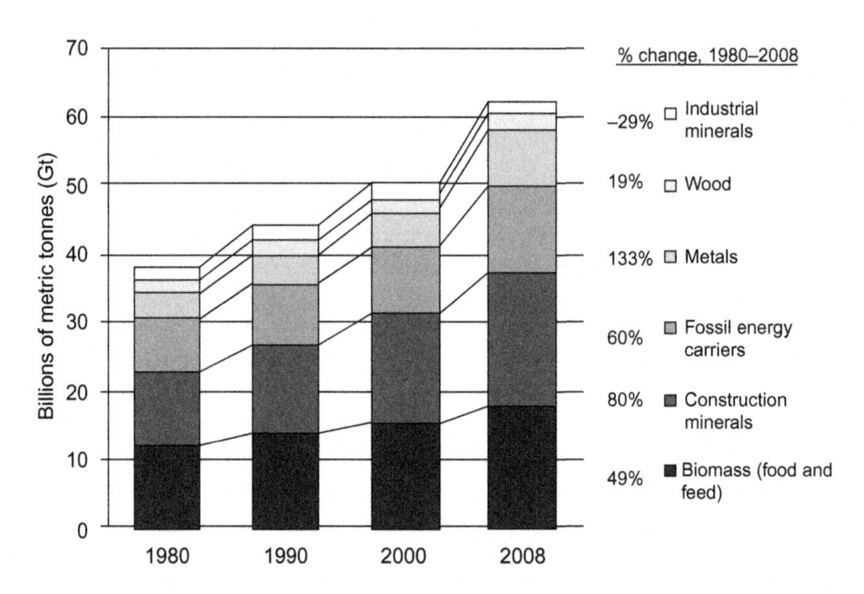

FIGURE 35.1 Global material resource extraction. Source: *OECD, 2015. Material Resources, Productivity and the Environment. OECD, Paris.*

BOX 35.1

PLANETARY BOUNDARIES

Recent research shows that four of the nine planetary boundaries have already been crossed as a result of human activity (Steffen et al., 2015). These four boundaries are climate change, change in biosphere integrity, land-system change, and altered biochemical nitrogen and phosphorus cycles. The suggested boundary of 350 parts per million (ppm) of carbon dioxide (CO_2) was crossed several years ago, and the current level is around 400 ppm. According to certain estimates, the planet was ice free at a CO_2 concentration above 450 ppm (Hansen et al., 2008). With the currently increasing CO_2 levels, there is already an increase in the shift in weather patterns, wherein drier regions are becoming even drier. Similarly, water availability and food security are already problematic in several regions. Ocean acidification and biochemical nitrogen and phosphorus cycle inflows into the oceans are reducing these carbon sinks' capacity to absorb the rising levels of CO_2. Global anthropogenic manipulations of freshwater resources have resulted in the drying of an estimated 25% of the world's river basins. Thus, climate change, ocean acidification, and the biochemical nitrogen and phosphorus cycles are considered as three different, yet interdependent, planetary boundaries. The land-system change that occurs primarily as a result of agricultural expansion and intensification contributes to global environmental change.

Figures (A) to (D) show the distributions and current status of the control variables for four of the boundaries where subglobal dynamics are critical: biogeochemical cycles, land-system change, and freshwater use.

(A) **Phosphorus**

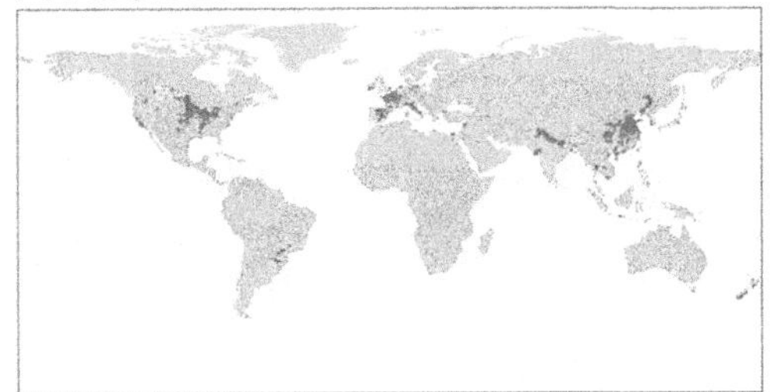

(B) **Nitrogen**

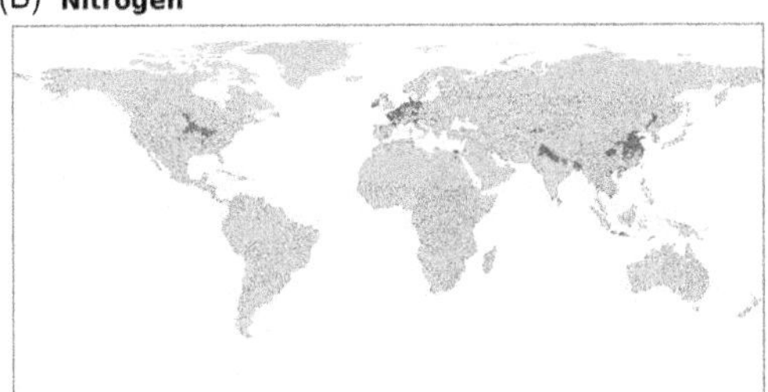

(C) **Land-system change**

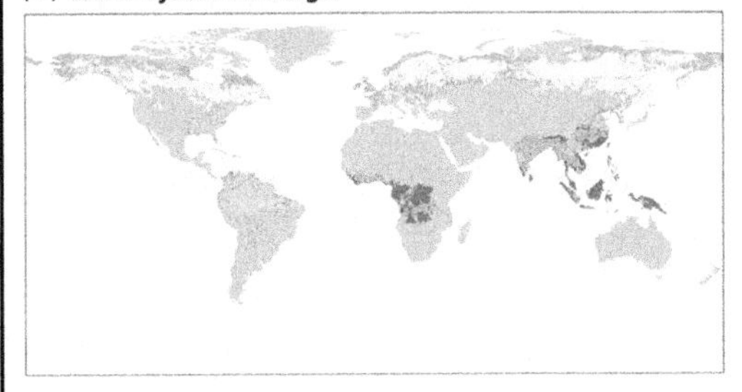

(D) **Freshwater use**

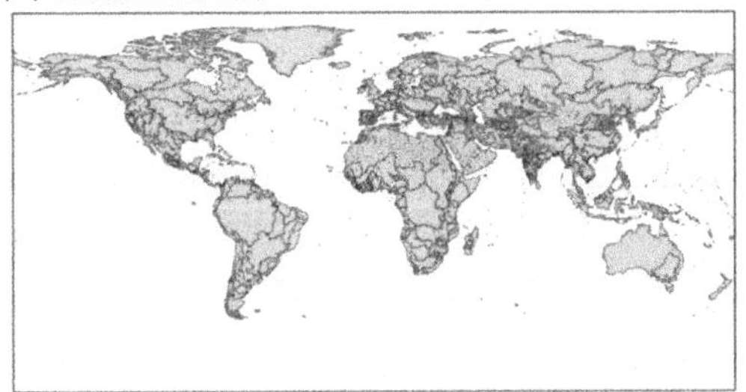

■ Beyond zone of uncertainty (high risk) In zone of uncertainty (increasing risk) Below boundary (safe)

BOX 35.1 (*cont'd*)

In each panel, green areas are within the boundary (safe), yellow areas are within the zone of uncertainty (increasing risk), and red areas are beyond the zone of uncertainty (high risk). Gray areas in (A) and (B) are areas where phosphorus and nitrogen fertilizers are not applied; in (C), they are areas not covered by major forest biomes; and in (D), they are areas where river flow is very low so that environmental flows are not allocated.

Operating spaces may appear to be safe at the broader scales because average values mask specific situations, and vice versa. Such a misinterpretation is revealed when processes are observed at higher scales. Examples of the negative case can be detected in Argentina when the analysis scales down from the national to the regional and the local scale (Viglizzo and Frank, 2006). While critical boundaries were not apparently transgressed, cases of irreversible shift can be detected at smaller scales in important farming areas of Argentina.

Cases of critical boundary transgression at different spatial scales in agroecosystems in Argentina

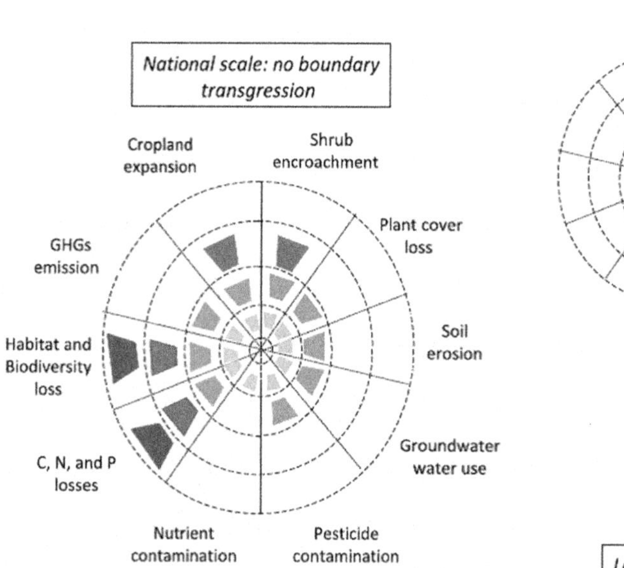

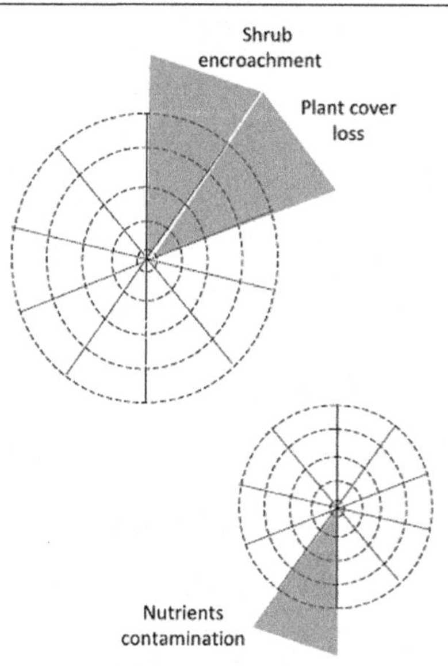

Source: *Derived from Steffen, W., Richardson, K., Rockstrom, J., Cornell, S.E., Fetzer, I., Bennett, E.M., et al., 2015. Planetary boundaries: Guiding human development on a changing planet. Science 347, 11, and UNEP, 2014. Assessing global land use: balancing consumption with sustainable supply. A Report of the Working Group on Land and Soils of the International Resource Panel. In: Bringezu, S., Schütz, H., Pengue, W., O'Brien, M., Garcia, F., Sims, R., Howarth, R., Kauppi, L., Swilling, M., Herrick, J. (UNEP, 2014). For interpretation of the references to color in this figure legend, the reader is referred to the web version of this book.*

volumes remain in-stream. Large lakes and inland seas have shrunk, and half the wetlands of Europe and North America no longer exist. Runoff from eroding soils is filling reservoirs, reducing hydropower and water supply. Groundwater is being intensively overpumped, and aquifers are becoming increasingly polluted and salinized in some coastal areas. Competition for water from domestic and industrial users is growing fast, and many regions face water scarcity with reduced quantities available for irrigation. In water-scarce regions such as the Middle East and North Africa, strong competition is reducing water available for agriculture, as observed in Saudi Arabia, for example (FAO, 2011a).

In addition, land and water use for food production regularly competes with other ecosystem services. Large parts of all continents are experiencing high rates of ecosystem degradation, particularly reduced soil quality, biodiversity loss, and harm to amenity and cultural heritage values. The steady increase in bioenergy production further contributes to the competition with AFS for biomass, land, and water resources (Box 35.2).

Poorly managed natural resources form a growing social and economic threat and foment instability. In the last 60 years, competition for natural resources has played a key

BOX 35.2

THE MULTIPLE DIMENSIONS OF RESOURCE SCARCITY

Demographic shifts, environmental pressures, and a rapidly changing global economy are exacerbating scarcities and shaping resource politics. So far, investment and technological advancements have attenuated the increasing competition for critical resources. Increasingly, there are efforts to address scarcity through improving resource use efficiency, closing nutrient loops, and developing of alternatives to scarce resources. Biofuels are an example of such a substitution.

Given the important linkages and feedback loops across multiple types of resources, resource scarcity can be seen as a "nexus." The food–energy–water nexus is a commonly cited example. High oil prices, for example, tend to lead to high food prices, as costs increase for fertilizer, for on-farm energy use, and for processing and transportation. They can also intensify competition for land, for instance, as biofuels become more cost-effective as an alternative to fossil fuels. And they can mean higher water prices because of the high energy requirements of water pumps, desalination plants, and purification systems. This dimension represents the "biophysical" interdependence of resources, which implies that it is critical to explore unintended consequences of different policy choices (such as biofuel subsidies) or support to the development and deployment of new technologies.

Beyond the biophysical linkages, natural resource scarcity is manifest through markets and trade flows of raw materials. Less attention has been paid to how resource scarcity is shaping trade patterns and intensifying economic interdependencies. The volume of natural resources traded globally has increased over 60% since the turn of the century (Chatham House, 2018). International trade

BOX 35.2 (*cont'd*)

redistributes resources across the globe, allowing some countries to export resources and to raise revenues and other countries to increase their supply of raw materials and products. Some countries are becoming increasingly reliant on foreign sources for resources such as land, water, and nutrients. Global markets offer the opportunity to trade "virtual resources," such as virtually embedded land and water, in traded commodities. This allows countries with scarce resources to effectively import from countries with abundant resources. In 2009, Saudi Arabia abandoned its policy on wheat self-sufficiency in favor of imports. The driver of this policy change was a strong concern over the depletion of the country's scarce fossil water reserves, as wheat production was entirely irrigated. FAOSTAT data on Saudi wheat production shows a gradual decline to 660,000 tons in 2013, from 1.2 million tons in 2009, and a corresponding increase in wheat imports by 75% over the same period.

A third dimension of resource scarcity is political. As the value and competition for natural resources increases, governments may use restrictive policy instruments (bans, taxes, quotas, and subsidies), trade restrictions, and investment policies to promote new technologies with the aims of protecting domestic production, protecting producers from competition, and protecting consumers from high prices. For example, in 2010, Russia's export ban on wheat in response to the heat wave and drop in grain production drove up international prices, giving rise to the initial protests in North Africa that became the Arab Spring. Export controls can suppress domestic agricultural investment, reduce output and have long-term consequences for sector growth and food security. Some policy support mechanisms may lead to further environmental degradation and depletion. This is the case in Argentina, which imposed taxes on agricultural products' exports: soy 35%, sunflower seed 32%, wheat 23%, corn and sorghum 20%, and beef 15% (Regúnaga and Tejeda Rodriguez, 2015).

role in at least 40% of all intrastate conflicts (UNEP, 2009). A changing climate is further exacerbating such tensions.

As competition for resources increases so do the costs of resource degradation. For example, the annual costs of degradation of arable and grazing land at the global level have been estimated at US$300 billion (Nkonya et al., 2016). Annual global loss of ecosystem services due to hypoxia, including damage to fisheries from coastal nitrogen (N) and phosphorus (P) pollution alone, costs US$170 billion (Sutton et al., 2013).

Production of food principally takes place in managed ecosystems (agroecosystems), both terrestrial and aquatic. Managed arable and grazing lands have become one of the largest terrestrial biomes on the planet, rivaling forest cover in extent and occupying 40% of the land surface (Foley et al., 2005).

Geographically, AFS are embedded in complex, diverse environments such as natural grasslands, forests, mountains, hills, wetlands, coastal areas, infrastructure, and human settlements of all sizes. Production occurs in a wide range of ecosystems, from those relatively undisturbed, such as seminatural forests, through food-producing landscapes with mixed patterns of human use, to ecosystems that are fully modified and managed by humans. These agroecosystems include polycultures, monocultures, aquaculture, rangelands and pastures, fallow lands, and mixed systems, such as crop-livestock and crop-aquaculture systems, agroforestry, and agrosilvopastoral systems (see Chapter 33, Agrifood Systems). Mobile pastoral livestock systems span a diversity of landscapes, from the dry rangelands of Africa to the steppes of central Asia. Artisanal fisheries are found along rivers, lakes, estuaries, coastal waters, and the open sea in temperate, tropical, and arctic zones.

Humanity has invested substantial effort into engineering ecosystems to produce services such as food, timber, and fodder, but commonly this has interfered with other ecosystem services such as flood control, regulation of soil and water quality, carbon sequestration, and biodiversity. The Millennium Assessment (MA, 2005) reported that approximately 60% (15 out of 24) of services were being degraded or unsustainably used as a consequence of agricultural management and other human activities.

The depletion and degradation of natural resources together with damage to ecosystem functioning undermine the productive capacity of AFS and their ability to provide food security. We are now drawing on natural stocks more than ever and degrading ecosystems, in some cases, irreversibly. The crucial role of AFS in the degradation or depletion of natural resources is evident:

- Thirty percent of total global land is degraded (Nkonya et al., 2016) and 33% of soils are moderately to highly degraded due to erosion, nutrient depletion, acidification, salinization, compaction, and chemical pollution (FAO, 2015). Each year, about 24 billion tons of fertile soils are lost due to erosion, and 12 million hectares (ha) of land are degraded through drought and the encroachment of the desert (UNCCD, 2011).
- Sixty-one percent of "commercial" fish populations are fully fished, and 29% are fished at a biologically unsustainable level and therefore overfished (FAO, 2014).
- At least 20% of the world's aquifers are overexploited, including those located in important production areas such as the upper Ganges Delta (India) and California (United States) (Gleeson et al., 2012).
- Sixty percent of global terrestrial biodiversity loss is related to food production (UNEP, 2016).
- Seventy-five percent of the genetic diversity of agricultural crops has been lost (FAO, 2004).
- Nearly 100 livestock breeds have gone extinct between 2000 and 2014, and 17% (1458) of the world's farm animal breeds are currently at risk of extinction (FAO, 2017).
- During the last century, about 50% of the wetlands, 40% of the forests, and 35% of the mangroves have been lost (TEEB, 2011).

Most AFS are not efficient resource users. Of the total input of N and P fertilizers, only 15%–20% ends up in the food that reaches consumers' plates, with the bulk of nutrients lost to the environment. Some regions have lower efficiency and higher losses (e.g., North America and East Asia), while in sub-Saharan Africa soil nutrient depletion (where nutrient extraction is higher than input) is common (UNEP, 2016).

Globally, agriculture, forestry, and other land uses (AFOLU) account for around 21.5% of global greenhouse gas (GHG) emissions. Carbon dioxide and methane account for 49% and 30%, respectively, of the emissions generated by AFOLU. Nitrous oxide emissions from AFOLU account for as much as 75% of global anthropogenic emissions of the gas (FAO, 2016c). Most managed agricultural lands, both arable and grassland, continue to lose carbon.

The impact of AFS on natural resources varies across regions. In some locations, land degradation and biodiversity losses are the major issues, while in others high nutrient losses leading to declines in air and water quality are of greater concern.

This overexploitation is alarming, not just in terms of degradation, but also in terms of loss of ecosystem productivity and resilience, which reduce production potential and the capacity to recover following major disturbances. These environmental impacts usually feedback harmfully on the renewable resources needed for both AFS and non-AFS activities. An example of the former is the impact of AFS on water quality, which makes water less suitable for irrigation purposes. An example of the latter is the effect of pollution from agricultural sources on drinking water quality. The feedbacks are sometimes very local with impacts occurring within a short time frame, for example, water pollution, whereas in other cases the feedbacks are through global systems with a time horizon of decades or centuries, for example, GHG emissions leading to climate change.

Over recent decades, efforts have been made to improve resource use efficiency in AFS. For example, since 1990 the rate of net forest loss went down by more than half (from 7.3 million ha per year in 1990 to 3.3 million ha per year in 2015 (FAO, 2015). One reason for the change is that countries in South and Central America and in Asia have better managed to enforce laws and monitor progress. Supply chain interventions for beef and soybean, such as the moratorium on soybean production in the Brazilian Amazon, have contributed to Brazil alone reducing annual losses of carbon (C) in above- and below-ground biomass from 240 million tons per year in the 1990s to about 80 million tons per year between 2010 and 2015. The Indonesian government has established a forest policy, renewing a moratorium that prohibits new licenses to clear forests in certain primary forests and peatlands. Slower deforestation reduced emissions from 4.6 to 3.8 gigatons of carbon dioxide equivalent ($GtCO_2$-eq) per year in the 1990s and 2000s to 3.7 $GtCO_2$-eq per year in 2010 (Tubiello et al., 2014).

35.2 THE ROLE OF AGRIFOOD SYSTEMS IN C CYCLES AND ENERGY USE

The C Cycle

The C cycle is the exchange of C among three reservoirs: land, oceans, and the atmosphere. Active C is that which stays within a reservoir less than a thousand years or so. More than 90% of global C is stored and cycled through aquatic systems, including oceans and coastal waters. The land, including its plants and animals, commonly referred to as the "terrestrial biosphere," is the next largest reservoir of active C (about 2500 gigatons of C). Up to 80% of total organic C in the terrestrial biosphere is stored in soils, with about

20% stored in vegetation (IPCC, 2000). Soils are critically important in determining global C cycle dynamics because they serve as the link between the atmosphere, vegetation, and the oceans. The atmosphere is the smallest pool of actively cycling C.

The C cycle enables life on Earth through photosynthesis and respiration, allowing the Earth to produce food and other renewable resources. Actively cycling C in its three reservoirs affects human life every day: C in the atmosphere, in the presence of sunlight, serves as food for plant growth; C in the soil provides energy for the growth of microbes; C in plants feeds humans and other animals; and C absorbed by the ocean is used by marine animals in several biological and chemical processes.

Solar energy is converted into chemical energy through photosynthesis in plants, a process that powers nearly all ecosystems. It is estimated that about 1% of solar energy reaching the Earth drives primary production, the rest being reflected by the atmosphere before reaching the Earth. Of this 1%, plants and other primary producers capture less than 3%. Plants and algae take C during photosynthesis and store it in their tissues and in the soil, which counteracts the accumulation of carbon dioxide (CO_2) in the atmosphere and maintains a stable climate.

Human activity has a large impact on the global C cycle, which has been perturbed by burning fossil fuels and clearing land. Burning fossil fuels releases large amounts of CO_2 and other GHGs into the atmosphere. Changes in land use have reduced the net capacity of ecosystems to sequester C and account for about 20% of anthropogenic C emissions to the atmosphere. Humans are currently emitting just under a billion tons of C into the atmosphere per year through land use changes (Riebeek, 2011).

Carbon-rich grasslands and forests have been extensively converted to cropland and pasture, which are poorer in C in most cases. Agriculture is considered to be the direct driver of around 80% of deforestation globally (FAO, 2016b). Forests are replaced with crops or pasture, which capture less C, and their exposed soils release more C from decayed plant matter into the atmosphere. Despite the large quantity of C stored as soil organic C (SOC), consensus is lacking on the sizes of global SOC stocks, their spatial distribution, and the C emissions from soils due to changes in land use and land cover (Scharlemann et al., 2014). In some cases, converting native vegetation to cropland has resulted in losses of 25%–50% of the SOC in the top 1 m of soil (Post and Kwon, 2000; Guo and Gifford, 2002; Lal, 2004), while conversion to pasture typically has resulted in smaller losses of SOC (Houghton et al., 2012). Forests absorb approximately 30% of anthropogenic CO_2 emissions (Pan et al., 2011). C sequestration in most forest systems levels off after a few decades. Both oceans and forests are critical buffers against climate change.

Oceans are an important sink for atmospheric CO_2 and play a key role in regulating the climate. Estimates vary, but between 30% and 50% of anthropogenic CO_2 emissions have been absorbed by the oceans. Coastal marine systems are phenomenally efficient at the continuous mitigation of C, with absorption rates up to 50 times greater than terrestrial forests. Furthermore, coastal marine systems can maintain these rates for centuries.

When coastal ecosystems are degraded, lost, or converted to other land uses, the large stores of blue C (C captured by the world's oceans and coastal ecosystems) are exposed and released as CO_2 into the atmosphere and/or oceans. Currently, on average, 2%–7% of blue C sinks (mangroves, seagrasses, and marshes) are lost annually, a sevenfold increase compared to only half a century ago (Nellemann et al., 2009). It is estimated that current rates of loss of these ecosystems may result in 0.15–1.02 billion tons of CO_2 being released annually. Although the combined global area of mangroves, tidal marshes, and seagrass meadows is only 2%–6% of the total area of tropical forest, degradation of these systems equates to 3%–19% of the C emissions from global deforestation (Nellemann, et al., 2009; Pendleton et al., 2012).

Unlike the other GHGs, carbon dioxide has a potentially beneficial effect for plant growth. Plants utilize the CO_2 in the air together with sunlight to manufacture carbohydrate for their energy and growth. There is a CO_2 fertilization effect with increasing concentrations of ambient CO_2. The climate assessments assume that this CO_2 benefit will be extensive under global climate change. From a quarter to half of Earth's vegetated lands have shown significant greening over the last 35 years largely due to rising levels of atmospheric CO_2 (Zhu et al., 2016). The CO_2 fertilization effect does, however, depend on optimal supplies of other nutrients and appears to be stronger at higher temperatures.

Biomass Appropriation

All AFS are a form of biomass appropriation. Biomass production depends on a range of environmental factors, including insolation, precipitation, temperature, humidity, and soil quality. About 68% of biomass is produced on land, whereas oceans are responsible for the remaining 32%. Estimates of the amount of biomass in the world's terrestrial ecosystems range from 773 to 1300 petagrams of carbon (Houghton et al., 2009).

Global biomass production has declined as a result of growing modifications of natural environments for agricultural and other purposes, such as urbanization. Decreases in the human-induced biomass production decline over 1980–2006 are estimated to range between one-quarter and one-third (Le et al., 2016). The human appropriation of net primary production (HANPP)—that is, the quantity of C in biomass that is harvested, grazed, burned, or lost as a result of human-induced land use change—has been estimated in the range of 15–20 gigatons of carbon (GtC) a year (Running, 2012; Krausmann et al., 2013).

The single most important driver of these changes is the shift of natural biomes to agricultural systems, which now cover about 40% of the terrestrial land surface, of which 12% is arable land and 26% pasture (FAOSTAT, 2017). Particularly striking is the increase in pasture from 3% of land cover in the year 1700 to 26% in 2000 (Ellis et al., 2010). Demand for fuels derived from biomass has created a new driver of land transformation, inducing pasture and cropland conversion from food production to fuel production. In total, crop and livestock production accounts for 78% of global HANPP, while the remaining 22% is made up of forestry, infrastructure, and human-induced fires (Haberl et al., 2007).

HANPP takes many forms: harvesting crops, residues and timber, forest slash, forages consumed by livestock, and biomass lost to human-induced fires, for example. HANPP amounts to 25% of overall production, a proportion that has doubled since 1900 (Fig. 35.2). HANPP rose from 6.9 GtC (13%) per year in 1910 to 14.8 GtC (25%) per year in 2005 (Krausmann et al., 2013).

Fig. 35.3 illustrates where on Earth, and how strongly, human activities alter ecological energy flows, showing the extent and intensity of human ecosystem use for human purposes. The green–yellow–red color gradient indicates increasingly positive HANPP

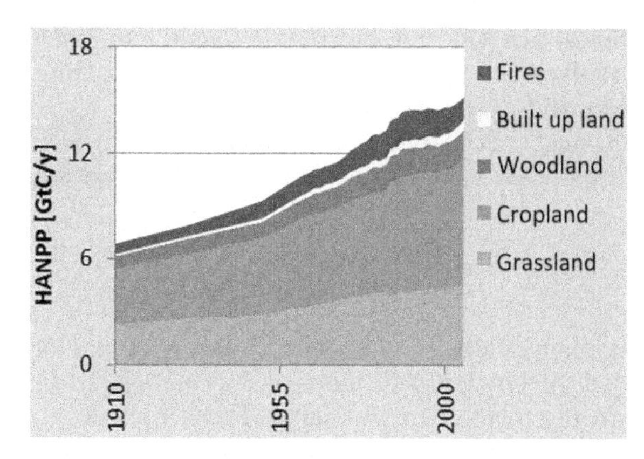

FIGURE 35.2 Development of global human appropriation of net primary production (HANPP) by major land use type and human-induced fires (1910–2005). Source: *Krausmann, F., Erb, K.H., Gingrich, S., Haberl, H., Bondeau, A., Gaube, V., et al., 2013. Global human appropriation of net primary production doubled in the 20th century. Proc. Natl. Acad. Sci. U.S.A. 110, 10324–10329.*

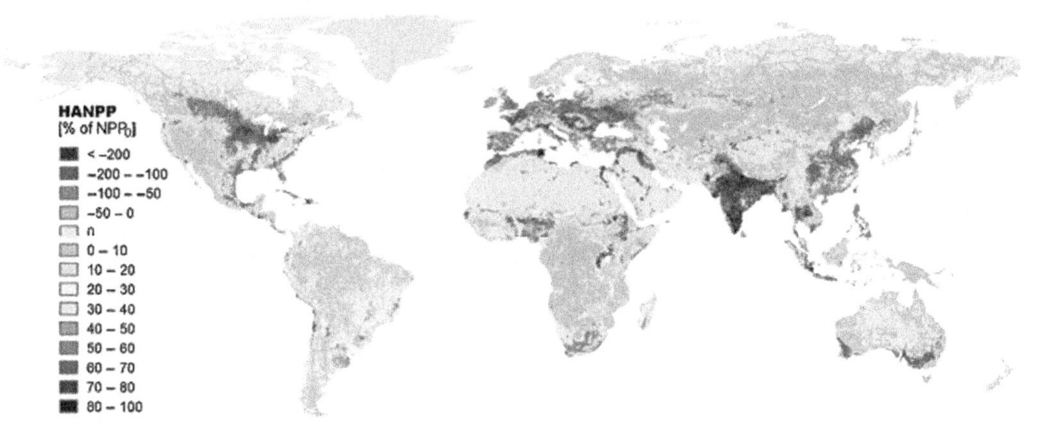

FIGURE 35.3 Human appropriation of net primary production (HANPP). Source: *Haberl, H., Erb, K.H., Krausmann, F., Gaube, V., Bondeau, A., Plutzar, C., et al., 2007. Quantifying and mapping the human appropriation of net primary production in earth's terrestrial ecosystems. Proc. Natl. Acad. Sci. U.S.A. 104, 12942–12945. For interpretation of the references to color in this figure legend, the reader is referred to the web version of this book.*

values, while blue colors show areas where HANPP is negative, that is, those ecosystems having higher HANPP than their natural NPP levels. These areas are predominantly highly irrigated in arid regions such as the Nile Delta, where agricultural activities increase NPP (Haberl et al., 2007).

A breakdown of global HANPP reveals considerably different patterns in various world regions. Aggregate HANPP may be as low as 11%–12% in central Asia, Russia, and Oceania, whereas land is used much more intensively than in other regions. For example, southern Asia has an overall HANPP of 63%, and land use intensity is also high in eastern and south-eastern Europe (52%). Land use-induced reductions in biomass productivity vary from 5% in eastern Asia to 27% in eastern and southeastern Europe (Krausmann et al., 2009).

HANPP varies from 1.3 tons C per capita per year in Asia, to 6–7 tons C per capita per year in the Americas. Latin America also has a high HANPP per capita (5.8 tons C per year) due to a high level of biomass consumption, moderate yields and large agricultural exports. Asia, on the other hand, has the lowest HANPP per capita (1.3 tons C per year) (Krausmann et al., 2009). This low figure results from prevailing intensive, high-yielding production systems, lower wood consumption, and a heavy reliance on imports.

HANPP not only reduces the amount of energy available to other species, but it also influences biodiversity, water flows, C flows between vegetation and atmosphere, and the provision of ecosystem services.

Fossil Fuel Use in Agrifood Systems

The previous discussion has shown that photosynthetic efficiency in terms of use of solar energy is quite low. In order to increase the efficiency of biomass production, additional energy is required, which is usually in the form of fossil energy. Fossil fuels (coal, oil, and gas) represent solar energy originally stored as biomass and eventually converted into energy stored in the C and hydrogen bonds of fossil fuels. These fossil fuels are being used rapidly. In the 20th century, the world population grew 4-fold, economic output 22-fold, and fossil fuel consumption 14-fold (UNEP, 2011). Of the 540 exajoules of energy consumed at the primary energy level in 2010, about 81% was provided by fossil fuels and 19% from renewable energy sources (Popp et al., 2014; Fig. 35.4).

Energy is not only an essential input to the production of food, but it also fuels the movements of inputs, products, and services that enable AFS to function at different scales. AFS, including input manufacturing, production, processing, transportation, marketing, and consumption, account for approximately 30% of global energy consumption, of which more than 70% is used beyond the farmgate (FAO, 2011b). Fossil fuel use has increased rapidly at all stages in the AFS. Fossil fuels power boats, ships, tractors, and other vehicles that transport food. Natural gas is used to manufacture chemical fertilizers and pesticides. Fossil fuels are combusted to generate electricity and heat for processing, refrigeration, and packaging. Both finite resources, oil and natural gas are the most widely used fossil fuels. Direct and indirect energy use by AFS emit around 3.4 $GtCO_2$-eq of GHG (FAO, 2011b).

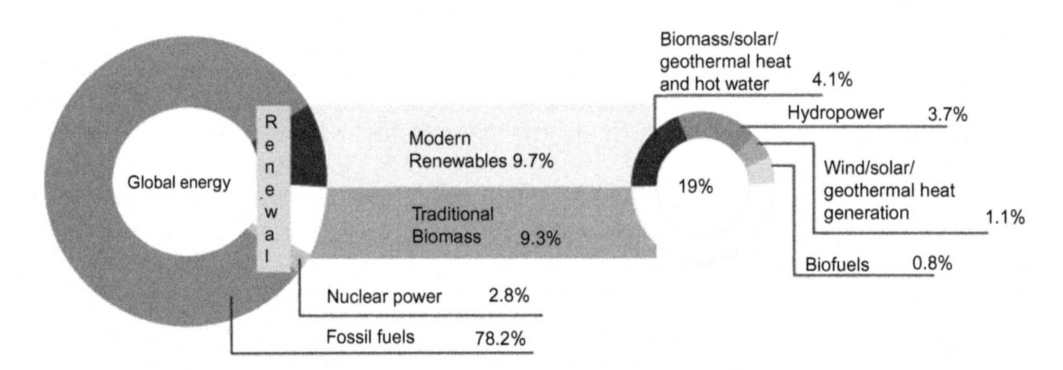

FIGURE 35.4 Estimated renewable energy share of global final energy consumption in 2011. Source: *Popp, J., Lakner, Z., Harangi-Rákos, M., Fári, M., 2014. The effect of bioenergy expansion: Food, energy, and environment. Renew. Sustain. Energy Rev. 32, 559–578.*

Bioenergy

Concerns over volatility in oil prices and attempts to replace fossil fuels with renewables have triggered the emergence of biofuels (ethanol and biodiesel). Biofuel production more than doubled from 60 billion liters in 2007 to around 130 billion liters in 2015 (IEA, 2016). Currently, around 80% of the global production of liquid biofuels is in the form of ethanol. In 2012, the United States was the world's largest producer of biofuels, followed by Brazil and the European Union (Figs. 35.5 and 35.6).

Bioenergy production must deal with competition for land from food and feed crops, which creates the risk of food prices increasing and affecting food security. Elsewhere, bioenergy has shown the potential to increase energy access, creating local employment and raising incomes (FAO, 2009). Attempts to bridge the large- and small-scale worlds of biofuels can be found in Brazil, where companies sell micro-distilleries for the production of 400, 1000, or 2000 L of bioethanol per day, using a variety of starch and sucrose-based crops (UNCTAD, 2014).

Competition between food and non food uses of biomass has increased interdependence between food, feed, and energy markets (FAO, 2017). Land diverted from food production needs to be compensated for by cropped land elsewhere unless abandoned agricultural land can be reclaimed for cropping and/or crop yields can be rapidly increased. However, when land lost from food production is replaced by bringing new areas under cultivation (indirect land use change), there may be damaging consequences in terms of GHG emissions if this land formerly stored more C than will its agricultural replacement.

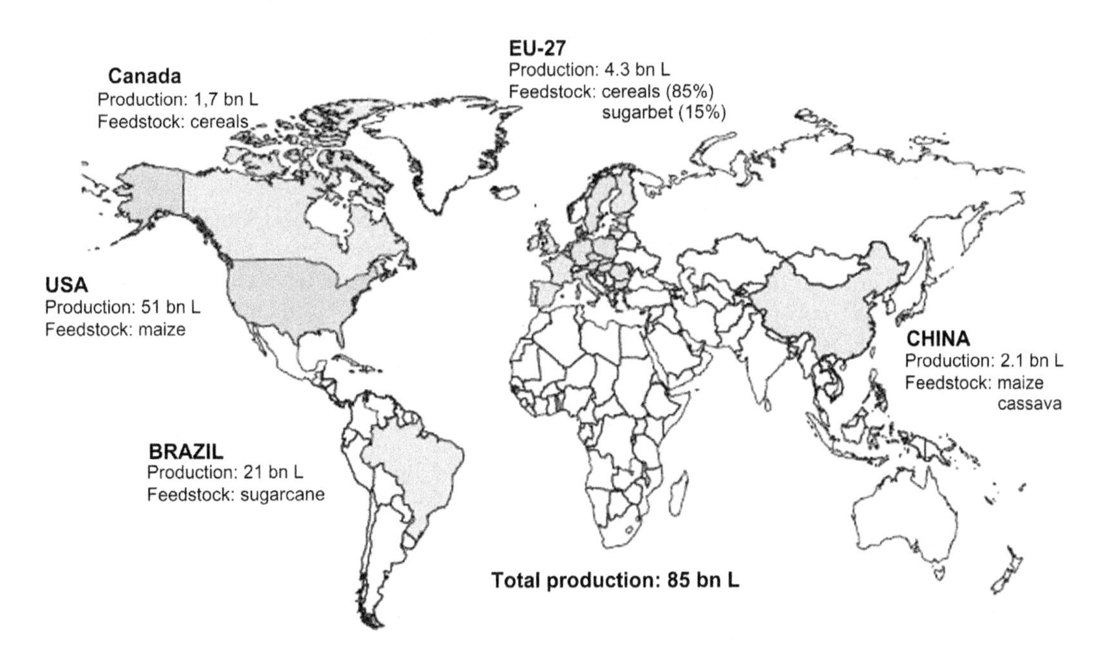

FIGURE 35.5 Global ethanol production (2012). Source: *IEA, 2016. Key World Energy Statistics. International Energy Agency (IEA), Paris.*

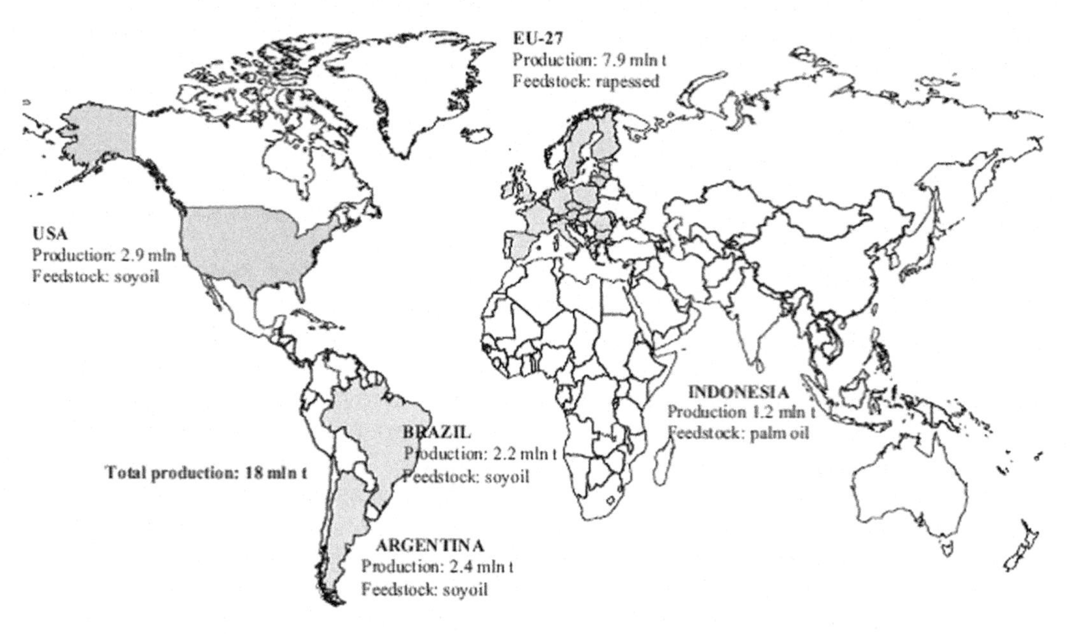

FIGURE 35.6 Global biodiesel production (2012). Source: *IEA, 2016. Key World Energy Statistics. International Energy Agency (IEA), Paris.*

35.3 LAND USE AND TRANSFORMATION

The management of land-based resources, including cropland, grazing land, forests, wetlands, and other land uses, has a major impact not only on the welfare of direct users of these resources, but also on other environmental services, such as prevention of erosion and runoff, removal of pollutants from water, sequestration of atmospheric carbon and other GHGs, and preservation of biodiversity.

Global land use plays a central role in determining supplies of food, materials, and energy. Global land area is estimated at 13.2 billion ha. Of this, 12% (1.6 billion ha) is currently in use for cultivation of agricultural crops, 31% (4 billion ha) is under forest, and 25% (3.3 billion ha) comprises grasslands (FAOSTAT, 2017). Aquaculture occupies an additional 45.2 million ha (Waite et al., 2014). The global area of cultivated land has grown by 19 million ha since 1961. Western Europe, Eastern Europe, and North America have shown a decline in cropland use, whereas more land has been brought into cultivation in Latin America, Africa, and Asia (Fig. 35.7).

Differences between countries and regions need to be interpreted against the background of global trends, such as the increase in international trade. For instance, the decline of cropland in Europe is in consequence of largely replacing domestic feed production by imports of soybean and soybean meal, mainly from Latin America. Trade is associated with "virtual resource use" embodied in traded commodities, such as water, biomass, and land use. This decoupling of production and consumption allows importing countries to enjoy the consumption of imported goods, while the exporting country bears the environmental consequences of producing those goods.

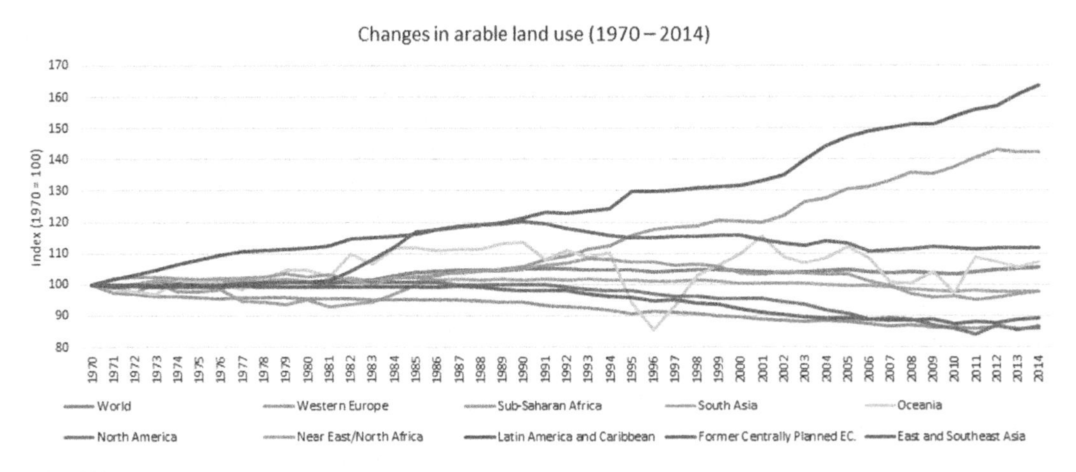

FIGURE 35.7 Regional trends in arable land use. Source: *FAOSTAT, 2017. FAOSTAT Database. FAO, Rome.*

Many changes in land use occur due to the increase in human population and the resulting demand for more resources—among them, minerals, soil, and water. Human activities have directly affected around 10 billion ha of the 13.2 billion ha of global land area, leaving about 30% of the land surface largely untouched. Between 2.3 and 3.8 billion ha (18%–29%) of affected land has been converted, mainly through deforestation, for agriculture, infrastructure, and urban use (Luyssaert et al., 2014; Fig. 35.8).

Land degradation implies lower productivity and compromised ecological function, including desertification, deforestation, overgrazing, salinization, and soil erosion. Evidence abounds of the potentially dramatic consequences of neglecting soil and land resources. The Dust Bowl years on the Great Plains of the United States in the 1930s were the result of rapid erosion caused by decades of continuous monocropping of shallow-rooted annual crops. In northwestern China, similar unsustainable practices led to widespread dust and sandstorms from the 1970s to 1990s. Almost 75% of cropland in Central America, 20% in Africa (mostly pasture), and 11% in Asia is seriously degraded (IFPRI, 2000), and 6% of India's agricultural land has been rendered unproductive as a result of salinization (Rosset et al., 2000).

Globally, about 75 billion tons of soil are lost every year to unsustainable practices. As a result, ecosystem services provided by fertile soil are diminishing fast, with devastating results. Between 10% and 20% of drylands are degraded and 24% of globally usable land on Earth is degraded at an economic loss of US$40 billion per year (Bai et al., 2008). It is estimated that 1 to 1.5 billion people in all parts of the world are already directly negatively affected by land degradation (Bai et al., 2008).

Agricultural expansion is a major driver of land use change, in particular through deforestation and forest degradation, contributing about 11% to total global GHG emissions net of reforestation; if reforestation and afforestation are excluded, the impact rises to nearly 20% of global GHG emissions. Annual net deforestation was 5.2 million ha/year during the period 2000 to 2010 (FAO, 2010), when the world lost an average of 13 million ha of forest (gross) each year to deforestation (5.2 million ha net of reforestation and

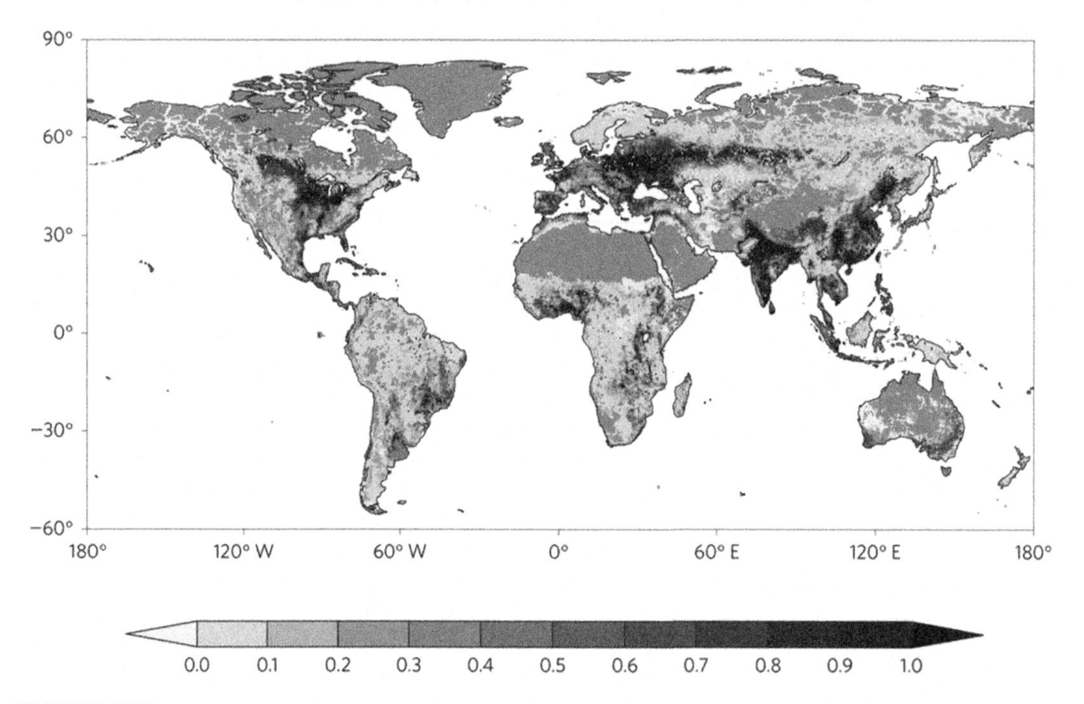

FIGURE 35.8 Spatial extent of land cover change, land management, wilderness, and nonproductive areas. Wilderness and nonproductive areas are shown in green and represent land largely unaltered by humans. The remaining land is used for producing food, fiber, and fuels, and for hosting infrastructure. The color scale represents the fraction of each grid cell for which the original plant cover was converted. Dark colors indicate regions where most of the original plant cover was converted; these regions are the subject of typical land cover change studies. The light colors show areas for which land cover change is low, but which are nevertheless under anthropogenic land management. Source: *Luyssaert, S., Jammet, M., Stoy, P.C., Estel, S., Pongratz, J., Ceschia, E., et al., 2014. Land management and land-cover change have impacts of similar magnitude on surface temperature. Nat. Clim. Change 4, 389–393. For interpretation of the references to color in this figure legend, the reader is referred to the web version of this book.*

afforestation). Africa and Latin America had the highest net annual loss of forests between 2010 and 2015, with 2.8 and 2 million ha, respectively. In contrast, net forest area has increased in temperate countries, while there has been relatively little change in the boreal and subtropical regions.

In tropical countries, a net 7 million ha of forest were lost per year over the period 2000–10 (FAO, 2016b and 6 million ha of agricultural land were added. Forests are cleared for both crop and livestock production, timber harvesting, extraction for fuelwood or charcoal, mining and road building. Large-scale commercial agriculture accounts for about 40% of deforestation in the tropics and subtropics, local subsistence agriculture for 33%, infrastructure for 10%, urban expansion for 10%, and mining for 7% (Kissinger et al., 2012). There are significant regional variations; for example, commercial agriculture accounts for almost 70% of the deforestation in Latin America, but for only one-third in Africa, where small-scale crop and livestock expansion is more important (Fig. 35.9).

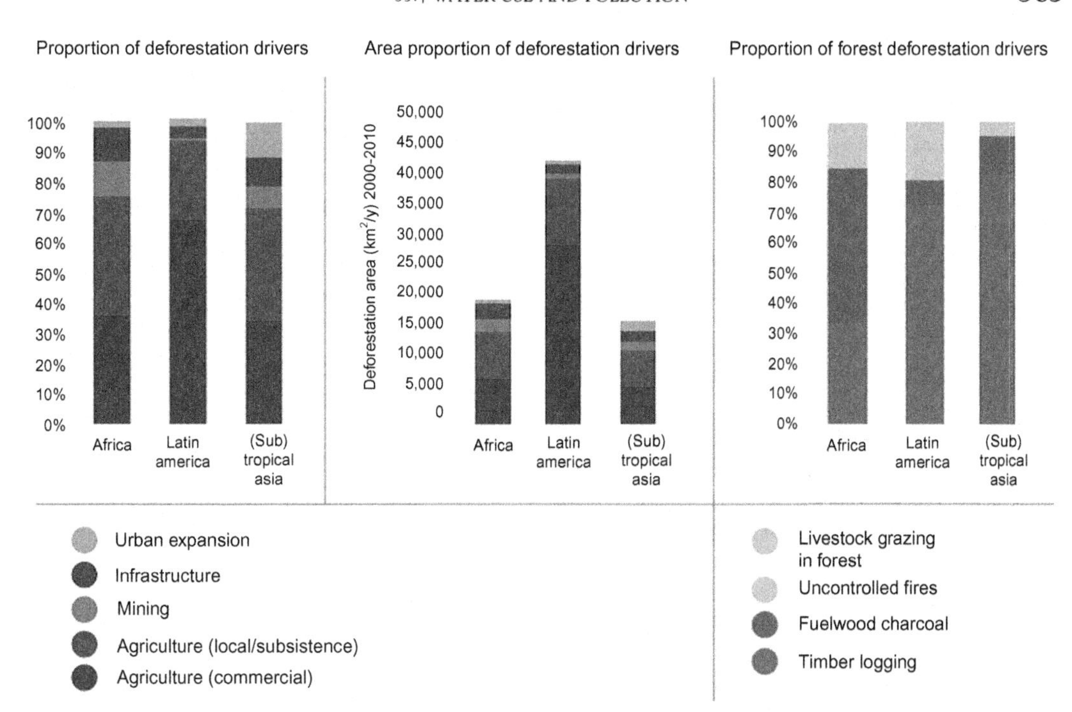

FIGURE 35.9 Key drivers of deforestation. Source: *Kissinger, G.M., Herold, M., De Sy, V., 2012. Drivers of deforestation and forest degradation: a synthesis report for REDD + policymakers. Lexeme Consulting, Vancouver.*

These land cover changes cause shifts in regional and global climates by adjusting biogeochemical processes, such as C and N cycling, and biophysical processes, such as surface albedo, C sequestration, and evapotranspiration. While each incident of land cover change occurs on a local scale, the aggregated impacts have consequences for Earth system processes. Land use and land use change directly affect the exchange of GHG between terrestrial ecosystems and the atmosphere. Annual GHG flux from land use and land use change activities accounts for approximately 4.3–5.5 $GtCO_2$-eq per year (Smith et al., 2014).

35.4 WATER USE AND POLLUTION

Most of the Earth's surface is covered with water: 97.5% of this is saltwater, leaving only 2.5% as freshwater. Nearly 70% of that freshwater is frozen in the ice caps of Antarctica and Greenland, and most of the remainder is present as soil moisture or lies in deep underground aquifers as groundwater not accessible to human use (Shiklomanov, 1997). The total quantity of global freshwater potentially available for human use has been estimated at about 475 million km^3, but much of this is inaccessible (Shiklomanov, 1997).

About 9% of global renewable freshwater resources are withdrawn for human uses, of which 70% goes to agriculture. In richer countries, this is 42%–44%, mainly because water use in other sectors is higher (OECD, 2010). In some areas such as the Middle East, North Africa, central Asia, India, and parts of China, 80%–90% of the water is used for agriculture (OECD, 2010). Water withdrawals for agriculture have been on the rise as the world's irrigated area has doubled over the last five decades (FAO, 2011a). Globally, some 38% of irrigated areas depend on groundwater (Siebert et al., 2013), and groundwater abstraction for irrigation has grown 10-fold over the last 50 years. However, losses of 50% of water are common. In many countries, relatively inefficient irrigation techniques such as flooding or high-pressure raingun technologies are still being used, which use considerably greater quantities of water than low pressure sprinklers and drip irrigation techniques (OECD, 2008). Inefficient use of water for crop production depletes aquifers, reduces river flows, degrades habitats, and has caused salinization of 20% of the global irrigated land area (FAO, 2011a).

Gleeson et al. (2012) assessed groundwater footprints of major aquifers that are important to agriculture (Fig. 35.10), showing that humans overexploit groundwater in many large aquifers that are critical to agriculture, especially in Asia and North America.

Use of nonconventional sources of water as an alternative to freshwater, although currently a minor source, is increasing in certain regions and countries. Globally, only 1% of

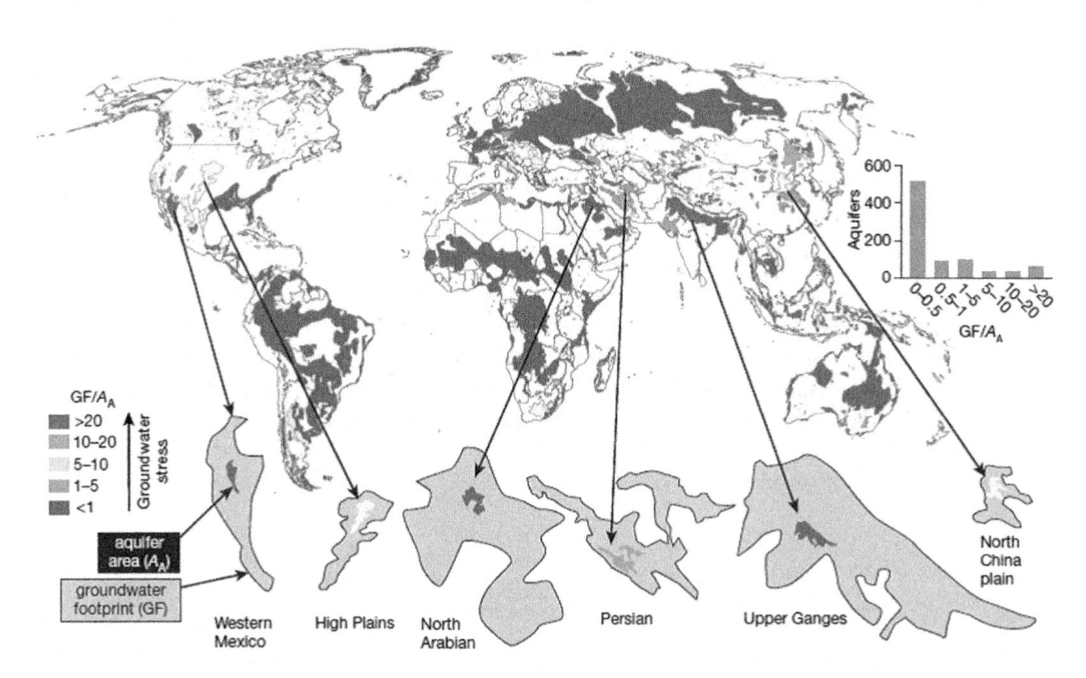

FIGURE 35.10 Groundwater footprints of aquifers that are important to agriculture are significantly larger than their geographic areas. Note: At the bottom of the figure, the areas of the six aquifers (Western Mexico, High Plains, North Arabian, Persian, Upper Ganges, and North China Plain) are shown at the same scale as the global map; the surrounding gray areas indicate the groundwater footprint proportionally at the same scale. The ratio groundwater footprint (GF)/aquifer area indicates widespread stress of groundwater resources and/or groundwater-dependent ecosystems. Inset, histogram showing that GF is less than the aquifer area for most aquifers. Source: *Gleeson, T., Wada, Y., Bierkens, M.F.P., van Beek, L.P.H., 2012. Water balance of global aquifers revealed by groundwater footprint. Nature 488, 197–200.).*

the water used in agriculture consists of treated wastewater or desalinated water. About 300 million people acquire freshwater from more than 17,000 desalination plants in 150 countries. In regions such as the Arabian Peninsula, water from unconventional sources constitutes more than 50% of the water used, 40% being desalinated water and 10% wastewater (FAO, 2011a). Without the energy-intensive process of desalination, the Arab states of the Persian Gulf and Israel cannot support their current populations. Bahrain, for example, has virtually no freshwater, its population completely reliant on desalination of seawater.

Water is degraded directly by chemical or nutrient pollution, and indirectly when land use and land use change increases soil erosion or reduces the capacity of ecosystems to filter water. A serious global problem, nutrient runoff from agriculture results in eutrophication and human health hazards in coastal regions. Impacts of nutrient loss from agroecosystems include groundwater pollution and increased nitrate levels in drinking water, eutrophication, increased frequency and severity of algal blooms, hypoxia and fish kills, and "dead zones" in coastal marine ecosystems (Bouwman et al., 2009). Globally, over 400 coastal dead zones exist—up from 49 in the 1960s—and these are expanding at the rate of 10% per decade (Diaz and Rosenberg, 2008). In the United States alone, agriculture accounted for around 60% of river pollution, 30% of lake pollution, and 15% of estuarine and coastal pollution in 2010 (OECD, 2012).

35.5 WATER AS A PRODUCTION ENVIRONMENT (FISHERIES AND AQUACULTURE)

Humans depend on aquatic ecosystems for critical goods and services, but human use has also altered them through direct and indirect means. Ocean-based activities extract resources, add pollution, and change marine species composition. Land-based activities are major sources of runoff of pollutants and nutrients into coastal waters and remove, alter, or destroy natural habitats. These human activities vary in their intensity of impact on the ecological systems and in their spatial distributions.

Ocean health is vital to food production; about 56% of fish harvested in 2014 came from wild catches with the remainder coming from aquaculture (FAO, 2016a). Globally, marine capture fisheries production has leveled off since the late 1980s (Pauly et al., 2005), as reduced harvests in coastal maritime regions were compensated for by fishing further out, at deeper sea levels, and harvesting other species (World Ocean Review, 2013). The Food and Agriculture Organization of the United Nations estimates that 85% of the world's fish stocks are either overexploited or exploited to their maximum potential (FAO, 2012). Other issues affecting ocean health include contamination from nutrients, pesticides, sediment, acidification, and growing levels of microplastics in aquatic environments.

Fish produced by aquaculture is a growing industry that meets the growing demand for fish supplies that can't be met by wild capture. In response to rapidly growing demand, world aquaculture production expanded by almost 12 times from 1980 to 2010 (FAO, 2012). However, aquaculture requires fishmeal and plant-based protein feed, creates pollution, and is involved in disease transmission to wild fish populations (McIntyre et al., 2009). Aquacutlure also draws on water resources and affects water quality (Box 35.3).

BOX 35.3

WATER USE IN AQUACULTURE

In 2010, aquaculture consumed an estimated 201 cubic kilometers (km^3) of freshwater, equal to approximately 2% of global agricultural water consumption. Freshwater inland aquaculture uses water to maintain pond levels, compensating for water lost through seepage, evaporation, and intentional discharge. More intensive systems use frequent water exchanges to aerate and filter ponds. Production of plant-based fish feed also consumes water. Freshwater is becoming increasingly scarce in many aquaculture-producing areas because of upstream dams and diversion of water for agriculture and urban uses. Intensification of production and greater recirculation of water are leading to increases in aquaculture's water use efficiency. Extensive pond aquaculture consumes more than 10,000 cubic meters of water per ton (m^3/t) of fish produced because of the need to drain and fill ponds and replace water lost through seepage and evaporation. More intensive operations consume much less (2000–5000 m^3/t), and cages and recirculating systems consume virtually no freshwater.

Aquaculture not only consumes freshwater, but also has an impact on water quality through water pollution. Discharges can contain excess nutrients from fish feed and waste, antibiotic drugs, other chemicals (e.g., pesticides, hormones), and inorganic fertilizers. In comparison to terrestrial livestock production, it is difficult to collect wastes from aquaculture production because they rapidly disperse into the surrounding water. Pollution associated with aquaculture can cause degradation of aquatic habitats and eutrophication of lakes or coastal zones, and is also a threat to the aquaculture operation itself.

Hall et al. (2011) used the life cycle assessment (LCA) approach to examine, quantify, and compare the environmental performance of major aquaculture production systems around the world. This particular LCA compiled data on inputs (e.g., land, water, feed, energy) and environmental releases (e.g., waste nitrogen [N] and phosphorus [P]), and evaluated the potential environmental impacts associated with each. The use of freshwater and impacts on water by various species groups is variable (see the following table).

Freshwater consumption and water pollution attributable to aquaculture

	Freshwater Consumption	**Water Pollution**	
Species Group	**m^3/kg Edible Protein**	**kg P/t Edible Protein**	**kg N/t Edible Protein**
Carps	61.4	97	329
Mollusks	0.0	−148	−136
Shrimps	4.4	104	422
Tilapia	15.9	82	349
Catfish	52.2	97	234
Salmonids	0.0	48	182
Average figures for world aquaculture	**40.4**	**76**	**273**

Based on Waite, R., Beveridge, M., Brummett, R., Castine, S., Chaiyawannakarn, N., Kaushik, S., et al., 2014. Improving Productivity and Environmental Performance of Aquaculture. Working Paper, Installment 5 of Creating a Sustainable Food Future. World Resources Institute, Washington, DC.

35.6 NUTRIENT FLOWS IN AGRIFOOD SYSTEMS

Biological production in natural and agricultural ecosystems is strongly determined by availability of N and phosphorus (P), and both are used heavily in AFS. N and P fertilizers have greatly increased the amount of N and P in the biosphere and have complex, often harmful, effects on natural ecosystems (Vitousek et al., 1997). Since the 1960s, human use of synthetic N fertilizers has increased ninefold, from 12 million tons to 109 million tons in 2014, while P use has increased fourfold, from 10 million tons to 45 million tons in 2014 (FAOSTAT, 2017).

Because of its scarcity in nature, reactive nitrogen (Nr) has long been the most limiting element to primary productivity in terrestrial and aquatic ecosystems. Although substantial amounts of N are fixed through naturally occurring processes such as lightning and biological N fixation by leguminous plants, the amounts are not sufficient to meet the food demands of an increasing world population. Globally, humans introduce 210 teragrams (Tg) of N per year of new N as synthetic fertilizer for plant production (Fowler et al., 2013), in addition to the 58 Tg N fixed biologically (Vitousek et al., 2013). The efficiency of N use is very low. Considering the complete food chain, of about 180 Tg N input through a combination of manufactured fertilizers and biological N fixation annually, only 28 Tg (16%) is available in food for human consumption, with only 19 Tg (11%) actually consumed.

The supply of Nr is not evenly distributed around the world. Regions of excess Nr include North America, Europe, and South and East Asia, especially China. N-limited regions include much of Africa, Latin America, and parts of Asia. A paradox of Nr is that when present either in excess or in deficit it causes negative environmental impacts. In areas of excess, it accumulates in and moves between air, soil, and water, causing environmental and human health-related problems. In areas where there is too little Nr, primary production falls short of its potential and agriculture fails to produce enough food to sustain the population. Insufficient Nr and other agricultural nutrients can also lead to land degradation, soil erosion, and desertification. In many African countries, N extracted in harvested crops is not replaced sufficiently by new N inputs, a process termed nutrient mining, resulting in lost fertility. Henao and Baanante (2006) found that in Africa 40% of farmland suffers from nutrient depletion rates greater than 60 kg/ha per year. Chapter 36, Molecules, Money, and Microbes, takes a detailed look at the N cycle, in the context of dairy farming.

Unlike Nr, P cannot be manufactured, and there is no substitute available. P is a limiting nutrient in crop growth and its deficiency limits crop yields. P moves in a cycle through rocks, water, soil, sediments, and organisms. Unlike the N cycle, the P cycle has no atmospheric component. Most of the Earth's P is locked up in sediments and rocks, unavailable for plants to use. P used in fertilizer comes predominantly from phosphate rock, a finite resource formed over 10–15 million years ago in the Earth's crust (Cordell et al., 2009). Ninety percent of the world's mined P rock is used in AFS, mostly as fertilizer but also as an animal feed additive. P is one of the most geographically concentrated commodities; 90% of the P rock reserves are located in just five countries: Morocco, China, South Africa, Jordan, and the United States. The world's annual rock P mining in 2011 amounted to 191 million tons (Scholz and Wellmer, 2013), which corresponds to about

25 Tg P per year. Currently, the quality of the available reserves is declining, while the cost of extraction and processing is increasing (Gregory et al., 2010).

Most P is wasted. Of 25 Tg P mined annually, only 3–4.9 Tg is consumed by humans (food waste not included), yielding an estimated full-chain P use efficiency of 12%–20%. Of the rest, the major losses occur in the steps from mining to preparation of mineral fertilizer and other P products, and from mineral fertilizers to crop and livestock production (Sutton et al., 2013).

These nutrient flows and the agricultural activities through which they are manipulated represent major disturbances to natural biogeochemical cycles. The AFS involved have relatively low production efficiency, besides leaking nutrients into water and air and generating large volumes of animal manure and food chain waste. The changes in global nutrient cycles have both positive and negative effects. The increased use of N and P fertilizers has allowed for the production of food necessary to support the rapidly growing human population (Galloway and Cowling, 2002), reducing pressures on other natural resources such as land and water. Regardless of the origin of these nutrients, they contribute to a cascade of spatial and temporal impacts at various scales. As scale increases, the relationship between these nutrients and ecosystem services becomes increasingly complex due to dependencies on the same biogeochemical cycles. Significant fractions of the anthropogenically mobilized N and P in watersheds enter groundwater and surface water and are transported to coastal marine systems. This has resulted in numerous negative human health and environmental impacts, such as groundwater pollution and increased nitrate levels in drinking water, loss of habitat and biodiversity, an increase in frequency and severity of harmful algal blooms, eutrophication, hypoxia, and fish kills (Vollenweider, 1992; Vollenweider et al., 1992; Howarth et al., 1996; Rabalais, et al., 2002; Turner et al., 2003; Diaz and Rosenberg, 2008; Bouwman et al., 2009).

35.7 BIODIVERSITY

Biodiversity is the quantity and variability of living organisms within species (genetic diversity), between species, and between ecosystems. Biodiversity is not itself an ecosystem service but rather underpins the supply of services. The value placed on biodiversity for its own sake is captured under the cultural ecosystem service called "ethical values" (TEEB, 2008).

Biodiversity is a crucial natural resource for food production in all its forms: crops, livestock, forestry, aquaculture, fisheries, and hunting and gathering (Le Roux et al., 2008; MA, 2005). It provides genetic material and variability for crop, tree, livestock and fish selection and breeding, chemicals for medicines, and raw materials for industry. Diversity of living organisms and the abundance of populations of many species are also critical to maintaining biological services, such as pollination and nutrient cycling. Less tangibly, but equally important, diversity in nature is regarded as providing both intrinsic value and a widely recognized recreational function for most people. Agroecosystems are both providers and consumers of ecosystem services.

According to evidence from geology and evolutionary science (Waters et al., 2016), the world is currently undergoing a mass extinction of plants and animals. Unlike past mass extinctions, caused by events like asteroid strikes, volcanic eruptions, and natural climate shifts, the current crisis is almost entirely caused by humans. According to the Center for Biological Diversity, 99% of currently threatened species are at risk from human activities, primarily those driving habitat loss, introduction of exotic species, and global warming.

Besides depending on biodiversity and ecosystem services, AFS also exert major pressures on biodiversity. The erosion of global biodiversity over the past century is alarming; as a key economic sector, food and agricultural exerts the largest impact on biodiversity, contributing 60%–70% of total biodiversity loss to date in terms of the "mean species abundance" (MSA) indicator in terrestrial ecosystems and about 50% of MSA in freshwater systems (PBL, 2014). Major losses have occurred in virtually all types of ecosystems, much of it through loss of habitat area (Box 35.4).

BOX 35.4

BIODIVERSITY TRENDS IN KEY ECOSYSTEMS

Agro-Biodiversity

Agricultural systems are considerably more simplified than natural ecosystems; they are multifunctional and, in addition to food, if managed well they can provide a range of regulating, supporting, and cultural ecosystem services. As much as 30% of the potential area of temperate and subtropical forests has been lost to agriculture through conversion. Intensification also displaces a great diversity of traditional crop varieties with modern high-yield, but genetically uniform, crops. More than 90% of crop varieties have disappeared, and half of the breeds of many domestic animals have been lost. In fisheries, the world's 17 main fishing grounds are now being fished at or above their sustainable limits, with many fish populations becoming extinct. Global trends and figures related to agrobiodiversity include the following:

- Since the 1900s, some 75% of plant genetic diversity has been lost, as farmers worldwide have abandoned their multiple local varieties and landraces for genetically uniform, high-yielding varieties.
- 30% of livestock breeds are at risk of extinction and six breeds are lost each month—making food systems unnecessarily vulnerable to climate change and disease.
- Today, 75% of the world's food is generated from only twelve plants and five animal species. Of the 4% of the 250,000–300,000 known edible plant species, only 150–200 are used by humans (Esquinas-Alcazar, 2005).

Coastal Biodiversity

Indicators of habitat loss—disease, invasive species, and coral bleaching—all show declines in biodiversity. Sedimentation and pollution from land are smothering some coastal ecosystems, and trawling is reducing diversity in some areas. Commercial species such as Atlantic cod, five species of tuna, and haddock are threatened globally, along with several species of whales, seals, and sea turtles. Invasive species are frequently reported in enclosed seas, such as

BOX 35.4 (*cont'd*)

the Black Sea, where the introduction of Atlantic comb jellyfish caused the collapse of fisheries.

Forest Biodiversity

Forests, which harbor about two-thirds of known terrestrial species, have among the highest species diversity and endemism of many ecosystems, as well as the highest number of threatened species. Many forest-dwelling large mammals, half the large primates, and nearly 10% of all known tree species are at some risk of extinction. Significant pressures on forest species include conversion of forest habitat to other land uses, habitat fragmentation, logging, and competition from invasive species.

Freshwater Biodiversity

The biodiversity of freshwater ecosystems is much more threatened than that of terrestrial ecosystems. More than 10,000 species, or 20% of the world's freshwater fish, have become extinct, threatened, or endangered in recent decades. Physical alteration, habitat loss, and degradation, water withdrawal, overexploitation, pollution, and the introduction of exotic species all contribute to declines in freshwater species.

Grassland Biodiversity

Natural grasslands across the globe suffer from conversion, overgrazing, simplification, and lower productivity with domino effects to their biodiversity. Since 1945, 680 million hectares out of 3.4 billion hectares of rangelands have been affected, while 3.2 million hectares are currently degraded every year. Over 50% of flooded grasslands, savannahs, and tropical and subtropical grasslands have been destroyed.

Source: *Adapted from World Bank, 2004. Responsible growth for the new millennium: Integrating Society, Ecology, and the Economy. World Bank, Washington, DC (World Bank, 2004) and TEEB, 2011. The Economics of Ecosystems and Biodiversity in National and International Policy Making. Earthscan, London (TEEB, 2011).*

Biodiversity has suffered as agricultural land, which supports far less biodiversity than natural habitats, has expanded primarily at the expense of forest areas. Biodiversity is also diminished by intensification, which reduces the area allotted to hedgerows or wildlife corridors and displaces traditional varieties of seeds with modern high-yielding but genetically similar and uniform varieties. Pollution, overexploitation, and competition from invasive species represent further threats to biodiversity.

35.8 CONCLUSIONS

The Earth is a single complex, integrated system. Natural ecosystems operate as an interdependent set, with any change having multiple impacts on resources and ecosystem services. This has profound implications for local and global sustainability, as it emphasizes the need to address multiple interacting environmental processes simultaneously. Stabilizing the climate system, for example, requires stable terrestrial and water ecosystems. As both the source of critical inputs and the recipient of the waste streams and by-products from AFS, the biophysical environment is an integral component of the AFS, with its condition being critical to its long-term sustainability.

Besides needing to be embedded into ecosystem functions to preserve them, AFS must become more productive and efficient. Functions to be restored and amplified include the protection of biodiversity, watersheds, C stores, and natural pollinators; microclimate regulation; and the conservation of soils. In many regions, agricultural lands are the main source of such ecosystem services, and a large potential source of income if such services were to accrue better rewards.

AFS will need to shift away from heavy dependence on nonrenewable inputs and chemical-based intensification toward sustainable intensification based on fostering ecological processes and conserving local natural resources. Agroecological methods, based on locally adapted practices and new technologies, such as low till and precision agriculture, and improved varieties that are more resilient to water and heat stresses, will increase the efficiency of inputs used and realize synergies among species and systems. Better management of ecosystems for benefits such as rainwater harvesting, flood control, natural pollination, and improved soil health will be sources of yield growth and stability.

The concept of closing the loop, as applied to waste and resources, expresses a desire to move away from a linear process of resource extraction, manufacture, consumption, and disposal toward a system where resources remain in long-term cycles, commonly referred to as a circular bioeconomy. In global terms, a relatively small proportion of resources is recycled, and a large proportion is in productive use for only a very short time. This process depletes non renewable resources, degrades habitats, landscapes, and biodiversity, and compromises environmental "sinks" needed to absorb the pollution we generate as we produce and consume. Recycling resources within AFS, rather than just passing them through once, reduces the environmental impacts associated with extraction of resources and disposal of waste. Nutrient loops can be closed in many different ways, for example by ensuring that nutrients are applied at times and places that best match with plant needs (precision cropping). Recycling also includes efforts to recover nutrients in usable form from places in the AFS where nutrients concentrate—including wastewater treatment plants, livestock production facilities, compost operations, and food processing plants—and reuse them to produce more food and feed. Recovering nutrients in forms that are more easily used than manure (e.g., bio-ammonium sulfate crystals, phosphorus-rich solids) can increase the amount of nutrients recycled, creating a more "closed" system.

In order to maintain their productive capacity, agricultural systems and landscapes must diversify. At the farm level, this means crop diversification, polycultures, multiple varieties, and appropriate integration of livestock to enhance resilience, manage pest and disease risks, cycle nutrients, adapt to climate change, and use inputs most efficiently. At landscape scale, this means including natural areas in and around farms to sustain biodiversity and ecosystem services. Locally developed and adapted solutions are almost always biodiverse; with the complementary application of modern science, intensification without simplification is both possible and desirable, as is the diversification of simplified agroecosystems to restore ecosystem services.

Being a dominant user of land globally, AFS must produce ecosystem services and host biodiversity as an essential complement to natural areas. This calls for a better understanding and management of competing uses of land, water, and ecosystem services, and coordinated management of farms and landscapes to protect watersheds and habitat values. An integrated approach will be essential to increase agricultural production, whilst sustaining and regulating ecosystem functions, protecting and preserving genetic pools, and enhancing terrestrial C stocks.

In this chapter we have shown that AFS are responsible for land degradation, depletion of resources, nutrient losses, impacts on air, soil and water quality, biodiversity losses, and GHGs emissions that contribute to climate change. With the expected population growth, increasing wealth, expansion of cities, and changing eating habits, AFS will continue to grow and evolve as they respond to new opportunities. This will create many benefits but also complex trade-offs, especially those relating to increased pressure on an already stressed natural resource base, and to climate change. These trade-offs need to be quantified and clearly communicated to policy makers, consumers, and other AFS actors, so that they can make appropriate, evidence-based decisions. The analytical framework, developed in Chapter 36, Molecules, Money, and Microbes, aims to provide insights into the impacts, interactions, and trade-offs that occur as natural resources are transformed into human benefits through food and agriculture.

References

Bai, Z.G., Dent, D.L., Olsson, L., Schaepman, M.E., 2008. Proxy global assessment of land degradation. Soil Use Manage. 24, 223–234.

Bouwman, A.F., Beusen, A.H.W., Billen, G., 2009. Human alteration of the global nitrogen and phosphorus soil balances for the period 1970–2050. Glob. Biogeochem. Cycles, GB0A04 23.

Chatham House, 2018. resourcetrade.earth. Available from: <http://resourcetrade.earth/>.

Cordell, D., Drangert, J.-O., White, S., 2009. The story of phosphorus: global food security and food for thought. Glob. Environ. Change 19, 292–305.

Diaz, R.J., Rosenberg, R., 2008. Spreading dead zones and consequences for marine ecosystems. Science 321, 926–929.

Ellis, E.C., Goldewijk, K.K., Siebert, S., Lightman, D., Ramankutty, N., 2010. Anthropogenic transformation of the biomes, 1700 to 2000. Glob. Ecol. Biogeogr. 19, 589–606.

Esquinas-Alcazar, J., 2005. Protecting crop genetic diversity for food security: political, ethical and technical challenges. Nat. Rev. Genet. 6, 946–953.

FAO, 2004. Building on Gender, Agrobiodiversity and Local Knowledge. FAO, Rome.

FAO, 2009. Making Sustainable Biofuels Work for Smallholder Farmers and Rural Households. Issues and Perspectives. FAO, Rome.

FAO, 2010. Global Forest Resources Assessment 2010—Main Report. FAO, Rome.

FAO, 2011a. Energy-Smart Food for People and Climate. Issue Paper. FAO, Rome.

FAO, 2011b. The State of the World's land and Water Resources for Food and Agriculture—Managing Systems at Risk. Food and Agriculture Organization of the United Nations, Rome and Earthscan, London.

FAO, 2012. The State of World Fisheries and Aquaculture. FAO, Rome.

FAO, 2014. The State of World Fisheries and Aquaculture. FAO, Rome.

FAO, 2015. Global Forest Resources Assessment 2015. FAO, Rome.

FAO, 2016a. State of the World Fisheries and Aquaculture. Contributing to Food Security and Nutrition for All, State of the World Fisheries and Aquaculture. FAO, Rome.

FAO, 2016b. State of the World's Forests 2016. Forests and Agriculture: Land-Use Challenges and Opportunities. FAO, Rome.

FAO, 2016c. The State of Food and Agriculture—Climate Change, Agriculture and Food Security. FAO, Rome.

FAO, 2017. The Future of Food and Agriculture—Trends and Challenges. FAO, Rome.

FAOSTAT, 2017. FAOSTAT Database. FAO, Rome.

Foley, J.A., DeFries, R., Asner, G.P., Barford, C., Bonan, G., Carpenter, S.R., et al., 2005. Global consequences of land use. Science 309, 570–574.

Fowler, D., Pyle, J.A., Raven, J.A., Sutton, M.A., 2013. The global nitrogen cycle in the twenty-first century: introduction. Philos. Trans. R. Soc. B Biol. Sci. 368.

Galloway, J.N., Cowling, E.B., 2002. Reactive nitrogen and the world: 200 years of change. Ambio 31, 64–71.

Gleeson, T., Wada, Y., Bierkens, M.F.P., van Beek, L.P.H., 2012. Water balance of global aquifers revealed by groundwater footprint. Nature 488, 197–200.

Gregory, D.I., Haefele, S.M., Buresh, R.J., Singh, U., 2010. Fertilizer Use, Markets, and Management. Rice in the Global Economy. Strategic Research and Policy Issues for Food Security. International Rice Research Institute, Los Banos, Philippines, pp. 231–263.

Guo, L.B., Gifford, R.M., 2002. Soil carbon stocks and land use change: a meta-analysis. Glob. Change Biol. 8, 345–360.

Haberl, H., Erb, K.H., Krausmann, F., Gaube, V., Bondeau, A., Plutzar, C., et al., 2007. Quantifying and mapping the human appropriation of net primary production in earth's terrestrial ecosystems. Proc. Natl. Acad. Sci. U.S.A. 104, 12942–12945.

Hall, S.J., Delaporte, A., Phillips, M.J., Beveridge, M., O'Keefe, M., 2011. Blue Frontiers: Managing the Environmental Costs of Aquaculture. The WorldFish Center, Penang, Malaysia.

Hansen, J., Sato, M., Kharecha, P., Beerling, D., Berner, R., Masson-Delmotte, V., et al., 2008. Target atmospheric CO_2: where should humanity aim? Open Atmos. Sci. J. 2, 217–231.

Henao, J., Baanante, C., 2006. Agricultural Production and Soil Nutrient Mining in Africa Implications for Resource Conservation and Policy Development: Summary. International Center for Soil Fertility and Agricultural Development, Muscle Shoals.

Houghton, R.A., Hall, F., Goetz, S.J., 2009. Importance of biomass in the global carbon cycle. J. Geophys. Res. Biogeosci. 114.

Houghton, R.A., House, J.I., Pongratz, J., van der Werf, G.R., DeFries, R.S., Hansen, M.C., et al., 2012. Carbon emissions from land use and land-cover change. Biogeosciences 9, 5125–5142.

Howarth, R.W., Billen, G., Swaney, D., Townsend, A., Jaworski, N., Lajtha, K., et al., 1996. Regional nitrogen budgets and riverine N & P fluxes for the drainages to the North Atlantic Ocean: Natural and human influences. Biogeochemistry 35, 75–739.

IEA, 2016. Key World Energy Statistics. International Energy Agency (IEA), Paris.

IFPRI, 2000. Pilot Analysis of Global Ecosystems. World Resource Institute, Washington, DC.

IPCC, 2000. Land Use, Land-Use Change and Forestry—Special Report. Intergovernmental Panel on Climate Change. Meteorological Office, Bracknell, UK.

Kissinger, G.M., Herold, M., De Sy, V., 2012. Drivers of deforestation and forest degradation: a synthesis report for REDD + policymakers. Lexeme Consulting, Vancouver.

Krausmann, F., Gingrich, S., Eisenmenger, N., Erb, K.-H., Haberl, H., Fischer-Kowalski, M., 2009. Growth in global materials use, GDP and population during the 20th century. Ecol. Econ. 68, 2696–2705.

Krausmann, F., Erb, K.H., Gingrich, S., Haberl, H., Bondeau, A., Gaube, V., et al., 2013. Global human appropriation of net primary production doubled in the 20th century. Proc. Natl. Acad. Sci. U.S.A. 110, 10324–10329.

Lal, R., 2004. Soil carbon sequestration impacts on global climate change and food security. Science 304, 1623–1627.

Le, Q.B., Nkonya, E., Mirzabaev, A., 2016. Biomass Productivity-Based Mapping of Global Land Degradation Hotspots, Economics of Land Degradation and Improvement—A Global Assessment for Sustainable Development. SpringerLink, Washington D.C., pp. 55–84.

Le Roux, X., Barbault, R., Baudry, J., 2008. Agriculture et biodiversite. Valoriser les synergies. Expertise scientifique collective, synthese du rapport. INRA, France.

Luyssaert, S., Jammet, M., Stoy, P.C., Estel, S., Pongratz, J., Ceschia, E., et al., 2014. Land management and land-cover change have impacts of similar magnitude on surface temperature. Nat. Clim. Change 4, 389–393.

MA, 2005. Ecosystems and Human Well-Being: Synthesis. World Resources Institute, Washington, DC.

Maddison, A., 2001. The World Economy. A Millennial Perspective. OECD, Paris.

McIntyre, B.D., Herren, H.R., Wakhungu, J., Watson, R.T., 2009. Agriculture at a Crossroads. The Global Report. IAASTD, Washington, DC.

Nellemann, C., Corcoran, E., Duarte, C., De Young, C., Fonseca, L., Grimsdith, G., 2009. Blue Carbon—The Role of Healthy Oceans in Binding Carbon. United Nations Environment Programme, GRID-Arendal.

Nkonya, E., Anderson, W., Kato, E., Koo, J., Mirzabaev, A., von Braun, J., et al., 2016. Global cost of land degradation. In: Springer (Ed.), Economics of Land Degradation and Improvement—A Global Assessment for Sustainable Development. SpringerLink, Washington D.C., pp. 117–165.

OECD, 2008. Environmental Performance of Agriculture in OECD Countries since 1990, Paris.

OECD, 2010. Sustainable Management of Water Resources in Agriculture, Paris, France.

OECD, 2012. Agriculture and Water Quality: Monetary Costs and Benefits across OECD Countries. In: Moxey, A. (Ed.), Pareto Consulting, Edinburgh, Scotland, United Kingdom.

OECD, 2015. Material Resources, Productivity and the Environment. OECD, Paris.

Pan, Y.D., Birdsey, R.A., Fang, J.Y., Houghton, R., Kauppi, P.E., Kurz, W.A., et al., 2011. A large and persistent carbon sink in the world's forests. Science 333, 988–993.

Pauly, D., Watson, R., Alder, J., 2005. Global trends in world fisheries: impacts on marine ecosystems and food security. Philos. Trans. R. Soc. Lond. B Biol. Sci. 360, 5–12.

PBL, 2014. How Sectors Can Contribute to Sustainable Use and cCnservation of Biodiversity. PBL Netherlands Environmental Assessment Agency, The Hague.

Pendleton, L., Donato, D.C., Murray, B.C., Crooks, S., Jenkins, W.A., Sifleet, S., et al., 2012. Estimating global "Blue Carbon" emissions from conversion and degradation of vegetated coastal ecosystems. Plos One 7 (9), e43542.

Popp, J., Lakner, Z., Harangi-Rákos, M., Fári, M., 2014. The effect of bioenergy expansion: Food, energy, and environment. Renew. Sustain. Energy Rev. 32, 559–578.

Post, W.M., Kwon, K.C., 2000. Soil carbon sequestration and land-use change: processes and potential. Glob. Change Biol. 6, 317–327.

Rabalais, N.N., Turner, R.E., Dortch, Q., Justic, D., Bierman, V.J., Wiseman, W.J., 2002. Nutrient-enhanced productivity in the northern Gulf of Mexico: past, present and future. Hydrobiologia 475, 39–63.

Regúnaga, M., Tejeda Rodriguez, A., 2015. Argentina's Agricultural Policies, Trade, and Sustainable Development Objectives, Issue Paper No. 55. International Centre for Trade and Sustainable Development, Geneva, Switzerland.

Riebeek, H., 2011. The Carbon Cycle: Feature Articles, Available from: <https://earthobservatory.nasa.gov/Features/CarbonCycle?src = features-recent>.

Rosset, P., Collins, J., Lappé, F.M., 2000. Lessons from the green revolution. Tikkun 15, 52–56.

Running, S.W., 2012. A measurable planetary boundary for the biosphere. Science 337, 1458–1459.

Scharlemann, J.P.W., Tanner, E.V.J., Hiederer, R., Kapos, V., 2014. Global soil carbon: understanding and managing the largest terrestrial carbon pool. Carbon Manage. 5, 81–91.

Scholz, R.W., Wellmer, F.W., 2013. Approaching a dynamic view on the availability of mineral resources: What we may learn from the case of phosphorus? Glob. Environ. Change 23, 11–27.

Shiklomanov, I.A., 1997. Comprehensive Assessment of the Freshwater Resources of the World. Stockholm Environment Institute (SEI), Stockholm, Sweden.

Siebert, S., Henrich, V., Frenken, K., Burke, J., 2013. Update of the Digital Global Map of Irrigation Areas to Version 5. FAO and University of Bonn, Germany.

Smith, P., Bustamante, M., Ahammad, H., Clark, H., Dong, H., Elsiddig, E.A., et al., 2014. Agriculture, forestry and other land use (AFOLU). In: Edenhofer, O., Pichs-Madruga, R., Sokona, Y., Farahani, E., Kadner, S., Seyboth, K., et al.,Climate Change 2014: Mitigation of Climate Change. Contribution of Working Group III to the Fifth Assessment Report of the Intergovernmental Panel on Climate Change. Cambridge University Press, Cambridge, United Kingdom, and New York.

Steffen, W., Richardson, K., Rockstrom, J., Cornell, S.E., Fetzer, I., Bennett, E.M., et al., 2015. Planetary boundaries: guiding human development on a changing planet. Science 347, 11.

Sutton, M.A., Bleeker, A., Howard, C.M., Bekunda, M., Grizzetti, B., de Vries, W., et al., 2013. Our nutrient world: the challenge to produce more food and energy with less pollution. NERC/Centre for Ecology & Hydrology, Edinburgh.

TEEB, 2008. The Economics of Ecosystems and Biodiversity—An Interim Report. A Banson Production, Cambridge, UK, pp. 64.

TEEB, 2011. The Economics of Ecosystems and Biodiversity in National and International Policy Making. Earthscan, London.

Tubiello, F., Salvatore, M., Cóndor Golec, R., Ferrara, A., Rossi, S., Biancalani, R., et al., 2014. Agriculture, Forestry and Other Land Use Emissions by Sources and Removals by Sinks. FAO, Rome, Italy.

Turner, R.E., Rabalais, N.N., Justic, D., Dortch, Q., 2003. Global patterns of dissolved N, P and Si in large rivers. Biogeochemistry 64, 297–317.

UNCCD, 2011. Land and soil in the context of a green economy for sustainable development, food security and poverty eradication. Submission of the UNCCD Secretariat to the Preparatory Process for the Rio + 20 Conference, November 18, 2011.

UNCTAD, 2014. The global biofuels market: energy security, trade and development. Policy Brief.

UNEP, 2009. From Conflict to Peacebuilding: The Role of Natural Resources and the Environment. UNEP/ Earthprint, Nairobi.

UNEP, 2011. Decoupling natural resource use and environmental impacts from economic growth. A Report of the Working Group on Decoupling to the International Resource Panel. In: Fischer-Kowalski, M., Swilling, M., von Weizsäcker, E.U., Ren, Y., Moriguchi, Y., Crane, W., Krausmann, F., Eisenmenger, N., Giljum, S., Hennicke, P., Romero Lankao, P., Siriban Manalang, A., Sewerin, S.

UNEP, 2014. Assessing global land use: balancing consumption with sustainable supply. A Report of the Working Group on Land and Soils of the International Resource Panel. In: Bringezu, S., Schütz, H., Pengue, W., OBrien, M., Garcia, F., Sims, R., Howarth, R., Kauppi, L., Swilling, M., Herrick, J.

UNEP, 2016. Food Systems and Natural Resources. A Report of the Working Group on Food Systems of the International Resource Panel. In: Westhoek, H., Ingram, J., Van Berkum, S., zay, L., Hajer, M.

Viglizzo, E.F., Frank, F.C., 2006. Land-use options for Del Plata Basin in South America: Tradeoffs analysis based on ecosystem service provision. Ecol. Econ. 57, 140–151.

Vitousek, P.M., Aber, J.D., Howarth, R.W., Likens, G.E., Matson, P.A., Schindler, D.W., et al., 1997. Human alteration of the global nitrogen cycle: sources and consequences. Ecol. Appl. 7, 737–750.

Vitousek, P.M., Menge, D.N.L., Reed, S.C., Cleveland, C.C., 2013. Biological nitrogen fixation: rates, patterns and ecological controls in terrestrial ecosystems. Philos. Trans. R. Soc. B-Biol. Sci. 368 (1621), 20130119.

Vollenweider, R.A., 1992. Coastal Marine Eutrophication: principles and control. In: Vollenweider, R.A., Marchetti, R., Viviani, R. (Eds.), Proceedings of the International Conference on Marine Coastal Eutrophication, first ed Elsevier, Bologna, Italy, pp. 1–20. March 21–24, 1990.

Vollenweider, R.A., Rinaldi, A., Montanari, G., 1992. Eutrophication, structure and dynamics of a marine coastal system: results of a ten years monitoring along the Emilia-Romagna coast (Northwest Adriatic Sea). In: Vollenweider, R.A., Marchetti, R., Viviani, R. (Eds.), Proceedings of the International Conference Marine Coastal Eutrophication, 1st ed Elsevier, Bologna, Italy, pp. 63–106. March 21–24, 1990.

Waite, R., Beveridge, M., Brummett, R., Castine, S., Chaiyawannakarn, N., Kaushik, S., et al., 2014. Improving Productivity and Environmental Performance of Aquaculture. Working Paper, Installment 5 of Creating a Sustainable Food Future. World Resources Institute, Washington, DC.

Waters, C.N., Zalasiewicz, J., Summerhayes, C., Barnosky, A.D., Poirier, C., Galuszka, A., et al., 2016. The Anthropocene is functionally and stratigraphically distinct from the Holocene. Science 351, 137.

Wentworth, J., 2011. Living with Environmental Limits. In: Brogden, G. (Ed.), Living Within the Limits of Natural Resource Systems: Select UK Analyses. Nova Science Publishers.

World Bank, 2004. Responsible Growth for the New Millennium: Integrating Society, Ecology, and the Economy. World Bank, Washington, DC.

World Ocean Review, 2013. The Future of Fish. The Fisheries of the Future, maribus gGmbH, Pickhuben, Hamburg, pp. 143.

Zhu, Z.C., Piao, S.L., Myneni, R.B., Huang, M.T., Zeng, Z.Z., Canadell, J.G., et al., 2016. Greening of the Earth and its drivers. Nat. Clim. Change 6, 791.

CHAPTER

36

Molecules, Money, and Microbes

36.1 INTRODUCTION

Frameworks to map and quantify interactions can help with understanding the complex links between human and natural systems (Liu et al., 2015; Bruckner et al., 2015). A variety of such frameworks have been proposed, including ecosystem services (Bennett et al., 2009), environmental footprints (Hoekstra and Wiedmann, 2014), planetary boundaries (Rockstrom et al., 2009), human-nature nexuses (Liu et al., 2015), telecoupling (Liu et al., 2013), and virtual resource use (Dalin et al., 2012). In this chapter, we draw on these approaches to develop a framework with which to explore the links between the natural resource base and human benefits in the context of food and agriculture. Two that are particularly relevant, beyond ecosystem services and planetary boundaries, which are already elaborated upon in this section in the context of agrifood systems (AFS), are telecoupling and virtual resource use.

Telecoupling is a framework proposed by Liu et al. (2013) to account for the effects on sustainability of spatially distant interactions between natural (environmental) and human (socioeconomic) systems. The telecoupling framework comprises five interrelated components: (1) coupled human and natural systems; (2) flows of material, information, and energy among systems; (3) agents that facilitate the flows; (4) determinants that drive the flows; and (5) effects that result from the flows. For example, in the context of globalization, changes in behavior such as consumption patterns in one location may give rise to displaced impacts in another: goods consumed cumulatively from 1990 to 2008 in the European Union contributed to approximately 9 million hectares of deforestation in other locations (European Commission, 2013, cited by Bruckner et al., 2015).

Similarly, virtual resource use (Dalin et al., 2012; Dalin and Rodríguez-Iturbe, 2016) relates products consumed in one place to the resources that were used in their production elsewhere. Agricultural production depends on resources such as water, land, and agroecological conditions that are largely immobile, thus restricting certain types of agriculture to suitable locations (Dalin and Rodríguez-Iturbe, 2016). Trade, however, has allowed people access to agricultural produce from different places (Porkka et al., 2013; D'Odorico et al., 2014). Virtual resource use is defined as the quantity of resources, such as land or water, required to produce a commodity. Though imported meat, for example, contains

Sustainable Food and Agriculture
DOI: https://doi.org/10.1016/B978-0-12-812134-4.00036-4

very little water as a product, its production required a considerable amount of water, giving rise to a considerable transfer of virtual water (Chapagain and Hoekstra, 2003).

In this section, we use the AFS framework presented in Fig. 32.1 to analyze the flows and linkages connecting the environment and natural resource base to human benefits through agricultural production and agro-industries. The dairy subsector functions as an example illustrating these flows and nodes from three different technical perspectives. A lifecycle assessment approach is used to determine the fate of nitrogen (N), a value chain approach to examine value flows, and a One Health approach to describe the transmission of microbes and their interactions with other organisms. Each of these takes into consideration the range of AFS that exist: extensive, labor-intensive, and capital-intensive. Nitrogen, money, and microbes are just three of many other perspectives that could be taken; for example, a focus on carbon (C) flows and climate change, or emerging diseases or a specific focus on gender and youth.

It is important, however, that such integrated analysis is built around some common concepts.

We have previously introduced the notion of AFS, which connect natural resources and the environment with human benefits, through agricultural production and the attendant agro-industries, both upstream and downstream of production. This provides us with the four interdependent domains, or nodes, described in Fig. 32.1, among which various flows take place. The nodes and flows take different forms. From a nutrient point of view, the flows are chemical reactions that take place in different environments (nodes). From a health perspective, nodes can be the host or vector, and flows can be disease transmission and buildup of resistance. In socioeconomic terms, nodes are people or "agents," and the different organisms engaged in production are connected by flows of value streams.

36.2 THE DAIRY SUBSECTOR

Milk is the single most valuable agricultural commodity, with an annual production (in 2013) of around 770 billion liters, valued at US$328 billion. The species contributing to the global supply of milk are cattle (82.6%), buffaloes (13.9%), goats (1.9%), sheep (1.3%), and camels (0.3%) (FAOSTAT, 2016). Milk, in powdered form, and other processed dairy products are heavily traded, accounting for about 14% of agricultural trade. Dairy farming is carried out in diverse contexts and AFS, reflecting different combinations of production factors and technologies. Extensive dairy production occurs in pastoralist systems, with herders keeping large herds of animals, ranging from a few dozens to hundreds. Dairying is extremely popular in labor-intensive AFS, usually as part of a mixed farming system with two or three animals, typified by smallholder production systems in the East African highlands and South Asia. Smallholder farms produce the largest share of milk production and dairying, even in industrialized economies, and are largely a family business. In Europe, for instance, the average herd of specialized dairy holdings, which derive at least two-thirds of their output from dairy, is about 30. However, a trend toward larger scales is present in almost all countries, and mega-dairy farms are becoming more common. In Ohio, for example, the number of dairy farms with 500 or more cows increased from 10 in 1997 to 50 in 2007 (Ohio Department of Agriculture, 2007); mega-dairies, on the other hand, can host upward of 2000 animals. The largest in the United States reportedly has a capacity of 36,000 animals, and the Mudanjiang mega-dairy farm, located in northeast China, has over 100,000 dairy cows, supplying milk and cheese to Russia.

This great diversity makes dairy a good example with which to illustrate the flows of money, microbes, and molecules. While relatively low economies of scale have enabled smallholders in lower middle-income countries (LMICs) to remain competitive, fierce competition and tight profit margins in high-income countries (HICs) are driving the expansion of mega-dairies. Milk plays an important direct role in nutrition and livelihoods and, with its complex processing chain and wide variety of dairy products, provides employment and many options for value addition. Even in the more automated systems, daily milking requirements create a large human-animal interface, presenting opportunities for microbial transmission in both directions. Dairy cows are big animals, which also creates a considerable environment-animal interface. They require large quantities of food and water, produce large quantities of excreta, and emit large quantities of greenhouse gases (GHGs). Additionally, the high volumes of trade make issues of telecoupling and virtual resource use highly relevant.

36.3 MOLECULES: THE FLOW OF NITROGEN

Achieving sustainable resource use and ensuring that the stocks and flows of resources are managed in an efficient manner throughout the AFS are critical, not only from an environmental standpoint, but also from an economic and social perspective. These measures help improve resource productivity, achieve efficiency gains, and secure adequate supplies of goods and services to the populace, while at the same time limiting adverse impacts on health and the environment associated with their extraction, processing, use, and disposal.

In order to assess all environmental impacts of a product or service, the whole cycle—from cradle to grave—must be assessed. These life-cycle approaches also capture flows that do not enter the economy as priced goods, but that are relevant from an environmental point of view. They can help identify hotspots of losses, accumulation of pollutants, and environmental pressure; they can also reveal how flows of materials shift among countries and within countries, and how this affects the environment within and beyond geographies over various timescales.

Natural Resources and Environment

Nitrogen is a necessary element for every living creature, essential to the nutrition of plants and animals alike and a constituent of all proteins and genetic material. But most atoms of N—which represent 78% of the atmosphere—are bound tightly in pairs as nitrogen gas (N_2). Most organisms cannot break the powerful bond in an N_2 molecule to access N. For plants to grow and animals to thrive, they need the element to be "fixed" into a reactive form (Nr) that can support life, that is, bonded to C, hydrogen (H), or oxygen (O), most often as organic N compounds (such as amino acids), ammonium (NH_4), ammonia (NH_3), or nitrate (NO_3). Plants obtain Nr from the soil or water. Animals get their Nr from eating plants.

Conversion of N_2 into Nr is the process of "nitrogen fixation." In nature, N_2 can be fixed nonbiologically or biologically (Galloway et al., 2003). N fixation through nonbiological processes requires large amounts of energy and accounts for relatively little Nr. N molecules in the atmosphere can be split by the energy generated by lightning strikes and volcanic action. Whenever lightning flashes in the atmosphere, some N combines with O and forms the gas nitric oxide (NO). This NO is converted to nitric acid (HNO_3), which is highly soluble in water and falls to the ground in rainwater to be absorbed by soils. Volcanic eruptions release stored Nr from the Earth's crust in the form of NH_3.

The most important source of Nr is biological nitrogen fixation (BNF), carried out by a few groups of bacteria and archaea. Three groups of bacteria are responsible for fixing 90% of the Earth's Nr: free-living bacteria, which live in the soil (*Azotobacter*); symbiotic and mutualistic bacteria (*Rhizobium*), which live in the root nodules of leguminous plants and which supply the photosynthetically derived energy needed to fix atmospheric N_2; and photosynthetic cyanobacteria, which are found most commonly in water. All of these fix N either as nitrate (NO_3) or ammonia (NH_3). Most plants can take up nitrate and convert it to amino acids.

Agricultural Production

Having enough Nr is fundamental to plant growth, which in turn determines the extent of world food supply. N fixed through natural processes corresponds to less than 0.1% of the N_2 pool and limits primary production in both terrestrial and aquatic ecosystems (Thamdrup, 2012). Pre-modern agriculturalists realized that they could increase soil fertility by including N-fixing crops such as legumes in rotations or adding naturally occurring fertilizers such as manure, guano, or fossil nitrate mineral deposits. But traditional recycling of organic waste can supply only a limited amount of the required Nr: the N content of these products is inherently low, and N losses before and after their application are high. BNF is an expensive process in terms of energy (Oldroyd and Dixon, 2014). Hence, plants fixing N with the help of partner bacteria have to pay a considerable price in the form of carbohydrates delivered to the bacteria per unit of N received, energy which could otherwise be used for growth and reproduction.

By the beginning of the 20th century, it was apparent that there would not be enough natural Nr available to produce the volume of food needed to feed a growing population; consequently, the development of the Haber-Bosch process enabled vast increases in food production. The Haber-Bosch process generates NH_3 from N_2 and hydrogen gas (H_2) (Smil, 2001), using a metal catalyst under high atmospheric pressure and temperatures. This became the mainstay of industrial N fertilizer production, resulting in huge amounts of anthropogenically fixed N being added to the environment. Human-induced production of Nr is now estimated to greatly exceed the entire N fixation achieved by natural processes (Rockstrom et al., 2009). In 2010, human activities created 210 teragrams (Tg) of Nr compared to 58 Tg of Nr by natural processes (Fowler et al., 2013; Vitousek et al., 2013).

The N used in the Haber-Bosch process is taken from the air, but H sources vary depending on location. In China, for example, most H comes from gasified coal, but elsewhere in the world most comes from natural gas. The creation of Nr is extremely energy intensive due to the high temperatures and pressures needed to break nitrogen's triple bond. Fertilizer production consumes about 2% of the global use of energy. Most of that, however, is used as feedstock (to produce H), not fuel. It has been estimated that without the additional Nr produced by the Haber-Bosch process, only 3 billion people—less than 50% of the current global population—would have enough food given current diets and agricultural practices (Smil, 2001; Erisman et al., 2008).

A second source of anthropogenic Nr is a by-product of energy production. Fossil fuel combustion emits N as a waste product (nitrogen oxide, or NO_x) from either the oxidation of atmospheric N_2 or organic N in the fuel (primarily coal) (Socolow, 1999; Galloway et al., 2002). The former creates new Nr, while the latter mobilizes sequestered Nr. Globally, around 75% of anthropogenic Nr production stems from industrial N fixation and 25% from fossil fuel and biomass burning (in the form of NO_x) (Galloway et al., 2008; Fowler et al., 2013).

Nitrogen has arguably the most complex cycle of all the major elements; it forms some of the most mobile compounds that have the ability to travel through a variety of environmental reservoirs through a number of processes. A single N-containing molecule can have a series of impacts on the environment because Nr moves so easily among air, soil, and water. In the air, it can contribute to higher levels of ozone in the lower atmosphere, causing human health problems and damaging vegetation and crops. From the atmosphere, it generally falls to the surface in atmospheric deposition, generating a series of effects: acidifying soils and waterbodies and fertilizing trees and grasslands, creating higher than natural growth rates and nutrient imbalances and impacting on biodiversity. Leaching out of the soils, Nr can pollute both groundwater and surface water, making it unfit for human consumption, and cause eutrophication in inland water and coastal ecosystems, with negative impacts on fish stocks and biodiversity. Eventually, most Nr is denitrified back to molecular N, but a portion is converted to nitrous oxide (N_2O), which is a potent GHG and contributes to stratospheric ozone depletion. This complexity makes tracking anthropogenic N through the natural environment a challenge.

Agricultural Production

To illustrate the many pathways of N in AFS, we use the examples of dairy production in Europe (capital-intensive) and Africa (labor-intensive), following the fate of N from its application as fertilizer to a field, through its uptake by feed crops, its harvest, uptake by the animal, the production and processing of milk and, ultimately, to the consumption of dairy products (Uwizeye et al., 2017). Figs. 36.1 and 36.2 illustrate the flow of N in the two contrasting dairy food systems, including the various issues concerning resource use efficiency and environmental impacts. Each stage has multiple inputs and outputs, which may be produced in different regions. N enters the AFS through synthetic fertilizers, manure and other organic fertilizers, BNF, and atmospheric deposition. It leaves the AFS in the form of meat, milk, and dairy products and losses to the air and water in the form of ammonia, nitrogen oxides, and N_2O.

Nitrogen enters the primary production stage of dairy AFS in various forms, depending on the production system. Nutrients are taken up by pasture and feed crops. In intensive

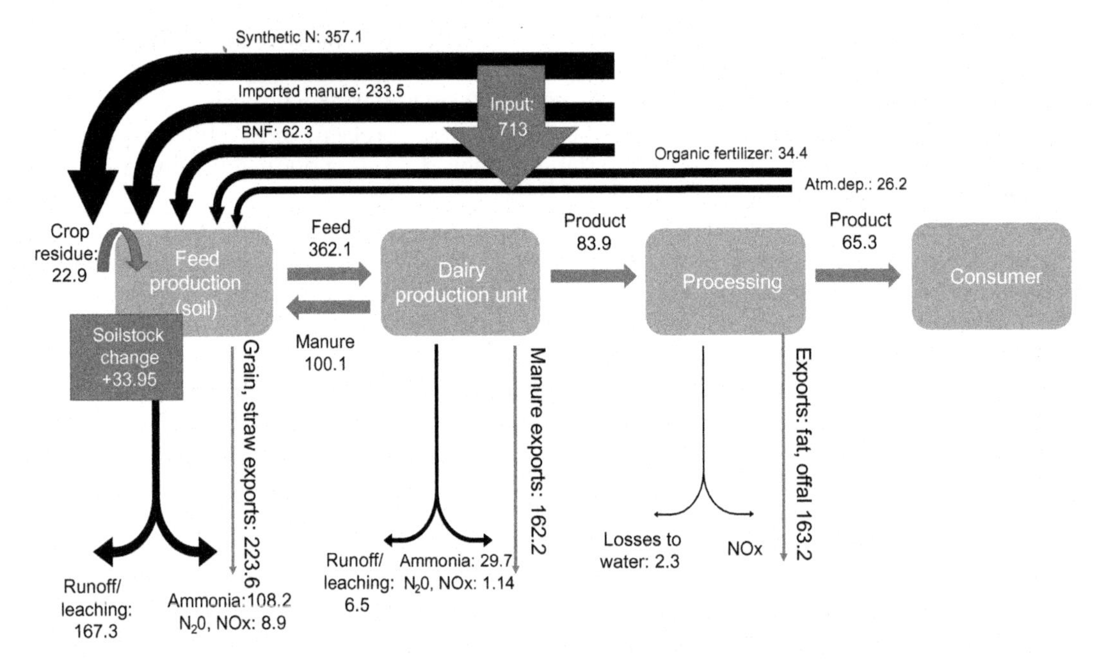

FIGURE 36.1 Nitrogen cycle in intensive dairy systems in Europe. Inputs into the dairy system are shown on the left and outputs of products and losses are shown on the right. Values are given in thousand tons. Source: *Adapted from Uwizeye, A., Gerber, P.J., Groen, E.A., Dolman, M.A., Schulte, R.P.O., de Boer, I.J.M., 2017. Selective improvement of global datasets for the computation of locally relevant environmental indicators: a method based on global sensitivity analysis. Environ. Model. Softw. 96, 58–67.*

systems, including fertilized pastures, most of the Nr is in the form of synthetic fertilizers, while in smallholder dairy systems recycled manure constitutes the main source (Figs. 36.1 and 36.2). There are internal exchanges of Nr within the feed system, with some nutrients in manure being reused and others lost. The fate of Nr in manure depends in part on how the manure is managed.

Figs. 36.1 and 36.2 show the flow of N and the losses to air, groundwater, and surface waters in intensive and smallholder dairy systems, respectively. The single largest source of N pollution is from feed crops and managed pastures and from dairy production. Synthetic fertilizer or manure applied to agricultural soil is not fully taken up by plants: some is lost to the environment, either by volatilization or runoff. In both systems, feed production is by far the largest source of gaseous N losses to air and water (87% and 92% of the N losses in intensive and smallholder dairy systems, respectively Figs. 36.1 and 36.2). About 45% and 56% of the initial N input into the system is lost at the production level in intensive and smallholder dairy systems, respectively.

Human-induced N inputs to agricultural ecosystems through N fertilizer and manure application are very unevenly distributed. In many developing countries, N inputs are lower than crop N uptake, thus mining the soil (see Chapter 35: Natural Resource and Environmental Dimensions of Agrifood Systems, of this section). In contrast, many

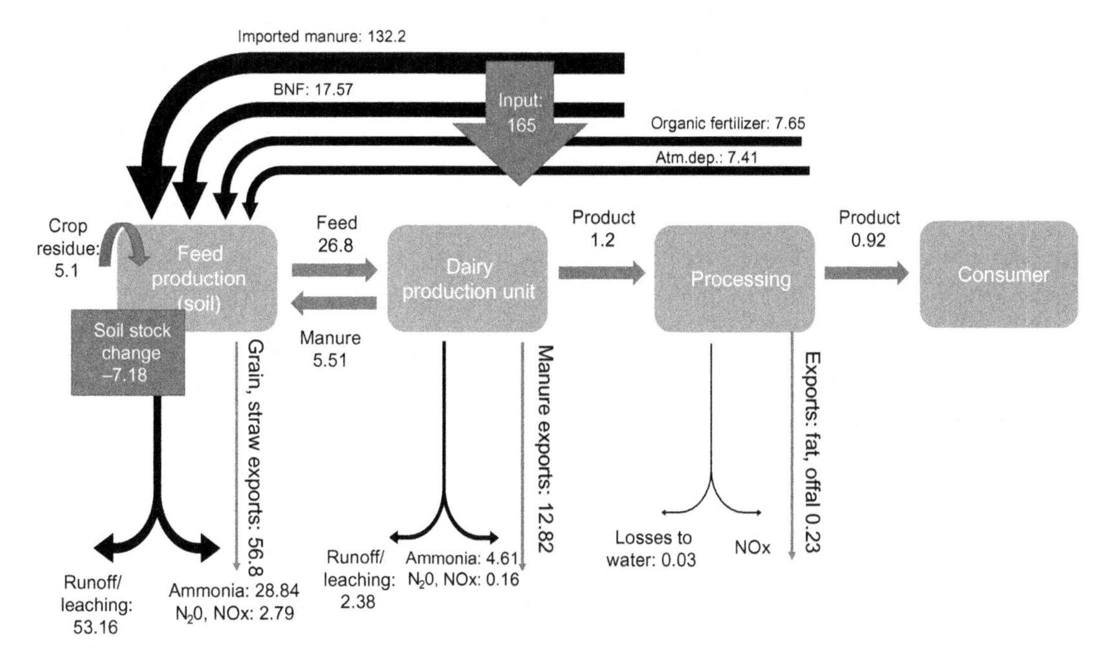

FIGURE 36.2 Nitrogen cycle in smallholder dairy systems in Africa. Inputs into the dairy system are shown on the left and outputs (products and losses are shown on the right. Values are given in thousand tons. Source: *Adapted from Uwizeye, A., Gerber, P.J., Groen, E.A., Dolman, M.A., Schulte, R.P.O., de Boer, I.J.M., 2017. Selective improvement of global datasets for the computation of locally relevant environmental indicators: A method based on global sensitivity analysis. Environ. Model. Softw. 96, 58–67.*

developed and emerging countries show N surpluses. The stock of N in the soil is a balance of N input (atmospheric N deposition, BNF, and synthetic fertilizer) and output (N uptake by plants and N emission to groundwater or the atmosphere). In many African countries, N extracted in the harvest is not fully replaced by N additions, which mines the soil and lowers fertility. Such low-input systems rely largely on manure, BNF, and organic fertilizer (Fig. 36.2), which are not always sufficiently available, are of low nutrient content, and are not optimally managed. In these systems, the soil nutrient status is already low and exported or lost nutrients are not replenished, leading to serious land degradation and low crop yields. In contrast, in large parts of China, the United States, and Europe, N application in excess of crop demand causes elevated N losses to air, soil, and water (Fig. 36.1).

Human activities have not only increased the supply, but also have enhanced the global movement of N. Applied N fertilizers pollute rivers and lakes, cause eutrophication of terrestrial and aquatic systems, and increase global acidification and stratospheric ozone losses (Galloway et al., 2003), in addition to the greenhouse effects of the N_2O released by denitrifying bacteria.

Though the first two distribution mechanisms are important, they are mostly local and regional in scope due to the short persistence of Nr in the atmosphere and to the inherently regional character of hydrological systems. Because of this increased mobility, excess N from human activities has serious and long-term environmental consequences. While passing through the AFS, considerable amounts of Nr are lost in fields, during manure management, in processing, and as waste. The subsequent fates of lost Nr are diverse. Denitrification converts most of the Nr back into unreactive N_2, but also releases ozone-depleting N_2O. If N is lost in reactive form through leaching or volatilization, it can unleash a cascade of effects on the environment, such as eutrophication and acidification.

Distribution via international trade accounts for much of the N that is transported. Lassaletta et al. (2014) observed that during the period 1961–2010, the amount of N traded in the form of agricultural commodities among countries increased eightfold (from 3 to 24 Tg N) and now accounts for one-third of the total N fertilizer used in crop production, with the largest part in the form of animal feed. International trade in agricultural products must therefore be considered an integral part of the biogeochemical processes, shaping the N cycle at the global scale.

The disconnection of crop and livestock farming, resulting from specialization, larger scales, and greater reliance on external inputs, has resulted in the opening of nutrient cycles in agroecosystems. Due to the spatial concentration of animal production, and its geographical separation from feed production, manure has become a form of waste, leading to nutrient overload and pollution instead of being used as fertilizer. This separation also means that crop residues cannot be utilized by livestock and are often burned instead. When N in exported feed is not returned to its place of origin or the excess deposited in its place of use, the cycle is broken, resulting in loss of N efficiency. Imbalances generated by these losses are negative externalities not considered in the price of the product.

Only 23% and 5% of the N in feed used in intensive and smallholder dairy systems, respectively (cf. Figs. 36.1 and 36.2), are captured in the products destined for consumption; the rest is excreted in manure. Much of that manure is reapplied to cropland, where its N is released into the air, leeched into groundwater, or stored in soils. Organic N in manure can be mobilized for crop growth, but timing of application is critical as the plants' demands for N vary seasonally.

The Agro-Industry

Produce leaves the dairy farm either as milk or as live animals destined for slaughter. In smallholder systems, milk is often bulked at milk collection points (dairy hubs) before processing, or may be consumed by the household or sold directly to consumers at the farm-gate or through informal markets. In industrialized countries, the bulk of the milk either undergoes primary processing into fresh livestock products or secondary processing into cheese, butter, and other products. Irrespective of the production system, the losses at the agro-industry node are relatively minor contributions to the N loading of the environment.

Human Benefits

Food produced from N fertilizers provide considerable benefits to consumption, in terms of nutrition and health. However, consumers are responsible for a major part of total waste along the food supply chain. One recent estimate puts the amount of N lost to the environment because of global consumer food waste at 2.9 million tons per year (Grizzetti et al., 2013). In Europe, about 10% of dairy products are wasted by consumers, whereas in sub-Saharan Africa, wastage at the consumer level is negligible (FAO, 2011).

Nitrogen pollution plays an important role in air quality and the resulting impacts on human health. N-related air pollution is linked to higher rates of cardiopulmonary ailments and overall mortality in urban areas. There is also concern about the potential health impact of high levels of nitrate in drinking water. Finally, ecological feedbacks to excess N may inhibit crop growth, increase allergenic pollen production, and possibly increase the prevalence of several parasitic and infectious human diseases, including, but not limited to, cholera, malaria, and West Nile Virus (Townsend et al., 2003).

Discussion

Nitrogen cycles are an integral part of AFS and critical to sustaining the world's population. They have considerable environmental and public health impacts. Environmental consequences of increases in reactive N can have negative effects on food supply and complex feedbacks on environmental systems. For example, N-fueled eutrophication of marine coastal waters contributes to harmful algal blooms and fish kills (Burkholder, 1998), as well as causing dead zones and reef degradation that can harm shellfish and fisheries (NRC, 2000). On land, high levels of tropospheric ozone caused by excess atmospheric N oxides have been shown to decrease crop yields and C sequestration in highly polluted areas. N-intensive agriculture can also reduce soil fertility through acidification, promote "weedy" species invasions, and elevate the risk of agricultural pests and diseases (Matson et al., 1997).

Nitrogen dynamics are closely coupled with C stocks and flows. Because N is limiting in most natural ecosystems, increased N levels usually lead to higher net primary production by stimulating C capture in those systems. Improved N management can increase C sequestration in both soils and plants; it has been estimated that deposited Nr increases carbon dioxide (CO_2) sequestration by 10–50 kg C per kg N (de Vries et al., 2009, 2014). The potential C sequestration benefits of CO_2 fertilization are likely to be constrained by nutrient availability (Wieder et al., 2015).

Developing effective policies to address excesses in Nr is hampered by the complex pathways of Nr, including air, soil and water, humans, plants, animals, and microbes. Considering that a policy to address one problem may aggravate another, policies targeted at reducing Nr emissions need to consider possible cascade effects and trade-offs among different ending points.

Drivers of anthropogenic N fixation, as well as its impacts and solutions to mitigate N losses, are subject to large spatial and temporal variations. In general, the spatial extent of Nr effects increases with time. For example, N_2O remains for more than 100 years in the atmosphere and contributes to climate change on a global scale. Over a shorter time period, Nr effects are mainly local or regional in nature, and its different forms (NH_3, NO_3, and NO_x) have distinct and specific effects.

The mobility of Nr through air, water, and traded commodities (human food, animal feed, and fertilizer) means that some of the impacts of excessive Nr extend to regional and global levels, implying the need for collective responses that span political and administrative borders. Therefore, policy responses to excesses and deficiencies of Nr will be required at all scales, from local to global.

Currently, most HICs as well as emerging economies practice high N inputs to food production, with associated negative environmental and health effects. In many developing regions, on the other hand, anthropogenic N fixation is still low. These regions are expected to show the largest increases in synthetic N inputs in the future. Policies that deal with regional variations will be key to addressing the challenge of maximizing the benefits of Nr for food production while containing its many environmental and health impacts.

Understanding how N cycling affects climate change and vice versa is essential. A changing climate and greater Nr release can interact in a variety of ways. Nr can play a critical role in all aspects of climate change considerations, including mitigation, adaptation, and impacts (Davidson et al., 2011). Several N-cycling processes have contrasting effects on GHG emission processes, including a possible net cooling effect by enhancing CO_2 uptake and sequestration in vegetation and soils. In agriculture, improvements in nutrient management such as proper timing and use of fertilizer will reduce N losses and could also provide some adaptive protection to crops from climate variability (Bruulsema et al., 2009). Yet, variability in weather will also make efficient nutrient management, such as matching N crop needs, more difficult for farmers. Climate change is significantly altering N cycling processes, affecting both terrestrial and aquatic ecosystems, as well as human health (Hristov et al., 2011).

36.4 MONEY: VALUE FLOWS

Dairy value chains link consumers with all of the actors involved in producing, transporting, processing, packaging, and storing milk and dairy products. In each segment of the value chain, a multitude of actors combine natural resources, labor, and capital with the objective of further adding value to milk and dairy products to satisfy consumers' demand. At each step, actors aim to increase value addition and reduce costs, with the consumer ultimately driving the value chain. Consumers include not only household individuals, but also businesses and institutions that collectively feed people, such as retailers, restaurants, and canteens.

Developed in the business world (Porter, 1985), the value chain approach has since been adapted and widely used as a framework within which to explore and analyze the economics and marketing of agricultural produce at local, regional, and global levels. Value chains are the various activities and actors involved in producing and marketing a product. A value chain approach describes how economic value is added and distributed at different stages, including inputs, production, processing, marketing, and sales. Distribution of the benefits to agents is determined by the structure and characteristics of the value chain. Market power affects distribution of benefits and costs across economic agents, such as producers, processors, and consumers. Movements in prices along the supply chain drive the allocation of resources and influence individual economic agents about production levels and use of production factors. Here, we explore the heterogeneity of the dairy value chain, with a focus on telecoupling, or on how benefits from the sector are distributed across actors along the segments of the value chain and across geographies.

Human Benefits

More than 80% of the world's population, or about 6 billion people, regularly consume liquid milk or other dairy products. In 2014, the global dairy market was estimated at US $330 billion (FAOSTAT, 2014). In India in 2011–12, the average person spent between US $32 and US$36 on milk and dairy products, that is, about 20% of all food expenditure and 10% of total individual expenditure (NDDB, 2012). In the United States, per capita expenditure on milk and milk products was between US$116 and US$305 in 2014, depending on income (Food Institute, 2015).

Liquid milk or minimally processed dairy products account for most purchases in LMICs, whereas in HICs, the largest share of expenditure is for highly processed dairy products. This is not surprising, given the price difference between milk and processed dairy products. In many European countries, for instance, a liter of pasteurized milk costs between 0.8 and 1.5 euro, while a liter of fermented milk or yogurt costs about double that amount. In the United States, 30%–35% of consumer expenditure on dairy products is on fresh milk and cream, and 65%–70% on processed dairy products. This latter share is increasing, following the growth of convenience foods.

People benefit from consuming milk and dairy products. They are nutrient-dense foods that supply energy and significant amounts of protein and micronutrients, including calcium, magnesium, selenium, riboflavin, and vitamins B5 and B12. They are the fifth largest provider of energy and the third largest provider of protein and fat for human beings and an important source of affordable nutrition to meet recommended levels. In the United States, at about US$0.23 per 100 kcal, milk and milk products cost less per kcal than meat, poultry, or fish (at US$0.41 per kcal on average) or fruit and vegetables. Their cost is similar to that of eggs, sugar, sweets, and other beverages, and scarcely more expensive than grains, dry beans, legumes, and nuts (Drewnowski, 2010). In Germany, the price of 100 kcal of milk and milk products is about 0.19 euro: only fats, including butter and margarine (0.08 euro per kcal), noodles, rice, and bread (0.11 euro per kcal), and sweets and nibbles (0.13 euro per kcal), are less expensive forms of dietary calories (Westenhöfer, 2013).

However, consumers do not derive the same nutritional benefits from expenditure on milk and dairy products in all global locations. These benefits depend on the average level of income and dietary patterns. For example, household expenditure on milk translates to a high level of per capita consumption ($>$150 L/capita per year) in Argentina, Armenia, Australia, Costa Rica, Europe, Israel, Kyrgyzstan, and North America. Per capita consumption is intermediate (30–150 L/capita per year) in India, Iran, Japan, Kenya, Mexico, Mongolia, New Zealand, northern and southern Africa, most of the Near East, and most of Latin America and the Caribbean. Levels of dairy consumption are low ($<$ 30 kg/capita per year) in most of Central Africa and in east and southeast Asia (FAO, 2017).

The Agro-Industry

Consumers' expenditure on milk and dairy products supports a vast network of retailers, wholesalers, processors and, more generally, the entire agro-industry supporting dairy. The agro-industry operates both upstream and downstream of the primary production unit. Upstream are the inputs to dairy production: the seeds, specialized feed, fertilizers, health care, genetic material, equipment, and financial services. Downstream are the handling, processing, preservation, transportation, and marketing of dairy products and other outputs such as animals for slaughter and hides for leather. The structure and functions of dairy AFS determine the number and types of jobs or livelihoods opportunities it generates, both on-farm and in the associated agro-industries.

In low-income countries, dairy AFS typically have a simple structure. The production of the dairy herd is closely connected to the natural resource base where the herd is raised, external inputs are minimal, milking is manual, and milk is either sold directly across the farmgate or traders collect milk from farmers and sell it to small-scale dairy processing units and consumers in local markets. This is a labor-intensive chain: estimates from Bangladesh, Kenya, and Ghana suggest that between 1.2 and 5.7 jobs are created for every 100 L of milk exchanged on the market, i.e., purchased and sold by local traders either as liquid milk or as processed dairy products (Omore et al., 2004). Liquid milk is the principal product traded, and food safety issues loom large in this type of value chain. For example, lack of cold-chain facilities or improper hygiene practices often lead consumers to purchase spoiled milk, which harms rather than benefits human health. Some developing countries, such as India and Kenya, are characterized by a well-structured dairy value chain, where the agro-industry largely comprises small- to medium-scale cooperatives. Dairy farmers are members of local or national cooperatives that process and sell to consumers either pasteurized milk or other dairy products, such as butter or ghee. India is a case in point: a number of federal and dairy cooperatives of states and union territories collaborate under the National Cooperative Dairy Federation of India to collect milk daily from millions of dairy farmers and sell either liquid milk or dairy products to consumers. In India, 50%–65% of milk is sold as liquid milk, with the rest being processed in products such as cheese, ghee, curd, and butter (FASAR, 2016). The quality of milk and dairy products is high, as milk quality tests are implemented at regular points along the dairy value chain. While relatively few jobs are created directly along the dairy value chain, many are created in the associated agro-industries: a multitude of companies and individuals provide equipment (e.g., milk chilling machines) and services (e.g., marketing) to dairy cooperatives and their members.

The dairy value chain in high- and middle-income countries is typically characterized by a combination of small, local enterprises and international dairy companies. For example, in Italy, there are about 4000 dairy companies, employing a total of less than 40,000 people. In the country, only 0.3% of these companies employ more than 250 people (De Gregorio and Tugnolo, 2015). Lactalis, for example, is one of the top 20 international milk processors by milk volume, which, according to the International Farm Comparison Network, jointly collects, transforms, and passes on to retailers over one-quarter of all milk produced worldwide. The Lactalis Group, with an annual turnover of over 17 billion euro, owns more than 230 industrial plants in 43 different countries and employs a total of 75,000 people (Lactalis, 2017). Another example is Fonterra, a New Zealand-based company owned by over 10,500 farmers. The company has a total revenue of about US$22 billion per year and employs about 22,000 people in more than 100 countries around the world (Fonterra, 2017). The strength of these companies resides in aggregating farmers at the local, national, and global level, improving their competitiveness and capacity to supply affordably priced dairy products to consumers. These companies are responsible for much of the trade associated with milk and dairy products, which accounts for about 14% of global agricultural trade. Whole milk powder and skimmed milk powder are the most traded agricultural commodities globally in terms of percentage of production traded, while fresh dairy products with less than 1% of production traded are the least traded agricultural commodity (OECD-FAO, 2016).

Agricultural Production

Dairy processors purchase liquid milk from a multitude of dairy farmers—dairy animals being a popular asset in rural areas. More than one-quarter of the 570 million farmholdings worldwide, or more than 150 million farmers, are estimated to keep at least one milk animal, including cows, buffaloes, goats, and sheep. There are estimated to be 133 million holdings keeping dairy cattle, 28.5 million with buffaloes, and 41 and 19 million with goats and sheep, respectively (FAO, 2016). Farmers often keep mixed herds with more than one species of dairy animal. Cows are by far the most common dairy animal, with farmers in developing countries usually keeping them in herds of two or three animals (FAO, 2016). In HICs and middle-income countries, however, herds are much larger than this: the average dairy farm in the United Kingdom and the United States manages 90 and 300 dairy cows, respectively (DEFRA, 2016; USDA, 2015). Farms with more than 100 cows represent less than 0.3% of all dairy farms globally (IFCN, 2015). In general, while in low-income countries dairy farming is a popular activity in rural areas—there are over 1 million smallholder dairy farmers in Kenya and over 75 million in India, for example (KDB, 2015; IFCN, 2015) —in HICs, there are few dairy farmers: 13,000 in the United Kingdom and 6400 in Australia, for example (DEFRA, 2016; Dairy Australia, 2015). India is the world's largest milk producer, accounting for 18% of global production, followed by the United States, China, Pakistan, and Brazil. In 2016, however, the European Union generated over half (52%) of global milk exports, followed by the United States and Oceania (USDA, 2017). Dairy animals provide a regular source of food and cash for farmers, who either consume or sell milk and dairy products every day to traders, cooperatives, or processors.

But in the LMICs, dairy animals are far more than a source of food or cash: they are a store of wealth and enhance resilience. Farmers can sell calves in time of need to generate cash; use animals as collateral for loans; and animals can be transported over long distances, thereby maintaining an important asset base even if farmers are forced to leave their homesteads. Dairy animals also generate dung, which is valuable as fertilizer, fuel, and construction material and which can be marketed at any time. They provide a valuable use for crop residues and patches of grass and confer social status and social capital, thereby facilitating networking, which is at the core of effective market and supply chains. In about 25% of cattle-keeping households, or in about 35 million farms, dairy cows are directly owned and/or managed by women (FAO, 2016). Women often feed, milk, and clean the animals and their stalls, compost manure, and take responsibility for breeding and maintaining animal health and for the sale of milk. Dairy often serves as a platform for rural women to consolidate a better place for themselves in their society. As about 22% of the world's women of working age are employed in agriculture and about one-fourth of agricultural holdings, headed by both men and women, keep milk animals, about 80 million women are to some extent engaged in dairy farming (FAO, 2016).

The Natural Resource Base

The dairy farmers' business relies on the use of natural resources, such as land, water, nutrients, and energy. Feeding dairy cows, sheep, goats, and buffaloes requires around 1 billion hectares (ha) of land, or 7% of the total emerged land on earth. The majority of this is grassland (pastures and rangelands), but dairy also uses about 150 million ha of arable land. The global dairy herd consumes about 2.5 billion tons of dry matter feed annually, accounting for about 40% of global livestock feed intake. Seventy-seven percent of this is grass and straw, meaning that the global dairy herd converts materials that are not edible to humans into high-quality protein and essential micronutrients (Mottet et al., 2017). Producing these feed materials requires significant amounts of nutrients and water, but in most cases, access and use of natural resources—with the exception of the cost of purchasing or renting land—come at either low or zero cost to farmers. For example, access to natural pastures and to water points is usually free, though regulated in some circumstances; when in place, taxes on agricultural land are generally low; and possible negative impacts of dairy farming on land, water, and genetic diversity are rarely, if ever taxed, though farmers in some HICs are rewarded for compliance with environment-friendly practices. The situation is even more complex with regard to GHG emissions. Milk animals produce around 3.1 gigatons of CO_2 equivalent per year, or 40% of global livestock emissions, with dairy cattle accounting for 75% (Gerber et al., 2013). In spite of this large contribution to GHG emissions, costs associated with these emissions are not borne by the dairy sector.

Discussion

The dairy value chain is characterized by flows of financial resources from consumers through the agro-industry to farmers, and a corresponding flow of products, from liquid milk to yogurt, cheese and ice-cream, from farmers through the agro-industry to consumers. Benefits are created throughout the chain, with millions of people being directly or indirectly employed by the sector. Additional benefits for society include, for example, nutrition and health, women's empowerment, and employment in the agro-industry.

The way these benefits are distributed is heterogeneous, with a multitude of institutional setups. At one extreme, there are farmers in extensive systems consuming unprocessed milk and selling it to their neighbors; at the other, capital-intensive dairy farms supplying milk under contract to large processors that, in turn, sell processed dairy products to affluent consumers thousands of kilometers away.

In a relatively low state of development and where mixed systems prevail, the dairy sector directly benefits about 30% of rural households, who keep milk animals, and a large number of consumers, who mainly purchase liquid milk and, to a lesser extent, lightlyprocessed dairy products. The average quantity of dairy products consumed is in many cases insufficient to contribute significantly to well-being and nutrition; when the quality of milk is poor, because of the presence of pathogens such as *Brucella*, consuming milk may cause illness.

As economic development progresses, dairy farming intensifies, and large processing companies emerge. Large numbers of people purchase and regularly consume affordably priced liquid milk and processed dairy products, which become staple foods. Intensification reduces livelihoods opportunities at the farm level, with relatively few jobs created along the value chain and many jobs lost at the farm level. In HICs, therefore, the dairy sector largely benefits consumers through the availability of a variety of affordably priced, nutritious, and healthy dairy products, from processed liquid milk to milk-based protein bars and snacks.

From a socioeconomic perspective, intensification of dairy farming allows for increased efficiency along the dairy value chain, which benefits producers, consumers, and all other actors engaged both within and outside of the chain. However, as in all transitions, there are potential losers. As dairy intensification progresses, there is a net reduction of direct livelihoods opportunities, as the number of jobs created in the agro-industry does not compensate for the number of jobs lost at the farm level. Creating alternative economic opportunities is important to absorb the labor surplus such dairy intensification generates. At the same time, telecoupling disconnects the different segments of the value chain; for example, when feed is transported to animals from a distance, and when households consume dairy products produced thousands of kilometers away from where they were produced. The increased disconnection between factors of production along the value chain and between animal agriculture and human ecosystems is a challenge in systems where access and use of natural resources is not only cheap but often unregulated. Eventually, it could lead to increased environmental and public health threats, further exacerbating challenges associated with the fast-growing population and rapidly changing environmental conditions. A key challenge will be to find ways to better regulate and cost the access and use of natural resources, in particular, environmental and health negative externalities: pricing previously uncosted resources has proven to be an effective way to maximize the benefits of telecoupling.

36.5 MICROBES

The One Health approach is defined as "...the collaborative effort of multiple disciplines—working locally, nationally, and globally—to attain optimal health for people, animals and our environment..." (AVMA, 2008). One Health recognizes the connection between the health of people, the health of animals, and the health of the environment. This approach to dealing with global health issues is, where appropriate, gaining considerable traction. It is increasingly accepted that a holistic approach is the only way to address many health issues, and particularly those that are zoonotic. Many microbes have important human, animal, and environmental aspects to their emergence, transmission, and persistence; dealing with the microbes as a whole requires that each aspect be addressed. One Health approaches are generally thought of as means of intervention: acknowledging the need to address issues simultaneously in the three

One Health compartments. We use the approach here to ensure completeness in looking at the evolution, persistence, and transmission of microbes within the AFS.

Natural Resources and Environment

Many components of the natural resource base—soil and water in particular—are teaming with microbes of all types—protozoa, bacteria, archaea, certain fungi and algae, and viruses. These are the most ancient of organisms that have coevolved over millions of years and are the mainstay of life on earth. They are essential components of primary production through symbiotic relationships with plants and are fundamental to the recycling of organic matter and nutrients. They have evolved to exist in a fine balance determined by synergistic relationships and competition for resources. In the oldest of arms races, they have acquired the ability to produce compounds that inhibit the growth of other strains and species—antimicrobial agents—but it is thought that these are also important signaling compounds for microbes. Importantly, gene flows among many of the bacteria occur not only vertically but also horizontally, via mobile genetic elements or plasmids, which means that traits evolved in one species can very quickly be transferred to other species.

Agricultural Production

Agricultural production, be it of crops, livestock, forestry, aquaculture, or fisheries, is heavily dependent on the microbes occurring in the natural environment. Microbes release C and nutrients for primary production, and crops and trees depend on their symbiotic relationship with the soil mycorrhizae to absorb water and nutrients effectively. Of particular importance to the dairy farm are the N-fixing rhizobia bacteria that occur in the root nodules of leguminous plants such as clover, an important component of pasture in temperate environments, which greatly reduces the need for synthetic N fertilizer.

Many tropical forage legumes perform a similar role in balancing the N cycle of grazed pastures. Even in the more intensive dairy production systems, with a higher dependence on concentrated feeds, BNF can play an important role in rotating feed crops with legumes, although much substitution with inorganic N fertilizers would likely be the norm in such systems.

Animals also host myriad microbes, largely within their guts, and particularly so in the case of ruminants, where the gut flora, or microbiome, plays an essential role in the digestion of plant cell walls. The gut microbiome is fundamental to productivity in dairy animals. As with the soil microbiome, there is a fine balance in its composition, which adjusts to the diet and prevailing environmental conditions and changes as animals grow and mature. With dairy cows, the gut microbiome varies also with physiological status—where they are in the lactation cycle—with microbial activity driving the fermentation processes that release the substrates required for producing milk. Any disruption to the gut microbiome therefore threatens not only the health and welfare of the animal, but also the production and quality of milk produced. Finally, dairy cattle account for 20% of livestock-derived GHG emissions through enteric methane production and the gut microbiome is an extremely important mediator of these emissions.

Besides depending for their survival on the naturally occurring microbiome, plants and animals come into contact with other microbes from the environment, some of which are pathogenic in nature. A dairy cow, for example, is exposed to microbes present on the grass where it grazes, its feed and water, the atmosphere and all of the equipment, and people and other living organisms with which it makes contact on the farm, including other members of the herd.

Mastitis, for example, a mammary gland infection usually caused by pathogenic bacteria entering the teat canal, is probably the costliest disease to the dairy industry globally. Yearly, mastitis is estimated to cost the United States dairy industry about US$1.7 to US$2 billion (Jones and Bailey, 2009). There is a long list of mastitis-causing bacteria, including very ubiquitous species such as *Escherichia coli*, *Klebsiella pneumoniae*, and *Staphylococcus aureus*.

Diseases of agriculture caused by microbial agents are treated with antimicrobials, the origins of which mostly lie in the naturally occurring microbes in the environment. These have been developed largely with the objective of treating antimicrobial infections in people, but the very same compounds are used to treat diseases of domestic animals, both terrestrial and aquatic, and of farmed trees and crops. The treatment of diseases with antimicrobials brings about the phenomenon of antimicrobial resistance (AMR) and the inappropriate use of antimicrobials, which includes incorrect dosing, use of antimicrobials that do not correspond to the target pathogen (often resulting from poor diagnosis), and the use of antimicrobials to prevent (rather than treat) disease and to promote growth, which all greatly exacerbate the problem. On the farm, there is considerable transfer of microbes, AMR genetic determinants, and antimicrobials residues to other animals, to the farm environment, and to people and equipment. AFS practices, and in particular on-farm, have a profound impact on these dynamics. In consequence, an emerging pattern is that increased reliance on antimicrobials follows the trajectory of intensification. Large numbers of high density, genetically homogenous animals in highly synchronized production cycles are more prone to the evolution and spread of virulent forms of pathogens. High investments and strong competition put pressure on farmers to speed up growth and prevent disease—using antimicrobials. But this phenomenon is not limited to large-scale farms, as when smallholders operate in high densities, similar pressures are created, for example, in smallholder poultry production in Vietnam.

In 2013, the global consumption of antimicrobials in food animals, estimated at over 130 tons, was almost three times that of human medicine and is projected to reach about 200,000 tons by 2030 (Van Boeckel et al., 2017). In the highly regulated livestock production environments of the HICs, there is growing awareness and public pressure to reduce antimicrobial use in livestock production, particularly with regard to growth promotion and disease prevention. In these hygienic environments, reducing antimicrobial use can be achieved with no loss to productivity. In the burgeoning industrial systems of the LMICs, however, regulation is more difficult and complex. Within these systems, antimicrobial consumption is expected to increase the most if remedial action is not urgently taken.

The Agro-Industry

As farm produce leaves the farmgate (in the case of dairy farms either as milk or as live animals destined for slaughter), it encounters the agrifood industry, whether it be highly developed and industrialized, as with the capital-intensive systems of the HICs, or smaller scale and less formal, as is more typical in the labor-intensive, smallholder systems of LMICs. Untreated milk contains generalist foodborne pathogens, of the types listed above, but may also contain more specific zoonotic pathogens such as *Brucella abortus*, an agent responsible for brucellosis, which remains an important human disease in many parts of the world (WHO, 2006). Milk production and processing usually involves bulking up so even clean milk can readily become contaminated with microbes from other contaminated milk. Live animals arrive at slaughterhouses where they and their attendant microbes come into contact with new environments and different groups of people, such as those working at the slaughterhouses and others who move inputs and products to and from it. The processing plant, slaughterhouse, and associated agro-industry provides a new environment for the transmission of microbes and AMR genetic determinants. It also provides a route for the contamination of the environment with microbes and AMR genetic determinants largely through the disposal of waste and of wastewater from cleaning. From the slaughterhouse the traditional foodborne pathogens—*E. coli*, *Klebsiella* spp., *Staphylococcus* spp., and *Campylobacter* spp., for example—can be transmitted into the human food system.

Also, under consideration here is the agro-industry associated with agricultural inputs: feeds, drugs, and other substances. Such inputs from the pharmaceutical industry enables and extends contamination with antimicrobials and derivatives into the environment.

The nature of the AFS is a very strong determinant of microbial transmission both through milk and meat emerging from dairy farms. In the low-input, labor-intensive systems, standards of hygiene are generally far lower, but so is the inherent risk of pathogen mixing due to the much smaller volumes involved. The higher throughput of the capital-intensive systems has created much greater risks for contamination, which, in conjunction with a far better informed and enfranchised customer base, has driven the development and enforcement of much higher sanitary standards in both milk processing units and slaughterhouses in these AFS.

While some microbes can render milk poisonous, others play critical roles in transforming milk into diverse products, both modern and traditional, and as a main ingredient of the food industry, with a strong presence in fast food. The origins of dairy products such as yogurt and cheese are in prolonging the lifespan of these most perishable of products. An important step in the production of yogurt and cheese, for example, is the bacterial fermentation of milk—converting the lactose in milk to lactic acid—using cultures of *Lactobacilli* and *Streptococci* bacteria.

Human Benefits

In environments where milk is used raw or is inadequately treated, microbes are transferred directly to people who drink it, resulting in infection with zoonotic or foodborne microbes, which may have genes conferring resistance to antimicrobials. If milk for consumption has been contaminated at some stage of the AFS chain, this may cause a "foodborne disease." The global burden of foodborne diseases is quite high (WHO, 2015), as they are the most direct and frequent mode of transmission of pathogens of agricultural origin—mostly with AMR genetic determinants—to humans. Once in the human population, microbes are passed readily from person to person through direct contact, aerosols, and indirect contact (e.g., contaminated surfaces). Transmission rates are particularly high in hospital environments. Transmission also occurs back to agri-food systems and to farms when people (consumers) are involved in these activities. Closing the One Health circle, transmission occurs back into the environment through waste disposal and human sewage.

Humans are treated for microbial infections with antimicrobial drugs, and this in turn gives rise to the development and amplification of AMR genetic determinants. These get back into the agricultural production node either directly through people working on or coming in contact with farms, or indirectly through the environment.

Discussion

Living in close association with microbes of many types, the humans depend on many of them directly or indirectly for its survival. Many other microbes have developed pathogenicity to humans or to the plants and animals that we farm. Some pathogens have jumped the species barrier from animals to people; these zoonotic pathogens are particularly worrisome and difficult to control as they often have an animal reservoir. In many agricultural settings, there is also transmission between domestic livestock and wildlife. The nature of our AFS has very profound implications for the dynamics of microbial flows (including AMR) and much can be done to control these flows for human benefit.

Using antibiotics that are critical to human medicine in livestock and aquaculture systems to prevent disease, cover up for poor sanitary standards, or promote fast growth, for example, is not sustainable. This is why the General Assembly of the United Nations, in September 2016, agreed to a set of resolutions to tackle the problem of AMR, specifically acknowledging the influential role of food and agriculture.

There is also a concerning equity issue around AMR. The poor of the world undoubtedly face the greatest burden of infectious diseases, that is, those with microbial causal agents, and at the same time have the least access to effective antimicrobials with which to treat these infections. The problem of AMR exacerbates this situation by further reducing the effectiveness and availability of antimicrobial drugs, particularly the second- and third-line antibiotics needed to treat AMR infections.

Antimicrobial resistance has been termed the quintessential One Health issue (Robinson et al., 2016), with clear links to the One Health domains of people, animals, and environment. The problem must be tackled on each of these fronts if it is to be resolved. Progress in human and animal health will be limited if we continue to pump antibiotic residues and AMR microbes into the environment, and AMR in human infections must be tackled hand in hand with AMR and the abuse of antibiotics in agriculture. The responsible use of antimicrobials in agriculture, particularly in livestock and aquaculture production, is a fundamental aspect of the sustainability of AFS.

Many of the problems are known and can be dealt with. In recent years, significant advances in genome sequencing technologies have been made, which now enables the elucidation of microbial transmission pathways—microbes can be tracked through the

AFS by genetic fingerprinting to identify hotspots of transmission that can be targeted by interventions. What is needed now is the will to act responsibly, to create incentives to act, and to put institutions and policies in place to facilitate such action.

36.6 POINTS OF CONVERGENCE, SYNERGIES, AND TRADE-OFFS

The preceding sections have taken three technical approaches to exploring the flows of diverse agents—molecules, money, and microbes—through AFS, taking the dairy sector as an example. These are very different agents, each relating to key aspects of sustainability: environment, equity, and health, respectively. The nature of these agents may be different, but they are interconnected. With each agent, there are both positive and negative aspects in relation to the sustainability of the AFS in which they operate, associated with different nodes and with different flows. This model helps identify likely points of intervention for improving sustainability—where we can amplify the positive and reduce the negative contributions. Such a perspective may help build synergies and address trade-offs, both within and between cycles, that need to be understood and accounted for.

For example, applying inorganic N fertilizers to crops and pastures clearly increases yields where N is deficient in the soil and can also result in increased C stocks, which are good for sustainability. The trade-offs, however, are that fertilizer production depends on GHG-emitting fossil fuel combustion and that excessive Nr can pollute air, soil, and water resources.

The intensification of dairy systems, for example, not only raises profits, but also results in cleaner milk and reduced GHG emissions. Intensification, however, is also associated with more distant market connections, such as sourcing protein feed from low-cost production regions such as the Americas. Intensification may benefit society by increasing the availability of affordably priced and healthy milk and dairy products for the population but will reduce on-farm job opportunities for smallholders and noncompetitive or less-efficient farmers, which is a cost to society. Since this trade-off is not peculiar to dairy, but rather characterizes the development of agriculture as a whole, when debating this issue, decision-makers need to consider the fallout on labor markets and the need to develop alternative employment opportunities. Increased efficiency in agriculture releases productive resources, including labor that can be used to satisfy the demand of nonagricultural goods and services of the population, which contribute to societal welfare.

Antibiotics are used to treat diseases of livestock, greatly improving animal welfare, health, and productivity—adding to human benefits. Their inappropriate usage, however, can give rise to the development of resistance in pathogens of medical importance, with potentially massive societal costs.

Another example of trade-offs and synergies in flows and stocks is the application of manure to fields and pastures as an organic fertilizer. In principle, though recycling the waste to enhance primary production of crops or pastures is good practice, there are associated risks, particularly if the manure originates from intensive pig or poultry units in which antibiotics have been used excessively in production. While increasing the flows and recycling of nutrients, the system must also try to curb the flows of antimicrobial residues and AMR genetic determinants. There are many examples linking the cycles of microbes with those of money. Livestock diseases such as mastitis come with enormous economic burdens that eat away at profits, for example. Assuming that the overuse of antimicrobials in agriculture contributes to the problem of AMR in human medicine, the economic costs are potentially enormous (O'Neill, 2016).

In a drive toward increased sustainability of AFS, some of these flows need to be enhanced, in particular those that close nutrient cycles, while others need to be curtailed, such as the transmission of pathogens, particularly those resistant to antimicrobials. Some of the options for working toward these goals are introduced in Chapter 37, Policy Orientations for Sustainable Food and Agriculture. These examples of synergies and trade-offs illustrate the nature of sustainability as an optimization process; aiming towards multiple and variable objectives across space and over time.

References

AVMA, 2008. One Health: A New Professional Imperative. American Veterinary Medical Association.

Bennett, E.M., Peterson, G.D., Gordon, L.J., 2009. Understanding relationships among multiple ecosystem services. Ecol. Lett. 12, 1394–1404.

Bruckner, M., Fisher, G., Tramberend, S., Giljum, S., 2015. Measuring telecouplings in the global land system: A review and comparative evaluation of land footprint accounting methods. Ecol. Econ. 114, 12–21.

Bruulsema, T., Lemunyon, J., Herz, B., 2009. Know your fertilizer rights. Crops Soils 42, 13–18.

Burkholder, J.M., 1998. Implications of harmful microalgae and heterotrophic dinoflagellates in management of sustainable marine fisheries. Ecol. Appl. 8, S37–S62.

Chapagain, A.K., Hoekstra, A.Y., 2003. Virtual Water Flows Between Nations in Relation to Trade in Livestock and Livestock Products. UNESCO-IHE Delft, The Netherlands.

Dairy Australia, 2015. Australian Dairy Industry in Focus 2015. Dairy Australia, Melbourne.

Dalin, C., Rodríguez-Iturbe, I., 2016. Environmental impacts of food trade via resource use and greenhouse gas emissions. Environ. Res. Lett. 11, 035012.

Dalin, C., Konar, M., Hanasaki, N., Rinaldo, A., Rodriguez-Iturbe, I., 2012. Evolution of the global virtual water trade network. Proc. Natl. Acad. Sci. USA 109, 5989–5994.

Davidson, E., David, M.B., Galloway, J.N., Goodale, C.L., Haeuber, R., Harrison, J.A., et al., 2011. Excess nitrogen in the US environment: trends, risks, and solutions. Issues Ecol. 15, 1–16.

De Gregorio, P., Tugnolo, A., 2015. Overview of the Italian Dairy Market. Unibanca - Private & Corporate Unity.

de Vries, W., Du, E., Butterbach-Bahl, K., 2014. Short and long-term impacts of nitrogen deposition on carbon sequestration by forest ecosystems. Curr. Opin. Environ. Sustain. 9–10, 90–104.

de Vries, W., Solberg, S., Dobbertin, M., Sterba, H., Laubhann, D., van Oijen, M., et al., 2009. The impact of nitrogen deposition on carbon sequestration by European forests and heathlands. For. Ecol. Manage. 258, 1814–1823.

DEFRA, 2016. Statistical Data Set: Structure of the Agricultural Industry in England and the UK at June. Department for Environment, Food and Rural Affairs, London.

D'Odorico, P., Carr, J.A., Laio, F., Ridolfi, L., Vandoni, S., 2014. Feeding humanity through global food trade. Earths Future 2, 458–469.

Drewnowski, A., 2010. The cost of US foods as related to their nutritive value. Am. J. Clin. Nutr. 92, 1181–1188.

Erisman, J.W., Sutton, M.A., Galloway, J., Klimont, Z., Winiwarter, W., 2008. How a century of ammonia synthesis changed the world. Nat. Geosci. 1, 636–639.

European Commission, 2013. The impact of EU consumption on deforestation: Comprehensive analysis of the impact of EU consumption on deforestation.

FAO, 2011. Global Food Losses and Food Waste. FAO, Rome.

FAO, 2016. The Global Dairy Sector: Facts. FAO, Rome.

FAO, 2017. Gateway to dairy production and products. Accessible at: <www.fao.org/dairy-production-products/en/>.

FAOSTAT, 2014. FAOSTAT Database. FAO, Rome.

FAOSTAT, 2016. FAOSTAT Database. FAO, Rome.

FASAR, 2016. Innovation in Cold Chain – The Dairy Value Chain Perspective. YES Bank, Mumbai.

Fonterra, 2017. Fonterra, About us.

Food Institute, 2015. Demographics of Consumer Food Spending. Upper Saddle River, USA.

Fowler, D., Pyle, J.A., Raven, J.A., Sutton, M.A., 2013. The global nitrogen cycle in the twenty-first century: introduction. Philos. Trans. R. Soc. B-Biol. Sci. 368 (1621), 20130165.

Galloway, J.N., Cowling, E.B., Seitzinger, S.P., Socolow, R.H., 2002. Reactive nitrogen: too much of a good thing? AMBIO J. Hum. Environ. 31, 60–63.

Galloway, J.N., Aber, J.D., Erisman, J.W., Seitzinger, S.P., Howarth, R.W., Cowling, E.B., et al., 2003. The nitrogen cascade. Bioscience 53, 341–356.

Galloway, J.N., Townsend, A.R., Erisman, J.W., Bekunda, M., Cai, Z., Freney, J.R., et al., 2008. Transformation of the nitrogen cycle: Recent trends, questions, and potential solutions. Science 320, 889–892.

Gerber, P.J., Steinfeld, H., Henderson, B., Mottet, A., Opio, C., Dijkman, J., et al., 2013. Tackling Climate Change Through Livestock: A Global Assessment of Emissions and Mitigation Opportunities. FAO, Rome.

Grizzetti, B., Pretato, U., Lassaletta, L., Billen, G., Garnier, J., 2013. The contribution of food waste to global and European nitrogen pollution. Environ. Sci. Policy 33, 186–195.

Hoekstra, A.Y., Wiedmann, T.O., 2014. Humanity's unsustainable environmental footprint. Science 344, 1114–1117.

Hristov, A.N., Hanigan, M., Cole, A., Todd, R., McAllister, T.A., Ndegwa, P.M., et al., 2011. Review: ammonia emissions from dairy farms and beef feedlots. Can. J. Anim. Sci. 91, 1–35.

IFCN, 2015. Dairy Report 2015. For a Better Understanding of the Dairy World, Kiel.

Jones, G.M., Bailey, T.L., 2009. Understanding the Basics of Mastitis, VCE Publications, 404/404-233 ed. Virginia Cooperative Extension, pp. 1–5.

KDB, 2015. Annual Report and Financial Statements for the Year Ended 30 June 2014. Kenya Dairy Board (KDB), Nairobi.

Lactalis, 2017. The Lactalis Group.

Lassaletta, L., Billen, G., Grizzetti, B., Garnier, J., Leach, A.M., Galloway, J.N., 2014. Food and feed trade as a driver in the global nitrogen cycle: 50-year trends. Biogeochemistry 118, 225–241.

Liu, J., Hull, V., Batistella, M., DeFries, R., Dietz, T., Fu, F., et al., 2013. Framing sustainability in a telecoupled world. Ecol. Soc. 18 (2), 26.

Liu, J., Mooney, H., Hull, V., Davis, S.J., Gaskell, J., Hertel, T., et al., 2015. Systems integration for global sustainability. Science 347 (6225), 1258832.

Matson, P.A., Parton, W.J., Power, A.G., Swift, M.J., 1997. Agricultural intensification and ecosystem properties. Science 277, 504–509.

Mottet, A., de Haan, C., Falcucci, A., Tempio, G., Opio, C., Gerber, P., 2017. Livestock: on our plates or eating at our table? A new analysis of the feed/food debate. Glob. Food Sec. 14, 1–8.

NDDB, 2012. Per Capita Monthly Consumption Expenditure in Milk & Milk Products. National Dairy Development Board.

NRC, 2000. Clean Coastal Waters: Understanding and Reducing the Effects of Nutrient Pollution. The National Academies Press, Washington, DC.

O'Neill, J., 2016. Tackling Drug-Resistant Infections Globally. Joint Programming Initiative on Antimicrobial Resistance (JPIAMR).

OECD-FAO, 2016. OECD-FAO Agricultural Outlook 2016–2025. OECD, Paris, p. 136.

Ohio Department of Agriculture, 2007. Ohio Department of Agriculture annual report and statistics. Accessible at: <http://www.agri.ohio.gov/divs/Admin/Docs/AnnReports/ODA_Comm_AnnRpt_2007.pdf>.

Oldroyd, G.E.D., Dixon, R., 2014. Biotechnological solutions to the nitrogen problem. Curr. Opin. Biotechnol. 26, 19–24.

Omore, A., Cheng'ole Mulindo, J., Fakhrul Islam, S.M., Nurah, G., Khan, M.I., Staal, S.J., 2004. Employment generation through small-scale dairy marketing and processing. Experiences from Kenya, Bangladesh and Ghana. ILRI and FAO, Nairobi and Rome.

Porkka, M., Kummu, M., Siebert, S., Varis, O., 2013. From food insufficiency towards trade dependency: a historical analysis of global food availability. PLoS One. 8 (12), e82714.

Porter, M.E., 1985. Competitive Advantage. Free Press.

Robinson, T.P., Bu, D.P., Carrique-Mas, J., Fevre, E.M., Gilbert, M., Grace, D., et al., 2016. Antibiotic resistance is the quintessential One Health issue. Trans. R. Soc. Trop. Med. Hyg. 110, 377–380.

Rockstrom, J., Steffen, W., Noone, K., Persson, A., Chapin, F.S., Lambin, E.F., et al., 2009. A safe operating space for humanity. Nature 461, 472–475.

Smil, V., 2001. Enriching the Earth. Fritz Haber, Carl Bosch, and the Transformation of World Food Production. The MIT Press, Cambridge.

Socolow, R.H., 1999. Nitrogen management and the future of food: Lessons from the management of energy and carbon. Proc. Natl. Acad. Sci. USA 96, 6001–6008.

Thamdrup, B., 2012. New Pathways and Processes in the Global Nitrogen Cycle. Annu. Rev. Ecol. Evol. Syst. 43 (43), 407–428.

Townsend, A.R., Howarth, R.W., Bazzaz, F.A., Booth, M.S., Cleveland, C.C., Collinge, S.K., et al., 2003. Human health effects of a changing global nitrogen cycle. Front. Ecol. Environ. 1, 240–246.

USDA, 2015. Agricultural Statistics 2015. National Agricultural Statistics Service, United States Department of Agriculture, Washington, D.C.

USDA, 2017. Dairy: World Markets and Trade. United States Department of Agriculture, Foreign Agricultural Service, Washington, D.C.

Uwizeye, A., Gerber, P.J., Groen, E.A., Dolman, M.A., Schulte, R.P.O., de Boer, I.J.M., 2017. Selective improvement of global datasets for the computation of locally relevant environmental indicators: A method based on global sensitivity analysis. Environ. Model. Softw. 96, 58–67.

Van Boeckel, T.P., Glennon, E.E., Chen, D., Gilbert, M., Robinson, T.P., Grenfell, B.T., et al., 2017. Reducing antimicrobial use in food animals. Science 357, 1350–1352.

Vitousek, P.M., Menge, D.N.L., Reed, S.C., Cleveland, C.C., 2013. Biological nitrogen fixation: rates, patterns and ecological controls in terrestrial ecosystems. Philos. Trans. R. Soc. B Biol. Sci. 368 (1621), 20130119.

Westenhöfer, J., 2013. Energy density and cost of foods in Germany. Ernaehrungs Umschau Int. 3, 31.

WHO, 2006. Brucellosis in Humans and Animals. World Health Organization, Geneva, Switzerland.

WHO, 2015. WHO Estimates of the Global Burden of Foodborne Diseases: Foodborne Disease Burden Epidemiology Reference Group 2007–2015. World Health Organization.

Wieder, W.R., Cleveland, C.C., Smith, W.K., Todd-Brown, K., 2015. Future productivity and carbon storage limited by terrestrial nutrient availability. Nat. Geosci. 8, 441–444.

CHAPTER

37

Policy Orientations for Sustainable Agrifood Systems

37.1 INTRODUCTION

In this final step, we combine the different approaches introduced in Chapter 33, Agrifood Systems; Chapter 34, Socioeconomic Dimension of Agrifood Systems; Chapter 35, Natural Resource and Environmental Dimensions of Agrifood Systems; and Chapter 36, Molecules, Money, and Microbes in an attempt to distill some general orientations that may guide the development of policies for sustainable food and agriculture.

Building on the five principles laid out in FAO (2014), such policies need to focus on how to: (1) manage flows of energy, matter, and information in agrifood systems (AFS); (2) protect and enhance natural resource stocks; (3) improve human benefits and their distribution; (4) manage risks; and (5) build institutions and governance.

The fifth of these principles, governance and institutions, is addressed in detail in Sections IV and V of this book. Here, it is covered as a cross-cutting issue in relation to the other four principles, in particular, regarding rules that govern resource allocation and overarching conditions of political authority.

37.2 MANAGE FLOWS OF ENERGY, MATTER, AND INFORMATION

Agriculture involves organic processes that transform natural resources (e.g., carbon, nitrogen, water, and genes) into a stream of human benefits. Natural flows are increasingly complemented with human interventions that provide additional and targeted flows of energy (e.g., fossil fuels), nutrients (e.g., synthetic fertilizer), and other inputs. Only a tiny fraction of the resources initially engaged in this transformation is ultimately retained in the product and consumed. Large losses occur, particularly at the production level, accentuated by the large variability of environmental conditions, especially in low- and middle-income countries (LMIC). These losses are particularly high in systems of low

Sustainable Food and Agriculture
DOI: https://doi.org/10.1016/B978-0-12-812134-4.00037-6

productivity. Low efficiency of resource use is also caused by large amounts of food waste in households, retail, and restaurants, mainly in high-income countries (HIC).

There are various ways to measure the efficiency of flows. Factor productivity considers the economic efficiency of a transformation process. Material flow analysis allows for stocks and flows to be accounted for and helps identify inefficient use of natural resources, energy, and materials in processes, chains, or the economy at large. Lifecycle analysis considers the fate of energy and materials from "the cradle to the grave," not only at the site of production, but also upstream and downstream of production. Indicators like emission intensity consider the amount of greenhouse gases (GHGs) emitted in relation to a unit of product or an amount or value. Measures can also be combined to reflect composite objectives.

Agriculture relies directly on energy, soil, water, and a variety of biological processes. Conditions of production and transformation are highly variable and not easily controlled. They involve complex associations among plants, animals, and other organisms such as microbes, all of which differ from place to place. The large variety of production conditions makes industrialization difficult. AFS "leak" in many important ways, from provision of inputs through production and processing to retail and consumption.

Only a tiny portion of the carbon (C) engaged in the production of food is actually captured in the final product and consumed. C is lost from soils through burning of biomass, methane emissions from livestock and rice, and burning fossil fuels. Global nitrogen (N) efficiency, which is the amount of N captured in food compared to N used during the production of food, is about 20%. N losses to the environment pose severe health and environmental risks and contribute to climate change. More than half of the total N supply to agriculture is now of synthetic origin, increasing and shifting N flows. N is lost from fertilizers, through animal waste, and all along the food chain.

While soil N is a common property resource that has no cost, fertilizer has a price, dependent on energy prices, infrastructure, and subsidies or taxes. Pricing usually does not consider the consequences of N losses, impacts of which occur both in the short and long terms. Some impacts are local, such as pollution of rivers, lakes, and coastal water, while nitrous oxides are important GHGs with global impact. Water quality is increasingly regulated, in particular, putting limits on pollution from nitrates.

Nitrogen efficiency varies widely across AFS. N efficiencies tend to be high in low-intensity, extensive systems; low in systems of intermediate intensity such as labor-intensive systems; and high again as techniques and inputs are refined in high-intensity, capital-intensive systems (Hirel et al., 2011; Ibarrola-Rivas and Nonhebel, 2016). Efficiency of natural resource use, with the exception of land, tends to be high in extensive systems that do not rely on external inputs, but their productivity is typically low. At initial stages of intensification, natural resource use efficiency tends to be rather low, as a wide range of improvements are required. Efficiency then gradually improves with further technological and structural changes that are characteristic of capital-intensive systems. This broad pattern of resource use efficiency is subject to location-specific variations across sectors.

Efficiency of global phosphorus (P) use is very low: on average, 70% of P consumed is lost to the environment (Van Dijk et al., 2016). P is supplied from soils, organic fertilizers, and as a mineral fertilizer. Loss of P, mostly from fertilizer runoff, is generally a local externality with a short- to long-term timeframe. P is priced as an input, but externalities are not costed.

Water efficiency and water depletion are complex location-specific issues. Losses occur through inefficient water storage and irrigation systems, and water use and losses in production and processing. Time frames of impact can be short to long term. While water depletion and degradation are local and regional issues, shortages can be addressed via imports of water-intensive commodities. Water, too, is often treated as a common property resource, with producers only charged for costs of extraction. However, in regions of scarcity, water prices can be high, particularly for consumers, and water access and use is increasingly regulated.

Efforts to optimize the management of flows in food and agriculture have been captured under the term "circular bioeconomy." While a linear economy uses external inputs to produce outputs and waste, a circular economy minimizes the leakage of energy and materials from the system by reusing them. A circular bioeconomy is made up of efficient AFS, and accounts for the fact that flow cycles have different spatial and temporal scales and extents.

Efficiency gains have been made by integrating the different subsectors of crops, trees, livestock, and fish in various combinations. Such mixed systems take advantage of the complementarities between components with integrated C and nutrient flows. Mixed systems are diversified, generating multiple benefits.

Intensification, by contrast, is built on specialization and scale across all sectors. Because of the large natural variability of agriculture, economies of scale are strongest where production conditions are rather uniform and controlled. In their extreme form, they are found in landless monogastric production, fish tanks, greenhouses, and laboratories. Such specialization, often geographically clustered, tends to disrupt nutrient cycles. Some new forms of integration can be found, for instance, in greenhouses that use carbon dioxide emissions from power plants to boost plant growth.

Different flows, of C, N, P, and water, for example, can be combined to optimize production conditions, resulting in synergies and substitution effects. As with all agroecological processes, the flows in AFS span a range of scales. Local flows, at the farm level, can be optimized through on-farm integration. Regional and landscape flows can be optimized through area-wide integration of specialized AFS (between specialized crop and livestock production, for example) and through integration with other sectors. International flows, usually associated with capital-intensive AFS, have greatly expanded their geographical reach, posing a challenge to the circular bioeconomy at global scale. Such flows include the transfer of N and water, typically from North and South America to consumption centers in Asia and Europe and to water-scarce regions of the Near East.

The large variation in resource use efficiency among producers and actors in agroindustries indicates that considerable gains can be made by closing efficiency gaps. These gaps can be found in all types of systems. Capital-intensive crop systems in South America (e.g., Argentina and Brazil) and in Eastern Europe (including Russia), for example, both have the potential to increase their contribution to the global food supply. Such intensification can be sustainable if managed carefully through interventions, such as precision farming and conservation agriculture. Efficiency gaps are usually greatest in labor-intensive, crop and mixed systems in sub-Saharan Africa and Asia, which are currently being held back by low levels of investment and infrastructure. Increasing efficiency requires new forms of integration and a focus on overall efficiency rather than that of a single operating unit. Food waste, for example, is fed to pigs and poultry in Japan and Korea, but in most other countries this practice is banned for fear of introducing pathogens into the food chain. By building a system based on sterilized food waste, South Korea recycled more than 45% of its food waste as animal feed in 2008 (Ju et al., 2016). Flows can also be considered at the microbial level in managing manure, where the benefits of nutrient cycling should not occur at the risk of spreading antimicrobial residues or genetic determinants of antimicrobial resistance. Appropriate manure management can mitigate this trade-off, reducing risks of transmission while also reducing GHG emissions.

While the integration of different flows among systems and subsystems is beneficial, related efforts are often hampered by the ongoing contamination of soils, water, and products (e.g., heavy metals, dioxins, mycotoxins, and pesticides). For example, untreated water contaminated with high levels of agricultural runoff is often unsuitable for modern aquaculture production.

Large gains can be made in natural resource use efficiency across AFS through sets of incentives and regulations, with a clear focus on efficiency and renewable natural resources. In agriculture, many resources can be used without cost, and overuse and pollution are not addressed. Rewards and fees, and implementation of best practices, can address these externalities.

37.3 PROTECT AND ENHANCE NATURAL STOCKS

Progress in efficiency reduces pressure on resource stocks, but it does not protect them from depletion. Food and agriculture critically depend on natural resource stocks and functioning ecosystems. These stocks include C, soils, water, forests, fish stocks, and genetic material. AFS need to protect the natural resource pool to enhance their productive capacity.

Most natural stocks used by agrifood systems have been depleted over centuries, with a dramatic recent acceleration in many parts of the world. Many of the natural resource stocks are held as common property resources (CPRs), which are transformed into private goods such as food. This leads to problems of overuse. Some are global, such as air, oceans, and availability of effective antimicrobials, while others, such as common grazing land, soil quality, communal forests and rivers, have local features. Due to additional demands, these already strained common pool resources have come under pressure everywhere. Adaptive management can work where there is a sense of ownership and effective local institutions. But more often than not, common pool resource use has become restricted, privatized, or turned into degraded and "club" goods, accessible only to a subset of the original users. Stock management also has a role to play in risk management, as healthy stocks act as a buffer in coping with risks.

Soil C stocks have been depleted ever since the advent of agriculture. Soils hold more C than plant biomass (or vegetation) and account for 81% of the world's terrestrial C (IPCC, 2000). Soil C stocks also vary by ecosystem, ranging, for instance, from 100 gigatons (Gt) in temperate forests to 471 Gt in boreal forests (IPCC, 2000). It is estimated that the current rate of C loss due to land-use change (deforestation) and related land-change processes (erosion, tillage operations, biomass burning, excessive fertilizers, residue removal, and drainage of peat lands) is between 0.7 and 2.1 Gt per year (World Bank, 2012).

Approximately 35% of soil used for agriculture around the world is considered degraded (Bai et al., 2008). Soil degradation and depletion leads to a reduction in soil nutrient stock, in particular N, P, and C, and some important micronutrients. For example, soil erosion is a major land degradation process that emits soil C. Because soil organic matter is concentrated near the soil surface, accelerated soil erosion leads to a progressive loss of soil C. The annual rate of soil loss ranges from 7.6 Gt for Oceania to 74 Gt for Asia. Globally, 201 Gt of soil is lost annually to erosion, corresponding to 0.8 to 1.2 Gt of emitted C per year. Africa, Asia, and South America emit between 0.60 and 0.92 Gt of C per year through soil erosion (World Bank, 2012).

Water stocks are being depleted both in quantity and quality. Underground aquifers supply 35% of the water used by humans worldwide. However, groundwater is being over exploitedin many locations, including in areas of major importance for global food security, such as northern China, the Middle East and North Africa, South Asia, the western United States, and Mexico. More than half of the Earth's 37 largest aquifers are being depleted; 21 of them have exceeded their sustainability tipping points, meaning more water is removed than is replaced. Water stocks are being salinized in many dry areas and polluted with effluents from agriculture and other sources.

Forest stocks are declining in terms of both area and forest types. Some estimates suggest that the global forest area has decreased by around 1.8 billion hectares (ha) in the past 5000 years, a decline equivalent to nearly 50% of the total forest area today. According to the Global Forest Resources Assessment 2015 (FAO, 2015a), the global forest area fell by 129 million ha (3.1%) in the period 1990–2015, to just under 4 billion ha. Deforestation, most prevalent in the temperate climates until the late 19th century, is now mostly occurring in the tropics. In those regions, the net annual loss of forest area from 2000 to 2010 was about 7 million ha, whereas agricultural land increased by more than 6 million ha each year. There are significant regional variations: Central and South America, sub-Saharan Africa, and south and southeast Asia all had net losses of forest and net gains in agricultural land.

Genetic resources for food and agriculture are of value to food security, nutrition, and livelihoods. However, biodiversity, and in particular genetic diversity, is being lost at an alarming rate: 75% of genetic diversity of agricultural crops has been lost; 75% of the world's fisheries are fully exploited or overexploited; of the 8800 livestock breeds, 7% are extinct and 17% at risk of extinction. Genetic resources are held both as CPR and as private goods in the form of privately owned genetic material.

AFS are characterized by extensive CPR assets and liabilities, which, in the absence of effective mechanisms of collective coordination, are subject to a range of dysfunctionalities, including conflicts, a deteriorating resource base, and environmental pollution. Without appropriate institutions and governance, these malfunctions undermine the workings of AFS and the socioeconomic benefits that depend on them. Because markets fail to regulate the use of CPR in the interests of economic efficiency and overall human benefits, agencies for collective action, together with broad regulatory frameworks that define resource access and use, are needed in addition to market instruments.

Stocks occur at scales from local to global, but even with global stocks, the rebuilding of compromised stocks depends on local action. Soils must be maintained and, where damaged, recovered. Stocks of soil organic carbon (SOC) must be managed by restoring vegetation, building organic matter, and preventing losses of SOC. There is considerable potential for growing terrestrial C stocks in forests, pastures, and arable land through sustainable forestry, agroforestry, silvopastoralism, conservation agriculture, fertilization, and other mechanisms to improve soil quality. Aquifers and surface water need to be protected from depletion, degradation, and pollution. Green and blue water resources should be managed together and based on integrated water resources management (Rockstrom et al., 2010). Forest stocks must be maintained and restored by slowing and reversing forest losses. Grasslands also must be protected, maintained, and restored where they have become degraded. Organic, terrestrial stocks of all kinds are dependent on pollinators, but the world's natural pollinators, especially bees, are under threat. While pesticide use, monocropping, and pests and diseases are implicated, research is needed to establish the nature of the problems so that interventions to protect them can be implemented. Fish stocks continue to decline largely due to overexploitation. The introduction of fishing quotas may have slowed, but has not reversed these declines. Biodiversity losses continue at an alarming rate, resulting mainly from habitat loss, exacerbated by pollution from fertilizers and pesticides. Recovering lost biodiversity will be difficult, but measures have been identified and implemented to conserve and sustainably use what remains (CGRFA and FAO, 2011).

Extensive systems are well-suited to preserving resource pools if human pressure on the ecosystem stays within the limits that it can absorb. Slash-and-burn agriculture, for example, is perfectly sustainable at the very low population densities that allow the regeneration of the forest between brief cultivation cycles. Payments for environmental services can help by rewarding protective practices, aiming at clean air and water and landscape integrity through noninvasive forms of management. Similarly to managing flows, realistic pricing of access and use of CPR may have an important role to play in their preservation and restoration, as well as regulations and institutional interventions. Pricing not only affects the incentives to conserve the resource, but also the ability to transport and trade it to offset local physical scarcity through imports. Freshwater, for example, is often underpriced for consumers, leading to overconsumption.

An important trade-off question is whether all systems should uniformly intensify, or whether selected areas, which are suitable for intensification, having high productive potential and low vulnerability, should be intensified preferentially. If so, areas where intensification would be difficult to sustain, such as fragile environments with less productive potential, can be spared and freed up for other purposes.

As natural resources will continue to decline at least for some time, these will need to be substituted and complemented by stocks of an anthropogenic nature: knowledge and human capital, physical stocks such as infrastructure, and financial resources, all of which are rapidly expanding. The development of new technologies has been one of the important factors and in allowing continued access to resources. Technological development allows for resource substitution, lowers the cost of development, increases yields and recovery rates, and enables the discovery and use of new resources.

37.4 IMPROVE HUMAN BENEFITS

AFS provide vital benefits to humans, including social, environmental, and health outcomes. They provide food and nutrition, livelihoods, and employment, and support poverty reduction; they also sustain health and contribute to social and political stability. Agriculture provides a number of environmental benefits, such as through protecting and maintaining the quantity and quality of soil, water, air (climate change), and biodiversity.

The benefits derived from agriculture have grown incessantly, and levels of malnutrition and food insecurity have declined dramatically. However, the benefits and costs originating from AFS remain unevenly distributed between and within countries, and often within households as well.

First, poverty is largely a rural phenomenon, with a significant share of the poor depending on agriculture for a considerable part of their livelihoods. This is indicative of poorly performing AFS in some places, notably in low-income countries (LIC). For example, even though the value of African agriculture has increased by more than 16% over the past 30 years, this growth has been largely achieved through cultivating more land and raising more animals, with few improvements in labor productivity, and little improvement to rural livelihoods (NEPAD, 2013). The expansion, particularly since 2000, of social protection programs in rural areas, is a response to malfunctioning markets and institutions that have been unable to support inclusive and efficient AFS, which reduce poverty and hunger and improve farmers' livelihoods (FAO, 2015b). Nowadays, in sub-Saharan Africa, about 40% of the population, or 390 million people, live below the US$1.90/day poverty line.

Second, in many countries there is coexistence of malnutrition and stunting alongside overweight and obesity. This is indicative of the unequal distribution of nutrients and unbalanced diets in a world that produces enough to satisfy the nutritional requirements of the entire populace. In the last 40 years, from 1975 to 2016, at the global level there have been small relative changes in moderate and severe underweight prevalence, from 9.2% to 8.4% in girls and from 14.8% to 12.4% in boys. Conversely, the global prevalence of obesity in children and adolescents has increased from 0.7% to 5.6% in girls and from 0.9 to 7.8 in boys (Abarca-Gomez et al., 2016). The long-term costs of these trends for society cannot be overstated. The most powerful intervention will be to create public awareness and to educate people about the importance of healthy diets. But there are also legislations that can be used, such as price signals that discourage overconsumption, for example, fat and sugar "taxes" and portion size restrictions. If currently free inputs to and externalities of agricultural production are more realistically costed, this measure will ultimately be reflected by increases in the price of food products. People will then tend to substitute these foods for other, less expensive foods. A quick convergence is urgently needed on the issue of diversified healthy diets. Steps to achieve that are described elsewhere in this publication.

Third, foodborne and zoonotic diseases, which jump from animals to humans, diminish the overall benefits AFS generate for society. For example, because of novel and growing interactions between domesticated animals, wildlife, and humans, the frequency of emerging infectious diseases (EIDs) has increased in the last five decades and is anticipated to further increase in the coming years. Outbreaks of EIDs caused by animal pathogens can result in influenza pandemics, whose cost has been estimated to range from US$374 billion (in 2014 US$) for a mild pandemic to US$7.3 trillion for a severe one, with GDP losses estimated at 12.6%, without considering an estimated 142 million deaths (Pike et al., 2014).

Finally, the capacity of AFS to provide environmental benefits is declining, as discussed in Chapter 35, Natural Resource and Environmental Dimensions of Agrifood Systems, of this section. In many settings, AFS contribute to soil and water pollution, GHG emissions, agricultural-driven deforestation, and biodiversity loss, which both reduce the long-term capacity of agriculture to provide social benefits and have broader negative impact on society.

There is ample evidence for development and adoption of technologies and practices that support sustainable AFS. The International Food Policy Research Institute documented a series of success stories for different times and locations, spanning interventions enhancing crop productivity, facilitating market access, eradicating animal diseases, and conserving natural resources (IFPRI, 2010). Elements common to these stories of success are the application of science, policy, multistakeholder partnership, and leadership.

There are three broad areas in which the effective combination of science, policy, multistakeholder partnership, and leadership is most needed for sustainable AFS. First is the design of policies and investments that target multiple objectives simultaneously, for example, economic growth, social inclusion, health, and emissions, and explicitly address their relationships.

Second is the inclusion of the employment dimension in the policy discourse around sustainable AFS. Available evidence shows that agricultural development is associated with intensification, that is, with a reduction of the number of farms and the concomitant increase in average farm size. This process can be sustainable and does not necessarily result in the establishment of mega-farms, as for many products, such as milk for example, contract farming is a viable option. In all cases, there will be a reduction of the total number of jobs available in AFS, as agricultural value chains are rarely labor intensive. Policies and investments that increase the efficiency of AFS, therefore, should go hand in hand with actions that support an increased demand for labor in non agricultural sectors to avoid exarcerbating unemployment and migration. As current technological advancements are making both manufacturing and services less and less labor intensive, this requires a rethinking of labor markets in the coming decades, including a focus on the type of manual, technical, and managerial skills that will be in high demand.

Third is a comprehensive assessment of the costs and benefits of AFS to society in order to formulate policies and investments that support their sustainable transformation. As Stiglitz et al. (2010) note: "We will not change our behavior unless we change the ways we measure our economic performance." Today, we value agriculture by estimating the value addition of agricultural production and the agro-industry, that is, without assessing the positive and negative non monetary contribution of AFS to society. These include supporting farmers' livelihoods and contributing to poverty reduction, quality education, women's empowerment, sustainable migration, and social stability, as well as costly epidemics, overweight and obesity, and environmental pollution and degradation. The value of AFS, as calculated in national accounts, therefore, does not properly measure the net benefits human beings derive from agriculture. This will become a major issue, as in the coming decades a larger and increasingly affluent and urbanized population will demand ever more food, which can, and must be produced sustainably. Devising and implementing methods to accurately assess the value of AFS for society, including valuing nontradable benefits and externalities, is a key step to channel investments that can optimize the multiple benefits that AFS provide.

The global food and agriculture sector is managing to feed the world's population, but with a great deal of inequity and many disservices. Policies governing AFS need to promote a better distribution of human benefits, such as food security, healthy diets, equity, economic growth, employment, and public health. The diverse AFS are intended for distinctive purposes and provide different mixtures of human benefits. Primarily, smallholder systems function to provide subsistence and livelihoods. Capital-intensive systems are effective in supplying growing, often distant, markets. It is important that these diverse and vital roles be recognized in tailoring policies to each situation.

37.5 MANAGE RISKS

Risks to AFS are on the rise. Risks come in the form of growing environmental threats, including droughts, flash floods, and inundations and extreme heat. They also come in the form of emerging diseases, such as avian influenza, that spread through markets and with travel. Risk exposure is growing because of AFS being pushed into marginal and fragile ecologies where the impact of climate variability is more severe.

Growing environmental risks stem from higher temperatures and greater climate variability. These make yields less predictable and require additional costs to make systems more resilient against stress and shocks. Environmental change adds to high resource pressure in the form of water, land, and nutrient scarcity, particularly in extensive and labor-intensive systems. Many dryland regions are affected by even higher average temperatures and prolonged drought, affecting vast areas in Africa, the Near East, Central Asia, Australia, and parts of western North America and Central America. Flooding and torrential rains are raising the risks of destruction and disruption in many low-lying areas, including tropical Asia (recent examples include northeastern India, Bangladesh, Pakistan, and the Philippines) and areas bordering the Gulf of Mexico and the Caribbean Sea. In 2016, insurance companies experienced the highest volume of claims ever from natural disasters (Prodhan and Sheahan, 2017).

Changing weather patterns open new opportunities and transmission pathways for plant pests (e.g., fall armyworm) and animal diseases (e.g., African swine fever, Avian influenza, SARS, and MERS), and some of these pathogens have the capacity to spread to humans. The scale of these disruptions has grown, as pathogens can move rapidly to distant places through trade and travel, which also have provided new opportunities to "established" diseases such as foot-and-mouth disease. Longer food chains also require complex food safety compliance protocols. The disruption of supply and the impact of disease are often large and growing, affecting producers, market supply, and consumer confidence. Corruption, negligence, and poor implementation capacity frequently result in "food scares" (such as *Salmonella* and *Escherichia Coli*) that can greatly affect AFS, with concomitant impact on incomes, employment, and prices, as well as those resulting directly from the pathogen.

Global connectivity both mitigates and accentuates these risks of disruption. The trade in fertilizer, agricultural commodities, and, increasingly, processed products takes advantage of substantial differences in comparative advantage when producing agricultural products and is key to balancing regional food shortages and keeping a relatively stable price regime. The disruption of trade, in the form of export bans or tariffs, has exacerbated food insecurity and social and political instability. With national egoism on the rise, there is a growing risk that trade embargoes will include food, such as the case of the trade restriction imposed in Qatar in 2017. Less confidence in reliable trade may lead countries to hedge against market exposure by more strongly supporting domestic production, reducing the efficiency and welfare gains that can be made from trade. High international demand for certain products has already contributed to price spikes in local markets, such as quinoa in the Andean countries and avocado in Mexico, long considered a staple food in these areas, and "Jamon Iberico" in Spain. Market risks have been changing and pose a particular challenge for small producers and poor consumers. Price transmission can be very rapid, and shortages in one region of the world can have repercussions in faraway places. Most extensive and labor-intensive AFS have poor coverage against risks; attempts to extend insurance schemes to these systems have been shown to work effectively.

As the costs of coping with risks and variability are growing, it is necessary to build resilience to equip AFS to function under a wider range of environmental and microbial parameters (physical structures, technology, and buffers) at each step (considering flows and nodes, compare Chapter 36, Molecules, Money, and Microbes, of this section). Surveillance and response capacity to deal with weather, markets, and disease threats need to be strengthened at all levels, from local to country to global, to deal with the risks of disruption in a proactive and timely manner. Access to insurance schemes for smallholders needs to be facilitated and, where appropriate, subsidized in combination with climate adaptation programs (FAO, 2017).

37.6 BUILD INSTITUTIONS AND GOVERNANCE

AFS operate in different institutional and governance settings that cover diverse aspects, such as land tenure, labor laws and regulations, environmental regulations, food safety requirements, markets and insurance, taxation, subsidies, and service provision, as well customary rights and socio-cultural traditions.

Institutions relevant to sustainable food and agriculture are often patchy and fragmented, and strengthening measures are often poorly implemented, caused by a lack of finance and human capacity. The 2030 Agenda for Sustainable Development (UN, 2015) provides an overall framework for assessing a country's current institutions and governance, and to close gaps, particularly with regard to newly arising challenges such as climate change, disease pressure, and market volatility.

Institutions need to build and facilitate the efficient flow of matter and energy in AFS, starting with agreeing on metrics and indicators that can be used across systems, and at the various steps of transformation. Technical and policy options need to be identified, supported by institutions that provide the necessary regulatory and financial support and that enhance the technical capacity of all actors in different AFS. In the Paris Agreement on climate change, most countries have committed to reducing their emissions, and the majority of countries have included agriculture in their climate action plans. However, implementation has only just started; governance frameworks need to be built at the country level to measure emissions, identify adaptation and mitigation options, and support implementation at all levels, from the natural resources and inputs to production and processing to consumption and waste.

Numerous regulations and restrictions help to protect natural resource stocks, including air, forests, waterbodies, soils, and genetic resources. Implementation is often an issue, particularly in low-income settings, as demographic pressures push smallholders and pastoralists into natural habitats and protected areas. However, implementation is also an issue in MIC and HIC because of the many actors involved in AFS that make compliance difficult to enforce. For example, illegal deforestation continues in many tropical countries despite strict laws, increased policing, and satellite surveillance. Air quality, and the contribution to air pollution through open burning or ammonia emissions, is often not addressed. In many countries, voluntary guidelines cover soil health and erosion, but rates of erosion have not slowed. The use of genetic resources is often poorly regulated, and benefit sharing between the hosts of agricultural biodiversity—farmers and herders—with the commercial users of such diversity remains largely elusive.

Many institutions deal with enhancing the human benefits derived from AFS—and address the human costs of failing AFS in terms of poverty, hunger, poor nutrition, and disease. For poverty and hunger, great improvements have been made through institutional reforms, including the introduction of social protection schemes and the right to adequate food, but efforts are hampered by lack of capacity and finances, corruption, and remoteness of beneficiaries. Many countries are poorly equipped to deal with the consequences of overweight and obesity, including lack of awareness and education, misleading food labeling, and lack of commitment on the side of food suppliers and retailers to offer and promote healthy food choices. Poor food safety, despite the existence of public and private standards, continues to cause widespread illness and death. Poor gender equity, dangerous and unhealthy working conditions, and child labor are persistent institutional problems in many countries, hampering the sustainability of AFS.

Institutions need to be built that allow AFS and their diverse actors and beneficiaries to cope with growing risks. These include functioning markets and supply buffers for inputs and products at different stages of AFS. They also include concerted multistakeholder platforms to reduce the risks of accelerated climate change and its impacts and to counter the risks of disease transmission, and the emergence and spread of antimicrobial resistance, for example. Further, insurance and compensation schemes are required to deal with the fallout of emergencies and disasters in order to limit the impact of disruptions.

At a more general level, environmental and social governance has become a criterion for selective investment and is rapidly growing in importance, based on new-found evidence that ethical investment actually provides higher returns. In agriculture, related efforts have been underpinned by the RAI (responsible investment in agriculture and food systems) principles (CFS, 2014).

37.7 CONCLUSIONS

The issues that we have discussed in this section revolve around the question of feeding a growing and increasingly urban and affluent population sustainably, in a way that does not compromise the ability of future generations to continue to do so. The provision of food, as well as many other human benefits, is entirely dependent on agriculture. Agriculture is, and will continue to be, the largest user of natural resources and is the economic sector that is most exposed to environmental change. With many natural resources already stretched to their limits, the increasing demand for food is clashing with the availability of natural resources and the health of the planet. Addressing the balance of demands for food and nutrition security, equitable livelihoods, and good health on the one hand, and natural resources and the environment on the other hand, is a complex problem. It touches on many aspects of society and the natural world, both present and future, and across geographical space. Addressing such a complex and multifaceted problem requires that it be broken down into more specific and accessible components. Throughout this section, we have taken a number of approaches to do this, providing different lenses through which to consider the sustainability of AFS.

First, we have developed a conceptual model of AFS (Fig. 32.1), comprising four nodes, which are each linked the others in a complex web of interactions and feedbacks. Two of the nodes represent the opposition of human needs and demands (benefits) with natural resources and the environment. Sitting at the interface between human benefits and the environment are two other nodes: agricultural production and the agro-industries that serve production both upstream and downstream of production.

But food and agriculture is extremely diverse, so we have distinguished different AFS for each of the broad subsectors of agriculture: crops, livestock, forestry, aquaculture, and fisheries. We have developed a systems typology based on the level of endowment with the three classical factors of production—land, labor, and capital—and mapped these globally for livestock-, crop- and forest-based AFS as examples. Such mapping approaches distinguish land-intensive (extensive), labor-intensive, and capital-intensive systems, as well as many intermediate combinations of these. With different issues for sustainability applying to the different AFS, locating the systems geographically shows where in the world different types of technical, policy, and institutional interventions will need to be implemented. This framework allows us to embrace the diverse roles of the different systems and the large variations among them. Furthermore, it lends itself to the contemplation of different development pathways and trajectories in terms of their consequences for sustainability.

We then considered the human benefits and the natural resource base in greater detail; selecting the issues that predominate in the different AFS for special attention. This highlights the complexity of the human benefits derived from AFS and of the interrelations between food and agriculture and the environment. This review exposes the large interface between food and agriculture and CPRs, such as clean water, fertile soils, fish stocks, a healthy climate, and effective antimicrobials. As is all too common with shared resources, free access leads to their overuse and depletion. Our review reveals that although human benefits from food and agriculture are large, they are unevenly distributed and include negative benefits such as poor nutrition and disease. It further reveals the major resource footprint of food and agriculture as resulting in considerable unnecessary damage and waste. However, both sides of the balance present opportunities to increase sustainability, for example, through improving equity and efficiency.

We then reverted to our conceptual AFS model, with its representation of nodes and flows, using the example of dairy, in a cross-disciplinary analysis of the movement of molecules (N), money, and microbes to illustrate how different strings of analysis can be constructed around a simple basic model to enable multiobjective decision making. The dairy example shows that there are many processes at work within the food system—biological, social, and economic—which results in feedbacks and interdependencies. We have shown that there are direct and indirect consequences of interactions across these domains, and over different time frames. The dairy case shows that in considering any changes, decision-makers need tools to analyze intended and unintended effects, enabling them to weigh potential effects and balance trade-offs.

This highlights the need to integrate models, particularly combining natural resources with socioeconomics and the health of people, animals, and the environment. Using this integrative approach can help identify likely synergies and trade-offs of different interventions among different components of AFS across geographical space and time. Such analysis points toward the growing dependence on trade over a range of distances and issues related to telecoupling and virtual resource use.

Finally, the five principles of sustainability are interpreted in light of these perspectives. Steering AFS toward current and future sustainability will require comprehensive investments in order to address the need for long-term AFS development, innovations to accelerate technical and organizational change, incentives to promote good practices that address external effects and reflect scarcities, and institutions to protect people and resources, manage risks and facilitate a smooth transition of AFS to accommodate growing demands.

If the current unsustainable AFS practices continue to be followed, the future of food and agriculture is heavily discounted. Unless realistic costing of CPRs and externalities of AFS is introduced and widely adopted, this will continue to be the case. Addressing the sustainability of AFS requires approaches that account for the diversity of systems, so that interdisciplinary interventions are tailored and targeted and embrace the concerns and aspirations of present stakeholders and future generations.

References

Abarca-Gomez, et al., 2016. Worldwide trends in body-mass index, underweight, overweight and obesity from 1975 to 2016: a pooled analysis of 2416 population-based measurement studies in 128.9 million children, adolescents and adults. Lancet 390 (10113), 2627–2642.

Bai, Z.G., Dent, D.L., Olsson, L., Schaepman, M.E., 2008. Proxy global assessment of land degradation. Soil Use Manage. 24, 223–234.

CFS, 2014. Responsible investment in agriculture and food systems, In: Commission of Food Security. (Ed.). FAO, WFP and IFAD, Rome.

CGRFA, FAO, 2011. Second Global Plan of Action for Plant Genetic Resources for Food and Agriculture. FAO, Rome.

FAO, 2014. Building a Common Vision for Sustainable Food and Agriculture, Principles and Approaches. FAO, Rome.

FAO, 2015a. Global Forest Resources Assessment 2015. FAO, Rome.

FAO, 2015b. The State of Food and Agriculture 2015 (SOFA): Social Protection and Agriculture. FAO, Rome.

FAO, 2017. Tracking Adaptation in Agricultural Sectors - Climate Change Adaptation Indicators. FAO, Rome.

Hirel, B., Tetu, T., Lea, P.J., Dubois, F., 2011. Improving nitrogen use efficiency in crops for sustainable agriculture. Sustainability 3, 1452–1485.

Ibarrola-Rivas, M.J., Nonhebel, S., 2016. Variations in the use of resources for food: land, nitrogen fertilizer and food nexus. Sustainability 8, 16.

IFPRI, 2010. Proven Successes in Agricultural Development. A Technical Compendium to Millions Fed. IFPRI, Washington D.C.

IPCC, 2000. Land Use, Land-Use Change and Forestry - Special Report. Intergovernmental Panel on climate change, Meteorological Office, Bracknell, United Kingdom.

Ju, M., Bae, S.-J., Kim, J.Y., Lee, D.-H., 2016. Solid recovery rate of food waste recycling in South Korea. Journal of Material Cycles and Waste Management 18, 419–426.

NEPAD, 2013. Agriculture in Africa. Transformation and Outlook. New Partnership for Africa's Development, Johannesburg.

Pike, J., Bogich, T., Elwood, S., Finnoff, D.C., Daszak, P., 2014. Economic optimization of a global strategy to address the pandemic threat. Proc. Natl. Acad. Sci. USA 111, 18519–18523.

Prodhan, G., Sheahan, M., 2017. Insurers Paid Out $50 Billion for Natural Disasters in 2016. Reuters.

Rockstrom, J., Karlberg, L., Wani, S.P., Barron, J., Hatibu, N., Oweis, T., et al., 2010. Managing water in rainfed agriculture - the need for a paradigm shift. Agric. Water Manage. 97, 543–550.

Stiglitz, J.E., Sen, A., Fitoussi, J.-P., 2010. Mismeasuring Our Lives. The New Press, New York.

UN, 2015. Transforming our World: the 2030 Agenda for Sustainable Development. United Nations, New York.

Van Dijk, K.C., Lesschen, J.P., Oenema, O., 2016. Phosphorus flows and balances of the European Union Member States. Sci. Tot. Environ. 542, 1078–1093.

World Bank, 2012. Carbon Sequestration in Agricultural Soils. World Bank, Washington, DC, pp. 1–118.

SECTION IV

OPERATIONALIZING SUSTAINABLE FOOD AND AGRICULTURE SYSTEMS

Lead Authors

Leslie Lipper and Jeroen Dijkman (CGIAR Independent Science and Partnership Council, c/o Food and Agriculture Organization of the United Nations (FAO), Rome, Italy)

Co-authors

Regina Birner (Institute of Agricultural Sciences in the Tropics (Hans-Ruthenberg-Institute), University of Hohenheim, Stuttgart, Germany), Andrew Hall (Agriculture and Food, Commonwealth Scientific and Industrial Research Organisation (CSIRO), Black Mountain, Canberra, ACT, Australia), Patrick P. Kalas (Food and Agriculture Organization of the United Nations (FAO), Rome, Italy), Rachid Serraj (CGIAR Independent Science and Partnership Council, c/o Food and Agriculture Organization of the United Nations (FAO), Rome, Italy), Kumuda Dorai (LINK limited, Canberra, ACT, Australia), Misael Kokwe (Food and Agriculture Organization of the United Nations (FAO), Lusaka, Zambia), Preet Lidder and Molly Conlin (CGIAR Independent Science and Partnership Council, c/o Food and Agriculture Organization of the United Nations (FAO), Rome, Italy).

Summary

Progress toward sustainable food and agriculture (SFA) in policy and practice has thus far been sporadic and inconsistent. This section proposes a conceptual framing for operationalizing SFA that identifies key social-ecological features of systems and uses evidence and dialogue to build knowledge and address trade-offs. It subsequently highlights a number of mechanisms and approaches to realize behavioral and practice change. The section concludes that while there has been progress in identifying entry points for action, achieving the changes required in existing production and consumption patterns will take time, capacity, and political courage.

CHAPTER

38

Conceptual Framing of the Operationalization of Sustainable Food and Agriculture

38.1 INTRODUCTION

Sustainable agriculture has been on the policy agenda for decades. Nevertheless, we still live in a world where the lack of sustainability in food and agriculture systems is widespread, posing grave threats to human well-being both now and in the future. The way in which different systems are unsustainable is as varied and complex as the concept itself. Environmental damage is perhaps most commonly associated with unsustainable systems, but lack of economic viability or negative social impacts are equally important features of unsustainable systems, given the multiobjective nature of the concept of sustainability. There is a vast amount of literature on the nature of sustainability in agriculture and food systems, which indicates considerable variation in the specific definition of the concept. Nonetheless, there is a common understanding that sustainability requires the balancing of economic, social, and environmental objectives to achieve "development that meets the needs of the present without compromising the ability of future generations to meet their own needs" (WCED, 1987, p. 41). van Kerkhoff and Lebel (2006) define sustainable development as "the process of ensuring all people can achieve their aspirations while maintaining the critical ecological and biophysical conditions that are essential to our collective survival." Clark et al. (2016) define sustainable development as "the promotion of inclusive human well-being; this is to say, well-being that is shared equitably within and across generations and is built on the enlightened and integrated stewardship of the planet's environmental, economic, and social assets."

Operationalization is defined as "put into operation or use" (Oxford Dictionary).

Sustainable Food and Agriculture
DOI: https://doi.org/10.1016/B978-0-12-812134-4.00038-8

Unsustainable systems then are ones where imbalances between multiple objectives exist and manifest themselves in a great variety of ways. In this section, we aim to identify different typologies of unsustainability in food and agricultural systems—and the actions that can be taken to enhance dynamic processes that underpin sustainability across a range of food and agricultural systems.

We should first step back, however, and consider why the lack of sustainability in agricultural systems continues to be an enduring problem for humankind. As Velten et al. (2015) notes. "Since the beginnings of the debate about sustainable agriculture, there has been a great variety of conceptions of the term. It has been claimed that this multitude of different and partially opposing definitions has made the realization of sustainable agriculture a fuzzy affair, and caused confusion by exacerbating differences in the views of different stakeholder groups."

The difficulty of operationalizing sustainability in food and agricultural systems lies in the nature of the problem, as well as our capacity to grapple with it. Sustainability is a "wicked" problem, meaning a "social or cultural problem that is difficult or impossible to solve for as many as four reasons: incomplete or contradictory knowledge, the number of people and opinions involved, the large economic burden, and the interconnected nature of these problems with other problems" (Kolko, 2012; Rittel and Webber, 1973). For example, DeFries and Nagendra (2017) point out that the regulation of ecosystem functions has increasingly become a serious problem since mechanisms related to self-regulation have been disrupted by the use of capital and technologies to substitute for ecosystem services. Wicked problems are not solved, but rather involve solutions that are judged better or worse—or good enough over time (Batie, 2008; adapting Kreuter et al., 2004). This fits well with the notion of sustainability as a continuing dynamic and adaptive process rather than an endpoint at which one can arrive.

Looking through the vast literature on how and why food and agricultural systems are unsustainable, much of which is covered in earlier sections of this book, we can see that two major issues consistently arise:

1. People lack the knowledge needed to understand the complex dynamics of food and agricultural systems and, consequently, management strategies have frequently led to unintended consequences.
2. There are significant costs, or trade-offs, involved in moving onto sustainable trajectories, which are often not recognized, or relate to imbalances of power among different social groups and are thus not addressed through mechanisms for reducing or resolving them.

At a very general level, operationalizing sustainability dynamics in food and agricultural systems requires addressing these issues through a combination of knowledge and awareness creation to inform stakeholders about the complexities of food and agricultural systems and the potential effects of different strategies. In addition, it should empower their actions to enhance the sustainability of the food and agriculture system in which they participate in—whether as a policymaker, farmer, food processor, retailer, or consumer. Ultimately, operationalizing sustainability involves changes in the behavior of actors operating throughout agrifood systems to create the required changes, but also to demand needed changes in social, ecological, and economic performance as central pillars of sustainability.

38.2 ELEMENTS OF AN OPERATIONAL FRAMEWORK

Using the conceptual framing of sustainable food and agriculture (SFA) presented in Section III, we can think of operationalizing sustainability as a dynamic process of changing the balance between the biophysical and socioeconomic dimensions of food and agricultural systems with multiple entry points, depending on linkages and flows in the systems. This gives us a starting point for identifying actions to stimulate desired change. Generally, given the nature of the problem, we know that injecting actions to stimulate sustainability into the current dynamic trajectories of food and agricultural systems requires a mixture of knowledge, capacity, and incentives to take action, identify, and resolve potential conflicts and adapt them in response to changing circumstances. We can, however, become more context-specific in identifying which entry points make sense under which conditions and how they might be pursued.

The FAO (2014) outlined four areas of action to instigate the transition to SFA: (1) building relevant, coconstructed, and accessible evidence; (2) engaging stakeholders in dialogue to build common understanding and joint action; (3) formulating tools and levers to enable and incentivize changes in food and agricultural systems; and (4) realizing practice change through innovative approaches and solutions. In this section, we revisit and expand upon these four action areas to derive guidance on how to approach the wicked problem of operationalizing sustainability across different types of agrifood systems. The definition of agrifood system that we will use for the remainder of this section is:

> An interconnected web of activities, resources and people that extends across all domains involved in providing human nourishment and sustaining health, including production, processing, packaging, distribution, marketing, consumption and disposal of food. The organization of agri-food systems reflects and responds to social, cultural, political, economic, health and environmental conditions and can be identified at multiple scales, from a household kitchen to a city, county, state or nation *Grubinger et al. (2010)*.

Priority actions for operationalizing sustainability will vary across different types of agrifood systems, depending on the specific characteristics or imbalances that generate unsustainable dynamics. Thus, though we cannot build a blueprint for sustainability action across all food and agriculture systems, we can develop typologies and systems to help identify priority actions to enhance sustainability dynamics. The analysis presented in Section III—as indeed in all the sections of the book and much of the literature on SFA systems—indicates that specific combinations and interactions between biophysical and socioeconomic processes drive the dynamics of sustainability, or lack of thereof.

Section III of this book relates the five sustainability principles for agrifood systems laid out by the FAO (2014) to four major action areas, as follows:

1. Manage flows of energy, matter, and information in agrifood systems.
2. Protect and enhance natural resource stocks.
3. Improve human benefits and their distribution.
4. Manage risks.

In the analysis presented, the fifth sustainability principle on governance and institutions is addressed as a cross-cutting issue in each of these four action areas; for example, governance and institutions are needed to realize all of them.

This analysis gives us a basis for characterizing the nature of the forces underlying unsustainability in different agrifood systems and identifying priority areas for operationalizing sustainability. The key features and drivers of unsustainability in agrifood systems involving extensive and largely subsistence-oriented pastoralist livestock production in sub-Saharan Africa, for example, are very different from those of a highly automated United States Midwest monocropping corn system producing for animal feed and high fructose corn to be processed into food products for domestic and international consumer markets. In the pastoralist case, the flows of energy and information are generally low. Natural resource stocks are being drawn down from processes occurring both within and external to the agrifood system, food insecurity is a major problem, and production risks from transboundary diseases and climate are high. Conversely, in the United States Midwest corn case, flows of energy and information are very high, natural resource stocks may be damaged through pollution, human health is at risk from consumption of high fructose corn syrup products, and the system faces major risks from changes in markets and trade patterns that are often higher than production-related risks. In both cases, there is a need for changes in behavior from different participants in the food system: policymakers, producers, traders, consumers, processors, and other value chain participants of the agrifood systems—but the nature of the changes required to enhance the sustainability of the system are radically different.

This brief example gives rise to a first principle for operationalizing sustainability: the need to understand the relevant context by identifying the nature of imbalances in the particular agrifood system under consideration. Essentially this requires a process of information gathering, discussion, and consultation to understand the setting and leverage points within a given agrifood system to help identify imbalances and the drivers in the system that are generating these unsustainable processes.

Clearly, just understanding the basis of unsustainability is not enough to effectuate changes toward a more sustainable agrifood system. A second operational principle involves generating useful knowledge to identify priority actions and their modes of implementation to support changes toward sustainability. Here, the key is making sure knowledge is useful, and that in itself requires interactions with key stakeholders who have differing levels of capacity and power to effectuate change, necessitating a somewhat different approach to knowledge generation than has generally been applied in the past.

The third basic operational principle is facilitating changes in behavior from a wide and very diverse range of actors in the food system: from consumers, processors, regulators, and marketers to farmers, herders, fishers, or foresters who are actively involved in transforming natural resources into agricultural products and services. Though most effective entry points and feasible actions to operationalize sustainability vary, they consistently involve a process of building knowledge among key stakeholders and utilizing a range of approaches to realize change on a meaningful scale.

These elements of action can be outlined in a simple framework consisting of three main elements: (1) identifying the relevant context; (2) generating knowledge for action; and (3) facilitating pervasive change in behavior to achieve major shifts in the dynamics of agrifood systems. The components under each element—and the two elements themselves—do not follow a linear progression, but often occur in repeating loops or by

FIGURE 38.1 SFA action framework.

jumping from one component to another. For this reason, the action framework takes the form of a circle with three major interacting components (Fig. 38.1).

In operationalizing SFA through agrifood systems, it is important not only to consider the wide scope of agriculture and food-related activities, contexts, and players, but also to recognize that collectively these operate as dynamic, complex systems that operate at farm to societal scales. Operationalizing SFA involves changes in these systems as much as it does changes in component parts. That is to say, it involves changes in values, networks, and behavior as well as technology, market, and policy responses and drivers.

The move toward sustainably and locally produced food, for example, reflects a societal change that is mirrored in and supported by policy and market changes, technological responses, and support networks.

CHAPTER

39

Identifying the Relevant Context: Key Features of the Social-Ecological System

Given the importance of balancing human and natural systems to achieve sustainability, it is essential to understand the interactions between the two in order to identify the relevant context in operationalizing sustainability for agrifood systems. The concept of social-ecological systems provides a means of doing just that. Social-ecological systems (SES) are defined as linked systems of people and nature, emphasizing that humans must be seen as a part of—and not apart from—nature (Berkes and Folke, 1998). SES are complex and adaptive, operating over different scales and that embody feedback mechanisms. Knowledge, institutions, and governance are all instrumental in linking social and ecological systems, thus constituting an essential feature of SES. According to Matson et al. (2016), SES "are complex adaptive systems—systems that have multiple interconnected components that interact in diverse ways. They exhibit positive and negative feedbacks, connections across space and time, and non linearities and tipping points that influence the way the system works and the way it changes with each new intervention."

There are different approaches to describing SES (Musters et al., 1998; Adger, 2000; Eakin and Luers, 2006; Matson et al., 2016). A set of 10 of the major approaches being used was identified and compared by Binder et al. (2013). The approaches vary in terms of their focus, but all consider various aspects of interactions between social, economic, and ecological systems. Binder et al. (2013) argue that deciding which framework is most applicable in any given situation depends on the type of questions that the analysis seeks to address—is the focus more on how the social system impacts the ecological or vice versa?

In the remainder of this section, we present three examples of SES approaches being applied to specific issues of sustainability in food and agricultural systems and the way they can help develop action plans for operationalizing sustainability in each context.

Sustainable Food and Agriculture
DOI: https://doi.org/10.1016/B978-0-12-812134-4.00039-X

Matson et al. (2016) propose an approach that uses the stock of five capital assets (natural, manufactured, social, human, and knowledge) and the flows in and out of the stocks driven by natural and social processes to characterize SES. This approach can be applied to the Midwest corn agrifood system example described below, based on information provided in an article raising issues about the sustainability of this agrifood system (Foley, 2013).

39.1 UNITED STATES MIDWEST CORN SYSTEM PRODUCING FOR ANIMAL FEED AND HIGH-FRUCTOSE CORN SYRUP

- *Natural capital*: In the United States, more land is devoted to corn than any other crop. Between 2006 and 2011, the amount of cropland devoted to growing corn in America increased by more than 5 million hectares, mainly in response to rising corn prices and the increasing demand for ethanol. US corn consumes a large amount of freshwater resources, including an estimated 23.34 km^3 per year of irrigation water withdrawn from America's rivers and aquifers.
- *Human capital*: The United States Midwest corn system, while very efficient in producing corn, is not very efficient in actually producing food for people since most of the corn is for animal feed or high-fructose corn syrup (Foley, 2013). The corn-fed livestock have a generally low rate of conversion of grain to meat, and high levels of consumption of meat and products made with high-fructose corn syrup are associated with increased risk of noncommunicable disease.
- *Social capital*: The corn system receives more subsidies from the United States government than any other crop, including direct payments, crop insurance payments, and mandates to produce ethanol. In all, US crop subsidies to corn totaled roughly US $90 billion between 1995 and 2010—not including ethanol subsidies and mandates, which helped drive up the price of corn.
- *Manufactured capital*: Over 5.6 million tons of nitrogen is applied to corn each year through chemical fertilizers, along with nearly a million ton of nitrogen from manure. Much of this fertilizer, along with large amounts of soil, washes into the nation's lakes, rivers, and coastal oceans, polluting waters and damaging ecosystems along the way.

Trends: There is a reduction in the diversity of the American agricultural landscape, with even more land devoted to corn monocultures. Roughly 0.5 million hectares of grassland and prairie were converted to corn and other uses in the Western Corn Belt between 2006 and 2011, posing a threat to waterways, wetlands, and species that reside there.

Vulnerabilities: The monolithic nature of corn production presents a systemic risk to US agriculture, with impacts ranging from food prices to feed prices and energy prices.

Trade-offs: The US corn system essentially involves a trade-off of depleting natural resources to deliver relatively little food and nutrition to the world. The US investment in natural and financial resources is not paying the best dividends to its national diet, rural communities, federal budget, or environment. For sustainability, the US corn agrifood system should consider not only just the production of corn (which is highly efficient), but also the interests of lobbyists, big business, and the government, who have largely created the system (Foley, 2013).

More detailed SES assessments have been done using the SES framework approach, which has been identified as one of the most integrative of the SES assessment techniques. An updated version of the framework is presented in McGinnis and Ostrom (2014). Leslie et al. (2015) applied the SES framework to small-scale coastal fisheries in the Mexican state of Baja California Sur. In this study, they identified 13 variables relating to four dimensions:

1. Governance system
   a. Operational and collective-choice rules
   b. Territorial use privileges
   c. Fishing licenses
2. Actors
   a. Diversity of relevant actors
   b. Number of relevant actors
   c. Migration
   d. Isolation
   e. Livelihood diversity potential
3. Resource units
   a. Diversity of targeted taxa
   b. Per capita revenue
4. Resource system
   a. System productivity
   b. System size
   c. System predictability

In their study, indicators for each of the 13 variables were identified and quantified on the basis of primary data and then used to create composite, quantitative measures of each of the four SES dimensions. The analyses provided several findings relevant to building an action plan for enhancing the sustainability of the system. The study indicated considerable spatial variation in the potential for social-ecological sustainability as expected, but also yielded some rather surprising results. The authors hypothesized that there would be a positive correlation between ecological and social dimensions—e.g., areas with greater ecological richness would also be more likely to score higher on social dimensions. A small correlation was in fact found, which opens the door for identifying different strategies to enhance sustainability across the various areas surveyed. For example, one highly taxon-rich location was found to have a very low score on actors, which could be interpreted as having either very good or weak potential for a sustainable fishery. The weak score on actors, however, pointed to the need for capacity building in this area, whereas in the resource units dimension, maintaining existing management capacity may be more important.

For the final example of how social-ecological characterization may be used to identify strategies to operationalize sustainability, we move to a micro-household level, with an example of a project that used extensive characterization of socio-ecological features of smallholder agricultural systems in sub-Saharan Africa to identify potential entry points for enhancing the use of biological nitrogen fixation.

39.2 THE N2AFRICA PROJECT IN SUB-SAHARAN AFRICA: UNDERSTANDING THE CONTEXT TO ENHANCE BIOLOGICAL NITROGEN FIXATION

Legumes can play a key role in developing new strategies for promoting sustainable increase in agricultural productivity without harming the environment (Serraj et al., 2004). In addition to their dinitrogen (N_2)-fixing capacity, legumes are extremely important in human and animal nutrition. Globally, they supply a third of the proteins in human diets, and some species such as soybean and groundnut are important sources of oil. Other legumes are also a valuable source of phytochemicals and essential minerals like iron and zinc. Legumes also promote soil carbon sequestration and, ultimately, reduce soil erosion when included in intercropping farming systems and/or used as cover crops. Legume intensification within the farming systems contributes to interseasonal food security and increases household health standards. Furthermore, due to their high nutritional value, pulses are also valuable allies in fighting hunger worldwide (FAO, 2016a).

Biological nitrogen fixation (BNF) is a key component of the biogeochemical nitrogen cycle, and particularly important for global agricultural productivity; it can be considered as one of the most important biological processes on the planet. It provides about 100 million tons of nitrogen, which leads to an annual saving of around US$10 billion in nitrogen chemical fertilizer. Furthermore, the reduced need for synthetic fertilizers indirectly reduces the amount of greenhouse gases released into the atmosphere.

The first oil crisis of the 1970s stimulated a global interest for BNF and for research on legume-rhizobium symbiosis. However, this research has mostly focused on specific aspects, such as microbial strain selection and inoculation for enhancing the role of BNF contribution in agriculture. Moreover, the focus was often restricted to a limited number of legume species and cropping systems. Little attention was given to the socioeconomic impacts of technologies or to the integrated management of soil fertility within farming systems. For instance, research on the adaptation of symbiotic N_2-fixing legumes to environmental constraints for promoting farmers adoption has been insufficient or inefficient. Furthermore, the use of BNF technologies has often been disincentivized by national policies that reduce the economic competitiveness of BNF options, such as subsidies for chemical fertilizers. Little has been done to assist countries in developing appropriate strategies and understanding the socioeconomic context for BNF technology adoption, or to identify robust marketing chains for legume grains. As a result, implementation of a holistic approach for sustainable crop productivity and soil fertility improvement has been challenging. Consequently, the average quantities of N_2 fixed by legumes annually remain low, and the yield gaps of leguminous crops across Africa and most developing countries are still substantial (Giller, 2001). Various environmental factors, such as drought, soil salinity, acidity or alkalinity, phosphorus deficiency, and excessive applications of nitrogen fertilizers affect potential N_2-fixing activity and limit the optimal contribution of BNF in cropping systems (Serraj and Adu-Gyamfi, 2004), in addition to political and socioeconomic factors (Shiferaw et al., 2004).

The N2Africa project is attempting to address the issue of low adoption and building context- specific strategies. N2Africa is an international large-scale and multidimensional "research-in-development" project, focused on harnessing the potential of BNF for enhancing productivity and sustainability of legume crops in smallholder farming systems in sub-Saharan Africa. The project aims at increasing the productivity of grain legumes and the use and efficiency of BNF, and contributes to using the legume-rhizobium symbiosis to enhance soil fertility in the cropping systems, improving household nutrition, and increasing income and livelihoods of smallholder farmers (Giller et al., 2013).

N2Africa has progressively evolved and adapted from a role of direct implementation to a role of knowledge provider and catalyst while building trust and legitimacy with local communities over the project period. During its initial phase, N2Africa project workers were responsible for the design and implementation of activities through subcontracting of local research and development institutions. N2Africa is now a successful, widespread, and well-known project; it has become well-connected with development partners and private companies as a knowledge provider in the field of sustainable intensification of smallholder farming systems through legume-based and BNF technologies. A strong investment in public-private partnerships has also led to promising results in upscaling of technologies to ensure project sustainability in the future.

A set of principles, criteria, and indicators to assess the sustainability of farming systems was developed based on N2Africa's objectives and additional literature study

TABLE 39.1 Household Level Principles, Criteria, and Indicators

Principles	Criteria	Indicators
Productivity	[...] greater food and nutrition security	Protein from legumes
	[...] closing the yield gaps	Food self-sufficiency legume yield gap
		Maize yield gap
Viability	To have a viable farm size	Farm size
	[...] increased incomes [...]	Farm gross margin
	[...] expand the area of legume production within the farm [...]	Legume intensity
		Valuable assets
	Increase or maintain farm assets ...	Livestock owners
Resilience	Increase or maintain the natural resource base to spread and reduce risks of crop failure	Nitrogen input from N_2-fixation
		Agro-diversity
		Price variability
		Yield variability
Social well-being	[...] introducing labor-saving technologies from which women benefit [...]	Share of women in labor
		Women's empowerment
	Empower women [...]	Food security
	[...] greater food and nutrition security [...]	Postharvest storage
	To reduce postharvest losses	Market access
	To be connected to markets	Frequency of extension services
	To receive regular extension advice	

From Marinus, W., Ronner, E., van de Ven, G.W.J., Kanampiu, K., Adjei-Nsiah, S. and Giller, K.E. 2018. The devil is in the detail! Sustainability assessment of African smallholder farming. In: Bell S. Morse S. (Eds.) Routledge Handbook on Sustainability Indicators, pp. in press. Routledge, London (Chapter 28).

(Table 1; Marinus et al., 2018). The principles of sustainability used in the assessment show a high correlation with SFA principles.

The role of legumes in farming systems sustainability was presented at the household level in spider charts with scores on a scale from 0 to 10 for principles and indicators (Table 39.1).

The average scores on the level of principles were close to the middle score for nearly all principles in Kenya and Ghana (Fig. 39.1), meaning that sustainability of interviewed households could be considered low or just sufficient. Outcomes on the level of principles were very similar between countries and between research sites within a country.

The methodologies used to identify key drivers of change influencing the current status of farming systems in the study areas included stakeholder interviews and systematic literature analysis, focusing on the past three decades. Stakeholder workshops were also used to verify and deepen the results of the study. Stakeholders were interviewed in both countries, including local lead farmers and farmer organizations, extension workers, non-governmental organizations, local government officials, and officials of the Ministry of Food and Agriculture, Furthermore, farmer interviews were used during the household surveys to gather a broader knowledge of local developments and to triangulate results from the other stakeholder interviews.

In Kenya, additional research for development priorities were the assessment of the economic viability of legumes and the role of the government to promote and institutionalize legume cultivation. In northern Ghana, the focus was more on the identification of optimal configurations for cereal-legume intercropping systems, increased availability and affordability of legume inputs, identification of options for value addition through small- or medium-scale processing enterprises, and on climate-resilient cropping practices and area-specific fertilizer recommendations.

The large variation in levels of indicators between households within the same site suggested that the averages within a research site or on the levels of principles might only give a rough indication of general trends. It is important for future interventions to measure and interpret the indicators for individual households, which could help to understand the variability of indicator scores. For instance, the highest scores in an area could be seen as the attainable score within the given socioeconomic and agroecological circumstances of that research site and be possibly used as a benchmark. A better understanding of indicator score variability and the reasons for high and low scores could lead to the identification of entry points for improvements on that particular indicator for sustainability.

The N2Africa project used the analysis of drivers of change and results of the household survey to identify priority areas for research during stakeholder workshops. It has distinguished priority areas for research interventions on the role of legumes in sustainable intensification from the work on sustainable intensification of farming systems in the case study areas. This distinction indicated that possible pathways for sustainable intensification, legumes in this case, should be assessed in the context of the farming systems – and the agrifood systems – to which they belong.

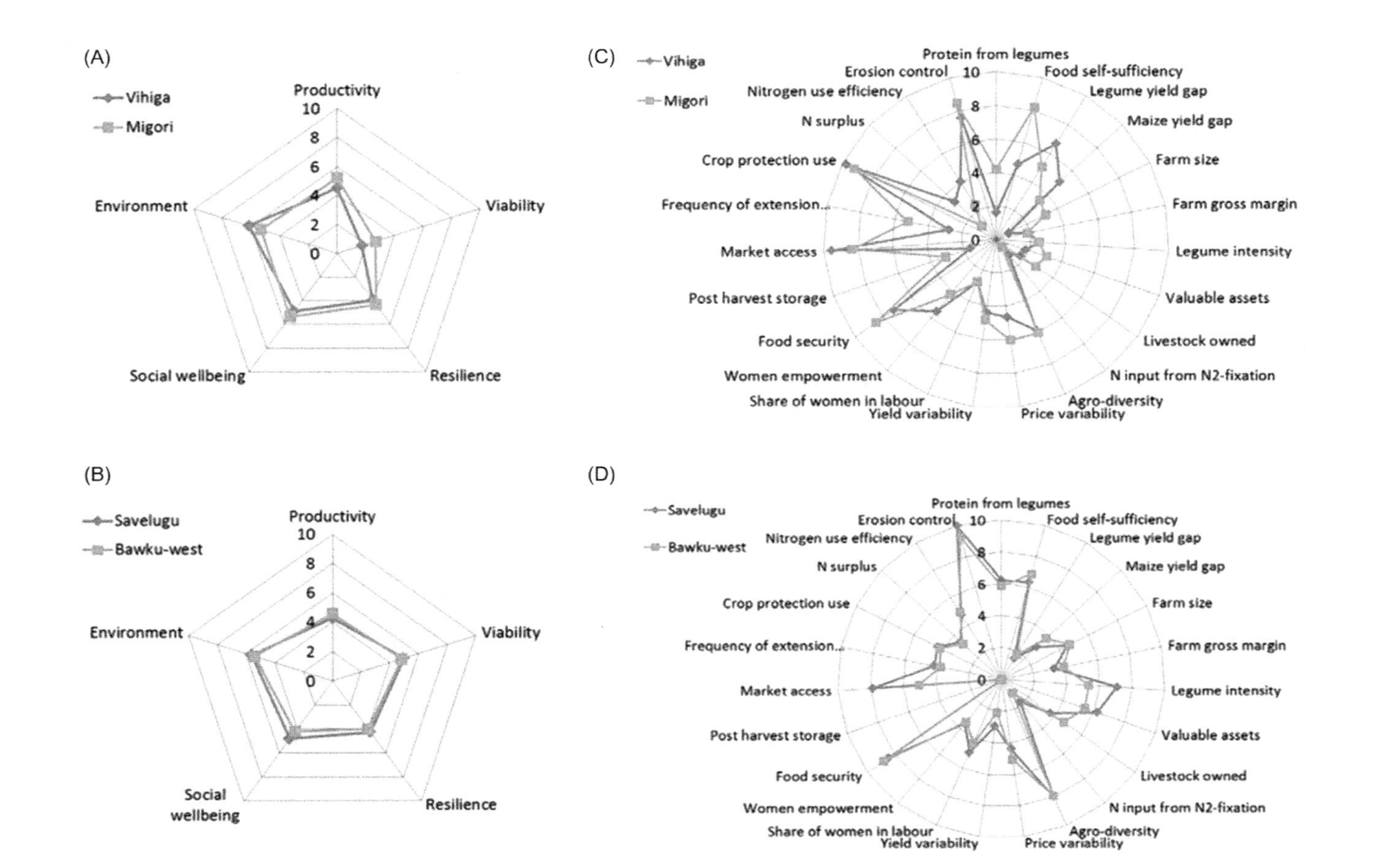

FIGURE 39.1 Average results across all households per research site in western Kenya (A, C) and northern Ghana (B, D). Indicators for sustainability were aggregated on the level of principles using equal weighing for the outcomes of each indicator (A, B). Outcomes for all indicators for sustainability are shown in C and D. Source: *From Marinus, W., Ronner, E., van de Ven, G.W.J., Kanampiu, K., Adjei-Nsiah, S., Giller, K.E. 2018. The devil is in the detail! Sustainability assessment of African smallholder farming. In: Bell, S., Morse, S. (Eds.) Routledge Handbook on Sustainability Indicators, pp. in press. Routledge, London (Chapter 28).*

CHAPTER

40

Using Evidence and Dialogue to Build Knowledge and Address Trade-Offs

In order to make changes that will enhance sustainability, people need to understand what is hampering sustainability in the current system. They also need to be aware of technically feasible options to address it and have the capacity to build agreement among the key participants, where often significant differences in power and agency are observable. Evidence—both science-based and practical—is an essential precondition to meeting both of these objectives. Evidence here refers to the available body of facts, data, and information that indicate whether a belief or proposition is true or valid (see http://pubs.iied.org/pdfs/G04132.pdf). In the development context, we can observe an increasing importance placed on "evidence-based" policy advice. However, what constitutes credible evidence, both politically and across disciplines, is often contested. Moreover, there has been a growth of "post-truth" driven dialogues in recent years. Hence, it is important to clarify how evidence can actually be used to enhance sustainability in food and agriculture systems. Detailing the use of evidence is particularly important for science-based knowledge organizations with sustainable development objectives, as they are actively involved in building knowledge from evidence.

In presenting a framework for pursuing sustainability, "knowledge capital" has been identified as a key asset needed by society to enhance sustainability processes (Matson et al., 2016). Knowledge capital encompasses both conceptual and practical knowledge and takes the form of a public good, which in principle can be used by anyone; one person's use of it does not reduce the potential for another's use. To be influential, knowledge must be trusted, and trust is built upon the perceived relevance, credibility, and legitimacy of the knowledge. Practical means of building such trust include using collaborative methods where the intended end users of the knowledge are involved in developing the agenda for producing knowledge (instead of being treated as a group of passive recipients), building a system for coordinating varying streams of knowledge development into a broader system of innovation, and enabling adaptive management through an explicit learning agenda as well as explicit links to actions (Matson et al., 2016).

Sustainable Food and Agriculture
DOI: https://doi.org/10.1016/B978-0-12-812134-4.00040-6

The perspective of knowledge capital by Matson et al. (2016) is consistent with the findings of Jenkins-Smith and Sabatier (1993), who studied natural resource management processes from a political science perspective. They found that groups that have different positions on environmental policy issues (in the framework developed by Sabatier and Jenkins-Smith they are referred to as "advocacy coalitions") often use research-based knowledge to justify their own position while criticizing the research-based knowledge that their opponents invoke as evidence. Advocacy coalitions use different strategies to discredit evidence of their opponents, such as challenging the validity of the data that were collected for a study, questioning the adequacy of the methods used, and challenging the validity of the analysis. Jenkins-Smith and Sabatier (1993) argue that analytical debates and policy-oriented learning across coalitions are required to build a broad-based consensus on what constitutes credible evidence.

40.1 DEVELOPING USEFUL EVIDENCE

In order to be effective, evidence needs to be considered as credible and it needs to be utilized by relevant stakeholders. This implies the need for a learning process whereby evidence is not only accepted, but also internalized by stakeholders. Considerable literature exists on epistemology—the study of knowledge—which is beyond the scope of this work to delve into in detail. However, some recent works on sustainable food and agriculture provide useful ways of conceptualizing knowledge for sustainability and the role of evidence in building it. van Kerkhoff and Lebel (2006) define knowledge as "justifiable belief," with different criteria for justification depending on the type of knowledge. For example, science-based knowledge must be credible according to science standards, whereas local knowledge must be justified based on experience and links to a specific place. Cash et al. (2003) argue that in order for science-based knowledge to be actionable for sustainability, it must be perceived as credible, salient, and legitimate in the eyes of prospective users. They also note the importance of "boundary work" between communities of experts and decision-makers, which involves a process of dialogue to identify various norms, beliefs, and expectations among the groups as to what constitutes justifiable beliefs and the role evidence can play therein.

Research communities are recognizing the need to better position research-based evidence to affect policy outcomes, particularly organizations that are being tasked with generating development outcomes and not just research outputs. A good example is the CGIAR, which is a global agricultural research partnership dedicated to reducing poverty, enhancing food and nutrition security, and improving natural resources and ecosystem services. As such, it is a key global provider of research-based evidence to support sustainability in food and agricultural systems. In recent years, CGIAR has adopted the agricultural research for development (AR4D) model, implying that within a multiscale and multi-objective innovation process, research will be planned and implemented to facilitate development processes. This shift in the objectives and operational modalities of the organization has given rise to a need for developing a new and commonly agreed upon approach for assessing the quality of the research for development (QoR4D). The analysis and debate involved in developing such an approach is highly relevant to the issue of generating evidence to support the operationalization of sustainability in food and agricultural systems. Box 40.1 describes the proposed frame of reference for assessing the quality of research in the CGIAR system, which incorporates several features of knowledge systems to support SFA outlined above.

The degree to which the standards in Box 40.1 will actually have an impact in terms of increasing the utility of the evidence generated for operationalizing sustainability depends on how they are implemented. The intent is for the CGIAR to integrate all four elements

BOX 40.1

QUALITY OF RESEARCH IN THE CONTEXT OF AGRICULTURAL RESEARCH FOR DEVELOPMENT

In 2016, the Independent Science and Partnership Council of the CGIAR launched a consultative process, involving representatives from entities across the system, for developing a common frame of reference for assessing the QoR4D in the CGIAR system. This process led to a consensus that QoR4D in the CGIAR context should be viewed as an integrated whole of four key elements: relevance, scientific credibility, legitimacy, and effectiveness (Belcher et al., 2016).

Relevance refers to the importance, significance, and usefulness of the research objectives, processes, and findings to the problem context and to society associated with CGIAR's comparative advantage to address the problems. It incorporates strategic stakeholder engagement along the AR4D continuum, original and socially relevant research aligned to national and regional priorities, as well as the CGIAR Strategy and Results Framework and the Sustainable Development Goals. It also recognizes the importance of international public goods.

Scientific credibility requires that research findings be robust and that sources of knowledge be dependable and sound. This includes a clear demonstration that data used are accurate, that the methods used to procure the data are fit for purpose, and that findings are clearly presented and logically interpreted. It also recognizes the importance of good scientific practice, such as peer review.

Legitimacy means that the research process is fair and ethical and perceived as such. This encompasses the ethical and fair representation of all involved and consideration of interests and perspectives of intended users. It suggests transparency/lack of conflict of interest, recognition of responsibilities that go with public funding, genuine recognition of partners' contributions, and partnerships built on trust.

Effectiveness means that research generates knowledge, products, and services with high potential to address a problem and contribute to innovations and solutions. It implies that research is designed, implemented, and positioned for use within a dynamic theory of change, with appropriate leadership, capacity development, and support to the enabling environment to translate knowledge to use and to help generate desired outcomes.

Source: ISPC (2017).

at all stages in research management and implementation. That is relatively straightforward at a project scale. In the theory of change for the project, who is expected to do what differently as a result of the project? Are the research and supporting activities designed/executed such that the intended users will use this knowledge in a way (i.e., change their behavior compared to what they would do in the absence of that knowledge and supporting activities) that will result in next-level outcomes?

In policy-oriented research, the outcome might mean informing and influencing not only policymakers, but also advocates and civil society groups and perhaps also influencing opponents and vested interests. Theory suggests that knowledge needs to be considered relevant, credible, and legitimate by various actors/stakeholders, and it needs to be positioned for use with adequate supporting activities (e.g., capacity building, networking, coalition building) to be effective.

At higher scales, especially when thinking about international public goods, it becomes more indirect, more abstract, and more challenging. But still, the focus is on relevance, scientific credibility, legitimacy, and effectiveness for the main intended users. That includes immediate users and secondary and tertiary users. For example, the Center for International Forestry Research (CIFOR) has done a great deal of work to support the United Nations collaborative process on Reducing Emissions from Deforestation and Forest Degradation (REDD +) process. CIFOR aimed to directly influence REDD + project implementers through collaborative activities and to indirectly influence REDD + implementation by developing and disseminating knowledge and recommendations and organizing knowledge events with various actors who can themselves influence national and international discourse, policy, and practice. The knowledge needs to satisfy quality criteria in each dimension in order to be useful and used by immediate users and intermediate users in the sphere of influence.

40.2 LINKING KNOWLEDGE WITH ACTION

Knowledge generation on its own is not enough and does not necessarily lead to action. In recent years, there has been considerable debate and discussion on how best to link research-based knowledge to action (policy outcomes). Johnson and Birner (2013) identified three main schools of thought on the links between research and policy formulation: (1) linear and logical; (2) iterative; and (3) approaches centered on discourse (Box 40.2).

Both the linear and iterative models of linking research policy are based on the notion of a distinct community of suppliers linking to a distinct community of users, while the discourse-based models allow for a wider range of both suppliers and users of research-based evidence. All three models involve some type of interactions between suppliers and users of research-based evidence, which often involve dialogue as well as institutional arrangements to facilitate interactions.

Since the policy formation process varies considerably among countries and even within them, the applicable model for building knowledge from research-based evidence that can contribute to effective policy making also varies: different models are applicable in different circumstances. However, across all circumstances, it is important to recognize that there is an underlying process by which policy is formulated—and it is necessary to

BOX 40.2

MODELS OF THE RESEARCH TO POLICY PROCESS

Linear or trickle down model of research production function. This approach is built on the notion of a linear transfer of research outputs delivered by the science community to policymakers, who then make use of the evidence provided. This model assumes that research is purely objective and that the main barrier for using evidence in policy making is the lack of awareness of decision-makers of the evidence.

Iterative models. In these models, policymakers are assumed to be muddling through a policy process in an incremental, iterative, and complex process that involves compromise. This group includes models of bounded rationality, satisficing, incrementalism, and "muddling" through. These models recognize the presence of multiple and competing policy objectives.

Models based on discourse. The third model of utilization of research outcomes is built upon the notion that research feeds into policy over time and through a learning process that involves multiple actors—not just producers and users of science-based evidence. Research-based evidence only gains importance as the demand for change is built through enhancement of knowledge among many different actors and through many different processes.

Source: *Johnson and Birner (2013).*

have an understanding of that process, as well as its possible entry points for research-based evidence in order to enhance the potential effectiveness of such evidence.

Process Net-Map

One tool that can be used to analyze a specific policy process is a mapping method called "Process Net-Map." This tool, which is well suited to study the role of evidence in policy processes, has been derived from Net-Map, a method that was originally developed by Schiffer to analyze multistakeholder governance structures for natural resource management (Schiffer and Hauck, 2010). Net-Map is "an interview-based mapping tool that helps people understand, visualize, discuss, and improve situations in which many different actors influence outcomes" (https://netmap.wordpress.com/about).

To conduct a Process Net-Map, a set of stakeholders and experts currently or formerly involved in a particular policy process are interviewed. Such an interview proceeds in three phases. In the first phase, the interviewer asks the respondent to identify all steps involved in the policy process under consideration, starting from the first initiative of developing a policy to its adoption or implementation. During the interview, each step is marked on a large sheet of paper in the form of an arrow. The actors involved in each step are marked on the paper sheet using "actor cards" (that is, sticky notes on which the name of the actor is written). In the second phase, the interviewer asks the respondent to rate the influence of each actor on the final outcome of the process. The rating is visualized by placing a stack of checker's game pieces next to the respective actor card. The height of the stack denotes the ranking of the actor's influence, for example, on a scale from 1 to 6. The respondent is asked to explain the reasons behind the influence ratings of the actors. In the third phase, the Process Net-Map is analyzed together with the participants of the interview, focusing on questions of interest, such as the ways in which research-based evidence was considered in the different steps of the policy process.

Fig. 40.1 displays the example of a Process Net-Map, which was constructed by Mockshell (2016) for the case of Ghana. The policy process shown in the map is related to

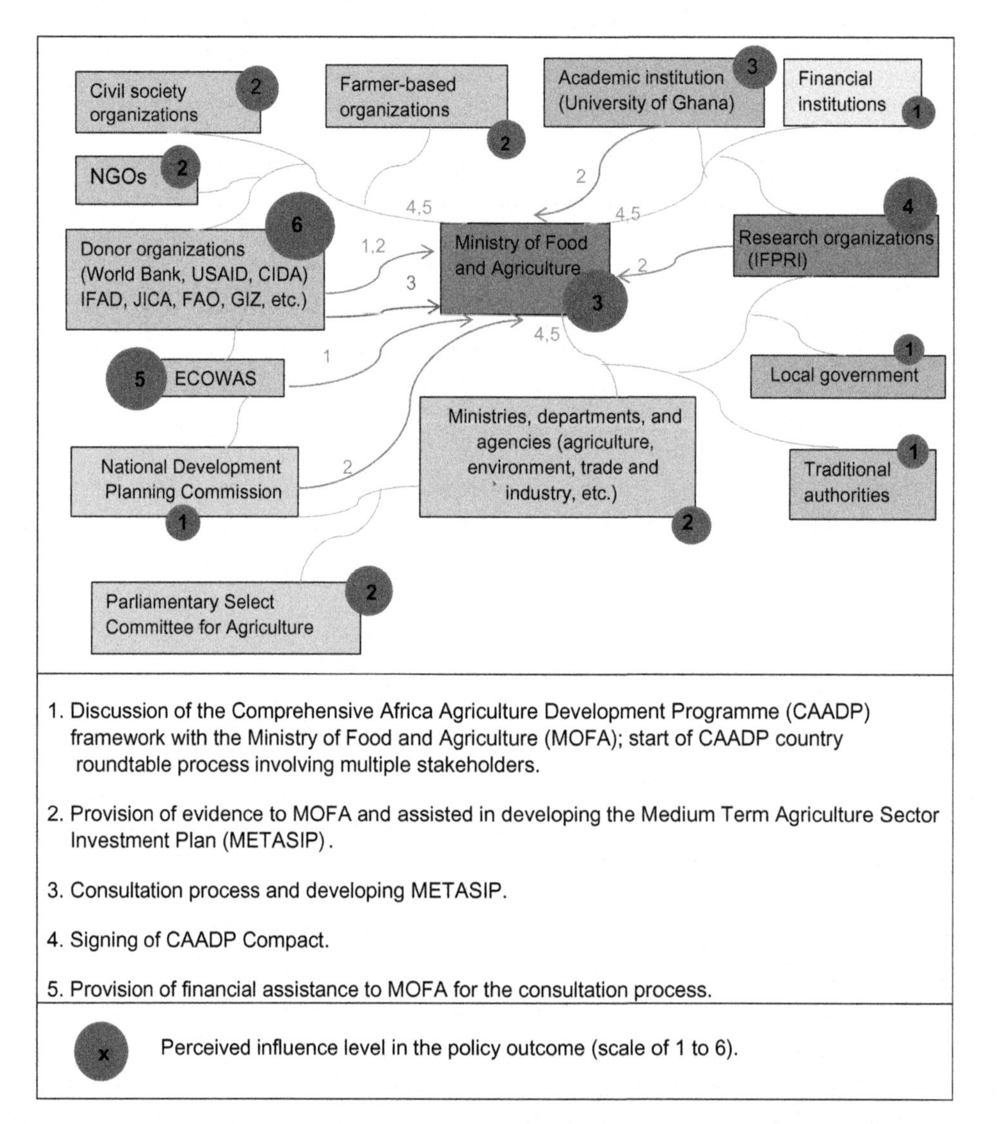

FIGURE 40.1 Example of a Process Net-Map. Source: *Adapted from Mockshell, J. 2016. Two Worlds in Agricultural Policy Making in Africa?: Case Studies from Ghana, Kenya, Senegal and Uganda, p. 100.*

the Comprehensive Africa Agriculture Development Program (CAADP), which aims at promoting sustainable agricultural development across the African continent. In Ghana, the CAADP policy process resulted in a Medium Term Agriculture Sector Investment Plan (METASIP). As can be seen from the Process Net-Map, the policy process involved a range of different stakeholders. Evidence for developing the METASIP was provided both by national and international research organizations (see Step 2). The influence rating shows that research organizations had an intermediate level of influence, which indicated that evidence did indeed play a significant role in the process. However, the influence rating also revealed that international organizations, specifically the Economic Community of West African States and the donors, had a rather high influence on the process, whereas the influence of national and local stakeholders, such as parliamentarians and local authorities, was rather limited. This finding indicates that the process may have been affected by limited country ownership. The finding was confirmed by further research, which involved the mapping of other policy processes for the purpose of comparison (Mockshell, 2016).

Boundary Work

"Boundary work" is a relatively recent concept that can help bridge different communities, from science to policy to a broader conceptualization between knowledge and actions (Clark et al., 2016). The main idea behind boundary work is actively managing the interface and interactions between communities with different views of what constitutes reliable and useful knowledge, as well as between communities of researchers (knowledge) and of policymakers (action) (Clark et al., 2016; Matson et al., 2016). On the one hand, a hard barrier between two communities means no communication or learning is taking place between them. On the other hand, a highly porous barrier could result in politics mixing with evidence—and alternative facts. Boundary work is thus a key tool for identifying and overcoming imbalances between social, environmental, and economic objectives.

Clark et al. (2016) provide a framework for boundary work that categorizes both the source of knowledge and its use to identify six different kinds of boundary situations with different approaches required for each. The matrix is presented in Table 40.1. The individual cells of the framework reflect how the particular combinations of knowledge sources and uses determine the challenges facing boundary work in particular contexts.

Clark et al. (2016) argue that there are three main uses for knowledge: general enlightenment, decision making, and negotiation; the approaches to linking science and policy—or, more generally, knowledge and action—are quite different in each case. Additionally, the degree to which a relatively homogenous community of expertise exists is a key factor in determining how best to manage the boundary. In the context of operationalizing sustainability for food and agricultural systems, the community of expertise is quite heterogeneous with, in some cases, starkly varying perspectives. The importance of overcoming trade-offs in the operationalization of sustainability indicates that we should expect knowledge to frequently be used for negotiation, although its use in enlightenment and decision making is also important, which is clearly highlighted in the case study of India below.

TABLE 40.1 Boundary Work Framework

		USE of Knowledge to Support....		
Boundary Work		**Enlightenment (U_0)**	**Decision (U_1)**	**Negotiation (U_m)**
SOURCE of knowledge...	Personal expertise (S_0)	$S_0 \leftrightarrow U_o$ *Contemplation*	$S_i \leftrightarrow U_j$ *Decision*	U_k $S_i \leftrightarrow \updownarrow$ U_l *Politics*
	Single community of expertise (S_1)	$S_1 \leftrightarrow U_o$ *Demarcation*	$S_i \leftrightarrow U_j$ *Expert advice*	U_k $S_i \leftrightarrow \updownarrow$ U_l *Assessment*
	Multiple communities of expertise (S_n)	S_i $\updownarrow \leftrightarrow U_o$ S_j *Integrative R&D*	S_i $\updownarrow \leftrightarrow U_j$ S_j *Participatory R&D*	S_i U_k $\updownarrow \leftrightarrow \updownarrow$ S_j U_l *Political bargaining*

Source: Clark, W.C., Tomich, T.P., van Noordwijk, M., Guston, D., Catacutan, D., Dickson, N.M., and et al., 2016. Boundary work for sustainable development: Natural resource management at the Consultative Group on International Agricultural Research (CGIAR). Proc. Natl. Acad. Sci. 113 (17), 4615.

Politically Contested Evidence: The Case of Groundwater Irrigation in India

Irrigated wheat and rice production in India is an important example of an agrifood system that faces major sustainability challenges and involves important trade-offs. As a consequence of the green revolution, wheat and rice production in India is characterized by high levels of productivity and plays a major role for economic prosperity and food security in the region (Hazell and Ramasamy, 1991). However, this highly intensive production system is confronted with major environmental challenges, one of which is the overuse of groundwater. Steadily declining groundwater levels in India's major production areas of wheat and rice are a striking indication of this sustainability problem (Government of India, 2007). While the evidence for this environmental problem is not contested, there is a major political controversy regarding the underlying causes of this problem and regarding appropriate policy responses.

Combining the Advocacy Coalition Framework developed by Sabatier and Jenkins-Smith (1993) with the discourse analysis approach developed by Hajer (1995), Birner et al. (2011) analyzed this controversy on groundwater overuse in India. The authors identified two discourse coalitions, which were defined as groups of actors who share similar policy beliefs that are reflected in similar discourses on the issue under consideration, in this case being the overuse of groundwater.

One discourse coalition, which the authors referred to as the "market-oriented coalition," identified the lack of a well-functioning price mechanism for groundwater as the major cause of its depletion. In most Indian states, electricity for groundwater irrigation is highly subsidized and, importantly, it is also provided at a flat rate. In some states, electricity for pumping groundwater is even provided free of charge. The market-oriented discourse coalition was comprised of international financial organizations, most notably the World Bank, officials in the Ministry of Finance, members of the Planning Commission, and a number of renowned Indian and international academics. The main storyline of this discourse coalition was that, due to the policy of providing electricity at a flat rate or free of change, farmers had no economic incentive to save electricity and, as a consequence, they had no incentive to save groundwater. A related argument was that the electricity subsidies prevented a reform of the power sector and resulted in low quality of electricity services, for example, due to lacking investment in the infrastructure for electricity transmission. Accordingly, this discourse coalition suggested the abolition of the flat rate and the reduction or elimination of subsidies as a major policy strategy to encourage both a sustainable use of groundwater and a reform of the power sector.

The second discourse coalition identified by the study was opposed to the elimination of electricity subsidies to farmers. The main storyline of this discourse coalition was that electricity subsidies were an important instrument to address the crisis of the agrarian sector, which was, according to this coalition, indicated by high levels of farmers' indebtedness and farmers' suicides (Birner et al., 2011, p. 140). This discourse coalition was labeled the "welfare-state-oriented coalition." Members of this coalition included, among others, farmer organizations, officials in the Ministry of Agriculture, and a number of renowned academics in Indian universities.

Both discourse coalitions used research-based evidence to support their arguments. The market-oriented coalition concentrated on evidence that was supposed to show that the current electricity subsidy policy had negative effects. A major effort to provide such evidence was a two-volume study on "Power Supply to Agriculture," which was commissioned by the Energy Unit of the World Bank's South Asia Regional Office (World Bank, 2001a, 2001b).

As stated in the introduction, the study was specifically designed to provide evidence that farmers would benefit from a reduction of electricity subsidies, as this measure would result in an improved quality of electricity services.

> After almost a decade of high-level effort to bring the charges (tariffs) that farmers pay for electricity more nearly into line with the costs of supply, India has barely made a dent in the longstanding and increasingly uneconomical practice of subsidizing power to agricultural consumers for irrigation. Progress has been slowed by the understandable but misplaced concern that higher tariffs would harm farmers – and that the injured parties would take political revenge on the reformers. This study seeks to dispel that anxiety. [...] Its central contribution to policy discussion is the detail in which it documents the costs – usually neither acknowledged nor clearly defined – to farmers in those states of subsidies that actually harm agricultural operations more than they help as well as the benefits that the farmers would get from improved quality of electricity services ***World Bank. 2001a. India - Power Supply to Agriculture Volume 1: Summary Report. Energy Sector Unit, South Asia Regional Office, World Bank, p. 1***

The study was based on an extensive survey of farmers in two states where groundwater irrigation plays an important role: Andhra Pradesh and Haryana. Thus, the study produced quantitative evidence in support of the market-oriented discourse. However, this and other evidence provided by the market-oriented coalition did not result in any policy change. Ten years after the publication of the World Bank study, the number of states that provided electricity free of charge had even increased (Birner et al., 2011). Less far-reaching reform proposals, e.g., moving from general to targeted electricity subsidies, were not implemented either.

Why was research-based evidence not more influential in stimulating reforms that could have resulted in increased sustainability of India's rice-wheat production systems? One reason was the fact that electoral politics played a major role in preventing the reduction of electricity subsidies. Political parties could gain electoral advantage in state-level elections if they promised free electricity to farmers. Therefore, even political parties that did not consider free electricity a sensible policy instrument adopted this policy (Birner et al., 2011; Lal, 2006; Srinivasulu, 2004).

In addition to electoral politics, the nature of the studies presented by the two discourse coalitions was also a reason why evidence did not play a more important role in the policy process. The World Bank study mentioned above illustrates the problems involved. The opponents of a subsidy reform did not contest the facts presented in this study, but rather disagreed with the conclusion. Farmers, in particular, did not trust that a reduction of electricity subsidies would encourage investments to improve the quality of electricity services. They feared that they would lose the subsidies without getting anything in return (Birner et al., 2011).

Another problem was the fact that relatively few empirical studies addressed the key question of the controversy: To what extent do the electricity subsidies indeed cause over-extraction of groundwater? Members of the welfare-state-oriented coalition pointed out that the relations between electricity pricing and groundwater extraction were far more complex than the storyline of the market-oriented discourse coalition suggested. The argument of the welfare-state-oriented coalition was that in most Indian states, electricity supply to agriculture was heavily rationed: electricity was supplied to farmers only for a very limited number of hours, so that farmers could not use more water than required by their crops. On this ground, members of this coalition argued that electricity pricing was not the major reason for overuse of groundwater.

Abolishing the electricity subsidies would, therefore, only reduce farmers' income without solving the groundwater problem According to the welfare-state-oriented coalition, other measures were required to reduce the overuse of groundwater, such as investing in water-saving technologies and promoting a switch to less water-intensive crops in regions affected by overuse of groundwater.

The evidence regarding this argument has remained ambiguous. Two studies illustrate this point: in 2005, a study was published that provided empirical evidence that volumetric pricing would lead to a more efficient use of groundwater (Kumar, 2005). In 2012, a study was published that provided evidence for the argument that, due to electricity rationing, farmers used less water than would have been required for optimal crop yields (Banerji et al., 2012). Both studies appeared in internationally renowned journals.

Thus, the policy process was dominated by two discourse coalitions espousing fundamentally different beliefs on electricity subsidies and overuse of groundwater. Both discourse coalitions referred to theoretical arguments as well as selected empirical evidence, but no efforts were made to enter into a dialogue. This situation prevailed for more than a decade. One strategy to strengthen the use of evidence in such situations is promoting "policy-oriented learning across coalitions" (Jenkins-Smith and Sabatier, 1993). Experience in similar situations has shown that it would be essential to create a high-level forum, convened by a neutral broker, where representatives of both coalitions could meet and agree on contested issues for which empirical evidence is missing. They could then engage in "analytical debates," that is, debates on research methodologies that are acceptable by members of both coalitions. In the Indian case, such a forum was never established, and at the time of writing this report (2017) electricity subsidies still prevail and the groundwater problem remains unresolved.

Addressing Divergent Stakeholder Preferences for More Sustainable Outcomes

Recognizing and engaging with a diverse and appropriate range of stakeholders is essential for navigating trade-offs. King et al. (2015) used an environmental economic framework, as a point of reference, to examine the nature of trade-offs between protecting tropical dry forests versus transforming them into pastures for cattle ranching in Mexico. Stakeholder values were mapped onto a production possibility frontier of woody biomass against herbaceous biomass—conservationists and peasants who conserved forests favored an outcome where most of the tropical dry forest area was preserved, while ranchers preferred one where pasture dominated for the provision of cattle fodder. Since mutually acceptable outcomes for ranchers and forest conservationists were excluded along the efficiency frontier, the potential impact of improved agroforestry practices to alleviate the severity of trade-offs was considered. This changed the shape of the efficiency frontier to allow higher woody biomass and high livestock forage production, thereby enabling mutual acceptability and a viable win–win option.

While the significance of boundary work cannot be disputed, an alternative approach that highlights the importance of hybridizing and blurring of boundaries is now emerging. "Antiboundary work" necessitates interdisciplinarity (better integration among disciplines), transdisciplinarity (a holistic approach involving nonacademic stakeholders), and solution-oriented research (De Pryck and Wanneau, 2017).

CHAPTER

41

Mechanisms and Approaches to Realizing Behavioral Change at Scale

41.1 BUILDING INNOVATION CAPACITY

Policy and practice on linking knowledge to actions that operationalize SFA have evolved significantly over the years. Early emphasis on solely transferring technology through building capacity in science and technology proved inefficient due to its weak demand orientation and insensitivity to social and environmental agendas. More recently, the concept of innovation systems has provided an empirical and conceptual basis for understanding innovation as a networked and socially embedded process. Consequently, the concept has emerged as a powerful tool for revealing the processes and capabilities associated with linking knowledge to actions to operationalize SFA.

This section explores the different pathways and policy instruments through which innovation processes that operationalize SFA occur. It presents a conceptual framework of typologies of modes of innovation, and the identification of mechanisms to drive desirable changes in organizational strategy and priorities to unlock agrifood system innovation toward SFA at different scales.

In developing a framework to better understand the relationship between different innovation configurations (partnerships, networks, and practices) and SFA impact, the starting assumption was that while configurations are contextually specific, broad patterns of practices and partnerships associated with innovation and impact would emerge. These patterns could then form the basis of a framework to explain how impact takes place, and to point to tools and practices that increase the likelihood of innovation and impact under varying circumstances.

Fig. 41.1 illustrates the different analytical lenses that innovation systems' thinking brings to bear. This framing was used explore a series of case studies using secondary sources, backed up with interviews where possible (ISPC/CSIRO, 2017).

Sustainable Food and Agriculture
DOI: https://doi.org/10.1016/B978-0-12-812134-4.00041-8

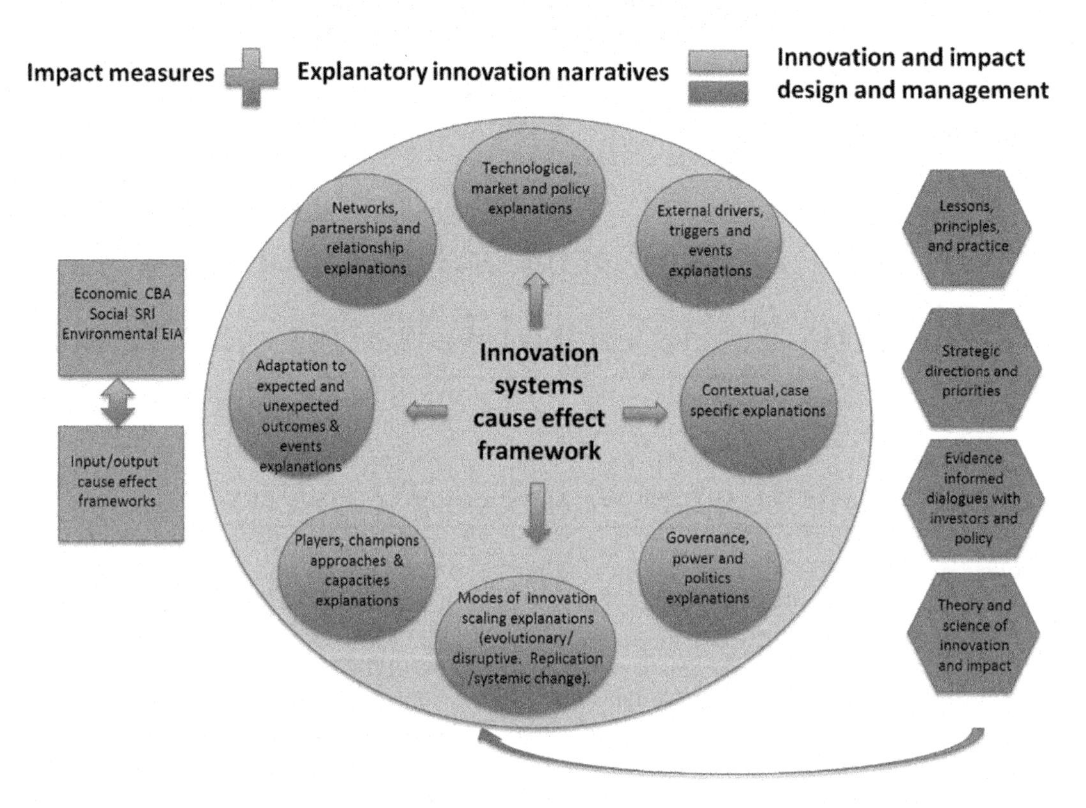

FIGURE 41.1 Conceptual framework for links between innovation configurations and impact.

Elements used for the case study analyses include the following:

1. initiating events and key turning points during the innovation process;
2. the role of research and technology in the wider process of change;
3. the range of players involved in the innovation process and their changing roles over time;
4. alliances and partnerships that were pivotal in the innovation process;
5. institutional arrangements and the certification, regulatory, pricing, and other policy measures that formed part of the innovation;
6. the nature of the innovation process (e.g., the commercialization of a [public] research technology by the private sector; the public policy, regulatory regimes, or governance arrangements that stimulated/facilitated technological and practice changes; and market disruptions arising from new business models and/or changing societal demands and values); and
7. evidence from independent evaluations and impact assessments about the current and future scale and nature of impacts.

The case study analysis indicates the importance of institutional development to accompany technological improvements and the power of linking public, private, and civil society in a joint effort to effectuate changes contributing to sustainability.

One case study covers the emergence and continuing development of the Chilean salmon industry. This industry is based on a nontraditional agricultural output and was developed over a relatively short period of about 35 years. At present, Chile is the second largest salmon exporter globally, with an annual export value of over US$4 billion and providing 30,000 jobs directly. Early public investment in the development of commercial salmon farming was part of government policy at the time of promoting scientific and technological innovations that added value to the country's natural resource endowment. However, the rapid expansion and economic success of the industry outpaced the development of socio-political institutions required for environmentally friendly and sustainable natural resource exploitation, social inclusiveness, and equitable outcomes for the local communities. The result was the lopsided growth of the industry, where economic success was favored over social and environmental considerations. Lack of knowledge about the resource "loading capacity" in different aquaculture locations, poor understanding of the ecological and biological equilibrium, and limited government regulation and law enforcement capacity of the public sector agencies resulted in the significant infectious salmon anemia crisis in 2007. This case illustrates that free market processes could not ensure long-term environmental sustainability, and the emergence of the institutions required for a socially inclusive process of development, without government intervention.

Another case study traces the development of a live vaccine against East Coast fever, a tick-borne disease of cattle caused by *Theileria parva* in Africa. Technological breakthroughs in the 1950s and 1960s led to a control measure, commonly known as the "infection and treatment" method (ITM). However, the widespread deployment of ITM was stalled for decades because key development and donor agencies focused resources and efforts on a frontier science research agenda rather on the delivery of an existing solution. This resulted in an effective and simple control measure being sidelined by the search for recombinant vaccine solutions in a newly established international research center that ultimately failed to produce the needed technological breakthrough. ITM finally received renewed attention when a public-private sector mechanism revisited the technology for field deployment and development impacts. The establishment of GALVmed in the early 2000s, as a platform to engage the private sector in the commercialization of existing public science solutions, eventually formed sufficient political alignment and incentives to create an effective vaccine production and delivery mechanism.

As a final example, one of the case studies examines the substantial investments that have been made in the development and promotion of biofortified orange-fleshed sweet potato (OFSP) as a complementary approach to reducing vitamin A deficiency in sub-Saharan Africa. Vitamin A deficiency is a serious public health concern in many countries, which can cause blindness and increase mortality. Breeding breakthroughs in high-yielding OFSP varieties combined up to fiftyfold increases in beta-carotene levels with drought tolerance and adaptation to local conditions. To combat vitamin A deficiency, these technological breakthroughs needed to be coupled with improved access to OFSP varieties, and education to build awareness about the nutritional and health benefits of OFSP in improving the adoption, production, and consumption of OSFP among rural households. Through the involvement of a major philanthropic foundation, significant investment has been directed to large advocacy and educational campaigns promoting household consumption of OFSP and associated value chain development in countries where the sweet potato is either the staple crop or an important secondary staple. To date, the primary evidence of OFSP impact comes from such interventions in Mozambique and Uganda, where the investigation of scaling up showed that the project led to OFSP adoption rates of 61%–68% among project households, improved vitamin A knowledge at the household level, and significantly increased (nearly doubled) vitamin A intake among targeted women and children. Recent evaluations point towards the need to ensure the timely availability of quality planting material, addressing regional differences in consumer preferences, and the creation of dual strategies to increase consumption of OFSP among the rural poor and the parallel development of value chains, among others, for OFSP to achieve the pervasive impact desired in sub-Saharan Africa.

The 27 case studies undertaken in this effort (ISPC/CSIRO, 2017) illustrate three broad patterns of innovation, each with distinctive configurations and processes, and each with different scope for scale of SFA impact. These are (1) incremental innovation and systems optimization; (2) radical innovation and subsystems transformation; and (3) transformative innovation and systems transformation. This innovation typology draws upon the work of Freeman and Perez (1988), Weterings et al. (1997), Geels (2002), and Scrase et al. (2009). A short description of each follows below.

Incremental Innovation or Systems Optimization

Key characteristics: Incremental improvement of existing products and services or incremental improvement of value chain efficiencies that deliver marginal social, economic, and environmental impact in specific production systems and value chains.

Key processes and enablers: Case studies illustrate the way research helps develop incremental improvements in existing farming systems and individual value chains. Demand-led research and collaborative action by local stakeholders are critical in defining and developing solutions. The scale of impact, however, is often restricted by the absence of policy, institutional, and market systems changes, as well as investments needed to spread and sustain these innovations.

Radical Innovation or Subsystems Transformation

Key characteristics: Technological and/or market "step changes" or discontinuities that open up new economic, social, and environmental impact opportunities in a specific subsector or market sector and open up new opportunities for incremental innovation.

Key processes and enablers: Case studies illustrate ways in which new types of products and services (e.g., animal health products, livestock insurance, and novel agricultural inputs) emerging from research organizations have created step change improvements in specific subsectors. Mission-focused research and novel forms of public–private sector partnerships have provided new solutions to generic subsector challenges, followed by incremental innovations to improve impact effectiveness. All cases demonstrate a degree of subsector transformation, including market disruption, collaboration between the public and private sector to create delivery and control systems, and infrastructure investment. All cases indicate the creation of new economic and other value-added opportunities, new incremental innovation opportunities in production and marketing systems, and new opportunities for the delivery of a wider range of products and services through the establishment of new delivery systems.

Transformative Innovation or Systems Transformation

Key characteristics: Deep systems changes underpinned by broad-based consensus that significantly advance the economic, social, and environmental frontiers of the agrifood sector as a whole and that open up opportunities for new waves of radical and incremental innovation.

Key processes and enablers: Case studies illustrate far-reaching types of innovation with implications for the entire agrifood sector. These cases are not demand driven per se, but instead emerge from a broad-based consensus concerning the need to pursue new directions or take advantage of new platform technologies. Nascent transformative cases associated with advances in information technology suggest that a key bottleneck is the lack of mechanisms to convene stakeholders to achieve political alignment needed to create a "joined up" approach to facilitate the economic, social, and environmental step changes that new technology promises.

This proposed typology of innovation modes allows for an analysis of the way clusters of policies, practices, and stakeholder interests can lock agriculture into incremental innovation and system optimization at a time when step changes are needed. The typologies also provide a framework for allocating scarce public and private sector resources in ways that open up new opportunities for innovation and SFA impact, as well as an alternative explanation of the way agrifood system innovation and impact occurs.

Insights for operationalizing SFA

Much of the received wisdom on innovation good practice and its links to impact is evidenced in the case studies: (1) client orientation and involvement in innovation processes; (2) various forms of partnerships and alliances and the importance of both public and private investment; and (3) the importance of science, technology, and research as both initiators and enablers of innovation.

However, the case studies also suggest that the overriding ingredient in innovation processes that lead to impact and transformational change does not relate to the fine-grained arrangements involved in the innovation processes per se (although these are critical implementation strategies). Rather, the main ingredient is macro-level alignment of public policy, private sector, and often civil society objectives. This is particularly important where larger societal issues such as environmental protection, health, and nutrition and food security are at stake.

Despite the critical role played by the private sector, purposeful and proactive public investment is evident in the radical and transformative modes of innovation. This involves responding to market failures through, for example, investing in research that creates opportunities for the private sector through commercialization. It also involves solving system failures through, for example, investment in the creation of a mechanism to bring industry, civil society, and research players together to tackle systemic challenges.

The three modes of innovation discussed all have a value in operationalizing SFA, albeit with different scales of impact. The results also suggest that public (but also industry body) investments giving primacy to demand-led, bottom-up processes and short-term impacts at the farm scale skew the allocation of resources towards the local optimization route at the expense of investment in transformative changes. These path dependencies have locked many development stakeholders into incremental innovation and change processes that are often out of step with rapidly evolving agrifood systems' trends. Some of the key lock-ins identified relate to short-term funding models with unrealistic impact expectations, evaluation traditions based on historical performance measures with weak learning orientation, and demand-led research and innovation that is looking for short-term quick impact and quick wins only, rather than long-term systemic change. Our results resonate with the original intent of innovation systems and indeed recent writing on the innovation policy—e.g., the "entrepreneurial" state (Mazzucato, 2015) and the politics of innovation (Mason et al., 2016). They also support the work of Meadows et al. (2004) on stages of transformation. However, much of this perspective has been lost in the implementation of these perspectives in agriculture, which has led to a misallocation of resource and policy attention towards local optimization at the expense of systems transformations needed to reinvent the agricultural sector sustaining many of the world's poorest.

The appropriate mix of public and private sector investments needed for transformation requires an agreement on what the critical challenges lie ahead; this in turn requires a strategic partnership between public, civil, and private sectors at the political level (Table 41.1).

41.2 THE ROLE OF AR4D PARTNERSHIP IN OPERATIONALIZING SFA

The innovation systems case studies (ISPC/CSIRO, 2017) show that innovation is often enabled by institutional and policy arrangements that support knowledge flows through partnerships and other forms of interaction. However, these same studies also show that for innovation to emerge and spread, the wider regime of institutions and policies often needs continuous adaptation. For example, a food processing company can develop an inclusive way of conducting business. The agribusiness sector, however, will not become equally inclusive until pervasive changes in rule sets and incentives lead to changes in value chains (including consumer preferences) and the wider policy enabling environment (e.g., modes of education, research financing, regulations).

This has important implications for the research and development sector's aspirations to create impact at scale. It suggests that scaling is not a task of replicating effective strategies at a local scale (although this may have value when the degree of systems' complexity is low). Rather, it suggests that scaling requires systemic change. In other words, it requires innovation in institutional and policy arrangements that shape relevant aspects of social and economic activity. This is not only required to sustain and spread innovation,

TABLE 41.1 Practice and Policy Considerations in Different Innovation Modes

	Incremental Innovation	Radical Innovation	Transformational Innovation
Realm of applications	Continuous upgrading and improvement of existing production and value addition processes	Defined subsector challenges where game-changing technological breakthroughs and other advances exist or are likely	Complex, contested concerns at the sector or societal level
Public investment rationale	Market failure	Market and systems failure	Systems failure and uncertainty
Tensions to be managed	Overinvestment in immediate improvements jeopardizes long-term opportunities	Reinforces position of incumbent market players at the expense of emergent players with strong innovation potential	Conflicts between emerging and incumbent stakeholders in reaching consensus and implementing joined-up action
Limiting factors	Local vested interests	Effective public–private sector partnerships	Lack of consensus at societal level
		Social license	Clarify on public and private sector roles and investments
Characteristics of tools and approaches	Need to bridge scales		
	Need a stronger political economy perspective		
	Need to support experimentation in both the technology sense and the impact effectiveness sense		
	Need to help navigate the transition between local optimization and transformation, including tools for integrated diagnostic analysis of systems to be transformed		
	Need to help with alignment of stakeholder agendas and consensus building		
Innovation capacity metrics	Rural innovation capacity	Ability of players to respond to subsector challenges and opportunities	Agricultural innovation systems health

Adapted from Hall, A., Dijkman, J., Taylor, B.M., Williams, L., Kelly, J., 2016. Synopsis: towards a framework for unlocking transformative agricultural innovation. DISCUSSION PAPER #1 – 13.07.2016. Hall et al. (2016).

but also to enable an evolving process of technical, organizational, institutional, and policy innovation in response to changing conditions.

Understanding innovation as a process of systemic change also provides a lens to explore the effectiveness of different modes of multistakeholder partnership. It does this by providing two analytical considerations:

- the degree of complexity and therefore the level of systems change needed to enable and spread innovation; and
- the range of stakeholders relevant to these different levels of systemic change.

As much of the analysis in previous sections of this book indicate, operationalizing SFA requires systemic change: transformative, systems-wide innovation involving interlinked technological, institutional, and policy change across scales. To make a coherent contribution to SFA—with the local to global dimensions implied by the Sustainable Development Goals (SDGs)—approaches to partnerships will thus need to evolve.

In this section, we synthesize emerging patterns of good practice in partnership to derive some general principles of engagement for operationalizing SFA. In doing this, we look at AR4D or SFA partnership practice and at global multistakeholder partnerships (MSPs) for development practice framed by systemic challenges.

Partnerships for SFA Impact

Building on a long tradition of the progressive adoption of partnership approaches in the international agricultural research community, a range of mechanisms has been used in recent years to better interface research with stakeholders involved in the innovation process (Adekunle and Fatunbi, 2012). This has found operational expression as innovation platforms (Nederlof et al., 2011). Innovation platforms have often been local level MSPs to help better use research products and expertise in local development processes. Innovation platforms have also been used to some degree at national and international scales with varying degrees of success. Their key feature is that they have been initiated and often led by agricultural research organizations as a way of addressing concerns about the impact of their research investments in SFA.

Over the past two decades, there has also been enormous growth in collective action for international development, much of which has been based on establishing new global partnership organizations and initiatives (Patscheke et al., 2014). The reasons given to explain their growth include recognition that the scale and complexity of major global challenges cannot be addressed successfully by single actors, a decline in confidence in established aid structures and business models, the rapid spread of new technologies, and, increasingly, well-organized and effective advocacy on specific issues by NGOs. The operational expression of these global MSPs varies from lofty platforms with little connection to on the ground realities to truly multiscale architectures that link global and local agendas and global initiatives to local expression and impact (Patscheke et al., 2014). Organized around development challenges rather than research problems, the initiators and leaders of these global MSPs are quite diverse.

MSPs represent a specific form of partnership involving structured alliances of stakeholders from public, private, and civil society sectors. These include companies, policymakers, researchers, NGOs, development agencies, interest groups, and stakeholders from local, national, regional, and international governance regimes. A key feature is the dissimilarity of partners. This is qualitatively and functionally different from research or business partnerships, where similar stakeholders pool together resources and actions to address goals within their collective control.

The rationale for forming MSP groupings varies considerably. It can, however, be categorized as follows:

- Economic efficiency: Value for money can be achieved by building alliances between stakeholders who can play to their comparative advantage (Byerlee and Echeverría, 2002).

- Inclusiveness and governance: Partnerships are mechanisms for ensuring that notions such as "inclusiveness," "participation," and "voice" are addressed in the design and implementation of interventions (Malena, 2004).
- Complexity, "wicked" problems, and systemic change: Many development challenges sit at the interface of professional, organizational, sectoral, and national boundaries and are systemic in nature. A wide range of stakeholders needs to be involved in systemic change, which involves technological, institutional, and policy innovation at multiple levels (Befani et al., 2015; Burns and Worsley, 2015).

The operational manifestation of MSPs also reveals different forms of practice. Peterson et al. (2014) argue that the form of partnership adopted depends on whether the goal of the MSP is more concerned with addressing defined problems or whether it is concerned with addressing systemic problems (Fig. 41.2).

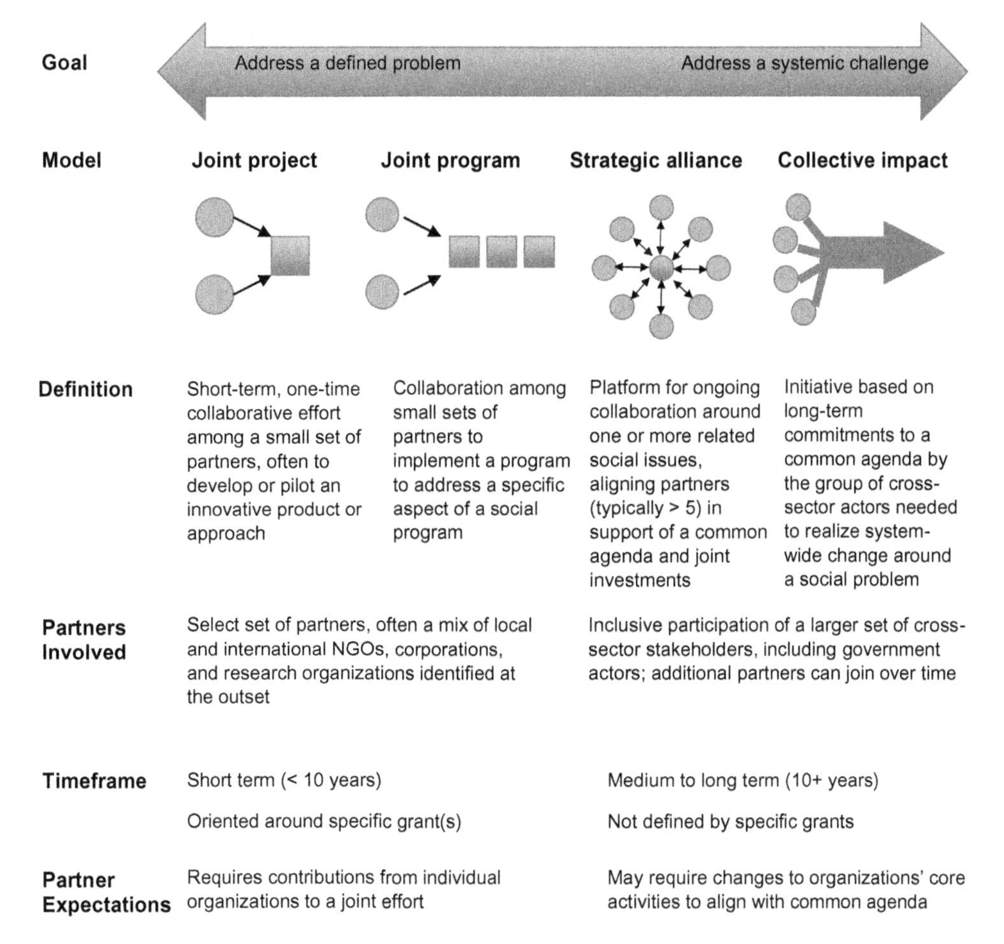

FIGURE 41.2 MSP typology. Source: *Peterson, K., Mahmud, A., Bhavaraju, N., Mihaly, A., 2014. The Promise of Partnerships: A Dialogue between INGOs and Donors. FSG: www.fsg.org.*

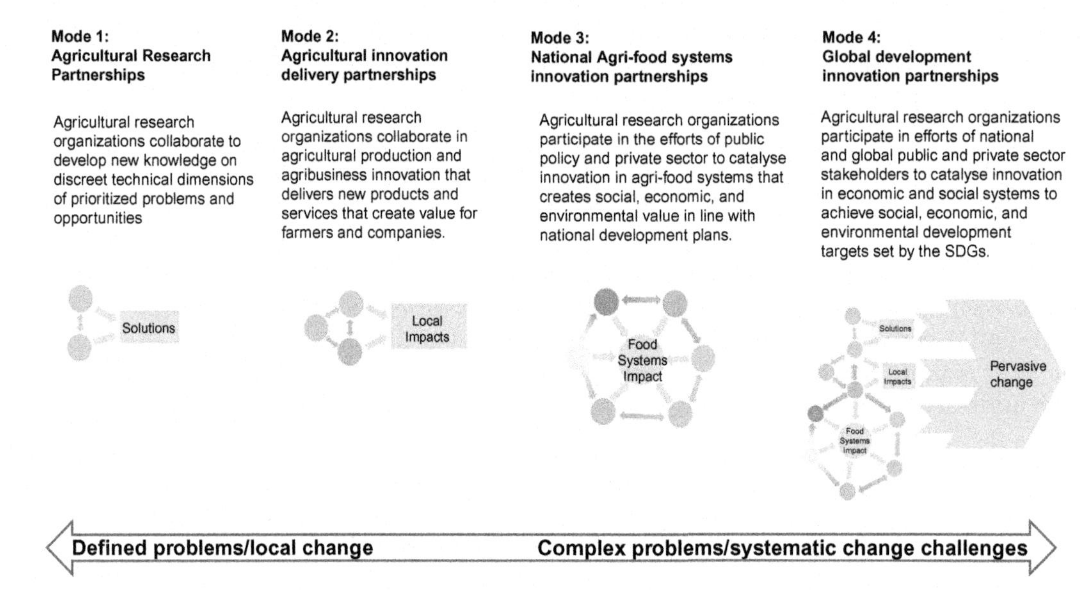

FIGURE 41.3 Modes of innovation and partnership. Source: *ISPC. 2015. Strategic study of good practice in AR4D partnership. Rome, Italy. CGIAR Independent Science and Partnership Council (ISPC), xiii + 60pp + annex 50pp.*

Partnership in SFA: A Framework for Analysis

Thinking of partnerships on a continuum from knowledge discovery to systemic change resonates with the innovation systems perspectives that suggest partnership needs to be understood along two axes: (1) the degree of complexity and, therefore, the level of systems change needed to enable innovation and SFA impact; and (2) the range of stakeholders relevant to these different levels of systemic change and scale of impact.

Using these innovation systems perspectives and building on the work of Peterson et al. (2014), Fig. 41.3 presents four modes of partnership and innovation that map onto different types of challenges, ranging from defined problem to complex challenges that require systemic change of the sort articulated by the SDGs and SFA.

Elements of Good Practice Across MSPs

Reviews of MSPs in recent years have almost been as numerous as the partnerships themselves (e.g., Bezanson and Isenman, 2012; Dodds, 2015; Hanleybrown et al., 2012; Hazlewood, 2015; Patscheke et al., 2014; Pattberg and Wilderberg, 2014; Peterson et al., 2014; Rajalahti et al., 2008; Severino and Ray, 2010; van Huijstee, 2012; Moench-Pfanner and Van Ameringen, 2012; Malena, 2004; Lele et al., 2007; Horton et al., 2009). These differ in terms of the nature of partnerships looked at (health, agriculture, nutrition, disease prevention, etc.), or the nature of the analytical lens used to critique and evaluate them (governance, development impact, value for money, etc.). Here, we draw on a "review of reviews" of MSPs to highlight elements of good practice drawn from comparative studies of different partnerships tackling varied themes and noting common problems.

A useful starting point is to recognize that while evidence of effectiveness of global MSPs is far from an exact science, there is enough evidence that some have worked better than others. Peterson et al. (2014) cite a study of 330 global MSPs from the Global Sustainability Partnership database, which suggests that only 24% are functioning effectively. However, the review does indicate that when functioning well MSPs can prove to be very effective in tackling global challenges, including SFA.

MSPs need an effective coordinating agency. The most important common feature of success across several MSPs is the existence of a coordinating agency, also referred to in the literature as an implementing agency, backbone structure, or even "broker." This agency acts as a glue, holding all the partners together, guaranteeing that conditions to maintain the partnership are met and ensuring progress towards the common goal that the partnership set out to achieve. This agency does not necessarily need to have the expertise to tackle a complex issue. Rather, its role is to understand the ultimate aim of the partnership, to identify the challenges and the gaps, to bring in the necessary expertise and resources where required, and possibly access funds when necessary. This role involves skills concerning building and maintaining networks, resolving conflicts, and sustaining working relationships between all actors in the partnership. According to Patscheke et al. (2014): "The backbone provides strategic coherence around the common agenda, establishes shared measurement and learning systems, supports the mutually reinforcing activities of the different partners, and facilitates continuous communication. It needs to provide strong leadership for the initiative while building ownership among the different partners like a conductor of a symphony, allowing each participating organization to bring their particular strengths to the joint effort."

Global partnerships cannot be formed simply around an agenda or theme; they need focused strategies and execution plans in order to succeed. The process of committing to a shared agenda can take place through an agenda-setting process involving all stakeholders, who can exchange their perspectives on the problem. This process can ensure buy-in from partners, legitimacy of the partnership, and can build understanding and trust among partners. For instance, the World Economic Forum's New Vision for Agriculture spent an initial six months meeting with governments, agribusinesses, investors, farmer groups, development agencies, and civil society groups to make the case for action and agree on the core issues to address. Once the boundaries of the issue were set, it took another year to develop a strategy to guide the partners' actions. The resulting agenda contains a three-pronged vision for change that encourages a holistic approach to agricultural development by addressing food security, environmental sustainability, and economic opportunity.

Careful consideration needs to be given to the regions in which partnerships operate. From the outset, new partnerships must assess countries and regions most in need of action when it comes to what the partnership aims to achieve. However, they should also consider the level of the national government's recognition of the problem and its willingness to act. What are the resources available, financial and otherwise to dedicate to the problem? Are there local champions to take on the issue? Partnerships work best when they align with country priorities and work through national and local planning, budgeting, and fund allocation systems in order to protect national sovereignty, build genuine ownership and strengthen capacity, and enhance the efficient and effective delivery of finance and other means of support.

Strong governance structures. Legitimacy of the coordinating agency is important, and it should be seen as representing the interests of all partners. Local ownership of the in-country implementing agency is critical to ensuring buy-in and the long-term viability of the partnership.

Managing MSPs for Operationalizing SFA

1. *Strengthen existing and emerging MSP platform architectures*. Architectures linking local to global scales are key to achieving impacts at scale and as a way of reconciling immediate and long-term development agendas. Often the building blocks of such architecture already exist. The priority is to ensure that efforts at different levels coordinate with and support each other rather than establishing new parallel and competing arrangements.
2. *Clarify roles within emerging architectures*. The principles of comparative advantage and subsidiarity are key, both in terms of effectiveness and in terms of capacity building. This is a particularly important consideration for international agencies. In many ways, these emerging global architectures represent a new world order wherein they need to locate an appropriate route of engagement, which in turn might mean a reframing of roles and responsibilities.
3. *Strengthen learning and capacity building*. Engaging with complexity means engaging with uncertainty. Arriving at modes of practice that are effective in addressing system challenges are, therefore, by their very nature necessarily experimental. A key priority for building capacity will be strengthening learning in and around MSP practice and translating this into practice change.

An example of a functional MSP is presented in the following case study.

The Global Agenda for Sustainable Livestock

Initiated by the Food and Agriculture Organization of the United Nations (FAO) in 2010, the Global Agenda for Sustainable Livestock is a partnership that brings together a variety of actors committed to sustainable development of the livestock sector, including business actors, NGOs, community-based organizations, social movement groups, governments, research organizations, intergovernmental agencies, and foundations.

The Agenda aims to achieve the following: (1) have an ongoing process where a shared understanding is created and constantly reaffirmed about views, ideas, solutions, expectations, problem definitions, actors involved, and required knowledge and data; (2) create coownership and commitments; and (3) energize stakeholders to take concrete actions to realize the goals that have been chosen (FAO, 2018).

The initial idea of the Agenda emerged from several consecutive reports and meetings of FAO's Committee on Agriculture (COAG), which is one of FAO's governing bodies. During its 20th session in April 2007, and based on the FAO publication *Livestock's Long Shadow: Environmental Issues and Options*, FAO's Animal Production and Health Division reported on the need for "managing livestock-environment interactions." At COAG's 22nd session (June 2010), it was agreed that FAO would engage in "consultations to establish a global dialogue" with a wide range of stakeholders to sharpen the definition of the livestock sector's objectives and to identify issues that could require intergovernmental action. In its document "Stakeholder Dialogue in Support of Sustainable Livestock Sector Development," FAO suggested a voluntary agenda, open to all stakeholders, targeting the "improvement of resource use efficiency in the livestock sector to support livelihoods, long-term SFA and economic growth, while safeguarding other environmental and public health outcomes." It also called for a "novel and functional governance system." During COAG 25, held in September 2016, 115 governments supported the Agenda.

The general aim of the Agenda is to offer a platform where ideas are exchanged and where stakeholders can reach out to other stakeholder platforms and organizations to distribute ideas about sustainable livestock. To this end, Agenda partners: (1) facilitate multistakeholder dialogue at the international and local level; (2) implement and support joint analyses and assessments, including the development of harmonized metrics and methodologies (e.g., Livestock Environmental Assessment and Performance, or LEAP); (3) identify and provide tools and guidance; and (4) promote and support innovation and local practice change (e.g., through the Climate and Clean Air Coalition agriculture initiative). A number of countries and international organizations, called the Dialogue Group, took the lead in moving the Agenda forward. They also provided the budget and decided that: "The Agenda should be built on broad based and voluntary stakeholder commitment, and act towards improved sector performance by targeting natural resource protection, while including poverty reduction and public health protection" (FAO, 2018).

The basis of the Agenda is the open MSP, which aims to meet once a year for stakeholders to discuss sustainable livestock issues. The guiding group and the support group of the Agenda make an effort to invite as many and as varied stakeholders as possible. Between 2010 and 2017, seven MSP meetings were held. They attracted between 100 and 200 attendees from across the globe. The MSP consists of seven different stakeholder clusters. During the second meeting in December 2011, stakeholders agreed on three focus areas: (1) closing the efficiency gap; (2) restoring value to grasslands; and (3) waste to worth. Different groups were then formed around these three focus areas. To take part in the focus area groups, stakeholders were asked to sign a consensus document to show commitment to the aims and intentions of the Agenda. During the third meeting (January 2013), stakeholders further refined the program for each of the three focus areas. Since then, the focus area groups have been meeting regularly to develop new ideas, pilots, and projects on specific goals of the Agenda in different regions, encompassing regional and national MSPs.

The governance of the Agenda continues to follow the basic organizational structure endorsed by the fourth MSP. Constituted and elected by the MSP, the guiding group is composed of representatives of the different stakeholder clusters. Consisting of three FAO staff members, the support team acts as process manager and conducts administrative tasks.

The Agenda aims to facilitate an open dialogue among stakeholders. The variety results in surprising interactions. For instance, representatives of pastoralists engage in debates with multinational meat processing organizations. While such diverse stakeholders express their framing of issues very differently, the MSP setting encourages them to explore where such different frames overlap, align, or can be linked.

During the meetings, the exchange of perspectives is organized through a wide variety of breakout sessions, for example, between stakeholder clusters, between focus area groups, or sessions that address one of the cross-cutting themes of the Agenda. To increase the participation of different stakeholders, MSP meetings are held in different regions of the world, employing a special mechanism to ensure the effective inclusion and participation of pastoralists (nomadic and sedentary), agricultural workers, smallholder farmers, and indigenous people in its processes.

The Agenda is constantly changing its internal structure to adapt to new circumstances. Initially, the Agenda was composed of one platform, but eventually evolved to seven different stakeholder clusters, which helped to identify the changing needs and circumstances of the stakeholder groups, to a setup that now includes seven action networks. An essential feature is the combination of a very open MSP and a set of more restricted focus area meetings. Through so-called twinning or mentoring mechanisms, countries are learning from each other. Learning is also facilitated by asking other parties to provide feedback and by setting up study portals.

The Agenda catalyzes the global dialogue into local action and uses local action to inform the global dialogue. It remains, however, a challenge to bridge between the local and global level. The focus seems mainly on the flow from global to local, containing shared ideas, knowledge, and innovative concepts about how to advance sustainable livestock. However, what goes upstream and finds its way into the focus area groups at the global level remains unclear. These could be observations about local barriers to change, such as technical conditions, infrastructure, and governance blockades, but also experiences about building trust for change, for example, knowledge about the role of local and national governmental organizations.

The action networks have already achieved tangible results in Latin America, Europe, Africa, and Asia through development of methodologies that have been adopted by a wide range of stakeholders (e.g., environmental assessment guidelines, efficiency analysis, guidelines for grassland management, knowledge transfer in Dairy Asia, and silvopastoral systems) and have supported several pilot projects worldwide. Networks around local projects may be more explicit and inclusive, involving end users, producers, local policymakers, and national governments. The Agenda aims to facilitate the dynamics between these global, regional, and local networks Breeman et al., 2015).

41.3 SYSTEM-WIDE CAPACITY DEVELOPMENT

The previous chapters in this section have argued that the transition to SFA is by nature a complex, multisector, multiactor, and multilevel process requiring systemic change, commitment, and participation from all stakeholders to ensure its success. How will the systemic transition to SFA materialize? Who will own, drive, and commit to this process at the field, district, and country levels? What are the national and subnational capacities of people, organizations, institutions, and the enabling policy environment that need to be enhanced in order for implementation to succeed? How will countries be supported in this process?

FIGURE 41.4 System-wide capacity development approach.

The thinking and practice of system-wide capacity development addresses these questions directly. Capacity development (CD) is defined as "the process whereby individuals, organizations and society as a whole strengthen, create, adapt and maintain capacity to set and achieve their own development objectives over time" (OECD, 2006). Therefore, an effective and system-wide CD approach aims to facilitate an endogenous development process rooted in development effectiveness principles (OECD, 2005) and national empowerment (Sen, 1999; Chambers, 1994). It aims to enable country stakeholders to own, lead, and drive their own development process and thus maximizes intended country ownership, impact, sustainability, and scale (FAO, 2010, 2016a; UNDP, 2003; Kalas et al., 2017).

Operationally this approach (Fig. 41.4; FAO, 2016b) enhances capacities interdependently across *individual capacities* (e.g., knowledge, skills, and competencies), *organizational capacities* (e.g., performance of organizations, cross-sectoral, multistakeholder coordination/collaboration mechanisms), and *enabling environment* (e.g., sound regulatory and policy frameworks, institutional linkages, and enhanced political commitment and will) based on jointly assessed needs, joint context-specific interventions identification, and joint tracking of results (FAO, 2010).

How to Apply System-Wide Capacity Development for SFA Transition at the Country Level?

As noted in the first chapter of this section, FAO's SFA approach identified four areas to operationalize transition towards SFA. These include (1) building relevant, coconstructed and accessible evidence; (2) engaging stakeholders in dialogue to build common understanding and joint action; (3) formulating tools and levers to enable and incentivize changes in food and agricultural systems; and (4) realizing practice change through innovative approaches and solutions.

The proposed system-wide CD approach is directly relevant across all areas. It provides an operational entry point to deepen the desired participatory and contextualized nature of the process. In particular, it fosters dialogue, common-understanding, trust, ownership, and subsequent commitment for action (i.e., practice change) at scale, recognizing the critical added-value of the participatory and inclusive process to affect the desired end result.

Three iterative and highly participatory phases are recommended to apply system-wide CD for SFA transition jointly with stakeholders:

1. assessing needs and identifying options for change;
2. designing and implementing contextualized interventions/solutions; and
3. defining and tracking the results through fostering learning and mutual accountability.

The first phase is to jointly assess capacities with stakeholders. Undertaking a careful assessment of strengths and needs to diagnose what and whose capacities need to be developed is a fundamental precondition for all successful and sustainable development projects. Such participatory assessments ensure the context is commonly understood and that existing capacities and needs are identified, allowing the project or program to be customized to the local situation. In addition, through the participatory nature of the process, stakeholder ownership and commitment are fostered, key components during the implementation to reach impact, sustainability, and scale.

The following case study for Rwanda illustrates how national capacities for SFA transition were "assessed" an inclusive and interactive process that fostered country ownership and commitment. In 2014, the Government of Rwanda expressed the interest to engage in a dialogue with FAO to progress on the transition towards SFA. Jointly with stakeholders, it was agreed to assess the national capacities for SFA in Rwanda in a participatory and inclusive process facilitated by FAO. The SFA assessment process in Rwanda included three steps.

Step 1—Joint scoping: This step consisted of a consultative scoping study conducted by a trained national expert. Through several individual meetings and a collective workshop, the aim was to raise awareness, galvanize commitment, and start identifying key sustainability issues related to food and agriculture in Rwanda.

Step 2—Joint assessment: From July to September 2014, issues pertaining to the sustainability of food and agriculture in Rwanda, ongoing actions undertaken for solving them, and possible solutions to be implemented were assessed. Two complementary sources of data and knowledge were used during the assessment stage. First, a bibliography of 208 documents was collected and analyzed, including governmental strategies and laws, scientific literature, and development partner reports. Second, a stakeholder analysis (FAO, 2015) identified 60 key stakeholders from which 42 were interviewed based on the following questions:

1. What are the three most important sustainability issues in food and agriculture (including land, water, forests, and fish stocks) in Rwanda?
2. How is agriculture (including land, water, forests, and fish stocks) impacted by these issues?
3. Are there horizontal (among stakeholders) and vertical (across administrative and geographic levels) organizational and institutional coordination mechanisms in place to deal with cross-sectoral sustainability issues? If yes, please describe them. If no, what coordination mechanisms should be in place to support sustainability in Rwanda and who should be involved in such a mechanism?
4. How do you currently manage these three issues? For example, do you have projects or programs that deal with them? Are there any gaps? Who are the key stakeholders involved in managing these issues?

5. How should these three issues be managed? Which stakeholders should be involved? Do the stakeholders have adequate capacities to manage these issues (e.g., clear mandates, organizational structures)? How would you think that FAO could support your effort to cope with them?
6. Can you recommend other sources of information and resource persons that would help to understand important sustainability issues in your country?

The findings culminated into an assessment report that identified ten key obstacles to SFA in Rwanda, namely land scarcity and fragmentation; land degradation, soil fertility, and deforestation; water resources management and utilization; climate change and climate variability; improved seeds and breeds, and conservation of genetic resources; fertilizer availability, access and affordability; postharvest handling, storage, and transformation; access to markets, finance, and investment; professionalization of the farmers and cooperatives; and institutional capacity enhancement.

Strong "cross-cutting" emphasis was placed on the need to improve organizational and institutional capacities of government, as well as nongovernmental actors (including farmers and cooperatives). For government, due to the degree of decentralization in Rwanda, adequate human and institutional capacities for planning, coordination, and implementation at the local level are critical to guarantee good performance in the agriculture sector.

Step 3—Jointly refined assessment, validation, and action planning workshop: This step consisted of an inclusive and interactive national 3-day workshop with representatives of the Government of Rwanda, NGOs, civil society, development partners, and members of the research community. The objective was to validate and deepen the findings from the first two steps (i.e., scoping study and joint-assessment interviews), identify and prioritize options for change, elaborate on a road map, and throughout the process strengthen country ownership and commitment for action. The basis for the interactive group discussions were the findings from the initial capacity assessment and a jointly developed SFA capacity assessment questionnaire. Tailored for the Rwanda SFA context, the structure identified the current state, desired future state, and actionable recommendations for capacity improvements across the different capacity development levels (individual, organizational, and enabling policy environment). The workshop further identified trade-offs and strategies on how to address them, as well as drafting a road map and detailed action plan, including roles, responsibilities, and funding requirements.

From the ten initial areas identified in step 2 (see above), four critical SFA areas for action were prioritized, in line with proposed FAO's SFA principles covering both biophysical as well as socioeconomic and institutional capacity issues (Table 41.2).

Besides the actual findings and recommendations (i.e., the "product"), the "process" mattered equally as country ownership was deepened through joint identification and common understanding of challenges and ways forward. In addition, through the process, high-level political commitment to act on identified entry points towards the transition to SFA was achieved. The findings of the SFA assessments were presented by the Minister for Agriculture during the FAO conference in Rome in 2015 (FAO, 2015), and culminated into a subregional workshop on transition towards SFA that included countries from Mali, the Democratic Republic of the Congo, Côte d'Ivoire, Chad, Cameroon, Rwanda, Mozambique, Kenya, and Zambia (FAO, 2016c).

TABLE 41.2 Critical Areas for SFA Capacity Development

Area of Recommendation	Detailed Recommendation
Institutional coordination (horizontal and vertical) Part of strengthening institutional capacities, effective planning, and implementation of sustainability measures demands close collaboration between partners who may have potentially conflicting agendas.	Improve horizontal and vertical institutional coordination through joint mechanisms across sectors and government institutions to harmonize and ensure effective and good implementation of policies on seeds, land, fertilizers, mining, forests, and rural development. Particular attention needs to be given to improving intersectoral governance mechanisms, as well as between administrative levels in order to foster more integrated and effective action.
Intensification Improving livelihoods of farmers depends on their capacity to participate in viable commercial activities.	Support crop diversification and integrated production systems, e.g., small livestock, aquaculture. Improve conservation of soil and water, e.g., agroforestry, composting, small-scale irrigation.
Finance and marketing Improving livelihoods of farmers depends on their capacity to participate in viable commercial activities.	Better access and affordability of loans to farmers. Improve postharvest handling and storage.
Ownership and professionalization Part of strengthening organizational capacities includes empowering farmers to protect and enhance their producing environments.	Awareness raising and targeted technical capacity development on contract farming. Enabling the testing of innovative approaches to conservation based on farmers' participation and empowerment. Strengthening organizational capacities, such as improved mandates, processes and procedures, and knowledge sharing for better performance.

The second phase of system-wide CD for SFA is to design and implement contextualized interventions/solutions that go beyond training. Alongside the delivery of training, other successful and perhaps more appropriate CD modalities across the three CD dimensions for SFA include coaching, South–South cooperation, policy support, support to organizational development, farmer field schools, strengthening of networks, multistakeholder platforms, convening for national/regional events, and strengthening institutional coordination.

The third phase to integrate system-wide CD for SFA following the assessment and design and implementing contextualized interventions/solutions is to define and track the results through fostering learning and mutual accountability. Although complex and challenging due to its nonlinear nature of assessing change, jointly monitoring national capacity-related activities with stakeholders across the three dimensions is critical to enable learning and track progress while fostering country ownership and commitment. As transition to SFA is a complex, knowledge-intensive process requiring multiple learning loops and change of practice, this calls for innovative approaches for tracking this complex change process across the three CD levels.

The transition to SFA is by nature a complex, multisector, multi-actor, and multilevel process requiring a systemic change, commitment, and participation from all stakeholders to ensure its success. Integrating a system-wide CD approach directly fosters country-ownership and commitment through a participatory process to jointly with stakeholders assess, design, and track CD for countries to embark on a country-driven pathways towards more SFA.

41.4 GOVERNANCE AND POLICY CHANGE

Governance is the set of formal and informal rules that determines the interactions of people within social-ecological systems. The concept covers a wide range of factors influencing the decisions people take, from personal values and knowledge to formal laws and regulations. Overall, governance determines "the way we do things" from the technologies we use, the markets we access, the natural resources we transform into products and the processes used to do so, and the distribution of the benefits and costs from any of these actions. Changing governance is thus an essential tool for operationalizing SFA systems.

By governing we mean "all those interactions and activities of social, political and administrative actors that can be seen as purposeful efforts to guide, steer, control or manage (sectors or facets of) societies," and governance refers to "the patterns that emerge from governing activities" (Kooiman, 1993, p. 2). Most literature on the governance of wicked problems focuses on how-to strategies, and how to cope, control, and manage wicked problems. However, when dealing with wicked problems, appropriate ways of observing are also key. A characteristic of a wicked problem is that actors define problems differently and constantly develop new problem definitions along the way, which calls for the need to observe these often-emerging problem definitions – using methods such as those described in Chapter 39, Identifying the Relevant Context: Key Features of the Social-Ecological System, of this section. Hence, governance capability refers to the ability of actors to observe wickedness, to take action, and the ability of governance institutions to enable such observations and actions.

Many scholars have already shown that wicked problems cannot be solved in a straightforward way with actions taken by a hierarchic or monocentric form of governance (Bitzer, 2012; Duit and Galaz, 2008; Head, 2008; Koppenjan and Klijn, 2004; Ostrom, 1999; Roberts, 2000). Rather, when dealing with wicked problems, new forms of innovative polycentric governance arrangements are proposed, such as roundtables (Schouten and Glasbergen, 2011), leadership networks (Nooteboom and Termeer, 2013), multilevel forums (Bates et al., 2013), public–private partnerships (Diamond and Liddle, 2005), communities of practice (Wenger, 1998), and multistakeholder dialogues (Dentoni et al., 2012; Warner, 2006). Some examples of these different types of arrangements have already been given in the analysis of innovation systems and partnerships above.

The analyses provided in this section—and indeed throughout this entire volume—indicate that there are many different ways in which governance of agrifood systems can be managed to enhance sustainability, such as by providing usable knowledge to people in the system that empowers them to change, by engaging in political processes and reforming institutions to change the set of incentives and enabling conditions people face, and by improving the capacity to take collective action to solve problems. Thus, the means and nature of governance changes to support the operationalization of sustainability vary depending on the root causes of unsustainability, as well as the capacity in place for change.

In analyzing transitions to sustainability in socio-technical regimes (Rip et al., 1998), Smith et al. (2005) develop typologies of specific contexts of governance transitions that can indicate potentially effective entry points for governance change for enhancing sustainability of agrifood systems. They argue that changes in sociotechnical regimes are driven by two major processes: selective pressures, and capacity to mobilize and coordinate use of resources within a regime to adapt to pressures. Selective pressures may be the result of major trends, such as demographic or environmental changes, and in the context of this chapter can be related to the drivers of unsustainability in any given agrifood system. The adaptive capacity of the system is a key concept for sustainability governance—and an important facet of the concept of resilience. According to Folke (2016), adaptive capacity "refers to the ability of the social-ecological system to learn and to adjust its responses to the impacts of external drivers and internal change." Adaptive capacity then is the ability of a system to undergo change while maintaining the main function and structure of the system. Folke (2016) also makes the point that when a system's adaptive capacity cannot cope with pressures, transformative change occurs where the basic functions, structures, and modalities of the system shift. As noted in the analysis of innovation systems in the first part of this chapter, innovation can take the form of incremental, radical, or transformative change with generally greater capacity and focus on incremental. In thinking about governance change to operationalize sustainability, it is thus important to consider the scale of change that is being aimed for, as well as the available capacity within the system to deliver it.

While there is considerable diversity in governance approaches to support sustainability, there are some basic underlying principles to guide us. Matson et al. (2016) identify three core concepts for governance of sustainability: fostering collective action, coping with externalities, and managing common-pool resources. Collective action involves coordination between individuals and groups to generate a good or service they could not individually produce. Externalities are unintended consequences of one person's or group's actions on another; for example, pollution from an upstream industry affecting downstream water users is an example of a negative externality. Common-pool resources are a specific type of collective action problem involving finite resources (e.g., soil, land, water, atmosphere, genetic resources) that are difficult to control access to, and thus can often be depleted in the absence of an effective system for sharing the resource. Operationalizing sustainability requires systems of governance that promote effective collective action, stimulate the generation of positive externalities, and regulate access and use of common-pool resources so as to avoid their depletion and degradation.

"Governance through goals," as epitomized by the 17 SDGs and the related 169 targets, is a novel approach to place nonlegally binding, inclusively set goals at the center of global policy and governance (Biermann et al., 2017). However, such universal goals linking social, economic, and environmental aspects amount to little unless concerted action is taken by governments and nongovernment actors. Understanding the integrated nature of the SDGs together with the range of interlinkages and interdependencies among them is key to unleashing their full potential, as well as for reducing the possibility of perverse outcomes. Further, there are inherent trade-offs in dealing with the interactions between goals across multiple scales of governance (global, national, and local). Therefore, political discussion is essential regarding how these trade-offs are handled to better support integrated governance and policy coherence; this of course presents challenges for governments in terms of overcoming entrenched institutional and sectoral behaviors.

In the final section of this chapter, we present three case studies on changes in governance to operationalize sustainability. The first case study is an example of how externalities associated with fishing were explicitly incorporated into value chains and the decisions of consumers in purchasing fish products, with the private and civil society sector playing a leading role. The second example describes a case where a national government took action to enhance the coordination of its climate and agricultural policies, with assistance from an external source. Finally, the third example illustrates how the California state government has dealt with the potential trade-off between reducing GHG emissions and enhancing the welfare of its poorest citizens through a process of dialogue and political mobilization.

Case Study 1: Innovation Through the Marine Stewardship Council (MSC): Labeling Fish for Sustainability

From the 1980s onward, potentially negative environmental aspects of fishing started to attract greater attention from both the fishing industry and environmental groups. In addition to causing severe decline of fish stocks, industrial and large-scale commercial fishing has been associated with the significant problem of bycatch—the capture of nontarget species, including fish, marine mammals, turtles, and seabirds. Some fishing methods can also be highly damaging to the environment, such as bottom trawling.

One way of addressing these problems is to empower consumers to make informed choices so they can confidently identify and buy fish and other seafood that is caught in well-managed and sustainable fisheries and avoid products from fisheries that use environmentally damaging practices. This was the opportunity that the MSC was established to address.

The MSC operates as an independent not-for-profit organization established to address the problem of unsustainable fishing and to safeguard wild-caught seafood supplies for the future. It does this through an assessment and certification process: seafood products that meet its Fisheries Standard are identified at the point-of-sale by the blue MSC logo. Seafood businesses that handle these certified products are identified with Chain of Custody certificates, which are designed to ensure traceability of these products back to MSC-certified fisheries.

The MSC considers that it is developing and revising its Fishery Standard based on the latest understanding of internationally accepted fisheries science and best practice management. It is based on three core principles: (1) sustainable fish stocks: the fishing activity must be at a level that ensures it can continue indefinitely; (2) minimizing environmental impact: fishery operations must be managed to maintain the structure, productivity, function, and diversity of the ecosystem; (3) effective management: the fishery must comply with relevant laws and have a management system that is responsive to changing circumstances.

Assessments and certification are undertaken by independent third-party conformity assessment bodies, which must meet best practice guidelines for standard setting agencies as set out by the ISEAL alliance (the global association for sustainability standards) and FAO. Conformity assessment bodies are typically specialist for-profit companies. Certification is voluntary and is open to all fisheries involved in the wild capture of marine or freshwater organisms.

The efforts of three organizations in the early to mid-1990s were critical to the establishment of MSC: World Wide Fund for Nature (WWF), Greenpeace, and Unilever. In the early 1990s, senior managers at Unilever, one of the world's largest multinational consumer goods companies, were concerned about the threat to dependable fish supplies for their highly successful Birds Eye and Iglo brands of fish products, notably fish fingers. They were especially concerned about the stability of North Sea whitefish stocks. At the same time, Unilever and other large companies were being targeted by Greenpeace to stop selling products containing or derived from fish oil.

They also wanted a scheme whereby the origin of all fish was labeled. Unilever considered this labeling scheme to be unworkable, but the campaign helped to further focus the company's attention on this issue.

The WWF entered into discussions with Unilever, which culminated, in February 1996, in a joint statement to establish "an independent Marine Stewardship Council which will create market-led economic incentives for sustainable fishing." The statement noted that the two organizations had "different motivations, but a shared objective: to ensure the long-term viability of global fish populations and the health of the marine ecosystem on which they depend."

In 1997, MSC was launched as an independent, nonprofit organization registered as a charity in the United Kingdom, without membership but with a board of trustees and chair, a technical advisory board appointed by the board of trustees, a stakeholder council, and a secretariat that coordinated a standards council. The 11-person board is currently made up of representatives of major seafood companies, a supermarket, an international organization, academics and researchers, a government fishery manager, and a nongovernmental conservation organization. Members are drawn from North America, Australasia, Africa, and Europe.

Between 2000 and 2015, out of 189 certified fisheries, 39 made one or more improvements specified in the full assessment: this was mandatory for them to retain their certified status. These mostly related to research improvements, such as mapping location/intensity of fishing and bycatch estimates, and technical actions, such as spatial closure and gear modifications. More generally, since 2000, 94% of MSC certified fisheries have been required to make at least one improvement, resulting in more than 1200 examples of change.

To assess whether MSC certified fisheries can be considered sustainable (https://www.msc.org/standards-and-certification/fisheries-standard), the 2017 report includes results of an analysis of 100 MSC certified fisheries using independent stock assessments taken from a public database. The MSC's conclusion was that:

> By comparing recent snapshots of stock health with values from 2000, when none of the stocks examined were MSC certified, we can see that stocks have higher biomass in years following certification in nearly all regions. This suggests that either a desire to obtain MSC certification incentivized better stock stewardship, or that the MSC label was sought as recognition of efforts made to recover stocks to healthy levels of biomass.

Case Study 2: Coordinating Climate Change and Agricultural Policies in Zambia

Around 2011, the Zambian Government and FAO, through the Economics and Policy Innovations for Climate-Smart Agriculture program, were engaged in a dialogue on the enhancement of the Zambian agricultural sector. This resulted in the collaborative development of a Climate-Smart Agriculture (CSA) project to support the Zambian Government's recognition of the fact that sustainable agricultural development to achieve food security, poverty reduction, and economic growth and responding to climate change are two of the greatest challenges currently facing the country and are closely linked. The Zambian CSA project was developed in the context of the Fifth National Development Plan and in line with the National Medium-Term Priority Framework.

The CSA project addressed three broad challenges by focusing on capacity building delivered through experiential learning to address the following specific problems and issues:

1. *Knowledge/data gaps*: evidence-based collection and generation for planning and making strategic choices about CSA defined as an approach to developing the technical, policy, and investment conditions to achieve sustainable agricultural development for food security under climate change.
2. *Fragmentation of policy, institutions, and financing*: despite recognition that the challenges of food security and climate change are closely linked within the agriculture sector, policy, institutional arrangements, and funding channels for climate change, food security, and rural development are poorly coordinated in the country.
3. *CSA strategy gaps*: the absence of national climate-smart strategies that can guide nationally led action and provide a basis for mobilizing international support.

The Theory of Change

Based on the theory of change designed for the project, the research component focused on a systematic process:

1. *Country-level institutional mapping and evaluation of existing key institutions related to agricultural sector development, food security, and climate change.* Two key findings from this mapping included the apparent gap in interministerial coordination of efforts between the Ministry of Agriculture and Livestock and the Ministry of Lands, Natural Resources and Environmental Protection (the focal ministry for the coordination of national climate change policy development and implementation) in addressing the impacts of climate change on the agriculture sector. Second, the two ministries were both in the process of reviewing and developing national policies in their respective areas.
2. *A review of major policy documents related to climate and agriculture focusing on complementarities and contradictions for policy harmonization.* The review highlighted the inherent complementarities and contradictions in the two draft policies. This review was augmented by the evidence that had been collected and analyzed by the project on

the state of climate change and agriculture in Zambia, with focus on areas such as the following:

a. climate variability and uncertainty in predictions;
b. statistical analysis on climate shocks, and productivity under different climate-related shocks; and
c. policy simulations using cost-benefit surveys of CSA "entry points" (entry points included conservation agriculture, livestock/crop mix, agriculture/forest interface, role of climate risk and uncertainty, role of legal and institutional environment, and input support efficiency).

Conceptual Framing of the Operationalization of the Evidence Base, Dialogue, and Policy Nexus

Based on the scenario outlined above, the project organized a dialogue between the Ministry of Agriculture and Livestock and the Ministry of Lands, Natural Resources and Environmental Protection to contribute to the maximum effectiveness of agricultural and climate change policies that were being prepared at that time. Facilitating this process was one of the main tasks of the CSA project Zambia with technical support from FAO. The dialogue was planned to feed into broader climate change and development coordination discussions, within which the Ministry of Agriculture and Livestock and the Ministry of Lands, Natural Resources and Environmental Protection were expected to play important roles.

Key Areas Considered for the Policy Harmonization

1. *Synergies and trade-offs between policy measures*

   There were considerable synergies between agricultural and climate change policy measures: both draft policies called for the development of agricultural systems that are more resilient, increased dissemination of climate change information to farmers and policy-makers, and increased use of sustainable land management practices. The trade-offs examined included, for example, the fact that reducing deforestation has important mitigation and environmental benefits but may reduce agricultural expansion, which can reduce food production or increase costs to farmers, depending on how intensification on existing agricultural lands is managed. Productivity-increasing agricultural practices/technologies were to be screened for their potential contribution to (or detraction from) needed adaptation, as well as for their potential mitigation benefits.

2. *Possible policy inconsistencies or conflicts*

   Policy measures that do not carefully consider the linkages between agriculture and climate change, especially any trade-offs, may result in policy conflicts or inconsistencies, which may negatively affect policy implementation. In some cases, perverse outcomes may also be generated (e.g., biofuel production in certain contexts may harm food security). Managing trade-offs requires some understanding of the projected climate change effects at the local level (e.g., including use of downscaled models), as well as analysis of the extent to which the range of potential options for technology/practice change in agriculture address the challenges posed by climate change and associated adaptation needs (such as increasing resilience, water use efficiency, crops to fit changed length of growing season). To prevent harmful outcomes, it is important to identify potential trade-offs and possible options for their management.

3. *Overlaps and gaps*

   Overlap may occur when two policies are advocating the same or similar action. While climate change is a cross-cutting issue and the national climate change policy is meant to enable overall coordination of climate change action, the draft National Policy on Climate Change included a large number of sector-specific measures relating to agriculture. This could create confusion on who should take the lead in implementation and incur duplication of effort. A rational distribution of measures across the two policies may be beneficial, based on who will ultimately implement the measures and taking into account the respective roles, mandates, competence, and outreach of the two ministries and other key stakeholders tasked with their formulation and implementation. Gap analysis should help to identify any missing measures (e.g., more efficient use of fertilizers, which can increase soil fertility and decrease nitrous oxide, and integrated production systems, such as crop-livestock-fisheries-agroforestry).

The Practical Approach Taken for the Harmonization of Dialogue

The dialogue between members of the two ministries was conducted using the following practical approach:

1. Using the comparative table, analyze specific measures in terms of their synergies/trade-offs, consistencies/inconsistencies, duplication/overlap or whether there are any related gaps.
2. Identify essential elements present in one of the policies, but missing from other documents (e.g., no explicit incorporation of climate change effects in agricultural technology choice in the agricultural policy or lack of consideration of need for agricultural growth in the climate change policy).
3. To the extent possible, identify options for managing/reconciling trade-offs.
4. Identifying the division of labor: Where should the two ministries work together and where should they work separately?
5. Identify concrete steps for cooperation across the two ministries, for example, pooling research findings to underpin policy development, consultation at the outset and at the final stages of policy formulation, and cooperation on policy implementation (strategy development, planning, programs, investment strategies, projects), using the Ministry of Agriculture and Livestock's outreach at the subnational level for joint project implementation.

From Dialogue to Policy

The result of the dialogue was a more harmonized national agricultural policy and climate change policy, and the follow-through thereafter has been the recognition and prominence of the CSA elements in key climate-change related policy documents in

Zambia, such as the Zambia REDD + Strategy (UNDP, 2015), Zambia's Intended Nationally Determined Contribution (UNFCCC, 2015), and Zambia's draft of the Seventh National Development Plan implementation plan 2017–2021 (Ministry of National Development Planning, 2017).

Case Study 3: Climate Policy and Environmental Justice in California (United States)

The state of California, in the United States, has taken a leading role in combating climate change through a mix of market and regulatory policies to achieve ambitious GHG emissions reductions. In 2006, the state passed the Global Warming Solutions Act, setting GHG reduction targets for 2020, which is on track to meet. In 2016, California passed Senate Bill 32 (and its companion bill AB 197), which commits the state to reduce GHG emissions to 40% below 1990 levels by 2030. The California Air Resources Board developed a plan to reduce state GHG levels that gives a prominent role to a state cap-and-trade program. This program involves the allocation of "allowances" to major GHG emitters such as industries and utilities through auctions, or in some cases, free allocations. Overall, the state's emissions cannot exceed the quantity allowed by the cap, but allowances can be traded among emitters, or achieved through the purchase of "offsets," which involves emissions reductions from a nonemitter on behalf of the emitter—for a price. As a market-based measure, cap-and-trade programs are considered to be a relatively efficient means of regulating GHG emissions since they can result in least-cost emissions reductions being prioritized (Stavins, 2008). However, there are concerns about the distributional impacts of the California cap-and-trade program, and these have been articulated by the state's environmental justice community, specifically in the context of California's climate change policies (London et al., 2013).

Growing disparities in income, wealth, and public health characterize California's population. An expanding body of research demonstrates that economic and racial inequality influences where people live, work, and play, and what health risks they are exposed to as a result of their location. For communities of color in the state, air quality is worse and health risks are higher, contributing to a "climate gap" in which the greatest effects of climate change may be felt by populations already challenged by economic and social disadvantage.

Advocates from environmental justice groups and labor unions have been actively involved in California's climate policy debates, bringing different perspectives to bear on climate policy and its implementation. These groups have pushed to make "climate equity" a guiding principle in climate policy, albeit with somewhat different interpretations of what the goals of climate equity should be. Environmental justice advocates have been deeply concerned about cap and trade exacerbating toxic hotspots in communities near polluting facilities. Cap-and-trade programs allow businesses to choose to trade allowances or buy offsets instead of reducing GHG emissions—and only the latter will lead to the reduction of copollutants (the toxic air pollution that accompanies GHG emissions). Given the state's commitment to cap and trade, environmental justice groups have fought to insure that some of the funds generated from the cap-and-trade program go to those very communities.

In 2012, the Legislature passed Senate Bill 535, directing that 25% of the proceeds from the Greenhouse Gas Reduction Fund be allocated to projects that provide a benefit to disadvantaged communities. The legislation gave the California Environmental Protection Agency (CalEPA) responsibility for identifying those communities.

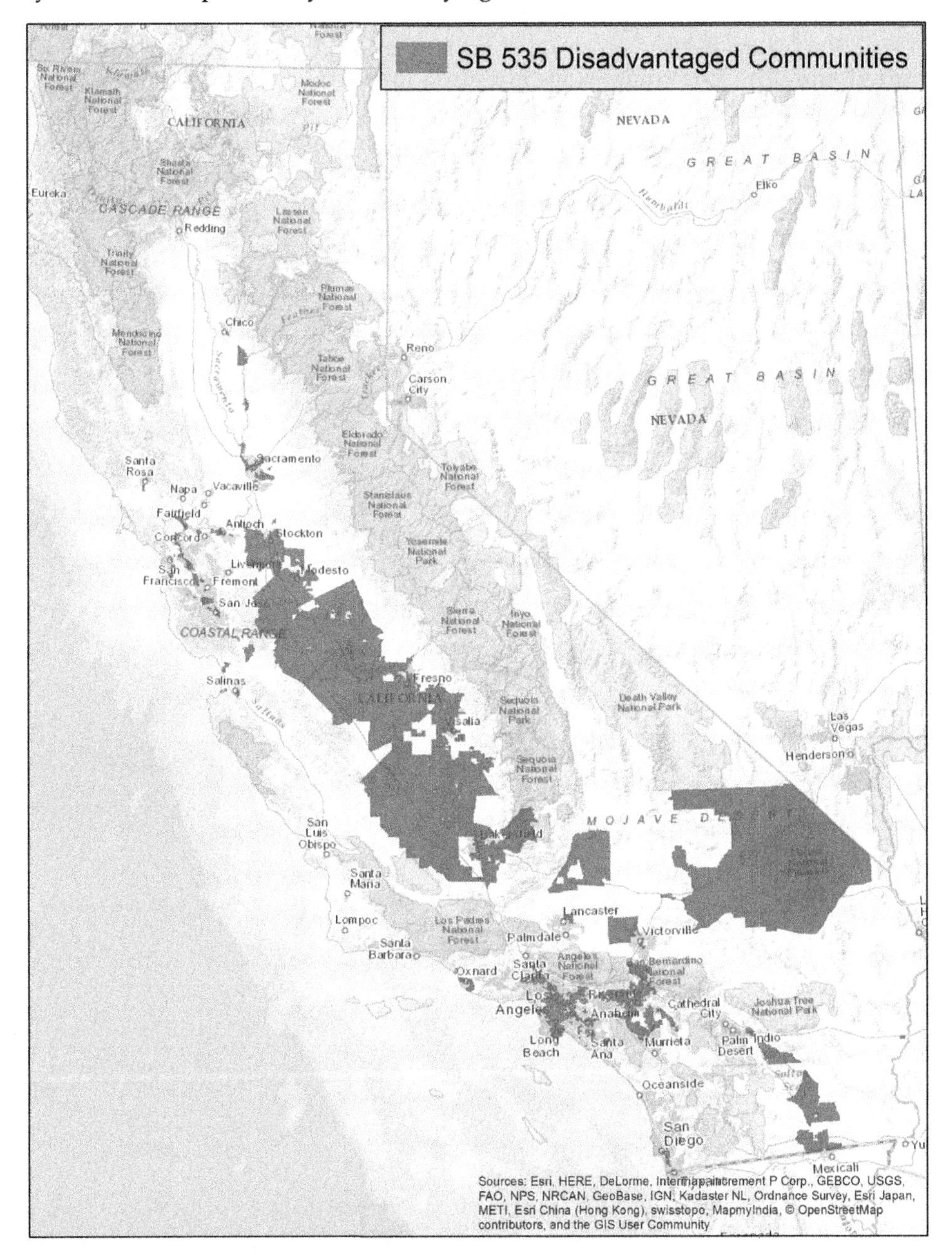

FIGURE 41.5 Map of disadvantaged communities in California, identified by CalEPA.

Following a series of public workshops in February 2017, CalEPA released its list of disadvantaged communities for the purpose of SB 535 (see Fig. 41.5). The agency used a tool called CalEnviroScreen to identify California's most pollution-burdened and vulnerable communities. This tool uses a quantitative method to evaluate multiple pollution sources and stressors, and vulnerability to pollution, in California's approximately 8000 census tracts. Using data from federal and state sources, the tool consists of four components in two broad groups. The four components are made up of environmental, health, and socioeconomic data from 20 indicators.

These communities, designated as disadvantaged through the application of the tool, are specifically targeted for investment of proceeds from the state's cap-and-trade program. These investments are aimed at improving public health, quality of life, and economic opportunity.

Thus, the state of California is seeking to achieve lower GHG emissions and improved welfare of its most disadvantaged citizens through innovative policy making, supported by strong political leadership and organization, as well as relevant and accessible evidence.

CHAPTER

42

Conclusions

The globally endorsed SDGs have set a development agenda that implies deep systemic changes in the social, economic, and environmental performance of agrifood systems and, indeed, of society as a whole. Admittedly, the radical changes implied by the SDGs will require time to take hold, as the underpinning processes needed to achieve them require pervasive changes in the way humans relate to nature and to each other across a wide range of interactions.

In this section, we have considered the operationalization of sustainability in food and agriculture systems as an ongoing process of the changing balance between human and natural interactions. The process occurs through a variety of means and entry points, depending on the specific nature of the imbalances that drive unsustainability in different contexts. Since SFA is a wicked challenge that does not have a simple or unique pathway, it requires a process of achieving improvements through dynamic and adaptive interaction involving different stakeholders, including from science, civil society, and political communities. The shift in the sustainability discourse from focusing on the features of the ideal sustainability state to one on processes that will result in an improvement in human/nature balancing is, thus, a major step forward in operationalizing SFA.

We used a simple conceptual framework for operationalizing SFA that consists of three major action areas: (1) identifying context and major imbalances in human/nature interactions; (2) using evidence and dialogue to build knowledge for action; and (3) facilitation of solutions at scale. We argue that operationalizing SFA requires actions in all three areas in a dynamic and nonlinear process of interaction and feedback.

We have also attempted to shed light on the way these three action areas have been translated into processes to operationalize SFA, drawing upon case studies. Table 42.1 provides a summary of the analysis provided in these case studies, identifying the actions taken across the three major areas we identified for a SFA operational approach.

The analysis indicates that there has been significant progress in a number of areas that are relevant to operationalizing SFA. We have, for example, seen the development of conceptual frameworks, analytical tools, and case studies for understanding in an integrated fashion the social and ecological context of agrifood systems. As these frameworks are applicable from micro- household scale to macro-global scale, it is observable that across

Sustainable Food and Agriculture
DOI: https://doi.org/10.1016/B978-0-12-812134-4.00042-X

TABLE 42.1 Summary of Case Studies

	Context	Key Imbalances	Evidence/Dialogue	Solution Approach
Case 1: N2Africa	Smallholder farm households in sub-Saharan Africa with low productivity, high levels of poverty and food insecurity	Lack of nitrogen to enhance productivity of agricultural system; expense and environmental impacts of using chemical nitrogen fertilizers	Participatory development of household level indicators of sustainability relating to economic, social, and environmental factors used to pinpoint key bottlenecks	Targeting research to address identified bottlenecks specific to local context
Case 2: Irrigation and electricity nexus in India	Irrigated wheat and rice farming in India depleting groundwater resources	Subsidized electricity rates for pumping water indicated by a "market-oriented" coalition of stakeholders as the main reason for groundwater overuse, whereas a "welfare state-oriented" coalition of stakeholder fiercely contests this	Evidence constructed by two conflicting interest coalitions to support their case; neither coalition, however, is convinced by the legitimacy of the evidence generated by the other	Create a high-level forum, convened by a neutral broker, where representatives of both coalitions could meet and agree on contested issues for which empirical evidence is missing; they could then engage in "analytical debates," that is, debates on research methodologies that are acceptable by members of both coalitions
Case 3: Global Agenda for Sustainable Livestock	Global community of livestock sector stakeholders responding to major sector controversies	Environmental, social, and health implications (both positive and negative) of increased demand for and consumption of livestock products	Relevant, co-constructed and accessible evidence; engagement of stakeholders in dialogue to build common understanding and joint action	Development of innovative approaches and solutions; and formulating tools and levers to enable and incentivize changes in food and agricultural systems
Case 4: Marine Stewardship Council (MSC)	Global commercial and environmental fishery interests leverage the opportunity of growing ethical consumer behavior	Unsustainable fishing and the safeguarding of wild-caught seafood supplies for future generations	Independent wild-caught fishery assessments, certification and labeling program, supported by science and technology advances and commercial interests under the aegis of the MSC, which includes representatives from all key stakeholder groups	Certification of seafood products as meeting sector-wide endorsed Fisheries Standards to signal to consumers on the sustainability of the product, and to incentivize fishing, processing, and retail businesses in making changes that contribute to sustainability

(Continued)

TABLE 42.1 (Continued)

	Context	Key Imbalances	Evidence/Dialogue	Solution Approach
Case 5: Climate Smart Agriculture (CSA) policy	Zambia's national level climate change and agricultural policymaking where a high number of food insecure and poor people depend on agriculture for their livelihoods, and are highly vulnerable to negative impacts of climate change	Need to increase agricultural productivity to enhance food security and reduce poverty, while addressing the need for adaptation to climate change and meeting national goals for GHG emission reductions	Country-level institutional mapping indicated apparent gap in interministerial coordination of efforts to address the impacts of climate change on agriculture. A review of major policy documents related to climate and agriculture focusing on policy harmonization developed to support a dialogue between the relevant ministries	Greater alignment of the Zambian national agriculture and climate change policies with the specific actions in the national development plan and nationally determined contributions to the UNFCCC process
Case 6: State of California climate change and environment justice policy	Economically dynamic and relatively rich state in the United States, but with large pockets of poverty concentrated in minority communities	New cap-and-trade programs allow businesses to choose to trade allowances or buy offsets instead of reducing GHG emissions, exacerbating toxic hotspots in poor communities near polluting facilities	Mapping of the location of disadvantaged communities identifying various sources of vulnerability; developed through a series of public workshops and expert analysis	Legislation directing that 25 percent of the proceeds from the Greenhouse Gas Reduction Fund go to projects that provide a benefit to disadvantaged communities

the various dimensions of SFA they highlight opportunities and gaps that provide a basis for the identification of entry points for action.

The analysis also indicates a fairly major shift toward evidence-based policymaking, and a more gradual understanding that a deliberate process is needed to actually ensure that the evidence will be useful, empowering, and used for positive change. Building evidence to support policymaking requires an understanding of the policymaking process and mapping of various players within that process in their various roles; we have tools and examples to support this effort. It also requires an understanding of how knowledge is legitimized—considering factors such as the source and the degree to which there is a single community of expertise versus multiple communities. In recent years, questioning concerning the legitimacy of science-sourced knowledge has increased the need to build positive engagement across a range of stakeholders to recognize sources of doubt and to address them through constructive dialogue that identifies areas of agreement and consensus actions.

The case studies are illustrative of the fact that transformation toward SFA is not something that happens at the scale of one company or among consumers on their own, or even in farming enterprises. Instead, it involves an interlocking set of adaptations from individuals to the entire "system of use." There are bottom-up processes of change driven by changing societal needs and top-down processes of change driven by changing policy incentives and regulation. Frequently, there is a dynamic interplay of drivers across scales, with new technology being brought into play, either as a disruptive force or as a response to new social, environmental, and economic imperatives. The consequence of this characteristic is that multiscale change processes involve multiple stakeholders, with transformation toward SFA involving numerous alliances and collaboration to orchestrate the integration of various institutional technological dimensions of the innovation process, as already alluded to above.

We emphasize that technological solutions are insufficient on their own to create the benefits sought without addressing the larger institutional and socioeconomic and political features of the agrifood system into which they are delivered. The analysis we provide indicates that step changes—for example, impact at scale—happen when technology is coupled with system-level changes in values, behavior, and networks that are also reflected in markets, policies, and institutions. Thus, good practice matters, but not as much as a favorable political economy. Strong alignment between public, private, and civil society interests and agenda at a macro level is a key ingredient in system change, and can be deliberately constructed through approaches illustrated in this section with the case studies.

Multistakeholder partnerships (MSPs) are particularly important for operationalizing SFA. MSPs play an essential role in linking local to global scales and are key to achieving impacts both at scale and as a way of reconciling immediate and long-term development agendas. Often, the building blocks of such structures already exist. The need is to ensure that efforts at different levels are synchronized in direction, rather than establishing new parallel and competing arrangements.

The principles of comparative advantage and subsidiarity are fundamental in MSPs, both in terms of effectiveness and in terms of capacity building. Finding an appropriate route of engagement might require a reframing of roles and responsibilities. A key priority for building capacity is strengthening learning in and around MSP practice. Also important is the development of appropriate (and widely accepted) evaluative and analytical frameworks to help assess partnership performance. System-wide capacity development is an essential tool for operationalizing SFA and requires (1) needs assessment; (2) design and implementation of contextualized solutions; and (3) defining and tracking results through accountability and building in learning processes.

Finally, we have seen that innovations in governance are needed at varying scales and including both formal and informal governance arrangements. Operationalizing SFA requires governance arrangements that enhance the capacity to act collectively, allowing for the explicit accounting of the potential external and unintended consequences of actions and building capacity to manage common-pool resources such as the atmosphere, waterways, and rangelands. Governance arrangements that facilitate dialogue and capacity to manage conflicts and trade-offs such as multilevel forums, public–private partnerships, communities of practice, and multistakeholder dialogues are essential to operationalizing SFA. In many contexts, key elements of such arrangements are already in place, and mobilizing the potential of the existing institutional infrastructure represents an important opportunity for operationalizing SFA.

Although this section has provided a range of concepts, analysis, and case studies related to operationalizing SFA, it does not cover the full richness of the emerging work related to this topic. Aside from a rapidly developing literature on sustainability science, there are very useful developing bodies of work on evidence-based decision-making approaches, new insights from behavioral economics on "nudging" to facilitate desired behavior changes and governance for adaptive change, and new governance approaches that offer additional insights on how to operationalize SFA. Though much of this material is not directly related to agriculture and food systems per se, it is highly relevant to their sustainability.

References

Adekunle, A.A., Fatunbi, A.O., 2012. Approaches for setting up multi-stakeholder platforms for agricultural research and development. World Appl. Sci. J. 16 (7), 981–988.

Adger, W.N., 2000. Social and ecological resilience: are they related? Prog. Hum. Geogr. 24 (3), 347–364.

Banerji, A., Meenakshi, J.V., Khanna, G., 2012. Social contracts, markets and efficiency: groundwater irrigation in North India. J. Dev. Econ. 98 (2), 228–237.

Bates, L.E., Green, M., Leonard, R., Walker, I., 2013. The influence of forums and multilevel governance on the climate adaptation practices of australian organizations. Ecol. Soc. 18 (4), 62–74.

Batie, S.S., 2008. Wicked problems and applied economics. Am. J. Agric. Econ. 90 (5), 1176–1191.

Befani, B., Ramalingam, B., Stern, E. (Eds.), 2015. Towards Systemic Approaches to Evaluation and Impact IDS Bulletin 46:1.

Belcher, B., Rasmussen, K.E., Kemshaw, M.R., Zornes, D.A., 2016. Defining and assessing research quality in a transdisciplinary context. Res. Eval. 25 (1), 1–17.

Berkes, F., Folke, C., 1998. Linking Social and Ecological Systems: Management Practices and Social Mechanisms for Building Resilience. Cambridge University Press, Cambridge.

Bezanson, K.A., Isenman, P., 2012. Governance of new global partnerships: challenges, weaknesses, and lessons. CGD Policy Paper 014. Center for Global Development, Washington DC. Available from: http://www.cgdev.org/content/publications/detail/1426627.

Biermann, F., Kanie, N., Kim, R., 2017. Global governance by goal-setting: the novel approach of the UN Sustainable Development Goals. Curr. Opin. Environ. Sust. 26–27, 26–31.

Binder, C.R., Hinkel, J., Bots, P.W.G., Pahl-Wostl, C., 2013. Comparison of frameworks for analyzing social-ecological systems. Ecol. Soc. 18 (4), 26.

Birner, R., Gutpa, S., Sharma, N., 2011. The Political Economy of Agricultural Policy Reform in India: Fertilizers and Electricity for Irrigation. International Food Policy Research Institute, Washington, D.C.

Bitzer, V., 2012. Partnering for change in chains: the capacity of partnerships to promote sustainable change in global agrifood chains. Int. Food Agribusiness Manage. Rev. 15 (Special Issue B), 13.

Breeman, G., Dijkman, J., Termeer, C., 2015. Enhancing food security through a multi-stakeholder process: the global agenda for sustainable livestock. Food Sec. 7 (2), 425–435. April 2015.

Burns, D., Worsley, S., 2015. Navigating Complexity in International Development: Facilitating Sustainable Change at Scale. Practical Action, Rugby.

Byerlee, D., Echeverría, R.G. (Eds.), 2002. Agricultural Research Policy in an Era of Privatization. *CAB International*, Wallingford, UK.

Cash, D.W., Clark, W.C., Alcock, F., Dickson, N.M., Eckley, N., Guston, D.H., et al., 2003. Knowledge systems for sustainable development. Proc. Natl. Acad. Sci. 100 (14), 8086.

Chambers, R., 1994. The origin and practice of participatory rural appraisal. World Dev. 22 (7), 953–969.

Clark, W.C., Tomich, T.P., van Noordwijk, M., Guston, D., Catacutan, D., Dickson, N.M., et al., 2016. Boundary work for sustainable development: Natural resource management at the Consultative Group on International Agricultural Research (CGIAR). Proc. Natl. Acad. Sci. 113 (17), 4615.

De Pryck, K., Wanneau, K., 2017. (Anti)-boundary work in global environmental change research and assessment. Environ. Sci. Policy 77, 203–210.

DeFries, R., Nagendra, H., 2017. Ecosystem management as a wicked problem. Science 356 (6335), 265–270.

Dentoni, D., Hospes, O., Ross, R.B., 2012. Managing wicked problems in agribusiness: the role of multi-stakeholder engagements in value creation. Int. Food Agribusiness Manage. Assoc. 15 (B), 1–12.

Diamond, J., Liddle, J., 2005. What are we learning from the partnership experience? Public Policy and Administration 20 (3), 1–3.

Dodds, F., 2015. Multi-Stakeholder Partnerships: Making Them Work for the Post-2015 Development Agenda. Global Research Institute, University of North Carolina, USA.

Duit, A., Galaz, V., 2008. Governance and complexity—emerging issues for governance theory. Governance 21 (3), 311–335.

Eakin, H., Luers, A.L., 2006. Assessing the vulnerability of social-environmental systems. Annu. Rev. Environ. Resour. 31 (1), 365–394.

FAO, 2010. Corporate Strategy for Enhancing Capacity Development in Member Countries. Rome.

FAO, 2014. Building a Common Vision for Sustainable Food and Agriculture: Principles and Approaches. (2014). FAO Books, Rome.

FAO, 2015. Rwanda and FAO acting for Sustainable Food and Agriculture. <www.fao.org>.

FAO, 2016a. Soils and pulses: Symbiosis for life. <http://www.fao.org/3/a-i6437e.pdf>.

FAO, 2016b. Towards productive, sustainable and inclusive agriculture, forestry and fisheries in support to the 2030 agenda for sustainable development. Synthesis report. Regional workshop for Africa in Kigali, Rwanda. FAO, Rome, Italy.

FAO, 2018. <http://www.livestockdialogue.org/> (accessed 26.01.18).

Foley, C. 2013. <https://www.scientificamerican.com/article/time-to-rethink-corn/>.

Folke, C., 2016. Resilience (Republished). Ecol. Soc. 21, 44.

Freeman, C., Perez, C., 1988. Structural crises of adjustment, business cycles and investment behaviour. In: Dosi, G., et al., (Eds.), Technical Change and Economic Theory. Francis Pinter, London, pp. 38–66.

Geels, F.W., 2002. Technological transitions as evolutionary reconfiguration processes: a multi-level perspective and a case-study. Res. Policy 31 (8–9), 1257–1274. December 2002.

Giller, K.E., 2001. Nitrogen fixation in tropical cropping systems, second ed. CABI Publishing, Wallingford, UK0851994172, 423 p. https://doi.org/10.1079/9780851994178.0000.

Giller, K.E., Franke, A.C., Abaidoo, R., Baijukya, F., Bala, A., Boahen, S., et al., 2013. N2Africa: Putting nitrogen fixation to work for smallholder farmers in Africa. In: Vanlauwe, B., van Asten, P.J.A., Blomme, G. (Eds.), Agro-ecological Intensification of Agricultural Systems in the African Highlands. Routledge, London, pp. 156–174.

Government of India, 2007. Report of the Expert Group on Ground Water Management and Ownership. Planning Commission. Government of India, New Delhi.

Grubinger, V., Berlin, L., Berman, E., Fukagawa, N., Kolodinsky, J., Neher, D., et al., 2010. Transdisciplinary Research Initiative Spire of Excellence Proposal: Food Systems. Proposal. University of Vermont, Burlington, 2010.

Hajer, M., 1995. The Politics of the Environmental Discourse: Ecological Modernization and the Policy Process. Clarendon Press, Oxford.

Hall, A., Dijkman, J., Taylor, B.M., Williams, L., Kelly, J. 2016. Synopsis: Towards a framework for unlocking transformative agricultural innovation. DISCUSSION PAPER #1 – 13.07.2016.

Hanleybrown, F., Kania, J. and Kramer, M. 2012. Channeling change: making collective impact work. January 26, 2012, Stanford Social Review, Leland Standford Jr. University, USA.

Hazell, P.B.R., Ramasamy, C. 1991. The Green Revolution Reconsidered - The Impact of High-Yielding Rice Varieties in South India. Baltimore and London: The Johns Hopkins University Press for the International Food Policy Research Institute.

Hazlewood, P. 2015. Global Multi-Stakeholder Partnerships: Scaling up public-private collective impact for SDGs, Background paper 4, Independent Research Forum, IRF 2015.

Head, B.W., 2008. Three lenses of evidence-based policy. Aust. J. Public Admin. 67 (1), 1–11.

Horton, D., Prain, G., Thiele, G., 2009. Perspectives on Partnership: A Literature Review. International Potato Center (CIP), Lima, Peru, Working Paper 2009-3. 111 pp.

ISPC, 2015. Strategic Study of Good Practice in AR4D Partnership. CGIAR Independent Science and Partnership Council (ISPC), Rome, Italy, xiii + 60pp + annex 50pp.

ISPC. 2017. Brief No. 62: Quality of Research for Development in the CGIAR Context. <https://ispc.cgiar.org/sites/default/files/pdf/ispc_brief_62_qord.pdf>.

ISPC/CSIRO, 2017. Agri-food system innovation. Reframing the conversation. ISPC/CSIRO Workshop Report. 27–29 June 2017. International Crops Research Institute for the Semi-Arid Tropics (ICRISTAT), Hyderabad, India, https://ispc.cgiar.org/sites/default/files/pdf/ispc_csiro_report_agri-foodsystemsworkshop.pdf.

Jenkins-Smith, H., Sabatier, P.A., 1993. The dynamics of policy-oriented learning. In: Sabatier, P.A., Jenkins-Smith, H. (Eds.), Policy Change and Learning: An Advocacy Coalition Approach. Westview Press, Boulder, pp. 41–56.

Johnson, M., Birner, R. 2013. Understanding the Role of Research in the Evolution of Fertilizer Policies in Malawi. IFPRI Discussion Paper No. 01266.

Kalas, P., Abubakar, A., Chavva, K., Gordes, A., Grovermann, C., Innes-Taylor, N., et al., 2017. Enhancing capacities for a country-owned transition towards climate smart agriculture, Climate-Smart Agriculture Sourcebook, Second ed. FAO, Rome, Italy. Available at: <http://www.fao.org/climate-smart-agriculture-sourcebook/en/>.

King, E., Cavender-Bares, J., Balvanera, P., Mwampamba, T.H., Polasky, S., 2015. Trade-offs in ecosystem services and varying stakeholder preferences: evaluating conflicts, obstacles, and opportunities. Ecol. Soc. 20 (3), 25.

Kolko, J. 2012. Wicked problems: problems worth solving (SSIR). Retrieved from: <https://ssir.org/articles/entry/wicked_problems_problems_worth_solving>.

Kooiman, J., 1993. Governance and governability: using complexity, dynamics and diversity. In: Kooiman, J. (Ed.), Modern Governance – New Government – Society Interactions. SAGE Publications, London, 280 p.

Koppenjan, J.F.M., Klijn, E.H., 2004. Managing Uncertainties in Networks. A Network Approach to Problem Solving and Decision Making. Routledge, London, ISBN 0415-36941-X/0-415-36940-1, 290 pp.

Kreuter, M.W., De Rosa, C., Howze, E.H., Baldwin, G.T., 2004. Understanding wicked problems: a key to advancing environmental health promotion. Health Educ. Behav. 31 (4), 441–454.

Kumar, M.D., 2005. Impact of electricity prices and volumetric water allocation on energy and groundwater demand management: analysis from Western India. Energy Policy 33 (1), 39–51. Available from: https://doi.org/10.1016/S0301-4215(03)00196-4.

Lal, S., 2006. Can Good Economics Ever Be Good Politics? Case Study of India's Power Sector. World Bank Working Paper No. 83, World Bank, Washington, DC.

Lele, U., Sadik, N., Simmons, A. 2007. The Changing Aid Architecture: Can Global Initiatives Eradicate Poverty? <http://www.oecd.org/dataoecd/60/54/37034781.pdf>. Agri-food innovation and impact. Discussion paper series. ISPC and CSIRO, 2016.

Leslie, H.M., Basurto, X., Nenadovic, M., Sievanen, L., Cavanaugh, K.C., Cota-Nieto, J.J., et al., 2015. Operationalizing the social-ecological systems framework to assess sustainability. Proc. Natl. Acad. Sci. 112 (19), 5979.

London, J., Karner, A., Sze, J., Rowan, D., Gambirazzio, G., Niemeier, D., 2013. Racing climate change: collaboration and conflict in California's global climate change policy arena. Glob. Environ. Change 23 (4), 791–799.

Malena, C. 2004. Strategic Partnership: Challenges and Best Practices in the Management and Governance of Multi-Stakeholder Partnerships involving UN and Civil Society Actors. Background paper prepared for the Multi-Stakeholder Workshop on Partnerships and UN-Civil Society Relations, Pocantico, United Nations, New York.

Marinus, W., Ronner, E., van de Ven, G.W.J., Kanampiu, K., Adjei-Nsiah, S., Giller, K.E., 2018. The devil is in the detail! Sustainability assessment of African smallholder farming. In: Bell, S., Morse, S. (Eds.), Routledge Handbook on Sustainability Indicators. Routledge, London (Chapter 28), (in press).

Mason, N., Doczi, J., Cummings, C., 2016. Innovating for pro-poor services. Why politics matter. ODIInsights March 2016.

Matson, P., Clark, W., Andersson, K., 2016. Pursuing Sustainability: A Guide to the Science and Practice. Princeton University Press, Princeton.

Mazzucato, M. 2015. The Entrepreneurial State (US Edition), Public Affairs. ISBN 9781610396134.

McGinnis, M.D., Ostrom, E., 2014. Social-ecological system framework: initial changes and continuing challenges. Ecol. Soc. 19 (2), 30.

Meadows, D.H., Randers, J., Meadows, D.L. 2004. Limits to Growth-The 30 year Update, hardcover ISBN 1-931498-51-2.

Ministry of National Development Planning (Zambia). 2017. Zambia's Seventh National Development Plan Implementation Plan. <http://www.mof.gov.zm/jdownloads/Development%20Plans%20and%20Reports/Seventh%20National%20Development%20Plan/7ndp.pdf>.

Mockshell, J. 2016. Two Worlds in Agricultural Policy Making in Africa?: Case Studies from Ghana, Kenya, Senegal and Uganda.

Moench-Pfanner, R., Van Ameringen, M., 2012. The Global Alliance for Improved Nutrition (GAIN): A decade of partnerships to increase access to and affordability of nutritious foods for the poor. Food. Nutr. Bull. 33 (4(supplement)), United Nations University.

Musters, C., De Graaf, H., ter Keurs, W., 1998. Defining socio-environmental systems for sustainable development. Ecol. Econ. 26 (3), 243–259.

Nederlof, S., Wongtschowski, M., van der Lee, F. (Eds.), 2011. Putting Heads Together: Agricultural Innovation Platforms in Practice. Development, Policy and Practice. Bulletin 396. KIT Publishers, Amsterdam, The Netherlands.

Nooteboom, S.G., Termeer, C., 2013. Strategies of complexity leadership in governance systems. Int. Rev. Public Admin. 18 (1), 25–40.

OECD. 2005. Paris Declaration on Aid Effectiveness and Accra Agenda for Action. Paris.

OECD. 2006. The Challenge of Capacity Development. Working Towards Good Practice. Paris.

Ostrom, E., 1999. Coping with tragedies of the commons. Annu. Rev. Polit. Sci. 2, 493–535.

Patscheke, S., Barmettler, A., Herman, L., Overdyke, S., Pfitzer, M., 2014. Shaping Global Partnerships for a Post-2015 World. Stanford Social Innovation Review. Leland Standford Jr. University, USA.

Pattberg, P., Wilderberg, O., 2014. Transnational Multi-stakeholder Partnerships for Sustainable Development: Building Blocks for Success. IVM Institute for Environmental Studies, Amsterdam: The Netherlands.

Peterson, K., Mahmud, A., Bhavaraju, N., Mihaly, A. 2014. The Promise of Partnerships: A Dialogue between INGOs and Donors. FSG: <www.fsg.org>.

Rajalahti, R., Janssen, W., Pehu, E., 2008. Agricultural Innovation Systems: From Diagnostics to Operational Practices. Agricultural and Rural Development Discussion Paper 38. World Bank, Washington, DC.

Rip, A., Kemp, R.P.M., Kemp, R., 1998. Technological change. In: Rayner, S., Malone, E.L. (Eds.), Human Choice and Climate Change. Vol. II, Resources and Technology. Battelle Press, Columbus, Ohio, pp. 327–399.

Rittel, H.W.J., Webber, M.M., 1973. Dilemmas in a general theory of planning. Policy Sci. 4 (2), 155–169.

Roberts, N., 2000. Wicked problems and network approaches to resolution. Int. Public Manage. Rev. 1 (1), 1–19.

Sabatier, P., Jenkins-Smith, H., 1993. Policy Change and Learning - An Advocacy Coalition Approach. Westview Press.

Schiffer, E., Hauck, J., 2010. Net-map: collecting social network data and facilitating network learning through participatory influence network mapping. Field Methods 22 (3), 231–249.

Schouten, G., Glasbergen, P., 2011. Creating legitimacy in global private governance: the case of the roundtable on sustainable palm Oil. Ecol. Econ. 70 (11), 1891–1899.

Scrase, I., Stirling, A., Geels, F.W., Smith, A., Van Zwanenberg, P., 2009. Transformative Innovation: a report to the Department for Environment, Food and Rural Affair. Science and Technology Policy Research (SPRU). University of Sussex, Brighton.

Sen, A., 1999. Development as Freedom. Oxford University Press, Oxford.

Serraj, R., Adu-Gyamfi, J., 2004. Role of symbiotic nitrogen fixation in the improvement of legume productivity under stressed environments. West Afr. J. Appl. Ecol. 6 (1), 96–109.

Serraj, R., Adu-Gyamfi, J., Rupela, O.P., Drevon, J.J., 2004. Improvement of legume productivity and role of symbiotic nitrogen fixation in cropping systems: Overcoming the physiological and agronomic limitations. In: Serraj, R. (Ed.), Symbiotic Nitrogen Fixation: Prospects for Enhanced Application in Tropical Agriculture. Oxford & IBH, New Delhi.

Severino, J.-M., Ray, O., 2010. The End of ODA (II): The Birth of Hyper-Collective Action. Center for Global Development, Working Paper 218, June 2010.

Shiferaw, B., Bantilan, M.C.S., Serraj, R., 2004. Harnessing the potential of BNF for poor farmers: technological policy and institutional constraints and research need. In: Serraj, R. (Ed.), Symbiotic Nitrogen Fixation; Prospects for Enhanced Application in Tropical Agriculture. Oxford & IBH, New Delhi.

Smith, A., Stirling, A., Berkhout, F., 2005. The governance of sustainable socio-technical transitions. Research Policy 34, 1491–1510.

Srinivasulu, K., 2004. Political Articulation and Policy Discourse in the 2004 Election in Andhra Pradesh. GAPS Working Paper. Center for Economic and Social Studies, Hyderabad, India.

Stavins, R., 2008. Addressing climate change with a comprehensive US cap-and-trade system. Oxford Rev. Econ. Policy 24 (2), 298–321.

UNDP, 2003. Ownership, Leadership and Transformation. Earthscan.

UNDP. 2015. Zambia National Strategy to Reduce Emissions from Deforestation and Forest Degradation (REDD+). <https://info.undp.org/docs/pdc/Documents/ZMB/Zambia%20REDD+%20Strategy%20%28FINAL%20ed.%29%20%282%29.pdf>.

UNFCCC. 2015. Zambia's Intended Nationally Determined Contribution (INDC) to the 2015 Agreement on Climate Change. <http://www4.unfccc.int/ndcregistry/PublishedDocuments/Zambia%20First/FINAL+ZAMBIA%27S+INDC_1.pdf>.

van Huijstee, M., 2012. Multi-Stakeholder Initiatives: A Strategic Guide for Civil Society Organisations. SOMO (Stichting Onderzoek Multinationale Ondernemingen), Amsterdam, The Netherlands.

van Kerkhoff, L., Lebel, L., 2006. Linking knowledge and action for sustainable development. Annu. Rev. Environ. Resour. 31 (1), 445–477.

Velten, S., Leventon, J., Jager, N., Newig, J., 2015. What is sustainable agriculture? A systematic review. Sustainability 7 (6), 7833–7865.

Warner, J.F., 2006. More sustainable participation? Multi-stakeholder platforms for integrated catchment management. Water Resour. Dev. 22 (1), 15–35.

WCED, 1987. Our Common Future. World Commission on Environment and Development (WCED). Oxford University Press, Oxford.

Wenger, E., 1998. Communities of Practice: Learning, Meaning, and Identity. Cambridge University Press, Cambridge.

Weterings R., Kuijper J., Smeets E., Annokkée G.J., Minne B. 1997. '81 mogelijkheden: Technologie voor duurzame ontwikkeling' ('81 possibilities: Technology for sustainable development'). TNO report for VROM (Ministry of Housing, Spatial Planning and Environmental Affairs), The Hague.

World Bank, 2001a. India - Power Supply to Agriculture Volume 1: Summary Report. Energy Sector Unit, South Asia Regional Office, World Bank.

World Bank, 2001b. India: Power Supply To Agriculture - Volume 2: Haryana Case Study. Energy Sector Unit, South Asia Regional Office, World Bank.

SECTION V

INNOVATIONS, POLICIES, INVESTMENTS, AND INSTITUTIONS FOR SUSTAINABLE FOOD AND AGRICULTURE SYSTEMS – AND THE WAY FORWARD

Lead Authors

Clayton Campanhola, Kostas Stamoulis and Shivaji Pandey (Strategic Program Leader (Sustainable Agriculture), Assistant Director-General (Economic and Social Development Department), and Special Advisor (Sustainable Agriculture), respectively, of the Food and Agriculture Organization of the United Nations (FAO), Rome, Italy)

CHAPTER

43

Innovations, Policies, Investments, and Institutions For Sustainable Food and Agriculture Systems: And the Way Forward

Since agriculture (crops, livestock, forests, fisheries, and aquaculture) provides the only source of food, humanity's survival depends upon it. So far, it has been successful in feeding the world's growing population and reducing world hunger from a little over a billion in 1990–92 to less than 800 million in 2014–16 (FAO, 2016). Agriculture makes an important contribution to employment, providing labor for over 1 billion people—one in three of all who are retained in the agricultural sector (http://www.fao.org/docrep/015/i2490e/i2490e01b.pdf; accessed March 2018). While global poverty declined from about one-third of the population in 1990 to just over one-tenth in 2013 (World Bank, 2014), agricultural development also significantly contributed to increasing incomes and reducing poverty (Timmer, 2010; Hazel, 2010). As the majority of the world's poor live in rural areas and derive their livelihoods from agriculture, the role of agricultural growth in future poverty reduction will remain large. Besides providing food and income, agriculture also supplies a myriad of other products and services, such as fuel, fiber, clean air and water, and climate change mitigation. Paradoxically, while hundreds of millions of people remain hungry and 2 billion suffer from micronutrient deficiencies, 40% of the population above 18 years is overweight (WHO, 2014). Thus, hunger persists in different forms, despite achievements in the food and agriculture sector.

As the largest user of natural resources, food and agriculture also exerts negative pressure on them. For example, one-third of agricultural land is moderately to highly

The views expressed here are the authors' own and do not necessarily reflect the views of Food and Agriculture Organization of the United Nations (FAO).

Sustainable Food and Agriculture
DOI: https://doi.org/10.1016/B978-0-12-812134-4.00043-1

degraded, agriculture uses 70% of all water withdrawals, and agriculture causes environmental pollution (FAO, 2011a). Food and agriculture is a major source of loss of biodiversity: approximately 75% of crop biodiversity is lost (FAO, 2004), 61% of commercial fish stocks are fully exploited (FAO, 2014a), and 17% of farm animal breeds are at risk (FAO, 2017). It is estimated that 60% of the loss in biodiversity is attributed to food production (UNEP, 2016). Inefficient use of inputs during production and processing, large amounts of loss, and waste of food already produced (FAO, 2011b) further increase pressure on natural resources and reduce social and economic benefits from agriculture. Agriculture also is a major contributor to climate change: it (including forestry and land use) releases 21% (see Chapter 2: Global Trends and Challenges to Food and Agriculture into the 21st Century) of the greenhouse gases, and food systems as a whole emit 29% of the global greenhouse gases (Vermeulen et al., 2012). Increase of temperature reduces agricultural productivity in already vulnerable areas and increases extreme weather events (Mann et al., 2017). In 2050, climate change is expected to add another 70 million people to those already hungry (see Chapter 4: Climate Change, Agriculture and Food Security: Impacts and the Potential for Adaptation and Mitigation). Wars and civil strifes also negatively impact food and agriculture. Increasing volumes of internal and international trade enhance the likelihood of propagation of pests and diseases.

The world's food and agriculture systems are at a crossroads. The systems must feed a world population of 9.73 billion by 2050, requiring 50% more food than was needed in 2013 (FAO, 2017), with extra food coming largely from increases in productivity from a resource base declining both in quantity and quality. In addition, about two-thirds of the world population will live in cities in 2050 (IFPRI, 2017), and higher incomes and urban food preferences (animal-sourced foods, fruits and vegetables, and processed foods) will exert a major influence on the world's food and agricultural systems.

FAO (2014b) proposed five principles for achieving sustainable food and agriculture in a way "that meets the needs of the present without compromising the ability of future generations to meet their own needs" (World Commission on Environment and Development, 1987): (1) improving resource use efficiency; (2) conserving, protecting, and enhancing natural resources that underpin food and agriculture; (3) protecting and improving the livelihoods and well-being of people; (4) enhancing resilience of people, communities, and ecosystems; and (5) ensuring responsible and effective governance.

A holistic and integrated approach that aligns national development goals and actions with global goals contributes to enhancing policy coherence and synergies among different ministries and agencies dealing with food and agriculture at the national level; increases on- and off-farm employment and income opportunities, especially in rural areas of developing countries; and increases investment in food and agriculture to address many of those challenges. Another set of supporting actions would include maximization of resource use efficiency, reduction in food loss and waste, and reuse for conversion of waste to wealth throughout the food chain, along with protection, conservation, and sustainable use of key natural resources that underpin food and agriculture.

However, no country has yet implemented such a system. Addressing these challenges requires a better understanding of the positive and negative impacts of food and agriculture on the global and national social, economic, and environmental systems. It also requires an understanding of the drivers of change and linkages between rural and urban economies. Many trade-offs emerge during production to consumption with impacts on efficiency, sustainability, and inclusiveness, and they must be balanced with policies that promote an integrated approach. Policies and technologies should especially prioritize the challenges and opportunities for the 500 million smallholder farmers (FAO, 2017) and their contributions to sustainable rural transformation and alleviation of hunger and poverty.

Fortunately, the international community has recognized these challenges and the interdependencies among them. The 2030 Agenda for Sustainable Development provides a vision for transformative change to put economies and agriculture and attendant food systems on sustainable footing. The second Sustainable Development Goal (SDG 2) explicitly aims at ending hunger, achieving food security and improving nutrition, and promoting sustainable agriculture simultaneously by 2030. The Addis Ababa Action Agenda on financing for development calls on countries to pursue policy coherence and establish enabling environments for sustainable development at all levels and by all actors. The 2015 Paris Agreement reflects global commitments for concerted actions to address the perils of climate change, and the United Nations Framework Convention on Climate Change agreed, in 2017, to include agriculture in its work program. The Sendai Framework for Disaster Risk Reduction also gives priority to agriculture.

While these developments at the global level are important, policy coherence and sustained efforts at the country level are critical as sustainable food and agriculture must be achieved by the countries themselves. Therefore, the proposals and recommendations included in this book address different stakeholders and the policies, strategies, technologies, and examples that have relevance for both the short and long term.

Section I of this volume aimed at increasing understanding of factors, issues, and challenges to the productivity and sustainability of global food and agriculture systems that currently and will continue to impact it well into the 21st century. Section II described policies, technologies, approaches, and systems that have proven to be successful in the recent past or are being tested and promoted for addressing many of these challenges faced in different parts of the world. Section III considered the growing conflict between escalating human demand for food and other benefits, with resource degradation and scarcity. It analyzed how different agrifood systems respond to this conflict and suggests cross-cutting approaches that can deal with their complexity and diversity. Section IV described the importance of identifying and balancing the trade-offs that occur because of the tension between natural and social systems discussed in Section III, and presented a general approach to implementing sustainable food and agriculture. It also provided examples of successful implementation of the approach at the global and national levels.

The following four chapters of Section V—research and innovation, policies and incentives, resource mobilization for agriculture and food security, and governance and institutions—are considered most critical to delivering sustainable food and agriculture through meaningful structural transformations. Depending on their relevance, previous sections have also dealt with these topics but, given their importance to the overall subject of food and agriculture sustainability, they are treated in some detail throughout.

References

FAO, 2004. Building on Gender, Agrobiodiversity and Local Knowledge. FAO, Rome.

FAO, 2011a. The State of the World's Land and Water Resources for Food and Agriculture. Managing Systems at Risk. FAO, Rome.

FAO. 2011b. Global food losses and waste. Extent, causes and prevention. <http://www.fao.org/docrep/014/mbo6oe/mbo6oeoo.pdf>.

FAO, 2014a. The State of World Fisheries and Aquaculture. FAO, Rome.

FAO, 2014b. Building a Common Vision for Sustainable Food and Agriculture. FAO, Rome.

FAO, 2016. Food Security Indicators. FAO, Rome.

FAO, 2017. The Future of Food and Agriculture – Trends and Challenges. FAO, Rome, 163 p.

Hazel, P., 2010. An assessment of the impact of agricultural research in South Asia since the green revolution. In: Pingali, P., Evenson, R.E. (Eds.), Handbook of Agricultural Economics. Elgar Publishing, Ltd., Amsterdam, pp. 3469–3530.

IFPRI, 2017. 2017 Global Food Policy Report. IFPRI, Washington, D.C., 137 p.

Mann, M.E., Rahmstorf, S., Kornhuber, K., Steinman, B.A., Miller, S.K., Coumou, D., 2017. Influence of anthropogenic climate change on planetary wave resonance and extreme weather events. Sci. Rep. 7, 10.

Timmer, P.C., 2010. A world without agriculture: the structural transformation in historical perspective. Washington, D.C.

UNEP, 2016. Food Systems and Natural Resources. United Nations Environment Programme.

Vermeulen, S.J., Campbell, B.M., Ingram, J.S., 2012. Climate change and food systems. Annu. Rev. Environ. Resour. 37, 195–222.

WHO, 2014. Global Health Observatory (GHO) Data. Obesity and Overweight. WHO, Geneva.

World Bank, 2014. World Development Indicators. Poverty Rates. World Bank, Washington, D.C.

World Commission on Environment and Development, 1987. Our Common Future: Report of the World Commission on Environment and Development. Oxford University, New York.

CHAPTER

44

Research and Innovation

Ren Wang[1], *Alexandre Meybeck*[2] *and Andrea Sonnino*[3]

[1]Beijing Genomics Institute, Shenzhen, P.R. China [2]Bioversity International (CIFOR), Maccarese, Italy [3]Agenzia Nazionale per le Nuove Tecnologie, L'Energia e lo Sviluppo Economico Sostenible (ENEA), Biotechnologies and Agroindustry Division, Casaccia Research Centre, Rome, Italy

44.1 INTRODUCTION

Feeding the growing world population and reducing poverty, in the context of an eroding natural resource base and of climate change, is a challenge of unprecedented dimensions and nature. Agriculture and food systems (AFS) have a key role to play to achieve the Sustainable Development Goals (SDGs) and the objectives of the Paris Agreement. A paradigm shift is needed to reposition the world's AFS from being an important driver of environmental degradation to being a key contributor for the global transition to sustainability. Such a transformation can only happen through both generation of new knowledge and enhanced translation of knowledge into use (FAO, 2017). This translates into three major needs: the generation, diffusion, and adoption of knowledge; the integration of diverse disciplines and perspectives; and the involvement of decision-makers at all levels, from government to farm, to address multiple objectives, decide on priorities, and identify and manage synergies and trade-offs.

This section identifies some main research needs for sustainable AFS, examines the conditions that will facilitate the transition toward sustainable AFS, highlights the prevailing hurdles that prevent the systems from fully benefiting from innovation, presents some examples, and makes recommendations for improving the contribution of research and innovation to sustainable AFS.

Sustainable Food and Agriculture
DOI: https://doi.org/10.1016/B978-0-12-812134-4.00044-3

44.2 RESEARCH AND INNOVATION NEEDS FOR SUSTAINABLE FOOD SYSTEMS

The complex and dynamic nature of food and agricultural systems and their interactions with the environment call for a more systemic way of doing research, embracing production systems and their determinants from ecosystems and natural resources to food chains and consumption drivers (Hammond and Dubé, 2012). For agriculture to respond to current challenges, innovation cannot be directed only to increasing the productivity of production factors at the farm level, but it must also be directed to conserve and even enhance the natural resource bases at the landscape level. The practices required should be grounded in a holistic approach based on ecosystem management, by closing production cycles, integrating environmental and productivity objectives, and contributing to resilience of landscapes. Scientific analysis, therefore, cannot be limited to the study of individual elements of AFS, but considers the interplay of factors within the production system and between the system and the environment as a key strategy to reconcile food security and environmental conservation and to solve, at least partially, the relevant trade-offs (European Union, 2015).

However, the three broad dimensions of field, landscape, and food chain, as well as rural-urban linkages and the food-health nexus, can help organize main research questions into four broad areas:

1. How to improve agricultural productivity and efficiency in the use of resources (land, water, nutrients). Improved (higher-yielding, more nutritious, and stress-tolerant) breeds and varieties, greater efficiency in the use of resources (especially water, nutrients, feed), and optimization of the use of resources by bringing the right amount at the right place at the right time through precision agriculture are some of the options.
2. How to optimize management of ecosystem services in field and landscape to improve productivity, sustainable management of natural resources, and resilience in the context of resource scarcity and climate change. This will require understanding the interactions between both the various elements in a landscape as well as between species, and the way these relationships are modified by factors such as climate change. Researchers must adopt systematic thinking, and move from sectoral approaches to integrated and holistic approaches. The most urgently needed research would focus on identifying drivers and trade-offs for the changes of the agricultural systems and establishing indicators for the changes.

   It covers such issues as the role of trees in regulating water cycle at local and regional levels; best ways for land restoration; diversification of production to optimize the use of land, water, and nutrients while increasing resilience; integrated systems at various scales; prevention, monitoring, and management of biophysical risks (changes in biophysical characteristics of ecosystems, and plant, pests, and animal diseases), as well as positive and negative social innovations. A powerful tool in this perspective can be the ecosystem approach, defined by the Convention on Biological Diversity as a strategy for the integrated management of land, water, and living resources that promotes conservation and sustainable use in an equitable way (CBD, 2017).

3. How to optimize food chains to improve resource use efficiency, including by reducing food losses and waste and using coproduction and byproducts, and to promote the access of smallholders to markets and of consumers to diverse and nutritious foods. It includes enabling the collection, use, transformation, and distribution of more diverse products, and improved conservation and transformation of food for food safety, quality and nutritional value, and greater convenience. Such issues require not only technological innovations but also social innovations, in particular information and data collection and sharing to facilitate organization and the sharing of risks, costs, and benefits.
4. How to improve the agriculture-food-nutrition and human health nexus and rural-urban linkages through social and economic studies and interventions. The fragmentation of academic disciplines (biological sciences, social research, etc.) and the tendency for overspecialization must be overcome (Rockström et al., 2017) and systemic thinking promoted. Consolidation of local, indigenous, and formal scientific knowledge is also required. The advances of information and communication technologies and artificial intelligence allow for the design of processes and mechanisms that efficiently gather, systematize, analyze, and share large amounts of data and knowledge and make substantive contributions to govern complexity.

44.3 INNOVATION POTENTIAL CAN BE FULLY EXPLOITED IF PREREQUISITES ARE MET

Reconcile Research and Users' Expectations

The lag between when research projects are started and the time when their outcomes are delivered is long, sometimes taking years or even decades to realize. Therefore, research results may reach the users too late, especially in times of rapid demographic, economic, climatic, and environmental change. This calls for strategic planning of scientific and technological research backed by adequate forward thinking and supported by methodologies, such as scenario development, foresight, and horizon scanning (European Union, 2015).

The prevailing contract between science and society expects science to produce new knowledge to recompense the public resources it receives. The only condition is that reliable discoveries are communicated to society (Gibbons, 1999). Unfortunately, this model fails to consider societal priorities, values, and concerns of the users, as demonstrated by recent controversies that prevented some scientific achievements from being accepted by the general public, limiting their adoption (genetically modified organisms or nanotechnologies for the food industry, for example). This contract should be replaced by a new arrangement that ensures that the new knowledge produced is "socially robust," besides being scientifically sound (Sonnino and Sharry, 2017). Society should be engaged from the beginning in knowledge production processes, so expectations and attitudes are fully incorporated in research. This would ensure the necessary political support for science and technology and facilitate the full economic exploitation of scientific results. Recently, policy efforts have been dedicated to improving citizen-science dialogue on the adoption of new and emerging technologies, for instance, through the responsible research and innovation approach (Stilgoe et al., 2013; Sonnino et al., 2016).

As the physicist Nikola Tesla wrote in 1919: "Science is but a perversion of itself unless it has as its ultimate goal the betterment of humanity." Science has no impact without application to practice or, in other words, if it does not generate innovation. Agricultural innovation is defined as the process whereby individuals or organizations bring existing or new products, processes, and forms of organization into social and economic use to increase effectiveness, competitiveness, resilience to shocks, or environmental sustainability, thereby contributing to food and nutritional security, economic development, and sustainable natural resource management (Tropical Agriculture Platform, 2016). Innovation is therefore not limited to adoption of new technologies, but includes market access, technology dissemination, establishment of partnerships and networks, and other changes in institutions and policies. In this context of increasing complexity and changed scope, the linear model of the transfer of technology, successfully adopted during the green revolution, has gradually been replaced by more comprehensive and participatory approaches (Table 44.1).

Agricultural Innovation Systems

Environmental and socioeconomic triggers—such as the depletion of natural resources, changing international trade and local market demand, urbanization, climate change, concentration of food production and distribution, integration along value chains, changing consumption patterns, and food safety standards—are shaping AFS and making them increasingly dynamic and inherently complex. Addressing this complexity requires that innovation in agriculture and rural development be based on multistakeholder interaction to include both conventional actors (research, extension, farmer associations, cooperatives) and nonconventional players (e.g., input suppliers, producer associations, nonprofit organizations, market operators, food industry, and credit agents) (Tropical Agriculture Platform, 2016).

Agricultural innovation systems (AIS) are networks of actors or organizations and individuals that bring existing or new products, processes, and forms of organization into social and economic use. They are arranged into three main groups: research and education; business and enterprises, including farmers and their associations; and bridging institutions, including extension services, brokering agencies, and contractual arrangements. The fourth component is composed of supporting policies and institutions (formal and informal), which determine the way these actors interact, reflect, generate, share, and use knowledge as well as jointly learn and adapt to the external changes, shaping the "enabling environment" (Tropical Agriculture Platform, 2016) (Fig. 44.1).

Innovation outcomes, such as increases in agricultural productivity or resource use efficiency, are determined more by the properties and capacity of the system in which organizations or individuals engage with each other. Individual actors are no longer considered the sole drivers, initiators, or owners of the process of agricultural innovation; their services have to be considered in relation to the roles of others that interact and learn within a dynamic network (Gildemacher and Wongtschowski, 2015; Sonnino et al., 2016). The

TABLE 44.1 Evolution of the Theoretical Perspectives on Agricultural Innovation

	Transfer of Technology	Farming System Research	Agricultural Knowledge and Information Systems	Agricultural Innovation Systems
Periods/era	Central since the 1960s	Starting in the 1960s and 1970s	From the 1990s	Since the 2000s
Research agenda	Centrally developed	Based on interviews	Based on consultations	Based on stakeholders' participation
Objective	Production increase	Productivity increase	Livelihood improvement	Sustainability of food systems
Model	Supply technologies through linear processes	Supply technologies through circular processes (farmers to farmers)	Collaboration research/education/extension (knowledge triangle)	Codevelop innovation involving multiactor networks
Scope	Farm	Farm	Farm-based livelihoods	Value chains, territory
Innovators	Scientists	Scientists and extension workers	Farmers, scientists, and extension workers together	Multiple actors
Role of farmers	Adopters and laggards	Source of information	Experimenters	Partners, entrepreneurs, and innovators exerting demands
Role of scientists	Innovators	Experts	Collaborators	Partners, one of the actors responding to demands
Key changes sought	Farmers' behavior change	Removing farmers' constraints	Empowering farmers	Institutional change, innovation
Market integration	Nil	Nil	Low	High
Capacity development	Technology adoption and uptake through development of technical skills and infrastructure	Technology adoption and uptake through development of technical skills and infrastructure and integration of ecological and farm-economic conditions	Enhancing communication between actors, coevolved technologies better fit with livelihood systems	Functional capacities (to collaborate, navigate complexity, reflect and learn, and engage in strategic and political processes)

Source: Modified from the Tropical Agriculture Platform, 2016. Common Framework on Capacity Development for Agricultural Innovation Systems. Conceptual Background. CAB International, Wallingford, UK.

prominent role of the enabling environment needs to be carefully taken into account and properly addressed in order to unleash the potential of innovation. In that respect, the capacity of AIS actors to engage in strategic and political developments and to influence decision making is of paramount importance.

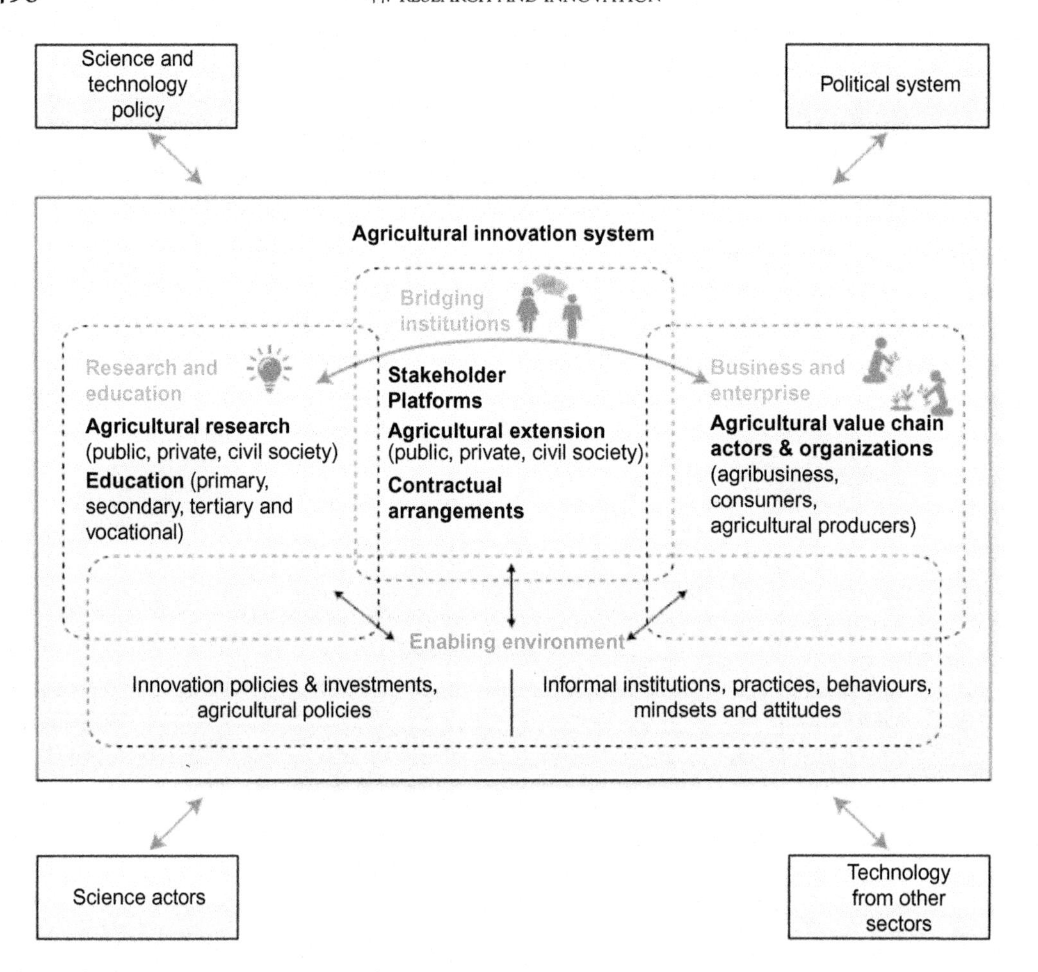

FIGURE 44.1 Conceptual diagram of an agricultural innovation system. Source*: Tropical Agriculture Platform. Common Framework on Capacity Development for Agricultural Innovation Systems. 2016. Conceptual Background. CAB International, Wallingford, UK.*

The AIS model attaches a high importance to access to markets as a fundamental driver for agricultural innovation (Sonnino et al., 2016). Technological innovation allows farmers to produce marketable surpluses, but functional capacities as well as physical and organizational infrastructure such as producer organizations and cooperatives are equally essential to gain access to markets (De Meyer, 2014).

Global Research and National Capacities

In a globalized world, no single country can afford to sustain full research and innovation programs able to address all its food and agriculture issues. All countries depend to a variable extent on knowledge generated elsewhere and adapted to the local contexts.

Many research projects are so vast that they can be technically or financially afforded only by international consortia, such as, for instance, in the deciphering of large genomes (e.g., the International Wheat Genome Sequencing Consortium and the African Orphan Crops Consortium [AOCC, 2017]).

Agricultural innovation is nevertheless an endogenous process and cannot rely only on spin-offs of research conducted abroad: rather it requires that domestic resources and capacities are present in each country to generate, evaluate, and adapt knowledge; develop, adopt, and scale up new technologies and organizational set-ups; and promote social change. Successful agricultural innovation requires technical expertise and experience in the relevant fields, as well as the functional capacities to navigate complexity, to collaborate, to reflect and learn, and to engage in strategic and political processes (Tropical Agriculture Platform, 2016).

In most low-income countries (LICs) and many lower-middle-income countries, however, individuals, organizations, and the systems as a whole lack these capacities and are therefore unable to develop effective AIS (Aerni et al., 2015). AIS are often disjointed, and behaviors, mind-sets, policies, and processes disarranged in a way that prevents innovations from being adopted. Promoting agricultural innovation requires upgrading investments and capacities of research and extension organizations and strengthening all components of AIS and, even more prominently, the way they interact. In this context, the capacity to collaborate assumes a pivotal role.

44.4 CURRENT TRENDS DO NOT FACILITATE THESE EVOLUTIONS

Lack of Sustained and Predictable Public Funding for Research

In spite of the recognized importance of innovation for the transition toward sustainable AFS, and of the high return rates of investments in agricultural research (Alston, 2010), public spending on agriculture research and development (R&D) suffered a neglect during the last part of the 20th century. This tendency seems to be now reversed, considering that public expenditure on R&D grew by an average of 3.1% during the first decade of the present century (Pardey et al., 2014), largely driven by China (+8.7%), India (+5.2%), and a handful of middle-income countries. Overall R&D investment in most lower-middle-income countries lagged behind (+2.3%). In addition, investments in agricultural R&D remain unstable in many LICs, especially in Africa (Nienke et al., 2012). Considering the ratio between the national public expenditure in R&D and the agricultural gross domestic product, commonly indicated as agricultural research intensity (ARI), LICs and middle-income countries remain well under the recommended level of 0.1% (IFPRI, 2017). As differences within each income group remain quite large, the investment dearth is particularly acute in some countries.

There has also been a change in the types of funding for public sector international agricultural research, particularly the CGIAR (Consultative Group for International Agricultural Research). In 2011, CGIAR received 28.5% of its total funding in an "unrestricted" type, that is, money that the system could allocate to its research centers to cover their staff and operational costs based on the CGIAR Strategy and Results Framework and/or the centers' strategic plans. The level of this "unrestricted" funding declined to 15.1% in 2015 (CGIAR System Management Office, personal communication, 2017). Such an "unrestricted" type of funding is particularly important for supporting CGIAR's activities, typically requiring long-term resource commitment and research, such as conservation and studies on plant genetic resources and breeding programs.

Arguably, total funding for the CGIAR has increased significantly since the 2009 CGIAR reform, from US$894 million in 2011 to US$945 million in 2015 (CGIAR System Management Office, personal communication, 2017). This was primarily due to increased targeted funding of the CGIAR Research Programs (CRPs) or toward specific programs/ projects at the center level. The shift in the type of funding has played a major role in making the CGIAR centers more "development oriented" rather than "science driven." On the other hand, CGIAR is now less motivated to, and less capable of, providing products and services that are public goods by nature, with the main purpose of enabling partner institutions, especially the developing country national agricultural research systems, to develop locally adapted crop varieties or improve the underutilized or "orphan" crops that are important for the food and nutrition security or cash income of smallholder farmers. With the significant shift of public funding from long term and "unrestricted" toward short-term, development-oriented programs, the CGIAR centers along with their developing country partners are more "driven" by donor agenda, often with high transaction costs.

Long-term, stable, and predictable funding for public-sector agricultural research is a prerequisite for developing new crop varieties and achieving genetic gains necessary for increasing crop and livestock productivity. Decline in such funding has led, at least partially, to the stagnant/slow growth of the yield of main food crops (wheat, rice, and maize) at about 1.2%–1.5% during the recent decade. It may have also contributed to the world's food supply increasingly relying on a few global crop commodities (Khoury et al., 2014) and to the lack of investment in the genetic improvement of "orphan" crops.

Many agricultural research organizations in LICs are also dependent on donor funding, with the result that their research priorities are not necessarily aligned with national priorities, and they do not respond to the needs of local smallholders (Beye, 2002). In addition, the level of foreign assistance to research and extension remains low and highly volatile (Angelico et al., 2015). Investments in agricultural research and innovation need time before giving returns. Instability and unsustainability of investments, both donors' funding and internal expenditures, can therefore prevent the innovation potential from being realized.

The Growing Importance of Private Research

Over the past 25 years, private investments grew faster than public agricultural research spending, increasing from US$5.1 billion in 1990 to US$15.6 billion in 2014 (Fuglie, 2016). The composition of private investments in agricultural R&D has also dramatically changed: in the 1990s, private investments were concentrated in R&D for farm machinery, agrochemicals and fertilizers, and animal health, nutrition, and genetics; currently, private investments in R&D for the crop seed and biotechnology sector have outpaced other

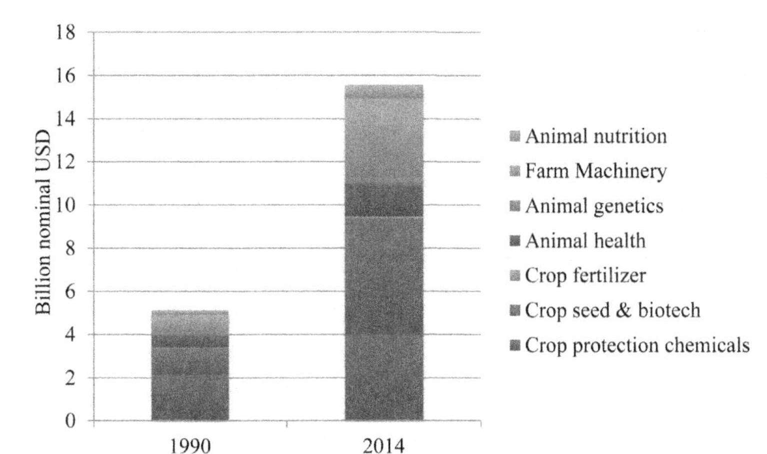

FIGURE 44.2 Private sector R&D expenditures for agriculture at the global level, 1990 versus 2014. Source: *Authors' elaboration of data from Fuglie, K., 2016. The growing role of the private sector in agricultural research and development world-wide. Glob. Food Sec. 10, 29–38.*

sectors and are now predominant (Fuglie and Toole, 2014) (Fig. 44.2). Investments for research in the food industry primarily come from the private sector.

The companies that account for most agricultural R&D spending are based in developed countries, but their technologies are increasingly being adopted in developing countries (Beintema et al., 2012; Pray et al., 2011). Privately funded research is focused on commodities, where large markets for agricultural inputs can potentially compensate high, long-term research investments, and tends to concentrate on product or process specialization. Over the 2008–09 period, private companies in India invested US$88 million in seed research, US$41 million in agricultural machinery research, and US$39 million in pesticides research (Rada and Schimmelpfennig, 2015). Private sector research depends to a large extent on using knowledge, methods, and technologies developed in the public sector (Byerlee, 1998). In India, for instance, the most common drought-tolerant rice varieties were developed by the public sector, and the varieties developed by multinationals and national seed companies are hybrids from public inbred lines (Rada and Schimmelpfennig, 2015). Private investment cannot, therefore, substitute public agricultural R&D, which covers the less commercially attractive innovation fields. Public and private agricultural R&D are, in this perspective, acquiring complementarity.

Technology Transfer

The growing gap between knowledge generated by publicly funded research and its use by agricultural producers has been called the "valley of death" (European Union, 2015). This gap is present worldwide; its causal factors range from research design that does not consider the farmers' real needs to inefficient or missing technology transfer mechanisms to lack of infrastructure and inadequate access to credit and markets. Bridging the "valley of death" and facilitating the translation of research into social and economic value requires greater linkages between research providers and end users. An important cause of nonadoption is lack of communication between researchers and the users, resulting in products that do not address the perceived needs of the users (Wildner et al., 1993; Fujisaka, 1994; Bassi and da Silva, 2014).

Trust between research institutions and enterprises, spatial proximity, efficiency of communication, and flexibility in intellectual right policies and mechanisms can facilitate technology transfer (Santoro and Gopalakrishnan, 2001). Bassi and da Silva (2014) suggest designing mechanisms to identify users' demands and to better integrate the technology transfer team within the research projects.

Five technology transfer mechanisms have been proposed (Rogers et al., 2001): contracts between industry and research institutions; scientific publications; licensing contracts; interaction between users and researchers; and academic spin-offs by which a technology is managed by a new company often consisting of former employees of the original institution. Embrapa, a Brazilian agricultural research corporation, uses several means of technology transfer (Gomes and Atrasas, 2005; Bassi and da Silva, 2014). For knowledge exempt from intellectual property rights, Embrapa uses extension services, technical assistance, networks, field days, training, and publication. It also commercializes technologies through licensing, direct selling of user rights, or consultancies. Bassi and da Silva (2014) note that "spin-offs" can facilitate integration with users and markets. Cornell University, United States, has developed a dedicated unit, the Center for Technology Licensing, to bring the university's scientific discoveries and technological innovations to the marketplace (Cornell University, 2016). Transfers from research institutions toward markets often rely on privileged relationships with local actors. For instance, fruit and nut tree varieties developed by UC Davis are field tested by farmers in the area and are licensed only to California producers during the first years following their release (UC Davis, 2016a, 2016b).

The linkages between researchers and users should extend from the codesign of research programs to the joint steering of research projects, passing through the farmers' direct involvement in development and validation of new technologies (Tropical Agriculture Platform, 2016). In this context, the roles of the players, including farmers and their organizations and cooperatives, research institutes and other academic institutions, the private sector, and governmental bodies, may need substantial reshaping (Sonnino et al., 2016). New functional capacities emerge as essential, such as the capacity to adapt and respond to changing circumstances and the development of appropriate professional skills, such as facilitation and communication. Studies on technology transfer in Brazil (Mendes, 2015; Gargenal, 2015) noted that farmers often adopt technological packages that enterprises propose, along with their financial and technical services, as part of a broader relationship. Technological innovations that are not part of such relationships are less likely to be adopted. Furthermore, this tendency risks ignoring smallholder farmers who are not integrated in most value chains.

44.5 PROMISING EXAMPLES CAN INSPIRE BROADER CHANGES

Interdisciplinary System Solution-Oriented Research Programs

In 2017, the CGIAR launched its new research portfolio with eleven CRPs and three platforms. The new portfolio aims at strengthening the multidisciplinary, cross-sectoral, and system-based approaches, as well as strengthening the research and application at scale linkage (CGIAR, 2017). The second-phase portfolio (2017–22) is structured around two interlinked clusters of challenge-led research. The first is innovation in AFS, which involves adopting an integrated approach through eight AFS research programs, focused on fish; forests, trees, and agroforestry; livestock; maize; rice; roots and tubers; bananas; and wheat. The second cluster consists of four cross-cutting programs in the areas of agriculture for nutrition and health; climate change agriculture and food security; policies, institutions, and markets; and water, land, and ecosystems. A final cluster includes research support platforms that underpin the entire research system, focused on big data in agriculture, excellence in breeding, and gene banks.

The Chinese Academy of Agricultural Sciences (CAAS) is China's, and one of the world's, largest public research organization in the agriculture sector. It has a network of 34 directly affiliated research institutes and nine joint institutes with local Chinese governments or international organizations and about 10,000 staff. Since 2011, CAAS has embarked on an "Agricultural Science and Technology Innovation Program" funded by the Chinese Government. Through the program, CAAS streamlined its research programs and teams to focus on strategic and interdisciplinary research, solving problems of national and global importance and developing enabling platforms for international and nationwide collaboration. As a result, CAAS, in collaboration with a broad range of partners in both the public and private sectors, has developed "Green Technology Packages and System Solutions" for major crops such as rice, oilseed rape, maize, and horticulture crops, as well as small ruminants that have been used on a large scale by family farmers in the country. The total funding of CAAS received from the government increased nearly threefold during 2011–15 (CAAS, 2013, 2017).

Science/Policy Interaction Mechanisms

At the national level, there are increasing efforts to better include stakeholders' concerns and perspectives in the definition of research orientations and priorities. In 2006–07, the French National Institute for Agricultural Research conducted CAP-Environment, a participative process based on the dialogue between researchers and stakeholders (firms, associations, elected representatives, media, farmer organizations, public bodies, etc.) to identify future issues and research priorities in the field of interactions between agriculture and the environment (INRA, 2017). In Australia, the National Climate Change Adaptation Research Facility developed National Climate Change Adaptation Research Plans for nine key sectors. The Plan for Primary Industries, developed in 2010 (Barlow et al., 2011), identified research required to provide decision-makers within government, industry, and communities with information they need to effectively respond and adapt to the impacts of climate change on primary industries. Updated in 2013, it involved the active participation of both the research community and adaptation stakeholders (Barlow et al., 2013).

At the international level, broad mechanisms for bringing science to decision-makers have also been created. Examples for the food and agriculture sector are the International Assessment of Agricultural Knowledge, Science and Technology for Development, the Millennium Ecosystem Assessment, the Intergovernmental Panel on Climate Change, the Intergovernmental Science-Policy Platform on Biodiversity and Ecosystem Services, and the High-level Panel of Experts on Food Security and Nutrition.

Innovative Technology Transfer Mechanisms

South–South cooperation is particularly active in agricultural research and development, led by a commitment to solidarity and an interest in "mutual benefits," and based on the premise of similarity of contexts (Shankland and Gonçalves, 2016). China has established over 40 agricultural demonstration centers in African countries (Amanor, 2016) that drive programs aimed at increasing yields and training African farmers in new low-cost techniques, particularly for rice and vegetable cultivation (Amanor and Chichava, 2016). China's support to the Food and Agriculture Organization of the United Nations (FAO) program on South–South cooperation provides additional opportunities.

In its approach to South–South cooperation in Africa, Brazil highlights the similarities of conditions between Brazil and Africa, its scientific and technical successes, and their transferability to Africa (Amanor, 2016). The African-Brazilian agricultural innovation marketplace collaborates with Embrapa, the Forum for Agricultural Research in Africa, and the World Bank to strengthen agricultural technology development in the region through the establishment of research partnerships (Amanor and Chichava, 2016). Between 2010 and 2013, it connected 103 research institutions from more than 20 African countries to various Embrapa centers, supporting 66 research projects in Africa (Freitas, 2015).

The "Science and Technology Backyard" (STB) approach, established by professors and students of China Agricultural University, in collaboration with local partners (Zhang et al., 2016), is an innovative model that not only illustrates, but also implements, how to effectively disseminate technical know-how to smallholder farmers at a large scale. The university professors and students live in villages among farmers to carry out participatory innovation and technology transfer. Outreach and scaling up were achieved through training of the leading farmers. Thus, the STB connects the scientific community with the local farming community and has helped smallholder farmers to successfully close the yield gap of wheat crops by 30%, with a yield increase from 67.9% of the attainable yield to 97.0%.

44.6 RECOMMENDATIONS

If the challenges in front of us are unprecedented, it is also true that we can count on unprecedented opportunities, including the following:

- Strong political will for sustainable development both at the global level, including through the adoption of the 2030 Agenda for Sustainable Development, and at the national and local levels.
- Recognition of innovation as an engine for sustainable development. Agenda 2030, for instance, explicitly recognizes the crucial role of innovation to achieve the SDGs, and calls for increased investments in science and technology and for strengthened developing countries' scientific, technological, and innovative capacities (UN, 2017).
- Rapid development and deployment of information and communication technologies that enable the management of large quantities of data and that connect people, producers, and suppliers, with consumers, humans, and machines.

- Exponential growth of scientific knowledge to achieve goals considered unattainable only a few years ago (FAO, 2017). For instance, gene-editing techniques are transforming the very promising potential of "omics" sciences into the reality of next generation plant and animal breeding.

However, if these opportunities are to be seized, and agricultural research and innovation are to substantially contribute to the transition of AFS to sustainability and to the achievement of the SDGs, a number of prerequisites need to be met.

As Kennedy (2014) notes, "The much-needed revolutions in agriculture can only come through an investment that we make now." Increased and steady investments in research, both at the national and international level, are the necessary and nonrenounceable condition for the transition toward sustainability of the global food and agriculture systems.

Increased levels of investment must be accompanied by adoption of a systemic research approach that considers the system as a whole and pays more attention to the interaction between system components or factors than to the isolated components or factors themselves is a crucial requisite for research planning. Reynolds et al. (2017) make the case for improving global integration of crop research, highlighting the need for a global network of controlled "field laboratories," harmonization of research practices, and data sharing. Agricultural research should be placed in the framework of AIS, where all stakeholders are actively engaged in collective reflection and learning processes. In this context, formal science outputs may be reconciled with traditional knowledge and integrated with other ingenuity sources (rural advisory services, farmer associations, the private sector, local governments, etc.) in order to unleash innovation potential. Within AIS, synergies and partnerships between public and private agricultural R&D should be promoted.

Support for innovation should therefore be focused on developing capacities (both technical and functional) of AIS at all the levels (individual, organizational, and institutional). The enabling environment is in fact crucial to ensure the effective development of sustainable AFS.

The needed paradigm shift, including transformation of behaviors, mindsets, policies, and processes, is summarized in Table 44.2.

As highlighted above, many of the promising areas for research and innovation rely on a considerable increase in data and information collection, management, and sharing, including by distant and or automatic means such as through remote sensing. Even more important are issues related to data ownership and intellectual property rights, many of which are only beginning to be raised, such as for big data. Some mechanisms that deal with such issues for plant genetic resources already exist. The extension of open-access data policies in the public sector and of the potential of data to generate value-added calls for a broader reflection. Progress will require a thorough consideration of the various means by which data collected from farms could be returned to farmers in a useful format and also benefit the researchers.

Leveraging the potential of research and innovation for sustainable AFS requires the following:

- ensuring long-term, higher, and predictable investment in research on priority areas that are also compatible with global objectives;
- promoting systemic approaches, valorizing diverse research objectives, and aligning disciplines and actions from the generation of research hypothesis to the effective development of research programs and innovation systems;

TABLE 44.2 Cultural Changes Made Necessary by the New Context of Agricultural Innovation

	From	To
Ultimate aim of research	Creation of knowledge	Social, economic, and environmental change
Social contract	Science for society	Science with and within society
Scientific approach	Reductionist	Systemic
Knowledge created	Scientifically sound	Scientifically sound and socially robust
Evaluation	Results' indicators (publications, patents)	Impact indicators (social, economic, and environmental change)
Relationships with society	Consultations with potential beneficiaries	Direct involvement of the interested parties in decision-making processes
Communication type	Top-down, unidirectional	Participatory
Communication tools	Scientific communication (conferences, and scientific and technical papers)	Facilitation, recording, management, and sharing of knowledge
Innovation site	Farm	Territory
Training type	Education	Collective learning
Type of working organization	Individual merit and competition between research institutes	Teamwork and collaboration within and between research institutes and between research institutes and other actors

Source: From Sonnino, A., Carrabba, P., and Iannetta, M., 2016. The role of research bodies, from leaders of the system to responsible partners. In: Petruzzella D., Di Mambro A., (Eds.), Innovation in the Mediterranean agrifood sector. Concepts, experiences and actors in a developing ecosystem. Bari: CIHEAM pp. 39–50. Available from: URL http://om.ciheam.org/om/pdf/b74/00007183.pdf.

- ensuring systematic coownership of research programs by all users and facilitating participatory research processes and interactions between researchers and users to develop and deliver products and services that respond to user demands;
- organizing effective data and knowledge gathering and sharing, fully exploiting the potential of information and communication technologies; and
- facilitating technology transfer and capacity development at all levels, including through South–South and triangular cooperation.

44.7 CONCLUSION

Research and innovation have a major role to play in the profound transformation that food and agriculture systems have to undergo to become both more sustainable and more productive. Research will need to integrate diverse disciplines and perspectives in order to address multiple objectives at the field, landscape, and value chain levels, with a systemic approach and oriented toward user needs, mobilizing both global research programs and strengthened national capacities. Innovation will be produced by agricultural innovation systems that include researchers, farmers, enterprises, and bridging institutions. Such a transformation will only happen with sustained and predictable public funding, as well as dialogue with the increasing private investment sector. Progress calls for effective data and knowledge gathering and sharing, fully exploiting the potential of information and communication technologies and for exploiting the potential of diverse forms of technology transfer, building upon successful experiences. Such a transformation has the potential not only of making food systems more sustainable, but also to benefit and attract young people, both men and women, provided that they are endowed with the necessary education and training to enable them to participate in the sector. It will be key in addressing and meeting the challenges of the future.

References

Aerni, P., Nichterlein, K., Rudgard, S., Sonnino, A., 2015. Making agricultural innovation systems (AIS) work for development in tropical countries. Sustainability 7, 831–850.

Alston, J.M., 2010. The Benefits from Agricultural Research and Development, Innovation, and Productivity Growth. OECD Food, Agriculture and Fisheries, Paris, Working Papers No. 31.

Amanor, K.S. 2016. South-South cooperation in context: perspectives from Africa. Future Agricultures, Working Paper 054.

Amanor, K.S., Chichava, S., 2016. South–south cooperation, agribusiness, and african agricultural development: Brazil and China in Ghana and Mozambique. World Dev. 81, 13–23.

Angelico, C., Grovermann, C., Nichterlein, K., Sonnino, A., 2015. Is Aid to Agricultural Innovation a Priority for the International Community? A Comprehensive Analysis of 2002 to 2012 OECD Data on Foreign Assistance to Research and Extension in Agriculture, Forestry and Fishing. FAO, Rome. Available from: http://www.fao.org/publications/card/en/c/90bfc23f-6e73-4387-9175-590f140e4794/.

AOCC (African Orphan Crops Consortium). 2017. AOCC website. <http://africanorphancrops.org>.

Barlow, S., Grace, P., Stone, R., Gibbs, M., Howden, M., Howieson, J., et al., 2011. National Climate Change Adaptation Research Plan for Primary Industries. National Climate Change Adaptation Research Facility, Gold Coast, Australia, p. 64.

Barlow, S., Eckard, R., Grace, P., Howden, M., Keenan, R., Kingwell, R., et al., 2013. National Climate Change Adaptation Research Plan Primary Industries: Update Report 2013. National Climate Change Adaptation Research Facility, Gold Coast, Australia.

Bassi, N.S.S., da Silva, C.L., 2014. As estratégias de divulgação científica e transferência de tecnologia utilizada pela Empresa Brasileira de Pesquisa Agropecuária (Embrapa). Interações, campo grande. 15 (2), 361–372. Available from: URL:https://www.alice.cnptia.embrapa.br/bitstream/doc/1003411/1/final7774.pdf.

Beye, G. 2002. Impact of foreign assistance on institutional development of national agricultural research systems in Sub-Saharan Africa. Research and Technology Paper, no. 10. FAO, Rome.

Beintema, N., Stads, G.J., Fuglie, K., Heisey, P., 2012. ASTI global assessment of agricultural R&D spending: developing countries accelerate spending. Agricultural Science and Technology Indicators Project, IFPRI, Washington, D.C.

Byerlee, D., 1998. The search for a new paradigm for the development of national agricultural research systems. Rural Development Department. World Bank, Washington, DC.

CAAS (Chinese Academy of Agricultural Sciences), 2013. The Chinese Academy of Agricultural Sciences: envisioning an innovative future. Sponsored Suppl. Sci. 340 (6136), 1122.

CAAS (Chinese Academy of Agricultural Sciences). 2017. CAAS website. <www.caas.cn>.

CBD (Convention on Biological Diversity). 2017. Ecosystem approach. CBD website. <https://www.cbd.int/ecosystem>.

CGIAR. 2017. CGIAR website. <http://africanorphancrops.org>.

Cornell University. Center for technology licensing - 2016 annual report. 2016. Available from: URL: <http://www.ctl.cornell.edu/news/annual-reports/2016-AnnualReport.pdf>.

De Meyer, J., 2014. Apple-Producing Family Farmers in South Tyrol: An Agriculture Innovation Case Study. FAO, Rome. Available from: URL: http://www.fao.org/documents/card/en/c/2927b768-aa16-4d17-8a36-4967b212090d/.

European Union, 2015. New Ways of providing knowledge to tackle food and nutrition security: what should the EU do? Expo 2015 EU Scientific Steering Committee. Available from: https://doi.org/10.2788/802016.

FAO. 2017. The future of food and agriculture – Trends and challenges. Rome. Available from: URL: <http://www.fao.org/3/a-i6583e.pdf>.

Freitas, A. 2015. Innovative partnerships for agricultural research and development: Examining the Africa-Brazil agricultural innovation marketplace. Briefing Note 82. Maastricht, ECDPM.

Fujisaka, S., 1994. Learning form six reasons why farmers do not adopt innovations intended to improve sustainability of upland agriculture. Agric. Syst. 46 (4), 409–425.

Fuglie, K., 2016. The growing role of the private sector in agricultural research and development world-wide. Glob. Food Sec. 10, 29–38.

Fuglie, K.O., Toole, A.A., 2014. The evolving institutional structure of public and private agricultural research. Am. J Agric. Econ. 96 (3), 862–883.

Gargenal, I. 2015. Da inovação à transferência de tecnologia. Jornal da Unicamp No. 644.

Gibbons, M., 1999. Science's new social contract with society. Nature 402 (Suppl), C81–C84.

Gildemacher, P., Wongtschowski, M. 2015. Catalysing innovation: from theory to action. KIT working paper 1.

Gomes, G., Atrasas, A.L., 2005. Diretrizes para trasferência de tecnologia: modelo de incubação de empresas. Embrapa Informação Tecnológica, Brasília.

Hammond, R.A., Dubé, L., 2012. A systems science perspective and transdisciplinary models for food and nutrition security. PNAS 109 (31), 12356–12363. Available from: https://doi.org/10.1073/pnas.0913003109.

IFPRI, 2017. Food Policy Indicators: Tracking Change: Agricultural Science and Technology Indicators (ASTI). International Food Policy Research Institute, Washington, DC. Available from: URL: https://www.asti.cgiar.org/sites/default/files/GlobalFoodPolicy-ASTI.pdf.

INRA (Institut National de la Recherche Agronomique). 2017. Projet CAP environnement. Available at: <https://www6.paris.inra.fr/depe/Projets/CAP-Environnement>.

Kennedy, D., 2014. Building agricultural research. Science 20154 346 (6205), 13.

Khoury, C.K., Bjorkman, A.D., Dempewolf, H., Ramirez-Villegas, J., Guarino, L., Jarvis, A., 2014. Increasing homogeneity in global food supplies and the implications for food security. PNAS 111 (11), 4001–4006. Available from: https://doi.org/10.1073/pnas.1313490111.

Mendes, C.I.C. 2015. Transferência de tecnologia da Embrapa: rumo à inovação. PhD paper, Unicamp. Brazil.

Nienke, M., Beintema, N., Stads, G.J., Fuglie, K.O., Heisey, P.W., 2012. ASTI Global Assessment of Agricultural R&D Spending: Developing Countries Accelerate Investment. International Food Policy Research Institute, Washington.

Pardey, P., Chan-Kang, C., Dehmer, S., 2014. Global Food and Agriculture R&D Spending, 1960–2009. University of Minnesota, Sant Paul, InSTePP Report.

Pray, C.E., Gisselquist, D., Nagarajan, L. 2011. Private investment in agricultural research and technology transfer inAfrica. ASTI/IFPRI-FARA Conference: Agricultural R&D: investing in Africa's future. Accra, Ghana, 5–7 December2011, IFPRI, Washington and FARA, Accra.

Rada, N.E., Schimmelpfennig, D.E., 2015. Propellers of Agricultural Productivity in India. ERR-203. US Department of Agriculture, Economic Research Service. Available from: URL https://www.ers.usda.gov/mediaImport/1957187/err-203.pdf.

Reynolds, M.P., Braun, H.J., Cavalieri, A.J., Chapotin, S., Davies, W.J., Ellul, P., et al., 2017. Improving global integration of crop research. Science 357 (6349), 359–360. Available from: https://doi.org/10.1126/science.aam8559.

Rockström, J., Williams, J., Daily, G., Noble, A., Matthews, N., Gordon, L., 2017. Sustainable intensification of agriculture for human prosperity and global sustainability. Ambio 46 (1), 4–17.

Rogers, E.M., Takegami, S., Yin, J., 2001. Lessons learned about technology transfer. Technovationm, Amsterdã 21 (4), 253–261.

Santoro, M., Gopalakrishnan, S., 2001. Relationship dynamics between University research centers and industrial firms: their impact on technology transfer activities. J. Technol. Transf. 26 (1-2), 163–171.

Shankland, A., Gonçalves, E., 2016. Imagining agricultural development in South–South cooperation: the contestation and transformation of ProSAVANA. World Dev. 81, 35–46.

Sonnino, A., Sharry, S., 2017. Biosafety communication: beyond risk communication. In: Adenle, A., Morris, J., Murphy, D. (Eds.), Genetically Modified Organisms in Developing Countries. Cambridge University Press.
Sonnino, A., Carrabba, P., Iannetta, M., 2016. The role of research bodies, from leaders of the system to responsible partners. In: Petruzzella, D., Di Mambro, A. (Eds.), Innovation in the Mediterranean Agrifood Sector. Concepts, Experiences and Actors in a Developing Ecosystem. CIHEAM, Bari, pp. 39–50. Available from: URL http://om.ciheam.org/om/pdf/b74/00007183.pdf.
Stilgoe, J., Owen, R., Macnaghten, P., 2013. Developing a framework for responsible innovation. Res. Policy 42 (9), 1568–1580.
Tropical Agriculture Platform. Common Framework on Capacity Development for Agricultural Innovation Systems, 2016. Conceptual Background. CAB International, Wallingford, UK.
UC Davis (University of California, Davis). 2016a. Annual report 2016a. Available from: URL: <http://interim-chancellor.ucdavis.edu/reports/2016/UC%20Davis%20Annual%20Report%202016_150DPI.pdf>.
UC Davis. 2016b. 2015–2016 annual report – technology management & corporate relations. Available from: URL: <http://research.ucdavis.edu/wp-content/uploads/2015-16-Technology-Management-Corporate-Relations.pdf>.
UN. 2017. Sustainable Development Knowledge Platform. Transforming our world: the 2030 Agenda for Sustainable Development. <https://sustainabledevelopment.un.org/post2015/transformingourworld>.
Wildner, L.P., de Nadal, R., Silvestro, M., 1993. Metodologia para integrar a pesquisa, a extensão rural e o agricultor. Agropecuária Catarinense, Florianópolis 6 (3), 39–47.
Zhang, W., Cao, G., Li, X., Zhang, H., Wang, C., Liu, Q., et al., 2016. Closing yield gaps in China by empowering smallholder farmers. Nature 537 (7622), 671–674.

Further Reading

FAO. 2014. State of Food and Agriculture: Innovation in family farming. Rome. Available from: URL: <http://www.fao.org/3/a-i4040e.pdf>.

CHAPTER

45

Policies for Sustainable Food Systems

Prabhu Pingali

Tata-Cornell Institute for Agriculture and Nutrition, Cornell University, Ithaca, NY, United States

Over the past 50 years, we have seen tremendous progress in poverty reduction and overall economic development across much of the developing world. The incidence of poverty in non-OECD (Organisation for Economic and Co-operation and Development) countries, as measured by the proportion of population living under US$2 a day, has dropped from 47% in 1960 to 14% in 2015. The United Nations reports that developing countries in aggregate have achieved the Millennium Development Goal poverty reduction target. However, there are significant regional differences, with sub-Saharan Africa (SSA) and South Asia lagging behind on poverty, hunger, and other welfare indicators.

Several dozen countries in Asia and Latin America have graduated to middle-income status, some well on their way to becoming developed countries. In most of these countries, agricultural growth has played an important role in "jumpstarting" overall economic development and in moving countries along the structural transformation pathway (Timmer, 2010). Agricultural growth, often associated with the green revolution, focused on enhancing smallholder staple food crop productivity, and has been credited with the high rates of poverty reduction (Hazell, 2010). These emerging economies are well on their way toward achieving agricultural modernization and structural transformation. The challenge for agriculture in these emerging economies is to maintain competitiveness in the face of global integration of food markets and to close the interregional income gap (Pingali, 2010). These countries are also facing negative environmental trade-offs associated with the intensification of agricultural systems, such as long-term soil fertility decline, degradation of water resources, and loss of agricultural biodiversity.

In the low-income countries of SSA, continued high levels of food deficits and reliance on food aid and food imports have reintroduced agriculture as an engine of growth on the policy agenda (Pingali, 2012). Rising population densities and a degrading agricultural resource base have focused attention on the need for a more sustainable approach to

Sustainable Food and Agriculture
DOI: https://doi.org/10.1016/B978-0-12-812134-4.00045-5

agricultural development. There is also increasing awareness of the detrimental impacts of climate change on food security, especially for tropical agricultural systems in low-income countries (Byerlee et al., 2009). These countries continue to be plagued by age-old constraints to enhancing productivity, such as the lack of technology, market infrastructure, appropriate institutions, and an enabling policy environment (Binswanger-Mkhize and McCalla, 2010). Political conflicts and civil strife have also contributed to poor growth performance.

A majority of the world's agricultural production takes place on small farms, and currently there are nearly 500 million smallholder farmers around the globe (FAO, 2017). In Asia and SSA, where the problem of hunger and poverty is the most severe, 80% of the food supply comes from smallholders (FAO, 2012). Therefore, assuring the viability of small farms is crucial to enhancing rural incomes, ensuring sustainable food security, and triggering the structural transformation process.

This chapter draws on two publications by its author: Abraham and Pingali (2017) and Pingali et al. (2016). It provides a review of agricultural development and food security and sustainability challenges as economies move along the structural transformation pathway, from their subsistence roots to increased market orientation and commercialization. The challenges of meeting the rising demand for food, including diversity in food, in the context of environmental and climate pressures are described. The first part of the chapter discusses the role of agricultural development in the process of economic growth and structural transformation. It also notes the changes to food and nutrition security and environmental consequences as countries transition from subsistence to modernizing agricultural systems. The second part of this chapter looks broadly at sustainable food systems policies from production to consumption. These policies address sustainable productivity improvement, diversification of food systems, smallholder participation in value chains, nutrition enhancement, gender empowerment, and policies for managing climate risks and reducing environmental trade-offs.

45.1 AGRICULTURAL DEVELOPMENT AND STRUCTURAL TRANSFORMATION

Agricultural development is central to the structural transformation process in all developing countries. Productivity growth in agriculture leads to surplus creation and increased market participation by small farms, resulting in rising household level incomes and welfare gains. This increased engagement with markets is referred to as commercialization (Carletto et al., 2017; Pingali and Rosegrant, 1995). Commercialization is essential for the transfer of surplus in the form of food, labor, and capital from the agrarian sector to the industrial and service sectors to enable structural transformation (Timmer, 1988). The economies of developing countries are at various stages of structural transformation and can be categorized as low-productivity agricultural systems, modernizing agricultural systems, and commercialized agricultural systems (Pingali et al., 2015).

Countries with low per capita incomes and a larger share of agricultural contributions to gross domestic product are referred to as low-productivity agricultural systems. Many of the countries in SSA are classified as such, and in these regions hunger and poverty remain high. Agricultural lands in these economies are also prone to high levels of environmental degradation, as farm productivity growth rates struggle to match population growth rates.

The emerging economies of Latin America, South East Asia, and South Asia are witnessing increasingly market oriented and modernizing agricultural systems. These regions successfully implemented green revolution technologies, gained from the resulting agricultural productivity increases, and have substantially reduced poverty and hunger. In these regions, however, high levels of income inequality and regional disparities in development persist. Their economies are also facing the negative consequences of food policy trade-offs that promoted productivity growth over environmental sustainability. The East Asian economies of Japan, Taiwan, and South Korea, which are dominated by small farms, are referred to as commercialized agricultural systems due to their high per capita incomes, low share of agriculture in gross domestic product, and high market integration of the agriculture sector. Rising values of environmental services have induced policy reforms that have reduced some of the negative environmental trade-offs in many of these countries, for example, the return of tree cover to low productive lands that have been released from agricultural production.

The challenges for sustainable food systems in each of these production systems are different; consequently, by assessing the major characteristics of small farm economies in different stages of structural transformation, we can better understand the economic, nutritional, social, and environmental challenges such systems face.

Low-Productivity Agricultural Systems

Several countries across the globe are beset with low-productivity agricultural systems, and often with low income levels. Most of these countries are in SSA, where the adoption of green revolution technologies in staple grains such as wheat, rice, and maize was low (unlike in Asian and Latin American countries). While 82% of the area under staples in Asia concentrated on modern high-yielding varieties in 1998, SSA had only 27% (Evenson, 2003). While yields in cereals doubled in SSA, they quadrupled in South Asia, Latin America, and Southeast Asia between 1970 and 2010. The difference in productivity primarily stems from the fact that agricultural production in low-productivity agricultural systems is carried out in marginal environments with constraining agroclimatic, socioeconomic, and technological or biophysical conditions, where input intensive green revolution technologies could not be adopted (Pingali et al., 2014). This—coupled as it is with poor access and provision of essential public goods such as research and development (R&D), factors such as seeds and fertilizers, and essential infrastructure such as irrigation, storage, and roads—affects production incentives at the farm level.

Other challenges to adoption are problematic governance, lack of institutional support (like extension services and markets), and the effects of conflicts plaguing several parts of Africa. Factors such as low and inelastic demand for agricultural products have also affected development (Pingali, 2010). In recent years, production increases have taken place in these regions via area expansion and not through yield increases (Binswanger-Mkhize and McCalla, 2010). Also, much of the increase has been in maize rather than in traditional staples, which often have higher nutrient content and are better adapted to agroclimates of SSA, such as millets, cassava, and beans.

Expansion of low-input, low-yield production systems across SSA has contributed to significant land erosion and soil degradation problems. Sanchez (2002) states that African soils over the last 30 years have lost, on average, 22 kg of nitrogen, 2.5 kg of phosphorus, and 15 kg of potassium per hectare of cultivated land—an annual loss equivalent to US$4 billion in fertilizers. Expansion of cultivation into previously uncultivated lands has also caused significant losses in biodiversity (Pingali, 2017).

Finally, in some parts of SSA, frequent incidents of water stress and drought have contributed to high volatility in food outputs and frequent shocks to food security and household welfare. To be effective in moving the needle on agroclimatic risks, investments in irrigation infrastructure ought to be complemented by other measures, such as promotion of drought-tolerant crop varieties and animal breeds, sustainable management of soils, conservation of ecosystem services, information services to empower farmers to anticipate and manage crises, and innovations in agricultural insurance. These measures are crucial for achieving sustainable small farm growth and development in low-productivity agricultural systems.

Modernizing Agricultural Systems in Emerging Economies

Emerging economies face a myriad of challenges that have implications for food system transformation and sustainability. First, rapid growth in incomes, urbanization, and the rise of the middle class led to fast diversification of diets and boost demand for higher value crops and livestock products (Pingali, 2007; Pingali and Khwaja, 2004; Reardon et al., 2009; Reardon and Minten, 2011). Second, despite significant gains in food supply and food access, interregional inequalities in income and nutritional status continue to persist at high levels, especially in the more marginal agroclimatic zones that were bypassed by the green revolution (Pingali, 2012). Third, reversing the negative consequences of the productivity environment trade-offs that were made during the green revolution is a major challenge for emerging economies as they endeavor to transition to a more sustainable food system. A common thread throughout all the above-mentioned issues is the need to reexamine the emphasis given to staple crop production systems in developing countries. Additionally, there is urgency to promote diversity across agroecologies and the food system and to enhance resource conserving technical change.

The green revolution has had an unquestionably positive impact on the calorie and protein consumption of the population due to its direct (access to food) and indirect (through enhanced real incomes) effects. Increased income due to the green revolution led to a rise in demand for nonstaple foods such as vegetables, fruit, meat, and dairy products. This rising demand for diet diversity as countries move along the structural transformation pathway is consistent with Bennett's law. However, the increased demand for nonstaples was not always matched by a corresponding increase in their supply. Hence, high relative prices of nonstaple food persisted. A large number of crops (such as legumes, fruits, and vegetables) whose relative prices compared to staple grains are high are especially rich in micronutrients. This limited the impacts of diet diversification on nutrition outcomes.

Despite the rising demand, policy and structural impediments as well as a weak private sector limited the supply responsiveness for vegetables and other nonstaples. Policies that promoted staple crop production, such as fertilizer and credit subsidies, price supports, and irrigation infrastructure (particularly for rice), tended to crowd out the production of traditional nonstaple crops, such as pulses and legumes in India (Pingali, 2015). The persistence of staple grain fundamentalism in agricultural policy hampers farmer incentives for the diversification of their production systems (Pingali, 2015).

Within emerging economies, regions left behind during the green revolution-led growth process face the dual problem of declining competitiveness with the more agroclimatic favorable zones in the country in terms of staple grain production and limited ability to diversify out of low productive staple grain agriculture. To add to the problem, in many of these regions, traditional crops rich in certain micronutrients, such as millets and pulses, have been crowded out by the push to promote the big three staples—rice, wheat, and maize (Pingali, 2015; Webb, 2009). While closing the interregional productivity gap remains critical, the focus ought to be on crops, livestock, and aquaculture production systems that are relatively more suitable to the marginal environments and enhance access to nutritious food. For marginal environments to respond to changing market demand for higher value crops and other nonstaples in ways that would benefit small farms, infrastructure and support resources are required to enable them to participate in the value chain.

The agricultural productivity-led growth strategy has often resulted in significant negative environmental externalities in terms of land and water resource degradation, agricultural biodiversity loss, and chemical run-off due to excessive fertilizer and pesticide use. The environmental consequences have been exacerbated by the policy environment that promotes injudicious and overuse of inputs and expansion of cultivation into areas that cannot sustain high levels of intensification, such as sloping lands. Output price protection and input subsidies – especially fertilizer, pesticides, and irrigation water—provide distorted incentives at the farm level for adopting practices that would enhance efficiency in input use and thereby contribute to sustaining the agricultural resource base.

45.2 POLICY AGENDA FOR SUSTAINABLE FOOD SYSTEMS

A policy agenda for sustainable food systems strives for simultaneous improvements in economic, social, and environmental welfare of rural ecosystems. It adds to the sustainable intensification principles by incorporating human and social capital dimensions. Pretty et al. (2011) define sustainable intensification as "producing more output from the same area of land while reducing the negative environmental impacts and at the same time increasing contributions to natural capital and the flow of environmental services." Also important here is agricultural intensification without increasing negative externalities of agricultural production, such as diminishing biodiversity, increased greenhouse gases, and land and water degradation, among others.

In addition, sustainable food systems policies explicitly address the welfare of producers, especially smallholders, the rural poor, and consumers, including considerations of nutrition and food safety.

R&D for Enhancing Food and Nutrition Security

Agricultural research is often cited as the single best investment in terms of increasing productivity and reducing poverty in the developing world (Fan, 2000; Fan et al., 2000; Fan and Pardey, 1997). Among many investments made in agricultural research during the past five decades, South Asia's green revolution—doubling the yields and output of major food staples between 1965 and 1985—is one of the most cited examples of this high payoff (Hazell, 2010; Pingali, 2012). But similar successes have also been achieved in Africa at different scales and with different crops and technologies (Haggblade and Hazell, 2010; Spielman and Pandya-Lorch, 2009). A shared characteristic of many of these high-return investments was the contribution of modern science, particularly plant breeding and cultivar improvement, which was supported by the donor community (Evenson and Gollin, 2003; Raitzer and Kelley, 2008; Renkow and Byerlee, 2010).

Continued high levels of investments are needed for enhancing the productivity of the major staple grains—rice, wheat, and maize—in order to meet their rising demand due to population and income growth. Research into crop breeding and genetics, with a targeted focus on varieties, which will thrive within a specific agroclimatic zone, can substantially influence the productivity of a smallholder and maximize yields. Additionally, productivity gains in traditional staples, such as cassava, millets, barley, and sorghum as well as fruits and vegetables, that were not the focus of the green revolution now need to be emphasized to improve diversity of diets and essential micronutrient availability. High returns have been demonstrated for programs focused on productivity gains for cassava cultivation in SSA (Binswanger-Mkhize and McCalla, 2010).

R&D is also essential for enhancing the efficiency of input use, with a particular focus on soil fertility, water-use efficiency, and pest-resistant cultivars. Modern information and communications technology tools, such as geographic information systems and remote sensing, could contribute significantly to sustainable use of inputs. Research on delivery systems for these intrinsically knowledge-intensive technologies is crucial, especially in developing country smallholder systems. Also essential is policy research for effective means of reducing incentive distortions in the adoption and use of efficiency-enhancing technologies.

Biofortification of staple and nonstaple food can be a sustainable means to reducing immediate concerns of micronutrient deficiency (Bouis et al., 2011). Essential micronutrients such as iron, zinc, and vitamin A can be accessed through biofortified foods and in a cost-efficient way (Asare-Marfo et al., 2013). Vitamin A-biofortified maize and iron-biofortified pearl millets have been successfully adopted in Zambia and Benin, respectively (Cercamondi et al., 2013). Biofortification has also been successfully carried out in nonstaples that can be easily grown and accessed in remote regions where commercially available fortified foods are scarce (Bouis et al., 2011). Examples of this include vitamin A-biofortified cassava in Nigeria and the Democratic Republic of the Congo, and vitamin A-biofortified orange flesh sweet potato in Uganda (Hotz et al., 2012) and Mozambique (Low et al., 2007). These measures may be an effective approach to remedy deficiencies until such time as production of nonstaple micronutrient-rich crops becomes more established.

Promoting Food System Diversification

Despite rising demand, the persistence of green revolution era policies and structural impediments, as well as a weak private sector, limited the supply responsiveness for vegetables, nonstaple food, and other sources of food, including livestock, aquaculture, and neglected or underutilized plant species and breeds. Creating a "level policy playing field" that corrects the historical bias in favor of staple crops would help improve the incentives for diversification of production into nonstaple foods. Agricultural policy of the past was focused on staple crop intensification. The need both today, and well into the future, is for a policy that is "crop neutral," removes distortions, and allows farmers to respond to market signals in making crop production choices (Pingali, 2015).

In addition to leveling the playing field, investments in road and transport infrastructure and cold storage systems are required for developing markets for perishable products. Investments in market information systems and farmer connectivity, especially through cell phones, could significantly cut transactions costs for market participation. Policies promoting food safety should be a priority for upgrading traditional markets and ensuring that human health is safeguarded (Pingali et al., 2015). In addition to reducing instances of foodborne illness and disease, food safety policies can make traditional markets a viable place for procurement by modern retail value chains.

Investments in general literacy and specialized training for farmers in meeting quality and safety standards for high value crops would help integrate smallholders into market value chains. Finally, institutional investments in establishing clear property rights to land and other assets, formalized contractual arrangements that depersonalize market transactions, and access to finance (that is not tied to particular commodities) are essential for diversifying production systems.

Linking Farmers to Modern Fresh Food Value Chains

Linking small producers to fresh food value chains requires particular attention given the rapidly growing urban demand for fruit, vegetables, and livestock products. The rise of supermarkets and their growing reach into rural areas to procure fresh food provides new growth opportunities for small producers. It has been noted that small farms participating in fresh food value chains have both direct and indirect gains (Swinnen and Maertens, 2007). The direct gains accrue through productivity increase, quality improvement, rise in household level incomes, and improved nutrition (Birthal et al., 2009; Dries et al., 2009; Ramaswami et al., 2009). The indirect effects include reduced risks in production, increased access to credit and technology, improved market participation, and the productivity spillovers to other crops (Bellemare, 2012; Swinnen and Maertens, 2007). Therefore, effective linkages to product markets play an important role in incentivizing production, diversification, and intensification in all production systems.

Policy interventions to create infrastructural public goods and mitigating locational disadvantages in low potential areas will help decrease regional disparity in market access. Increased investment by the state to expand storage facilities, cold chains, and improved connectivity is also vital to reduce wastage and increase marketing options for smallholders. These interventions in infrastructure are often needed for a private sector response to engage in markets and enable the successful emergence of vertical coordination where farms can directly connect with retail.

Institutional interventions such as producer organizations and cooperatives have helped provide inputs, reduce transaction costs, and also form market linkages (Barrett et al., 2012; Bellemare, 2012; Boselie et al., 2003; Briones, 2015; Reardon et al., 2009; Schipmann and Qaim, 2010). Promoting these institutions will help smallholders mitigate some of the transaction costs associated with entry into fresh food value chains, as this measure addresses problems associated with economies of scale. These production systems also require an incentive to attract public-private partnerships and collaborate with civil society organizations to enable such linkages.

Growth That is Inclusive of Rural Women

Women are among the largest group of landless laborers and the largest group dispossessed or with restricted access to land (Agarwal, 1994; Deere and Leon, 2001). They also make up two-thirds of the livestock keepers (Thornton et al., 2002) and 30% of labor in fisheries (FAO, 2011). Despite having an important role in production, studies have also shown that women face high costs in accessing capital, engaging in entrepreneurial activities (Fletschner and Carter, 2008), and adopting technological inputs and mechanization (FAO, 2006). Therefore, in many developing countries, woman-headed households have lower yields and incomes due to poor access to markets and productive resources (Croppenstedt et al., 2013), affecting their contributions to agricultural productivity (FAO, 2011). Women also provide nonmarketable goods and services at the household level, such as gathering water and fuel, child health and nutrition, and subsistence food production, which is essential for household welfare (Floro, 1995). Time savings in this context is relevant for reducing women's workloads, income, and household-level welfare.

In low-productivity agricultural systems, women's participation in the agricultural labor force is higher than the global average (Croppenstedt et al., 2013). Therefore, closing the gender gap and addressing gender-specific transaction costs and constraints to agricultural production is crucial to increase agricultural productivity and women's empowerment.

Improving access to the factors of production such as cultivable land and institutional credit is central to providing women with control over productive resources. Better access to tapped water and clean fuel for household use, helps improve women's health, reduce drudgery, and free up labor for more productive activities. Agricultural policies related to natural resource management, input, and technology access and production affect man- and woman-headed households differently and therefore necessitates a more gendered policy focus in agriculture (FAO, 2011). Promotion of women's self-help groups for education, information dissemination, microcredit access, provisions of essential public goods, and supporting production-based activities are essential.

The two major interventions needed to address gender-specific challenges in agriculture are improved access to product markets and labor savings for rural women. With regard to access to product markets, studies have shown that women involved in both traditional and modern crop production and marketing face considerable disadvantages and risks (Cabezas et al., 2007). A more gender-sensitive value chain is required to address access problems in markets (Rubin and Manfre, 2014; Nakazibwe and Pelupessy, 2014; Quisumbing et al., 2015).

Policy initiatives to promote women's organizations and build capacity to make them self-sustaining are important in tackling gender-specific challenges in production and marketing. Gender-sensitive value chains that facilitate women's participation in high value markets are essential. Collaboration with state and civil society organizations is vital for promoting and empowering women's producer organizations and self-help groups.

As women are often involved in agricultural labor and nonmarketed household labor, measures to improve labor efficiency and productivity of women will enable cost savings and free up time. Labor-saving technology through mechanization in agriculture is needed to reduce drudgery. Mechanization, like marketing, is scale sensitive and, again, collective action to enable joint access of labor-reducing machinery is vital. Targeting mechanization in women-dominated activities in agriculture, such as transplanting, harvesting, and postharvest operations, must take precedence in modernizing agricultural systems.

Managing Climate Change Impacts

Mitigating the effects of climate change and the need to increase yield simultaneously pose a major challenge to growth and development of the agricultural sector. This challenge could be particularly important for crops that are important to the poor, such as millets and cassava. Little is known about the long-term climate impacts on crops beyond the major staples. To offset the current impact of climate change, investments in R&D must be made to promote heat- and drought-resistant crop technologies and infrastructure investment like microirrigation systems. Making these technologies easily accessible to smallholders is also crucial. Policy interventions to promote sustainable agricultural intensification are necessary to manage the dual challenge of climate change and productivity growth (Matson et al., 1997; Pretty et al., 2011). For instance, supplementing agricultural productivity programs with agroforestry for carbon sequestration, soil conservation, and watershed management programs to limit land degradation and promote water conservation will prove to be essential in agricultural policy formulation (Lipper et al., 2006; Pretty et al., 2011). Promoting market mechanisms for carbon sequestration through conservation agriculture could contribute to small-farm income improvements while addressing climate mitigation. Finally, insurance systems, disaster relief programs, and safety net programs need to be specifically developed for dealing with the welfare costs associated with increased incidence of extreme events, such as more frequent droughts, floods, and other disasters.

45.3 CONCLUDING REMARKS

Today, the challenges for agricultural development and food security improvement are as formidable as they were 50 years ago. Given the multiple and concurrent threats that food systems face, from unabated growth in food demand to intensification pressures on the agricultural resource base and the growing threat of climate-related risks, the complexity of the task ahead is significantly greater than that faced in the past.

Because of the divergent growth trajectories of developing countries, the future pathways to agricultural growth and food security will differ according to the stage of economic development of each country. A "one-size-fits-all" approach is no longer appropriate in the design of agricultural development programs. While the least-developed countries are still facing chronic conditions of low productivity and high levels of food insecurity, emerging economies are rapidly moving toward market integration and agricultural commercialization. Feeding the cities with a diverse food basket provides new growth opportunities for these economies.

While much of the initial focus of R&D for the green revolution was on the major staples, such as rice, wheat, and maize, the R&D needs for the future are more diverse. This needs to be reflected not only in terms of a diversity of crops and livestock species, but also in terms of the push toward sustainable natural resource management and adaptations related to climate change.

The need today is not just for enhancing the yields of the primary staple grains, but also for addressing the genetic improvement of the so-called "orphan or neglected staples," such as sorghum, millets, and tropical tuber crops, in addition to furthering the diversification of agricultural production by emphasizing the role of livestock and fisheries in our agricultural ecosystem. Such investments could provide new opportunities for growth in the marginal production environments, as well as enhancing the supply and accessibility of micronutrient rich food to the rural poor. There is also an urgent need for R&D investments in making food crops climate resilient, especially in marginal production environments.

Sustainable intensification of food systems can help address multiple societal goals, enhancing food and nutrition security while minimizing the damage to the environment and helping societies deal with climate risks. As in the past, continued and sustained donor support for agricultural development and food security is absolutely crucial, particularly to help the least-developed countries achieve their Sustainable Development Goals.

References

Abraham, M., Pingali, P., 2017. Transforming smallholder agriculture to achieve the SDGs. In: Riesgo, L., Gomez-Y-Paloma, S., Louhichi, K. (Eds.), The Role of Small Farms in Food and Nutrition Security. Springer.

Agarwal, B., 1994. A Field of One's Own: Gender and Land Rights in South Asia. Cambridge University Press, Cambridge.

Asare-Marfo, D., Birol, E., Gonzalez, C., Moursi, M., Perez, S., Schwarz, J., et al., 2013. Prioritizing Countries for Biofortification Interventions Using Country-Level Data (HarvestPlus Working Paper No. 11). Washington D.C.

Barrett, C.B., Bachke, M.E., Bellemare, M.F., Michelson, H.C., Narayanan, S., Walker, T.F., 2012. Smallholder participation in contract farming: comparative evidence from five countries. World Dev. 40 (4), 715–730. Available from: https://doi.org/10.1016/j.worlddev.2011.09.006.

Bellemare, M.F., 2012. As you sow, so shall you reap: the welfare impacts of contract farming. World Dev. 40 (7), 1418–1434. Available from: https://doi.org/10.1016/j.worlddev.2011.12.008.

Binswanger-Mkhize, H., McCalla, A.F., 2010. The changing context and prospects for agricultural and rural development in Africa. In: 4th ed. Pingali, P.L., Evenson, R.E. (Eds.), Handbook of Agricultural Economics, vol. 4. Elsevier B.V, Oxford, pp. 3571–3712.

Birthal, P.S., Jha, A.K., Tiongco, M.M., Narrod, C., 2009. Farm-level impacts of vertical coordination of the food supply chain: evidence from contract farming of milk in India. Indian J. Agric. Econ. 64 (3), 481–496. Retrieved from http://search.proquest.com/docview/201483208?accountid = 10267.

Boselie, D., Henson, S., Weatherspoon, D., 2003. Supermarket procurement practices in developing countries: redefining the roles of the public and private sectors. Am. J. Agric. Econ. 85 (5), 1155–1161. Retrieved from http://www.jstor.org.proxy.library.cornell.edu/stable/1244887.

Bouis, H.E., Hotz, C., McClafferty, B., Meenakshi, J.V., Pfeiffer, W.H., 2011. Biofortification: a new tool to reduce micronutrient malnutrition. Food. Nutr. Bull. 32 (1), S31–S40.

Briones, R.M., 2015. Small farmers in high-value chains: binding or relaxing constraints to inclusive growth? World Dev. 72, 43–52. Available from: https://doi.org/10.1016/j.worlddev.2015.01.005.

Byerlee, D., de Janvry, A., Sadoulet, E., 2009. Agriculture for development: toward a new paradigm. Annu. Rev. Resour. Econ. 1 (1), 15–31. Available from: https://doi.org/10.1146/annurev.resource.050708.144239.

Cabezas, A., Reese, E., Waller, M., 2007. In: Cabezas, A., Reese, E., Waller, M. (Eds.), The Wages of Empire: Globalization, State Transformation, and Women's Poverty. Paradigm Publishers, Boulder CO.

Carletto, C., Corral, P., Guelfi, A., 2017. Agricultural commercialization and nutrition revisited: empirical evidence from three African countries. Food Policy 67, 106–118. Available from: https://doi.org/10.1016/j.foodpol.2016.09.020.

Cercamondi, C.I., Egli, I.M., Mitchikpe, E., Tossou, F., Zeder, C., Hounhouigan, J.D., et al., 2013. Total iron absorption by young women from iron-biofortified pearl millet composite meals is double that from regular millet meals but less than that from post-harvest iron-fortified millet meals. J. Nutr. 143 (9), 1376–1382. Available from: https://doi.org/10.3945/jn.113.176826.

Croppenstedt, A., Goldstein, M., Rosas, N., 2013. Gender and agriculture: inefficiencies, segregation, and low productivity traps. World Bank Res. Observ. 28 (1), 79–109. Available from: https://doi.org/10.1093/wbro/lks024.

Deere, C.D., Leon, M., 2001. Empowering Women. Land and Property Rights in Latin America. University of Pittsburgh Press, Pittsburgh.

Dries, L., Germenji, E., Noev, N., Swinnen, J.F.M., 2009. Foreign direct investment, vertical integration, and local suppliers: evidence from the Polish dairy sector. World Dev. 37 (11), 1742–1758.

Evenson, R.E., 2003. In: Evenson, R.E., Gollin, D. (Eds.), Crop Variety Improvement and Its Effect on Productivity: The Impact of International Agricultural Research. CABI, Cambridge, MA, pp. 447–472.

Evenson, R.E., Gollin, D., 2003. Assessing the impact of the green revolution, 1960 to 2000. Science 300 (5620), 758–762. Available from: https://doi.org/10.1126/science.1078710.

Fan, S., 2000. Research investment and the economic returns to chinese agricultural research. J. Prod. Anal. 14 (2), 115–137.

Fan, S., Pardey, P.G., 1997. Research, productivity, and output growth in Chinese agriculture. J. Dev. Econ. 53 (1), 115–137. Available from: https://doi.org/10.1016/S0304-3878(97)00005-9.

Fan, S., Hazell, P., Thorat, S., 2000. Government spending, growth and poverty in rural India. Am. J. Agric. Econ. 82 (4), 1038–1051. Retrieved from http://www.jstor.org/stable/1244540.

FAO, 2006. Agriculture, Trade Negotiations and Gender. Rome, Italy.

FAO, 2011. The Role of Women in Agriculture (ESA Working Paper No. 11–2). Retrieved from <www.fao.org/economic/esa>.

FAO, 2012. Sustainability pathways. Smallholders and family farmers. Available at: <http://www.fao.org/fileadmin/templates/nr/sustainability_pathways/docs/Factsheet_SMALLHOLDERS.pdf>.

FAO, 2017. State of Food and Agriculture. Food and Agriculture Organization, Rome, Italy.

Fletschner, D., Carter, M.R., 2008. Constructing and reconstructing gender: reference group effects and women's demand for entrepreneurial capital. J. Socio-Econ. 37 (2), 672–693. Available from: https://doi.org/10.1016/j.socec.2006.12.054.

Floro, M.S., 1995. Economic restructuring, gender and the allocation of time. World Dev. 23 (11), 1913–1929. Available from: https://doi.org/10.1016/0305-750X(95)00092-Q.

Haggblade, S., Hazell, P., 2010. In: Haggblade, S., Hazell, P. (Eds.),) *Successes in African Agriculture: Lessons for the Future.* Johns Hopkins University Press, Baltimore.

Hazell, P., 2010. An assessment of the impact of agricultural research in South Asia since the green revolution. In: Pingali, P., Evenson, R.E. (Eds.), Handbook of Agricultural Economics. Elgar Publishing Ltd, Amsterdam, pp. 3469–3530.

Hotz, C., Loechl, C., Lubowa, A., Tumwine, J.K., Ndeezi, G., Masawi, A.N., et al., 2012. Introduction of β-carotene-rich orange sweet potato in rural uganda resulted in increased vitamin A intakes among children and women and improved vitamin a status among children. J. Nutr. 142 (10), 1871–1880.

Lipper, L., Pingali, P., Zurek, M., 2006. Less-favoured areas: looking beyond agriculture towards ecosystem services. In: Ruben, R., Pender, J., Kuyvenhoven, A. (Eds.), Sustainable Poverty Reduction in Less-Favoured Areas: Problems, Options and Strategies. CABI, Wallingford, UK, pp. 442–460.

Low, J.W., Arimond, M., Osman, N., Cunguara, B., Zano, F., Tschirley, D., 2007. A food-based approach introducing orange-fleshed sweet potatoes increased vitamin a intake and serum retinol concentrations in young children in rural Mozambique. J. Nutr. 137 (5), 1320–1327. Retrieved from http://jn.nutrition.org/content/137/5/1320.abstract.

Matson, P.A., Parton, W.J., Power, A.G., Swift, M.J., 1997. Agricultural intensification and ecosystem properties. Science 277 (5325), 504–509. Available from: https://doi.org/10.1126/science.277.5325.504.

Nakazibwe, P., Pelupessy, W., 2014. Towards a gendered agro-commodity approach. J. World Syst. Res. 20 (2), 229–256. Retrieved from https://search.proquest.com/docview/1561499952?accountid = 10267.

Pingali, P., 2007. Agricultural growth and economic development: a view through the globalization lens. Agric. Econ. 37, 1–12. Available from: https://doi.org/10.1111/j.1574-0862.2007.00231.x.

Pingali, P. 2010. Chapter 74 Agriculture Renaissance: Making "Agriculture for Development" Work in the 21st Century. In: B. T.-H. of A. Economics (Ed.), (Vol. Volume 4, pp. 3867–3894). Elsevier, doi: 10.1016/S1574-0072(09)04074-2

Pingali, P., 2012. Green Revolution: Impacts, limits, and the path ahead. Proceedings of the Natl. Acad. Sci. 109 (31), 12302–12308. Available from: https://doi.org/10.1073/pnas.0912953109.

Pingali, P., 2015. Agricultural policy and nutrition outcomes – getting beyond the preoccupation with staple grains. Food Security 7 (3), 583–591. Available from: https://doi.org/10.1007/s12571-015-0461-x.

Pingali, P., 2017. The green revolution and crop biodiversity. In: Hunter, D. (Ed.), Handbook of Agricultural Biodiversity. Routledge Press.

Pingali, P., Khwaja, Y. 2004. Globalisation of Indian Diets and the Transformation of Food Supply Systems," ESA Working Paper No. 04-05. (FAO: Rome, Italy, 2004) (ESA Working Paper No. 04–05). Rome, Italy.

Pingali, P., Rosegrant, M.W., 1995. Agricultural commercialization and diversification: processes and policies. Food Policy 20 (3), 171–185. Available from: https://doi.org/10.1016/0306-9192(95)00012-4.

Pingali, P., Schneider, K., Zurek, M., 2014. Poverty, agriculture and the environment: the case of sub-Saharan Africa. In: von Braun, J., Gatzweiler, F.W. (Eds.), Marginality: Addressing the Nexus of Poverty, Exclusion and Ecology. Springer, Dordrecht Heidelberg New York London, pp. 151–168.

Pingali, P., Ricketts, K., Sahn, D.E., 2015. The fight against hunger and malnutrition: the role of food, agriculture, and targeted policies. In: Sahn, D.E. (Ed.), Agriculture for Nutrition. Oxford University Press, Oxford, UK.

Pingali, P., Witwer, M., Abraham, M., 2016. Getting to zero hunger: learnin from the MDGs for the SDGs. In: Lalaguna, P.D.Y., Barrado, C.M.D., Liesa, C.R.F. (Eds.), International Society and Sustainable Development Goals. Thomson Reuters Aranzadi, Pamplona, Spain, pp. 175–202.

Pretty, J., Toulmin, C., Williams, S., 2011. Sustainable intensification in African agriculture. Int. J. Agric. Sustain. 9 (1), 5–24. Available from: https://doi.org/10.3763/ijas.2010.0583.

Quisumbing, A.R., Rubin, D., Manfre, C., Waithanji, E., van den Bold, M., Olney, D., et al., 2015. Gender, assets, and market-oriented agriculture: learning from high-value crop and livestock projects in Africa and Asia. Agric. Human Values 32 (4), 705–725. Available from: https://doi.org/10.1007/s10460-015-9587-x.

Raitzer, D., Kelley, T., 2008. Benefit–cost meta-analysis of investment in the international agricultural research centers of the CGIAR. Agric. Syst. 96, 108–123.

Ramaswami, B., Birthal, P.S., Joshi, P.K., 2009. Grower heterogeneity and the gains from contract farming: the case of Indian poultry. Indian Growth Dev. Rev. 2 (1), 56–74. Available from: https://doi.org/10.1108/17538250910953462.

Reardon, T., Minten, B., 2011. Surprised by supermarkets: diffusion of modern food retail in India. J. Agribusiness Dev. Emerg. Econ. 1 (2), 134–161. Available from: https://doi.org/10.1108/20440831111167155.

Reardon, T., Barrett, C.B., Berdegué, J.A., Swinnen, J.F.M., 2009. Agrifood industry transformation and small farmers in developing countries. World Dev. 37 (11), 1717–1727. Available from: https://doi.org/10.1016/j.worlddev.2008.08.023.

Renkow, M., Byerlee, D., 2010. The impacts of CGIAR research: a review of recent evidence. Food Policy 35 (5), 391–402.

Rubin, D., Manfre, C., 2014. Promoting gender-equitable agricultural value chains: issues, opportunities, and next steps. In: Quisumbing, A.R., Meinzen-Dick, R., Raney, T., Croppenstedt, A., Behrman, J.A., Peterman, A. (Eds.), Gender in Agriculture and Food Security: Closing the Knowledge Gap. Springer and FAO, New York.

Sanchez, P., 2002. Ecology. Soil fertility and hunger in Africa. Science 295 (5562), March 15.

Schipmann, C., Qaim, M., 2010. Spillovers from modern supply chains to traditional markets: product innovation and adoption by smallholders. Agric. Econ. 41 (3–4), 361–371. Available from: https://doi.org/10.1111/j.1574-0862.2010.00438.x.

Spielman, D.J., Pandya-Lorch, R., 2009. In: Spielman, D.J., Pandya-Lorch, R. (Eds.), Millions Fed: Proven Successes in Agricultural Development. International Food Policy Research Institute, Washington DC.

Swinnen, J.F.M., Maertens, M., 2007. Globalization, privatization, and vertical coordination in food value chains in developing and transition countries. Agric. Econ. 37, 89–102. Available from: https://doi.org/10.1111/j.1574-0862.2007.00237.x.

Thornton, P.K., Kruska, R.L., Henninger, N., Kristjanson, P.M., Reid, R.S., Atieno, F., and et al. 2002. Mapping Poverty and Livestock in the Developing World. Nairobi, Kenya.

Timmer, P.C., 1988. Chapter 8 The agricultural transformation. Handb. Dev. Econ. 1, 275–331. Available from: https://doi.org/10.1016/S1573-4471(88)01011-3.

Timmer, P.C. 2010. A World Without Agriculture: The Structural Transformation in Historical Perspective. Washington D.C.

Webb, P., 2009. Cultivated capital: agriculture, food systems, and sustainable development. Dimens. Sustain. Dev. 2 (74).

CHAPTER

46

Resource Mobilization for Agriculture and Food Security

Kartika Bhatia and Hafez Ghanem

Middle East and North Africa Office, World Bank, Washington, DC, United States

46.1 INTRODUCTION

Agriculture faces a major challenge: to produce sufficient food in a sustainable manner to reduce hunger and malnutrition in the world today, as well as feed the growing population that is expected to reach 9.7 billion by the year 2050 (Brookings Institution, 2015). The Food and Agriculture Organization of the United Nations (FAO) estimates that as many as 815 million people in the world today are undernourished. Moreover, about 155 million children under 5 years of age suffer from stunted growth, while more than 50 million suffer from wasting (FAO et al., 2017). What is also worrisome is that both the prevalence of undernourishment and the number of the undernourished have increased between 2015 and 2016, reversing a downward trend.

Agriculture is important for growth and poverty reduction in most developing countries. Moreover, in sub-Saharan Africa and parts of Asia, agriculture provides employment and supports the livelihoods of nearly 60% of the total population (Gollin, 2010). Gollin has reviewed the theoretical and empirical evidence on the role of agricultural productivity in fostering economic growth. Typically, in countries in early stages of development, where a large proportion of the world's poor and undernourished live, agriculture has a large share of gross domestic product (GDP), and agricultural labor is an even greater share of the total labor force. This points to low agricultural labor productivity and hence low returns to labor. In those countries, agricultural growth is important not only for food security, but also for overall growth.

The views expressed here are the authors' own and do not necessarily reflect the views of the World Bank.

Sustainable Food and Agriculture
DOI: https://doi.org/10.1016/B978-0-12-812134-4.00046-7

Development is historically associated with the structural transformation of economies, from agriculture oriented originally to industry, manufacturing, and services more recently. Declining shares of agriculture in employment and GDP, and productivity increases in all sectors, typically are part of this transformation. As a result, the contribution of agriculture to overall growth during development is likely to decline. On the other hand, agriculture is widely considered to be vital for spurring poverty reduction, as the vast majority of the extreme poor in developing countries live in rural areas and depend, whether directly or indirectly, on agriculture for their livelihoods (see also Sections I and II). Christiaensen et al. (2006) find that improving agricultural productivity is key to designing effective poverty reduction strategies. Cervantes-Godoy and Dewbre (2010) analyzed data from 25 developing countries and found growth in agricultural incomes to be an important determinant for poverty reduction.

Urbanization, and the increased demand for diversified diets and processed and packaged goods, offers a unique opportunity for domestic production to meet the increasing but also more complex demand for food. However, for farmers (especially smallholder farmers) to meet the challenge and turn it to their benefit, a substantial increase in investments is necessary. Investments in agriculture and the broader food system in general will be key to the development strategies, especially in the case of "late transformers" for which a rapid increase in nonagricultural sectors (manufacture and services) is slow or unfeasible. In such cases, growth in agriculture, and especially the food system, will be critical in absorbing the rapidly increasing labor force associated with high population growth.

Raising agricultural output will require an increase in investment both in agriculture and the broader food system. A 2015 joint report by the Food and Agriculture Organization of the United Nations (FAO), the International Fund for Agricultural Development, and the World Food Program estimates that an additional investment of US$265 billion a year is needed to eradicate hunger by 2030 (FAO et al., 2015). Estimates by the United Nations also highlight that the average annual cost of incremental investment in climate change mitigation and adaptation, sustainable agriculture development, and food security in low-income countries will amount to US$1452 billion (constant 2010 US$) (FAO, 2017a, Table 15.1).

To meet these investment goals, we require a new strategy that involves increasing public expenditure, encourages the private sector to invest in agriculture and the broader food system, and supports smallholder farmers to better link to agricultural value chains and produce sustainably. Governments need to increase public expenditure in rural areas, especially on agricultural research and development (R&D), infrastructure, and enhancing food security. Governments can also play a strategic role in encouraging the private sector to invest in agriculture and in upstream and downstream activities. Private investment in agriculture and the value chains should move beyond mere land ownership toward supporting and fostering a relationship with local growers. Both public and private investment initiatives should focus on enhancing the agricultural productivity of smallholders and women farmers considering that 2.5 billion people currently live and work on small farms (Brookings Institution, 2015).

46.2 SOURCES OF INVESTMENT IN AGRICULTURE

The majority of investment in agriculture comes from farmers themselves, followed by government expenditure and external private investment. Fig. 46.1 plots the investment in agriculture in low- and middle-income countries by source for the period 2005–07.

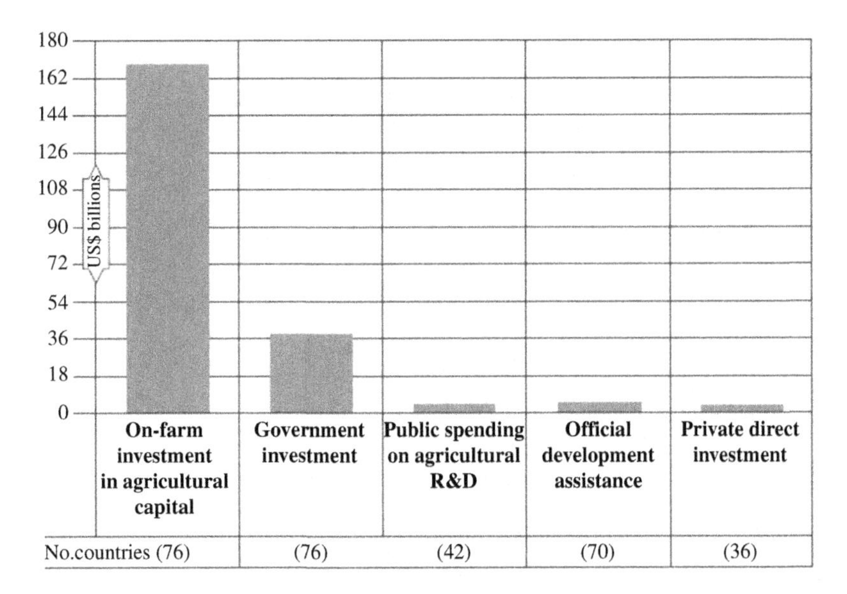

FIGURE 46.1 Investment in agriculture in low- and middle-income countries by source, 2005–07 (annual average).

Column 1 represents on-farm investment in capital stock by farmers (i.e., land development, livestock, machinery and equipment, plantation crops, and structures for livestock).[1] Columns 2 and 3 are public spending on agricultural development (agrarian reforms, extension services, irrigation, subsidies, loans to farmers, etc.) and agricultural research. Column 4 is official development assistance (ODA), which includes more than investment flows, and column 5 represents foreign direct investment (FDI). Investment by farmers is more than four times greater than the investment made by governments. Both ODA and FDI represent much smaller shares of total investment in agriculture. To meet the rising demand for food in a sustainable way, private sector participation in agriculture must increase together with public spending on agricultural R&D and other public goods such as infrastructure and communications.

Increasing the effectiveness of the use of funds channeled to agriculture is just as important as increasing the volume. Experience shows that too often publicly financed projects fail to achieve their objectives due to weak capacity and governance, which means that investing in capacity building in the public sector, as well as governance reforms, is also needed. Additionally, new delivery methods should be explored, for example, increasing reliance on nongovernmental organizations (NGOs) for implementation or for providing credits directly to farmers to implement or supervise projects. New technologies, such as intelligent cards, mobile phones, video-taking drones, and social media, provide new and exciting possibilities for project implementation and monitoring. Finally, a policy framework conducive to agriculture and food systems development—including, pricing, exchange rates, trade and

[1] Since it does not include investments made by farmers in training and education, it represents a lower bound.

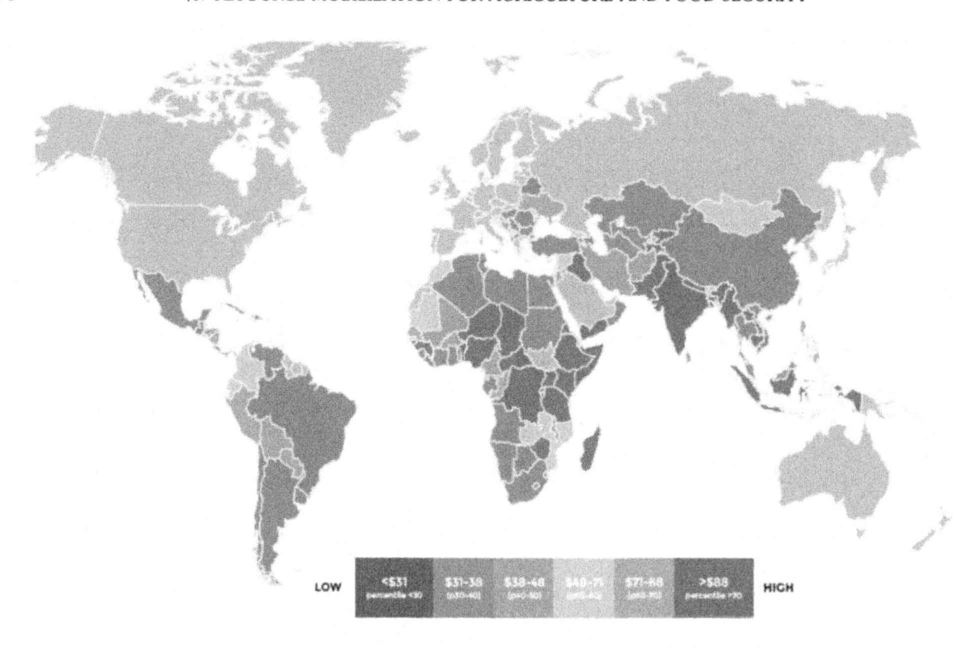

FIGURE 46.2 Investment in food and nutritional security (total resources, constant 2013 US$ rural per capita). Source: *Brookings Institution, 2015. Ending Rural Hunger: Mapping Needs and Actions for Food and Nutrition Security., Figure 3.7, p. 45.*

access to markets, and access to land—will affect not only the volume, but also the efficacy of private investments.

Wide disparity also persists in agricultural investment across different countries (Fig. 46.2; Brookings Institution, 2015). Total resources include public investment, ODA, FDI, spending by US-based NGOs and philanthropy, and exclude investments by farmers. Malaysia, Cyprus, Uruguay, Turkey, Belarus, and Croatia spend from US$350 to US$532 per rural inhabitant as domestic public spending, whereas governments in Myanmar, Mauritania, Yemen, Ghana, the Republic of the Congo, and the Democratic Republic of the Congo spend less than US$4 per rural inhabitant. The differences seem to reflect differences in per capita income as well as priority given to agriculture by governments. ODA is mostly directed at sub-Saharan Africa and Latin America. Private external investment, which comprises FDI, US NGOs, and philanthropy, is concentrated in Latin America, the Caribbean, and the Gulf countries.

46.3 PUBLIC INVESTMENT IN AGRICULTURE

Government expenditure is the second largest source of investment in agriculture. The economic rationale for public investment in agriculture stems from the inefficiencies brought about by market failures. Goods and services,[2] such as education, public health,

[2] Investment in public goods, such as health, education, and public infrastructure, have broader scope and benefit everyone, but in late transforming and agricultural dependent countries, such investments are directly linked to agriculture, which dominates rural areas.

roads, R&D, and clean environment, are underprovided by private agents due to the inherent "public good" nature of these goods, that is, goods that are nonrivalrous and nonexcludable. Many of these goods also produce positive or negative externalities that are often not reflected in their market prices, leading to socially inefficient production levels. In addition, poverty and equity concerns also drive governments to invest in agriculture, as the majority of the world's poor live in rural areas (Fan and Saurkar, 2006).

Trends in Public Agriculture Expenditure

Over the past decades, there has been an increase in public spending with the gap closing between high-income and low-income countries. Table 46.1 draws on data from the Statistics on Public Expenditure for Economic Development (SPEED) database developed and maintained by the International Food Policy Research Institute (IFPRI), as reported by Yu et al. (2015). Per capita total expenditure averaged US$400 in 2005 purchasing power parity (PPP) international dollars in the 1980s for developing countries, almost 19 times

TABLE 46.1 Total and Agriculture Expenditure, 1980–2010

	1980–89	1990–99	2000–10	1980–2010
PER CAPITA TOTAL EXPENDITURE (2005 PPP)				
World	1894	1969	2452	2116
Developing countries	400	533	936	633
Developed countries	7534	8382	10,041	8697
RATIO OF TOTAL EXPENDITURE TO GDP (%)				
World	30.7	27.3	26.9	28.3
Developing countries	21.1	18.6	20.9	20.2
Developed countries	33.8	31.5	31.0	32.0
PER CAPITA AGRICULTURAL EXPENDITURE (2005 PPP)				
World	49	41	54	48
Developing countries	23	23	42	30
Developed countries	145	121	114	126
SHARE OF AGRICULTURE IN TOTAL EXPENDITURE (%)				
World	2.6	2.2	2.3	2.4
Developing countries	6.1	4.6	4.4	5.0
Developed countries	1.9	1.5	1.2	1.5

Source: Yu, B., Fan, S., and Magalhães, E., July 2015. Trends and composition of public expenditures: a global and regional perspective. European Journal of Development Research, Palgrave Macmillan; European Association of Development Research and Training Institutes (EADI), 27 (3), 353–370.

less than the developed world. This gap has diminished over the years with high-income countries spending US$10,041 per capita in 2000–10, about 10 times the level in developing countries. Total government expenditure, as a percentage of GDP, averaged at 26.9% in the 2000s for the world, with developed countries spending 31% and developing countries spending 21% relative to the size of their economy.

However, this increase in public spending has not been translated into an increase in agriculture's share in total expenditure. Even if agriculture is not the largest sector in many developing countries, the share of agricultural labor in the total labor force far exceeds agriculture's share in the economy. Hence, spending on agriculture is an important policy tool for supporting livelihoods in rural areas. Average agriculture expenditure in developing countries has grown from US$23 per person in the 1980s to US$42 per person in the 2000s, but the share of agriculture in total expenditure has steadily declined from 6% in the 1980s to 4.4% in the 2000s (Table 46.1). The developed world has also seen a decline in the per capita agriculture expenditure during the same period, with agriculture spending representing less than 2% of total public expenditure.

The situation remains the same even when we break down public investment in agriculture by region (Figs. 46.3 and 46.4). While public expenditure in agriculture has risen steadily since the 1980s in all regions, the share of agricultural expenditure in total expenditure has fallen except for South Asia and East Asia and the Pacific, regions that have seen a moderate increase in the share of agricultural expenditure in recent years. Declining agriculture shares reflect the shift in government priorities from agriculture to other sectors such as defense, health, education, and social protection (Yu et al., 2015). A shift from agriculture to nonagriculture sectors is, by historical trends, a defining element of the structural transformation processes also reflecting strategic policy choices. Nonetheless, together with the transformation of the sector, maintaining a high intensity of investment in agriculture is crucial for smooth and inclusive development in rural areas.

Agricultural spending intensity is defined as the share of agriculture expenditure in agriculture GDP: Fig. 46.5 shows that developed countries typically have higher agricultural spending intensity than developing countries, with the Euro zone averaging above 20%. In sub-Saharan Africa, agricultural spending was 67.7% of agricultural GDP in 1980 and has declined dramatically to 2.4% in 2012. Unfortunately, this decline reflects a rapid

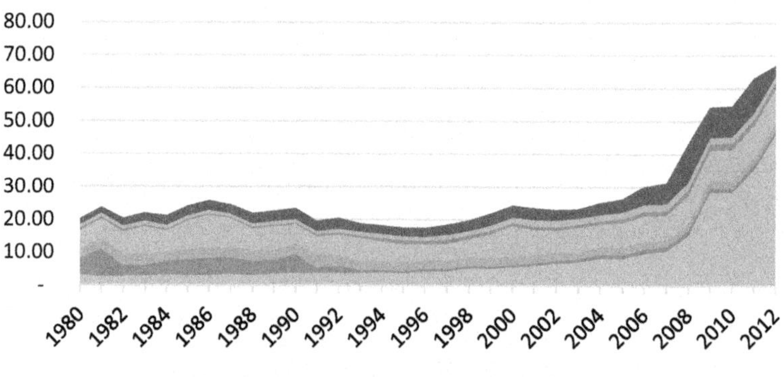

FIGURE 46.3 Agriculture expenditure in 2005 US$ PPP. Source: *SPEED, IFPRI.*

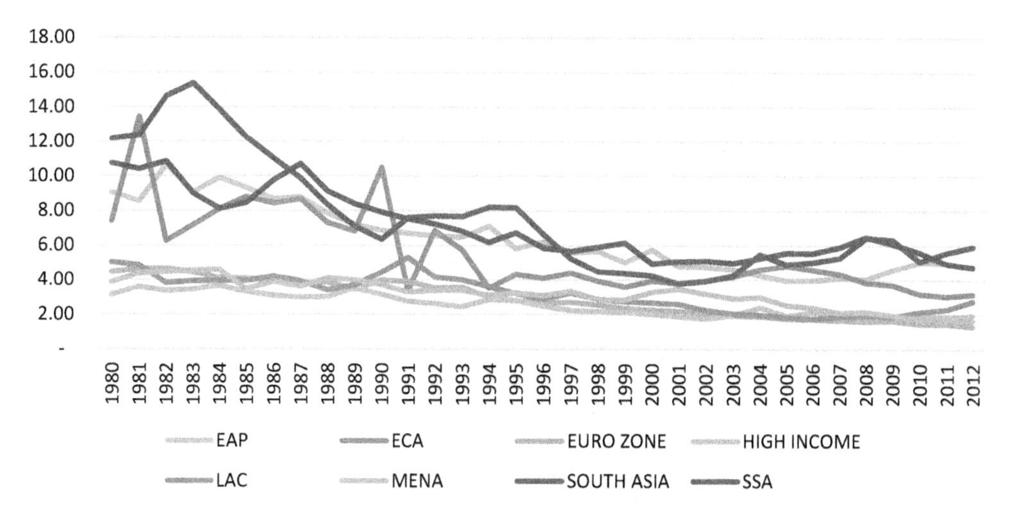

FIGURE 46.4 Percentage of agriculture expenditure in total expenditure. Source: *SPEED, IFPRI.*

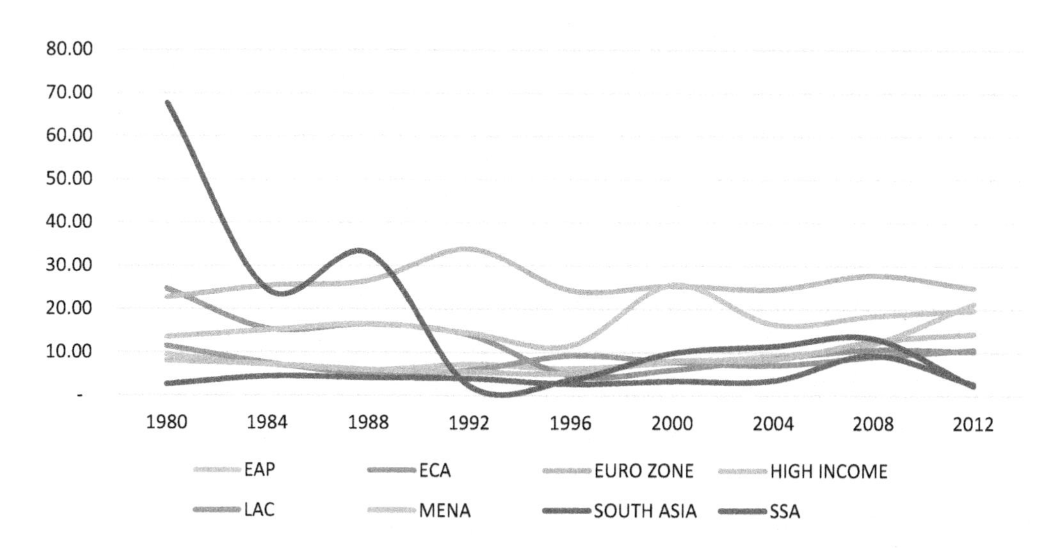

FIGURE 46.5 Ratio of agriculture expenditure in agriculture GDP (%), 1980–2012. Source: *SPEED, IFPRI.*

drop in public spending rather than a sharp increase in agricultural GDP. The East Asia and the Pacific (EAP) region, however, has increased its agricultural spending intensity during the same period. South Asia started with an agriculture expenditure intensity of 2.6% in 1980, moved up to an average of 4% during the 1980s and early 1990s, but then fell back to 2.7% in 2012.

To summarize, governments in developing countries have reduced not only their share of agricultural investment in total expenditure, but are also spending less intensively on agriculture. This is cause for concern, to the extent that increasing government

TABLE 46.2 Total Public Agricultural Research Expenditures by Country/Region (Million 2005 PPP Dollars)

Developing Country/Region	1981	1990	2000	2008
Africa south of the Sahara (45)	1207	1218	1314	1745
China	658	1055	1907	4048
India	451	779	1487	2121
Asia and the Pacific (26)	1863	2897	4736	7725
Brazil	972	1218	1244	1403
Latin America and the Caribbean (28)	2328	2464	2819	3297
West Asia and North Africa (13)	na	na	1544	1848
Eastern Europe and the former Soviet states (21)	na	na	514	963

Source: ASTI database, as published in Beintema, N., Stads G., Fuglie K., and Heisey P., 2012. ASTI Global Assessment of Agricultural R&D Spending. IFPRI Beintema et al. (2012).

expenditures in agriculture results generally in reduced incentives for farmers and the private sector to invest in the sector.[3] However, it is important to point out that aside from investments in agriculture, spending in other sectors such as education, health, and social protection could have important benefits for food and nutrition security in the medium or long term.

The public spending data discussed above do not include public expenditure on R&D. Government support and promotion of investment in agricultural research is essential for developing new technologies, inputs, and tools to produce sustainably. According to data compiled by the Agricultural Science and Technology Indicators, which is managed by IFPRI, total public expenditure on agricultural R&D increased by 22% between 2000 and 2008, with China, India, and Brazil accounting for a quarter of the global spending (Table 46.2). These three major developing countries have shown remarkable progress in R&D investment. In the period between 2000 and 2008, China doubled its spending on R&D, reaching US$4 billion (in 2005 constant PPP), with the private sector contributing 16% of total agricultural research expenditure in 2006 (Hu et al., 2011). Brazil has the largest agricultural research system in Latin America and employs the largest number of PhD-qualified agricultural researchers in the region. India has invested consistently in agricultural research during the past few decades, spending close to 0.4% of its agricultural GDP on public agricultural R&D in 2008. This ratio is close to the average for the developing world, but far from the average (2.4%) for high-income countries. However, India's total number of agricultural researchers declined by 10% from 2000 to 2008 due to stagnant recruitment by the country's universities. Other developing countries are also lagging behind in their R&D investment.

[3] Actually, this is a disincentive for farmers to stay in the sector, as they are to some degree "pushed" to leave and seek opportunities elsewhere.

Impact of Public Investment in Agriculture

Agriculture and rural development demands public investment not only in goods directly linked to agriculture such as irrigation services, R&D, and power, but also in key public goods such as education, health, rural roads, and telecommunications, which have a bearing on reducing poverty and food insecurity. Given tight budgets, policy-makers need to decide which investments will have the most impact. Tewodaj et al. (2012) conducted a comprehensive literature review of the impact of public investment in and for agriculture. Their analysis shows that public spending in R&D has high social returns, with the majority of studies finding an internal rate of return to investment greater than 20% (see also the chapter on research and innovation in this section).

The impact of public investment in R&D on agriculture production and productivity is greater than investment in other agricultural activities, such as irrigation, extension, and input subsidies (FAO, 2012a). Returns to R&D investment are also higher when compared with investments in rural road infrastructure, education, electrification, health, and telecommunication. In spite of all the evidence, governments continue to subsidize inputs and invest poorly in R&D. Allcott et al. (2006) analyze a panel dataset from Latin America and show that income inequality and political and institutional factors determine the structure of rural public expenditure and constrain governments from allocating funds efficiently.

No longer adequate as the only metric for the impact of public agricultural investment, income growth should be complemented by measures of social inclusion and environmental sustainability. Today, most low- and middle-income countries cope with very high rates of youth unemployment. While economic and social exclusion is bad in and of itself, it also threatens peace and stability in many parts of the world as one of the drivers of violence and conflict.[4] Moreover, despite many positive contributions, agriculture also contributes to climate change and natural resource degradation (see also Sections I and III). Thus, it is essential to analyze the ex-ante impact of agricultural projects and investments on livelihoods and natural resources.

Climate change is an important global threat. Hence, agricultural projects need to be also assessed by their climate co-benefits. Most of the poorest countries in the world are in areas that will be severely impacted by climate change owing to rising temperatures and increasing water scarcity that will impact agricultural productivity. Therefore, in addition to impact on mitigation, projects need to be assessed in terms of their contribution to climate change adaptation.

46.4 OFFICIAL DEVELOPMENT ASSISTANCE

Development financing is an important complementary source of investment in agriculture, especially for countries lacking adequate financial resources. A decline in aid to agriculture is traceable since the mid-1980s, reflecting the neglect of the sector by donors. However, the decline seems to be reversing after the 2007–08 food crisis with a renewed interest in promoting sustainable agriculture (Fig. 46.6). Bilateral aid commitments to

[4] For example, see Bhatia and Ghanem (2017).

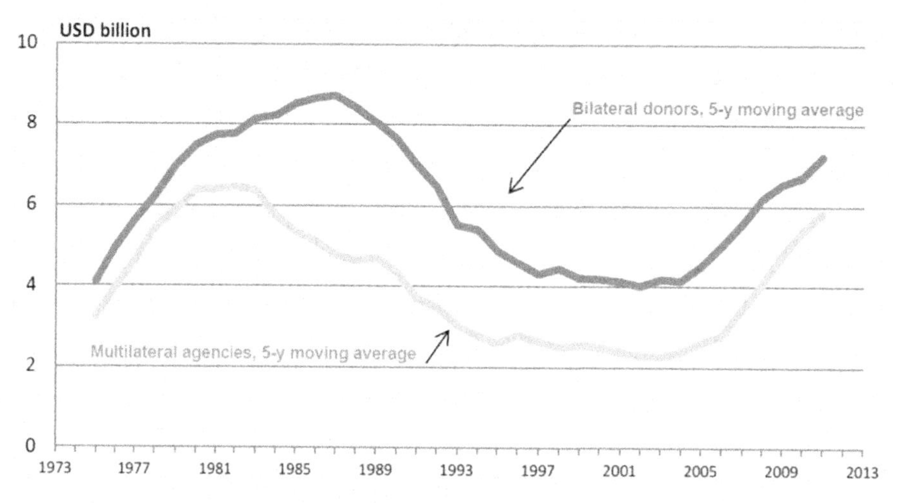

FIGURE 46.6 Trends in aid to the agricultural and rural development sector, 1973–2013 (5-year moving average ODA commitments, constant 2013 prices). Source: *Organization for Economic Co-operation and Development (OECD). 2017. List of OECD databases. Available at: <https://www.oecd.org/statistics/listofoecddatabases.htm>.*

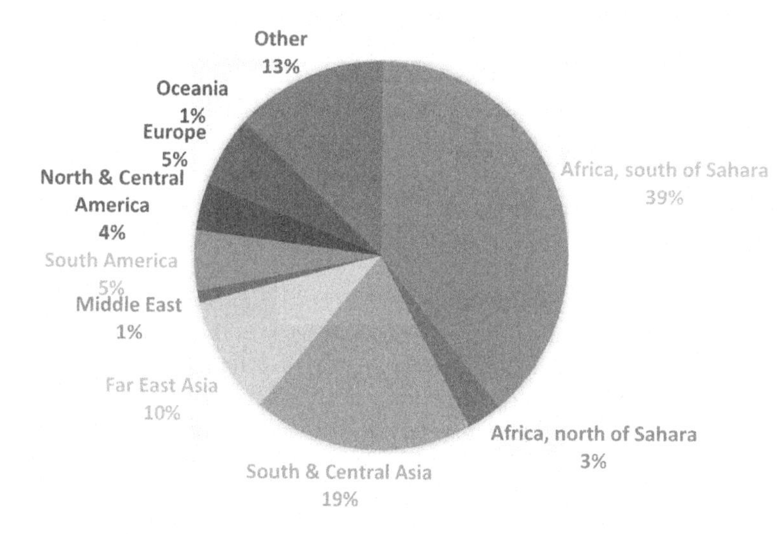

FIGURE 46.7 Regional breakdown of aid to agriculture and rural development (2012–13, bilateral and multilateral ODA commitments at constant 2013 prices). Source: *Organization for Economic Co-operation and Development (OECD). 2017. List of OECD databases. Available at: <https://www.oecd.org/statistics/listofoecddatabases.htm>.*

agriculture are driven by a few large donors, namely, the United States, Japan, France, and the United Kingdom. On the multilateral side, the International Development Association is the most important agency, followed by European Union institutions.

In 2012–13, aid flows to agriculture and rural development primarily went to sub-Saharan Africa and to South, Central, and East Asia (Fig. 46.7). The two regions together receive around 70%–80% of total development assistance. While poverty rates play a factor in aid commitments, geopolitical concerns also determine the magnitude and direction of aid. Afghanistan was the biggest recipient of agricultural aid in 2012–3, followed by Ethiopia, Turkey, Bangladesh, the Philippines, and Brazil (OECD, 2017).

In 2010, the Global Agriculture and Food Security Program (GAFSP), a multilateral mechanism, was created to improve incomes and food and nutrition security in low-income countries. The GAFSP supports both public and private sector investment across the entire food supply chain, from agricultural inputs to storage to processing and financing. The private sector window of the GAFSP initiative has been especially innovative in providing loans and offering guarantees to agribusinesses through risk-sharing facilities with financial institutions. Its direct investment in agribusinesses has supported the growth and expansion of africaJUICE, sub-Saharan Africa's first certified Fairtrade company to grow and process tropical fruit in Ethiopia. Also, it is investing in hazelnut production to boost incomes and create jobs for smallholder farmers in Bhutan, increasing production in the poultry sector in Madagascar, and expanding the reach of agri-inputs to farmers in Africa. As of December 2015, GAFSP's public sector window has supported projects worth US$1 billion, and the private sector window has allocated US$217 million (GAFSP, 2015). Initiatives, such as GAFSP, facilitate private and public investment to support smallholders and agribusinesses.

Wars and civil conflict are key drivers for undernourishment and malnutrition in the world today (FAO, 2017a). ODA has mostly focused on providing humanitarian assistance in zones such as the Horn of Africa, Yemen, and Syria. This is clearly a priority. However, there is also a need to invest in resilience, where ODA through international organizations and NGOs could play a major role. As conflicts are becoming more protracted, it is important to develop a nexus between humanitarian and development assistance. Relief activities through food or cash distribution need to be complemented by development projects that enhance resilience and build the basis for postconflict recovery and reconstruction. The work of the international community during the Yemen conflict provides an example of how humanitarian and development activities can work hand in hand (Box 46.1).

46.5 INVESTING IN FAMILY FARMS

Family farms occupy around 70%–80% of farmland and produce more than 80% of the world's food in value terms (FAO, 2014). Three-quarters of the world's farms are located in Asia, of which 60% are located in just two countries, China and India. The majority of the farms operate on less than 2 ha of land, but there is significant regional variation in farm sizes across regions. In Asia and Africa, the average farm size is 1–2 ha, whereas in the Americas the average is 74–118 ha (Shenggen et al., 2015).

Family farms are key to sustainable productivity growth in agriculture.[5] Smallholders, including women farmers, are big contributors to the regional food supply, but they are most likely to experience poverty, undernutrition, and food insecurity. They have poor access to financial services and are unable to secure the required capital to invest in sustainable practices, market integration, and increased productivity.[6] As a result, they rely

[5] A comprehensive portrait with indicators on smallholder family farmers' characteristics, contributions, and constraints for more than 20 developing countries can be found in FAO (2017b).

[6] Financial institutions meet less than 3% of the total smallholder financing demands (Zook et al., 2013). Of the formal financing, almost 80% is done via public policy banks—that is, state and agricultural development banks. The rest is covered by local branches of commercial banks and microfinance banks.

BOX 46.1

YEMEN: SMALLHOLDER AGRICULTURAL PRODUCTION, RESTORATION AND ENHANCEMENT PROJECT (SAPREP)

Background. The project has been prepared in partnership with the Food and Agriculture Organization of the United Nations (FAO) as an emergency response to the deteriorating food security situation in Yemen, which is currently facing an unprecedented food crisis. The agricultural sector is critically important to overall economic performance, food security, and poverty alleviation in Yemen. More than 17 million Yemeni are food insecure, making this the largest food security emergency in the world. Before the outbreak of the conflict, the sector employed more than half of the workforce and was the main source of income for 73% of the population. The sector faces many challenges that have been exacerbated by the current conflict, which has severely disrupted agricultural production. Locally grown food supply in 2016 was only 62% of the precrisis level. Being the main employment sector in Yemen, agriculture has also been the sector most affected by the crisis, with a loss of almost 50% of its workers.

Project interventions. The project focuses on two main areas:

1. providing support to poor households and smallholders to increase agricultural production, income, and nutrition; and
2. helping the most vulnerable and conflict-affected farmers to re-engage in the crop and livestock sectors to restore their livelihood and provide income for their basic needs.

The project invests in increasing smallholders' production, income, and nutrition through the following: (1) strengthening community land and water management, including water harvesting for production and consumption; (2) improving animal husbandry, livestock production, and animal health services; and (3) increasing value added of selected agricultural products. The project also provides urgently needed support to farmers affected by the conflict, internally displaced people, returnees, and other vulnerable groups to resume crop and livestock production through provision of livelihood kits and farm restoration start-up packages. Investments selected and implemented through community-based and participatory approaches will help affected communities to reclaim their livelihoods by reinforcing their resilience.

The project is funded by a grant from the Global Agriculture and Food Security Program (GAFSP) and implemented by FAO in collaboration with the Yemen Social Fund for Development (SFD). FAO and SFD have established institutional and implementation mechanisms for the delivery of the project-relevant activities in the country. As a specialized agricultural agency, FAO is well positioned in Yemen to support and lead on issues related to agriculture given its technical expertise and presence on the ground. Partnership with these institutions proved to be effective already at the start-up phase of project implementation.

BOX 46.1 (*cont'd*)

Project beneficiaries and outcomes. The project targets poor and food insecure households within the seven most food insecure governorates in Yemen. About 90,000 households (about 630,000 persons) will directly benefit from SAPREP investments. This includes about 35,000 conflict-affected households, including internally displaced people and returnees, that will be provided with start-up packages to resume agricultural production. In addition, animal vaccinations and treatment activities will benefit about 200,000 livestock owners. Some 18,000 households will benefit from improved water supply, as water infrastructure for production and harvesting on 440 hectares will be rehabilitated.

Implementation of community-based small infrastructure subprojects through cash-for-work schemes provides temporary work and income opportunities for participating communities, while the project support to the livestock sector and selected value chains would also generate seasonal and permanent employment opportunities. The project will also provide opportunities for private sector participation, as Yemeni private seed growers and livestock farmers are contracted to supply seeds and livestock. The main benefits expected from the project are the following: (1) prevention of further deterioration in households' food security and reduction of malnutrition; (2) increased food production and income generation; (3) restoration of agricultural and livestock production at the farm level through agricultural inputs; (4) livestock asset protection and increased livestock production; and (5) increased water supply resulting from the construction of water-harvesting structures.

on informal lenders such as relatives, local moneylenders, input suppliers, and traders, who charge exorbitant interest rates. Family farms are also vulnerable to climate change and extreme weather events and have limited access to formal safety nets. Food price volatility also affects smallholders disproportionately more than large landholders (Karfakis et al., 2011). Additionally, smallholders need access to modern markets, which requires the presence of multiple factors such as rural infrastructure (such as road access and irrigation), membership in cooperatives, education and information regarding price, supply, and demand, and quality standards.

Many smallholder family farmers are women, who face even greater constraints in accessing resources and services needed for farming.[7] Women farmers have little or no access to credit, and they are less likely than men to own land or livestock and receive extension services or advice on technology adoption (FAO, 2011). They are also more likely to hold smaller and less productive land. Closing the gender gap in agriculture could raise total agricultural output in developing countries by 2.5%–4% (FAO, 2011).

Given the variety of constraints faced by smallholders, public policy should consider the characteristics of smallholders. Profitable and viable smallholders should receive aid to move toward business-oriented farming, and governments should promote the creation of

[7] Women make up 43% of the agricultural labor force in developing countries.

nonfarm employment opportunities in rural areas for farmers who are unable to sustain their livelihoods by practicing only agriculture. Public and private funds can also help alleviate supply-side constraints. Banks interested in serving smallholders need assistance in developing expertise to assess smallholder loans and design appropriate financial products. Absence of land titles (required as a collateral) and credit bureaus (assessing customers risk profile) make lending to smallholders less attractive. Public policy directed toward securing property and tenure rights, developing rural infrastructure, incentivizing banks to increase their rural coverage, public procurement of agricultural products, and providing social protection and agricultural insurance can play a catalytic role in improving the incomes of smallholder farmers.

46.6 PRIVATE INVESTMENT IN AGRICULTURE

Private sector participation and investment in agriculture is almost negligible when compared to investment by farmers themselves and the public sector. To increase agriculture production sustainably, there is a need to bring in private investments from both industries and institutional investors to fill the investment gap. Large private investments in agriculture and the food system can assist in unleashing structural changes in low-income countries by supporting smallholders, developing and fostering adoption of climate smart technologies, and improving efficiency in storage and financing. Attracting private investment requires complementary public investment in key public goods, adequate agriculture-related infrastructure, well-defined tenancy rights, agricultural R&D, competitive financial markets, and transparent legislative policies. In addition, policies aimed at increasing private investment should ensure that these investments are sustainable and secure benefits not only to investors and recipient countries, but also to host communities and local populations.

In recent years, large-scale land acquisitions of farmland in low- and middle-income countries by international investors have generated a debate, with concerns being raised on the impact of such investments on the environment, local economy, and host communities. In 2010, the Organization for Economic Co-operation and Development, or OECD, contracted a private consulting firm to conduct a survey of private financial sector investment in agriculture (HighQuest Partners, 2010). The survey covered large-scale financial institutions, hedge funds, real estate investment trusts, and private-public companies. Of the 25 companies surveyed, 60% were based in Western countries, 24% in South America, 12% in Asia and the Pacific, and 4% in the Middle East and North Africa. South America (led by Brazil) and Africa were the principal destinations of investments by these firms.

Firms were primarily interested in investing in the production of major raw crops (soft oilseeds, corn, wheat, and feed grains) and livestock production (feed, and pastures for beef cattle, dairy, sheep, and swine). Investment in infrastructure, irrigation, storage, distribution, or inputs was of secondary importance for the firms. Firms preferred to build capacity by hiring and training local staff and indicated that building a good relationship with the local community was essential. Sometimes they undertook activities such as financing schools, building health centers, and promoting cultural activities in return for concessions on leasing farmland from the government. Cotula et al. (2011)

reviewed the literature on land acquisition in Africa and found that the impact of purchasing land was heterogeneous, depending on how the deals were structured. In cases where environmental and social impact assessments are conducted, property rights are respected and local communities are involved in decision making, as such types of agricultural investments can deliver local benefits. Arezki et al. (2011) conducted an empirical analysis of the factors driving transnational land acquisition. They found that in contrast to the general literature on FDI, good governance in the host country was not a determinant of acquisition. Countries with weak tenure system and poor land governance could be attractive to investors more interested in land grabbing than in long-term productive investment.

Outright purchases of land by international companies may lead to accusations of land grabbing, and accordingly do not represent the ideal form of private sector involvement in agriculture. Large firms can invest in agriculture by fostering a partnership with the farmers themselves through outgrower schemes or contract farming. Contract farming refers to long-term supply agreements between farmers and agribusiness processing/marketing companies/buyers, which bring mutual gains and normally include price and supply arrangements (FAO, 2012b, p. 1). The arrangement guarantees a secure market for the product, thereby allowing farmers to earn more money and buyers to secure a return on their investment. By linking smallholders to national and export markets and mobilizing FDI in agriculture, the outgrower schemes can alleviate some of the bottlenecks faced by smallholders. However, governments need to ensure that outgrower schemes also include small farmers and do not create social tensions through land grabbing. The schemes can also have adverse gender effects, as women are less likely to have access to contract farming than men (Schneider and Gugerty, 2010).

In Andhra Pradesh, India, farmers contract with large poultry input suppliers, where the firms provide day-old chicks, feed, and medicines, while the farmers supply land, labor, and other minor variable inputs. At the end of the cycle, farmers receive a net price, which is pegged to an industry price that is more stable than the retail price. The World Bank analyzed the benefit of this outgrower scheme and found that farmers benefited from access to credit and insurance schemes. The contract farmers were also more efficient than noncontract producers. While the whole value chain generates more money, most gains accrued to buyers. In the end, the outgrower scheme was successful in improving access to credit, insurance, and technology for small farmers.

The Rubber Outgrower Plantation Project in Ghana is a tripartite agreement between the financial operators—Agricultural Development Bank and the National Investment Bank, Technical Operator—Ghana Rubber Estates Limited (GREL), and farmers represented by the Rubber Outgrowers and Agents Association. Under the agreement, GREL provides technical assistance, advisory services, and commits to purchasing the rubber. Farmers obtain a bank loan on concessional rates, commit to technical assistance conditions of GREL, and produce rubber. FAO conducted a study of the outgrower scheme and found that farmers were able to increase their productivity because GREL was directly involved in production, processing, and marketing of the product, providing constant monitoring, technical assistance, and good quality inputs to the farmers (FAO, 2012b). The initiative also helped to mitigate climate change through the carbon sequestration capacity of rubber plantations.

Other examples for an inclusive model of corporate investment in agriculture are tenant farming and sharecropping. In tenant farming, small farmers lease land from large agribusinesses, whereas in sharecropping, farmers use land from agribusinesses in return for a share of the crop produced. In recent years, investment funds for agriculture have sprung up as a means to pool resources from different investors to provide capital to agricultural stakeholders. Agricultural investment funds are being used by both the public and private sector to invest in developing world agriculture. Many of these funds are set up as partnerships between public and private sector investors. African Agricultural Capital Limited (AAC), established in 2005 in East Africa, is an example of an agricultural investment fund. AAC observed that the agriculture sector in East Africa was dominated by smallholders who faced a lack of access to finance and technical assistance. AAC invests in early stage businesses, such as plant breeding, seed production, cereal crop handling and marketing, agribusinesses that contract small farmers as outgrowers, and aquaculture. AAC's investments target small and medium enterprises that benefit smallholders, thereby promoting agricultural development in East Africa.

Private investment in all forms can make a positive impact, especially if it supports small-scale and women producers. Private sector investment can provide the much-needed capital to boost agricultural production to meet the continuously rising demand of food. However, such investment needs to move beyond land acquisition and adhere to some key principles, such as targeting local growers, working with producer organizations, and respecting the rights of small farmers, workers, and communities.

In addition to appropriate policies and legal and regulatory frameworks, attracting more private investment in developing country agriculture will require interventions to lower the risk faced by investors. Agriculture being an inherently risky sector, farmers are used to dealing with risks related to weather and price fluctuations. Moreover, increasingly, they can purchase private insurance to deal with those types of risks. Domestic farmers and cross-border investors incur additional risks related to changes in government policies, or violence and civil unrest, while cross-border investors may assess the additional risk that their properties may be expropriated. Those types of "sovereign" risks discourage FDI from flowing to developing country agriculture. International organizations could play a role to reduce such risks, for example, by expanding the work of the World Bank's Multilateral Investment Guarantee Agency to cover more agricultural investments, particularly in fragile and conflict-affected countries.

46.7 CONCLUSIONS

More efficient and better investment in agriculture and food systems is needed if the sector is to play a fundamental role for food security and nutrition, for rural poverty reduction and economic inclusion, as well as for overall growth and sustainable development. This paper has focused on ways to increase resource mobilization for those investments as well as ways to improve their impact. Governments need to prioritize public investment in agriculture, especially in high payoff activities such as R&D, aimed at improving productivity, improving nutrition, and enhancing resilience of food and agricultural systems to climate change (see also Section I, and the chapter on research and innovation in this section). At the same time, it is important to note that investments in other types of public goods such as rural roads, power, water, social protection, health, and education could sometimes have an even greater impact on agricultural productivity, and food and nutrition security, than investments in the agricultural sector itself. Governments also need to change the delivery mechanisms for their agricultural investment and other spending. It is often more efficient to channel investments through NGOs or the private sector than to use public sector implementing agencies who may suffer from capacity and governance problems.

Experiences from countries like Afghanistan, Syria, Yemen, Somalia, and the Democratic Republic of the Congo indicate that protracted conflict, violence, and fragility are the major causes of increased undernourishment and malnutrition in our world today. International donors are adjusting their assistance strategies to reflect this new reality. There is an increased focus on financing projects that support resilience to shocks. It has also become apparent that humanitarian and development actors need to work closely together in conflict and protracted crises situations. The old approach of working on humanitarian relief during conflict, and only to start development work at the end of the conflict, is not viable when the conflict/crisis extends over several years.

Family farmers, including smallholder farmers, produce most of the world's food supply. Moreover, smallholders are among the poorest in nearly all developing countries. Hence, supporting them is key to food and nutrition security and to poverty reduction. Governments and their development partners need to prioritize programs that support family farmers, especially smallholders. Programs that promote adoption of sustainable and innovative technologies and practices to farmers and those that link them to markets and the value chain and support their transformation into entrepreneurs have been among the most successful over the years. It must also be noted that because women are heavily represented among poor smallholders, programs should emphasize the special needs of women farmers.

Government and donor resources are not sufficient to meet all of agriculture's investment needs. Farmers themselves and other private sector actors, including foreign investors, are and will be the most important source of increased resources for agriculture. Moreover, private investment tends to be more efficient than public investment. Governments need to put in place the right legal and regulatory frameworks that encourage private investments, while protecting local communities, smallholder farmers, and natural resources. Such a framework could encourage partnerships between local farmers and the private sector through programs such as outgrower schemes or sharecropping. Experience indicates that outright purchase of land could sometimes lead to accusations of land grabbing and hence be counterproductive. It is also important to consider ways of reducing the risk faced by foreign investors through insurance and guarantee schemes. Simply measuring the change in agriculture productivity and output is insufficient to analyze the impact of agricultural investment. Investment should also aim at creating jobs and contributing to inclusive growth (e.g., increasing opportunities for youth and women), and to conservation of ecosystem services (e.g., clean water, pollinators, natural enemies of diseases and pests, nutrient cycling in soils). In addition, agriculture needs to contribute to dealing with the challenge posed by climate change. Therefore, the impact of agricultural investment should also be measured in terms of its climate cobenefits and how it helps farmers adapt to changing temperatures and increased water scarcity.

Finally, investment in primary agriculture, forestry, and fisheries is unlikely to absorb the significant increase in the labor force in countries that are late transformers and characterized by a high growth of population. Today, hundreds of millions people are unemployed, and many more are underemployed. The growth in industry and services sectors can generate employment opportunities and raise household incomes. Evidence shows, however, that in some developing regions industrialization is slow and services are not necessarily the most productive and remunerative activity. At the same time, the middle class in urban centers, small cities, and rural towns influences diets and preferences, including the physical appearance of goods or how they should be packaged.

For those regions and countries, investment in agriculture should be considered in conjunction with investments in the broader food system, including in increasing linkages between production with upstream and downstream sectors. These parts of the agricultural and food systems can create jobs and growth opportunities by integrating small firms and farms into a chain of interdependent links that are able to capture and generate more value (FAO, 2018). Henceforth, investments in agriculture should consider the preferences for food of urban consumers, who demand more processed, packaged, and convenience food as well as organic, fresh, and novelty foods. A harmonized approach requires that agriculture is prepared to meet the demands from agroindustry, the role of which is increasing with urbanization.

References

Allcott, H., Lederman, D., Lopez, R. 2006. Political institutions, inequality, and agricultural growth: the public expenditure connection. Policy Research Working Paper Series 3902, The World Bank.

Arezki, R., Deininger, K., Selod, H. 2011. What drives the global land rush? IMF Working Papers 11/251. Washington, DC, International Monetary Fund.

Beintema, N., Stads, G., Fuglie, K., Heisey, P., 2012. ASTI Global Assessment of Agricultural R&D Spending. IFPRI.

Bhatia, K., Ghanem, H. 2017. How do education and unemployment affect support for violent extremism: Evidence from MENA countries, Brookings Working Paper 103, March 2017.

Brookings Institution. 2015. Ending Rural Hunger: Mapping Needs and Actions for Food and Nutrition Security.

Cervantes-Godoy, D., J. Dewbre. 2010. Economic Importance of Agriculture for Poverty Reduction, OECD Food, Agriculture and Fisheries Working Papers, No. 23, OECD Publishing

Christiaensen, L., Demery, L., Kuhl, J. 2006. The role of agriculture in poverty reduction an empirical perspective. Policy Research Working Paper Series 4013, The World Bank.

Cotula, L., Vermeulen, S., Mathieu, P., 2011. Agricultural investment and international land deals: evidence from a multi-country study in Africa. Food Sec. 3 (Suppl 1), 99. Available from: https://doi.org/10.1007/s12571-010-0096-x.

Fan, S., Saurkar, A. 2006. Public spending in developing countries: trends, determination and impact. (mimeo).

FAO, 2011. The State of Food and Agriculture 2010–11. Women in agriculture: closing the gender gap. Rome.

FAO, 2012a. The State of Food and Agriculture 2012. Investing in agriculture for a better future. Rome.

FAO, 2012b. Outgrower schemes: advantages of different business models for sustainable crop intensification. Ghana case studies. FAO Investment Centre.

FAO, 2014. The State of Food and Agriculture 2014. Innovation in family farming. Rome.

FAO, 2017a. The future of food and agriculture – Trends and challenges. Rome. Available at: <http://www.fao.org/3/a-i6583e.pdf>.

FAO, 2017b. Smallholders Dataportrait, weblink: <http://www.fao.org/family-farming/data-sources/dataportrait/en/>, Rome, FAO.

FAO, 2018. The challenge of employment in the 21st century: The potential for jobs in the agricultural and food systems, FAO Agricultural Development Economics, Policy Brief No. 6, Rome, FAO.

FAO, IFAD, WFP, 2015. The State of Food Insecurity in the World 2015. Meeting the 2015 international hunger targets: taking stock of uneven progress. Rome, FAO.

FAO, IFAD, WFP, 2017. The State of Food Insecurity in the World 2017: Building Resilience for Peace and Food Security. Rome, FAO.

Global Agriculture and Food Security Program (GAFSP), 2015. Smart Growth and lasting Results. Annual Report.

Gollin, D., 2010. Agricultural Productivity and Economic Growth. Handbook of Agricultural Economics. Elsevier.

HighQuest Partners (United States), 2010. Private financial sector investment in farmland and agricultural infrastructure. OECD Food, Agriculture and Fisheries Papers, No. 33. OECD Publishing, Paris.

Hu, R., Liang, Q., Pray, C., Huang, J., Jin, Y., 2011. Privatization, public R&D policy, and private R&D investment in China's agriculture. J. Agric. Resour. Econ. 36 (2), 416–432.

Karfakis, P., Velazco, J., Moreno, E., Covarrubias K. 2011. Impact of Increasing Prices of Agricultural Commodities on Poverty. ESA Working Paper 11–14 (Rome: FAO, Agricultural Development Economics Division).

Organisation for Economic Co-operation and Development (OECD). 2017. List of OECD databases. Available at: <https://www.oecd.org/statistics/listofoecddatabases.htm>.

Schneider K., M. Gugerty. March 2010. Gender and Contract Farming in Sub-Saharan Africa. Literature Review, University of Washington, pp. 1–2

Shenggen, F., Joanna, F., Tolulope, O., 2015. The business imperative: helping small family farmers to move up or move out. 2014–2015 Global food policy report. Chapter 4. International Food Policy Research Institute (IFPRI), Washington, D.C., pp. 25–31.

Tewodaj, M., Bingxin, Y., Shenggen, F.,Linden, M. 2012. The impacts of public investment in and for agriculture: Synthesis of the existing evidence. IFPRI discussion papers 1217, International Food Policy Research Institute (IFPRI).

Yu, B., Fan, S., Magalhães, E., July 2015. Trends and composition of public expenditures: a global and regional perspective. Eur. J. Dev. Res. 27 (3), 353–370. Palgrave Macmillan; European Association of Development Research and Training Institutes (EADI).

Zook, D., Deelder, W., Denny-Brown, C., Carroll, T., 2013. Local Bank Financing for Smallholder Farmers: A $9 Billion Drop in the Ocean. The Initiative for Smallholder Finance, France.

CHAPTER

47

Governance and Institutions: Considerations From the Perspective of Sustainable and Equitable Food Systems

Louise O. Fresco

Wageningen University and Research, Wageningen, Netherlands

When discussing sustainable agriculture and food production, we must address a wide variety of intertwined developments—technological, economic, as well as institutional. By 2050, over two-thirds of the world's population is projected to live in cities (United Nations, 2015). That single fact has major institutional implications: urban planning must reflect the design of the food supply, for example, how food will be grown in or in the periphery of cities, or on the contrary, be imported and stored, and how its distribution will be organized. This is not as simple as it sounds. Raising livestock close to cities requires awareness of the risk of zoonotic diseases and their control, including early warning systems. Similarly, urban horticulture may influence water quality, while the use of chemicals may need to be controlled to encourage careful usage. Issues like these require coordination in policy design and implementation among different government departments, such as agriculture, health, water, and finance. Also required is coordination between national, local, and urban authorities, simultaneously encouraging the population to express their priorities in shaping their environment. Furthermore, the private sector is increasingly assuming new roles in providing services and conducting applied research, for example, through the command of large data sets on climate, soils, and prices (see also Section IV).

All this indicates that we stand at the threshold of new thinking about the governance of food systems. The greatest paradigm shift in thinking about governance and institutions derives from the fact that urban demand will drive sustainable and equitable food supply.

Sustainable Food and Agriculture
DOI: https://doi.org/10.1016/B978-0-12-812134-4.00047-9

This paradigm shift assumes an additional significance given the fact that in the future poverty and malnourishment will recede slowly from rural areas, where the majority of the poor live today, to urban centers, where the majority of the needy will live.

One additional dimension makes the institutional design even more complicated and urgent: the fact that sustainability is rarely a matter of win–win situations and, most frequently, is a matter of a choice between diverging and partly incompatible objectives. This means that governance systems must make the choices and options transparent in order to obtain public support. Consider, for example, the supply of tomatoes to a city. Should they be produced locally, which would reduce the carbon footprint and transportation losses, or be produced in a remote province where production would provide income and employment but may require subsidies? Or should tomatoes be imported since that option would be cheaper and would allow poor urban consumers access to nutritious food? Or should a sustainable and equitable food system leave all options open and expect a natural balance to occur over time when market forces determine the outcomes?

As the FAO (2016a) concluded, "Designing such approaches to respond to the multiplicity of challenges will not be easy given past trajectories of mostly sector-specific policy-making and given major deficiencies in global and national governance mechanisms, regulatory systems, and monitoring and accountability frameworks."

Given the degree of complexity outlined above and the diversity of countries and cultures, how can we design governance systems that allow a better integration of these developments? This is all the more urgent because institutions and governance are weakening in terms of agriculture. Either cooperatives are less powerful or have proved a failure; dedicated departments of agriculture are increasingly integrated into other departments that do not include food; and extension services are downsized or have become obsolete. Poor governance, poor quality of civil servants, and downright corruption have eroded confidence in food system governance. The green revolution has shown that, notwithstanding its successes, the lack of a unified and integrated approach to technology and policies, especially in the environmental and social domains, has caused many negative side effects.

As the FAO (2016b) recognized, "A core element of a proposed common approach of agriculture, forestry and fisheries for a transformative change toward sustainability relates to how these sectors are governed. Building a common approach requires the development of a common understanding and better dialogue across sectors. It requires involving different stakeholders, including private sector and civil society, and developing partnerships for effective action at different levels."

So what approaches could be considered? It is useful to distinguish between the food chain, that is, the technical and economic process of transformation by humans of sunlight into a diversity of foods through plants and animals, and the food system that includes the web of all actors, regulatory frameworks, and decision-makers. Obviously, there is not one blueprint for governance of the food system, but instead a number of elements.

47.1 INTEGRATED POLICY MAKING

The first step to unlock the potential of any country is to acknowledge that the basis of a sustainable and equitable food supply lies in the food chain, that is, the complex relations between input suppliers (seeds, machinery, fertilizers, biocides, etc.) to farmers, processors of raw materials, food producers (such as bakers) to retailers and, ultimately, consumers. Second, there needs to be acknowledgment of the need to take a systematic approach to govern the food chain as a food system. No country has yet developed such an integrated food system policy. Although some countries have linked health and agricultural concerns in an attempt to cover both, in no country are the incentives in place for a truly integrated approach to involve all actors and sectors, including the environment. Nor are staff trained to think in a cross-sectoral manner. Integrated food system policy-making would include national, cabinet-level decision making, allocation of funds to provide incentives, training and monitoring of the entire food chain, and the articulation of these policies at subnational levels. An important aspect here is devolution, encompassing what needs to be governed at national or international levels (e.g., food safety legislation) and what can be undertaken at subnational or local levels (e.g., distribution, opening hours of sales points of foods, and other elements of consumer food environments).

47.2 MAPPING ALL THE ACTORS

By definition, food systems involve a multitude of actors, not only those along the chain (from input suppliers to consumers), but also those at supranational, national, regional, and local levels. Mapping the actors involved in specific food systems, such as the provision of staples like rice or wheat, or in meat or dairy production, is a necessary initial exercise. As an exercise that would create awareness at all levels, it would provide a useful starting point for a food system governance framework. In this process, care should be taken to look also at nongovernmental institutions, for example, consumer or farmer organizations and the multitude of private sector actors. Mapping the actors entails analyzing not just their roles, but also the constraints they may be facing, such as lack of credit and unnecessary regulations.

47.3 DEFINING THE GOALS AND ALIGNING THE ACTORS

Overall goals for the food system must be set at the national level, but they require translation into goals for different actors or subsectors. For example, the national government may wish to increase the consumption of vegetables or fruits, though what that exactly means remains to be defined: promoting local horticulture through loans or subsidies, reducing horticultural waste through better transport and storage, promoting fruits and vegetables in schools through education and vegetable gardens, and so on. Even this simple example shows the broad multitude of actors that need to be aligned, from schoolteachers to vegetable seed companies.

Whatever the precise goals, they will encompass the entire set of effects of food on human, animal, and ecosystem health, environment and nature, equity, and economic growth. The global One Health approach is the companion to an integrated food system approach (Frenk et al., 2016).

47.4 BUILDING CAPACITY AND APPLIED RESEARCH EXPERTISE

Few vocational schools or universities exist that teach the food system as an integrated part of their curricula. Creating cross-disciplinary courses will help to develop a class of civil servants and experts who can help implement a food system approach. Furthermore, applied research is needed to identify and remedy bottlenecks in the system or in decision making. For example, if a city council wants to promote local production by imposing a carbon or value added tax on imported foods, the exact footprint of each food item needs to be assessed, along with the intended and unintended side effects on food availability. This kind of assessment should be conducted for any government intervention, for example, on pricing, but also regarding the methods of monitoring the effects of food systems on the wider environment, such as water quality.

Successful systems of capacity building and applied research for food systems are found in countries where universities, government departments, and the private sector are fully aligned behind the same goals and are in constant interaction. The increasing role of nongovernmental organizations in functioning as critical watchdogs is highly beneficial. The agricultural and food sectors can learn from the medical sector in this respect, where academic and teaching hospitals, the pharmaceutical sectors, and patient organizations are involved in close, and often critical, dialogues.

47.5 CREATING A NATIONAL FOOD SYSTEM COORDINATION UNIT

Implementing a national food system strategy requires continuous monitoring and target setting to ensure that the overall goals are met or need adjustment. Again, no country to date seems to have installed a truly integrated national coordination, but elements are in place in several countries. A useful endeavor would be to take the lead in an inventory of integrated food system coordination at the country level in order to list and disseminate best practices.

47.6 GLOBAL INSTRUMENTS IN TRADE AND FOOD AND AGRICULTURE

For decades, areas of food and agriculture are among the most regulated at the international level. Most issues involve government support to the sector, including through barriers to free trade. Food safety, trade, and subsidies are all areas governed by extensive global, regional, and bilateral agreements. The FAO and other United Nations agencies play major roles in many of these areas, such as sanitary and phytosanitary measures (SPS agreements). International treaties, such as on climate, biodiversity, genetic resources, and desertification, touch the agricultural and food sectors directly. Likewise, even if not in a binding manner, many voluntary agreements exist on responsible fisheries, pesticide reduction, and various industry-led initiatives such as roundtables on oil palm or soybean.

The major problem with international instruments is that their focus is nearly always economic in nature and requires heavy monitoring. They rarely allow the private sector to develop new guidelines, even if policy exists in the area, for fear of provoking accusations of unfair trade, as in the case of the reduction of salt in processed food. Government policy exists in many countries to reduce salt levels, but the private sector hesitates to act because to be the prime adopter might mean that consumers, who are used to high salt levels, will seek another, possibly imported, brand. Also, if private sector companies agree among themselves to reduce salt levels together, they may be accused of collusion. Further complicating the issue is the fact that drastic reductions of salt levels may entice consumers to add salt at home, thereby rendering the entire policy useless. The only way to move forward on this subject is to obtain simultaneous agreement across the entire food system to reduce salt levels everywhere, across the sector and countries, and to enact this measure over a precisely defined period of time. Similar efforts are needed in the area of the environment. The introduction of a carbon tax in the food system can only be done collectively across countries and subsectors; otherwise, both the level playing field and equal opportunities for all actors are at risk. International organizations—such as FAO, the World Health Organization, the World Trade Organization, the United Nations Framework Convention on Climate Change, and others—should make an inventory of the areas that require special attention in the implementation of national and international food system policies.

47.7 INVOLVING THE PRIVATE SECTOR

At every step in the food system, the private sector plays a crucial role. In fact, most of the actors in the food system are private, not public actors. The primordial role of both government and other public actors is to enable private actors to fulfill their responsibilities to contribute to the ultimate goal of providing safe and nutritious food in a sustainable way. This enabling role entails many different dimensions, from guaranteeing the security of land title and rule of law to ensuring food safety and animal welfare and setting standards for labor conditions in the food chain. Clarity in regulation and implementation, lack of corruption, transparency of all steps, and appeal procedures are essential for a healthy private sector.

In recent years, private actors in many areas, especially large multinational companies, have taken the initiative to contribute to the Sustainable Development Goals directly. For many chief executive officers, sustainability is becoming part of a sound business model. Best practices are exchanged at forums like the World Economic Forum, the Consumer Goods Forum, and the Sustainable Business Council. The public actors are sometimes slow to follow, and more dialogue is needed to get all actors aligned on implementation and the identification of barriers to sustainability and fair trade.

47.8 EDUCATING CONSUMERS

With rising urbanization, consumers will increasingly be removed physically and psychologically from the actual process of food production. At the same time, rapidly changing food patterns, especially the huge increase in ready meals and fast food, may have major effects on health. While awareness of the relationship between diet and health is growing, the lack of understanding of the food chain seems to lead to a great sense of confusion about what is nutritious food and how the food must be produced. There is a growing sense among the urban middle classes that processed food is not natural, but is artificial and therefore dangerous. As the distance in time and space to the rural populations grows, the perception starts to hold that the past, with its manual work performed "close to the crops and the animals," was somehow better. Anybody who knows the reality of trying to plant one hectare of rice by hand knows that the past was not necessarily better; nor is it necessarily safer to have an animal slaughtered just around the corner instead of at a distance. This perception is already very strong in western countries and increasing everywhere in emerging markets. Public food systems policy must account for these perceptions, however erroneous they may sometimes be, when clarifying concepts such as the "naturalness," authenticity, and quality of food. Any food system will include niches for locally produced organic foods, even if the world's cities will depend for the majority of their calories on food produced elsewhere, either within the same country or internationally.

47.9 EQUITABLE DEVELOPMENT BETWEEN URBAN AND RURAL AREAS

International trade is known to have helped food availability for the urban poor, even if it sometimes occurred to the detriment of poor farmers and the rural poor. The rural out-migration to cities is such that fewer and fewer young people are willing to remain in farming or food processing. It is logical that people will leave agriculture, because agricultural work can be harsh and young people do not consider it to be attractive. At the same time, intensification of agriculture asks for larger farms with less labor. One of the great challenges for the future is to foster a class of young entrepreneurs who will farm in a professional way. This means that new employment, or differently put, value-added jobs must be found in the food chain for those abandoning agriculture. Usually for agriculture to be competitive with imported foods, major reforms are needed in the agricultural sector. This requires a policy of modernization, investment, and entrepreneurship, which in most countries must be a subset of an overall food system policy.

47.10 ANTICIPATING MAJOR TECHNOLOGICAL TRENDS IN FOOD AND AGRICULTURE

Probably the single most important change in the food system is the accelerated digitalization that occurs in every sector, whether nationally or internationally. Digital databases exist of land, water, weather conditions, agricultural prices and markets, crop varieties, animal breeds, and pest and disease outbreaks, which are increasingly linked to one another and to modeling tools. The result is an emerging "Internet of Food," which will, theoretically, provide total transparency, monitoring, and the potential to anticipate events such as droughts and diseases. Digitalization is the basis for the development of precision agriculture. We can now actually measure the soil water status or the nutrient status of leaves and fields to one square centimeter, which means that the monitoring of things like fertilizer and water can be far more precise, with all the negative effects of emissions being managed in a much better way. Accordingly, the crop is nurtured in a far better way than before, with every plant receiving a special treatment.

This digitalization extends further to affect the entire food chain and the consumer. Individual food products can be equipped with digital chips to monitor for origin and food safety, for example, observing meat as it comes out of the slaughterhouse and travels down the retail chain to the consumer. Such a chip not only describes the origin—and hence allays consumer fears on "where does my food come from?" —but also traces, for example, the temperature regime during transportation, all the way down to a refrigeration station or the supermarket. In the near future, consumers will be equipped with similar tools to monitor their refrigerators and even their individual health.

Information and communication technologies (ICT) enable food chains to be organized with much more precision and detailed data capturing. In the coming years, we will face a deluge of data, as sensors, satellites, robots, and other machines (including drones) enter the farm and the rest of the food chain. A revolution comparable to the introduction of the tractor or to chemical products in the 1950s is in the making. This may raise productivity, improve food security, make farming more climate smart, solve environmental issues, and help consumers opt for more healthy and sustainable personal diets in their smart kitchen. ICT can make policies more targeted and address inequality issues, for instance, by focusing on farms with a low income that can be mapped in great detail.

At the same time, developments in ICT are not neutral. Depending on who owns the data and how the exchange of data is organized, the food chain can be governed in many different ways. This can shift the balance of power and lead to more central decision making, turning farmers into mere franchise takers for a centrally organized food chain. As in previous rounds of mechanization, more technology could also lead to greater inequality in agriculture, as those who adopt it grow faster than those who do not. Its effects on the labor market remain to be seen, but are unlikely to be positive for low-skilled workers. An advantage is that the Internet is not necessarily scale dependent: small farmers can benefit as much as large farmers. However, digital skills are essential to all (Fresco and Poppe, 2016).

As developing countries are rapidly catching on to this technology, soon we will see the same quantum jump directly to drone technology as is occurring with the full potential of mobile phones. These new issues call for new regulations and institutions, concerning data management, transparency of data, and ownership. It is of utmost importance that public actors, nationally and internationally, whose awareness of ICT has been lagging behind, catch up rapidly and start preparing the governance models that are required to deal with these technological developments.

Last, but not least, urbanization and the concentration of people will promote the shift toward a truly circular economy, where waste does not exist anymore—where every single element from the food system can be retrieved and utilized, such as proteins and enzymes from sewage systems and from food waste, and brought back into the food chain. This requires another major institutional change, as it should be understood that waste and sewage are not negative, but are valuable resources for food, pharma, and energy. To manage this safely, however, governance needs to become adjusted and cross-sectoral.

The food systems, globally and nationally, will need major changes in order to deliver safe, sustainable, and nutritious food for all in the coming decades. Current governance models are not geared toward an integrated approach of the complexities of the challenges. Whatever governance models result, reforming food systems requires a balance between top-down government-led and bottom-up participatory approaches. The digitalization revolution makes this more feasible than ever before. The key to new food systems resides in the hands of the ultimate beneficiaries who need to understand and support the major changes that lie ahead.

References

FAO, 2016a. Reviewed Strategic Framework and Outline of the Medium Term Plan 2018–21. Document CL155/3. FAO, Rome. Available at: www.fao.org/3/a-mr830e.pdf.

FAO, 2016b. Agriculture and the 2030 Agenda for Sustainable Development. FAO, Rome.

Frenk, J., Fresco, L.O., Hong, P.K., Kuri-Morales, P., 2016. Global Health Challenge. Brookings Institution Press, 80 p.

Fresco, L.O., Poppe, K.J., 2016. Toward a Common Agricultural and Food Policy. Wageningen University and Research, 58 p.

United Nations. 2015. World population prospects: the 2015 revision. Accessed November 2016 from <https://esa.un.org/unpd/wpp>.

CHAPTER

48

Sustainable Agriculture and Food Systems: The Way Forward

Clayton Campanhola, Kostas Stamoulis and Shivaji Pandey

Strategic Program Leader (Sustainable Agriculture), Assistant Director-General (Economic and Social Development Department), and Special Advisor (Sustainable Agriculture), respectively, of the Food and Agriculture Organization of the United Nations (FAO), Rome, Italy

The following is a partial list of actions that, if taken at the global, national, and local levels, will enhance the productivity and sustainability of agriculture and food systems. At all levels, the recommendations must be adapted to best suit the short- and long-term goals and realities. It is important to perform a broad *ex ante* analysis of the impact produced by different actions, including trade-offs and synergies for selecting the most appropriate intervention(s). It is also desirable to develop a strategic plan with an associated work plan showing road maps, targets, and timelines that will allow better convergence and coherence of different interventions at different levels.

Ideally, all strategies and interventions that would contribute to sustainable agriculture and food systems in a given situation would be adopted and implemented together for maximum benefit. However, an "incremental approach," based on prioritization of critical challenges and matching them with appropriate interventions with greatest impact potential, would probably be more practical (Table 48.1).

Sustainable Food and Agriculture
DOI: https://doi.org/10.1016/B978-0-12-812134-4.00048-0

TABLE 48.1 Partnerships, Innovations, Policies, Investments, and Institutions for Sustainable Agriculture and Food Systems: Some Recommendations for Global, National, and Local Level Actions

Recommendations	Global-Level Actions	Nation-Level Actions	Local-Level Actions
Maximize synergies and align strategies and actions between global frameworks and goals (e.g., the 2030 Agenda, Paris Agreement, 2nd International Conference on Nutrition, Sendai Framework for Disaster Risk Reduction, Addis Ababa Action Agenda) and national development action plans	Support countries and strengthen their capacities with knowledge, capacity development, and financial resources to participate in the formulation of global initiatives and relevant agendas	Align national strategies and actions with relevant global goals and frameworks Develop methodologies and mechanisms for assessing and reporting on the implementation of global goals	Enhance awareness among local stakeholders, including the government, producers, processors, retailers, traders, financing institutions, and consumers about global and national efforts aimed at increasing the productivity and sustainability of agriculture and food systems, their impact on them, and their role in achieving them
Strengthen and encourage agricultural research and development (R&D) institutions to focus on policy and technological innovations for more productive, sustainable, and inclusive food systems	With financial resources, knowledge and technologies, support national efforts to pursue interdisciplinary and intersectoral research and innovation that encompasses the entire food system	Prioritize economic and social innovations that increase rural employment, expand rural enterprises and increase rural income, especially of smallholders, women and youth Prioritize R&D that increases productivity and climate change resilience especially of smallholder production systems, and increases landscape-level conservation, restoration, and sustainable use of natural resources Prioritize economic, social, and technological innovations that would provide diverse, safe, and nutritious foods, and contribute to healthy diets Prioritize innovations that reduce food loss and waste, increase efficiency in the use of natural and financial resources, and promote synergistic reuse throughout the food production and consumption chain Use modern ICTs for efficient and timely dissemination of agriculture and food system relevant information to producers, processors, and consumers, especially targeting small family farms, women and youth Reward development, promotion, and adoption of intersectoral and interdisciplinary technologies and practices for the entire food system	Using ICTs and other communication tools, use and promote participatory approaches to enhance scientist–producer–consumer interactions and partnerships in the development of innovations aimed at enhancing productivity and sustainability agriculture and food systems

(*Continued*)

TABLE 48.1 (Continued)

Recommendations	Global-Level Actions	Nation-Level Actions	Local-Level Actions
Prioritize, develop and implement policies and strategies that focus on elimination of hunger, malnutrition, and poverty	Support regional and national efforts aimed at poverty reduction and improvement of food security and nutrition, by sharing knowledge and experience, and provision of financial resources	Design and implement policies considering the entire value chain (i.e., from farm to plate) that simultaneously enhance economic, social and environmental benefits to societies by increasing productivity, efficiency in the use of natural and financial resources, and synergy and integration, and reducing waste	Working with producers, leaders, and development agencies, identify income-enhancing opportunities (e.g., contract farming and producing for cities), at the local level, and establish mechanisms to implement them
		Include monetary and nonmonetary valuation of benefits (e.g., food, health, nutrition, income, ecosystem services) and negative externalities (e.g., degradation of land and water, and loss of biodiversity) of agriculture and food systems in designing and prioritizing policies	Adapt and adopt most appropriate system(s) and framework(s) described in sections I, II and III for enhancing productivity and sustainability throughout the food chain
		Raise rural on- and off-farm employment and income opportunities (e.g., food processing, services, and agri- and ecotourism) by encouraging entrepreneurship and expand and strengthen social protection especially in rural areas to improve and secure livelihoods and reduce forced migration to cities	
		Link smallholders to markets by improving their access to financial and other services including appropriate infrastructure and institutions	
		Enhance awareness among producers, processors, and consumers regarding the benefits of diverse, safe, and nutritious food and healthy eating habits on health, productivity, and quality of life; its dependence on SFA systems, and their role in securing it	
Increase public, private, and ODA investments for more productive and sustainable agriculture and food systems	Increase financial support to countries so they can better participate in and contribute to related global frameworks, initiatives, and agendas	Allocate at least 1% of agricultural GDP to R&D for productive and sustainable agricultural and food system	Strengthen community capacities to adequately use their own and external financial resources to sustainably enhance their productivity and income, protect and use their natural resources, and improve their livelihoods

(Continued)

TABLE 48.1 (Continued)

Recommendations	Global-Level Actions	Nation-Level Actions	Local-Level Actions
	Increase and focus ODA support to national efforts for enhancing productivity and sustainability of their agriculture and food systems	Increase public sector investment in development and improvement of rural infrastructure (e.g., roads, transport, energy, storage, etc.) and rural agroindustries and enhance rural-urban linkages	
		Develop and implement transparent and conducive policies and strategies to leverage and increase private sector investment in the food and agriculture sector in provision of financial and market services, inputs, and in capacity development	
Build, strengthen, and improve institutions and governance for enhancing productivity and sustainability of agriculture and food systems, especially targeting smallholders, women, and youth	Support regional and national efforts that promote access to and sustainable use of food and agriculture-related natural resources, with evidence-based advice and examples	Build, support, and encourage institutions to pursue holistic and systems approaches to food and agriculture sector development, integrating the entire food system	Ensure that implementing communities also participate in the development of relevant policies, regulations, and solutions
		With transparent policies and their implementation, strengthen the rights and capacities of all stakeholders involved in the sector, particularly smallholders, women and youth, to own, access, and sustainably use natural resources (e.g., land, water, biodiversity, and ecosystem services)	Create awareness about distortions caused by perverse policies and actions that hinder access and use of natural resources, contribute to their deterioration, and thereby contribute to poverty and hunger
		Increase coherence and synergy in food and agriculture-related actions and investments across relevant global, regional, and national (public, private, and civil society) institutions	Build capacity of rural institutions and communities to identify opportunities for shared actions and to develop and implement strategies and practices for conservation and sustainable use of natural resources
		Harmonize policies and actions across relevant Ministries (e.g., Food, Agriculture, Fisheries, Forestry, Finance, Environment, Planning, Trade, Labor) that impact on food systems	Enhance access to ICT facilities especially for rural communities and the poor, build their capacities to use them, and facilitate their access to the opportunities from rural-urban linkages
		Reward initiatives and actions that conserve and enhance natural resources, including through payments for ecosystem services and penalties for abuse	

(*Continued*)

TABLE 48.1 (Continued)

Recommendations	Global-Level Actions	Nation-Level Actions	Local-Level Actions
		Establish a consultation and coordination mechanism to enhance synergy of action among public, private and civil society stakeholders on issues related to productivity and sustainability of agriculture and food systems	
Enhance institutional governance and technical capacities for: (1). reducing impact of climate change on agriculture and food systems; (2). reducing impact of agriculture and food systems on climate change; and (3). increasing resilience of nations and communities to disasters, conflicts, and natural hazards, and to build-back better in their aftermaths	Support and promote sharing and development of policies, strategies, and measures for agriculture and food systems to adapt to climate change and help mitigate it	Strengthen country capacities to implement the Nationally Determined Contributions to minimize impact of climate change and reduce emissions of greenhouse gases through synchronization of policies and actions across relevant public-, private- and civil society organizations on the nation's food, nutrition, and health sectors, prioritizing the poor and most vulnerable	Strengthen local capacities to adapt to and mitigate climate change by providing information, building capacities, and facilitating and encouraging use of sustainable practices, approaches, technologies, and innovation
	Provide early warning systems and support tools to enhance national capacities in disaster risk management and governance to reduce impacts of climate change and other calamities on food security, natural resources, and livelihoods	Strengthen national institutional capacity, mobilize public and private investments, and foster partnership to minimize impact of climate change and other disasters on agriculture and food systems and livelihoods	Strengthen capacities of local institutions and communities to increase their resilience to climate change, natural disasters, and other hazards, and building-back better by utilizing available knowledge and resources

Author Index

H

I

J

K

L

Subject Index

Note: Page numbers followed by "*f*," "*t*," and "*b*" refer to figures, tables, and boxes, respectively.

B

C

D

I

Printed in the United States
By Bookmasters